Biochemistry and Molecular Biology

Fourth edition

**William H. Elliott
& Daphne C. Elliott**

UNIVERSITY PRESS

OXFORD

UNIVERSITY PRESS

Great Clarendon Street, Oxford OX2 6DP

Oxford University Press is a department of the University of Oxford.
It furthers the University's objective of excellence in research, scholarship,
and education by publishing worldwide in

Oxford New York

Auckland Cape Town Dar es Salaam Hong Kong Karachi
Kuala Lumpur Madrid Melbourne Mexico City Nairobi
New Delhi Shanghai Taipei Toronto

With offices in

Argentina Austria Brazil Chile Czech Republic France Greece
Guatemala Hungary Italy Japan Poland Portugal Singapore
South Korea Switzerland Thailand Turkey Ukraine Vietnam

Oxford is a registered trade mark of Oxford University Press
in the UK and in certain other countries

Published in the United States
by Oxford University Press Inc., New York

British Library Cataloguing in Publication Data
Data available

Library of Congress Cataloging in Publication Data
Data available

Typeset by Graphicraft Limited, Hong Kong
Printed in China through Asia Pacific Offset

ISBN 978–0–19–922671–9

3 5 7 9 10 8 6 4 2

10 0645484 7

Preface

This book is designed for undergraduates taking science and health-related courses in biochemistry and molecular biology. With the rapid development of the subject, course structures in these areas have become complex but our broad aim is to cover most of the requirements of many BSc courses and those in health-related ones of a similar standard. The reason for writing this book continues to be to provide a text suitable for students meeting these subjects for the first time, but of sufficient depth to be intellectually satisfying. The book is written in an approachable style, with a lot of explanation to promote understanding. New terms encountered in the text by students are explained. Biological context to the topics is included so that students can see the relevance of the molecular devices which constitute life. We have also tried to explain what is not yet known so that students can appreciate the interest of a subject in which a good deal remains to be discovered. In general, we hope that the book can take students to the level at which they could handle many of the reviews of the type found in *Trends in Biochemical Sciences* and related journals as well as the mini-reviews found in some of the major journals.

There is a seemingly never ending flood of new developments in the subject, which is exciting, but the rapid increase of information poses problems for students and teachers alike. To achieve our aims and keep the book of a size that students can cope with, we have made subjective decisions to omit certain aspects, mainly in some metabolic areas, which may be less likely to be taught in today's courses. As an example we do not give the metabolic pathways for the synthesis and breakdown of all individual amino acids but rather concentrate on amino acid metabolism of more general relevance or of special interest or medical importance. This policy allows the metabolic and molecular biology areas of widest current interest to be covered more fully without increasing the book's size.

About the book

The book is organized into five main parts:
1. Basic concepts of life;
2. Structure and function of proteins and membranes;
3. Metabolism;
4. Information storage and utilisation;
5. Molecular biology in health and disease.

The chapters are arranged to give a seamless progression through the subject but teachers usually want to decide in what order topics should be taught. To facilitate this there is extensive cross-referencing between chapters.

What is new in this edition?

- All the chapters have been reviewed to update them and to include a number of new figures and references. This varies in the different chapters from extensive rewriting and major additions to minor changes. A number of complex areas, which some students find difficult, have been rewritten for increased clarity. These include mass spectrometry (Ch. 5), control of glycogen metabolism, (Ch. 16) and the role of histocompatibility complexes (MHCs) in the immune response (Ch. 30). In particular, students sometimes find protein synthesis a difficult topic, possibly because the elongation procedure is counter-intuitive. We have included a preliminary treatment of the chemistry of polypeptide synthesis so that students, having mastered this, can then focus on the complications of what happens on the ribosome more easily (Ch. 24).

  The exercise introducing students to the use of a molecular graphic program to obtain 3-D images of proteins using data from the Protein Data Bank (Appendix to Ch. 5) has also been completely rewritten using a more approachable viewer program.

Among important new developments during the past few years that have been added in this edition are the following:

- A new chapter (25) on the RNA world. This includes micro RNA genes and the revolutionary new concept of RNA controlling the expression of protein coding genes. It includes a description of the international ENCODE (Encyclopaedia of DNA Elements) project to document the functional elements of the genome, whose preliminary findings already have somewhat surprising implications.

- RNA interference, its mechanism, and its medical potential including control effects on cancer metastasis by a microRNA gene (Ch. 25).

- New material on the role of the mediator in eukaryotic gene transcription, and its possible mechanism of action; the role of DNA methylation in transcription regulation and imprinting; DNA insulators which restrict the effects of enhanceosomes to their intended targets are a new addition (Ch. 23).

- Control of gene expression by riboswitches and attenuation are also new items (Ch. 23).

- The mechanism of selenocysteine (the '21st amino acid') incorporation into proteins (Ch. 24).

- The apoptosis chapter has been largely rewritten to include more recent developments in the mechanism and control of this important area The last three chapters which were combined into one previously are now split into three new chapters. (The cell cycle and its control (Ch. 31), Apoptosis (Ch. 32), and Cancer (Ch. 33).)

Important technological developments include the following:

- The role of single nucleotide polymorphisms (SNPs) coupled with microarrays for the location of disease producing genes and individual genotype determination for prediction of disease susceptibility (Ch. 28).

- Developments in the reprogramming of human skin cells into pluripotent stem cells by treatment with transcription factors (Ch. 28). The potential of this may, if it proves to be clinically applicable, cope with immunological problems in stem cell therapy, and, since no human or animal eggs are involved, possibly ethical questions as well.

- The advent of synchrotron radiation in X-ray diffraction of proteins and the use of selenomethionine-labelled proteins is described. This facilitates 3-D determination of proteins especially since micro crystals can be used (Ch. 5).

- The production and use of humanized monoclonal antibodies in therapy (Ch. 30).

In metabolism, new material includes:

- Fructose metabolism and its control, and its repercussions on diabetes (Ch. 16).

- Some new factors in appetite control and fat metabolism.

- New boxes: Smoking, elastin, emphysema, and antiproteinases (Box 4.2); *Trans* fatty acids (Box 7.1); Repetitive DNA sequences (Box 28.1); Red wine and cardiovascular health (Box 29.1).

Using the book

This book includes a number of features to help make it easy to use, and to make learning from it as effective as possible.

- **Index of diseases.** A separate index of diseases and medically relevant topics helps students on health-related courses to identify relevant topics.

- **Medical boxes.** These illustrate the direct relevance of biochemistry and molecular biology to medicine and health-related issues. A separate list of these boxes is shown on the Contents pages.

- **Questions and answers.** Questions at the end of each chapter (with answers at the back of the book) are designed to support student learning.

- **Chapter summaries.** Summaries at the end of each chapter highlight the key concepts presented and aid revision.

- **Further reading references.** References at the end of each chapter direct the reader mainly to review articles of the shorter type found in Trends journals.

Online Resource Centre

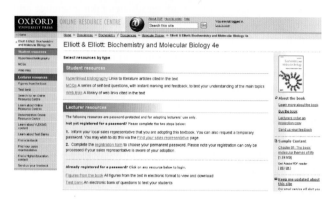

The Online Resource Centre has been developed to provide students and lecturers with ready-to-use teaching and learning resources. It is free-of-charge, designed to complement the textbook, and offers additional materials which are suited to electronic delivery. Visit **www.oxfordtextbooks.co.uk/orc/elliott4e** to find out more.

Lecturer resources

The following resources are password protected to ensure only lecturers can access them. Instructors may find these useful for preparing lectures and course assessment.

- **Illustrations.** All of the illustrations in the textbook available to download electronically. Ideal for the preparation of lecture slides.

- **Testbank of questions.** A customizable testbank featuring questions of different format, with feedback keyed to the book. Ideal for use in both formative and summative assessment.

Student resources

Accessible to all, with no registration or password required.

- **Web links.** A selection of annotated general web links as well as links relevant to specific topics and specific chapters in the book.

- **Hyperlinked bibliography.** A fully hyperlinked version of the further reading lists included in the text, providing links to electronic versions of literature articles cited in the text (where available). (Note that these are just hyperlinks: institutional subscription will be required to obtain full-text access in most cases, though abstracts for linked articles will usually be available.)

- **Online self-test questions.** A selection of multiple-choice questions for readers to check their understanding of each chapter. Each question has feedback keyed to the book to help the reader review concepts relevant to the questions posed.

- **Molecular structures.** A library of three-dimensional, interactive versions of key biological structures featured in the book.

The testbank and online self-test questions have been kindly compiled by Dr Tony Bradshaw, Oxford Brookes University, Dr Shirley McCready, Oxford Brookes University, Racheline Rogers, University of Adelaide, and Dr John Wright, University of East London.

The library of molecular structures has been prepared by Dr Keith Johnstone and Ms Katy Jordan, Department of Plant Sciences, University of Cambridge.

Adelaide   WHE
2008   DCE

Acknowledgements

We are greatly indebted to the following colleagues, who have reviewed sections or given valuable advice:

Professor David Catcheside, School of Biological Sciences, Flinders University, South Australia

Dr Chris Bagley, Protein Laboratory, Hanson Institute, Institute of Medical and Veterinary Science, Adelaide, South Australia

Dr Grant Booker, School of Molecular and Biomedical Science, University of Adelaide, South Australia

Professor John Bowie, Department of Chemistry, University of Adelaide, South Australia

Professor Peter Colman, Walter and Eliza Institute of Medical Research, Parkville, Victoria

Dr Michael Elliott, Immunology, Centocor Research & Development Inc., Malvern, Pennsylvania

Laura Frank, School of Paediatrics and Reproductive Health, University of Adelaide, South Australia

Dr Joan Kelly, School of Molecular and Biomedical Science, University of Adelaide, South Australia

Professor John Mattick, Institute of Molecular Bioscience, University of Queensland, St. Lucia, Queensland

Associate Professor Robert Richards, School of Molecular and Biomedical Science, University of Adelaide, South Australia

Professor George Rogers, School of Molecular and Biomedical Science, University of Adelaide, South Australia

Lynn Rogers, School of Molecular and Biomedical Science, University of Adelaide, South Australia

Dr Neil Street, Dept of Anaesthesia, The Children's Hospital at Westmead, Sydney, New South Wales

Our special thanks are due to Professor Leigh Burgoyne, School of Biological Sciences, Flinders University, who has acted as consultant advisor to the book. He has participated in many stimulating discussion meetings in planning content, contributed valuable suggestions, and critically reviewed every chapter.

Finally we would like to thank the staff at Oxford University Press for their help and cooperation in the development and production of this edition, in particular Leonie Slowman, Development Editor; Sara Miller, Publishing Editor, Science; and especially Jonathan Crowe, Publisher, Science, who has personally guided all aspects of this edition in a most friendly and considerate way.

OUP and the authors would like to thank the following members of an extensive advisory and reviewing panel who gave advice on the third edition, and reviewed draft chapters of the fourth edition.

Dr Alison Caswell, Leeds Metropolitan University, UK

Dr Peter Coussons, Anglia Ruskin University, UK

Dr Tuomo Glumoff, University of Oulu, Finland

Dr Simon Hardy, University of York, UK

Dr Robert Learmonth, University of Southern Queensland, Australia

Dr Pete Lund, University of Birmingham, UK

Dr Soren Mathiesen, University of Aarhus, Denmark

Dr Peter Nunn, University of Portsmouth, UK

Dr Caroline Owen, Kingston University, UK

Lynn Rogers, University of Adelaide, Australia

Dr Martin Ryan, University of St Andrews, UK

John Sadlier, University of Western Australia

Dr Alison Snape, Kings College London, UK

Dr Daniel Ungar, University of York, UK

Dr David Woodward, University of Tasmania, Australia

Dr John Wright, University of East London, UK

Dr Patrick Young, Stockholm University, Sweden

It may metaphorically be said that natural selection is daily and hourly scrutinising, throughout the world, the slightest variations; rejecting those that are bad and adding up all that are good; silently and insensibly working, whenever and wherever opportunity offers, at the improvement of each organic being in relations to its organic and inorganic conditions of life.

Charles Darwin

Evolution is a tinkerer

Francois Jacob

Brief contents

Contents

Part 1 Basic concepts of life

Part 2 Structure and function of proteins and membranes

Part 3 Metabolism

Part 4 Information storage and utilization

Part 5 Molecular biology in health and disease

Chapter 29 Special topics: blood clotting, xenobiotic metabolism, reactive oxygen species | 483

Chapter 30 The immune system | 493

Chapter 31 The cell cycle and its control | 507

Diseases and medically relevant topics

Abbreviations

A	adenine	COP	coat protein of transport vesicles
AA	aminoacyl group	CRE	cAMP-response element
ABC	ATP-binding cassette	CREB	CRE-binding protein
ACAT	acyl-CoA:cholesterol acyltransferase	CSF	colony-stimulating factor
ACP	acyl carrier protein	CTD	C-terminal domain
ACTH	adrenocorticotrophic hormone (also called corticotrophin)	CTP	cytidine triphosphate
ADH	antidiuretic hormone (also called vasopressin)	d-	deoxy
ADP	adenosine diphosphate	DAG	diacylglycerol
AgAp	agouti-related appetite stimulant	dd-	dideoxy
Akt	protein kinase (*see also* PKB)	DHAP	dihydroxyacetone phosphate
AIDS	acquired immunodefiaquired immunodeficiency syndrome	DNA	deoxyribonucleic acid
		DNase	deoxyribonuclease
ALA	5-aminolevulinic acid	DS	double-stranded (DNA, RNA)
ALA-S	aminolevulinate synthase		
AMP	adenosine monophosphate	E_o'	redox potential value at pH 7.0
AMPK	AMP-activated protein kinase	EF	elongation factor
AP	apurine or apyrimidine	EF-G	*E. coli* ribosomal translocase
APC	antigen-presenting cell	EFtu	elongation factor temperature unstable
Apaf1	apoptopic protease mediating factor	EGF	epidermal growth factor
ATCase	aspartyl transcarbamylase	eIF	eukaryote initiation factor
ATP	adenosine triphosphate	ELISA	enzyme-linked immunoabsorbent assay
AURE	AU-rich responsive element	ENCODE	Encyclopoedia of DNA elements (project)
AZT	azidothymidine	ER	endoplasmic reticulum
		ES cell	embryonic stem cell
BAC	bacterial artificial chromosome	ESI	electrospray ionization
Bp	base pair		
BPG	2:3-bisphosphoglycerate	F_o	membrane rotary subunit of ATP synthase
BSE	bovine spongiform encephalopathy (mad-cow disease)	F_1	catalytic subunit of ATP synthase
		F	Faraday constant (96.5 kJ V^{-1} mol^{-1})
C	cytosine	FAD	flavin adenine dinucleotide
c-	'cellular', denotes protooncogene (c-*ras*, c-*myc*, etc.)	$FADH_2$	reduced form of FAD
		Fd	ferredoxin
CAM	calmodulin	FFA	free fatty acid
cAMP	adenosine-3′, 5′-cyclic monophosphate	FH_2	dihydrofolate
CAP	catabolite gene-activator protein	FH_4	tetrahydrofolate
CBP	CREB-binding protein	fMet	formylmethionine
CD	cluster of differentiation (proteins)	FMN	flavin mononucleotide
Cdk	cyclin-dependent protein kinase	FSH	follicle-stimulating hormone
cDNA	complementary DNA		
CETP	cholesterol ester transfer protein	G_1 *or* G_2	'gap' phases of cell cycle
CoA	coenzyme A (A = acyl)	G	guanine
CoQ	ubiquinone (*see also* Q, UQ)	*G*	free energy (Gibbs)
COX	cyclooxygenase	$G^{0'}$	standard free energy (Gibbs)
CDP	cytidine diphosphate	G-1-P	glucose-1-phosphate
CKI	Cdk inhibitor protein	G-6-P	glucose-6-phosphate
		GAG	glycosaminoglycan

GAP	GTPase-activating protein
GLUT	glucose transporter (Glut 1 to GLUT 5)
GroEL	Hsp 60
GroES	lid of groEL
GRB	growth receptor-binding protein
GRK	G-protein receptor kinase
GSH	reduced glutathione
GSK3	glycogen synthase kinase 3
GSSG	oxidized glutathione
H	enthalpy
HAT	histone acetyltransferase
Hb	haemoglobin
HbO_2	oxyhaemoglobin
HDL	high-density lipoprotein
HGPRT	hypoxanthine–guanine phosphoribosyltransferase
HIF	hypoxia-inducible factor
HIV	human immunodeficiency virus
HLH	helix-loop-helix
HMG-CoA	3-hydroxy-3-methylglutaryl-CoA
hnRNP	hetero-ribonucleoprotein complex
HPLC	high-pressure liquid chromatography
Hsp	heat-shock protein
HTH	helix–turn–helix (DNA-recognition motif)
I	inosine
IAP	inhibitor of apoptosis
IDL	intermediate-density lipoprotein
IF	initiation factor (e.g. If1, If2, If3)
IF	intermediate filament
Ig	immunoglobulin (IgG, IgG1, IgA, etc.)
IGF	insulin-like growth factor (IGFI, IGFII)
Il	interleukin
Inr	initiator
IP_3	inositol triphosphate
IRE	iron-responsive element
IRS	insulin receptor substrate
JAK	type of tyrosine kinase (*Janus* *k*inase)
K_a	acid dissociation constant
K_{eq}	equilibrium constant of a reaction
K'_{eq}	equilibrium constant at pH 7.0
K_m	Michaelis constant:the substrate concentration at which a Michaelis–Menten enzyme works at half-maximal velocity
kb	kilobase
LCAT	lecithin:cholesterol acyltransferase
LDL	low-density lipoprotein
LINES	long interspersed repeated sequences

MALDI	matrix-assisted laser-desorption ionization
MAP	mitogen-activated protein (kinase)
MHC	major histocompatibility complex
miRNA	microRNA
mRNA	messenger RNA
MS	mass spectrometry
MTOC	microtubule-organizing centre
N	unspecified base in a nucleotide
NAD^+	nicotinamide adenine dinucleotide (oxidized form)
NADH	reduced form of NAD
$NADP^+$	nicotinamide adenine dinucleotide phosphate (oxidized form)
NADPH	reduced form of NADP
N-CAMS	nerve cell adhesion proteins
NES	nuclear export signal
NF-kB	nuclear factor family of eukaryotic transcription factors
NK	natural killer cells
NLS	nuclear localization signal
NPY	neuropeptide Y; appetite stimulant
NTF	nuclear transport factor
Ⓟ	high-energy phosphoryl group
P450	cytochrome P450
PBG	porphobilinogen
PC	phosphatidylcholine (lecithin)
Pc	plastocyanin
PCNA	proliferating cell nuclear antigen
PCR	polymerase chain reaction
PDGF	platelet-derived growth factor
PDH	pyruvate dehydrogenase
PDI	protein disulphide isomerase
PE	phosphatidylethanolamine (cephalin)
PEP	phosphenolpyruvate
PEP-CK	PEP carboxykinase
PFK	phosphofructokinase (PFK1, PFK2)
PG	prostaglandin
3-PGA	3-phosphoglycerate
PH	pleckstrin homology (domain)
P_i	inorganic phosphate
PIAS	protein inhibitor of activated STATS
PIC	preinitiation complex
PI 3-kinase	phosphatidylinositide 3-kinase
PK	protein kinase (PKA, PKB, PKC, etc.)
PK	pyruvate kinase
pK_a	the pH at which there is 50% dissociation of an acid
PKB	mammalian homologue of Akt
PKU	phenylketonuria
PLC	phospholipase C
PLP	pyridoxal-5′-phosphate
Pol	DNA polymerase

POMC	pro-opiomelanocortin; appetite repressor
PP_i	inorganic pyrophosphate
PPI	peptidylproline *cis–trans*-isomerase
Pq	plasto-quinone
PrP^c	prion protein (constitutive)
Prp^{sc}	prion protein (scrapie)
PRPP	5-phosphoribosyl-1-pyrophosphate
PS	phosphatidylserine
PS	photosystem (PSI, PSII)
PTGS	posttranscriptional silencing (plants)
PTS	peroxisome-targeting signal
PYY-3–36	neuropeptide appetite inhibitor
Q	ubiquinone (*see also* CoQ, UQ)
R	gas constant ($8.315\,J\,mol^{-1}\,K^{-1}$)
Rb	retinoblastoma
RFLP	restriction fragment length polymorphism
RF	release factor
RISC	RNA-induced silencing complex
RNA	ribonucleic acid
RNase	ribonuclease
RNAi	RNA interference
ROS	reactive oxygen species
R-5-P	ribose-5-phosphate
rRNA	ribosomal RNA
Rubisco	ribulose-1:5-bisphosphate carboxylase
S	Svedberg unit
S	'synthesis' phase of cell cycle
S	entropy
SAM	*S*-adenosylmethionine
SCNT	somatic cell-nuclear transfer
SECIS	selenocysteine insertion sequence
SDS	sodium dodecylsulphate
SH2	Src homology region 2
SINES	short interspersed repeated sequences
siRNA	small interfering RNA
Sn	stereospecific numbering
Snp	single nucleotide polymorphism
snRNAP	small nuclear RNA particles
SnoRNA	small nucleolar RNA
SOCS	suppressors of cytokine signalling
SOS	'son of sevenless'
SR	sarcoplasmic reticulum
SRP	signal-recognition particle
SSB	single-strand binding protein

STAT	signal transducer and activator of transcription
STR	short tandem repeats (microsatellites)
T	thymine
T_3	triiodothyronine
T_4	thyroxine
TAF	TBP-associated factor
TAG	triacylglycerol
TBP	TATA-box-binding protein
TCA	tricarboxylic acid
TCR	T cell receptor
TF	transcription factor
TFIID	transcriptional factor D for polymerase II
TIM	translocator of the inner mitochondrial membrane
TNF-α	tumour necrosis factor-α
TOF	time of flight
TOM	translocator of the outer mitochondrial membrane
TPA (t-pa)	tissue plasminogen activator
TPP	thiamin pyrophosphate
TRNA	transfer RNA
$tRNA^{phe}$	tRNA specific for phenylalanine (by analogy, $tRNA^{Leu}$, $tRNA^{Met}$, etc.)
$tRNA_f$	tRNA formyl (bacterial translation initiation)
$tRNA_i$	tRNA for eukaryote translation initiation
TSH	thyroid-stimulating hormone
U	uracil
UDP	uridine diphosphate
UDPG	uridine diphosphoglucose
UDP-Gal	uridine diphosphogalactose
UTP	uridine triphosphate
UQ	ubiquinone (*see also* CoQ, Q)
UTR	untranslated region (of mRNA)
UV	ultraviolet (light)
v-	'viral', denotes oncogene (v-*ras*, v-*myc*, etc.)
V	velocity of reaction
V_{MAX}	maximum velocity of reaction
VLDL	very-low-density lipoprotein
VNTR	variable number of tandem repeats
X-5-P	xylulose-5-phosphate
YAC	yeast artificial chromosome

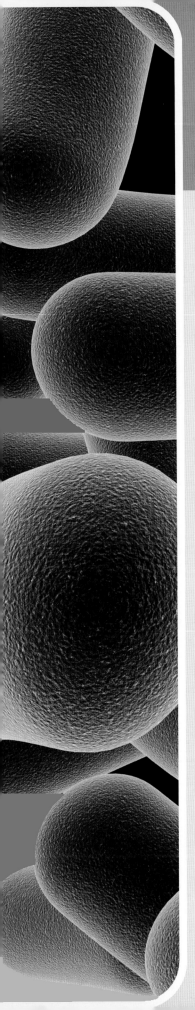

Part 1

Basic concepts of life

Tyler Boyes / iStockphoto

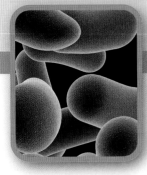

Chapter 1

The basic molecular themes of life

Living systems have the seemingly almost magical property of being able to reproduce themselves by a process of self-assembly from nonliving materials in the environment. Life is a molecular process in which ordinary chemical compounds are able to achieve this. It is a complex process in detail but is based on a few basic molecular themes common to all life forms on this planet.

A reviewer commented that all writers of biochemical texts face the problem that everything should come first because most aspects are interdependent. This chapter is as near as we could get to achieving this. It gives a preliminary survey of the concepts and of the laws of the universe that determined them. All of the topics discussed here are dealt with at greater length in later chapters to which page references are given. It is hoped that this preliminary survey will help students to understand where each topic fits into the overall picture as they come to them in more detail in later chapters.

Biochemistry and **molecular biology** are the disciplines concerned with understanding the mechanism of life in molecular terms. Biochemistry is the name for the earliest studied aspects in which the metabolism of food and small molecules is a principle focus. Molecular biology started later in about the 1950s and was the name given to the study of biological macromolecules, particularly proteins and DNA, and how these function in the genetic mechanism. The distinction between biochemistry and molecular biology has become blurred because the two are interdependent, using much the same techniques, but the terms are still used as convenient broad labels. Many biochemical departments and biochemical societies have added 'molecular biology' to their titles. The joint subject is of ever increasing importance in medicine, agriculture and all aspects of biology. It is an exciting subject with seemingly never-ending discoveries of molecular mechanisms by which life operates. Research progresses at an exhilarating pace in almost all areas of biological science; the medical potential of discoveries at the molecular level has given rise to the biotechnology boom.

All life forms are similar at the molecular level

All living organisms at the molecular level have a basic unity; whether this will extend to life on Mars or wherever it might be discovered is a fascinating question since the same physical and chemical laws that have governed the development of life on earth apply throughout the universe. Little green people might have molecular mechanisms similar in concept to those of life here, however different they may be in external form.

To return to earth, a famous dictum of the French Nobel prize winner, Jacques Monod, is that 'what holds for the Coli bacterium is true for an elephant', meaning that the similarities between a bacterium living in the human gut and an elephant far exceed the differences between the two organisms when viewed at the molecular level.

A mass of evidence shows that all current life had a single origin. Initially life, or its precursor, must have been extremely simple but it would have had to involve a molecule with the capability of directing its own replication and of determining the characteristics of 'offspring' of the replicating unit. This is where **nucleic acids** come into the picture. **DNA** and its close relative, **RNA**, are nucleic acids. DNA is the basis of all cellular life in that it carries the information necessary for reproduction. It carries the 'genetic code' that determines the characteristics of offspring. RNA has the same role in certain viruses. The essential structure of DNA and RNA, lends itself to self-directed replication, as explained later. It is believed that life originated with RNA as the 'genetic' material; this has been replaced in all cellular life with DNA. DNA is chemically more stable and therefore more suitable for storing genetic information. But, as explained in Chapter 25, RNA may be more important for life today than hitherto believed. At the origin of life, once the primordial form of a self-replicating cell-like 'unit' had been developed, many of the fundamental biochemical processes must have already been established and life was locked into these. This has given all life forms that evolved from this initial replicating unit their basic unity.

Mistakes in replication of the DNA are inevitable; small random variations continually occur in the form of mutations, so that offspring have slight variations from the parents. Any variation that increases the chances of the organism reproducing itself is preserved and any that reduces it is eliminated by natural selection. This leads to the evolution of new life forms better adapted to the environment, but their underlying molecular processes remain much the same.

This explains why, in biochemical research, a variety of organisms is often used to elucidate a given biochemical process. To understand how a process works, for example in humans, the best strategy may be first to study the bacterium, *Escherichia coli*, or a virus where the basic information might be more easily obtained from these simpler systems. The yeast cell is a eukaryote and has become an important model system for investigating human diseases such as cancer. There are differences in molecular processes between bacterial and human cells but they are more matters of detail rather than of principle. Most biochemical knowledge is applicable to all life forms.

The energy cycle in life

Living cells obey the laws of physics and chemistry. To grow and reproduce, cells take in simple molecules such as sugars, nitrogenous and other compounds from the external medium and build them up into the large organized complex molecules of cells. The synthesis of complex molecules from simpler ones involves an increase in energy content of the cell so chemical work must be done (Chapter 3). A living cell is at a higher energy level than the random collection of molecules in its external environment from which it was assembled. It is far from being in thermodynamic (energetic) equilibrium with its surroundings; this is achieved only by decomposition after the death of the cell. It is somewhat analogous to a flying aeroplane; the high gravitational potential energy state is maintained by constant fuel oxidation, which releases energy. If this stops, the aeroplane crashes to a minimum energy state on the ground.

Most cells get their fuel from food molecules, in which the energy initially came from the sun and was converted into chemical energy in the form of sugars by photosynthetic plants. For some organisms such as bacteria living around hydrothermal vents in the ocean the 'food' that supplies their energy are chemicals such as hydrogen sulphide from the earth's crust. They are known as **chemotrophs**.

The laws of thermodynamics deal with energy

Thermodynamics is the study of energy transformations. It is a daunting subject to some, or many, students partly because it usually has to deal with systems in which temperature and pressure changes are mechanistically involved, such as steam engines. This complicates the subject. However, living organisms are constant temperature and pressure systems. They do not depend on temperature and pressure gradients within the organism. This greatly simplifies thermodynamics as applied to life processes and indeed involves little more than common sense. It is probably true for example that most biochemists seldom use a thermodynamic equation but they need to understand the simple concepts presented here and in Chapter 3.

The **first law of thermodynamics** states that energy can be neither created nor destroyed – the total energy content of the universe remains constant. We all believe we know what energy is but it is difficult to define. A useful concept is that it is what makes things happen. Although energy cannot be destroyed, not all energy is capable of being useful in the sense of performing work. When you burn fuel, the liberated heat may be made to perform work as in a steam engine but part of it becomes heat that cannot be used to drive the engine. The heat in a hot car engine block is still energy but it cannot be used to propel the car. An important concept is that if you consider the total amount of energy released by any process, such as the oxidation of food, only part of it can be used to do useful work. The rest increases the total **entropy** of the universe. Entropy is the degree of randomness in any system or as was once expressed, the degree of mixedupness. A high entropy equals relatively low energy and a low entropy relatively high energy. Heat increases the random motion of molecules and hence increases entropy. When a molecule breaks down into smaller ones or anything increases the number of particles in a system this also increases entropy. To anyone unfamiliar with the concept it will probably seem odd that the vitally important **second law of thermodynamics** specifies that all processes must increase the total entropy of the universe. It seems, however, that increasing entropy is a major driving force in the universe and all processes must contribute to it. That is why no process can ever be energetically 100% efficient. It would seem that the ultimate fate is a dark, silent universe of infinite entropy, and maximum stability where nothing whatsoever can happen. The universe appears to have a relentless 'drive' to achieve this state.

At first sight, living cells appear to be magical and defying the second law. Cells reproduce by dividing into two after doubling in size. They have to convert small randomly arranged compounds of high entropy from the environment into the large highly organized structures of life (low entropy). They thus might appear to be unique islands in the universe where the drive to increasing randomness is reversed and the second law defied, since a living cell is at a lower entropy and higher energy level than the starting materials from which it was built. The answer to the paradox is that in oxidizing foodstuff molecules to release energy to drive the process, there is a large increase in entropy due to the liberation of heat and the breakdown of molecules to smaller ones such as CO_2, which escape into the cell surroundings. This entropy increase exceeds the decrease in entropy that occurs in the production of cells, so that if we consider the cell plus its surroundings there is a net increase in total entropy and the second law is obeyed.

Energy can be transformed from one state to another

A familiar example is the conversion of **kinetic energy** (movement) into heat. The kinetic energy of a rock crashing to the ground is converted into heat, by friction on the way down, and completely when it hits the bottom. There is **potential energy**; a rock perched up on the cliff has **gravitational potential energy**, which converts to kinetic energy when it falls. There is also **chemical potential energy**. Each molecule has a certain amount built into it depending on its structure. The molecules of the food that you eat are rich in potential chemical energy, while molecules such as H_2O and CO_2 have, in this context, none. When glucose is oxidized to carbon dioxide and water, its potential energy is released. Food is taken in by organisms where it is oxidized back to carbon dioxide and water and the energy so released is used to drive all the reactions of a living cell. This is summarized in the energy cycle of life shown in Fig. 1.1.

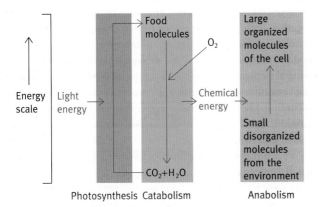

Fig. 1.1 The energy cycle in life. Catabolism is the breakdown of complex molecules releasing energy in the cell. Anabolism is the energy-requiring transformation of simple molecules into more complex ones.

ATP (adenosine triphosphate) is the universal energy currency in life

As already emphasized, so far as life is concerned, heat from food oxidation is 'waste' energy; it keeps you warm but that is different from it doing work. It has, however, the important role of increasing the entropy of the universe as required by the second law. Life is a chemical process and cells must harness the energy in a form of chemical energy. There are several different classes of food molecules to be oxidized – carbohydrates, fats, proteins, and a variety of other compounds and even alcohol. There are also different uses to which the energy must be coupled – chemical work, osmotic work, mechanical work, electrical work. It would be impossibly complicated to inflexibly link all the different type of foodstuff oxidations to all the individual uses of the energy.

This problem has been solved by an ingeniously simple concept. All processes releasing energy from all food molecules trap it in a single compound, **adenosine triphosphate** (**ATP**), shown in diagrammatic form in Fig. 3.3 (page 33). With trivial exceptions, all processes needing energy use ATP to supply it. ATP is the universal energy currency of life. To give a simple illustration, when you contract a muscle, ATP breaks down to adenosine diphosphate (ADP) and releases inorganic phosphate. To state an elementary point that may be obvious, for ATP to supply energy to a process, its breakdown must, in some manner, be tightly coupled to the process. If ATP were simply hydrolysed to ADP and phosphate, the energy would simply be released as heat. You will, in this book, come across a great many examples of the ways in which enzymes couple ATP breakdown to the performance of work. The breakdown supplies the requisite energy for muscular activity, for example, by the mechanism described in Chapter 8 (page 130). Food breakdown processes then immediately jump into action and replenish the ATP reserve by resynthesizing it from ADP and phosphate. This one molecule is the immediate source of energy that drives all living creatures. Dinosaurs roamed on it, electric fishes generate electricity on it, fireflies flash on it (to emphasize its universal role). There is nothing magical about ATP; it is an ordinary chemical with a structure that allows its remarkable role (Chapter 3, page 34). It is obtainable as a white powder that can be stored in the freezer. ATP cannot penetrate into cells, so they must generate it themselves. It cannot be supplied to them from the outside.

Trapping the energy released by food oxidation rather than dissipating it as waste heat is a complex task in detail and is a subject that occupies a large proportion of Part 3 of this book.

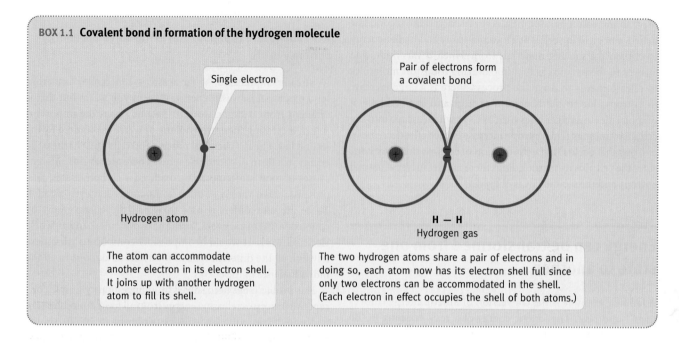

BOX 1.1 Covalent bond in formation of the hydrogen molecule

Single electron

Pair of electrons form a covalent bond

Hydrogen atom

H — H
Hydrogen gas

The atom can accommodate another electron in its electron shell. It joins up with another hydrogen atom to fill its shell.

The two hydrogen atoms share a pair of electrons and in doing so, each atom now has its electron shell full since only two electrons can be accommodated in the shell. (Each electron in effect occupies the shell of both atoms.)

Types of molecules found in living cells

Molecules are formed by the chemical bonding of atoms. There are several types of chemical bonds (Chapter 3, page 36); the most familiar and the strongest one is the **covalent bond**. It is formed by two atoms sharing a pair of electrons, a simple example being the formation of a hydrogen molecule from two hydrogen atoms (see Box 1.1). Each electron is attracted to the positive nucleus of both atoms and this holds them together. Hydrogen atoms react in this way to form molecules so that the proportion of hydrogen atoms in hydrogen gas is negligible. The reaction is energetically strongly downhill; energy is liberated as heat. Chemical systems tend to achieve the lowest possible energy state – to achieve the lowest chemical potential energy. The lower this is, the less reactive chemicals are – they are more stable.

Atoms are maximally stable when their outer electron shells are filled, as is the case with the inert noble gases such as helium, neon, argon and krypton. Chemical reactions appear to achieve atomic structures as close as possible to those of the noble gases closest to them in the periodic table. They might be regarded as part of the tendency of the universe to achieve the most stable state and maximum entropy.

Most biochemical reactions do not involve free atoms but rather rearrangements or exchange of chemical groupings between or within molecules. The same energetic principle applies in that a chemical reaction can occur only if it liberates energy and increases entropy as required by the second law of thermodynamics as described.

Biological molecules are based on the carbon atom bonded mainly to carbon, hydrogen, oxygen and nitrogen atoms. These four atoms constitute something like 99% of the total number

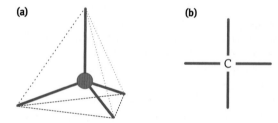

Fig. 1.2 (a) The tetrahedral arrangement of the four single bonds of the carbon atom. (b) As represented in structural formulae.

in the body. They all can form strong covalent bonds, the strength of these being inversely proportional to the weights of the atoms involved. The four bonds of carbon are tetrahedrally arranged allowing branching structures to form (Fig. 1.2). This together with its ability to form long chains by C—C bonds enables it to form a wide variety of structures of different shapes and properties. In this respect carbon is unique among the 92 natural elements. Silicon can form chains but has nowhere near the versatility of carbon. Other elements are important in life, including phosphorous and sulphur, which are also strong covalent bond formers. Sodium, potassium, and calcium are also of great importance.

We arbitrarily divide cellular molecules into two categories, **small molecules** and **macromolecules.** Small molecules are things like glucose, fats and amino acids with molecular weights in the range of a few hundred Daltons or less. (A **Dalton** or Da is a unit of **atomic or molecular mass** formally defined as one twelfth of the mass of a carbon 12 atom, effectively equal to the mass of a hydrogen atom. The size of molecules is commonly referred to as **molecular weight**).

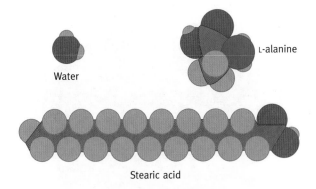

Fig. 1.3 Space-filling models of water, the amino acid L-alanine, and a lipid, stearic acid. The colours of the atoms are: carbon, dark grey; oxygen, red; hydrogen, blue; nitrogen, dark blue. The computer program that generates the models represents the size of the electron cloud of atoms, which is affected by the nature of their attached atoms. In the case of hydrogen with a single electron, the represented size is greatly reduced when attached to an electrophilic atom such as oxygen or nitrogen.

Small molecules

Water is the most prevalent of the small molecules, constituting about 70% of a typical cell. Although the water molecule has no charged groups and is overall electrically neutral it is a polar molecule (one that has opposing electrical charges). This is because the two hydrogen atoms are bonded to the oxygen at an angle as shown in Fig. 1.3, giving the molecule an asymmetric shape with the oxygen at one end and the hydrogens at the other. The oxygen atom is electronegative – it is greedy for electrons and takes more than its share from the O—H covalent bonds. The hydrogens therefore have **partial** positive charges and the oxygen a partial negative one giving an overall polarity to the molecule (see structure in Chapter 3, page 37).

Hydrogen bonds have a central role in life

The polarity of water has important repercussions because it allows the molecules to be linked by **hydrogen bonds** causing bulk water to have a cohesive structure, without which the world would be a different place. Hydrogen bonds are also important in cellular structures. They belong to the class of chemical bonds known as **noncovalent or weak secondary bonds**. Despite their description as 'weak' and 'secondary', these bonds lie at the heart of life processes. While covalent bonds give molecules stability, the molecules interact in life processes by noncovalent bonds, as will become evident. In water they are due to weak electrostatic attractions between the partial positive charges on the hydrogen atoms of water and the partial negative charge on the oxygen of adjacent water molecules. They can form between appropriate hydrogen atoms and oxygen or nitrogen atoms of other molecules (page 37) and play a vital role in biological structures. The genetic apparatus of all organisms is dependent on them.

Water is a good solvent for polar compounds

The polar nature of water makes it an excellent solvent for polar substances such as sugars. Molecules of sugars in solution become separated by weakly attached water molecules. Such soluble molecules are known as **hydrophilic** (water loving). Nonpolar **hydrophobic** (water hating) molecules such as benzene cannot form bonds with water and so do not dissolve (see page 38). The tendency of water to reject hydrophobic molecules has a profound effect on the structure of macromolecules such as proteins (page 51).

Water molecules in the pure state are only minutely dissociated into hydrogen ions and hydroxyl ions both being at 10^{-7} M.

Most of the rest of small molecules are monomers (single molecules) such as sugars, fats and amino acids mainly derived from foodstuffs. There are also thousands of other, different, molecules resulting from the chemical activities going on in the cells.

The sugars such as glucose and sucrose are called carbohydrates because they have the elements of carbon and water in equal proportions. They are important energy stores. Amino acids are short carbon chains with basic amino groups and acidic carboxyl groups (page 46). Lipids or fats have various roles, the two most prominent being the formation of cell membranes and as the major storage of energy in an animal. Fig. 1.3 shows molecular models of water, L-alanine (a typical amino acid) and stearic acid (a typical lipid), which has a long hydrocarbon chain.

Macromolecules are made by polymerization of smaller units

Glycogen, starch and **cellulose** are very large **polymers** known as **polysaccharides** formed by joining together glucose units in a slightly different manner in the three cases. Glycogen and starch are for energy storage in animals and plants respectively and cellulose is for structural strength in plants. Cellulose is broken down by bacteria in ruminants and is their major food source. It cannot be used by other animals. Only glucose monomers are involved in the synthesis of these polysaccharides (though more complex ones exist); all that is needed for their synthesis is a mechanism to link them together in the appropriate way. There is therefore no information content in them. They resemble long strings of a single letter.

Protein and nucleic acid molecules have information content

Proteins, DNA and RNA are macromolecules built up from a variety of monomers that must be linked together in the correct order in each specific molecule. They have information content based on the sequence. To achieve this, the cell must contain template instructions on the correct sequences for each protein, RNA, and DNA molecule. They should be compared with meaningful messages composed from alphabets.

The flow of information is DNA → RNA → Protein.

Proteins

It has been said that proteins may be the most remarkable molecules in the universe. The word protein is derived from the Greek meaning 'primary'; proteins are of primary importance in life; genes and the genetic system exist to make their production possible. As a generalized statement, proteins do everything in the mechanism of life. They are built up from a menu of 20 different species of amino acids, a large number of which are polymerized into long chains, known as **polypeptides** (Fig. 1.4). The DNA of the genes has coded information for the amino acid sequences for each protein, one gene coding for each polypeptide chain. Life depends on these sequences being correct. A single incorrect amino acid among hundreds or thousands, in a protein molecule (equivalent to one letter being incorrect in a word of that length), may result in a genetic disease. After synthesis, the chains fold up into three-dimensional compact shapes determined by the particular sequence of amino acids. Fig. 1.5 shows a space filling molecular model of human **deoxyhaemoglobin**, an average-sized protein of 574 amino acids and molecular weight of 64,500. Proteins range in size from the small insulin molecule (molecular weight 5733

Fig. 1.5 Space-filling model of haemoglobin. The CPK (Corey-Pauling-Koltun) colour scheme is used: carbon, light grey; oxygen, red; nitrogen, blue; sulphur, yellow. The Protein Data Bank accession code for haemoglobin is 1A3N.

Da), which is comprised of 51 amino acids linked together, to large ones of several thousand amino acids.

Catalysis of reactions by enzyme proteins is central to the existence of life

Enzymes are catalytic proteins. Thousands of different chemical reactions occur in a living cell even though the mild conditions there are not such that would facilitate chemical reactions – almost neutral pH, low temperature, no especially reactive substances and chemicals present in dilute aqueous solution. In the chemistry laboratory, reactions are commonly brought about by high temperatures, extreme pHs and high concentrations of reactants. A sugar such as glucose is stable at body temperature and left in air in a bowl will undergo no change for many years. If you eat the sugar, it is involved in rapid chemical reactions in the cell due to enzymes combining with the molecules and catalysing the reactions. Enzymes are specific protein catalysts – usually one enzyme, one reaction. Without the ability of proteins to bind precisely with their target molecules (in this case, enzyme substrates) and catalyse specific reactions, life would be impossible. Since there are thousands of different reactions in a cell there are thousands of different enzymes catalysing them. They are efficient catalysts. One molecule of the enzyme carbonic anhydrase, important in red blood cells (page 67), catalyses the conversion of 600,000 molecules of substrate per second. The mechanisms by which proteins are so catalytically efficient are described in Chapter 6, page 94.

Why are enzymes needed?

We have earlier described how the second law of thermodynamics determines whether a reaction may be permitted to

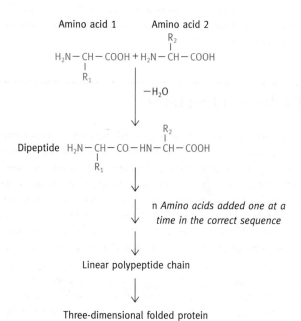

Amino acid 1   Amino acid 2

$$H_2N-\underset{R_1}{\overset{}{CH}}-COOH + H_2N-\underset{}{\overset{R_2}{CH}}-COOH$$

$$-H_2O$$

Dipeptide $\quad H_2N-\underset{R_1}{\overset{R_2}{CH}}-CO-HN-\underset{}{\overset{}{CH}}-COOH$

n Amino acids added one at a time in the correct sequence

Linear polypeptide chain

Three-dimensional folded protein

Fig. 1.4 Outline of protein synthesis. Note that although peptide synthesis involves overall the removal of a water molecule, the process in the cell is not a direct condensation. Protein synthesis is carried out by cellular bodies called ribosomes. The sequence of amino acids added to form the polypeptide chain is specified by a molecule of messenger RNA, which is a copy of the base sequence of the gene coding for the protein.

proceed, not whether a given reaction *will* actually take place. This is a separate question involving the nature of chemical reactions. There is a barrier to chemical reactions occurring. If this were not so everything that could react would have done so long ago. Everything combustible would burst into flames in the presence of oxygen were it not for this barrier. Enzymes cannot change the thermodynamics of a reaction – they cannot tinker with the laws of thermodynamics. But they can lower the barrier to chemical reactions. Enzyme catalysis is one of the wonders of life. It is explained in Chapter 6.

Proteins work by molecular recognition

As stated, to act catalytically on specific substrates, enzymes must 'recognize' their target molecule and combine with it. They do this by virtue of noncovalent bonds between the protein and the target molecule. This can be exquisitely specific (page 87). The ability of proteins to recognize other molecules is central to almost everything in life. Cellular structures, muscle contraction, nerve impulses, hormone action, chemical signalling, and regulation of metabolism all depend on it. Proteins are very versatile, ranging from delicate enzymes and exquisite molecular machines to the tough proteins of cartilage, hair, and horses' hooves.

Life is self-assembling due to molecular recognition by proteins

Life is based on a one-dimensional linear coding system (the DNA) and yet the information is translated into three-dimensional organisms. The linear code initially is translated into a linear, one-dimensional polypeptide chains but these fold up (page 48) to form three-dimensional proteins which have uniquely specific recognition sites on them (page 87). The folding is determined by the amino acid sequences of the polypeptides, which themselves are determined by nucleotide sequences in the genes. The specific abilities of proteins to recognize other molecules is therefore determined by evolution of genes, which therefore determines the automatic assembling of all living organisms. If there are rabbit genes, rabbit proteins are produced and a rabbit will assemble itself during embryonic development from a fertilized egg because of a multitude of molecular recognitions between those proteins. The proteins, as it were, click together as evolution has designed them to do.

Many proteins are molecular machines

We come now to one of the most remarkable properties of many or most proteins. They can undergo microscopic conformational changes in their three-dimensional shape on combining with their cognate molecules. (The term 'cognate' means 'having affinity to'. A cognate molecule is one that it is

appropriate for the protein to bind to.) When most or all enzymes bind to their substrate they change very slightly; it is part of the catalytic mechanism (Chapter 6). Also, enzymes must be regulated in their activities. You need to metabolize fuels more rapidly during exercise than when asleep. For example, key control enzymes change their shape on detecting that ATP supplies are adequate and switch off the oxidation of food (Chapter 12). Muscle contraction depends on vast numbers of protein molecules undergoing conformational change and interaction with protein fibres. Other proteins literally walk along special protein tracks in the cell pulling loads. They are molecular motors (Chapter 8). Haemoglobin does not just carry oxygen; it responds to signals so as to maximize oxygen pick up in the lungs and surrendering of it in the tissues (Chapter 4, page 42). A vast network of cell–cell communication regulates cell activities. This is dependant on protein conformational changes (Chapter 27). Gene control (Chapter 23), cell division control (Chapter 31), in fact just about everything depends on it.

How can one type of molecule do so many tasks?

If we regard amino acids as an alphabet of 20 letters, proteins are, as explained, typically 'words' several hundred or more letters long, the number of theoretically possible different amino acid sequences is virtually infinite. There is no theoretical limitation to the number of different polypeptide chains that evolution can 'try out'. The sequences of the proteins existing today have been evolved over billions of years of random mutation and natural selection and stored in the genes. It is a colossal information storage process.

Evolution of proteins

As stated, evolution is a process in which natural selection preserves those mutations that increase the chance of progeny reaching reproductive age. Deleterious mutations are eliminated. Since genes code for proteins, evolution depends on the synthesis of new proteins that give a selective advantage. Important recent discoveries of genes that do not code for proteins (Chapter 25) but which probably affect the expression of protein-coding genes, mean that this statement has to be qualified somewhat, but its essential correctness remains. The chances of random changes in the amino acid sequence of a protein being advantageous are small so that evolution is a slow, chancy business, but vast timescales are involved. Natural selection is infallibly efficient in assessing the 'experiments'.

Development of new genes

The evolution of proteins requires the development of new genes. The problem of how you can change an essential gene into a different one without eliminating the function of the original one can often be explained by accidental gene duplication. One of the duplicates can be mutated into new genes that results in new proteins, while the other continues to code for the original protein. There is much evidence in the base sequences of genes indicating that this is what often happened; families of related genes exist that have evolved from common ancestors (see Chapter 4, page 54).

DNA (deoxyribonucleic acid)

It was established in the 1940s that DNA (**deoxyribonucleic acid**) is the substance of genes. A complete DNA molecule is a **chromosome**, with protein components present as structural support. The protein–DNA complex is called **chromatin**. The *E. coli* chromosome has a molecular weight of 12 million daltons and the largest human chromosome is several billion daltons. As already explained, the DNA carries the chemical message that signals to the cell how to assemble the amino acids in the correct sequence to produce the protein for which each gene is 'responsible'. The information is contained in the sequence of the monomers called **nucleotides**, which make up DNA. A nucleotide has the structure:

base – sugar – phosphate

There are four different nucleotides in DNA, differing in the base components linked together forming a 'backbone' of alternating sugar-phosphate residues, with the bases projecting from the sugar residues. The sequence of different bases carries the information of the gene.

DNA exists in the form of a double strand held together by secondary bonds of which hydrogen bonds are critical, as illustrated in Fig. 1.6, Two of the four species of bases in DNA, **adenine** and **thymine** (A and T) automatically hydrogen bond together because their shapes are complementary. The same is true for the other pair, **guanine** and **cytosine** (G and C). This pairing is known as **complementarity** or **Watson–Crick base pairing**, after its discoverers. It is specific; base pairing in this way in the double helix occurs only between G and C, and A and T respectively. The two strands in a DNA molecule are not parallel as indicated in Fig 1.6 for simplicity, but rather wind around each other to form the well-known double helix, shown more realistically in the space-filling model shown in Fig. 1.7. Chapter 21 deals with DNA structure.

DNA directs its own replication

The central requirement of any genetic system is that the hereditary information can be replicated and passed on to daughter cells. The reason nucleic acids carry the genetic information is that they have the capacity to direct their own replication as well as performing their function of directing protein synthesis.

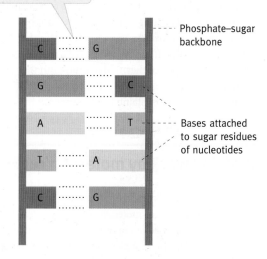

Noncovalent (hydrogen) bonds are critical to holding the two chains together.

Phosphate–sugar backbone

Bases attached to sugar residues of nucleotides

Fig. 1.6 Diagram of the structure of double-stranded DNA. The backbone consists of alternating sugar–phosphate residues to which the four types of base are attached. The base pairs are always between G and C or between A and T. Note that each base pair always includes one larger and one smaller base so that all base pairs are of the same size. The two chains are held together by noncovalent bonds (page 3); there are three between G and C, and two between A and T. The two strands are shown as being parallel for clarity but in fact they form a double helix as shown in the model in Fig. 1.7.

The two strands of DNA are separated and each strand acts as a template for the assembly of new partner strands. An A on the template strand matches a T on the new strand, G is matched to a C and *vice versa*. Fig. 1.8 illustrates the principle of this. This results in two new double helices identical to the original one. The synthesis is discussed in Chapter 22.

Genetic code

Triplets of bases in the DNA known as codons specify amino acids of polypeptide chains. The 'dictionary' of which triplet represents which amino acid is known as the **genetic code**. With 4 different bases, 64 different triplet combinations (**codons**) are possible $(4 \times 4 \times 4)$.

The DNA of the human genome (the complete collection of **genes**) contains 3.2 billion nucleotide pairs, The complete sequencing of these has been achieved by the **Human Genome Project** completed in 2003.

Genes are part of the continuous threadlike chromosomal DNA molecule. Each gene is distinct from the next, separated by spacer sequences between them.

Proteins are synthesized on cellular structures known as **ribosomes**. These take instructions (indirectly) from the gene.

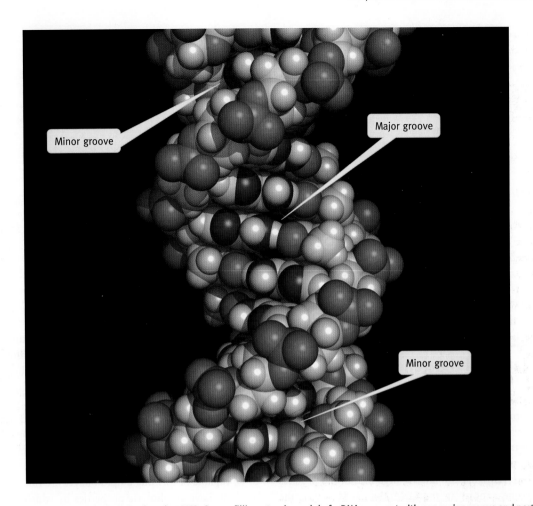

Fig. 1.7 A model of B DNA, Protein Data Bank code 1BNA. Space-filling atomic model of a DNA segment with one major groove and part of two minor grooves.

	Double-stranded, parental DNA, the two strands held together by Watson–Crick base pairing.
	The two strands are progressively separated by breaking the base pairing. The separated sections act as templates for synthesis of new strands.
	Nucleotide monomers from solution automatically base pair with the separated parent strands and a DNA polymerase enzyme joins them together forming new strands.
	The process continues to the end of the piece of DNA thus producing two identical copies of the original parent DNA with the same base sequences. Note that each copy has one strand from the parent DNA and one newly synthesized strand. This is known as **semi-conservative replication**.

Each gene is independently copied into a different relatively short nucleic acid called **messenger RNA (mRNA)**, which delivers the message of coded instructions from the gene to the ribosome. There is a separate mRNA for each protein. RNA has almost the same structure as a single strand of DNA with small chemical differences. The bases in RNA specifically base pair much as in DNA. The sequence of information flow is as shown below.

(+ RNA monomers)   (+ Amino acids)
         ↓   ↓
DNA of gene 1 → messenger RNA 1 → polypeptide 1 → folded protein 1
DNA of gene 2 → messenger RNA 2 → polypeptide 2 → folded protein 2

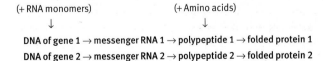

Fig. 1.8 (*opposite*) Principle of DNA replication. The two strands of the double helix are held together by hydrogen bonds between bases A and T, and G and C respectively. When the strands are separated the single strands are now available for base pairing by incoming monomer nucleotides. The nucleotides thus lined up are linked together to give two identical daughter double helices.

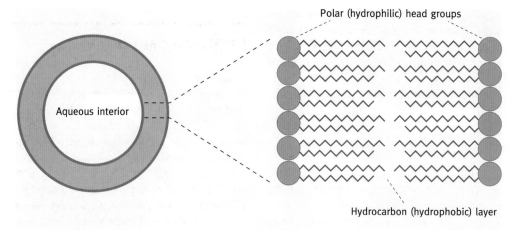

Fig. 1.9 A synthetic liposome made of a lipid bilayer structure.

Organization of the genome

The completion of the monumental task of sequencing the human genome delivered some surprises. The genes are scattered about within the linear DNA thread and over half the DNA is seemingly useless junk that couldn't be discarded and was given the name of 'junk DNA'. Bacteria have very little junk DNA and look more orderly in their organization, with genes adjacent to one another in the DNA thread. However, the assessment of junk DNA has been thrown into turmoil by the discovery in it of large numbers of noncoding **microRNA genes**, which have been conserved over long periods of evolution. They code for short **microRNAs** that are not for protein-coding purposes. It looks possible or probable that ideas on genetic inheritance and genome function will have to be modified. It also has revealed a new process of gene control by microRNAs, which promises to be one of the great medical advances. This area will be dealt with in Chapter 25.

How did life start?

Living organisms consist of one or more cells. Each cell is surrounded by a cell membrane, a thin sheet composed mainly of lipid (fatty) molecules that is necessary to hold the contents of the cells together (along with other functions).

As already stated, at some time in the establishment of life there must have been a primordial self-replicating molecular system from which living cells developed. Hypotheses have been formulated of how such systems might have been established on a mineral surface or in a drop of liquid or sea pool, but, at an early stage, it had to be contained by a membrane or it

presumably would have been dispersed. A striking fact is that when molecules of a suitable substance are simply agitated in water they form small spherical vesicles (Chapter 7). The boundary of these vesicles is made of a structure known as a **lipid bilayer,** which is virtually identical to the basic structure found in the membranes of modern cells (Fig. 1.9). Such vesicles may have enclosed a drop of the first self-replicating system. From such a primordial cell-like structure all life is postulated to have originated. The requirements for a molecule to be capable of forming a lipid bilayer are not demanding; it needs to have amphoteric properties by which we mean one part of a molecule is water-insoluble (hydrophobic) and the other water-soluble (hydrophilic) and of a roughly suitable shape as illustrated in Fig. 1.10.

What was the source of the molecular building blocks needed to produce the components of living cells? Experiments have been done in which electrical discharges were passed through a mixture of gases (hydrogen, methane and ammonia, in the presence of water) intended to resemble the atmosphere believed to exist on the primitive earth, as suggested by geologists and astronomers. A mixture of potential precursors of

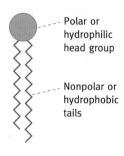

Fig. 1.10 An amphipathic molecule of the type found in cell membranes.

biomolecules including some amino acids was produced. The postulated primordial self-replicating cell must have taken in molecules from the environment to produce new cellular material. Diffusion through the containing membrane before the development of transport mechanisms would have been slow, and replication likewise slow, but vast time scales were involved.

The RNA world

A more difficult problem in the establishment of a self-replicating system is to identify the initial catalysts and the primitive 'genetic system' to ensure faithful replication. In short a chicken and egg problem; which came first, proteins to catalyse reactions or nucleic acids to direct the synthesis of primitive proteins? This dilemma received a possible answer with the discovery that RNA can catalyse some chemical reactions including conversion of short polynucleotides into longer sequences. Such molecules were given the name of 'ribozymes' (not to be confused with ribosomes). It was the first time that biological molecules other than proteins had been found to catalyse specific reactions. RNA has the same potentiality for acting as a template in its own replication as explained for DNA. RNA was probably both the catalyst and the primitive 'genetic system' for self-replication in the origin of life, thus avoiding the chicken and egg dilemma. It may be speculated that the first short polynucleotides were formed from nucleotide monomers by heat chemically condensing the nucleotides together by driving off water.

From this stage, evolution of more efficient catalysts, namely proteins, to replace RNA catalysts are postulated to have occurred, though the first 'proteins' must have been primitive and presumably were short peptides of low catalytic efficiency. The concept of an RNA-based biological world that preceded the DNA world is generally accepted for there is much supporting evidence. In modern cells, although protein enzymes bring about almost all catalysed reactions, the displacement of RNA from this role is not complete. What might be regarded as a few fossil catalysts – hangovers from the RNA world – exist in cells as ribozymes. One of these in ribosomes is involved in the synthesis of all proteins, providing an interesting link between one type of catalytic system (RNA) to a more efficient one (proteins). Ribosomes are giving us a glimpse into the ancient RNA world, somewhat akin to astronomers viewing the past universe through long-distance telescopes. Moreover, as already alluded to, the status of RNA has recently escalated (Chapter 25).

Proteomics and genomics

From what has been said in this chapter, it will be clear that sequences of amino acids in proteins and those of the nucleotides in DNA underlie just about everything in life. As these sequences were determined, it was realized that the flood of molecular information would be of little avail without an efficient retrieval system. To this end, protein and DNA computer databases were established in various centres around the world in which information on proteins and genes is recorded. Details of the sequences of thousands of genes and proteins together with the three-dimensional structures of many proteins are now available. Software in the public domain is available to search the databases and analyse the information contained in them. This area of science is known as **bioinformatics**, which has become of immense importance in biochemistry and molecular biology. These developments have made computers with their remarkable software programs essential tools in molecular research.

Parallel to this there have been developments of methods for the automatic sequencing of DNA that have resulted in the completion of the human genome project and the sequencing of the genomes of other species, such as those of the mouse, the rice plant, and Drosophila, the fruit fly, to cite only a few. Other technical developments, using DNA microarrays (Chapter 28), allow the simultaneous study of which genes are active. In the protein field, the relatively recent application of mass spectrometry to proteins, a development of immense importance, makes it feasible to investigate many proteins at once.

These developments are sometimes referred to informally as the 'omics revolution'. The entire collection of proteins in a cell (in any one state, for it varies from time to time) is called the **proteome** and that of genes, the **genome.** The collective studies of these are called **proteomics** and **genomics** respectively. The essential aspect is that large numbers of proteins are studied simultaneously and similarly for genes. An apt analogy has been put forward to illustrate the meaning of these terms to the effect that many of the instruments in an orchestra (genes, proteins and metabolites) have been identified. The next stage is to listen to the music the orchestra plays with them. In other words, the proteins and genes in a cell function as a collective whole and a full understanding of life and abnormalities will need to consider them as such and their interactions to be understood.

The collective study of the copying of genes into messenger RNA could have important practical uses. As a simple illustrative example, they make it possible to compare the collection of proteins and gene activities for example in cancer cells compared with their normal neighbours. The differences between them could give leads on what has caused the cancer and possibly suggest therapeutic strategies.

Summary

Unity of life

Despite the diversity of life forms, at the molecular level all life is basically the same, variations being modifications of the same theme. It suggests a single origin of life.

Living cells obey the laws of physics and chemistry

Energy is derived from breaking down food molecules (ultimately produced by plants using sunlight energy). The energy must be released in a form that can drive chemical and other work. Heat cannot do work in the cell.

ATP is the universal energy currency in life

The energy is used to synthesize ATP (adenosine triphosphate) from ADP (adenosine diphosphate) and phosphate, ATP breakdown is coupled to biochemical work.

Molecules found in living cells

These include small molecules such as water, food molecules and their breakdown products. Macromolecules, among which proteins and DNA are pre-eminent, are large molecules formed by polymerization of smaller units.

Proteins

These are the cell's workhorses and the basis of most living structures. They are long chains of amino acids, typically hundreds long, but folded up into a precise three-dimensional structure. There are 20 different amino acids in proteins; each protein is a unique sequence of these.

Enzyme catalysis

Enzymes are proteins that catalyse virtually all the thousands of chemical reactions of life One enzyme, one reaction. Relatively recently, however, it has been discovered that RNA (ribonucleic acid) can have catalytic activity.

DNA

The cell must have a blueprint of the sequence of each of the thousands of proteins it synthesises. This is the function of DNA in the form of genes, each gene specifying the amino sequence of one polypeptide.

DNA consists of two strands of polynucleotides in a double helix. A nucleotide has the structure: base–sugar–phosphate. The bases are paired by hydrogen bonds, the base A linked to T, and G to C. This automatic pairing is the basis of self-directed replication. The base sequences act as a code, specifying individual amino acids, three bases, known as codons each representing an amino acid. The genetic code is the table correlating codons to the amino acids they specify (though the term is commonly used in popular reporting to mean the total genetic information of an organism).

Ribosomes translate the base sequences of genes into proteins. The mechanism of this is that each gene is copied into messenger RNA (a polymer resembling DNA), which attaches to and instructs a ribosome.

Evolution of genes and proteins

DNA is the record of sequences needed to synthesize proteins, the information having been acquired by billions of years of evolution. Mistakes in replicating DNA inevitably occur. These are mutations that result in faulty amino acid sequences in proteins, which may in turn result in genetic diseases. The random mutations are also the material on which evolution, *via* natural selection, develops new genes.

Molecular recognition by proteins

Apart from recognition of substrates by enzymes, proteins recognize (bind to) other molecules such as hormones and growth factors, thus directing development, growth and metabolic processes. The binding is by multiple weak bonds whose formation depends on atoms being close enough for the bonds to form. This means that only molecules closely complementary to one another bind. It is the basis of biological specificity. The use of weak bonds in molecular recognition confers flexibility and reversibility.

How did it all start?

An RNA world is believed to have preceded DNA and proteins. Life presumably must have originated by the spontaneous formation of a molecule capable of self-replication without the aid of proteins. It is generally believed that life originated with RNA, which has the information to direct its own replication. The RNA world is still seen in the genes of some viruses, and in all cells in the form of ribosomes, which have a high content of RNA. DNA replaced RNA because it is a chemically more stable repository of genetic information.

The new 'omics' phase of biochemistry and molecular biology

In the past decade an explosion of new technologies has revolutionized biochemistry and molecular biology. Prominent among these are automated DNA sequencing, mass spectrometry for the study of proteins and DNA micro arrays for gene studies. They are having enormous effects on biological science, medicine and agriculture. The branches of science using these are described as proteomics, genomics, metabolomics, which are collective terms to specify that large numbers of proteins, genes and metabolites respectively, can be examined together.

Further reading

Cech, T. R. (1986). A model for the RNA-catalysed replication of RNA. *Proc. Natl. Acad. Sci. U.S.A.*, **83**, 4360–3. *Describes the formation of polycytidylate.*

Gilbert, W. (1986). The RNA world. *Nature*, **319**, 618.

Joyce, G. F. (1989). RNA evolution and the origins of life. *Nature*, **338**, 217–24.

Orgel, L. E. (1994). The origin of life on earth. *Sci. Am.*, **271**(4), 52–61. *Growing evidence supports the idea that the emergence of catalytic RNA was a crucial early step.*

Lafcano, A. and Miller, S. L. (1996). The origin and early evolution of life: prebiotic chemistry, the pre-RNA world and time. *Cell*, **85**, 793–8.

Junk DNA and microRNA genes

Gibbs, W. W. (2003). Hidden genes. *Sci. Am.*, **289**, 28–33.

Mattick, J. S. (2003). Challenging the dogma: the hidden layer of nonprotein-coding RNAs in complex organisms. *BioEssays*, **25**, 930–9.

Problems

1   How is the entropy level related to the energy level of a system?

2   What is energy?

3   List the various forms of energy.

4   What do we mean by saying that proteins and DNA have information content?

5   What is a messenger RNA in broad terms?

6   Why were nucleic acids 'chosen' to be the basis of life?

Chapter 2

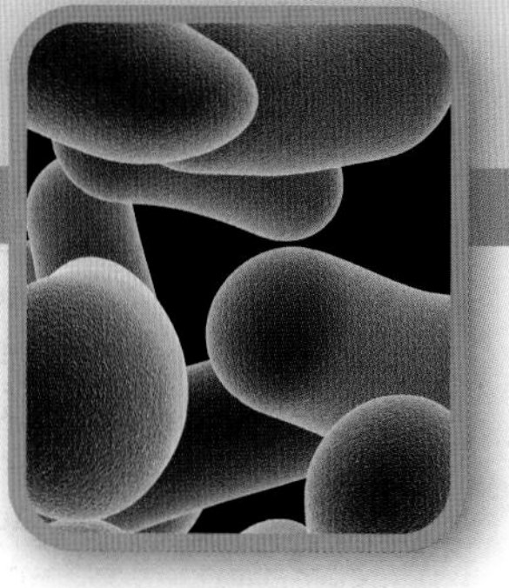

Cells and viruses

This chapter offers a broad survey of the structures and properties of cells and viruses, the aim being to provide a biological background for the more detailed mechanisms of cell biology, biochemistry and molecular biology that follow in subsequent chapters. Viruses, it should be noted, are not living organisms in the sense that cells are. They are incapable of replication themselves and have virtually no biochemical activities of their own. However they possess genomes and have the ability to enter living cells, where they cause the host's molecular biology machinery to replicate them, as a result of which they cause diseases. We will deal first with cellular life and then with viruses.

Cells are the units of all living systems

All living organisms consist of one or more cells and, with rare exceptions, cells are microscopic in size. Bacteria are the smallest; *Escherichia coli (E. coli)*, is rod-shaped with dimensions of about 2 μm in length and 1 μm in diameter. Animal and plant cells typically have dimensions about 10 times larger.

Bacteria are single cell organisms that live independently of each other (with the exception of sexual exchange of genes) or occasionally in small aggregations. Their strategy of life is, essentially, to grow as rapidly as possible and outstrip the growth of competitors in the struggle for survival. At the other end of the biological scale are multicellular organisms which, in the more complex ones such as mammals, are composed of vast numbers of cells. In a human it is estimated that there are somewhere around 10^{11}–10^{13} cells. In such organisms there are many types of different cells specializing in different functions, and this requires mechanisms by which their activities are controlled so that they are appropriate to the needs of the organism as a whole.

Why are cells microscopic in size?

There are compelling physical reasons for the small size of almost all cells, the surface area/volume ratio being probably the most important. Cells take in molecules from the environment, such as food, and release unwanted molecules such as carbon dioxide, so that there is constant molecular traffic across the membrane for which adequate surface area is needed to allow a rate of exchange sufficient to support the needs of the cell. In addition, incoming molecules and ions have to reach all parts of the cell. Diffusion in solution is a relatively slow process and would be inadequate for the cell's needs over more than tiny distances. The advantages of small cells are that the ratio of surface to volume is maximized, and the length of diffusion paths, within the cells, minimized. Macromolecules synthesized in one part of a cell often have to reach their functional sites in other parts. Bacteria are small enough to rely on diffusion for this but in animal and plant cells, 1000 times larger in volume, it would be inadequate, and energy-requiring transport systems are used. The necessity of accommodating the essential quota of macromolecules, particularly the DNA and proteins, places a minimum size on cells.

Classification of organisms

There are three main evolutionary branches of organisms, **bacteria, archaeobacteria**, and **eukaryotes**. (Slight variations in precise names, and their spellings are common.) The bacteria and archaeobacteria are **prokaryotes**. Bacteria include the photosynthetic cyanobacteria and the typical *E. coli* cell. Archaeobacteria include **extremophiles**, which are found in hostile environments such as acid hot springs and around hydrothermal vents in the ocean floor, possibly representing conditions on primitive earth. They live on chemical 'food' such as hydrogen sulphide emanating from the earth's crust. Possibly they represent survivors from life on primitive earth. Some of them have enzymes of unusual thermal stability, which have found important applications in recombinant DNA work (Chapter 28).

Prokaryotic cells have no nuclear membrane unlike eukaryotic cells

Eukaryotes are the so-called higher plant and animal cells which range from unicellular yeasts and protozoa to humans. 'Karyon' in Greek refers to a nucleus. Prokaryotes ('before nucleus') do not have a distinct nucleus bounded by a membrane. Eukaryotes ('true nucleus') have a nucleus bounded by a membrane. Although having a membrane or not around the DNA of a cell might seem like a trivial difference, in fact it represents a major evolutionary divide. In eukaryotes, the separation of genes issuing instructions to the cytoplasm where those instructions are carried out has profound repercussions on the molecular biology of the cell, as will be explained later in the book. We will now describe the characteristics of both classes of cells.

Prokaryotic cells

E. coli is a typical prokaryotic cell, which lives in the human gut. It is structurally simple being a single compartment with no **internal** membranes or, put in a different way, it has no **organelles**, these being membrane-bounded 'small organs' such as mitochondria, which are so important in eukaryotic cells (Fig. 2.1).

The *E. coli* cell is surrounded by the plasma membrane containing systems which transport molecules into the cell. It also houses the vitally important system for generating ATP. The membrane is constructed of lipid molecules (collectively called membrane or polar lipids) into which proteins are inserted (Chapter 7) for various functions. Outside of the membrane is a protective strong cell wall made of a meshwork of filaments of

polymerized amino sugars (**glycans**) cross-linked by short peptides. The structure resembles a seamless string bag, the whole cell wall being a single molecule. (**Penicillin** inactivates the enzyme catalysing the final step in formation of the peptide cross-links. A weakened cell wall results, causing the cell to burst from the high osmotic pressure inside. Penicillin works only on growing cells that are actively synthesizing new cell wall, since a nongrowing cell has already produced a sound cell wall.) A second membrane exists outside the cell wall in *E. coli* but not all bacteria have this.

The prokaryotic chromosome resides in the cell as a nucleoid, a diffuse tangled thread visible in the electron microscope (Fig. 2.2). The main chromosome of *E. coli* is a circle of double-stranded DNA, though in its tangled form in which it is packed into the cell it cannot be discerned to be circular. The *E. coli* chromosome is estimated to have just over 4000 genes. The cell is **haploid** – there is only a single copy of the main chromosome but, in addition, there are often small circles of DNA, variable in number, known as **plasmids**. These contain a variety of genes that often include those conferring resistance to antibiotics. They replicate independently of the main chromosome. Plasmids play a major role in recombinant DNA technology (genetic engineering), dealt with in Chapter 28.

Cell division in prokaryotes

Some prokaryotes such as *E. coli* can grow in simple media containing inorganic salts and a source of energy such as glucose; some can replicate in about 20 minutes under ideal conditions, which means that a single cell can, overnight, multiply to hundreds of millions.

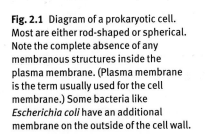
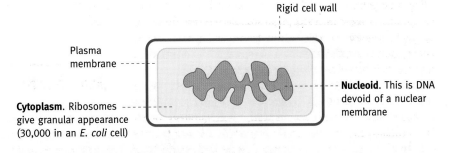

Fig. 2.1 Diagram of a prokaryotic cell. Most are either rod-shaped or spherical. Note the complete absence of any membranous structures inside the plasma membrane. (Plasma membrane is the term usually used for the cell membrane.) Some bacteria like *Escherichia coli* have an additional membrane on the outside of the cell wall.

Plasma membrane

Cytoplasm. Ribosomes give granular appearance (30,000 in an *E. coli* cell)

Rigid cell wall

Nucleoid. This is DNA devoid of a nuclear membrane

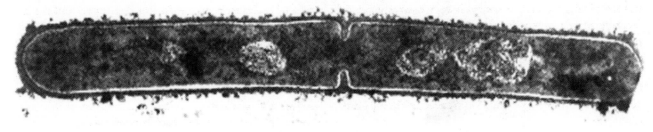

Fig. 2.2 Electron micrograph of vegetative *Bacillus subtilis* cell undergoing septum formation. Septum formation occurs at the middle of the cell and involves ingrowth of the cell membrane and cell wall layers. Scale bar is 1 µm. Photograph courtesy Dr E. J. Harry, School of Molecular and Microbial Biosciences, University of Sydney, Australia.

The prokaryotic cell cycle is uncomplicated in that the replication of the chromosome goes on continuously while the cell enlarges, and once a critical size is reached, cell division occurs by binary fission (Fig. 2.2). In this, a cell wall and new membranes begin to form around the circumference of the cell in the mid line, which, when complete, divides the cell into two. Each daughter cell must receive a copy of the chromosome, so the two DNA molecules arising from the chromosome replication have to be segregated into what will become the two daughter cells. In the most intensively studied bacteria (*E. coli* and *Bacillus subtilis*) the favoured model links segregation to chromosome replication. The replication apparatus (**replisome** or **segrosome**) remains at the mid-cell position for most of the cell cycle. As the circular chromosome is replicated, the two copies of DNA are extruded to either side of the mid-cell line and in the direction of the two ends. The newly replicated DNA molecules are continuously condensed (captured) through interactions with proteins homologous (see page 54) with the eukaryotic **condensins** that 'package' the chromosomes in preparation for mitosis (see Chapter 21). This is referred to as the **extrusion–capture model** for chromosome segregation in bacteria. The process is physically complex and not completely understood.

E. coli cells also have sexual conjugation, the purpose of which in all cells is to exchange genes between two genomes (as the total collection of genes is known); this recombination increases genetic variability, which is important for evolution. A 'male' cell is one that contains a single 100 kb (kilobase) circular plasmid known as the **F plasmid** while the 'female' lacks this. The process of conjugation is complex but, in simple terms, a section of the male chromosome is transferred into the female cell via a temporary bridge and it replaces the corresponding section in the female chromosome. The female thus receives new genes, creating genetic diversity. The mechanism is such that the male chromosome remains unchanged by the process because only single-strand transfer occurs. In both cells DNA copying restores the double-strand forms.

Eukaryotic cells

A **plasma membrane** surrounds the eukaryotic cell. It is a lipid structure, as in bacteria, with proteins inserted into it for molecular transport, receipt of cell signals and other functions (Chapter 7). In contrast to prokaryotic cells, however, the eukaryotic variety contains membrane-enclosed structures, known as **organelles** (Fig. 2.3).

The cell **nucleus** is the most prominent structure inside the cell and contains the chromosomes. The typical eukaryotic **genome** is **diploid**, there being two copies of each chromosome, one from each of the two parents (Chapter 21). The nucleus is

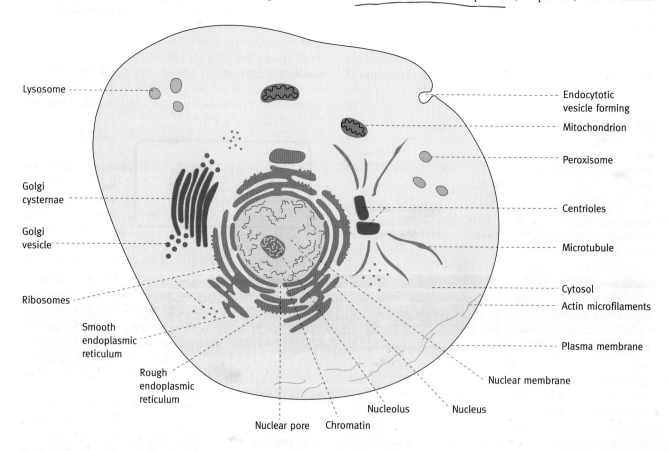

Lysosome
Golgi cysternae
Golgi vesicle
Ribosomes
Smooth endoplasmic reticulum
Rough endoplasmic reticulum
Nuclear pore   Chromatin
Nucleolus   Nucleus
Nuclear membrane
Plasma membrane
Actin microfilaments
Cytosol
Microtubule
Centrioles
Peroxisome
Mitochondrion
Endocytotic vesicle forming

Fig. 2.3 Diagrammatic representation of a generalized animal cell, showing different types of organelles.

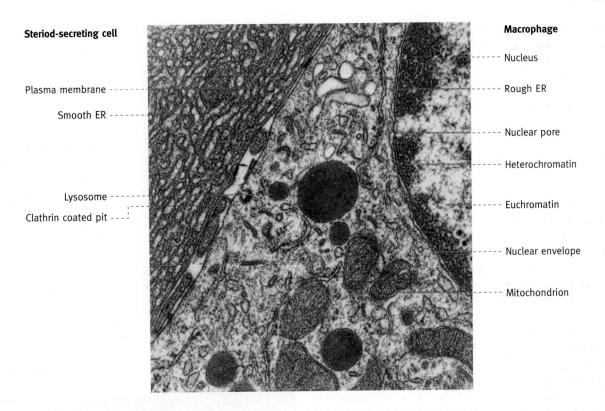

Fig. 2.4 Electron micrograph of part of sectioned testis showing a steroid-secreting cell (upper left) packed with smooth endoplasmic reticulum (ER), alongside a macrophage (right). Photograph kindly provided by Professor W. G. Breed, Department of Anatomy, University of Adelaide, Australia.

surrounded by a membrane supported on the inside by a layer of filamentous proteins (Chapter 8). The membrane is studded with **nuclear pores** (see Fig. 26.15) through which there is both diffusion and active traffic of specific molecules in both directions. An elaborate mechanism controls the movement of molecules through the pores.

The **endoplasmic reticulum (ER)** is a membranous structure in the form of a completely enclosed convoluted tubular sac. It surrounds the nucleus whose membrane is continuous with that of the ER. It forms a separate compartment, inside the ER being the **lumen**, with the cytoplasm outside. In some cells it is present in large amounts and in cross section (Fig. 2.4) almost fills the cell. In others there is much less (or none in mature red blood cells). It is involved in the synthesis and delivery of newly synthesized proteins to their various cellular destinations or, in the case of plasma proteins and digestive enzymes, to the plasma membrane for secretion. Part of the ER membrane known as the **rough ER**, is studded with **ribosomes**, the protein–RNA complexes that synthesize proteins. Other sections (continuous with the rough ER) have no ribosomes and constitute the **smooth ER**, which processes and packages proteins into vesicles, for transport to the Golgi vesicles (below). The smooth ER has other metabolic roles also. It is important to note that only those ribosomes that happen to be synthesizing proteins destined for secretion are attached to the rough ER

membrane. These detach and join the general pool of ribosomes in the cytoplasm when synthesis of each protein molecule is complete, and re-attach when they initiate a secretary protein. Additionally, however, large numbers of ribosomes are free in the cytoplasm synthesizing cytoplasmic proteins and proteins destined to be transferred into organelles.

The **Golgi vesicles**, named after their discoverer, play the role of a post office mail sorting room, the mail in this case being newly synthesized proteins. They consist of four to six (more in plant cells) closed membranous flattened structures enclosing spaces, the **Golgi cisternae** (Fig. 2.5). They resemble in cross section a stack of large platelike vesicles placed near the nucleus. Membrane-bounded transport vesicles carrying newly synthesized proteins from the lumen of the smooth ER are budded off, and travel to fuse with the Golgi membranes and deliver their contents into the cisternae where proteins are given 'destination labels', packaged into appropriately labelled vesicles and released into the cytoplasm to be carried to their cellular destinations. The complex process is described fully in Chapter 26 on protein delivery, or targeting as it is called. It is an astonishing sorting and delivery mechanism.

Mitochondria are the powerhouses of the cell. They are present in multiple copies in all eukaryotic cells except the mature red blood cells; they are about the size of an *E. coli* cell. They have a double membrane enclosing a dense collection of proteins.

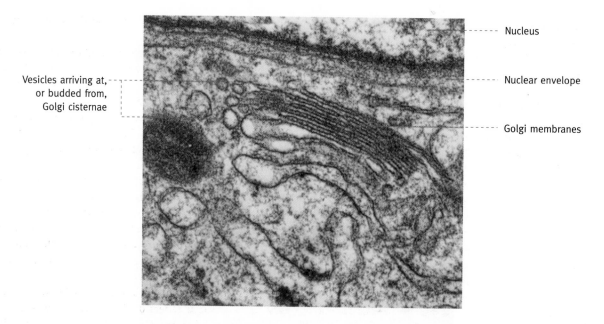

Vesicles arriving at, or budded from, Golgi cisternae

Nucleus

Nuclear envelope

Golgi membranes

Fig. 2.5 Higher magnification electron micrograph of Fig. 2.4 in which a prominent Golgi apparatus is evident. Photograph kindly provided by Professor W. G. Breed, Department of Anatomy, University of Adelaide, Australia.

Mitochondria are responsible for most of the ATP production of eukaryotic cells, the energy being derived from the terminal oxidation of food molecules such as carbohydrates and fat. The inner membrane is the site of ATP generation. Its area is increased by invagination to form **cristae** whose extent reflects the level of ATP demand (see Frey and Mannella review). Heart muscle mitochondria, for example, have closely packed cristae in keeping with the high ATP demand (Fig. 2.6).

It is generally accepted that mitochondria originated in the evolutionary sense from prokaryotic cells being engulfed by the precursor of eukaryotic cells and becoming established inside the cells. In modern eukaryotes, all aerobic (oxygen-dependent) energy release from food is performed by mitochondria. The acquisition of aerobic metabolism by eukaryotic cells was of very great evolutionary importance since the yield of ATP from complete glucose oxidation is 18–19 times greater than from its anaerobic breakdown.

The mitochondria have their own circular DNA molecule and undergo replication so that daughter cells have the full complement as they grow, but they have surrendered much of their genetic independence. Most of the proteins in a mitochondrion are coded for by genes in the nucleus of the cell, synthesized in the cytoplasm and transported into the organelle. The ribosomes of mitochondria are prokaryotic in type. (Prokaryotic ribosomes are smaller than those of eukaryotes and have certain different characteristics, but are basically the same molecular machines – see Chapter 24).

Chloroplasts are present in plant cells. They contain the photosynthetic apparatus (Fig. 18.2). It is believed that these originated from photosynthetic prokaryotes (cyanobacteria) engulfed by precursors of eukaryotic cells, and became established inside the cells much as occurred with mitochondria.

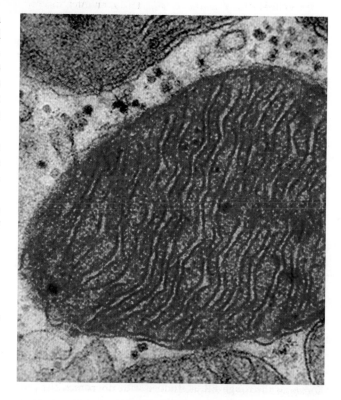

Fig. 2.6 Coloured transmission electron micrograph of a heart muscle mitochondrion, showing cristae. © Robert Harding.

They are analogous to mitochondria in having their own circular DNA and limited protein synthesis, and replicate by simple division.

Lysosomes are small spherical membrane vesicles, present in the cytoplasm of eukaryotic cells in large numbers. They contain a collection of about 50 hydrolytic enzymes capable of destroying biological molecules. They do not have their own genome. Their function is to destroy some material taken in from outside of the cell by endocytosis (page 107) or cellular material due for destruction. It is essential that destruction occurs within closed membrane vesicles to protect the cell itself from the destructive enzymes. There are several lethal diseases associated with lysosomal abnormalities (Box 26.1).

Peroxisomes are small membrane-enclosed vesicles in which a number of molecules not metabolized elsewhere are oxidized by enzymes that use molecular oxygen directly and produce hydrogen peroxide. Hydrogen peroxide is destroyed by further reactions catalysed by catalase and peroxidases. Oxidation in peroxisomes does not generate ATP as it does in mitochondria. Peroxisomes have no genetic system and do not synthesize their enzymes, which are transported into them from the cytoplasm.

Glyoxysomes are small membrane-bounded vesicles found in plants. They contain enzymes not present in animals. They make it possible for fat reserves to be converted to carbohydrate, which is important for example in the germination of seeds, the main energy reserve of which is in the form of fat. Importantly, animals cannot do this.

The **cytoskeleton** (Chapter 8, page 133) is an internal structure of protein fibres that, in most eukaryotic cells, pervade all of the cytoplasm. Many of the fibres are attached to the cell membrane and influence the shape of the cell. However the term 'skeleton' gives a misleading impression of its nature. The fibres are mainly tracks for molecular motors to operate on. They are involved in cell movement, separation of chromosomes at cell division, and transport of molecules within the cell. They are also disconcertingly ephemeral, being set up and demolished rapidly, as needed. Prokaryotes, being about 1000 times smaller in volume, do not have a cytoskeleton as most transport is by simple diffusion.

Cytoplasm is the term for all the contents of a cell but excluding membrane bounded organelles. It contains particulate structures such as ribosomes, the cytoskeleton and proteasomes.

The **cytosol** is not precisely defined but usually is taken to mean the cytoplasm from which particulate structures such as ribosomes have been removed by centrifugation.

Eukaryotic cell growth and division

Single cell eukaryotes such as yeast can be grown in simple media, as anyone who has home-brewed beer will know. Mammalian cells can be cultured in the laboratory but they are more demanding. They are usually grown in Petri dishes (flat circular plastic dishes with lids) on the surface of nutritive gels. In the whole animal, cells mutually control proliferation; that is to say one cell's divisional activity is strongly influenced, even controlled, by other cells, often its neighbours. (Independent growth of mammalian cells is characteristic of cancer.) To achieve this coordination, cells communicate by chemical signals known as **growth factors** and **cytokines**, which are produced by many types of somatic cells (see below). They are proteins or large peptides. To culture mammalian cells in the laboratory, these chemical signals are added, often as a mixture in foetal blood serum.

When cells of tissues such as liver are cultured, after a certain number of cell divisions (perhaps 40 in the case of skin cells known as fibroblasts) normal proliferation ceases. It seems that each cell has a limitation on the number of normal divisions that it can undergo and this has led to speculation that it may be a factor in ageing. The limitation is due to a structure, the **telomere**, at the end of each chromosome, which acts like a cell division counter, shortening every time DNA is replicated (Chapter 22) and beyond a critical shortening, normal cell division ceases. However, cells may mutate and escape this control, for example in cancerous cells, which can divide and be cultured indefinitely, establishing immortal cell lines; their telomeres are continually replenished. This problem does not apply to prokaryotes because the DNA is typically circular and there are no ends to shorten.

Cell division in eukaryotes is more complex than in prokaryotes, as might be expected. The process in somatic cells is called **mitosis** in which the resultant daughter cells are diploid. In **germ line cells** the process is called **meiosis,** which produces the haploid sperm and eggs. The mechanism of these cell divisions is described in Chapter 21.

Basic types of eukaryotic cells

If you consider an animal such as a mammal, its development starts with a single fertilized egg – a single diploid cell. From this, embryogenesis produces all the different types of cells found in the adult body. Most of these are known as **somatic cells**, which are all the cells of the body except the **germ cells** that give rise to eggs and sperm. Somatic cells of an adult are mainly **differentiated cells** – they have acquired the characteristics of their special nature. For example a liver cell has properties quite different from those of nerve cells and so on, even though these cells all have the same DNA sequences. The process of cells changing in this way is known as **differentiation** and this is a naturally irreversible process. A liver cell if it divides can only produce liver cells. In fact somatic cells are in a nondividing state most of the time. They divide only to maintain a constant number of cells appropriate to the organ they belong to, to replace dying cells or repair wounds. This keeps the organ size appropriate to the size of the organism. Most somatic cells can, however, divide quite rapidly if given the signal to do so (muscle cells, erythrocytes [red blood cells] and possibly some nerve cells are exceptions). If you experimentally remove two thirds of a mouse liver it will regain its full size in days and then abruptly stop growing. Control of cell division is elaborate and vital (Chapter 31). In cancer, the control has gone wrong (Chapter 33).

However, certain groups of cells are constantly renewed throughout the life of an organism. The skin is an example. The lining of the intestine is the most rapidly replaced of all. Blood cells are renewed at a great rate. The life span of an erythrocyte is 120 days in a human, where 2–3 million red blood cells are made per second to replace those old cells destroyed in the liver and spleen by macrophages.

You can see that since differentiated cells can only replace themselves and since all cells arise from a single one there must be a class of cells that can divide but give rise to other types of cells. These are known as **stem cells**. They are cells that divide, and the two daughter cells can do one of two things: they can remain as stem cells or they can differentiate into somatic cells. This means that there is a constant basis of stem cells supplying somatic cells. There are different types of stem cells. **Embryonic stem cells** are those initially produced from the fertilized egg during embryogenesis. They are **pluripotent** meaning that they can give rise to any cell type. **Adult stem cells** are a halfway house between embryonic stem cells and somatic cells in their capacity to form different cell types. If you consider for example renewal of skin, there are various types of skin cells that have to be replaced, so underlying skin there is a pool of stem cells that can give rise to all of these. But they cannot give rise to non-skin cells. There are adult stem cells giving rise to the various blood cells but only to these, and so on. The process by which all blood cells have their origin in bone marrow stem cells that continually divide is known as haemo-poiesis (Fig. 2.7). Stem cells, as everyone now knows are of major medical interest. The differentiation of a stem cell into a somatic cell is in mammals irreversible so far as natural processes go. However technologies exist that can reverse this, to reprogram the nucleus back to its pluripotent state as was shown by the celebrated Dolly the sheep work. We will deal further with this aspect later in the book.

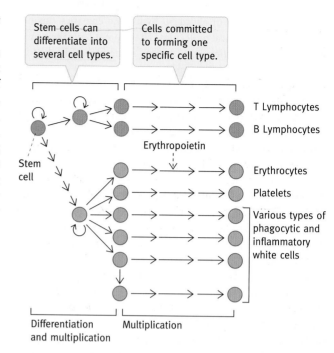

Fig. 2.7 Simplified diagram of haemopoiesis. At each arrow, cell multiplication can occur, and this may be controlled specifically by various protein cytokines – the colony-stimulating factors, interleukins, and erythropoietin. For example, erythropoietin stimulates erythrocyte production at the step indicated. Differentiation of stem cells into committed cells is likewise controlled. Embryonic stem cells can give rise to all types; adult stem cells to a limited range; somatic cells only reproduce themselves.

BOX 2.1 Some of the organisms used in experimental biochemical research

In Chapter 1 (page 4) we briefly alluded to the fact that it is best when trying to elucidate a molecular process to choose an organism that gives you the best chance of success. The unity of basic life processes in all life forms often means that discoveries in one organism are to a considerable degree relevant to other life. A brief survey of some of the most popular ones may help to make scientific literature more comprehensible.

Viruses often can be easily grown both in bacteria and animal cells. Their genomes are very small and reproductive times are measured in minutes. Phage lambda (λ) (see page 23) has been used extensively in gene cloning (Chapter 28).

Bacteria: Although a variety of bacteria is used for specific purposes in research, **E. coli** (non-pathogenic strain) was for a long time the main workhorse of biochemists. It is small, easily cultivated and replicates in 20 minutes, and has only 4000 genes as compared with the 25,000 of human cells. It is still a key organism in genetic engineering work (Chapter 28).

Yeast is a single cell eukaryote and is much used for genetic studies. It has most of the molecular characteristics of more complex eukaryotes but with a very much smaller genome.

C. elegans (*Caenorhabditis elegans*) is a small easily grown nematode roundworm that has become the simplest model in studies on animal cells. There are only 959 somatic cells plus germ cells. The lineage of every somatic cell back to the fertilized egg has been traced in a remarkable piece of work that earned the Nobel Prize in 2002 for Sydney Brenner, Robert Horvitz, and John Sulston. It is important for animal developmental studies and for the detection and isolation of genes. Genes identified in it have been found to occur widely in animals.

Drosophila melanogaster, the fruit fly, is one of the most used models for animal developmental studies and has been important in the development of molecular biology in general. It has a short reproductive time and the existence of a larval stage makes it a convenient model for identifying genes and their function. Work on it has revealed developmental and other control genes that exist throughout vertebrates.

The **mouse** (and also the **rat**) are the most used mammalian models for studies on vertebrates.

Cultured mammalian cells are also important for molecular biology studies since they give direct access to cells under controlled experimental conditions.

Arabidopsis thaliana is a small cress plant that is the workhorse for genetic and developmental studies on plants.

Viruses

Quite apart from the biological and medical importance of viruses, they are important in biochemical research; they provide simpler models for studying processes in gene function. They are also important research tools for many purposes, including roles in recombinant DNA technology, described in Chapter 28. The chief (but not the only) protection mammals have against virus infections is the immune system, which produces antibodies to combine with viruses, leading to their destruction, or it causes the destruction of infected cells (Chapter 30) and aborts viral replication in them. Viruses have often developed strategies to outwit the immune defense systems and knowledge of these strategies can be an important guide in research for unraveling the highly complex immune strategies.

Viruses are much smaller than cells, an electron microscope being needed to see them rather than the light microscope that is adequate for most bacteria. They have no metabolism by themselves – a virus particle or **virion**, as it is called, on its own does nothing. It is an inert, organized, complex of molecules that may sometimes be crystallized. Nonetheless, they are reproduced when they infect living cells. Different viruses infect specific animal and plant cells, and also bacteria (viruses that infect the latter are known as **bacteriophages**). Their strategy is to get their genetic material into cells and use the host cell machinery for their own replication.

A virus particle comprises a small amount of nucleic acid surrounded by a protective shell of protein molecules. The total number of genes may be about a hundred for a large virus, such as **vaccinia** (used for smallpox vaccination), or three or four for the smallest. The protein shell surrounds the genome and is called the **nucleocapsid**. In some viruses there is an additional membrane coat.

Viruses enter the cells they infect and release their genetic material (bacteriophages inject this without the whole assembly entering). The viral genes then direct the synthesis of the components required for the assembly of new virus particles which escape from the cell.

The membrane of a cell is a barrier to virus entry but there are ways past it. For example, animal cells have mechanisms for engulfing molecules (endocytosis, page 107) and some viruses exploit this; they hitch a ride on a normal cellular import mechanism. A second route, available to viruses with a membrane is direct fusion with the plasma membrane of the host cell. **HIV** uses this route.

In **bacteriophages**, a different route is followed. The bacterial cell has a rigid cell wall around it. **Phage lambda (λ)** has a tadpole-like shape (Fig. 2.8). The **head** is a capsule formed by protein molecules in which resides the viral DNA molecule. The phage attaches to the cell wall by its tail fibre and injects the DNA into the bacterial cell, almost like the action of a hypodermic syringe.

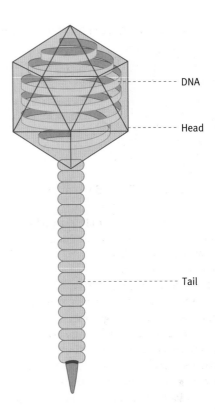

Fig. 2.8 A phage λ particle. The λ chromosome – some 50,000 base pairs of DNA – is wrapped around a protein core in the head.

Genetic material of viruses

The genetic material of viruses may be DNA, or RNA. RNA occurs in different viruses in single-stranded and double-stranded form. A single-stranded DNA also occurs in a known bacteriophage, but the replicative form is double-stranded.

Why has DNA superseded RNA as the medium for storing genetic information in all cells? The answer almost certainly is that DNA is chemically more stable than RNA. If a mistake is made in the synthesis of a DNA molecule, or it is damaged in some way, enzymes exist to repair it (Chapter 22). RNA is still the genetic material of many viruses. RNA damage is not repaired (as occurs with DNA) and RNA viruses therefore mutate rapidly. By constantly changing the proteins that the immune system recognizes (Chapter 30), new viral strains escape immune attack. So primitive molecular instability, coupled with lack of repair, is an advantage even in the modern world where most viruses are in fact RNA ones: human immunodeficiency virus (HIV), influenza, poliomyelitis, mumps, foot and mouth, measles, and rubella viruses to name a familiar few. The same applies to plant viruses.

Some examples of viruses of special interest

Influenza is an RNA virus consisting of a membrane with proteins inserted in it and enclosing its genetic material which is split into eight separate packages (Fig. 2.9). It mutates rapidly so that new strains develop each year requiring annual vaccinations

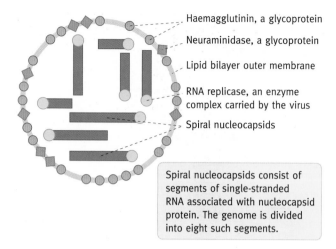

Haemagglutinin, a glycoprotein

Neuraminidase, a glycoprotein

Lipid bilayer outer membrane

RNA replicase, an enzyme complex carried by the virus

Spiral nucleocapsids

Spiral nucleocapsids consist of segments of single-stranded RNA associated with nucleocapsid protein. The genome is divided into eight such segments.

Fig. 2.9 Structure of the influenza virion. A layer of matrix protein underlying the outer membrane is not shown.

for protection against the current strain. Immune protection is directed against the surface proteins but these change rapidly because of the mutations, so protection is usually gradually lost as new strains appear (known as **antigenic drift**), but residual protection means that attacks may be comparatively mild. However, on fortunately rare occasions, the coat protein may be changed completely from a recombination between two strains of virus infecting the same cell. This is known as **antigenic shift**. In this situation, immunological protection of the populace due to previous exposure to the influenza virus is almost completely lost and a **pandemic** (a widespread, possibly worldwide, infection) may result. Pandemics of the past have killed many millions of people. The 1918 outbreak is believed to have originated from a reassortment of blocks of genes (recombination) between human and bird strains of the virus when an animal (possibly a pig) became infected by both types.

Antibiotics are not effective against viruses

Antibiotics are molecules produced by microorganisms and fungi, which they release to kill competing organisms in their neighbourhood. It is a chemical strategy in the battle of natural selection. Although all cells are to a large extent basically the same, there are differences, often subtle ones, that make processes susceptible to molecular spanners thrown into the works. From a human standpoint, the ideal antibiotic is one that attacks biochemical systems of infectious organisms that have no counterpart in human biochemistry. Penicillin (page 17) is the classic for it prevents susceptible bacteria making their cell walls, which animal cells lack. As resistance to antibiotics develops, it has been necessary for the pharmaceutical industry to create new antibiotics by modifying natural ones.

Human viruses present a different problem because they have essentially no biochemical machinery of their own, and reproduce using host cell machinery. It is difficult, therefore, to inhibit the virus reproduction without damaging the host.

No viral 'antibiotics' have evolved. The main human protection against viral infection is the immune system. This relies on distinguishing the amino acid sequence of viral proteins from human ones. This is a highly sophisticated task and makes the immune system almost incredibly complex (Chapter 30). However, antiviral, chemically made drugs have been successful in controlling a number of viruses, such as influenza and HIV (see below), by exploiting subtle differences or detecting loopholes in the viral processes.

The influenza virus has in its outer shell, molecules of an enzyme called **neuraminidase**, which play some part in the release of virus from infected cells but it may have other roles. A therapeutic strategy has been designed in the form of commercially available drugs – '**Relenza**' was the first – that bind to the active site of this enzyme with high affinity and block its action. The active site of an enzyme (page 87) is less likely to mutate and to become drug resistant because any change may interfere with its catalytic activity. The unchanging active centre structure may seem to expose the virus to antibody attack but it is not accessible to the relatively large antibody molecules, whereas the drugs are small enough to enter the site. The drugs are reported to lessen the severity of the attacks.

While antibiotics may be useful in controlling secondary bacterial infections following a viral disease, they are no use in combating viral infection themselves.

Retroviruses

Retroviruses are RNA viruses of immense current interest quite apart from the fact that HIV is of this type. The retrovirus particle carries within itself a few molecules of an enzyme, the discovery of which caused initial disbelief, to be followed by the award of a Nobel Prize in 1975 to its discoverers David Baltimore, Renato Dulbecco, and Howard Temin; it is called **reverse transcriptase**. Before its discovery it was dogma that DNA could direct RNA synthesis, but not the reverse. However, viral reverse transcriptase copies the viral RNA into DNA. Another enzyme, carried in the virus, causes integration of the DNA into the host chromosome where it becomes, in essence, an extra set of genes carried by the cell. Once in, it is replicated along with host DNA for cell division. For the production of new retrovirus particles, the proviral genes (that is, viral genes in the host chromosomes) are transcribed (copied) into RNA transcripts. The retroviral life cycle is described later (page 348). This viral genome integration is of relevance to cancer (Chapter 33), because retroviruses, on leaving a host, may pick up a gene fragment, or gene control segment, which, when inserted into a new host may be oncogenic (cancer-producing). Several examples of cancers in animals, and a single leukaemia in humans are known to have this origin.

HIV, like influenza, mutates very rapidly due to its error-prone replication mechanism, so that any immunity that is developed by the host is quickly bypassed. Also new mutant strains develop even during a single infection, giving little chance of any significant immunity developing. In addition the virus attacks helper cells of the immune system. These are essential in developing antibodies against the virus so that the patient's antibody production against the virus and any other pathogen is severely impaired leading, potentially, to multiple infections.

The drug **AZT** (**azidothymidine**), a nucleoside analogue, is used as one component of therapy against the AIDS virus (HIV); it works by inhibiting the viral reverse transcriptase by becoming incorporated in, and so terminating the synthesis of DNA chains. Other components of an antiviral drug 'cocktail' may be a non-nucleoside compound that binds directly to reverse transcriptase and prevents RNA conversion to DNA, and a protease inhibitor that binds to, and blocks, an essential HIV protease.

Viroids

Viroids are the smallest infectious particles known – even smaller and simpler than viruses. They are very small, naked RNA molecules without any protein or other type of coat. Infection of plant cells by them is dependent on mechanical

damage of the host plant. The effect of viroid infection varies from some impairment of the health of the plant to certain death of, for example, palm trees. The really remarkable feature is that no genes coding for proteins are present and it is not certain how viroids produce disease.

BOX 2.2 Structure of the drug azidothymidine

Azidothymidine (AZT)

Summary

Cells are the units of living systems (excluding viruses, which are arguably not living), each surrounded by a lipid membrane. They are with rare exceptions microscopic in size to maximize the surface area to volume ratio.

Prokaryotes include modern bacteria. They lack a nuclear membrane or any internal membrane-bounded organelles. Their small size makes it possible for all molecules, including macromolecules, to diffuse to parts of the cell where they are needed. *E. coli*, the typical prokaryote, has a single circular chromosome. It can replicate in about 20 minutes.

Eukaryotic cells are about 1000 times larger in volume and replicate in about 24 hours. They have a nuclear membrane containing the nucleus and other membrane-bounded organelles.

The **endoplasmic reticulum (ER)** is an extensive continuous membranous structure enclosing a lumen, separated from the rest of the cell. The ER modifies newly synthesized proteins and buds off transport vesicles to deliver them to the Golgi. The ER also synthesizes new lipid membrane.

Golgi vesicles are a series of enclosed sacs, involved in delivery of newly synthesized proteins to their destinations in the cell.

Mitochondria are the site of most of the ATP generation from oxidation of food molecules and are believed to have originated in evolutionary terms from engulfed prokaryotic cells. Chloroplasts are the site of photosynthesis in plants.

Lysosomes are bags of enzymes needed for the destruction of selected cellular material.

Peroxisomes oxidize certain molecules forming peroxide but do not generate ATP.

The **cytoplasm** is defined as the content of the cell from which membrane-bounded organelles have been removed. The cytosol refers to the soluble constituents of the cytoplasm.

The **DNA** of prokaryotes resides in the cytoplasm as an extended tangled thread. For cell division this is duplicated and the threads are pulled apart, followed by separation of the daughter cells. In eukaryotic cells the process is more elaborate, involving mitosis in which highly condensed chromosomes are segregated into daughter cells. The eukaryotic cell cycle is more highly controlled than that in prokaryotes.

Cells contain large numbers of **ribosomes**. These are not surrounded by a membrane but are large complexes of RNA and proteins. They translate mRNA into polypeptide chains – they synthesize proteins.

In animals, **cell growth** is controlled by growth factors and cytokines. Normal cells can only divide for a limited number of times because their telomeres shorten at each replication and eventually become so short that replication stops. In cancerous cells the telomeres are replaced continually.

Most of the **differentiated cells** in the body are known as somatic cells. These can only give rise to their own type of cell. The germline cells are those that give rise to sperm and eggs.

Embryonic stem cells are pluripotent and can give rise to all cell types. Adult stem cells can give rise to a limited number of cell types.

Viruses have genetic material that can be either DNA or RNA. When they infect a cell their genetic material is released into the cell and they utilize the metabolism of the host to reproduce.

Influenza virus is an RNA virus. It mutates rapidly and its antigenic properties change, making animals less resistant. Occasionally recombination generates a totally new strain, giving rise to a pandemic.

Retroviruses (HIV is of this type) are RNA viruses that after infection copy their RNA to form DNA. This then integrates into the host DNAs and the genes can be copied into new RNA viruses.

Viroids are smaller than viruses, but have no protective coats. They are naked RNA molecules that enter cells through sites of mechanical damage. They occur in plants.

 ## Further reading

Frey, T. G. and Mannella, C. A. (2000). The internal structure of mitochondria. *Trends Biochem. Sci.*, 25, 319–24.
Reviews the evidence, based on recent three-dimensional electron microscope tomography, that cristae structures are more complex than usually depicted in the 'baffle' model.

Hayes, F. and Barilla, D. (2006). Assembling the bacterial segrosome. *Trends Biochem. Sci.*, 31, 247.
Short article on bacterial cell division.

Viruses

Cullen, B. R. (1991). Human immunodeficiency virus as a complex prototype retrovirus. *J. Virology*, 65, 1053–6.
A minireview.

Frankel, A. D. (1998). HIV-1: Fifteen proteins and an RNA. *Ann. Rev. Biochem.*, 67, 1–25.
Describes the function of the components in the life cycle of the virus.

Ptashne, M. (2004). *A genetic switch: phage lambda revisited.* 3rd edn. Cold Spring Harbor Press.
A classic book on DNA structure and function, as well as phage molecular biology.

Moscona, A. (2005). Neuraminidase inhibitors for influenza. *N. Engl. J. Med.*, 353, 1363–73
Has an excellent diagram of virus release from infected cell.

 ## Problems

1 Influenza epidemics sweep the world at long intervals. Most are mild but occasionally, as in 1918, a highly lethal one occurs. Why is this so?

2 Comment on the nature of somatic cells and stem cells.

3 Briefly summarize the main characteristics of prokaryotic and eukaryotic cells.

4 Why are cells, with a few exceptions microscopic in size?

5 What sets the lower size limit of cells?

6 Why do biochemists often use many different types of organisms to elucidate a problem?

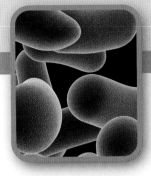

Energy considerations in biochemistry

If your theory is found to be against the second law of thermodynamics, I can give you no hope; there is nothing for it but to collapse in deepest humiliation.

Sir Arthur Eddington

In Chapter 1 there is a more general survey of the basic energy concepts that apply to life (pages 4 to 5), which it will be helpful to read. As explained there, because life operates at constant temperature and pressure, the relevant thermodynamic considerations are greatly simplified.

The consideration underlying everything in life is energy. A printed chemical equation on its own lacks a vital piece of information and that is the energy change involved. It may not be obvious what is meant by this, for energy change in chemical reactions occurring in solution is not something that is easy to see or detect, unless very violent, and such reactions do not occur in biology. But energy considerations in biochemical reactions determine whether a reaction is possible on a significant scale and whether the reverse reaction can occur to a significant degree. They are of prime importance in determining the chemical activities in cells.

An example of how this aspect applies directly to living processes is the conversion in muscle, during vigorous exercise, of glycogen (a polymer of glucose) to lactic acid (page 180) by the metabolic pathway called **glycolysis**. This pathway consists of a series of a dozen chemical reactions. After the exercise, during the ensuing rest period, accumulated lactic acid is converted back to glycogen (page 193) but not by the reverse of exactly the same set of chemical reactions by which it was formed. There are important differences between the forward and reverse pathways. This can seem to be pointlessly complicated chemistry but a knowledge of simple energy considerations makes it at once apparent why this strategy has to be followed by the cell. An understanding of simple energy considerations throws a flood of light onto the biochemistry of cells.

Energy considerations determine whether a chemical reaction is possible in the cell

As already implied, energy considerations permit certain chemical reactions to occur while others are not allowed. Note that in biochemistry we are concerned with reactions occurring to a *significant* extent. In principle, energy considerations do not completely prevent a reaction occurring (unless reactants and products are precisely at equilibrium concentrations; see below), but if the reaction ceases after a minute amount of conversion has taken place, then, for the practical purposes of life, it has not occurred.

First, what do we understand by energy change in a chemical system, by which we mean an assemblage of molecules in which chemical reactions can take place, such as exists inside a cell? This is not self-evident as is the case with, say, gravitational potential energy change in a weight falling. A chemical system involves a huge number of individual molecules each of which contains a certain amount of energy, dependent on its structure. This energy can be described as the heat content or **enthalpy** of the molecule. Enthalpy values for many compounds are available in physical chemical tables. When a molecule is converted to a different structure in a chemical reaction, its energy content usually changes; the change in the enthalpy is written as ΔH (delta H), which stands for the *change* in heat content. The ΔH may be negative (heat is lost from molecules and released, so raising the temperature of the surroundings) or positive (heat is taken up from the surroundings, which, correspondingly, cool down).

At first sight it may seem surprising that reactions with a positive ΔH can occur at all since it might seem to represent an energy uptake analogous to a weight simply raising itself from the floor and cooling the surrounding air as it does so. This is the point at which physical analogies such as weights falling become inadequate as models for chemical reactions. In chemical reactions, a negative ΔH favours the reaction and a positive ΔH has the opposite effect, but ΔH is not the final arbiter as is gravitational energy with a weight system. **Entropy change** (the drive to randomness), known as ΔS, also has a say in the matter.

Entropy is the degree of randomness of a system (see page 4 for a simple explanation of entropy if you are not familiar with the term). In a chemical system, this can take three forms: first, a molecule is not usually rigid or fixed – it can vibrate, twist around bonds, and rotate. The more this occurs the higher the entropy. Secondly, the vast numbers of individual molecules may be scattered in a random manner (higher entropy) or they may have a more ordered arrangement (lower entropy). A living cell has a more ordered arrangement of molecules than that of molecules in the outside-cell surroundings and therefore a lower entropy level. A finished house has a lower entropy level than that of the materials from which it was constructed. Thirdly, the number of individual molecules or ions may change as a result of a chemical change. The greater the number of individual particles (molecules and ions), the greater the randomness; the greater the randomness, the greater the entropy. These three factors can all contribute to entropy change. Thus a chemical reaction may change the entropy of the system. Increasing entropy lowers the energy level of the system; decreased entropy (increased order) increases the energy level.

As already stated above, both ΔH and ΔS have a say in determining whether a chemical reaction may occur. A negative ΔH and a positive ΔS both reinforce the 'yes' decision; a positive ΔH and negative ΔS both reinforce the 'no' decision. A negative ΔH and negative ΔS disagree on the decision, as do a positive ΔH and a positive ΔS, and whether the outcome is yes or no depends on which is quantitatively the larger.

The driving force of increasing entropy is illustrated by the example of the melting of a block of ice in warm water. Heat is taken up as the ice melts (the ΔH is positive) but the scattering of the organized molecules of the ice crystal, as it dissolves, increases the entropy so that the process proceeds. If energies are similar, the disordered state occurs more readily, and is more probable, than the ordered state, which is an everyday experience. In a collection of molecules, or any objects, there are vastly more possible random arrangements than organized ones. It is not easy, however, to intuitively accept that entropy change can affect processes in the way one can visualize gravitational energy causing a weight to fall or a car to roll downhill. If you find it difficult, rather than have a mental block, it would be best to accept the concept and it will become familiar as you progress through the book. (As pointed out earlier on page 4, despite entropy perhaps seeming a nebulous concept, achieving maximum entropy appears to be an irresistible tendency of the universe.)

The situation we have described between ΔS and ΔH is not convenient; we have two terms of variable sizes that may reinforce or oppose each other in determining whether a reaction is possible. Moreover, in biological systems it is difficult or impossible to measure the ΔS term directly. The situation was greatly ameliorated by the **Gibbs free energy** concept (see below

for an explanation of the term free energy). He combined the two terms into a single one. The change in free energy (ΔG after Gibbs) is given by the famous equation

$$\Delta G = \Delta H - T\Delta S$$

where T is the **absolute temperature** (page 30). This equation applies to systems where the temperature and pressure remain constant during a process, which is the case for biochemical systems. Note that we are concerned with the **change of free energy** in a reaction; ΔG is a term that applies to reactions. The ΔG of a reaction is an all-important thermodynamic term; in its application to chemical reactions the rule specified by the **second law of thermodynamics** is as follows: the products have less free energy than the reactants have.

The term 'free' in free energy means free in the sense of being *available* to do useful work, not free as in something for nothing. ΔG represents the *maximum* amount of energy available from a reaction to do useful work. It is somewhat like available cash being 'free' for you to make purchases with. Useful work includes muscle contraction, chemical synthesis in the cell, and osmotic and electrical work. ΔG values are expressed in terms of calories (cal) or joules (J) per mole (1 cal = 4.184 J); joules is now the official term (and the one used in this book) but calories is frequently used in biochemical texts. Since the values are large, the terms kilocalories (kcal) or kilojoules (kJ) per mole are used. Kilojoules per mole is usually abbreviated to kJ mol^{-1}.

Another wording of the second law given on page 4 specifies that all happenings must increase the total entropy of the universe. For example the released heat of a reaction does this (page 4).

The fact that a chemical reaction, if it were to occur, would comply with the second law does not mean that the reaction actually does occur though noncompliance guarantees that it will not. A negative ΔG is a necessary but not sufficient condition for a reaction to occur. We have explained earlier (page 8) that enzyme catalysis is also needed for biochemical reactions to take place. The reason briefly is due to the fact that there is an energy-barrier to chemical reactions occurring. Molecules must be raised to the transition state before they can react (see Chapter 6, page 88). This is why sugar in the bowl on the table does not burn despite the fact that the energy considerations of the sugar oxidizing to CO_2 and water are highly favourable.

Reversible and irreversible reactions and ΔG values

Strictly speaking, *all* chemical reactions are reversible. This might imply that the ΔG value must be negative in both directions – remembering that a reaction cannot occur unless the ΔG value is negative. The answer to this apparent paradox is that the ΔG of a reaction is not a fixed constant but varies with the reactant and product concentrations. The relationship is given later in this chapter. Thus in the reaction A$\rightleftharpoons$B, if A is at a high concentration and B at a low one, the ΔG may be negative in the direction A→B and, of course, positive from B→A. Reverse the concentrations and the ΔG can be negative for the reverse direction. A reaction will proceed to the point at which A and B are in concentrations at which the ΔG is zero in both directions and then no further net reaction can occur. This is the **chemical equilibrium point.**

If the ΔG of a biochemical reaction A $\rightleftharpoons$ B is small, significant reversibility may be possible in the cell because changes in reactant concentrations may be sufficient to reverse the sign of the ΔG of the reaction. If it is large, for all practical purposes the reaction is irreversible. In cellular reactions there is relatively little scope for concentration change in the metabolites (as reactants and products are called). The concentrations are always relatively low: 10^{-3}–10^{-4} M would be typical of many. The net result is that reactions with large negative ΔG values are irreversible because concentration changes are insufficient to reverse the sign of those values. (We later explain on page 191 why certain reactions may appear to contradict this.)

As a general guide, hydrolytic reactions in the cell – reactions in which a bond is split by water – are irreversible, in the sense that synthesis of substances in the cell does not occur by reverse of the hydrolytic reactions. Conversely reactions in which molecules become linked together by the elimination of a water molecule from between them will require input of energy from adenosine triphosphate (ATP). As you proceed through the chapters on metabolism you will become familiar with which reactions are reversible and which are irreversible.

To summarize, a reaction with a small ΔG is likely to be reversible in the cell, the direction being determined by small changes in reactant and product concentrations. A reaction with a large ΔG value will, in cellular terms, proceed in one direction only and, moreover, will proceed to virtual completion because the equilibrium point is so far to the side of the reaction. Put in another way, in the latter case, the ΔG of the reaction does not become zero until virtually all of the reactant(s) have been converted to product(s).

The importance of irreversible reactions in the strategy of metabolism

The major chemical processes of the cell usually involve not single reactions, but series of reactions organized into metabolic pathways in which the products of the first reaction are the reactants of the next and so on. In the example of the glycogen→lactic acid conversion in muscle mentioned earlier, a dozen successive reactions are involved.

An important general characteristic of metabolic pathways is that they are, as a whole, irreversible. Many of the individual reactions in a pathway may be freely reversible but it virtually always contains one or more reactions that cannot be directly reversed in the cell. Such irreversible reactions act as one-way valves and ensure that from a thermodynamic viewpoint the pathway can proceed to completion. Metabolic pathways are decisive – they go with a bang, as it were.

This is not the same as saying that overall physiological chemical processes are irreversible. Lactic acid formed from glycogen *is* converted back to glycogen in the body but by a different reaction pathway to that which produced it and, while many steps in the process are the simple reversal of those in the forward process, there are steps that cannot be reversed directly and alternative reactions are necessary. These involve the input of energy, which makes the alternative reactions also irreversible, but in the opposite direction. So the forward pathway (glycogen→lactic acid) is directly irreversible as in the reverse pathway (lactic acid→glycogen). A typical metabolic situation is:

$$A \rightleftharpoons B \;\circlearrowleft\; C \rightleftharpoons D \rightleftharpoons E \;\circlearrowleft\; F \rightleftharpoons \text{products.}$$

The red arrows represent irreversible reactions with large negative ΔG values.

A general biochemical principle emerges. *Whenever the overall chemical process of a metabolic pathway has to be reversed, the reverse pathway is not exactly the same as the forward pathway – some of the reactions are different in the two directions.*

Why is this metabolic strategy used in the cell?

There are basically two reasons. The alternative to the strategy we have outlined is that *all* the reactions of a metabolic pathway are reversible, that is,

$$A \rightleftharpoons B \rightleftharpoons C \rightleftharpoons D \rightleftharpoons E \rightleftharpoons F \rightleftharpoons \text{products.}$$

The major drawback of this arrangement is that the whole process is subject to mass action. If the concentration of A increases (perhaps due to something you ate), the reaction would swing to the right and more products would be formed. If substance A decreased in amount, some of the products would revert to A to maintain the equilibrium. Imagine that the pathway is for the synthesis of molecules such as the DNA of genes or of vital proteins, etc., and you can see how impossible this scenario is. It would be rather like building the walls of a house with the laying of bricks being a freely reversible process. The walls would rise and fall to maintain a constant equilibrium with the number of bricks lying on the ground.

There is a second, important, and related reason for having dissimilar reactions in the forward and reverse direction of metabolic pathways. Metabolism must be controlled. As already stated, during vigorous exercise muscles convert glycogen to lactic acid. During rest, the lactic acid is converted back to glycogen and the forward conversion switched off. To independently control the two directions (that is, to switch one on and the other off) there must be separate reactions to control; otherwise it would be possible only to switch both directions on and off together. Thus, the irreversible reactions are usually the control points. Metabolic control is a major subject that we will deal with in Chapter 16.

How are ΔG values obtained?

The ΔG value of a reaction, as explained, is *not* a fixed constant but the ΔG value of a reaction *under specified standard conditions* is a fixed constant. The standard conditions are with reactants and products at 1.0 M, 25°C, and pH 7.0; that is, the free energy difference between separated one-molar solutions of reactants and products. The ΔG value, *under these conditions*, is called the **standard free energy change of a reaction.** It is denoted as $\Delta G^{0'}$, the prime being used in biochemical systems to indicate that the pH is 7.0 rather than the value at pH 0, which is used in physical sciences.

The $\Delta G^{0'}$ value may often be calculated from readily available standard free energies of formation. For many reactions, the $\Delta G^{0'}$ value can be calculated by adding up the free energies of formation of the reactants and (separately) those of the products. The difference is the $\Delta G^{0'}$ value. An alternative is to determine experimentally the equilibrium constant for the reaction and from this the $\Delta G^{0'}$ value is easily calculated (see below).

There is a simple direct relationship between $\Delta G^{0'}$ values and ΔG values at given reactant and product concentrations. If, therefore, we know the relevant metabolite concentrations in the cell, the ΔG value for the reaction in the cell is readily determined (see Box 3.2, for an illustrative calculation). This has been done for a good many biochemical reactions so that such values are often quoted.

There is a snag – determining the actual concentrations of the thousands of metabolites in a cell is a nontrivial matter. They are present at low concentrations and are changing anyway due to metabolic activities in the cell. So, for many biochemical reactions, we do not have this data and therefore do not have the ΔG values. However, it is found that the more easily obtained $\Delta G^{0'}$ values usually correlate very well with known cellular happenings so that they are a useful guide in understanding metabolic reactions. Thus, although such values are not directly applicable to cells, because metabolite concentrations are never 1.0 M, they are frequently quoted to explain why certain reactions behave as they do. It is a compromise but a very useful one.

Standard free energy values and equilibrium constants

A particularly useful aspect of knowing the $\Delta G^{0'}$ value of a reaction is that, from it, the **equilibrium constant** of that reaction is readily determined. The equilibrium constant K'_{eq} of a reaction represents the ratio at equilibrium of the products to reactants. (The prime indicates that it is the K_{eq} at pH 7.0.) Thus, if one considers the reaction $A + B \rightleftharpoons C + D$, the K'_{eq} is calculated from the concentrations of A, B, C, and D present after the reaction has come to equilibrium – that is, when there is no net change in their concentrations,

$$K'_{eq} = \frac{[C][D]}{[A][B]}$$

The relationship between the value of the $\Delta G^{0'}$ and the K'_{eq} for a reaction is

$$\Delta G^{0'} = -RT \ln K'_{eq} = -RT\, 2.303 \log_{10} K'_{eq}.$$

R is the gas constant (8.315 J mol^{-1} K^{-1}) and T is the absolute temperature in degrees Kelvin (298 K = 25°C). At 25°C, RT = 2.478 kJ mol^{-1}. Thus, if the K'_{eq} is determined, the $\Delta G^{0'}$ for the reaction can be calculated and *vice versa*. Table 3.1 shows the relationship between $\Delta G^{0'}$ value and the K'_{eq} value for chemical reactions.

Table 3.1 Relationship of the equilibrium constant (K_{eq}) of a reaction* to the $\Delta G^{0'}$ value of that reaction.

Approximate $\Delta G^{0'}$ (kJ mol^{-1})	K'_{eq}
+17.1	0.001
+11.4	0.01
+5.7	0.1
0	1.0
−5.7	10.0
−11.4	100
−17.1	1000

* For a reaction $A + B \rightleftharpoons C + D$, the equilibrium constant is the molar concentration of $C \times D$ divided by that of $A \times B$.
$\left(K_{eq} = \dfrac{[C][D]}{[A][B]} \right)$; K'_{eq} is the K_{eq} at pH 7.0.)

The release and utilization of free energy from food

The conversion of simple precursor molecules to larger cellular molecules (such as DNA and proteins) involves increases in energy, and therefore cannot occur without energy input. The required energetic assistance comes ultimately from food breakdown. Chemical conversions involving positive free energy changes – the synthetic or 'building-up' processes – are collectively called **anabolism** or anabolic reactions. (The anabolic steroids of sporting ill-repute promote increase in body mass, hence their name.) The other half of metabolism consists of the 'breaking-down' reactions with negative free energy changes – the catabolic reactions, or collectively, **catabolism. Metabolism** is comprised of catabolism and anabolism. Catabolism of food liberates free energy, which is used to drive the synthesis of ATP from adenosine diphosphate (ADP) and inorganic phosphate. ATP is used to drive the energy-requiring processes of anabolism by the mechanism explained on page 34.

We can summarize the overall situation as in Fig. 3.1. Food oxidation releases free energy, captured in ATP, which is then used to drive energetically unfavourable processes. To keep the system going on a global scale, CO_2 and H_2O are reconverted during photosynthesis to food molecules (such as glucose or its derivatives) using light energy. This is converted by organisms

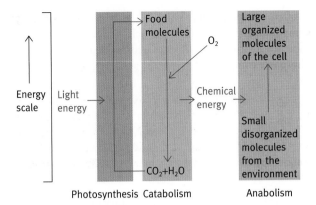

Fig. 3.1 The energy cycle in life.

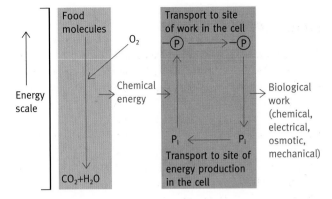

Fig. 3.2 The role of the phosphoryl group in the energy economy of a cell. —Ⓟ represents the phosphoryl group in a molecule whose hydrolysis to liberate P_i is associated with a $\Delta G^{0'} > 30$ kJ mol^{-1}.

to other food molecules such as fat. Although the assembly of large cellular structures involves a decrease in entropy (unfavourable), the oxidation of food molecules involves a greater entropy increase (favourable). The entropy change of the total system (cell and surroundings) is positive, and so the second law of thermodynamics is obeyed.

ATP is the universal energy intermediate in all life

As already explained in Chapter 1, oxidation of energy-supplying food molecules in the cell without appropriate coupling to the energy-requiring reactions would simply liberate heat and this cannot be used to do chemical or other work in the cell. Instead, the free energy change involved in food breakdown must be coupled to the energy-requiring processes. It does so by converting ADP plus inorganic phosphate to ATP, which is the universal energy intermediate of life.

ATP is a 'high-energy phosphate compound' (defined below). It is transported to wherever work is to be performed in the cell, where the attached phosphates, known as high-**energy phosphoryl groups**, are converted back to inorganic phosphate ions, with the liberation of the free energy that went into the formation of the groups in ATP. This does not mean that *direct* hydrolysis of the ATP occurs, as this could only liberate the energy as useless heat. The mechanisms by which the energy is harnessed for work will be described shortly. We now have to deal with why ATP has this central role. Before we get to its structure in detail we will first explain something about the chemistry of phosphates.

High and low energy phosphates

If we consider cellular compounds containing a phosphoryl group, they can be divided into two categories: **low-energy phosphate** compounds, the hydrolysis of which to liberate inorganic phosphate (P_i) is associated with negative $\Delta G^{0'}$ values

in the range of about 9–20 kJ mol^{-1}, and **high-energy phosphate** compounds with corresponding negative $\Delta G^{0'}$ values larger than about 30 kJ mol^{-1}. The concept of a high-energy phosphate compound used to be described in terms of the 'high-energy phosphate bond', but this use has been abandoned because a high-energy bond in chemical terms refers to a bond where the breakage requires a large input of energy, the reverse of the intended biochemical concept. The high-energy phosphoryl group can be regarded as the universal energy currency of the living cell. This concept is illustrated in Fig. 3.2 which is a refinement of Fig. 3.1.

What are the structural features of high-energy phosphate compounds

Phosphoric acid, H_3PO_4, is an oxyacid of phosphorus with three dissociable protons as shown below.

BOX 3.1 Henderson–Hasselbalch equation calculation

$$pH = pK_{a_2} + \log\frac{[\text{salt}]}{[\text{acid}]};$$

$$7.4 = 7.2 + \log\frac{[HPO_4^{2-}]}{[HPO_4^-]}.$$

Therefore,

$$\log\frac{[HPO_4^{2-}]}{[HPO_4^-]} = 0.2$$

so that

$$\frac{[HPO_4^{2-}]}{[HPO_4^-]} = 1.58.$$

At normal cell pH, a mixture of both singly negatively and doubly negaively charged phosphate ions exists in the solution. In biochemistry, this mixture of phosphate ions is symbolized as P_i, the i indicating 'inorganic'; it represents the lowest energy form of phosphate in the cell and can be regarded as the ground state of phosphate when considering its energetics. The degree of ionization of the molecule is a function of the three dissociation constants. These are known as pK_a values; a pK_a is the pH at which there is 50% dissociation of the group. An explanation of pK_a values and buffers is given in the appendix to this chapter; this should be studied if you are not already familiar with the subject. At physiological pH (say 7.4) the group with a pK_a of 2.2 will be fully ionized and that with a pK_a of 12.3 will be undissociated. The group with a pK_a value of 7.2 will be partially dissociated. The actual degree of the latter can be calculated from the **Henderson–Hasselbalch equation** (Box 3.1).

Inorganic phosphate, esterified with an alcohol, is a **phosphate ester.**

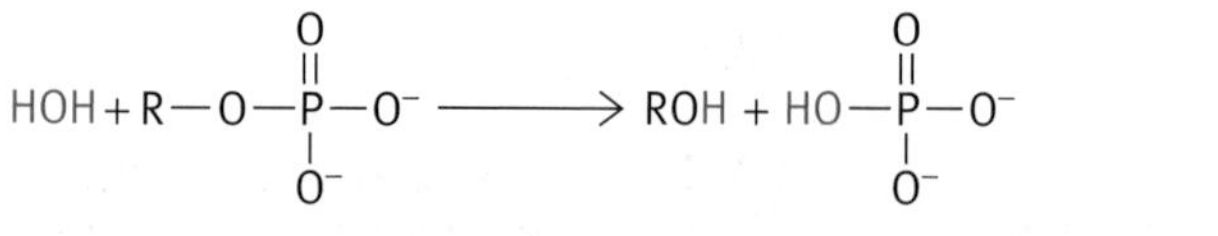

Phosphate ester Alcohol Inorganic phosphate (P_i)

The ΔG^0 for the hydrolysis of this class of ester is roughly of the order of -12.5 kJ mol^{-1}, resulting in the equilibrium for the hydrolytic reaction being strongly to the side of hydrolysis. In the cell the reverse reaction does not occur. However, relative to the pyrophosphate bonds in ATP, it is a low energy phosphate compound.

As well as forming phosphate esters with alcohols, Pi can form a **phosphoric anhydride** called **inorganic pyrophosphate** (*pyro* means fire; pyrophosphate can be made by driving off water from P_i at high temperatures) and in biochemistry it is often written as $\mathbf{PP_i}$. The $\Delta G^{0'}$ for the hydrolysis of this compound is -33.5 kJ mol^{-1}, which is much higher than the value for hydrolysis of the ester phosphate. It is a high-energy phosphate compound.

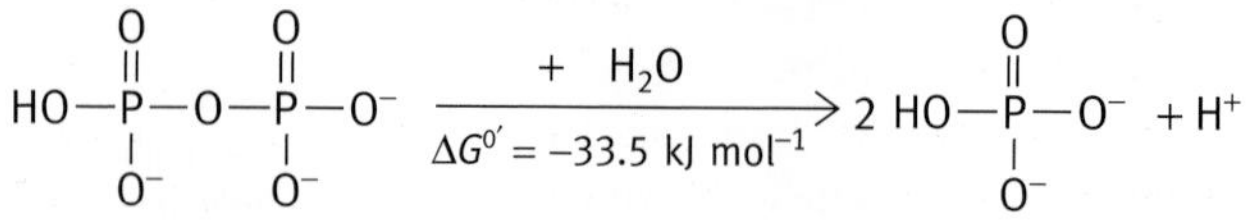

Inorganic pyrophosphate (PP_i) Inorganic phosphate (P_i)

We will explain the high-energy nature of the anhydride by referring to inorganic pyrophosphate because it makes the explanation simpler. (PP_i is not commonly an energy donor in the sense that ATP is). The very high free energy release associated with the hydrolysis of the phosphoric anhydride group is due to several factors. When the pyrophosphoryl group is hydrolysed, the electrostatic repulsion between the two negatively charged phosphoryl groups is relieved. This factor is made clear by the reflection that, in the reverse reaction to synthesize a pyrophosphoryl bond, the electrostatic repulsion of the two ions has to be overcome in bringing them together. The phosphoric anhydride bond formation might be likened to compressing a spring.

Secondly, the products of the reaction (in this case, two P_i ions) are **resonance stabilized** – they have a greater number of possible resonance structures (see below) than has the pyrophosphate structure. This increases the entropy and therefore decreases the energy level of the products so that on breaking the bond there is a bigger yield of energy.

In an inorganic phosphate ion, all of the P—O bonds are partially double-bonded in character rather than the proton being associated with any one oxygen, resulting in an increase in entropy and a lowering of their energy level. The main resonating forms of the phosphate ion are shown below:

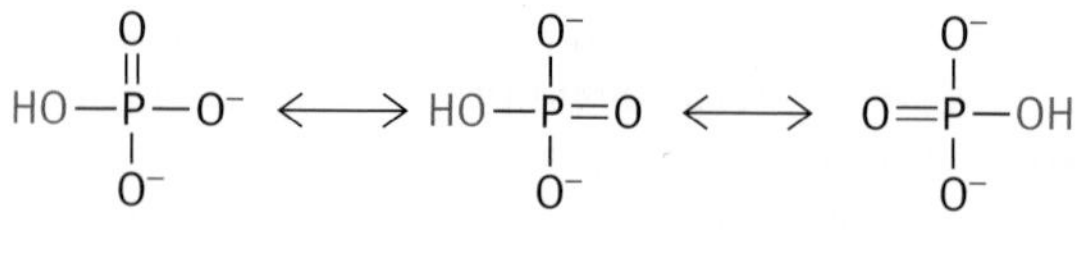

Note that the ↔ symbol has a special meaning in chemistry; it is not the same as ⇄. It does not imply that the different ionized forms are interconverting but that the structure which exists is not any of the forms shown but is an intermediate one in which all oxygens have a partial negative charge and the proton is not associated with any one form. The same comments apply to the structures of the resonating forms of carboxylic acid and guanidino compounds described below.

These factors mean that phosphoric anhydrides have equilibrium constants decisively in favour of P_i formation (equals a large negative $\Delta G^{0'}$ value). These considerations apply to any phosphoric anhydride group and in particular to those in ATP.

However, the factors we have discussed above do not apply to hydrolysis of ester phosphates, which explains why the energy release from hydrolysing them is much less.

The phosphoric anhydride structure discussed above is not the only biological high-energy phosphate compound (though it is the predominant one). Three other types are found.

The first is the mixed anhydride between phosphoric acid and a carboxyl group (often called **acyl-phosphates**), hydrolysis of which by the reaction shown has a very large negative $\Delta G^{0'}$ value (-49.3 kJ mol^{-1} in a typical case). The large free energy change is associated with the resonance stabilization of both products, namely, P_i and the carboxy acid (the latter illustrated below). This type of molecule occurs in metabolic reactions.

$$R{-}C\overset{\displaystyle O}{\underset{\displaystyle O}{\big\|}}\; \overset{O}{\underset{O^-}{\overset{\|}{\underset{|}{O{-}P{-}O^-}}}} + H_2O \longrightarrow R{-}C\overset{\displaystyle O}{\underset{\displaystyle O^-}{\big\|}} + HO{-}\overset{O}{\underset{O^-}{\overset{\|}{\underset{|}{P}}}}{-}O^- + H^+$$

Acylphosphate
anhydride

$$R{-}C\overset{\displaystyle O^-}{\underset{\displaystyle O}{}}$$

Resonance
stabilization

The second structure is that of **guanidino-phosphate**, the hydrolysis of which to produce P_i also has a very large negative $\Delta G^{0'}$ value (-43.0 kJ mol^{-1} in a typical case). Again this is due to resonance stabilization of both products. An example is creatine phosphate in muscle (page 126).

$$\overset{+}{N}H_2{\underset{\underset{\underset{R}{|}}{\underset{NH}{|}}}{\overset{\|}{C}}}{-}NH{-}\overset{O}{\underset{O^-}{\overset{\|}{\underset{|}{P}}}}{-}O^- + H_2O$$

Guanidino-
phosphate

$$HO{-}\overset{O}{\underset{O^-}{\overset{\|}{\underset{|}{P}}}}{-}O^- +$$

$$\overset{+}{N}H_2{\underset{\underset{R}{|}}{\underset{NH}{|}}}{\overset{}{C}}{-}NH_2 \longleftrightarrow NH_2{\underset{\underset{R}{|}}{\underset{+NH}{\|}}}{\overset{}{C}}{-}NH_2 \longleftrightarrow NH_2{\underset{\underset{R}{|}}{\underset{NH}{|}}}{\overset{}{C}}{=}\overset{+}{N}H_2$$

Resonance stabilization

A third type is in a different category – an enol-phosphate structure. This is found in the metabolite, **phosphoenolpyruvate**. This looks an unlikely candidate for high-energy status, but, on removal of the phosphate group, the **enol pyruvate** structure so formed spontaneously rearranges into the keto form of pyruvate, the equilibrium of this being far to the right.

$$\overset{CH_2}{\underset{\underset{O}{\overset{C}{\big\|}}\overset{}{\underset{O^-}{C}}}{\overset{|}{C}}}{-}O{-}\overset{O}{\underset{O^-}{\overset{\|}{\underset{|}{P}}}}{-}O^- + H_2O \longrightarrow \overset{CH_2}{\underset{\underset{O}{\overset{C}{\big\|}}\overset{}{\underset{O^-}{C}}}{\overset{|}{C}}}{-}OH + HO{-}\overset{O}{\underset{O^-}{\overset{\|}{\underset{|}{P}}}}{-}O^-$$

Phosphoenolpyruvate
(Enol phosphate)   Unstable

$$\overset{CH_3}{\underset{\underset{O}{\overset{C}{\big\|}}\overset{}{\underset{O^-}{C}}}{\overset{|}{C}{=}O}}$$

Stable

As a result of this the conversion of phosphoenolpyruvate to P_i and the keto form of pyruvate has a $\Delta G^{0'}$ value of -61.9 kJ mol^{-1}. Phosphoenol pyruvate is a component of the glycolytic pathway which harvests energy during the breakdown of glucose.

The structure of ATP

We have so far spoken of the phosphoryl groups present in high-energy phosphate compounds in general terms. Thus, in Fig. 3.2, P_i is shown as being elevated to a 'high-energy phosphoryl group' and then transported around the cell to where it is needed to supply the energy for work. Clearly, the phosphoryl group (—Ⓟ) must be covalently attached to another molecule that acts as its carrier within the cell. Which brings us to the next question.

What transports the —Ⓟ around the cell?

The general carrier is **adenosine monophosphate (AMP)** (the structure is shown in Fig. 3.3). AMP is, by itself, a relatively low-energy phosphate ester. This figure gives a diagrammatic representation of adenosine and its derivatives without any detailed structures since they are not needed in this context. (However,

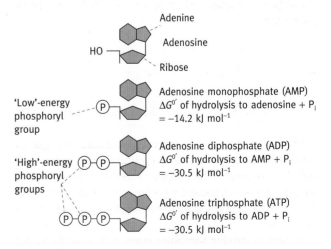

Fig. 3.3 Diagrammatic representation of adenosine and its phosphorylated derivatives. Adenosine is a nucleoside with ribose and adenine as its components.

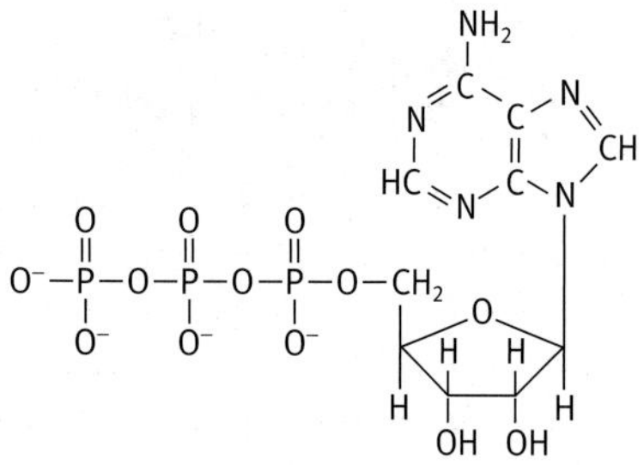

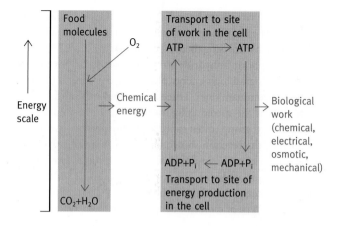

Fig. 3.4 The structure of adenosine triphosphate (ATP).

Fig. 3.5 The role of ATP in the energy economy of a cell. Note that some types of work involve breakdown of ATP to AMP, but, as described in the text, this does not change the concept given here.

for reference purposes, Fig. 3.4 gives the full structure of ATP.) AMP carrying a single —Ⓟ group is called ADP or **adenosine diphosphate**. AMP carrying two —Ⓟ groups is ATP or **adenosine triphosphate**. ATP therefore can be written as AMP—Ⓟ—Ⓟ. The two terminal phosphates are attached to AMP by phosphoric anhydride linkages of the same type as already met in PP$_i$, and the reasons given earlier for the large $\Delta G^{0'}$ for hydrolysis off of these two groups apply equally here. Note that, as stated, AMP itself is a low-energy ester phosphate and does not itself directly participate in the energy cycle except as carrier for the two —Ⓟ groups. You will see from Fig. 3.3 that the $\Delta G^{0'}$ of hydrolysis of each of the two terminal groups is -30.5 kJ mol^{-1}; that is, of the high-energy type. When food is oxidized or otherwise broken down, the free energy released, as explained, is coupled to the conversion of ADP (or AMP—Ⓟ) to ATP:

$$\text{ADP} + \text{P}_i \xrightarrow[\text{+ energy}]{} \text{ATP} + \text{H}_2\text{O}$$

Each cell has only a small quantity of ATP in it at any one time. The amount would last only a very short time and it cannot get any from the outside; ATP, ADP, or AMP cannot diffuse through the cell membrane because they are highly charged (Chapter 7). Each cell has to synthesize the entire molecule itself. ATP thus 'turns over' or cycles very rapidly, by which we mean it breaks down to ADP and P$_i$ and resynthesized to ATP. We can modify the diagram given in Fig. 3.2 to include this fact (Fig. 3.5).

How does ATP drive chemical work?

Suppose the cell needs to synthesize X—Y from the two reactants, X—OH + Y—H, and the $\Delta G^{0'}$ change involved in the conversion is 12.5 kJ mol^{-1}. The simple reaction XOH + YH $\rightarrow$ XY + H$_2$O cannot occur to any significant extent because the $\Delta G^{0'}$ is positive and the equilibrium is far to the left. The usual solution is to use **coupled reactions** involving ATP breakdown, as shown below. Coupled reactions are two or more reactions in which the product of one becomes the reactant for the next. No physical coupling is necessarily involved – they simply have to be present in the same chemical system, for example the same organelle in the same cell. Note that all reactions in the cell

involving ATP must be enzymically catalysed – the fact that hydrolysis of each of the two phosphoric anhydride groups is strongly exergonic (giving out energy) does not mean that ATP is an unstable or highly reactive molecule. For clarity ATP will be represented as AMP—Ⓟ—Ⓟ below. The coupled reactions to synthesize X—Y are as follows:

Reaction 1:   XOH + AMP—Ⓟ—Ⓟ $\rightarrow$ X—P + AMP—Ⓟ
Reaction 2:   X—P + YH $\rightarrow$ X—Y + P$_i$
Sum of 1 + 2:   XOH + YH + AMP—Ⓟ—Ⓟ $\rightarrow$ X—Y + AMP—Ⓟ + P$_i$

In reaction 1, a phosphoryl group is transferred from ATP to X—OH forming X—P. In the second reaction the phosphoryl group is replaced by Y, liberating inorganic phosphate and forming X—Y. Both reactions may occur on a single enzyme without the X—P ever being free, but in other cases it may leave the first enzyme and diffuse to a second enzyme, which carries out the second reaction. From an energetic viewpoint the two are the same. The overall $\Delta G^{0'}$ for coupled reactions is the arithmetic sum of the $\Delta G^{0'}$ values of the component reactions. The $\Delta G^{0'}$ for the XOH + YH $\rightarrow$ X—Y + H$_2$O we take to be 12.5 kJ mol^{-1}; the $\Delta G^{0'}$ for the reaction ATP + H$_2$O $\rightarrow$ ADP + P$_i$ is -30.5 kJ mol^{-1}. Therefore, the $\Delta G^{0'}$ of the overall process is -18.0 kJ mol^{-1}, a strongly exergonic process that will proceed essentially to completion – the equilibrium constant will be about 10^3 – which means that the reactions can proceed to approximately 99.9% to the side of X—Y formation (see Table 3.1). It is to be noted that the use of ATP here involves phosphoryl group transfer from ATP to one of the reactants and only then is P$_i$ liberated. *Direct* hydrolysis of ATP to ADP and P$_i$ is not occurring. In the cell, the actual ΔG value of the ATP breakdown to ADP and P$_i$ is considerably larger than the $\Delta G^{0'}$ value for this, because ΔG values are affected by reactant concentrations, and cellular levels of ATP, ADP, and P$_i$ are, of course, nowhere near 1.0 M (the standard concentrations specified in $\Delta G^{0'}$ determinations). If the actual cellular concentrations of

BOX 3.2 Calculation of ΔG value

$$\Delta G = \Delta G^{0'} + RT2.303\log_{10}\frac{[\text{products}]}{[\text{reactants}]}$$

where R is the gas constant (8.315×10^{-3} kJ mol^{-1} K^{-1}); and T is the temperature in degrees Kelvin (298 K = 25°C).

The concentrations of ATP, ADP, and P_i will vary from cell to cell. ATP 'levels' generally fall in the range 2–8 mM, ADP levels are about one-tenth of this, and P_i values are similar to those of ATP. However, for simplicity let us assume a situation in which all three are at 10^{-3}M. (For reactions in dilute aqueous solution, water has the value of 1.) Substituting these values into the equation:

$$\Delta G = \Delta G^{0'} + RT2.303\log_{10}\frac{[\text{ADP}][\text{P}_1]}{[\text{ATP}]},$$

we obtain:

$$\Delta G = -30.5 \text{ kj mol}^{-1} + (8.315 \times 10^{-3})(298)$$

$$\times 2.303\log_{10}\frac{[10^{-3}][10^{-3}]}{[10^{-3}]}$$

$$= -30.5 - 17.1 = -47.6 \text{ kj mol}^{-1}.$$

these components are known, the ΔG value for the hydrolysis of ATP to ADP + P_i can be calculated (see Box 3.2).

The reaction sequence given here is used for many biochemical reactions coupled to ATP breakdown. But, more commonly, for the synthesis of molecules such as nucleic acids and proteins, the cell uses an even more energetically effective trick to guarantee direct irreversibility of the reaction. Instead of breaking off one —Ⓟ of ATP to P_i it releases two. The negative $\Delta G^{0'}$ value that results from this is so large that the equilibrium is such that the reaction is completely irreversible in the cell. The way this is done is as follows, again taking XOH + YH → X—Y + H_2O as the example:

Reaction 1: XOH + AMP—Ⓟ—Ⓟ → X—AMP + PP_i
Reaction 2: X—AMP + YH → X—Y + AMP
Reaction 3: PP_i + H_2O → 2P_i
Sum: XOH + YH + AMP—Ⓟ—Ⓟ + H_2O → X—Y + AMP + 2P_i

In the first reaction, instead of a phosphoryl group being transferred to X—OH, the AMP group is attached, displacing the two terminal phosphoryl groups as inorganic pyrophosphate. The X—AMP reacts in a second reaction with X—OH, displacing the AMP and forming the desired X—Y. The two reactions occur on the surface of a single enzyme. So far this is energetically little different from the first scheme in which X—OH is phosphorylated, for only one phosphoric anhydride group has been broken. However a different enzyme hydrolyses the inorganic pyrophosphate. Assume that the $\Delta G^{0'}$ for the reaction from XOH + YH → X—Y + H_2O is 12.5 kJ mol^{-1}: the $\Delta G^{0'}$ for the reaction ATP + H_2O → AMP + PP_i is −32.2 kJ mol^{-1}; that for PP_i hydrolysis is −33.5 kJ mol^{-1}.

The $\Delta G^{0'}$ for the reaction ATP + H_2O → AMP + 2P_i is −65.7 kJ mol^{-1}. The overall process of synthesizing X—Y at the expense of ATP breakdown thus has a negative $\Delta G^{0'}$ of 65.7 − 12.5 = 53.2 kJ mol^{-1}, a very large value indeed.

The mechanism depends on the fact that inorganic pyrophosphate, PP_i, is broken down rapidly in the cell. We have stated earlier that ATP hydrolysis must be coupled to a reaction to drive it, and that simple direct hydrolysis would be useless. The fact that PP_i hydrolysis helps to drive the formation of X—Y is that it removes product (PP_i) of the reaction. This has the effect of swinging the reaction to the right. It is an equilibrium mass action effect. This explains why cells contain an enzyme, called **inorganic pyrophosphatase**, that catalyses the reaction (already shown above).

This enzyme, once regarded as an unimportant one, is a driving force in biochemical syntheses and is widely distributed. As already implied, this enzyme simply has to be present in the same chemical system (in the same cell) without physical association with the synthesis of X—Y. It is the free energy change for the overall conversion which is important. The total free energy change is the arithmetic sum of that of all the reactions.

How does ATP drive other types of work?

As well as performing chemical work, ATP breakdown powers muscle contraction, the generation of electrical signals, and pumping of ions and other molecules against concentration gradients, and much else. The mechanisms are, in principle, the same. Whatever the process, provided ATP breakdown is coupled into the mechanism the resultant free energy liberation will drive it. Many of these processes will be dealt with in subsequent chapters.

High-energy phosphoryl groups are transferred by kinase enzymes

Many ATP-requiring chemical syntheses in the cell produce AMP + 2P_i, not ADP + P_i. AMP cannot, itself, be converted to ATP by the foodstuff-oxidation system; only ADP is accepted as shown in Fig. 3.5. However, AMP is rescued by an enzyme that transfers —Ⓟ from ATP to AMP. The reaction is:

$$\text{AMP} + \text{AMP—Ⓟ—Ⓟ} \rightleftarrows 2\text{AMP—Ⓟ}$$

or, as written more conventionally:

$$\text{AMP} + \text{ATP} \rightleftarrows 2\text{ADP}.$$

The enzyme is called **AMP-kinase.** Kinase is the term for carrying a phosphoryl group from ATP to elsewhere; AMP-kinase means it transfers the group to AMP (AMP is also called adenylic acid; hence AMP-kinase may also be called adenylate kinase). Note that, in this case, the —Ⓟ group is transferred directly from one molecule to another without hydrolysis or significant release of energy. You will find as you go through the book that such freely reversible 'shuffling' at the 'high-energy level' occurs frequently. The ADP (AMP—Ⓟ) so produced is now accepted by the food-oxidation system and converted to ATP.

In addition to these transfers at a high energy level a multitude of specific kinases transfer phosphoryl groups to other molecules resulting in the formation of relatively low-energy phosphate esters. Such transfers are not reversible because of the large negative ΔG value involved. You will come across these kinases in almost every aspect of the subject.

To give perspective to where we are, we have not in this account dealt with the mechanisms by which ATP is synthesized from ADP and P_i at the expense of food catabolism. This is a very large topic that forms a substantial part of the chapters on metabolism later in this book. Also, we have dealt with the utilization of ATP energy to perform work only in general terms so far – as you progress through the book you will encounter example after example of this in later chapters, for ATP utilization is involved in virtually all biochemical systems. We come now to a change of topic though it is still concerned with free-energy changes in chemical processes.

Energy considerations in covalent and noncovalent bonds

The chemistry we have been discussing so far in this chapter involves covalent bonds, which are strong. They are formed by two atoms sharing a pair of electrons, a simple example being the formation of a hydrogen molecule from two hydrogen atoms:

$$H\cdot + H\cdot \rightarrow H:H$$

Each electron is attracted to the positive nucleus of both atoms, which holds the two together. When any chemical bond is formed, energy is liberated as required by the second law. To break a chemical bond the same amount of energy has to be provided as was liberated in its formation. Covalent bonds are needed to form stable molecules such as glucose. They are very stable because large amounts of energy are liberated in their formation. To give an example, in forming a molecule of O_2 from two oxygen atoms the negative standard free energy value is about 460 kJ/mol^{-1} so that the equilibrium is such that oxygen gas has a negligible number of free atoms. Collisions between molecules in solution can provide the energy required to break some chemical bonds. However, the average kinetic energy of molecules in solution at 25° is only in the range of about 4–30 kJ/mol^{-1}, far below the range needed to *destroy* a covalent bond. In biochemical reactions, the breakage of a covalent bond is accompanied by the simultaneous formation of another bond via the transition state (page 88) so that the net energy change involved is far less than the very large value required to *destroy* a covalent bond.

Now we come to a different type of molecular interaction involving **noncovalent** bonds, also referred to as secondary or weak bonds. The latter two terms belie their importance in life. There are few, if any, areas in the mechanism of life that do not depend on them.

Noncovalent, weak, or secondary bonds do not involve sharing of a pair of electrons between atoms; they are electrostatic attractions of several types. Their free energies of formation are about 0.5–40 kJ/mol^{-1} ; this means that the kinetic energy of thermal motion of water molecules is sufficient to disrupt them so that they are continually being formed and disrupted spontaneously. Their energies are so small that there is not the energy barrier to molecules interacting *via* noncovalent bonds so there is (usually) no need for enzyme catalysis to make or break them.

Noncovalent bonds would not be any good at forming stable discrete single molecules such as glucose. However, if a sufficient number of such bonds are present they can hold molecules together to form larger molecular structures.

Noncovalent bonds are the basis of molecular recognition and self-assembly of life forms

In Chapter 1 (page 9) we explained that one of the fundamental themes essential to life was that protein molecules can recognize other chemical structures and bind to them. There is virtually no biochemical process in which this is not a vital part. It is, for example, the reason why life is a self-assembling process. Produce the correct proteins inside a cell in the correct order and quantities and they will interact to assemble the organism. As we phrased it earlier, produce rabbit proteins and a rabbit will assemble. This is the quasi-incredible way in which a linear one-dimensional code in DNA can produce such a variety of three-dimensional living organisms.

How is this achieved? Noncovalent bonds form between appropriate atoms of molecules provided they are close enough to each other to do so. Because of their weakness many such bonds must be formed to hold them together.

Two different protein molecules can get close enough only if there are patches of structural complementarity; in other words only if they have complementary shapes so that they can fit together so that noncovalent bonds can form at the attachment points. There must be large enough numbers of bonds formed to hold the molecules together. Thirdly the attachment points must have the appropriate chemical groups to favour weak bond formation. Put all these together and the chance of random protein molecules being joined is remote. Only proteins that have evolved the appropriate structural complementarities will associate. Thus weak bond-dependant associations can be highly specific. Only appropriate associations will occur and therefore only molecular assemblies appropriate to the particular organism can form. Life assembles itself correctly in this way given the correct proteins. The same applies to enzymes recognizing their substrates, to gene control proteins recognizing their genes (Chapters 6 and 23).

Noncovalent bonds are also important in the structures of individual protein molecules and other macromolecules

So far we have been discussing the role of weak bonds in causing specific associations of protein molecules but they are also important in the structures of individual proteins. Their basic structure is the covalent linking of a long sequence of amino acids (Chapter 4) to form the long polypeptide chain. However, an extended polypeptide chain is rarely a functional protein. The chains fold up into compact three-dimensional shapes. The precise folding and therefore the external functional groups and their spatial arrangement on a protein is determined entirely in most cases, or mainly in some, by noncovalent interactions between amino acid residues of the polypeptide (see page 52 of Chapter 4). A few proteins which must face the rigours of the extracellular world such as insulin released into the blood or digestive enzymes released into the gut are stabilized in their folded state by a small number of covalent S—S bonds (page 54) resembling a few steel rivets in the structure.

There is a good reason for using weak bonds to produce the final compact structures, the reason being that they are weak. Many or most proteins are molecular machines that need to change their shape somewhat as they function. They have to undergo conformational change in response to ligand binding. Weak bonds allow this. To give another illustration, DNA depends on hydrogen bonding (see below) of the two strands; these must be separated for replication and gene activity. The weak bonds are sufficiently strong to hold the strands together and sufficiently weak to be broken as needed.

Types of noncovalent bonds

The three relevant types of bond or forces are shown in Table 3.2. All are electrostatic attractions between appropriate atoms of covalent molecules, each with a different strength.

Table 3.2 Noncovalent bonds and their characteristics

	Bond strengths (kJ/mol^{-1})
Wan der Waals interactions	0.4–4
Hydrogen bonds	12–30
Ionic bonds	20

For comparison covalent bonds typically have bond strengths in the region of a few hundred. The value for a C—C bond is about 350 kJ/mol^{-1}.
Additionally hydrophobic forces or interactions are not bonds but are important in structural determinations. See text.

Ionic bonds

An ionic bond is the attraction between negative and positive groups of ions. A typical case is the attraction between a negatively charged carboxylate ion and a positively charged amino group. The bond has no intrinsic directionality but since the groups are present in defined locations in the molecule(s) they can be involved in specific molecular recognition.

$$—COO^- — — —H_3^+N—$$

Hydrogen bonds

These are electrostatic attractions between atoms of polar molecules, but the attractions are not due to fully separated ionized groups but to a much weaker form of positive and negative charge separation, the electric dipole moment. Although a covalent bond is overall electrically neutral, the bond can have a weakly polar nature. An example is the water molecule.

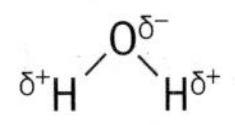

$$_{\delta^+}H \diagdown O^{\delta^-} \diagup H_{\delta^+}$$

It is asymmetric in shape because the hydrogen atoms are attached to the oxygen with an angle of 104.5° between them, giving the molecule an asymmetric shape with the oxygen at one end and the two hydrogens at the other. Although the covalent —O—H bonds are the result of a shared pair of electrons, the electrons are not shared exactly equally because oxygen is an electronegative atom. It is greedy for electrons and takes more than its half share. This gives the hydrogens a partial positive charge (δ^+) and the oxygen a partial negative (δ^-) one so that a weak attraction occurs between the O and H atoms of adjacent water molecules, known as hydrogen bonds. The partial charges on water molecules have a profound effect on electrical attractions between other polar molecules because they partially neutralize the charges. Each water molecule is bonded to four other molecules because the oxygen atom can form two hydrogen bonds and the hydrogen atoms one each. This means that bulk water has a cohesive structure.

A hydrogen atom of any molecule can participate in hydrogen bond formation provided it is attached to an electronegative atom. The main relevant electronegative atoms in biochemistry are nitrogen and oxygen. The bond must involve also an electronegative acceptor atom – also usually nitrogen or oxygen.

The types of hydrogen bond common in biochemistry are illustrated here with hydrogen bonds shown as dashed lines.

$$—O—H— — —O—$$
$$—O—H— — —N—$$
$$—N—H— — —O—$$
$$—N—H— — —N—$$
$$—N^+—H— — —O—$$

In contrast to ionic bonds, *the hydrogen bond is highly directional* and of maximal strength when all of the atoms are in a straight line.

van der Waals forces

This is the collective term used to describe a group of weak interactions between closely positioned atoms. Atoms are electrically neutral overall but very weak transitory polarities exist. The electrons of any atom are in constant motion; at any one time they may not be evenly distributed so that transitory fluctuations in electron density around the nucleus of the atoms occur. This means that the negative charge distribution in the atom can fluctuate, which causes, at any instant, one part having a slight positive charge and the other a slight negative one. This in turn can affect electron distribution in neighbouring atoms. A negative charge on one atom will tend to repel electrons in its neighbour, thus inducing a local positive charge resulting in an electrostatic attraction between the two atoms.

Van der Waals forces can operate between *any* two atoms, which must be positioned close together so that their electron shells are almost touching. The attraction between them is inversely proportional to the sixth power of the distance, so precisely close positioning is essential. If atoms tend to become too close together, their electron shells overlap and a repulsive force is generated. In short, precise positioning is a prerequisite for van der Waals attractions to arise. This means that for them to form between, say, two different proteins these must have precise complementarity of structure at contact points for van der Waals forces to be generated. Although the attractions are extremely weak they can exist in large numbers.

Hydrophobic force

This is also called **hydrophobic interactions**. The force does not involve bonds between hydrophobic molecules (Chapter 1, page 7) but it causes hydrophobic molecules to associate together. A hydrophilic (polar) molecule such as sugar that can form hydrogen bonds is soluble in water because, although it disrupts hydrogen bonding between water molecules (energy-requiring), it can itself form bonds with the latter (energy-releasing), making the process energetically feasible. Salts such as NaCl are soluble because the Na^+ and Cl^- ions become surrounded by hydration shells in which the ion-water attractions exceed those between the ions themselves and, when separated, there is a large increase in entropy from the crystalline state. If by contrast an attempt is made to dissolve a nonpolar substance such as olive oil in water, the oil molecules get in the way of hydrogen bonding between water molecules. Since hydrogen bonding is highly directional, the water molecules around the oil molecules rearrange themselves so that none of their bonding sites are aimed at the oil molecules. This more highly ordered arrangement is at a lower entropy level than that of randomly arranged water molecules (a higher energy level) and so the solubilization of the oil is opposed by the second law of thermodynamics. The nonpolar molecules are forced to associate together so as to present the minimum oil/water interface area. The olive oil forms droplets and then a separate layer, which is the minimum free energy state. Hydrophobic groups occur in proteins, DNA, and other cellular molecules. Hydrophobic forces play a crucial role in the structure of these molecules, as will be seen later in this book; it is remarkable how the necessity of 'hiding' these hydrophobic groups from water determines so much in living cells.

With that introduction to energy considerations in life we come to the next five chapters, in which protein structure and function are the main themes – protein structure, methods in protein investigation, enzymes, membranes, molecular motors, and the cytoskeleton. These topics are necessary for understanding metabolism, which is the major area covered in the middle section of the book.

Appendix: Buffers and pK_a values

It is very important in biochemistry to understand what buffers and pK_a values are. We have placed this is an appendix to avoid disrupting the text and because many will have dealt with it in their chemistry studies.

The pH of a living cell is maintained in the range 7.2–7.4. (pH is the negative logarithm of the H^+ concentration expressed in moles per litre). Special situations occur, such as in the stomach, where HCl is secreted, and in lysosomes (page 416) into which protons are pumped to maintain an acid pH, but, otherwise, the pH of cells and of circulating fluids is maintained within narrow limits. This is despite the fact that metabolic processes, such as lactic acid and acetoacetic acid formation and CO_2 conversion to carbonic acid (H_2CO_3) in the blood, occur on a large scale.

This pH stability is largely due to the buffering effect of weak acids. An acid in this context is defined as a molecule that can release a hydrogen ion, and a base as a proton acceptor.

A carboxylic acid dissociates, liberating a proton:

$$R\ COOH \rightleftharpoons R\ COO^- + H^+$$

Written in a more general form, the equation for acids is:

$$HA \rightleftharpoons H^+ + A^-$$

Acids vary in their tendency to dissociate. Stronger acids do so more readily than weaker ones – this is why, say, a 0.1 M solution of formic acid (HCOOH) has a lower pH than a 0.1 M solution of acetic acid (CH_3COOH). The tendency to dissociate is quantitated as a dissociation constant, K_a for each acid: the larger the value of K_a, the greater the tendency to dissociate and the stronger the acid:

$$K_a = \frac{[H^+][A^-]}{[HA]}.$$

For acetic acid $K_a = 1.74 \times 10^{-5}$. This value is not much used, as such, by biochemists, for there is another way of expressing the strength of the acid by a much more convenient term – the pK_a value. The two are related by the equation:

(a) for the molecule RCOOH with a pK_a of 4.2:

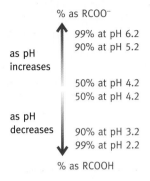

(b) for the molecule RNH_2 with a pK_a of 9.2:

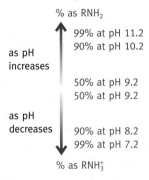

Fig. 3.6 Effect of pH on the ionization of **(a)** —COOH and **(b)** —NH_2 groups. Kindly provided by R. Rogers.

$$pK_a = -\log K_a$$

Thus, acetic acid has a pK_a of 4.76 and formic acid a pK_a of 3.75. These values represent the pH at which an acid is 50% dissociated. As the pH increases, the acid becomes more dissociated; as it decreases, the reverse. This is because the $HA \rightleftharpoons H^+ + A^-$ equilibrium is affected by the H^+ concentration as would be expected (Fig. 3.6).

An amine base also has a pK_a value because the ionized form can dissociate to liberate a proton and, in this sense, is an acid:

$$R\,NH_3^+ \rightleftharpoons R\,NH_2 + H^+$$

In this case, as the pH increases, the dissociation decreases; with both carboxylic acids and amine bases, an increase in H^+ concentration causes increased protonation. The difference is that protonation of a carboxylic acid reduces the amount of ionized form while, with amine bases, the proportion in the ionized form increases (Fig. 3.6).

pK_a values and their relationship to buffers

If you take a 0.1 M solution of acetic acid and gradually add 0.1 M NaOH, measuring the pH at each step, the curve shown in Fig. 3.7 is obtained.

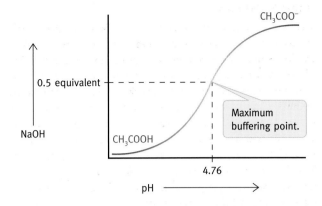

Fig. 3.7 Titration curve for acetic acid ($pK_a = 4.76$).

At the beginning of the titration, the added OH^- neutralizes existing H^+ (the acetic acid is slightly dissociated) and the pH rises rapidly. However, as the pH begins to approach the pK_a value of acetic acid, as more NaOH is added, the acetic acid dissociates into the acetate ion liberating hydrogen ions, which neutralize the hydroxide ions so that the pH change is relatively small until the acetic acid is all ionized. The reaction is:

$$CH_3COOH + OH^- \rightarrow CH_3COO^- + H_2O.$$

(Note that sodium acetate is fully dissociated.)

Similarly, from the other side of the pH scale, additions of H^+ cause little change to the pH because of the reaction:

$$CH_3COO^- + H^+ \rightarrow CH_3COOH.$$

This is the pH-buffering effect of acetic acid. It is maximal at the pK_a so that an equimolar mixture of acetate and acetic acid gives its maximum buffering effect at that pH. As a rule of thumb, in biochemistry, a useful buffer covers a range of about 1 pH unit on either side of the pK_a. The pH of a solution containing an acid-base conjugate pair can be calculated from the Henderson–Hasselbalch equation:

$$pH = pK_a + \log \frac{[\text{proton acceptor}]}{[\text{proton donor}]}$$

or, as often expressed:

$$pH = pK_a + \log \frac{[\text{salt}]}{[\text{acid}]}.$$

The pH of any mixture of acetic acid and sodium acetate can be calculated from this equation so that the composition of a buffer of any desired pH can be determined.

As an example for a solution containing 0.1 M acetic acid and 0.1 M sodium acetate:

$$pH = 4.76 + \log\frac{[0.1]}{[0.1]}$$

$$= 4.76 + 0 = 4.76.$$

If the mixture contains 0.1 M acetic acid and 0.2 M sodium acetate:

$$pH = 4.76 + \log\frac{[0.2]}{[0.1]}$$

$$= 4.76 + \log 2 = 4.76 + 0.30 = 5.06.$$

Suppose that to the buffer containing 0.1 M acetic acid and 0.1 M sodium acetate you added NaOH to a concentration of 0.05 M; the pH of the resultant solution would be:

$$4.76 + \log\frac{[0.15]}{[0.05]}$$

(since half of the acetic acid would be converted to sodium acetate)

$$= 4.76 + \log 3 = 4.76 + 0.48 = 5.24.$$

Acetic acid/acetate mixture has no significant buffering power at physiological pH values, but compounds with pKa values close to pH 7 exist in the body, and these are effective buffers. Among the most important are the phosphate ion and its derivatives. Phosphoric acid has three dissociable groups (page 31), the second one ($H_2PO_4 \rightleftarrows HPO_4 + H^+$) has a pK_a value of 6.86 so that phosphate is an excellent buffer in this region.

Another buffering structure is the imidazole nitrogen of the histidine residue in proteins with a pK_a around 6. The dissociation involved is shown below.

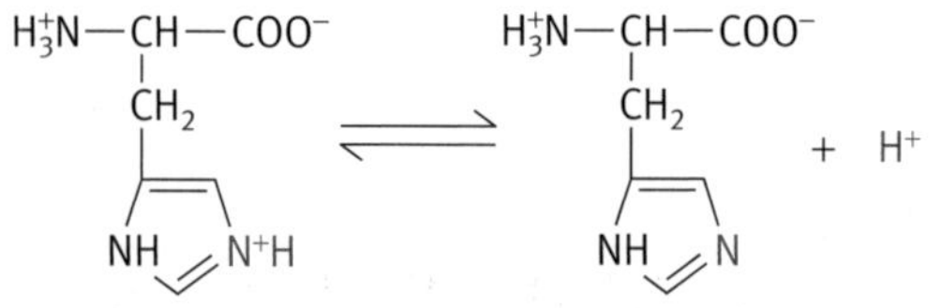

The phenomenon of buffering can be illustrated by a very simple observation. If you take a test-tube containing distilled water and adjust the pH to 7.0 (dissolved CO_2 makes it slightly acid) with NaOH and then add a few drops of 0.1 M HCl, there will be a precipitous drop in pH. If you take a test-tube containing 0.1 M sodium phosphate, also adjusted to pH 7.0, and add the same amount of HCl, the pH will hardly change, due to the buffering action of the phosphate ion.

The buffering action of compounds with pK_a values near 7 thus protects the cell and body fluids against large pH changes.

Summary

Energy changes in chemical reactions are reliable guides to the biochemistry of cells. The most useful value is the free energy change (ΔG), which is an expression of the amount of energy change in a reaction available to perform useful work. The ΔG value can be used to determine the equilibrium constant of a reaction and whether it is likely to be reversible in cells.

Free energy (a technical term meaning useable energy) made available from food breakdown is used to synthesise ATP the universal energy carrier of life.

Catabolism (breakdown) of food molecules drives anabolism (synthesis of molecules) with ATP being the energy-carrying go-between. ATP is called a high-energy phosphate molecule; the release of the two terminal phosphate groups liberates large amounts of free energy, which is used to perform work of all kinds in coupled reactions. There are other high-energy compounds in the cell that can transfer their phosphate groups to ADP to form ATP.

Weak noncovalent chemical bonds play a crucial part in life. These are hydrogen bonds, ionic bonds and van der Waals forces. Unlike covalent bonds, their formation and breakdown involves only small amounts of free energy change and occur without the need for enzyme catalysis. They are important because binding between molecules is dependent on a number of weak bonds forming. This will occur only if the atoms involved are sufficiently close, which means that only molecules with complementary shaped binding sites will bind by these forces. It is the basis of molecular recognition (biological specificity) by proteins on which all life depends.

Hydrophobic forces or hydrophobic interactions are not bonds but the phenomenon causes hydrophobic molecules to associate together.

Noncovalent bonds are essential for specifically recognizing (binding to) other molecules and are essential in most biochemical activities of cells. They also are crucially involved in biochemical structures such as proteins, DNA, and in formation of protein complexes. They have the advantage that because they are easily broken they confer flexibility on structures; for example the two DNA strands of a double helix are held together by hydrogen bonds. To replicate DNA the strands have to be separated.

Further reading

Nelsestuen, G. L. (1989). A partial remedy for the nonrelationship between reversibility of a reaction in the cell and the value of $\Delta G^{0'}$. *Biochem. Edu.*, 17, 190–2.
Discusses why the aldolase reaction with a large $\Delta G^{0'}$ value is freely reversible in the cell.

The high-energy phosphate compound concept

Lipmann, F. (1941). Metabolic generation and utilization of phosphate bond energy. *Adv. Enzymol.*, 1, 99–162.
The famous review that initiated the concept of ATP driving cellular processes. Very old, but the concepts still apply.

Covalent and noncovalent bonds

Kollman, P. A. (1999). Hydrogen bonding. *Curr. Biol.*, R501.

Problems

1 Suppose that a 70 kg man has a food intake for his energy needs of 10,000 kJ per day. Assume that the free energy available in his diet is used to form ATP from ADP and P_i with an efficiency of 50%. In the cell, the ΔG for the conversion of ADP + P_i is approximately 55 kJ mol^{-1}. Calculate the total weight of ATP the man synthesizes per day, in terms of the disodium salt (molecular weight = 551 D_d).

2 The $\Delta G^{0'}$ for the hydrolysis of ATP to ADP and P_i is 30.5 kJ mol^{-1}. Explain why, in Problem 1, a ΔG value of 55 kJ mol^{-1} was suggested as the amount of free energy required to synthesize a mole of ATP from ADP and P_i in the cell.

3 The hydrolysis of ATP to ADP + P_i and that of ADP to AMP + P_i have $\Delta G^{0'}$ values of −30.5 kJ mol^{-1}, while the hydrolysis of AMP to adenosine and P_i has a value of −14.2 kJ mol^{-1}. What are the reasons for the large difference?

4 Explain
   (a) why benzene will not dissolve in water to any significant extent
   (b) why a polar molecule such as glucose is soluble in water
   (c) why NaCl is soluble in water.

5 In the cell, ADP is converted to ATP using the energy derived from food catabolism, but AMP is not utilized by the ATP-synthesizing system; many synthetic processes convert ATP to AMP. How is AMP brought back into the system?

6 An enzyme catalyses a reaction that synthesizes the compound XY from XOH and YH, coupled with the breakdown of ATP to AMP and PP_i. The $\Delta G^{0'}$ of the reaction XOH + YH→XY + H_2O is 10 kJ mol^{-1}.

Determine the $\Delta G^{0'}$ of the reaction: (a) in the cell; and (b) using a completely pure preparation of the enzyme. Explain your answer. (You are told that the $\Delta G^{0'}$ values for ATP hydrolysis to AMP and PP_i, and for PP_i hydrolysis are −32.2 and −33.4 kJ mol^{-1}, respectively.)

7 What are the different types of weak bonds of importance in biological systems and their approximate energies. Why is it that their formation does not require enzymic catalysis? If they are so weak, why are they of importance?

8 The pKa values for phosphoric acid are 2.2, 7.2, and 12.3.
   (a) Write down the predominant ionic forms at pH 0, pH 4, pH 9, and pH 14.
   (b) What would be the pH of an equimolar mixture of NaH_2PO_4 and Na_2HPO_4?
   (c) Suppose that you wanted a buffer fairly close to physiological pH and all you had available were the 20 amino acids found in proteins, which one would be the most suitable?

9 Virtually all biological processes involve specific interactions between proteins and other molecules (which may also be proteins). Explain how the specific interactions are achieved.

10 The use of the term 'high-energy phosphate bond' (indicated by a squiggly bond) by Lipmann in 1940 has been of great importance in development of the concept of biological energy. Although sometimes still used, because it is a convenient shorthand notation, it has fallen out of favour because it is not chemically correct. Discuss this.

11 What is the fundamental driving force that causes chemical reactions to occur under appropriate

conditions? In other words why should reactions ever occur?

12 The second law of thermodynamics specifies that all processes must increase the total entropy of the universe and yet living cells are at a lower entropy level than the randomly arranged molecules in the environment from which the living cells are assembled. Does this mean that living cells are islands in the universe exempt from the second law? Explain your answer.

13 What is meant by 'free energy?

14 What is entropy? Discuss its significance briefly.

Part 2

Structure and function of proteins and membranes

Anthrax lethal factor protein
Lethal factor (LF) is one of the toxins produced by spores of the bacterium *Bacillis anthracis*. Spores enter the body and release their toxins, including LF which cleaves an enzyme vital for cell signalling. This renders the cells unable to function normally. If inhaled, anthrax is often fatal. The disease most commonly affects cattle, sheep and goats, but can also infect humans.
LAGUNA DESIGN/SCIENCE PHOTO LIBRARY

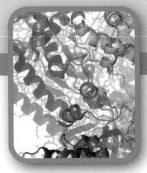

Chapter 4

The structure of proteins

It has been said, with considerable justification, that it is improbable that there exists in the universe any type of molecule with properties more remarkable than those of proteins. As stated in Chapter 1, life depends on thousands of different proteins whose structures are fashioned so that individual protein molecules combine, with exquisite precision, with other molecules. Chemical reactions in the cell and just about all cellular activities depend on them. Proteins essentially 'do' everything in life. They are the workhorses and the basis of most biological structures. The elaborate genetic machinery based on DNA coding is there so that the correct proteins are there at the right time and in the correct quantities. Given that, the fertilized egg develops into the 'right' organism.

Proteins are, in basic chemical structure, more complex molecules than DNA, though they do not reach the immense size of DNA. They are made of long strings of 20 different species of amino acids, while DNA has only four variable 'building blocks.'

This chapter deals with the structures of these giant molecules which carry out such a range of activities. Their versatility is based on the fact that a virtually unlimited number of different protein molecules can theoretically exist. A remarkable property of proteins is that while they have definite structures and shapes, they often, or usually, are flexible molecules that can change their conformations – their arrangements in three-dimensional space – in the process of performing their tasks. Many, probably most, proteins are to be regarded as molecular machines rather than inert chemical molecules. Haemoglobin is a classic example of this. Muscle contraction (Chapter 8) is an extreme example of a biological function resulting from protein molecules changing their conformation. It is remarkable that the miniscule atomic movements of groups of atoms can result in the massive contractions of animal muscles. In this chapter we will deal with the basic structures of proteins in general and how the structures of various types of proteins are engineered to fulfill their particular functions. Membrane proteins have special characteristics and we will discuss these in Chapter 7 on membranes.

Proteins are **polypeptide** chains. A protein molecule may have more than one such chain. A polypeptide chain consists of a large number of amino acids linked together – anything from 50 to several thousand. Twenty different amino acid structures are used for the assembly of the polymers.

Structures of the 20 amino acids used in protein synthesis

Learning to recognize 20 different structures of **amino acids** may be less daunting if you keep in mind what the different side groups are for. They are the building blocks with which evolution juggles to produce the vast variety of different proteins needed for the life of organisms. The same 20 are used for all life forms on earth. A variety of amino acids with different shapes, sizes, chemical properties, and polarity characteristics gives evolution flexibility in 'trying out' different protein structures, with natural selection relentlessly assessing whether variants should be 'adopted' *via* the genetic system. More than 20 different amino acids are found in different organisms, but only what Francis Crick has called the 'magic 20' are used for protein synthesis. (An exception is that, by a 'freak' mechanism, a few proteins have selenocysteine in polypeptide linkage – see page 393).

An *a*-amino acid, written in the nonionized form has the structure (a), but in aqueous neutral solution it exists as the zwitterionic form (b).

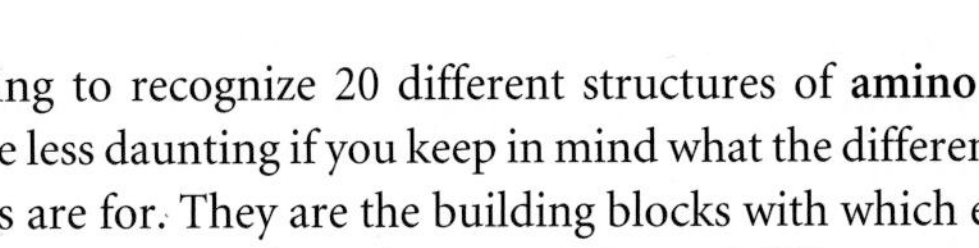

(a) $H_2N—\overset{\alpha}{CH}—COOH$ **(b)** $^+H_3N—\overset{\alpha}{CH}—COO^-$
$\qquad\qquad\;\;|$ $\qquad\qquad\qquad\qquad\;\;|$
$\qquad\qquad\;\;R$ $\qquad\qquad\qquad\qquad\;\;R$

Every amino acid, with the exception of proline which is actually an **imino acid** (see below), has the same $H_2N—CH—COOH$ part – only the R group attached to the α carbon atom varies. The R group, the '**side chain**' is shown below in red.

With the exception of glycine, which has no asymmetric carbon atom, amino acids in proteins are of the L-configuration. Fig. 4.1 gives the structures of L and D amino acids, the two

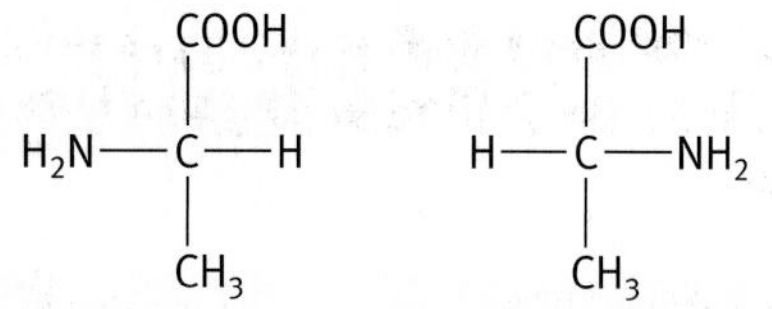

L-Alanine D-Alanine

Fig. 4.1 Stereoisomers of L- and D-alanine. In this projection, vertical lines represent bonds that project below the plane of the paper, and horizontal lines bonds that project outward from the paper. Note that the two stereoisomers are mirror images of one another.

being mirror images of one another. It is not necessary to specify that an amino acid in any biological context is of the L-configuration since D- amino acids are only rarely encountered (usually in things like antibiotics and microbial structures) and where they occur these are always specified.

Symbols for amino acids

There are two types of abbreviations for amino acids as shown in Table 4.1. The old three-letter system is often still used when short sequences are represented because they have the advantage of being self-evident. More usually now and for longer sequences, the single letter abbreviations are less cumbersome and save a lot of storage space in databases, though less easy to remember.

Aliphatic amino acids

The aliphatic amino acids (aliphatic means open noncyclic structures) are **glycine, alanine, valine, leucine,** and **isoleucine.**

Table 4.1 Single-letter and three-letter symbols for amino acids.

Amino acids	One letter	Three letter
Alanine	A	Ala
Arginine	R	Arg
Asparagine	N	Asn
Aspartic acid	D	Asp
Cysteine	C	Cys
Glutamine	Q	Gln
Glutamic acid	E	Glu
Glycine	G	Gly
Histidine	H	His
Isoleucine	I	Ile
Leucine	L	Leu
Lysine	K	Lys
Methionine	M	Met
Phenylalanine	F	Phe
Proline	P	Pro
Serine	S	Ser
Threonine	T	Thr
Tryptophan	W	Trp
Tyrosine	Y	Tyr
Valine	V	Val
Unspecified or unknown	X	Xaa

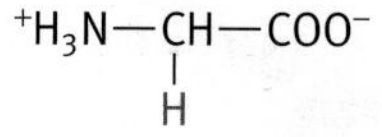

Glycine

Glycine with H as its side group is the smallest amino acid and has no marked hydrophilic or hydrophobic properties. **Hydrophilic** means affinity for water and tending to dissolve in water. **Hydrophobic** means low affinity for water, tending not to dissolve in it.

Alanine, valine, leucine, and isoleucine, have side chains of increasing hydrophobicity. The latter three are known as the **branched-chain aliphatics. Methionine** is a rather special hydrophobic aliphatic amino acid (special more because of its role in methyl group metabolism than its role in protein structure.

Hydrophobic amino acids

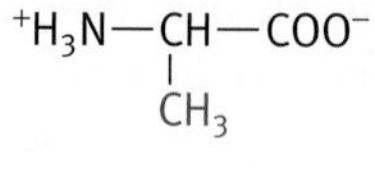

Alanine

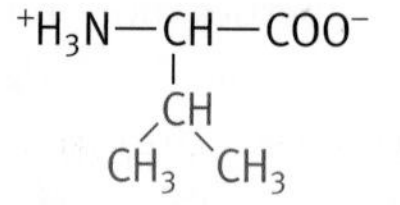

Valine

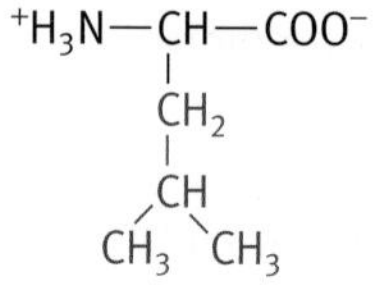

Leucine

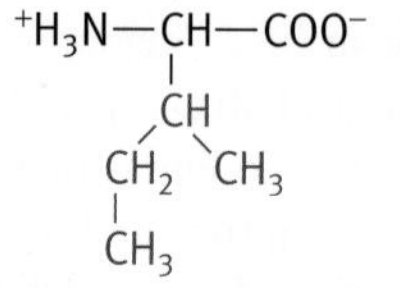

Isoleucine

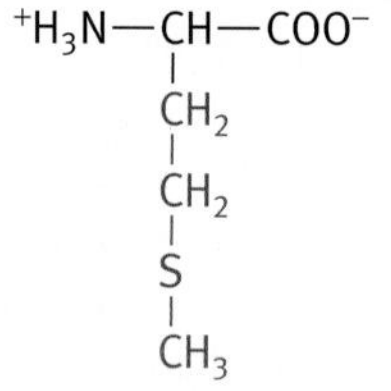

Methionine

Then we come to the large hydrophobic side chains – aromatic ones (those with cyclic planar rings):

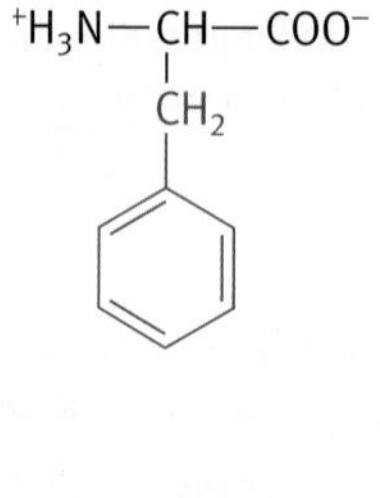

Phenylalanine

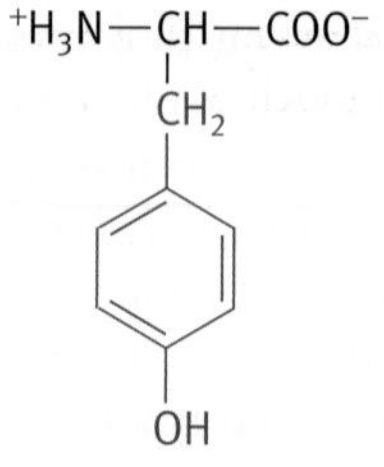

Tyrosine

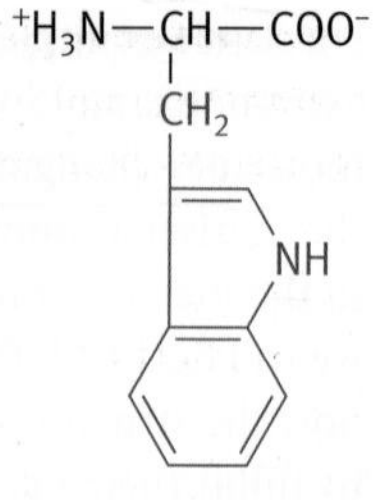

Tryptophan

Tyrosine with its —OH group can form a hydrogen bond, but its aromatic ring is large and hydrophobic so it is of somewhat mixed classification. Tryptophan can also form a hydrogen bond with its —NH group.

Ionized hydrophilic amino acids

An ionized group is hydrophilic. Acidic amino acids (with an extra —COO⁻ on the side chain) are negatively charged at the pH of the cell, and thus hydrophilic. Basic amino acids have an extra positive charge and are also hydrophilic. (The term basic means the opposite of acidic or that it will have a positive charge at cellular pH). **Aspartic** and **glutamic acids** both have acidic side chains, and are called **aspartate** and **glutamate** when negatively charged at physiological pH. **Lysine** and **arginine** have strongly basic side chains, and histidine is a weakly basic amino acid. The ring structure of histidine is known as the **imidazole ring**. The side chains of the these amino acids can form hydrogen bonds and salt bridges, the latter being any ionic bond between positive and negative groups.

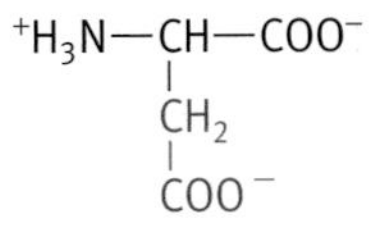

$$^+H_3N-CH-COO^-$$
$$|$$
$$CH_2$$
$$|$$
$$COO^-$$

Aspartate

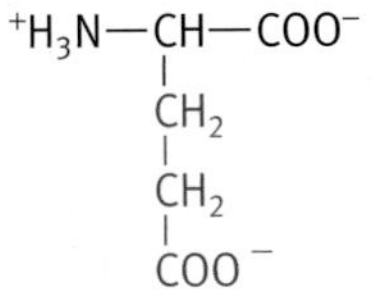

$$^+H_3N-CH-COO^-$$
$$|$$
$$CH_2$$
$$|$$
$$CH_2$$
$$|$$
$$COO^-$$

Glutamate

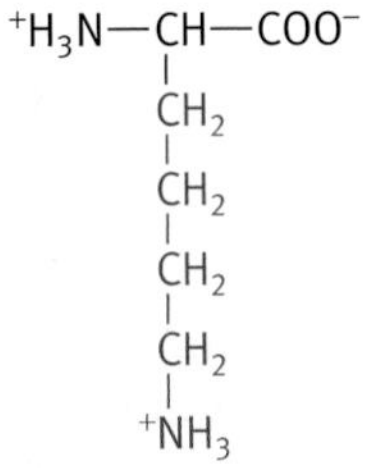

$$^+H_3N-CH-COO^-$$
$$|$$
$$CH_2$$
$$|$$
$$CH_2$$
$$|$$
$$CH_2$$
$$|$$
$$CH_2$$
$$|$$
$$^+NH_3$$

Lysine

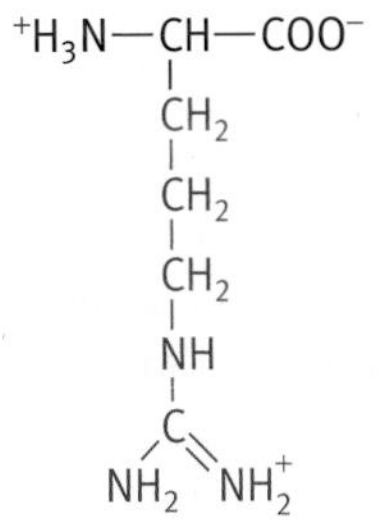

$$^+H_3N-CH-COO^-$$
$$|$$
$$CH_2$$
$$|$$
$$CH_2$$
$$|$$
$$CH_2$$
$$|$$
$$NH$$
$$|$$
$$C$$
$$NH_2 \quad ^+NH_2$$

Arginine

$$^+H_3N-CH-COO^-$$
$$|$$
$$CH_2$$
$$|$$
$$C-NH^+$$
$$||\quad\quad CH$$
$$HC-NH$$

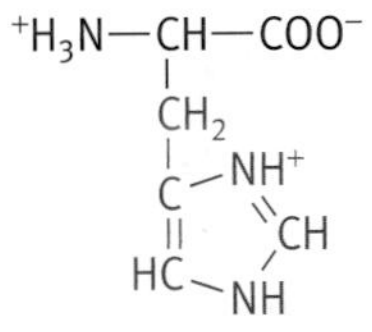

Histidine

The two acidic amino acids, aspartic acid and glutamic acid, exist also as the amides, **asparagine** and **glutamine**, respectively. Since the amide group does not ionize, these two amides are less hydrophilic and form much weaker hydrogen bonds than the parent compounds.

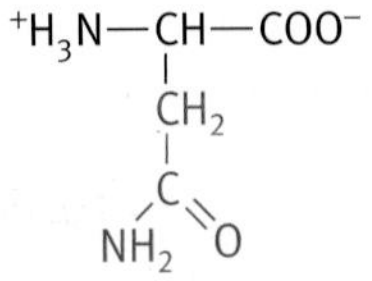

$$^+H_3N-CH-COO^-$$
$$|$$
$$CH_2$$
$$|$$
$$C$$
$$NH_2 \quad O$$

Asparagine

$$^+H_3N-CH-COO^-$$
$$|$$
$$CH_2$$
$$|$$
$$CH_2$$
$$|$$
$$C$$
$$NH_2 \quad O$$

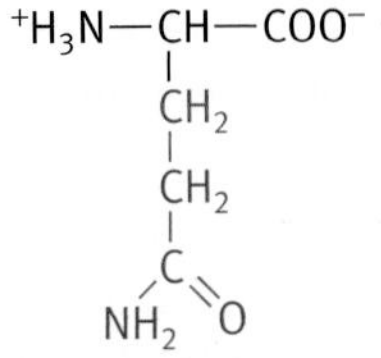

Glutamine

Serine and threonine side chains are nonionized and weakly hydrophilic. They can both form hydrogen bonds with their —OH groups.

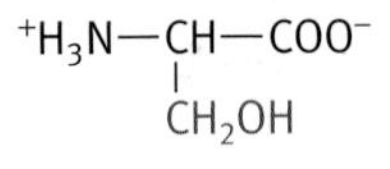

$$^+H_3N-CH-COO^-$$
$$|$$
$$CH_2OH$$

Serine

$$^+H_3N-CH-COO^-$$
$$|$$
$$CHOH$$
$$|$$
$$CH_3$$

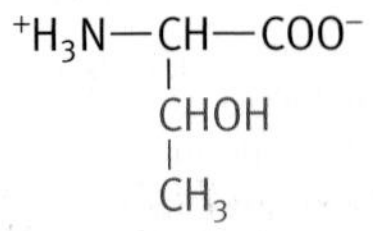

Threonine

Two amino acids with unusual properties

Cysteine is similar to serine but with —SH instead of —OH. It plays two special roles in protein structure – supplying external —SH groups such as in the active centres of enzymes, and forming covalent —S—S— bonds or disulfide bonds internally (see below). It is hydrophobic and its side chain cannot form strong hydrogen bonds or salt bridges.

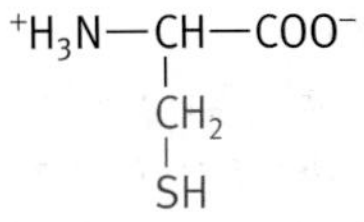

$$^+H_3N-CH-COO^-$$
$$|$$
$$CH_2$$
$$|$$
$$SH$$

Cysteine

Proline is the oddity – literally a kinky amino acid in that it puts a kink into the conformation of polypeptide chains (described later). It has an imino (—NH) group rather than an NH₂ amino group. If you have trouble in memorizing proline – which can happen – remember that it is an ordinary amino acid except that it is an —NH$_2^+$ imino acid rather than an —NH$_3^+$ amino acid. Its structure, that forbids rotation about the N—C bonds, causes it to have a large effect on protein structure.

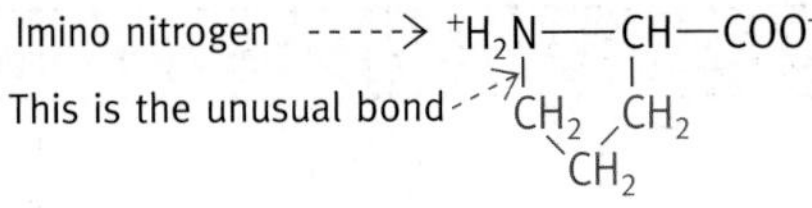

Imino nitrogen ----> ^+H_2N——$CH-COO^-$
This is the unusual bond $CH_2 \quad CH_2$
$$CH_2$$

The different levels of protein structure – primary, secondary, tertiary, and quaternary

We first give a brief overview of this topic, to be followed by a more detailed treatment.

The sequence of amino acids that are linked together covalently into a polypeptide chain is its **primary** structure (Fig. 4.2(a)). It says nothing about how that polypeptide is arranged in a three-dimensional space, just the order of the amino acids. The polypeptide backbone itself is arranged in a particular conformation known as the **secondary** structure (Fig. 4.2(b)). The secondary structure is folded yet again to give

(a) Primary

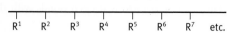

R¹   R²   R³   R⁴   R⁵   R⁶   R⁷   etc.

This structure will be represented below as a simple line.

(b) Secondary

The polypeptide **backbone** exists in different sections of a protein either as an α helix, a β-pleated sheet, or random coil. (These have yet to be explained.)

α helix   β-pleated sheet   Random coil or loop region

(c) Tertiary

The secondary structures above are folded into the compact globular protein.

This protein will be represented below as:

(d) Quaternary

Protein molecules known as subunits assemble into a multimeric protein held together by weak forces.

Fig. 4.2 Diagrammatic illustration of what is meant by primary, secondary, tertiary, and quaternary structures of proteins. The structures referred to will be described shortly in the text.

$$H_2N-CH-COOH \qquad H_2N-CH-COOH$$
$$\qquad | \qquad\qquad\qquad\qquad |$$
$$\qquad R \qquad\qquad\qquad\qquad\ R'$$

Two amino acids

Peptide bond

$$H_2N-CH-\overset{\overset{\displaystyle O}{\|}}{C}-NH-CH-COOH$$
$$\qquad |\qquad\qquad\qquad\quad |$$
$$\qquad R\qquad\qquad\qquad\quad R'$$

A dipeptide

The dipeptide is the simplest 'peptide unit' as in the **poly-peptide** structure shown:

$$H_2N-CH-\overset{\overset{O}{\|}}{C}-NH-CH-\overset{\overset{O}{\|}}{C}\left[-NH-CH-\overset{\overset{O}{\|}}{C}\right]_n -NH-CH-COOH$$
$$\quad\ |\qquad\qquad\quad |\qquad\qquad\qquad |\qquad\qquad\qquad |$$
$$\quad R\qquad\qquad\ R^2\qquad\qquad\quad R^n\qquad\qquad\quad R^3$$

A polypeptide

The order in which the 20 different amino acids are arranged in the polypeptide is the **amino acid sequence**. The sequence, as mentioned, is the **primary structure** of a protein. Determining the sequence is referred to as **protein sequencing**, or amino acid sequencing, dealt with in the next chapter. The naming of peptides of different lengths is an arbitrary matter but as a rough guide a short chain of a few amino acids, perhaps a dozen or so, is referred to as an **oligopeptide** (*oligo* – few), a polypeptide has many, and the terms polypeptide and protein are often used interchangeably. A protein may have several polypeptide chains, or just be a very long polypeptide. A few terms: the $-NH_3^+$ end of a protein is referred to as the **amino terminal** or **N-terminal** end and the other as the **carboxy terminal** or **C-terminal** end. The central chain, without the R groups, $(-CH-CO-NH-CH-CO-NH-CH-,$ etc.) is called the **polypeptide backbone**. An amino acid in a protein is referred to as an **amino acid residue** or **amino acyl residue**; the R groups are variously referred to as amino acid side chains, protein side chains, or simply as **side chains**.

By convention the amino acid sequence is read from the N-terminal end to the C-terminal end. Polypeptide chains have direction. Consider a given amino acid sequence such as ala-gly-leu-phe. If the N-terminal amino acid is ala, the molecule ala-gly-leu-phe is biologically different from its inverse, phe-leu-gly-ala.

Ionization of amino acids in polypeptide chains

As already mentioned, free amino acids in aqueous solution have the zwitterionic structures in which both the *a*-amino and *a*-carboxyl groups are ionized. (See page 38 if you wish to be reminded their pK_a values). However, when incorporated into a polypeptide chain, these groups are no longer ionizable,

the **tertiary** structure (Fig. 4.2(c)). The complete molecule so formed by the primary, secondary, and tertiary structures may be the final functional protein or it may be a **protein monomer** or **subunit**, which associates with other protein monomers (which may be the same or different) to form a functional protein. This is the **quaternary** structure (Fig. 4.2(d)).

With this overview, we will now deal in more detail with the different levels of protein structure.

Primary structure of proteins

Two amino acids can be linked together by the removal of H_2O, but note carefully that it doesn't happen by such a direct process in the cell. This would be thermodynamically impossible. The result is a **dipeptide**, as shown below (structures are written in the nonionized form for clarity). The $-CO-NH-$ bond is the **peptide bond** or link.

except for the terminal amino and carboxyl groups. The ionized state of a polypeptide, and therefore of a protein, is therefore almost entirely dependent on the side groups of aspartic and glutamic acid, lysine, arginine, and histidine residues, since their ionizable side chain groups are not blocked by peptide bond formation, as shown for glutamic acid and lysine in the illustration below.

```
—CO—NH—CH—CO—NH—CH—Polypeptide backbone
             |              |
             CH₂            CH₂
             |              |
             CH₂            CH₂
             |              |
             COO⁻           CH₂
                            |
                            CH₂NH₃⁺

        Glutamic        Lysine
        acid            side chain
        side chain
```

The side chain —COO⁻ groups of aspartic and glutamic acids have pK_a values around 4, so they are virtually fully dissociated at pH 7.4 – these provide the negatively charged side groups of proteins. The basic amino acids lysine and arginine, with pK_a values for their basic side chain groups of 10.5 and 12.5 respectively, are fully ionized at physiological pH. The third basic amino acid, histidine, has an imidazole ring as the side chain group whose pK_a value is near neutrality (around 6 in proteins), so that histidine is often found in active sites of enzymes where movement of a proton is involved in the reaction catalysed; the imidazole group can accept or donate a proton at a pH near that existing in the cell (see page 95 for an example of this).

The distribution of charged amino acid residues in a protein has an important effect on the conformation that a polypeptide chain can adopt. Charges of the same sign close to each other repel each other. Closely positioned positive and negative charges will attract each other.

The peptide bond is planar

The simple polypeptide structure above does not convey an important feature of a polypeptide chain Although the CO—NH peptide bond is written as an ordinary single bond (about which rotation might be expected), in fact the peptide bond is a hybrid between two structures: (1) in which the bond between the carbon and nitrogen atom is a single bond; and (2) in which it is a double bond.

```
(1)  —C      H        (2)  —C      H
        \   /                 \   /
         C—N                   C=N⁺
        /    \                /     \
       O      C—             O⁻       C—
```

The electron density is between the two, giving the C—N bond about 40% double-bonded character. This is sufficient to prevent rotation about it, making the polypeptide chain more rigid.

The architecture of a polypeptide chain is shown in Fig. 4.3. The successive α carbon atoms lie above and below the plane of

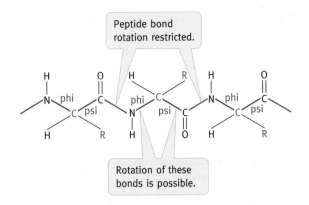

Fig. 4.3 Section of a polypeptide. Successive α carbon atoms (shown in red) of amino acid residues lie above and below the plane of the paper as also do the R groups (shown in blue). The peptide —CO—NH— bond (shown in red) has a partial double-bonded character that prevents rotation and gives the group a rigid planar structure. However, the adjacent bonds (shown in green) are capable of rotation with a single-bonded structure. The angle adopted by these bonds are known respectively by the Greek letters phi (Φ) and psi (ψ). The conformation of a polypeptide chain is determined by the value of these angles for each amino acid in the chain.

the paper. As shown, there is limited rotation possible in the bonds between the C—N and C—C bonds adjacent to the α carbon atoms whose angles of rotation determine the configuration of the chain. These angles are known by the Greek letters phi and psi as shown.

Ramachandran showed that the freedom of the chains to adopt certain combinations of these angles is restricted, presumably because of steric hindrances (atoms trying to share spaces too small for them both). In this famous piece of work, he plotted the pairs of phi and psi angles adopted by each amino acid in a large number of proteins. They clearly formed two tight clusters, indicating that most pairs of angles rarely or never occur because of steric hindrances.

Secondary structure of proteins

Most proteins have a compact globular shape rather than the extended configuration of a polypeptide, though fibrous proteins exist. The interior of globular proteins is a strongly hydrophobic environment. This imposes limitations on the possible types of secondary structures because, to fold the polypeptide backbone into a compact shape, it has to criss-cross the hydrophobic interior. The problem is that a polypeptide has polar groups capable of hydrogen bonding – two bonds per amino acid unit, since the C=O and N—H of the peptide bond are capable of participating in hydrogen bond formation. This bonding potentiality must be satisfied by bond formation as far as possible to produce a stabilized structure. The backbone groups in the interior of the molecule cannot hydrogen bond to the hydrophobic side chain groups. So what can these hydrogen atoms bond with? The answer is with groups on the same, or an adjacent, polypeptide backbone.

There are two main classes of secondary structures – the α **helix** in which the backbone is arranged in a spiral-like coil and the β-**pleated sheet** in which extended polypeptide backbones are side by side. These structures are stable; they can occur at the exterior of proteins with appropriate hydrophilic side chains or in the hydrophobic interior of proteins, with appropriate hydrophobic side chains. (Many proteins are located in cell membranes; in this case the requirements are different in that hydrophobic residues are exposed on the outside of the molecule where they contact the hydrophobic membrane structures, and hydrophilic groups are hidden away from them, page 109).

The α helix

In the α helix, the polypeptide backbone is twisted into a right-handed helix which, for L-amino acids is more stable than a left-handed one (Fig. 4.4(a)). You can visualize the direction of

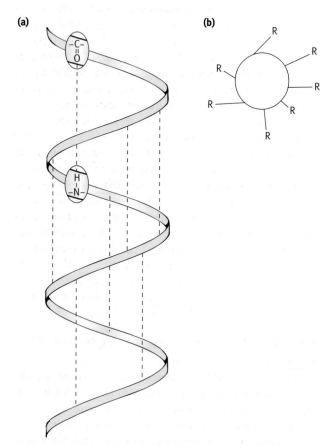

twist of the right-handed helix; if you look down the axis either way the helix turns clockwise. Its shape is shown in Fig. 4.4(a).

Linus Pauling deduced the α helix structure with a pitch of 3.6 amino acid units per turn, a seminal discovery reportedly made in Oxford, while he was in bed with a cold, using folded paper models. This results in the C=O of each peptide bond being aligned to form a hydrogen bond with the peptide bond N—H of the fourth distant amino acid residue. The C=O groups point in the direction of the axis of the helix and are nicely aimed at the N—H groups with which they hydrogen bond, giving maximum bond strength. Every C=O and N—H group of the polypeptide backbone are hydrogen bonded in pairs, forming a cylindrical, rod-like structure (Fig. 4.4(a)). In cross-section an α helix is a virtual solid cylinder, with all the R groups projecting outwards (Fig. 4.4(b)).

Not all of the polypeptide chain of a globular protein is in the α helix form. The sections that are, average 10 amino acid residues in length, but the range of lengths varies a great deal from this average in different proteins. Some proteins, like haemoglobin, are composed almost entirely of α helices, others have very little, but all variations can be found.

Amino acids vary in their tendency to form α helices

Some amino acids, such as leucine and methionine, are excellent helix formers, some indifferent, and a few are α helix breakers or terminators – a proline residue in particular forces a bend in the structure. With the other amino acids, the polypeptide chain can rotate around two bonds per residue as explained earlier (page 49). With proline in peptide linkage, there is no hydrogen atom on the nitrogen available for hydrogen bonding, and the structure of the residue restricts rotation, so that it cannot assume the conformation needed to fit into an α helix.

$$-NH-CH-\overset{\overset{\displaystyle O}{\|}}{C}-N-CH-\overset{\overset{\displaystyle O}{\|}}{C}-NH-$$

Imino nitrogen in peptide linkage not able to hydrogen bond

We will return shortly to how runs of amino acids in α helices fit into protein structures but, before that, we must deal with the alternative to the α helix, namely the β-pleated sheet. Proteins often contain mixtures of the two, with each constituting different sections of the polypeptide chain, but some have only one of these two structures.

Fig. 4.4 The α helix form of a polypeptide chain. **(a)** Hydrogen bonding between C=O and N—H groups of the polypeptide backbone (side chains not shown). The hydrogen bonds (broken lines) are shown in approximate positions only. **(b)** Looking down the axis of an α helix, with the amino acid side chains projecting from the cylindrical structure (each at a different distance below the plane of the paper). The R groups are not drawn in their exact orientation from the axis. Since there are 3.6 residues per turn, each residue occurs every 100° around a circle (360°/3.6 = 100°).

The β-pleated sheet

The β-pleated sheet also forms a stable structure in which the polar groups of the polypeptide backbone are hydrogen bonded to one another, thus forming a stable structure that is often in the hydrophobic interior of a globular protein.

The principle is simple. The polypeptide chain lies in an extended or β form with the C=O and N—H groups hydrogen

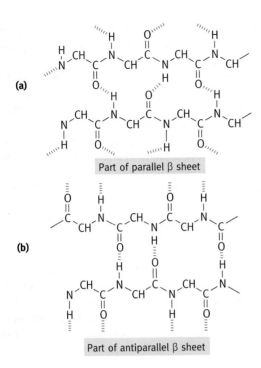

Fig. 4.5 (a) Hydrogen bonding between two polypeptide chains running in the same direction forming a parallel β sheet. The R groups attached to the —CH— groups are above and below the plane of the paper. **(b)** Adjacent polypeptide chains running in opposite directions can also mutually hydrogen bond forming an antiparallel β sheet. This enables a single chain to form a β sheet by folding back on itself. Hydrogen bonds are denoted by dotted lines.

bonded to those of a neighbouring chain (which may be formed by the same chain folding back on itself, or may be a separate chain lying alongside, Fig. 4.5). Several chains can thus form a sheet of polypeptide. It is pleated because successive β carbon atoms of the amino acid residues lie slightly above and below the plane of the β *sheet* alternately (see Fig. 4.3). The adjacent polypeptide chains bonded together can run in the same direction (parallel) or opposite directions (antiparallel). In the latter case, the polypeptide makes tight 'β turns' or hairpin bends to fold the chain back on itself (see Fig. 4.6 (b)). Four residues constitute the turn; a hydrogen bond between the C=O and N–H of residues 1 and 4 stabilize the hairpin bend. Parallel β sheets are connected by a longer motif, such as an α helix or connecting loop (see Fig. 4.7 (c)).

Connecting loops

In a protein, the α helices and β sheet sections are connected together by unstructured polypeptide known as **connecting loops**. The connecting loop may be in any conformation (other than a recognizable α helix or β sheet) as determined by the various group interactions in the protein structure. Since the structure of a loop may not satisfy the hydrogen bonding potentials of the C=O and N—H groups of the backbone, or those of the side groups, such loops are often found at the

exterior of proteins, in contact with water. In a given protein, connecting loops will have a conformation determined by their interactions with other groups.

Tertiary structure of proteins

An important factor in determining the tertiary structure is the hydrophobic force (page 38) which ensures that, as far as possible, hydrophobic side chains are sequestered in the middle of the protein, which maximizes van der Waals interactions (page 38) and hydrophilic groups on the outside in contact with water where they can maximize their hydrogen and ionic bonding.

A single polypeptide chain in a protein can be arranged as a mixture of the various secondary structures constituting different parts of the chain, which themselves are folded up and packed together to form the protein molecule. This arrangement of the various secondary structures into the compact structure of a globular protein is referred to as the tertiary structure.

To simplify structural diagrams of the tertiary structure of proteins, conventions have been adopted to depict the arrangements of the secondary structures of which they are composed. An *α helix* is represented either as a solid cylinder, sometimes with a helix inside it, or alternatively as a helical ribbon (Fig. 4.6(a)). The individual sections of polypeptide chains or β strands which participate in β-pleated sheet formation are represented as broad arrows (Fig. 4.6(b)). An extended polypeptide chain in the β configuration has been shown to twist slightly to the right so that the arrows are usually drawn in protein structures with this twist. Connecting loops are shown as any sort of line. It should always be remembered, however, that proteins are solid structures with tightly packed atoms, not the open structures of springs and wires these convenient diagrammatic motifs might suggest. Space-filling models (such as that of the haemoglobin molecule shown in the first chapter (Fig. 1.5)) representing the solid object with atoms depicted, are more realistic, but their internal structures cannot be seen. They are particularly useful where, as often is the case, we are particularly interested in interactions of protein surfaces with other molecules.

How do the three motifs – the α helix, the β-pleated sheet, and connecting loops – make up a protein?

In principle, the answer is that any combination of the three can be used to assemble a globular protein provided that the side chains can be folded and packed together in ways appropriate to their various affinities and repulsions, the unstructured sections having their backbone polar groups satisfactorily bonded – usually to water – and exposure of hydrophobic groups to water or other polar groups minimized. The folding of the polypeptide, and thus the protein structure, is determined by the primary structure, the sequence of amino acid residues. These are formidable provisions, which probably only one in countless numbers of random amino acid sequences can fulfill. Large numbers of proteins have had their three-dimensional structure determined by X-ray diffraction studies (Chapter 5).

(a)

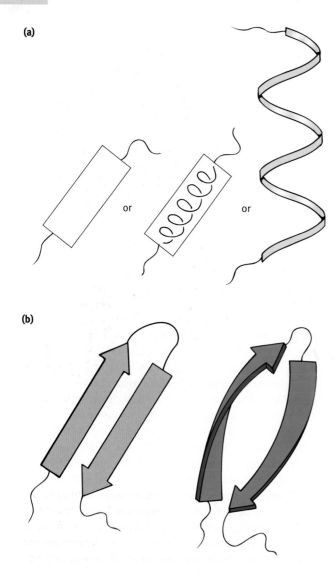

(b)

Fig. 4.6 Symbols used to indicate **(a)** α-helix structures; or **(b)** β-pleated sheets. In **(b)** a pair of polypeptide stretches are shown making an antiparallel sheet. Because of the slight right-handed twist the arrows are represented as on the right. The single lines connecting structures represent random coil or loop sections.

In Fig. 4.7, representations of a few illustrative protein structures are given. Myoglobin, (Fig. 4.7(a)) described later (page 62), has only α helices connected by short loop sections with its haem group inserted into a cleft. The staphylococcal nuclease (an enzyme hydrolysing nucleic acid) has a mixture of an antiparallel β sheet and three α helices connected by unstructured polypeptide (Fig. 4.7(b)). A common arrangement is the so-called α/β **barrel**. In this, there is a central core of β strands arranged like the staves of a wooden barrel except that they are twisted. The barrel encloses tightly packed hydrophobic side chains. Surrounding this barrel are α helices. The diagram of the triosephosphate isomerase in Fig. 4.7(c) illustrates this. The barrel is formed by a β sheet curving round to form a cylinder.

The enzyme pyruvate kinase, shown in Fig. 4.7(d), shows another type of structure with three sections.

When diagrams of numbers of proteins were constructed, it emerged that certain patterns of secondary structure organization are frequent. Families of proteins exist, different in detail but with the same basic 'recipe' of secondary structures evident. Myoglobin illustrates the 'globin fold', a structure of eight α helices which in this case forms a pocket into which the haem group fits. This fold is found in a large family of proteins. It seems that, often, old structural patterns are preserved as new proteins are evolved.

Noncovalent bonds are mainly responsible for stabilizing the tertiary structures of proteins

You can illustrate this yourself by frying an egg. Egg white is largely composed of a solution of the protein egg albumin, a water-soluble colourless protein. This can be purified and crystallized as sharp, shiny crystals. Albumin is a globular protein. Its polypeptide chain in its natural state, as explained, is folded up in a specific way into a compact structure. The three-dimensional structure of globular proteins – the folding of the polypeptide chain – is usually destroyed by heat. The folded form is called the **native** protein – the unfolded form, a randomly disarranged polypeptide chain, is called a **denatured** protein and the conversion from the native to the denatured state is known as **denaturation**.

When an egg is fried, the mild heat disrupts weak noncovalent bonds so the polypeptide folding comes apart, the extended chains of different albumin molecules get tangled up in to a white opaque insoluble mass (Fig. 4.8). This illustrates that the folding of the native protein is held together completely or largely by weak noncovalent bonds (page 36), for covalent peptide bonds are not disrupted by boiling at neutral pH.

The biochemical function of proteins, such as enzyme catalysis, is almost always destroyed during the heating process (though a few heat-stable proteins exist). When a polypeptide chain folds up into a globular configuration, the free energy drop (ΔG) is small so that the native folded protein structure is only marginally stable and easily denatured by mild heat in most cases.

The folded structure of a protein is determined by the amino acid sequence of the polypeptide chain

This was proved in the classical experiment of Anfinsen. He inactivated the enzyme ribonuclease by exposing it to high concentrations of urea which is a hydrogen bond breaker. He also reduced the —S—S— bonds with a thiol reagent (see below). This treatment denatures the protein. Its polypeptide chain is unfolded. On removal of the urea by dialysis, the protein refolded itself and enzyme activity was restored. It showed that proteins can spontaneously fold into the correct way, and the amino acid sequence is sufficient in itself to determine the final form. This is of very great importance; all life is dependent on this fact. It establishes that the simple one-dimensional linear

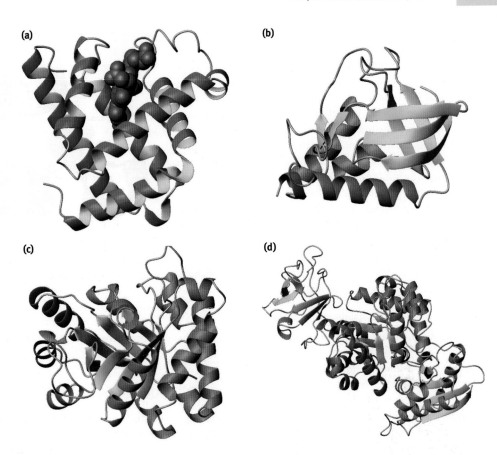

Fig. 4.7 Ribbon diagrams of the structures of different proteins: **(a)** myoglobin (1MBO), showing the haem molecule in blue; **(b)** staphylococcal nuclease (1A2T); **(c)** triosephosphate isomerase (1AG1); **(d)** pyruvate kinase (1A3W). (b)–(d) Colours indicate α helices in red, β strands in yellow, and all else in grey. (See text for description.)

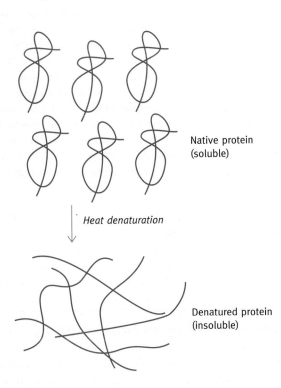

Native protein (soluble)

Heat denaturation

Denatured protein (insoluble)

Fig. 4.8 Hypothetical representation of egg albumin denaturation. (The folded configuration is drawn arbitrarily.)

code of genes is sufficient to specify the folded functional form of proteins and hence the three-dimensional form of thousands of different proteins, and hence of all life forms. It is difficult to see how life could exist if this were not the case. What makes this even more remarkable is that it is still not fully understood how it happens, as we will now discuss.

How does protein folding take place?

In the cell a newly synthesized polypeptide chain folds up in a minute or two. The seemingly obvious mechanism is that the polypeptide tries different conformations until the lowest energy state is found. This is not tenable for it has been calculated there are so many possible conformations even for a moderately sized protein of one or two hundred residues, it would take billions of years to fold correctly. The earth has not existed long enough for a single protein molecule to have folded up by this mechanism.

It is not fully understood how proteins in the cell fold up so rapidly. A postulated mechanism is that as the polypeptide is synthesized, sections of the polypeptide rapidly assume their secondary structures. It is a stepwise mechanism in which the series of secondary structures occur and finally arrange themselves in the correct form. However incorrect (nonprofitable) conformations inevitably occur and special proteins have been developed to cope with this problem. They prevent improper associations and like their Victorian human counterparts are called **chaperones** or **molecular chaperones.** There are various types. One way is to put the unfolded molecule in an isolated box of its own and given every encouragement to fold properly and then let out. This chaperone in a sense does what Anfinsen did for ribonuclease, allowing it to refold under favourable conditions, and is therefore sometimes described as an 'Anfinsen cage'.

The Anfinsen experiment was with a small protein and present in low concentration so that incorrect interactions between chains would be minimized. Conditions in a cell are very different, with tightly packed large polypeptides giving every chance of incorrect associations. It is important to note, however, that chaperones cannot give *directions* on how to fold; they can simply make it more likely that they will fold according to their amino acid sequences by stopping them interacting with other unfolded proteins. We will return to a fuller discussion of chaperones in the chapter on protein synthesis (page 395).

Covalent —S—S— bonds stabilize some proteins

Although their tertiary structures are largely the result of non-covalent bonds, some protein structures are 'locked' or strongly stabilized by **disulfide** (S—S) **bridges** which, being covalent, are very strong. Examples are proteins liberated into the blood (insulin is one example see Fig. 4.9), or the intestine (digestive enzymes). This stabilization is achieved by pairs of thiol groups of the cysteine side chains, brought together by polypeptide folding. An oxidase enzyme forms the S—S link between them by the following reaction:

$$2RSH + O_2 \rightarrow RS\text{—}SR + H_2O_2$$

This provides a very strong cross-linking bond – more or less like a steel rivet in the structure. A few of these between different sections of a chain, makes the folded shape more stable. Insulin has three disulfide or S—S bridges (Fig. 4.9). Proteins with disulfide bonds are often less easily denatured by heat. Few intracellular proteins have disulfide bonds in their structure, possibly because the interior of cells are strongly reducing and might be sufficient to disrupt them; most S—S cross-linked proteins are extracellular.

An extreme example of stabilization of a protein by disulfide bridges is the α-**keratin** protein of hair. The long polypeptides are interlinked by many disulfide bonds, which are important in locking in the configuration of the hair. In permanent waving, these are broken by thiol reduction, heat and moisture being used to disrupt the original hydrogen bonding, followed by setting the hair into a new configuration. On cooling, hydrogen bonds reform the α-helical structure, and then the new configuration is made permanent by the 'neutralizer', which reoxidizes cysteine —SH groups to reform disulfide bonds between the multiple polypeptide chains of the hair structure. These bonds are new ones between —SH groups that have been brought together in the new stable (permanent) curled configuration of the hair. Finally, the trivial name for two cysteine molecules

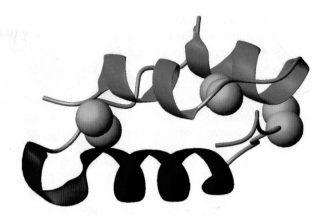

Fig. 4.9 Ribbon model of insulin (4INS) showing the two polypeptide chains joined by two disulphide bridges, with a third disulphide bridge internally in the A chain. The S—S bonds are visible in yellow.

which have become linked by an S—S bond is **cystine** and the symbol for it is Cys-Cys.

Quaternary structure of proteins

With the tertiary structure, we now have a protein molecule, and for many proteins that is the final stage. Many functional proteins, however, have more than one such protein molecule or protein monomer (see page 48) in them that remain as discrete molecules held together in a single complex by secondary bonds. These monomers may be the same or different, but the molecules have to be structured so as to fit *via* complementary surface patches, so that only the correct subunits complex together. The resultant multi-component molecules are variously called **oligomeric**, **polymeric**, or **multi-subunit proteins.** Allosterically regulated enzymes are mostly of this type, and so is haemoglobin (see page 64). This arrangement of subunits into a single functional complex is referred to as the quaternary structure of a protein (Fig. 4.2(d)).

Development of quaternary structures leads on to the self-assembly of the complex structures of life. It also greatly increases the number of functional proteins. For example, if the active form is a dimer of subunits, by forming heterodimers in which different but related subunits combine, many different functional proteins are possible. This strategy is very commonly used in DNA binding proteins, but not limited to those.

Protein homologies and evolution

Evolution involves the development of new protein structures and therefore modification of existing genes by mutation. It may seem puzzling that an existing gene coding for an essential protein can be modified into a different one without losing the original. Gene duplication followed by evolution of one copy avoids elimination of existing genes.

The amino acid sequence of a protein is a consequence of its evolution and can be used to gain insights into the past. Amino acids essential for the function of the protein tend to be **conserved** in evolution – they are not substituted for, or only by very similar amino acids. Mutational changes in a particular amino acid residue resulting in another one with similar properties are said to be **conservative** changes. Less important amino acids tend to change, and this is known as **nonconservative** substitution. It is observed repeatedly that different proteins with different functions have such close amino acid sequence similarities that they must have had a common ancestral protein. Such proteins that have evolved from a common ancestral protein are said to be **homologous**.

When the amino acid sequence of a protein has been determined, protein databases (page 8) can be searched for all known proteins recorded. Methods are available for aligning protein sequences to allow for deletions and insertions of residues, so that sequence similarities can be quantitatively assessed and *the statistical probability of resemblances being due to chance determined*. In this way gene families that evolved from common ancestors can be identified and information on evolutionary relationships obtained.

Tertiary structural resemblances between proteins can also be used to detect homologies, for example the number and arrangement of helices and β sheets (the tertiary structures) can be compared. These protein structures tend to be **conserved** (not changed throughout evolution) because they are most intimately related to function, and can be more diagnostic of evolutionary relationships between proteins than sequences.

The analysis of proteins in this way is referred to as **bioinformatics** (page 8).

Protein domains

If you consider a protein molecule consisting of a single polypeptide chain, its folded structure may be a single compact entity, no part of which could exist on its own, for the folding of the protein would be disrupted. However, especially in proteins larger than about 200 amino acids in length, it is often seen that there are two or more regions which form compact 'islands' of folded structure usually linked together by unstructured polypeptide. In a number of cases, they have been obtained separately and their integrity of structure and, in some cases, catalytic activity confirmed. A suggested definition of a protein **domain**, as these sections are called, is that it is a sub-region of the polypeptide that possesses the characteristics of a folded globular protein. A domain must be formed from a discrete section of polypeptide chain – it can't involve the folded structure being formed by the chain leaving the domain and then doubling back into it. It must be self-contained. The structure of the enzyme pyruvate kinase illustrated in Fig. 4.7(d) illustrates three domains joined together to form a single protein.

Protein domains are often associated with different partial activities of a protein that are involved in the function of the protein. Many enzymes with a single catalytic function combine with at least two substrates. The NAD^+ dehydrogenases (page 178) are a typical case. Different NAD^+ dehydrogenases all bind NAD^+, but each binds a different oxidizable substrate and all catalyse the reaction:

$$AH_2 + NAD^+ \rightleftarrows A + NADH + H^+$$

(where A is any substrate molecule).

It is found that several enzymes catalysing such reactions have separate domains for binding NAD^+ and the substrate (AH_2). The two binding sites together form the active centre of the enzyme. However, the NAD^+ binding domains of the different enzymes examined have a similar basic structure, suggesting homology and common ancestry, while the AH_2 binding domains are different. This suggested that once evolution had successfully developed an NAD^+ binding domain, it was used repeatedly, in combination with different substrate-binding domains evolving, specific for each substrate. Many examples are now known, that points to it being a general strategy used in the evolution of proteins. One particularly striking example is the SH3 domain of a protein involved in signal transduction (Chapter 27); in the human genome sequence over 300 DNA sequences exist that would be translated into amino acid sequences homologous with that of the SH3 domain.

In some cases, however, structural similarities may be the result of convergent evolution in which a particular sequence of amino acids, being the most efficient, is independently evolved in proteins of different ancestry. An example is the catalytic triad of proteinases (page 95), which occurs in a family of eukaryotic proteinases with clearly homologous structures, but also in the bacterial subtilisin, whose structure is ancestrally unrelated to the eukaryotic variety.

Domain shuffling

Domain shuffling (or domain swapping) is the name given to the evolutionary process in which new genes are assembled from sections of DNA coding for pre-existing protein domains. It leads to the synthesis of novel proteins made up of new mixtures of domains. Modular construction of enzymes and other proteins permits more rapid evolution of new functional proteins than would occur from single amino acid substitutions, somewhat like assembling variants of instruments using standard plug-in boards.

Protein domains are often (though not always) coded for in eukaryotes by specific separate gene subsections, called exons (described in Chapter 23). The localization within the gene of the DNA sections coding for domains, and the divided structure of eukaryotic genes, facilitates the chance of successful recombination of domains (see page 326).

Membrane proteins

We have so far described globular proteins as being water-soluble, but many membrane proteins have two external sections that are water-soluble at each end and a hydrophobic section in the middle (see page 109), that is in contact with the hydrocarbon layer in the centre of a membrane. The structure of membrane proteins is dealt with in Chapter 7.

Conjugated proteins and post-translational modifications of proteins

Many proteins are simply as already described – they need nothing but the folded polypeptide chain(s) for their function. However, many enzymes require a metal ion for activity; some have attached a complex molecule called a **prosthetic group**, which forms part of the active centre. The protein part in such cases is called an **apoenzyme** (*apo* = detached or separate) and the complete enzyme a **holoenzyme**.

Other proteins have carbohydrate attachments and are called **glycoproteins**. Most membrane proteins have oligosaccharides attached to the sections of the polypeptide on the outside surface of the protein via the —OH of serine or threonine side chains (*O*-linked) or to an asparagine side chain (*N*-linked). The latter attachment method is illustrated below.

Polypeptide backbone / Asparagine side chain

Sugar unit

Many secreted proteins such as blood and saliva proteins are glycoproteins. The degradation of carbohydrate attachments to serum proteins and to erythrocyte membrane proteins marks them for uptake and destruction by liver cells – the carbohydrates are acting as indicators of age of the components. Glycosylation of some proteins makes them effective lubricants, such as in saliva or protects them from proteolytic attack. In some cases the different sugars are involved in recognition. They play this role in the sorting of proteins by the Golgi apparatus (Chapter 26).

Post-translational modifications of proline residues are important in extracellular proteins (see below).

Extracellular matrix proteins

In structural terms we now come to a different class of proteins. The free soluble proteins we have discussed so far have been globular. Proteins of the extracellular matrix are mainly elongated fibrous proteins that are usually partly immobilized by being bound into larger structures. They play important organizational roles not previously appreciated. There is great medical interest in the area.

A general description of the extracellular matrix may be helpful here to give perspective on its role. All cells are embedded or bathed in it; even in tissues such as liver when the cells are in close contact and the layer is thin.

Between tissues there are spaces filled with **connective tissue** which, as the name implies, connects tissues together. It contains extracellular matrix proteins and cells producing them.

The **dense connective tissues** include bone and tendons, the latter linking muscles to bone transmit the tension of contraction. Both are predominantly **collagen**, the most plentiful protein in the mammalian body; in the case of bone, the tissue is calcified. At the other end of the spectrum are the **loose connective tissues** found under all epithelial layers; wherever there is a bodily cavity, such as the intestine and the blood vessels, there is a layer of epithelial cells lining the cavity. This has no mechanical strength but is supported by a layer of protein fibres known as the **basal lamina**. Underneath this is a layer of loose connective tissue as illustrated in Fig. 4.10 where skin is used as an illustrative example. The connective tissue joins up the epithelial layer to the underlying tissue. It is flexible and resists compression. The background substance that resists compression is a soft, highly hydrated gel formed by proteoglycans (described below), which on its own has no mechanical strength against

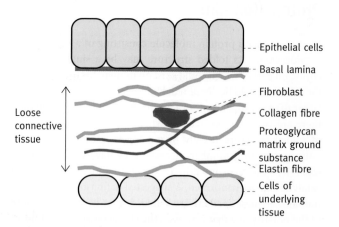

Fig. 4.10 Components of loose connective tissue that underlies epithelial cell layers (e.g. of skin or intestinal lining). The structures of the collagen and elastin fibres and of the proteoglycans are described later in the text.

tension but is reinforced with collagen and **elastin** fibres. Some fibres link the basal lamina to the epithelial cells above it and to the connective tissue below. Further links join the underlying tissue cells to components of the connective tissue so that the whole is a stable structure. There are also adhesive proteins that help to interconnect everything, the best known of which is **fibronectin**.

The components of connective tissue are secreted by cells known as **fibroblasts** dotted around in the background matrix and occupying little of the volume of connective tissue. Bone and cartilage have special fibroblasts known as osteoblasts and chondroblasts respectively.

With that general introduction we will now describe the structures of the reinforcing proteins, collagen and elastin, the structures of proteoglycans which form the jelly like ground substance of connective tissue, and then the adhesion proteins which bind components together. Finally we will describe the proteins (integrins) which connect the extracellular matrix to intracellular components; these are currently viewed with intense interest.

Structure of collagens

Collagen occurs outside cells. The protein from which it is assembled is secreted by cells in the form of **procollagen,** which is subjected to a variety of chemical changes catalysed by enzymes, resulting in the mature collagen. Procollagen consists of a triple superhelix – three helical polypeptides twisted around each other (see Fig. 4.11(a)). Each of the polypeptides in the triple superhelix of procollagen is an *unusual left-handed helix* not the right-handed α helix of globular proteins. However, *the three polypeptides are twisted around each other in a right-handed manner* to form the triple helix. About one in three amino acid residues is proline and every third residue is glycine. This gives an extended configuration to the polypeptide chains. At the ends are extra peptides which, after secretion

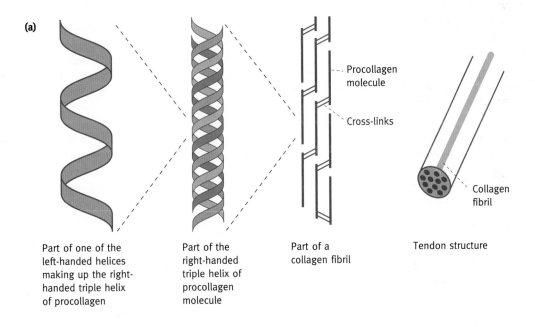

(a)

Part of one of the left-handed helices making up the right-handed triple helix of procollagen

Part of the right-handed triple helix of procollagen molecule

Part of a collagen fibril

Procollagen molecule

Cross-links

Collagen fibril

Tendon structure

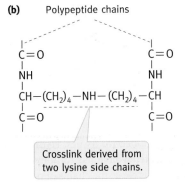

(b) Polypeptide chains

$$C=O \quad\quad C=O$$
$$NH \quad\quad NH$$
$$CH-(CH_2)_4-NH-(CH_2)_4-CH$$
$$C=O \quad\quad C=O$$

Crosslink derived from two lysine side chains.

Fig. 4.11 (a) Arrangement of collagen fibrils in collagen fibres. The bonds in red are covalent links formed between lysine residues. They exist also within the triple helix. **(b)** One type of cross-link formed between two adjacent lysine residues. Note that several specific types of collagen exist for individual functions.

of the procollagen, are cleaved to give **tropocollagen** molecules from which collagen fibrils are assembled.

After synthesis of the molecule, many of the proline and also lysine residues are hydroxylated to form **hydroxyproline** and **hydroxylysine** in the polypeptide. These hydroxylated amino acids are formed after the parent amino acids are in polypeptide form. Hydroxylation of proline in the polypeptide requires ascorbic acid (vitamin C), which keeps an essential Fe^{2+} atom in the enzyme prolyl hydroxylase in the reduced form. In deficiency of this vitamin, connective tissue fails to be properly formed, resulting in the painful consequences of scurvy – bleeding gums and failure of wound healing.

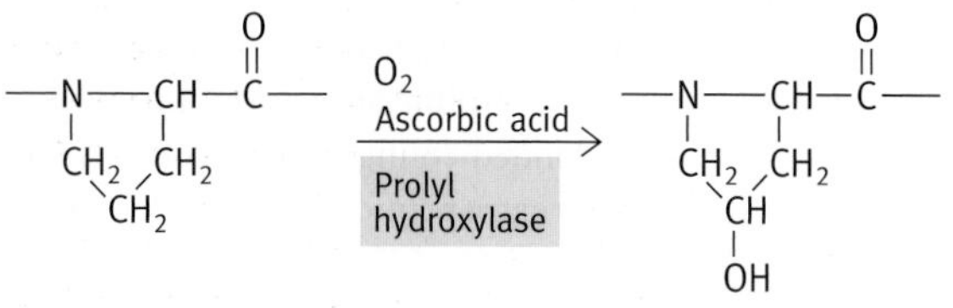

Proline residue in polypeptide

Hydroxyproline residue in polypeptide

(Note that the hydroxylation reaction is more complex than shown here and also involves α-ketoglutarate.)

The three strands of the triple superhelix are in close association, forming a very strong structure. The side chains of polypeptide chains would normally prevent such close association. In long stretches of collagen polypeptides the helical structure is more extended than in an α helix. There are three amino acid residues per turn. Every third residue is glycine and, in the triple superhelix, the contacts between the chains occur always at glycine residues, the side 'chain' of which is only a hydrogen atom that does not get in the way of close contact. The hydroxylated lysines and prolines form hydrogen bonds between the three chains, thus stabilizing the super helix.

The procollagen molecule so far described has about 1000 amino acid residues. These assemble into collagen fibrils by staggered head to tail arrangement as shown in Fig. 4.11(a). The 'holes' in this structure are believed to be sites where crystals of hydroxyapatite $(Ca_{10}(PO_3)_6(OH)_2)$ are laid down to initiate bone mineralization. The structure described so far would not have the required strength for a tendon. This is achieved by the formation of unusual covalent links between ends of the tropocollagen units; two adjacent lysine side chains are modified to form the link, one type of which is shown in Fig. 4.11(b). The collagen fibrils so formed aggregate to form tendons by a parallel arrangement. In skin, it is more of a two-dimensional network. Different subclasses of collagen exist, dependent on the precise structure of the three polypeptides in the triple superhelix. Variations between the different types of chains include the content of hydroxylysine and hydroxyproline and the degree of glycosylation of these residues.

Structure of elastin

The elastin molecule has a unique structure, different from that of collagen. It is a major component of connective tissue such as the lung and arteries. It is a hydrophobic insoluble protein which forms a three-dimensional elastic network which can reversibly stretch in any direction with a structure more elastic than rubber. The network is assembled from a soluble protein unit called proelastin, which is secreted from cells and then cross-linked to form the elastic network. Proelastin is rich in glycine, alanine and lysine. Every few residues are lysine and in between are short stretches of helical or possibly random coil sections with no recognizable secondary structure. It is these sections that can reversibly stretch in an elastic manner. The proelastin is assembled into the three-dimensional network of elastin by the formation of covalent links between lysine residues of four polypeptide chains forming desmosine, as shown in Fig. 4.12. Desmosine formation requires the enzymic oxidation of lysyl residues.

Structure of proteoglycans

The proteoglycans provide the background jelly-like matrix substance of the connective tissues. Hydrated jellies in nature are based on negatively charged carbohydrate polymers. The molecular strategy is clear; the chains of polar sugars are highly

BOX 4.2 Smoking, elastin, emphysema, and antiproteinases

We will see in Chapter 9 (page 146) how the digestive system escapes the actions of its own proteolytic enzymes. However, proteases exist elsewhere in the body. A particularly important one, in the present context, is the **elastase** of **neutrophils**. Neutrophils are phagocytic white cells attracted to sites of infection or irritation. When activated at such sites, they secrete elastase, which clears away connective tissue from the site. Elastin is the elasticity-conferring connective tissue protein from which elastase gets its name.

In the lung, air passages lead to minute pockets, the alveoli, which result in the lung having the very large surface area needed for the diffusion of gases between blood and air. Neutrophils attracted to the lung liberate elastase. In normal circumstances this is prevented from destroying the lung structure by α_1-**antitrypsin** (α_1-**antiproteinase**), a protein that is produced by the liver and secreted into the blood. The α_1-antitrypsin inhibits several proteases, including trypsin as the name implies, but is especially effective on elastase which, *in vivo*, is probably the most important of the proteases in this context. It inhibits by combining tightly with the enzyme and blocking its catalytic site. So tightly that it is known as a 'suicide inhibitor' for once it binds, both the enzyme and the inhibitor molecule are not recoverable.

α_1-Antitrypsin in adequate levels in the blood is essential for the protection of the lung. The molecule diffuses from the blood into the alveoli. If, due to a genetic defect, the level of α_1-antitrypsin is subnormal, neutrophil elastase may destroy alveoli, resulting in much larger pockets in the lung structure and consequent reduction of surface area available for gaseous exchange. The result is emphysema, a symptom of which is extreme shortness of breath.

Smokers are prone to emphysema for two reasons. The smoke irritation attracts neutrophils to the lungs, with a consequent increased release of elastase. Secondly, oxidizing agents in the smoke destroy α_1-antitrypsin; they chemically oxidize the sulphur atom of a crucial methionine side chain to a sulphoxide group ($S \rightarrow S{=}O$). This prevents the α_1-antitrypsin from inactivating elastase, and may result in proteolysis of lung tissue, resulting in emphysema.

This is not the only physiological role of antiproteinases in the body. Trypsin, chymotrypsin, and elastase are secreted by the pancreas into the intestine in an inactive zymogen form. They are activated by proteolysis in which an initial small activation becomes an autocatalytic cascade. This makes even a small amount of premature activation in the pancreas cells potentially dangerous, because any active proteinase may activate all of the zymogen in a proteolytic cascade and cause pancreatitis. The three digestion proteases all depend on an active serine residue (page 97) and the antiproteinases are collectively called **serpins** (**serine protease inhibitors**). As with α_1-antitrypsin they work by very tightly associating with the active centres of the enzymes and blocking their activity.

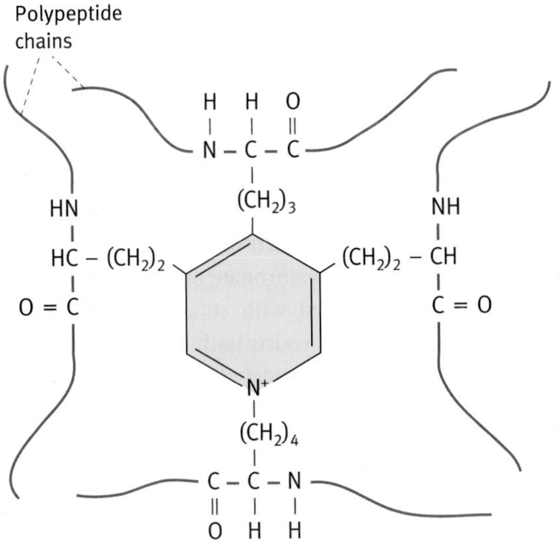

Polypeptide chains

Fig. 4.12 Desmosine cross-link between four polypeptide chains of elastin. The structure is formed by enzyme modification of four lysine residues.

hydrophilic and the mutual repulsion of the charged groups on them ensures that the chains are fully extended and occupy a large volume, thus entrapping a lot of water. A proteoglycan consists of chains of charged sugars attached to the serine side groups of **core protein** molecules (Fig. 4.13). The protein chain is fully extended, as are the carbohydrate chains. The negative charges localize a cloud of cations, which contribute to the osmotic pressure of the matrix, which is important in resisting compression.

The carbohydrate chains are all made of repeating disaccharide units each of which has either *N*-**acetylglucosamine** or *N*-**acetylgalactosamine** (Fig. 4.14) as one component so that the polysaccharides are known as **glycosaminoglycans** (GAGs). The second sugar usually has a carboxyl group and a sulfate group (though exceptions with only one of these groups occur). The general pattern of the repeating disaccharide is shown in Fig. 4.15.

There are many different types of proteoglycan; the basic design is extremely flexible. In different proteoglycans the length of the core protein varies from about 1000 to 5000 amino acid residues, and the number of polysaccharide chains attached to the core protein varies up to about 100; the length of the polysaccharide chain varies but typically is about 80–100 sugar residues long. Finally, the number and type of charged groups can vary. The main GAGs are known as **chondroitin sulphate**, **dermatan sulphate**, **heparan sulphate**, and **keratan sulphate**, which differ in the ways described above. (Heparin is structurally similar to heparan sulphate, but the latter has more sulphate groups. It has a different role, however, in blood vessels where it exists as free GAG, and is important in controlling blood clotting; see page 485). The GAGs are used in the assembly of different types of proteoglycan for different functions. The chief proteoglycan of cartilage provides an example to

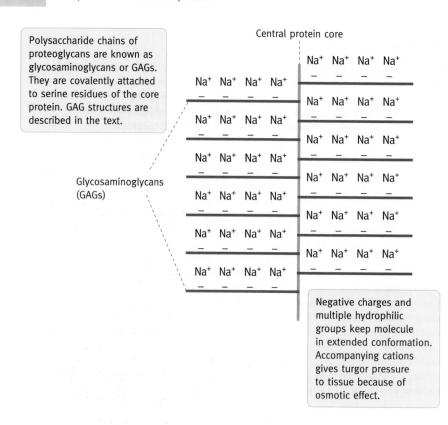

Polysaccharide chains of proteoglycans are known as glycosaminoglycans or GAGs. They are covalently attached to serine residues of the core protein. GAG structures are described in the text.

Glycosaminoglycans (GAGs)

Central protein core

Negative charges and multiple hydrophilic groups keep molecule in extended conformation. Accompanying cations gives turgor pressure to tissue because of osmotic effect.

Fig. 4.13 The design of proteoglycans. Note that there can be many variations on the common theme: the size of the central protein core can vary; the number of glycosaminoglycan (GAGs) attached to it can vary; the GAGs can vary in length, in the number, position, and nature of the charged groups, and in the details of their chemistry. We show the accompanying cloud of cations as Na^+ but other cations could be involved. The structures of the GAGs are described in the text.

N-Acetylglucosamine   *N*-Acetylgalactosamine

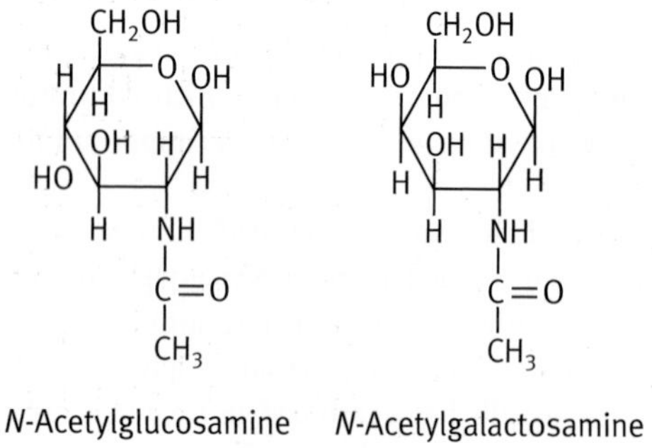

Fig. 4.14 *N*-Acetylglucosamine and *N*-acetylgalactosamine.

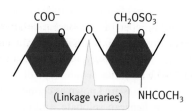

(Linkage varies)

Fig. 4.15 A disaccharide unit of glycosaminoglycan (GAG). For simplicity all bonds and substituents except those characteristic of GAGs have been omitted. The polysaccharide portion of proteoglycans are made of long unbranched chains of these disaccharides. Different GAGs vary in the number and positions of sulphate and carboxyl groups and in other details such as the nature of the glycosidic link between the sugars.

illustrate this. Cartilage has to withstand very large compressive forces and be very tough. In knee joints, for example, it is needed to prevent direct contact of leg bones and has to withstand enormous pressures. The proteoglycan in such cartilage made up of a protein core with two different GAGs attached to the serine side groups, and large numbers of these proteoglycan molecules are complexed noncovalently to yet another long GAG called hyaluronan, forming a gigantic molecule (Fig. 4.16) resembling a bottle brush as seen in the electron microscope. Hyaluronan is also called hyaluronate or hyaluronic acid. This matrix is heavily reinforced with collagen fibres that resist tension so that the cartilage resists both compression and tearing. The GAG hyaluronan is widely found in soft extracellular matrices and synovial fluid, where it exists free rather than associated with protein.

Adhesion proteins of the extracellular matrix

The best known of the adhesion proteins is **fibronectin**. It consists of two flexible elongated protein chains joined at one end by two disulphide bridges. It has several different sites on it that bind to proteoglycans, collagen, and cell surfaces, respectively, which serve to bind the entire structure together (Fig. 4.17).

Integrins are important signalling proteins

Integrins are transmembrane proteins that integrate the attachment of cells to extracellular matrix proteins (Fig. 4.18). A very

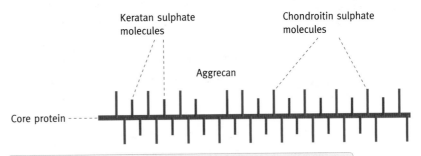

Aggrecan is a proteoglycan consisting of many copies of two GAGs attached to a long core protein via serine residues.

Keratan sulphate molecules

Chondroitin sulphate molecules

Aggrecan

Core protein

In cartilage many molecules of aggrecan are attached noncovalently to a third GAG (hyaluronan) via link protein molecules to form a huge complex.

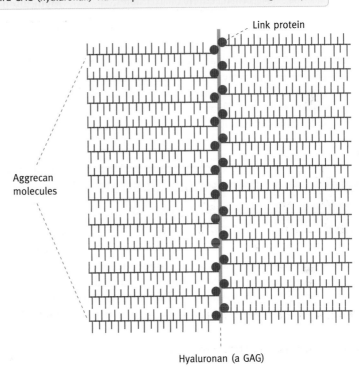

Link protein

Aggrecan molecules

Hyaluronan (a GAG)

Fig. 4.16 The main proteoglycan found in cartilage. Note that hyaluran is simply a glucosaminoglycan (GAG). It forms a huge noncovalent complex with multiple copies of the proteoglycan aggrecan, the attachment being via link protein molecules. Hyaluronate is also widely found in soft extracellular matrices where it exists free, not linked to proteins – often called hyaluronic acid in this context.

large family of different integrins exist. There are both homo-dimers and hetero-dimers, and they exist as surface proteins on cells. Different cells have different specific integrins to which specific matrix proteins or other cells can selectively attach to the extracellular domains of the integrins. The intracellular domains are attached to actin fibres of the cytoskeleton and so form a link between the extracellular matrix and the inside of the cell (Fig. 4.18).

The functions of the integrins are very complex. To illustrate their importance with a few examples, they are the attachment sites by which the neutrophils recognize other cells. Thus, at the site of an infection, neutrophils localize themselves from the blood stream and attach to integrins of cells at the infection site. They can then fight the infection.

Diffferent ligands can specifically attach to different integrins and can affect the activities of the cells. It is believed that integrin signalling may have profound effects on many aspects of cell life (See review by Hynes).

Cell adhesion proteins

Cell adhesion proteins, known as **CAMS**, are present on all vertebrate tissue cells. They are responsible for cell–cell attachments necessary for the formation of tissues. One very important class is known as **cadherins**. There is a very large superfamily of these with specificities for the different classes of tissues. The cadherins are cross-membrane proteins, which on the external

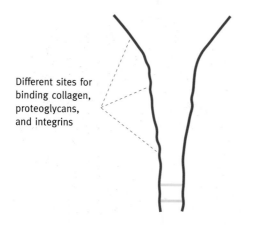

Fig. 4.17 A fibronectin molecule consisting of two polypeptides linked by two disulphide bridges (yellow). The integrins are transmembrane receptors on cells. The binding of fibronectin to collagen fibres, proteoglycans, and integrins links together the extracellular matrix structure.

Different sites for binding collagen, proteoglycans, and integrins

surface interlock with cadherins on adjacent cells through complementary patches of noncovalent bonds between cadherins on the cells. On the inside of the cell they connect via intermediate linker proteins with actin filaments of the cytoskeleton. Fibronectin is an important example.

Myoglobin and haemoglobin illustrate how protein structure is related to function

Myoglobin and haemoglobin have been studied intensively and form the classic illustration of the way in which knowledge of protein structure leads to an understanding of biological function.

Tissues of the body have to be supplied with oxygen. The most primitive animals and some cold-water fish transport oxygen in solution in their blood, but the solubility of the gas is too low for this to be adequate for more active animals. Much the same is true for the removal of CO_2 from the tissues.

Myoglobin is the red pigment found in striated muscle (page 126), where it has the role of an oxygen store to be used during intense muscular activity that results in the consumption of more oxygen than the blood can deliver. Haemoglobin is the oxygen carrier in blood. The two proteins are closely related, both in structure and ancestry, but myoglobin is a relatively simple molecule while haemoglobin is a sophisticated molecular machine superbly evolved for its complex roles. A comparison of the two proteins is helpful in understanding the latter.

Myoglobin

The myoglobin protein is monomeric – the single polypeptide chain of 153 amino acid residues is arranged entirely in the form of α helices connected by loops (Fig 4.23 (a)); inserted into a pocket formed by the helices is a molecule of haem. Haem is a ferrous iron – tetrapyrrole (Fig. 4.19) with hydrophobic side groups on three sides and the fourth with hydrophilic carboxyl radicals. The molecule is buried in the hydrophobic interior of the molecule with the hydrophilic side oriented to the exterior. Haem has an intense red colour resulting from its **conjugated double bond system** alternating single and double bonds around the molecule. The iron in haem can bond to six ligands, four being taken up by the pyrrole nitrogen atoms, the fifth by attachment to a histidine residue of the protein, and the sixth is available for the reversible attachment of oxygen (Fig. 4.20).

Fig 4.21 shows the percentage saturation of myoglobin at increasing oxygen tensions. The curve is hyperbolic, much the same as that of a 'classic' enzyme-substrate-binding response (see page 90). The molecule is almost fully saturated at the low oxygen tension of 20 torr present in capillaries. (A torr is a unit

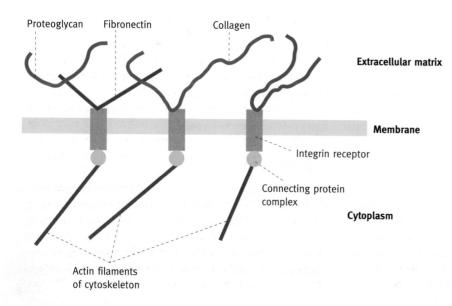

Proteoglycan   Fibronectin   Collagen

Extracellular matrix

Membrane

Integrin receptor

Connecting protein complex

Cytoplasm

Actin filaments of cytoskeleton

Fig. 4.18 The role of integrins in connecting components of the extracellular matrix to the cytoskeleton. A family of fibronectins exist. The fibronectin is represented as a single line rather than its dimer shape (Fig. 4.17) for simplicity. The cytoskeleton is described in Chapter 8.

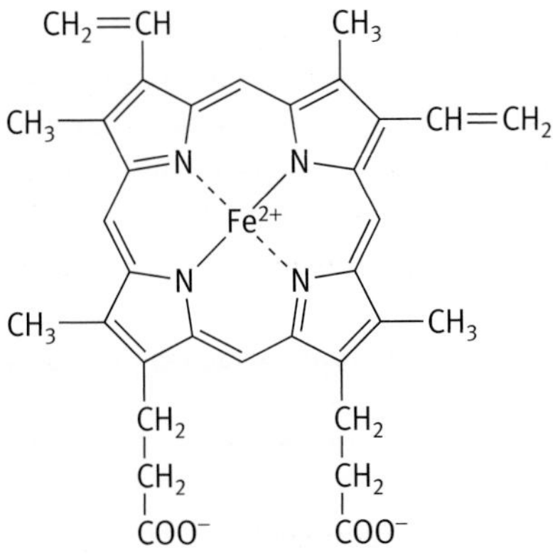

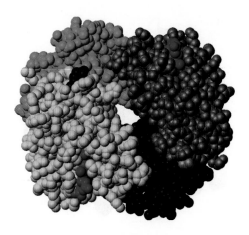

Fig. 4.19 The structure of haem. This form is found in haemoglobin; other haems differing in their side chains are found in cytochromes. Note that, at one side of the molecule, there are two hydrophilic propionate groups ($-CH_2CH_2COO^-$) while the remaining side groups are all hydrophobic. In myoglobin, haem sits in a cleft of the molecule with the hydrophilic side pointing out towards water and the hydrophobic groups buried into the nonpolar interior of the protein.

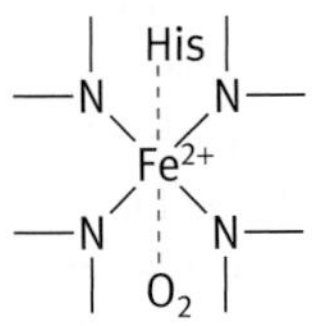

Fig. 4.22 Space-filling model of haemoglobin (1A3N). Model of the subunits and their arrangement in haemoglobin showing the central cavity into which 2:3-bisphosphoglycerate fits in the deoxygenated state. In sickle cell anemia the glutamic acid residues at position 6 in the β side chains are mutated to valines, creating hydrophobic patches on the molecule.

Fig. 4.20 Binding ability of Fe^{2+} in haem. The iron atom in haem can bond to six ligands in total: four bonds to the flat pyrrole nitrogen atoms as shown and the other two above and below the plane of the page. One of the perpendicular bonds is bound to the nitrogen atom of a histidine residue, The other is the binding site for an oxygen molecule.

from the blood to top up its store as required. Since it does not surrender oxygen at the levels normally found in tissues it would be unsuitable as the oxygen carrier in the blood. This is the role of haemoglobin.

Structure of haemoglobin

Haemoglobin is a tetramer of protein subunits (Fig. 4.22) each of which has a haem molecule capable of binding oxygen. The two α subunits are identical as are the two β ones; they are known as α_1 and α_2, and β_1 and β_2. The subunits closely resemble myoglobin in structure, being composed of α helices, and each subunit has a haem unit similarly located between the helices. Myoglobin and the β-haemoglobin subunit are virtually identical in structure, as shown in Fig. 4.23 (a) and (b).

Binding of oxygen to haemoglobin

In the body, the binding of oxygen to haemoglobin and its release in the tissues is not a passive reversible process. The haemoglobin molecule functions in such a way that the molecule almost seems to have intelligence.

A molecule of haemoglobin binds four molecules of oxygen, 1 per subunit. When an oxygen-saturation curve is plotted, instead of the hyperbolic curve seen with myoglobin, there is a sigmoid curve that is well to the right of that of myoglobin (Fig. 4.21). Higher oxygen concentrations are required to 50% saturate haemoglobin than is the case for myoglobin, indicating, as already stated, that myoglobin has a higher affinity for oxygen.

However, haemoglobin needs to readily pick up as much oxygen as it can in the lungs but then readily surrender it in the tissue capillaries. The sigmoidal oxygenation curve means that it is most steep (that is, surrenders the most oxygen) at oxygen pressures encountered in the capillaries (see Fig. 4.21) but,

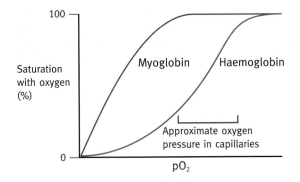

Fig. 4.21 Oxygen–haemoglobin saturation curve. The higher affinity for oxygen of myoglobin as compared to haemoglobin means that myoglobin in the muscles readily accepts oxygen from the blood.

of pressure named after Torricelli). This means that it has a high oxygen affinity and does not give it up at normal oxygen tensions. It does so only when intense muscular activity further lowers the oxygen tension in the tissue so that it is acting as a reserve store of oxygen in times of need. It has a higher affinity for oxygen than has haemoglobin and so can extract oxygen

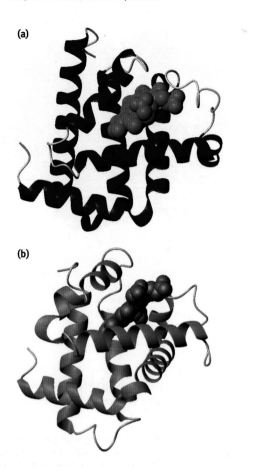

(a)

(b)

Fig. 4.23 Models of myoglobin (1MBO) (a) and a single chain (β chain) of haemoglobin (1HHO) (b). Computer-generated diagrams showing the folding of the polypeptide in myoglobin and haemoglobin and the positioning of haem in the molecule.

nonetheless, it is still capable of becoming virtually saturated at the oxygen pressures encountered in the lungs.

How is the sigmoidal oxygen saturation curve achieved?

Haemoglobin is an **allosteric protein** – a term that we have not introduced previously. Allosteric proteins, of which there are many, are multisubunit proteins that have more than one combining sites for ligands. Haemoglobin has four, each capable of binding an oxygen molecule. If you look at the sigmoid oxygen-binding curve of haemoglobin in Fig. 4.21 it means that binding of the oxygen at low pressures when few of the sites are occupied is more difficult than when more are occupied. There is a progressive increase in affinity of haemoglobin for oxygen as site occupancy increases, so that at the higher oxygen pressures the affinity is increased 20-fold. Although the haem molecules in haemoglobin are distant from one another, the initial binding of oxygen to one subunit facilitates the binding of further molecules of oxygen to the other subunits. This is known as a **homotropic positive cooperative effect** (homotropic, because only oxygen is involved). It is this that causes the sigmoidal curve.

It is possible to estimate cooperative interactions by a graphic method which gives the **Hill coefficient.** For a protein with a single binding site such as myoglobin, or where there is no cooperativity between sites if there are more than one, the value is 1. Values greater than 1 indicate the degree of cooperativity. For haemoglobin the value is 2.8.

Theoretical models to explain protein allosterism

What is the mechanism of this cooperative effect of oxygen binding? There are two theoretical models. In both of these, increase in substrate binding results in more and more subunits in a given collection of haemoglobin molecules being in a high affinity state, so that the initial binding of oxygen facilitates the further binding of more oxygen.

In the **concerted model** (also known as the MWC model after its discoverers, Monod, Wyman, and Changeaux) either *all* the subunits of haemoglobin bind substrate with low affinity (known as the tense or T state) or *all* with high affinity (known as the relaxed or R state), the two forms being in spontaneous equilibrium (Fig. 4.24), but with the equilibrium to the low affinity side. Binding of oxygen to a single subunit swings this equilibrium towards the high affinity state. As oxygen pressure increases more of the binding sites become occupied and swings the equilibrium further to the high affinity state. There is no precise number of oxygen molecules that need to be bound to cause the change but increased binding increases the statistical probability of the change and with it a progressive increase in the affinity.

The **sequential model** (Fig. 4.25) differs in that it assumes that, in the absence of substrate, *all* haemoglobin molecules are in the low affinity state; there is no equilibrium with high affinity state molecules. When one molecule of substrate binds to one of the subunits, this single unit changes its conformation from tense (T) to relaxed (R) so that unlike the postulated situation in the concerted model a molecule can have mixtures of subunits in the two states This has the effect of facilitating a

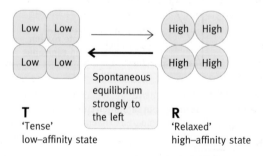

Fig. 4.24 Concerted model of cooperative binding of substrate (S). In this model the enzyme exists in two forms, T and R, the two being in spontaneous equilibrium. Binding of a substrate molecule to the R state swings the equilibrium to the right, thus increasing the affinity of all of the subunits in that molecule. 'High' and 'low' refer to affinities of the enzyme for its substrate.

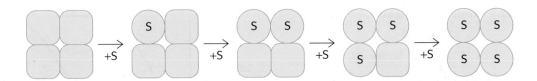

Fig. 4.25 The sequential model for the combination of an allosteric enzyme with its substrate. The binding of a single substrate molecule to a subunit causes a conformational change in the subunit. This facilitates the conformational change of a second subunit when it combines with a substrate molecule and so on for the next subunit. The net effect is to make it 'easier' for successive subunits to undergo the conformational change, which is seen as increasing affinity for the substrate by successive subunits.

similar change in an adjacent subunit so that a second molecule of substrate binds more easily. Binding of the second molecule of substrate increases still further the ease with which a third adjacent subunit can make the conformational change with a resultant increase of affinity for the substrate of that subunit and so on. Since, in a study of the oxygenation of haemoglobin, large numbers of individual molecules are involved, as more and more oxygen binds, more molecules of haemoglobin are in the relaxed state and, therefore, the observed oxygen affinity in the solution increases.

The concerted model explains much of the observed facts about haemoglobin, as also does the sequential model. It may be that the actual mechanism is somewhere in between the two models and individual allosteric proteins may differ in this respect. There are large numbers of allosteric proteins many of which are enzymes. As will be described in Chapter 16, allosteric proteins play a tremendously important role in virtually all aspects of biochemical regulation.

Mechanism of the allosteric change in haemoglobin

By means of X-ray diffraction studies, the conformation of the haemoglobin tetramer has been determined in the oxygenated and deoxygenated states. A useful way to view haemoglobin is that the α_1 and β_1 subunits are firmly associated as a dimer and, similarly, that the α_2 and β_2 form a dimer. It is the interaction between the two dimers in the tetramer that undergoes rearrangement in the T $\rightleftarrows$ R conversion. Figure 4.26 illustrates the relative rotation between the dimers' conversion from the T to R state caused by binding of oxygen.

What causes this allosteric change to the haemoglobin molecule when oxygen molecules bind? On binding of oxygen to haem, the iron atom moves slightly and, in doing so, brings about the T$\rightarrow$R conversion. Although haem, as usually written, appears as a planar molecule, in the deoxygenated state the Fe atom lies above the plane of the molecule because it is too big to fit into the tetrapyrrole (Fig. 4.27 (a)) and the tetrapyrrole is not quite flat. The Fe^{2+} atom itself is bonded to a histidine residue in one of the α helices of which the subunit is composed – designated F, the histidine group being identified as F8.

On binding of oxygen, the iron atom of haem becomes effectively smaller in diameter and moves into the plane of the

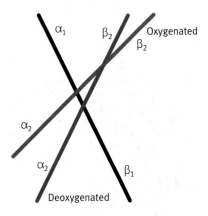

Fig. 4.26 The relative subunit positioning in deoxygenated and oxygenated haemoglobin molecules, the axes of which are represented by straight lines. The α_1, β_1 pair and the α_2, β_2 pair should be regarded as single dimer units. On oxygenation, these dimers rotate and slide, relative to each other, by about 15°. The black and blue lines represent the relative positions of the dimers in the T ('tense') state. In the oxygenated (R, 'relaxed') state the red line shows the rotation of the α_2/β_2 dimer relative to the α_1/β_1 dimer, which is represented as fixed. The change alters the contacts between the dimers.

tetrapyrrole, thus flattening the molecule. The protein rearranges itself as a result of the movement of the Fe atom (Fig. 4.27(b)). This causes a relative movement at the point where the α unit of one dimer interacts with the β unit of the other dimer (α_1–β_2/α_2–β_1 interfaces), and thus the relative rotation shown in Fig. 4.26. This movement results in the T$\rightarrow$R change.

The essential role of 2:3-bisphosphoglycerate (BPG) in haemoglobin function

$$COO^-$$
$$|$$
$$CH-O-P-O^-$$
$$|$$
$$CH_2-O-P-O^-$$

2:3-Bisphosphoglycerate

(a)   α helix of globin subunit

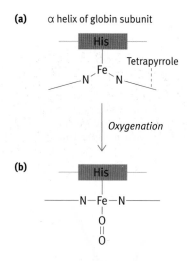

(b)

Fig. 4.27 The changes in the haem of haemoglobin upon oxygenation. (a) The haem molecule in deoxyhaemoglobin with the tetrapyrrole structure strained into a slightly domed shape. (b) Attachment of haem in oxyhaemoglobin. The movement makes the iron atom a microswitch which, by its attachment to an α helix of the haemoglobin, alters the conformation of the protein.

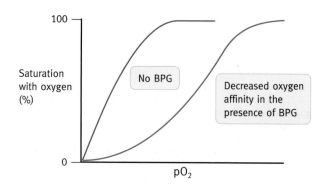

Fig. 4.28 Oxygen saturation curves for haemoglobin, illustrating the effect of 2:3-bisphosphoglycerate (BPG).

2:3-Bisphosphoglycerate (BPG) plays an important physiological role in oxygen transport by lowering the affinity of haemoglobin for oxygen, and thus increases the unloading of oxygen to the tissues; it moves the dissociation curve of oxyhaemoglobin to the right.

The haemoglobin tetramer, looked at from the appropriate viewpoint, has a cavity running through the molecule (see Fig. 4.22). Projecting into this cavity are amino acid side chains with positive charges. The BPG molecule has five negative charges at the pH of the blood, and has just the correct size and the configuration to fit into the cavity of deoxygenated haemoglobin and make ionic bonds with these positive charges. This helps to hold haemoglobin in its deoxygenated position; in effect it crosslinks the β units in that position. In the deoxygenated (T) state, the haemoglobin can accommodate a molecule of BPG. However, on oxygenation, because of the conformational change in the protein, the cavity of the R state becomes smaller and is unable to accommodate BPG. If we consider oxyhaemoglobin in the capillaries, the ability of BPG to strongly bind to, and stabilize the deoxygenated state favours unloading of the oxygen. In effect, the process is:

(a) Hb—oxygen $\rightleftharpoons$ Hb + oxygen
    (Relaxed)   (Tense)
(b) Hb + BPG $\rightleftharpoons$ Hb—BPG
    (Tense)   (Tense)

Reaction (b) with BPG will tend to pull the equilibrium of reaction (a) over and favour oxygen release.

If blood cells are stripped of all of their BPG, the haemoglobin remains virtually saturated with oxygen even at

oxygen concentrations below that encountered in the tissue capillaries. It would be incapable, therefore, in that state, of delivering oxygen to the tissues efficiently. The effect of BPG on oxygen binding by haemoglobin is illustrated in Fig. 4.28.

The normal molar concentration of BPG in the blood is roughly equivalent to that of tetrameric haemoglobin. The higher the concentration of BPG, the more the deoxygenated form is favoured. This constitutes a regulatory system – if oxygen tension in the tissues is low, synthesis of more BPG in the red blood cells favours increased unloading of oxygen. Acclimitization at high altitudes involves, in part, the establishment of higher BPG levels in red blood cells. It is to be noted that, while BPG causes greater delivery of oxygen to the tissues, the decreased oxygen affinity has little effect on the degree of oxygenation in the lungs. There is yet another refinement of the BPG-based regulatory system that illustrates how small changes to proteins can have major physiological effects. For a mother to deliver oxygen to a foetus, it is necessary for the foetal haemoglobin to extract oxygen from the maternal oxyhaemoglobin across the placenta. This requires the foetal haemoglobin to have a higher oxygen affinity than that of the maternal carrier. This is achieved by a foetal haemoglobin subunit (called γ) replacing the adult β chains, each of which lacks one of the positive charges of the β subunit. The missing charges are those that, in adult haemoglobin, line the cavity into which BPG fits; therefore foetal haemoglobin has two fewer ionic groups to bind BPG, the latter therefore being held less tightly. BPG is therefore less efficient in lowering the oxygen affinity, giving foetal haemoglobin a higher O_2 affinity than that of maternal haemoglobin. Thus, the maternal haemoglobin readily transfers its oxygen load to the foetus.

Effect of pH on oxygen binding to haemoglobin

Haemoglobin in the deoxygenated state has a higher binding affinity for protons than has oxyhaemoglobin. Put in another way, the oxygenated form is a stronger acid than is the deoxygenated form, resulting in dissociation of protons from the molecule when oxygen binds, a phenomenon known as the Bohr effect:

$$(1)\ Hb + 4O_2 \rightleftharpoons Hb(O_2)_4 + (H^+)_n$$

(where n is somewhere around 2; the number depends on a complex set of parameters).

The protons are released from histidine side groups of the protein. (The pK_a of a dissociable group can be altered somewhat by its immediate chemical environment.) The release of protons is caused by the T—R conformational change on oxygenation affecting the ionization (pK_a) of certain histidine groups and a terminal amino group.

Role of pH changes in oxygen and CO$_2$ transport

The Bohr effect, described above, has important physiological repercussions. In the tissues, CO_2 is produced and must be transported to the lungs. It enters the red blood cell where the enzyme **carbonic anhydrase** converts it to H_2CO_3, which dissociates into the bicarbonate ion and a proton,

$$(2)\ CO_2 + H_2O \rightleftharpoons H_2CO_3 \rightleftharpoons H^+ + HCO_3^-$$

The latter will drive the equilibrium shown in equation (1) above, to the left, causing the HbO_2 to unload its oxygen in the red cells, the effect thus being in harmony with physiological needs.

The HCO_3^- in the red cells passively moves out via the anion channel down the concentration gradient into the serum. The HCO_3^- movement is not accompanied by H^+ movement for there is no channel allowing passage of this across the membrane of the red cell. To electrically balance the exit of HCO_3^-, Cl^- moves into the cell via the same anion channel. The dual movement is known as the chloride shift (Fig. 4.29). The HCO_3^- travels in solution in the serum of the venous blood back to the lungs. Here, the proton concentration changes help to achieve the physiologically desirable results too. The release of protons from haemoglobin on oxygenation produces H_2CO_3 from HCO_3^- by a simple equilibrium effect:

$$HCO_3^- + H^+ \rightleftharpoons H_2CO_3$$

This permits carbonic anhydrase to form CO_2 (which is expired) because H_2CO_3 (not HCO_3^-) is the substrate of the enzyme:

$$H_2CO_3 \rightleftharpoons H_2O + CO_2$$

The destruction of HCO_3^- in the red blood cell causes HCO_3^- in the serum to enter the cell, down the concentration gradient, and Cl^- exits, that is, a reverse chloride shift occurs in the lungs, resulting in CO_2 expiration.

A small amount of CO_2 is transported in simple solution in the blood, but by far the greatest amount (about 75%) is transported as HCO_3^-. About 10–15% of the CO_2 is bound to the

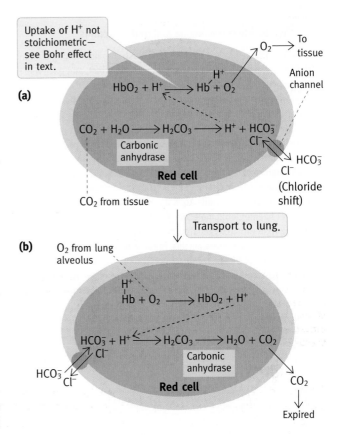

Fig. 4.29 Transport of CO_2 in the blood. (a) Reactions in the tissue capillaries. (b) Reactions in the lungs. The diagrams omit transport of CO_2 as carbamino groups of haemoglobin.

haemoglobin itself, as CO_2 chemically and spontaneously reacts with uncharged —NH_2 groups of the globin to form carbamino groups:

$$RNH_2 + CO_2 \rightleftharpoons RNHCOOH \rightleftharpoons RNHCOO^- + H^+$$

The RNH_2 groups available are mainly the terminal amino groups – lysyl and arginine side groups have too high a pK_a to be significantly uncharged.

pH buffering in the blood

(Buffers are described on page 38.)

From the above, it is clear that there are major changes in the hydrogen ion concentration of blood associated with CO_2 and oxygen transport.

When acid is produced in the tissues the buffering power of HCO_3^-, phosphates, and haemoglobin itself is important in maintaining a physiological pH. The Bohr effect, described above, also buffers. When oxygen is unloaded, haemoglobin takes up protons. This carries roughly half of the H^+ ions generated by CO_2 in the tissues, and this again helps to prevent a drop in the pH of the red blood cell to unphysiological levels.

BOX 4.3 Sickle cell anaemia and thalassaemias

Sickle cell anaemia is a disease that illustrates how a single amino acid change in a protein can have a profound effect, and gave rise to the concept of a molecular disease. In the normal β chains of the human haemoglobin tetramer (haemoglobin A), amino acid number 6 is glutamic acid, whose side group is negatively charged and highly hydrophilic. In the haemoglobin of sickle cell anaemia patients (haemoglobin S), this is replaced by the hydrophobic valine residue. This requires only a single base mutation from T to A in the triplet of DNA coding for this particular glutamic acid residue of haemoglobin. The abnormal hydrophobic patch on the haemoglobin S caused by valine, by chance, binds to a particular hydrophobic pocket on another haemoglobin molecule and so on, resulting in the formation of long rigid rods. Oxygenated haemoglobin, because of its different conformation, does not permit this. The long deoxyhaemoglobin rods distort the normal biconcave disc into sickle-shaped cells that tend to block capillaries and also to break up, causing anaemia. If both chromosomes are affected, the disease may be lethal, especially at low oxygen tensions such as occur at high altitude, causing unusually high deoxygenation of the haemoglobin.

The disease is prevalent in geographical areas with a malignant form of malaria (ignoring the effects of migration), where the high incidence can be explained only by positive selection of the diseased genome. The sickled red blood cell is unfavourable for the development of the malarial parasite and thus protects against malaria. Since malarial infection of normal persons has a higher death rate than does the heterozygous sickle cell state, the latter increases survival of persons with the genetic trait, and is therefore preserved by natural selection.

The thalassaemias are a family of genetic diseases in which haemoglobin production is abnormal. In α and β thalassaemias there is a deficiency of the corresponding subunits of haemoglobins. The disease is prevalent around the Mediterranean Sea (*thalassa* is the Greek for sea) and as with sickle cell anaemia appears to give some protection against malaria, thus accounting for its prevalence in malarial areas. Different types of mutations are known to cause the diseases. The β thalassaemias occur in varying degrees of severity.

Summary

Proteins are polypeptides (long chains of amino acids linked by peptide bonds) constructed from 20 species of amino acids, each polypeptide being of the length and sequence as specified by its gene. A protein has one or more polypeptides in it. The peptide bond is the CO—NH linkage between two amino acids. The 20 different amino acids are of differing sizes and degrees of hydrophobicity, hydrophilicity, and electrical charge. They present the possibility of a vast variety of different proteins.

The primary structure is the linear amino acid sequence. Secondary structure involves folding of the polypeptide backbone. The main motifs are the α helix and the β-pleated sheet, which satisfy hydrogen-bonding requirements of the backbone. Proteins are built up of various combinations of these structures linked by connecting loops.

Tertiary structure formation involves the folding of the basic motifs into the three-dimensional form of the protein. Proteins tend to conform to one of a number of 'recipes' made up of helices and pleated sheets. The folded structure is determined by the amino acid sequence of the polypeptide chains so that all proteins have unique structures. The association of protein molecules to form multi-subunit proteins produces the quaternary structure.

Larger proteins contain domains, which are sections of polypeptide chains folded into three-dimensional structures that could exist independently if they were separated from the rest of the chain. Proteins have evolved via mutations and also by domain shuffling, in which the coding regions of genes have

been reassembled to code for new combinations of domains to produce new proteins.

In globular proteins, hydrophobic residues are mainly inside the molecule and hydrophilic ones outside in contact with water. Membrane proteins have hydrophobic amino acids on the outside in contact with the hydrophobic membrane interior. (These are discussed further in Chapter 7 on membranes.)

Some proteins have groups such as ions or non-protein molecules firmly attached to them. When the protein is an enzyme and the attachment is required for activity it is a prosthetic group. In this case the active complex is a holoenzyme and the protein alone an apoenzyme.

Other proteins have carbohydrate attachments. The attachment is to one of the amino acids serine or threonine (via the —OH group of the side chain) or to asparagine (via the —N atom of the side chain).

Secreted proteins are often glycoproteins. The carbohydrate group may be important as a marker for either protecting the protein from degradation or marking it for degradation, depending on the nature of the carbohydrate.

Extracellular matrix proteins include collagens, which are in tendons, cartilage, and bone to confer toughness. Post-translational modification of proline residues are important in extracellular proteins. There are many genetic diseases of collagen. Elastin is the elastic protein of lungs. Elastase, liberated by neutrophils, clears away unwanted connective tissue from infection sites. This is kept in check by the antiproteinase α_1-antitrypsin. If the latter is inactivated by tobacco smoke components, the excess protease activity destroys alveoli. This produces cavities in the lungs, which reduce the surface area available for gaseous exchange, causing emphysema. Proteoglycans form the ground substance of loose connective tissues such as the flexible layer underlying skin.

Fibronectin is the best known of the adhesion proteins of the extracellular matrix. It has different sites which bind proteoglycans, collagen and cell surfaces binding the entire structure together.

Integrins are transmembrane proteins. Externally they are receptors for signalling molecules and internally they bind to actin microfilaments. This allows signal transmission across the membrane.

Cell adhesion proteins such as cadherin are responsible for cell–cell attachment necessary for the formation of tissues. They are also involved in transmitting signals to the interior of cells.

Myoglobin and haemoglobin are the most studied examples of protein structure related to function. Myoglobin is a monomeric protein that acts as an oxygen reserve in muscles. It has a haem prosthetic group to which the oxygen attaches. It surrenders its oxygen when the muscle has a low oxygen tension such as occurs in vigorous exercise.

Haemoglobin is the oxygen carrier of red blood cells. It is a tetramer of two α chains and two β chains, each subunit of which resembles myoglobin in structure; each has a haem prosthetic group.

Haemoglobin undergoes conformational changes during the attachment and release of oxygen, which are essential for its function. Rather than being a passive carrier of oxygen it is an amazing molecular machine designed to optimally perform its complex physiological functions, which require it to pick up the maximum amount of oxygen in the lungs and surrender as much as possible in the capillaries. It also carries CO_2 back to the lungs.

The binding of oxygen molecules to haemoglobin makes it easier for successive ones to attach, known as cooperative binding. The concerted model can account for this. It gives a sigmoid oxygen saturation curve as compared with the hyperbolic one of myoglobin, which does not undergo conformational change.

Proteins that change their shape on the binding of ligands are known as allosteric proteins. Large numbers of these exist and are of central importance to life. (See Chapter 16 for an account of allosteric enzymes.) The concerted model can explain the kinetics of haemoglobin's cooperative binding to oxygen.

2:3-Bisphosphoglycerate (BPG) plays an important physiological role in oxygen transport by lowering the affinity of haemoglobin for oxygen and thus increases the unloading of oxygen to the tissues.

Sickle cell anaemia is the classic genetic disease in which a single amino acid substitution causes aggregation of the protein into rods. The prevalence of this inherited trait in malarial areas of the world is explained by the fact that the sickle cell disease confers some protection on the more lethal malarial disease and thus has been selected.

 ## Further reading

Branden, C. and Tooze, J. (1999). *Introduction to protein structure*, 2nd edn. Garland Publishing.

Lesk, A. M. (2001). *Introduction to protein architecture*. Oxford University Press.
An advanced treatise on the structural biology of proteins.

Protein size and structure

Chothia, C. (1984). Principles that determine the structure of proteins. *Annu. Rev. Biochem.*, **53**, 537–72.
Excellent for the basics of protein structure.

Goodsell, D. S. (1991). Inside a living cell. *Trends Biochem. Sci.*, **16**, 203–6.
Presents a picture of the distribution of molecules on a proper scale in a cell of E. coli.

Goodsell, D. S. (1993). Soluble proteins: size, shape and function. *Trends Biochem. Sci.*, **18**, 65–8.
A survey of protein structures – their size and shape. Why are they so big? Why are they oligomeric?

Gerlt J. A. and Babbitt P. C. (2001). Divergent evolution of enzymatic function: mechanistically diverse superfamilies and functionally distinct suprafamilies. *Annu. Rev. Biochem.*, **70**, 209–46.

Söding, J. and Lupas, A. N. (2003). More than the sum of their parts: On the evolution of proteins from peptides. *BioEssays*, **25**, 837–46.

Protein modules or domains

Doolittle, R. F. (1995). The multiplicity of domains in proteins. *Annu. Rev. Biochem.*, **64**, 287–314.
A discussion of domains, domain shuffling, exons, and introns. (The latter two topics are dealt with in Chapter 24 of this book.)

Proteins of hair and connective tissue

Francis, M. J. O. and Duksin, D. (1983). Heritable disorders of collagen metabolism. *Trends Biochem. Sci.*, **8**, 231–4.
A relatively simple account of this very complex and medically important field.

Hulmes, D. J. S. (1992). The collagen superfamily – diverse structures and assemblies. *Essays Biochem.*, **27**, 49–67.

Arcangelis, A. D. and Georges-Labouesse, E. (2002). Integrin and ECM functions: roles in vertebrate development. *Trends Genet.*, **16**, 389–94.
Reviews signalling roles (ECM = extracellular matrix).

Mercer, J. F. B., *et al.*, (1993). Isolation of a candidate gene for Menkes disease by positional cloning. *Nat. Genet.*, **3**, 20–5.

Hymes, R. D. (2002). Integrins: bidirectional, allosteric signalling machines. *Cell*, **20**, 673–87.

Wheelock, M. J. and Johnson, K. R. (2003). Cadherins as modulators of cellular phenotype. *Ann. Rev. Cell. Dev. Biol.*, **19**, 207–35 .
A comprehensive review. Probably suitable for instructors rather than students.

Nelson W. J. and Nusse, R. (2004) Convergence of Wnt, β-Catenin, and Cadherin pathways. *Science*, **303**, 1483–7.
An advanced review describing the cellular arrangements and control of cell migration by cadherins. Suitable for instructors rather than students.

Problems

1   What is the primary structure of a protein?

2   What is meant by denaturation of a protein?

3   Write down the structure and name of an amino acid with each of the following side chains:
    (a) H
    (b) aliphatic hydrophobic
    (c) aromatic hydrophobic
    (d) acidic
    (e) basic.

4   Give the approximate pK_a values of:
    (a) acidic amino acid side chains
    (b) basic amino acid side chains
    (c) the histidine side chain.

5   Which amino acids are the major determinants of the charge of a polypeptide chain containing all 20 amino acids?

6   What are the four levels of protein structure?

7   What is the peculiar structural feature of elastin that gives it its elastic properties?

8   In collagen, sections of the polypeptide chains have glycine as every third amino acid residue. What is the significance of this?

9   The activities of proteins in general are readily destroyed by mild heat. The peptide bond is quite heat stable.
    (a) Why are proteins so inactivated by mild heat?
    (b) A few proteins, particularly extracellular ones, are more stable than usual. What structural feature is probably responsible for this?

10   What is a protein domain and why are they of interest?

11   In a globular protein where would you statistically expect to find most of the residues of (a) phenylalanine? (b) aspartic acid? (c) arginine? (d) isoleucine?

12   Explain why the molecular structure of proteoglycans is very suitable for creating mucins and gels with a very high water content.

13   The α helix and β-sheet structures are prevalent in proteins. What is the common feature that they have which makes them suitable for this role?

14   Compare the oxygen dissociation curves of myoglobin and haemoglobin. Discuss the rationale for the differences.

15   Binding of an oxygen molecule to haemoglobin causes a conformational change in the protein. Describe the mechanism of this.

16   Explain how the foetus is able to oxygenate its haemoglobin from the maternal carrier.

17   Explain the significance of the chloride shift in red blood cells.

18   In the context of protein tertiary structures it has been commented that when nature is onto a good thing it sticks with it. Discuss this briefly.

19   Sickle cell anaemia is so prevalent in certain areas of the world that there must be an explanation for its prevalence. Explain the nature of the disease and discuss why it is so prevalent in these areas.

20   Why does smoking cause emphysema?

21   The peptide bond is said to be planar. Explain briefly what is meant by this, give the structural basis of it and state the consequences of it.

22   Which of the following is out of place? Isoleucine, alanine, phenylalanine proline, leucine.

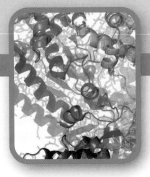

Chapter 5

Methods in protein investigation

Methods for investigating proteins have undergone a revolution in the past few years, though the more traditional methods, especially in protein separation, are still important and will be described in this chapter.

The main changes have resulted from the coming together of several technologies. One of the major developments has been the application of **mass spectrometry** to proteins, which allows investigations to be carried out with what a few years ago would have been regarded as unimaginable speed and sensitivity. This technology has been used for a long time in organic chemistry but was applicable only to volatile molecules. At the end of the last decade new methods allowed its application to proteins, with spectacularly successful results. A second major factor has resulted from DNA research. The amino acid sequence of proteins is determined by triplets of nucleotides in the genes (Chapter 24). Since we know which triplets specify each amino acid (page 381), if we know the nucleotide sequence of a gene for a protein it is a simple matter to deduce the amino acid sequence for that protein. The developments in automated DNA sequencing culminating in the sequencing of the entire human genome means that potentially the amino acid sequence for any human protein can be obtained.

The accumulated information on protein and DNA sequences increased explosively and would have been overwhelming had it not been for the development of protein and DNA databases into which the information is deposited as it is obtained. The databases established by international cooperation are freely available, as is computer software that permits the retrieval and analysis of the information in many different ways (page 81). This area of databases and their computer-assisted use is referred to as bioinformatics and has become a major tool in most protein research. The direct linking up of technologies such as mass spectrometry and DNA sequencing with bioinformatics, with its synergistic effects on research, has resulted in what is sometimes called the 'omics' revolution (page 13) in which large numbers of proteins and genes can be studied at once, the two areas being known as **proteomics** and **genomics** respectively.

We will now describe both the 'traditional' methods of studying proteins and the more recent advances. The complementary studies on DNA are dealt with in Chapter 28.

Purification of proteins

Most biochemical and molecular biology investigations require proteins to be isolated and, except for extracellular proteins, this means beginning by breaking open the cells usually by homogenization or more vigorous methods for bacteria because of their robust cell walls. Crude protein extracts contain thousands of different proteins and the amount of the protein you are interested in will be a very small fraction of the total. A cell extract has organelles (Chapter 2) and large protein complexes in them and if your protein is soluble the first step is to remove these unwanted items by centrifugation. If the opposite is the case, and it is present in an organelle, it may be purified by differential centrifugation followed by methods to solubilize the protein, such as by the use of detergents.

A method is needed to measure the amount of the protein of interest and preferably rapidly, for protein purifications can involve many steps. Total protein can be measured by UV absorbance of aromatic acids at 280 nm. If your protein is an enzyme, its amount can be determined by its catalytic activity, or some other specific property, such as colour for haem proteins. From this, and the total amount of protein, you can see if what you have done has increased the purity (specific activity) of your wanted protein. Specific activity is the ratio of the amount of enzyme (or other protein) to the total protein.

The method of purification adopted depends on the amount of the pure protein you need. If this is minute then it might be obtained in a day or so by electrophoretic methods (see below), which can separate hundreds of proteins as 'spots' provided you can identify the wanted protein. The importance of mass spectrometry (described later) is that it can handle minute, nanogram amounts and for many purposes this may be enough. There are situations, however, in which relatively large amounts of the pure protein are necessary. X-ray crystallography (below), for example, requires protein crystals, and to produce these, much larger amounts of pure protein may be needed.

For the latter situation, the more traditional methods are used based on size, charge and solubility of the protein, and the purification can take a long time. A common preliminary step is to precipitate proteins by increasing amounts of a highly soluble salt like ammonium sulphate, which drives proteins out of solution so that different proteins precipitate at different salt concentrations. Fractions are collected by centrifugation, redissolved and assayed for the wanted protein. Alternatively, selective precipitation of proteins by increasing alcohol concentrations (at low temperature to avoid denaturation) may be successful. Occasionally the protein may be relatively heat stable and this permits selective precipitation of undesired proteins by heat denaturation, often at about 55° for a short time. These methods are reminiscent of 'hit or miss' cookery and often achieve only moderate degrees of purification but they have the advantage of being cheap, easily done on a large scale, and valuable as preliminary steps to partially purify and concentrate the wanted protein into a smaller volume. This facilitates the application of more sophisticated methods. After such procedures, dialysis removes salt and puts the proteins into suitable buffers for further purification. Dialysis involves putting the solution of proteins into a cellophane 'sausage' immersed in a large volume of whatever solution you wish to have the protein in. Cellophane is a semi-permeable membrane which allows salts and other small molecules to freely pass but retains the protein molecules.

Most proteins are only marginally stable and are easily denatured so that conditions are often designed to avoid foaming and usually performed at low temperatures.

More sophisticated methods then become practical. Proteins may be separated on the basis of their differing sizes, electrical charges or chemical binding properties. Column chromatography is commonly used for these even in quite large-scale purification.

In those cases where a wanted protein is produced in *Escherichia coli (E. coli)* by the recombinant DNA procedures described in Chapter 28, an easily selectable 'tag' can be engineered into the protein. For example, a group of six histidine residues added to the C-terminal end of a protein has a very high affinity for nickel. This makes it possible to isolate the pure protein from the cell homogenate in a one-step or two-step procedure. Fusion proteins are used in a similar manner. This involves DNA engineering so that the wanted protein is produced fused to another easily selectable protein. The latter, after isolation, is removed proteolytically using a specific endopeptidase.

Column chromatography

The term chromatography is widely used for analytical and preparative techniques in which components of a mixture are separated on a solid matrix through which flows a mobile phase, usually liquid when separating protein. In column chromatography, the immobile material used to effect the separation is packed into a column, the sample to be treated applied to the top and washed through by the mobile phase. Fractions of the eluate are collected as they emerge from the column and analysed for the wanted material.

A common method is to separate on the basis of molecular size using **gel filtration**. A gel filtration column is packed with beads of an inert gel with known pore sizes (obtainable commercially) such as agarose or Sephadex. The protein solution is applied to the top of the column and then washed through with water or appropriate buffer solution. Protein molecules too large to enter the pores of the beads flow unimpeded around the beads but those small enough to enter the beads are retarded (Fig. 5.1). When the fractions from the column are quantitated and plotted on a graph, the individual components are seen to emerge as a series of 'peaks' in order of decreasing molecular weight. The fractions with the highest purity, that is to say with the highest ratio of specific properties as to total protein, can be selected for further purification.

Alternatively the column packing may be a polymer with ionized groups on it so that proteins with different charged groups bind with different affinities and are eluted sequentially with buffers of increasing ionic strength. This is known as **ion exchange chromatography**. Another variation is to use **affinity chromatography**. Suppose your protein of interest is known to specifically bind strongly to compound X; if an inert column packing material has X attached to it by chemical means, then the protein you want will be retarded by binding to the column while the rest are washed through. The desired protein can then be eluted, perhaps with solutions of different ionic strengths or by a solution of component X. **Reverse phase chromatography** involves using a packing material (often silicic acid with hydrophilic groups masked) with attached hydrophobic groups. Proteins with exposed hydrophobic groups are adsorbed; differential elution is effected by solutions of increasing hydrophobicity or different ionic strengths.

The speed and efficiency of column chromatography separations may be increased by using **high performance liquid chromatography (HPLC)**, also sometimes known as **high pressure liquid chromatography**; in this, column material is packed in a steel tube and the liquids forced through at high pressure. The method allows a more finely divided immobile phase material to be used, which increases the surface area and efficiency of separations. Usually a variety of methods are needed to purify a protein completely in significant amounts and it can be very time consuming. If a highly purified preparation of a protein is obtained, crystallization may be possible by adjusting a solution

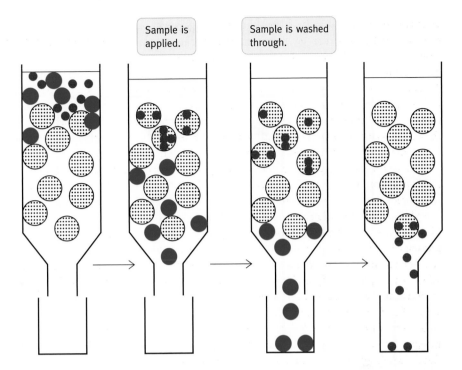

Sample is applied.

Sample is washed through.

Fig. 5.1 Protein separation by gel filtration. The column is packed with beads of a gel that has pores of defined size. A mixture of large and small protein molecules is allowed to enter the column. The green molecules are too large to enter the beads but the small ones can do so. The column is washed with a suitable buffer to move the molecules down the column. The blue molecules are retarded behind the green ones because they enter the beads and so emerge from the column later than the green ones. Pure samples of each can be collected as separate fractions.

almost to the point at which the protein precipitates, and then leaving it to stand. Crystallization is still an art, though robotic crystallization apparatus is proving useful.

SDS polyacrylamide gel electrophoresis

For analytical separation, when only small amounts of protein, in the microgram range, are involved, **gel electrophoresis** is often used. The principle is that molecules with a net charge migrating along an electrical field to the opposite pole through a gel are sorted according to size and charge. A polyacrylamide gel is usually used as the solid gel using a recipe giving the desired pore size. A common variant is to use a **denaturing gel** which contains a detergent, **sodium dodecylsulphate (SDS)**. This has a hydrophobic tail and a negatively charged sulphate group $(CH_3 (CH_2)_{10} CH_2 OSO_3^- Na^+)$ and the protein sample is dissolved in a solution of this, which denatures it. Disulphide bonds are disrupted by including a reducing agent. The SDS inserts by its hydrophobic tail into the proteins, which are thus covered with negative charges. Large amounts of the SDS attach, roughly one molecule per two amino acid residues, that swamp whatever charge the native protein had so that all proteins have a strong negative charge. The detergent also solubilizes water-insoluble hydrophobic membrane proteins so that these can be studied on gels, a major advantage of the technique. The polyacrylamide gel is polymerized *in situ* between two glass plates. The gel is made to a recipe giving suitable porosity so that as the SDS-proteins migrate in the electric field towards the anode, they are mainly separated by **molecular sieving**; small ones moving fastest. The proteins appear as

bands in the gel when visualized by staining, for example with a dye, Coomassie blue. The resolving power of this technique is very great. Such gels have many uses. Several samples can be run as separate 'tracks' on a single gel and if one of the tracks has a mixture of pure proteins of known molecular weight, known as 'markers', the gel is 'calibrated' so that the molecular weight of proteins in the samples can be roughly estimated. Or the variation of the amount of a given protein in cells may be studied.

In practical terms the apparatus is simple (Fig. 5.2). The plates with the gel between them are held vertically with the top and bottom edges of the gel exposed to tanks of the SDS-buffer solution and a voltage applied across the two tanks. When the gel is cast between the glass plates, before polymerization, a plastic 'comb' with teeth about 1 cm wide, is inserted into the top edge so that, when this is removed following solidification, the gel has separate wells into which different samples are introduced. The gel is set up in the apparatus with buffer in the tanks and the samples are applied with a pipette into the wells under the buffer. The samples contain glycerol to make them dense and settle into the wells without mixing with the buffer; a blue dye in the sample allows loading to be observed. A typical result is shown in Fig. 5.3, which also shows the purification of a protein achieved by selective elution from an ion exchange column.

Nondenaturing polyacrylamide gel electrophoresis

Nondenaturing gels do not use SDS, so that proteins are not denatured. In this case the separation is based partly on the net charge and varying electrophoretic mobilities of the proteins and partly on size. Positively charged particles will go in the

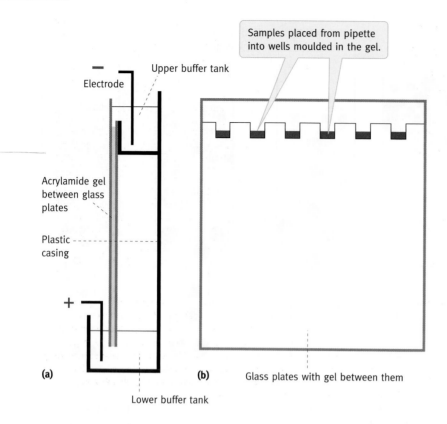

Samples placed from pipette into wells moulded in the gel.

Upper buffer tank

Electrode

Acrylamide gel between glass plates

Plastic casing

(a)

Lower buffer tank

(b) Glass plates with gel between them

Fig. 5.2 View of polyacrylamide gel electrophoresis apparatus (a) and a front view of the gel between the plates (b). The samples are injected into the wells through the buffer solution with a syringe or pipette. To prevent mixing of the sample in the wells with the buffer, the samples contain glycerol to make them dense. A blue dye makes it easy to see what is happening in the loading.

opposite direction to negatively charged particles. Nondenaturing gels have the advantage that separated proteins can be tested for a biological activity such as enzyme activity. Another variant is **isoelectric focusing**. The **isoelectric point** of a molecule with different ionizing groups (—COOH and —NH$_2$ in the case of proteins) is the pH at which the positive and negative charges exactly balance so the net charge is zero. The gel first has a pH gradient established electrophoretically with commercially available mixtures of many small polymers, known as **ampholines** or polyampholytes. These have varying charges and hence different isoelectric points. Native proteins migrate electrophoretically in such a gel towards that point in the gel at which the pH is such that the net charge on the protein is zero (its isoelectric point). The molecule then remains stationary.

Analytical ultracentrifugation can also be used to obtain pure nondenatured proteins but this is applicable only on a small scale and for special needs.

Two-dimensional gel electrophoresis

For maximum resolving power both isoelectric focusing and SDS gels are used. This is called **two-dimensional (2-D) gel electrophoresis**. In this, a single sample is first separated by isoelectric focusing on a gel strip with an established pH gradient (available commercially) and the 'strip' is transferred to an SDS gel and electrophoresed at right angles to the first direction. A crude cell extract so treated can give rise to many hundreds of separate protein spots spread in two directions on a single gel (Fig. 5.4).

Immunological detection of proteins

Antibodies are produced in the blood of animals against specific proteins. This process is described at length in Chapter 30. The antibodies have exquisite specificity for the protein it was raised against and to which it tightly binds. This can be used to detect proteins in minute amounts in gels and in solution. For electrophoresis gels, a method known as **Western blotting** is used in which the proteins on a gel are first transferred to a plastic sheet and the 'print' treated with an antibody against a given protein. The antibody can be directly labelled with radioactivity or a fluorescent compound to enable detection of binding.

Another use of immunoglobulins in protein chemistry is **ELISA** (enzyme-linked immunosorbent assay), which is widely used in clinical biochemistry (see page 503 for a description).

The principles of mass spectrometry

Mass spectrometry has become important in so many aspects of protein structure investigation that it will be best if we describe the various related methods in some detail before we get to its applications.

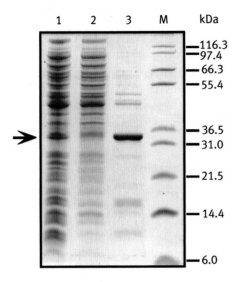

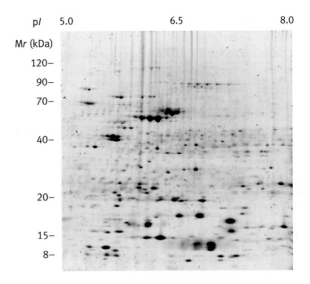

Fig. 5.3 SDS polyacrylamide gel electrophoresis. The figure also shows the purification of a protein from *Escherichia coli*. Samples from each stage of the purification were reduced and denatured, then electrophoresed on a 10% polyacrylamide gel in the presence of sodium dodecylsulphate (SDS). Separated proteins were visualized by staining with Coomassie brilliant blue. Lane 1 shows proteins extracted from *E. coli*; the arrow indicates the protein of interest in this experiment. The mixture was chromatographed on a column of cation-exchange resin (see page 72). Lane 2 shows the proteins that did not bind to the resin and were recovered in the flowthrough. Lane 3 shows the protein of interest in a fraction eluted from the column. Lane M (for markers) shows proteins of known molecular weights. The protein of interest had a molecular weight of 35.3 kDa. (kDa is the molecular weight unit, kiloDaltons). Photograph courtesy Dr Anne Chapman-Smith, Department of Molecular Biosciences, University of Adelaide, Australia.

Fig. 5.4 Representative two-dimensional electrophoresis gel of whole-cell proteins from the gut pathogen *Helicobacter pylori*. Proteins were separated using immobilized pH gradient isoelectric focusing in the first dimension and a second-dimension slab gel containing SDS buffer. Gels were stained with fluorescent Sypro Ruby. p*I* is the isoelectric point (see text). M*r* is the relative molecular weight. Image courtesy Dr Stuart Cordwell, Australian Proteome Analysis Facility, Sydney, Australia.

Mass spectrometry (MS) is now by far the best way of analysing the covalent structure of proteins because of its speed and extreme sensitivity. The amount of protein in a single spot on a 2-D gel is often sufficient for an analysis, the results of which permit rapid searching of protein databases for entries that match the 'unknown'. Computer software enables this to be done using data directly fed in from mass spectrometry. The main uses of the technique are as follows (you may be unfamiliar with some of the terms used but they are explained in due course):

- A protein can be identified in a database by **peptide mass analysis** (fingerprinting) or by **limited sequence analysis**.
- The molecular weight of a protein can be determined rapidly with an accuracy of 1 Da in 10,000 Da.
- A protein can be **sequenced**, partly or fully.
- Post-translational modifications of proteins can be investigated.

Mass spectrometers consist of three principal components

These are:

- an ion source
- one or more mass analysers that separate ions on the basis of their mass-to-charge ratios (m/z)
- an ion detector.

Ions are separated in the analyser(s), and are collected in the detector. A computer then displays a spectrum of ion intensity *versus* mass-to-charge ratio. At its simplest, ions are singly charged (z = 1) so that m/z values give molecular ion masses. Spectra containing multiply charged ions such as those often produced from proteins by electrospray ionization (see below) can be converted into mass values from inspection, or automatically by computer.

Ionization methods for protein and peptide mass spectrometry

The mass spectrometer deals with ions in the gas phase. Proteins and peptides are large nonvolatile molecules, but there are two 'soft' methods for generating gas phase ions from them. One is **matrix-assisted laser-desorption ionization** (MALDI). The material to be analysed (the analyte) is mixed with a UV-light-absorbing matrix (dihydroxybenzoic acid is an example) and deposited on a solid target surface. The target is placed in the mass spectrometer and pulsed with UV laser light, which is absorbed by the matrix molecules causing an explosive ejection of matrix molecules and carrying with them vaporized protein or peptide molecules, usually singly charged by accepting ions such as H^+ from the matrix.

The second method is the **electrospray ionization (ESI)** technique. In this, the analyte is dissolved in a solvent and the solution, raised to a high electrical potential (4 kV) is sprayed from a capillary. Fine droplets containing the analyte ions travel to the inlet of the mass analyser along a potential gradient, becoming desolvated, which reduces them to single peptide molecules suspended in a vacuum. These molecules are usually multiply charged ions that enter the high vacuum of the spectrometer. Typically, ESI is used for the direct ionization of the eluate from reversed phase liquid chromatography of proteins or peptides.

Types of mass analysers

This is a brief description of the different ways by which ions are separated to allow a spectrum of ions with different m/z values to be recorded.

Quadrupole (Q)

A quadrupole is a type of mass analyser (it has four pencil-like steel rods) in which ions are separated by a combination of direct voltage and radio-frequency fields. The quadrupole can be tuned to permit only ions of specific bands of selected m/z values to pass through. By tuning the quadrupole to different m/z values, and recording the numbers of ions that pass through at each step, a spectrum is generated. The mass accuracy of the spectrum is typically 1 in 10,000. A quadrupole may also be used to select a single ion for fragmentation analysis, such as in peptide sequencing (see below).

Time of flight (TOF)

In a TOF analyser, ions are created, often by MALDI, from material to be analysed. They are propelled from the ion chamber by a high voltage applied to a grid through which the ions move into the flight tube. The flight tube has no electrical or magnetic field – the ions simply move passively to the detector. During the flight, the ions are separated solely on the basis of the m/z values. All the ions with the same z charge have the same kinetic energy, so that heavier ions move more slowly than light ones and take more time to reach the detector. A mass accuracy of 1 in 100,000 is commonly achieved.

Ion-trap

An ion-trap mass analyser uses a combination of voltage and frequency fields to collect ions from an ion source. By applying increments to the field strength, ions of increasing mass-to-charge ratios are ejected one at a time from the trap towards the detector to produce a spectrum.

Types of mass spectrometers

The above types of analysers are assembled in various combinations by manufacturers to produce different types of mass spectrometers designed for different applications.

Single-analyser mass spectrometers

Single-analyser mass spectrometers are the simplest types, which produce the spectra of peptides and proteins using either ESI or MALDI ionization methods. Examples are ESI-single-quadrupole and MALDI-TOF spectrometers. The MALDI-TOF (Fig. 5.5(a)), is now extremely important in proteomic studies (page 80) because of its accuracy and relative simplicity of operation.

Tandem mass spectrometers

A tandem mass spectrometer (MS/MS) consists of two mass analysers in series separated by a central collision cell or fragmentation chamber, as shown in Fig. 5.5(b). Common types of tandem mass spectrometer use MALDI or ESI ion source, a fragmentation chamber and a TOF analyser. The importance of this type is that it gives amino acid sequences. The first analyser can be set by the computer to allow only a single peptide ion species of a given m/z ratio to progress into the collision cell. Other ions are scattered to the sides of the first analyser. The selected ions collide with argon gas molecules in the collision cell and are fragmented. The fragment (daughter) ions move into the second mass analyser which gives their m/z spectrum. The data so generated are sufficient for sequence of the peptide to be automatically obtained and any post-translational modifications to be localized (see below for an explanation of how the latter is done).

Identification of proteins using mass spectrometry without sequencing

The simplest method to identify an 'unknown' protein is to see if the protein already is recorded on a protein database, in which case a lot of information about it will be readily available. The method is to use peptide mass analysis or 'fingerprinting' as the identification method. The protein (which may be a spot from a 2-D gel) is enzymically digested into peptides by trypsin. MALDI-TOF MS analyses the mixture and records the m/z values of individual peptide ions as peaks, as shown in Fig. 5.5(a) (z is usually 1). Trypsin is specific in cutting a polypeptide at certain amino acid residues (lys and arg) *so the peptides produced from a given polypeptide are predictable.* The computer software allows every protein in the database to be theoretically digested by trypsin. The pattern of peptides obtained in the mass spectrometry experiment is compared with the theoretical (notional) digest of all proteins in the database. Any protein that would produce a set of peptides on trypsin digestion corresponding to ones found in the analysis is reported. Not all of the peaks (the peptides) need to be matched in the database but typically five peptide mass matches are required for unambiguous identification. The process can be completed in about one minute. About a few hundred femtomoles of protein are adequate. (One fmol equals 10^{-15} moles; a fmol of a protein of 100,000 molecular weight therefore equals 10^{-10} g.)

(a) Peptide mass analysis by MALDI-TOF

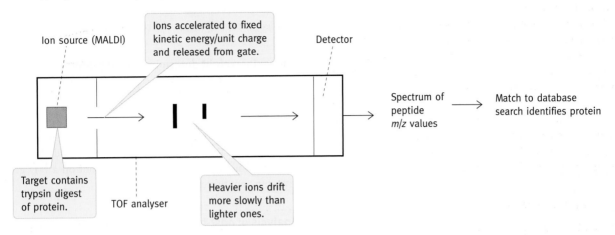

(b) Amino acid sequencing by tandem mass spectrometry (MS/MS)

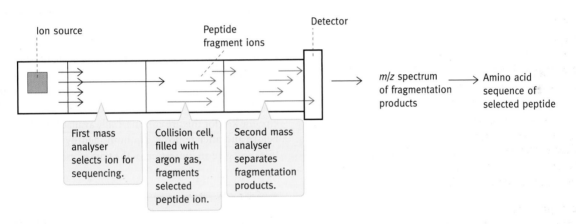

Fig. 5.5 Simplified diagram of the methods of peptide mass analysis and peptide sequencing. **(a)** Peptide mass analysis of a protein. The protein is trypsin-digested and the peptides subjected to matrix-assisted laser-desorption ionization time of flight (MALDI-TOF) 'fingerprinting'. Matching the mass analysis pattern to databases can identify a protein. **(b)** Amino acid sequencing. The protein is digested by, say, trypsin and then ionized, often by electrospray ionization. The first analyser selects a peptide ion of defined m/z value and releases it into a central collision cell where it is fragmented in a collision with argon gas. The fragmentation products pass into the final mass analyser and the m/z spectrum of the fragments recorded. This can be used to directly deduce the sequence of the selected peptide. The key to sequencing is that peptide fragmentation occurs in a predictable manner so that the 'fragmentation ladder', as the spectrum is called, can be interpreted in terms of amino acid sequences. The method is often coupled to high-pressure liquid chromatography separation of the tryptic digest and the peptides sequentially fed into the mass spectrometer.

Identification of proteins by limited sequencing and database searching

This differs from peptide mass analysis described above in that it produces some sequence data with which to match the 'unknown' to a protein in databases. MS/MS is used. (The method of protein sequencing is described below.) Even small patches of amino acid sequences from an unidentified protein is sufficient to identify the protein with one in the database with a high degree of confidence. Partial, but not exact matches of

sequences between the unknown and database entries can be of interest because they may be obtained from homologues of the unknown protein present in the database.

Methods of sequencing protein

Classical methods

The original method pioneered by Fred Sanger of Cambridge, UK was to cut up the protein into peptides and label the free amino groups with a detectable group. It is very laborious and requires a lot of protein. It took Sanger years to sequence for the first time a small protein, insulin, which earned him his first Nobel Prize. The method is not used now. Edman advanced the process by using the principle of deducting amino acids one by one from the ends of peptides, so that proteins could be automatically sequenced in a 'sequenator'. Runs were limited to about 30 amino acids. It took months to sequence a whole protein and is not used today except perhaps for special situations. With rapid new methods available it would never be used to sequence a protein as a major task.

Sequence analysis of proteins from gene DNA sequences

DNA of genes codes for the amino acid sequences of proteins. Each amino acid is coded for by a triplet of nucleotides in the DNA; the genetic code (page 381) relates the triplets to amino acids in proteins so that DNA nucleotide sequences can be interpreted as amino acid sequences of the proteins coded for by given genes. Advances in gene isolation and DNA sequencing (Chapter 28) means that the DNA sequences of many proteins are known and it is often easier to isolate the gene for a protein and determine its nucleotide sequence than to directly sequence the protein. The human genome project has determined the 3.2 billion base pairs of human DNA so that if the gene for any human protein can be identified in the base sequence print-out, the amino acid sequence of its protein is easily deducible. The genomes of several other species have been determined.

Sequencing by mass spectrometry

For sequencing by MS, a MS/MS is used. This is illustrated in Fig 5.5(b). In MS/MS, the individual peptides from a tryptic (or other protease) digest of the protein are analysed by the first mass analyser of a tandem mass spectrometer which is set so that only one selected peptide species of given m/z value is allowed to proceed into the central collision cell. There the selected peptide molecules are fragmented into a pattern of ions, which are separated in the second mass analyser, giving a spectrum of peaks known as a fragmentation ladder. The key to the method is that *the fragmentation occurs in a predictable manner (not described here) that is dependent on the amino acid sequence of the peptide* so that the fragmentation spectrum can be translated into sequence. This may be done by visual inspection but the computer also does this from the data directly fed into it. It is necessary to cleave the protein in different ways so that overlapping peptides are obtained which permits assembly of the complete sequence.

A refinement is to couple the MS/MS to reverse phase HPLC of peptides to be analysed; the separated or partially separated components are fed into the mass spectrometer as they emerge from the column. ESI is the most suitable for this. This permits sequencing of many or all the peptides obtained from a small sample of protein. About 100 fmoles of a protein from a gel are required for the analysis. A limitation of the MS approach is that the isoleucine and leucine residues are not readily distinguished, except under special conditions.

Molecular weight determination of proteins

The MS method determines molecular weight of a protein in a minute or so with an accuracy of 1 Da in 10,000 Da or 100 ppm. Molecular weight determinations are important in industrial biotechnology for quality control of proteins produced by recombinant DNA technology (Chapter 28) and also help to identify proteins in databases. A single spot on a 2-D gel can give sufficient protein for this.

Analysis of post-translational modification of proteins

Many proteins are covalently modified after synthesis such as by the addition of phosphoryl or glycosyl groups. The mass spectrometer can measure modifications at the level of mass changes to the whole protein or a peptide fragment of it. Within a peptide, the specific modified amino acid residue may be identified using MS/MS. Techniques are available which detect characteristic 'reporter' fragmentation products from specific modifications (described in the **Wilm** reference). Alternatively phosphoryl and glycosidic groups can be enzymically removed; enzyme kits for this are commercially available and MS analysis before and after removal can both identify the modified peptides and give an indication of the type of modification.

Determination of the three-dimensional structure of proteins

The linear sequence of amino acids tells us little about the protein as a functional unit because an unfolded protein has no biological activity. Biological activity depends on the folding of the polypeptide into a three-dimensional (3-D) structure, as specified in the amino acid sequence of the protein. Determining the 3-D structure of proteins is really the ultimate goal, for these permit elucidation of molecular mechanisms by which they function. There is a practical importance too; therapeutic drugs combine with specific sites on proteins, only recognizable in the 3-D structure, which are therefore of prime importance to developing new therapies. 3-D structures of many proteins have been determined and are available in the protein databases, but for novel proteins, the amino acid sequence is obtained and then two other methods are used. These are X-ray diffraction and nuclear magnetic resonance.

X-ray diffraction

In simplistic terms, X-ray crystallography 'photographs' the electron densities around the atoms of a protein.

The determination of protein structures by **X-ray diffraction** requires that the protein first be crystallized. A crystal is needed because the reflections (diffractions) from a single molecule would be too weak to observe; but the large numbers of identical molecules in a crystal reinforce the reflections so that they can be detected. Crystallization can be one of the most difficult phases of the work, though robotic methods for screening large numbers of crystallization recipes have a high success rate. The protein used for crystallization experiments is usually produced by recombinant DNA technology. The crystal is mounted in an apparatus that bombards it with X-rays with a wavelength of the same order as the distance between atoms (~1 Å). The scattered (diffracted) X-rays are measured directly because there is no lens suitable for focusing them into an image, as one does in a camera. To overcome this, protein crystals can be reacted with several different heavy metals and the data recollected, allowing a mathematical reconstruction of the image by a method known as multiple **isomorphous replacement**.

Experiments to measure the X-ray diffraction pattern from crystals are now commonly performed with **synchrotron radiation**. The extreme brightness of this X-ray source allows measurements to be made from very small crystals (~10 micron in diameter), an enormous advantage if these are the only crystals available. Further, the tuneability of synchrotron radiation allows the measurements to be made near the absorption edge of heavy elements such as selenium (in selenomethionine-labeled crystals) or the metal centres of metallo-proteins.

In ordinary X-ray diffraction the radiation is scattered by atoms without absorption of the incoming radiation. However, at certain wavelengths absorption effects (particularly for heavy metals) are sufficiently strong that the scattering is measurably affected. The advantage of collecting these anomalous effects is that one can then very easily reduce the experimental data to an image of the protein crystal without the need for serially labeling the crystals with different heavy atoms. Synchrotrons are currently the only source of tuneable X-rays for obtaining the required wavelengths. It is common now to obtain selenomethionine-labeled proteins for 3-D structure determination in *E. coli* expression systems (page 468). The selenomethionine is competetively incorporated during the protein synthesis instead of methionine.

Nuclear magnetic resonance spectroscopy

A second method of protein 3-D structure determination, based on the magnetic properties of certain atomic nuclei, and of increasing importance, is **nuclear magnetic resonance spectroscopy (NMR)**. Some atomic nuclei, of which protons are the most important in the present context, have a property known as spin and behave as minute magnets, which can be affected by a constant powerful magnetic field that causes the magnets to be aligned. Pulses of selected radiofrequency applied at the same time can cause the nuclei to jump into a slightly higher activated state as they absorb the radiation. This sets up a resonance between the two states and produces an emission spectrum. *The signals are sensitive to the surrounding atomic environment of the nuclei.* The technique gives a spectrum of peaks, which can be correlated with amino acid residues in a protein primary sequence and most importantly, gives information on the distance between pairs of atoms. It can identify atoms that are close together in the folded protein but distant from each other in the primary structure, and thus enables secondary and tertiary structures to be inferred. A major advantage of NMR over X-ray diffraction is that it is performed in solution, thus removing the necessity to crystallize the protein, which is frequently difficult or unachievable, but high concentrations of protein are needed. The method is most easily used on proteins below 100 amino acid residues in size but, with extra refinements in methodology, has been used for proteins three times larger than this. Another feature of NMR is that in solution structures such as connecting loops may be free to move and this can be observed; the protein is observed in its 'natural' state, rather than in a fixed crystal array.

Homology modelling

A large amount of work on theoretical prediction of 3-D protein structures has been done but it is still not possible, from theoretical principles, to predict the 3-D structure of a protein from its amino acid sequence. However, if there are two different proteins of similar sequence, the 3-D structure of one being known from, say, X-ray crystallography, then it may be possible to make a fairly reliable estimate of the structure of the other. The method, known as homology modelling, requires at least a 50% sequence homology (see page 55) between the two proteins for useful predictions.

An exercise in obtaining a 3-D structure from a protein database

The use of protein databases is important in searching for proteins for which the 3-D structure has been determined. In the appendix to this chapter we give step-by-step instructions on how to obtain the 3-D structure of human haemoglobin from a database, if you feel interested in trying it. It requires only a computer connected to the internet.

Proteomics and mass spectrometry

As mentioned in Chapter 1, in a living cell or organism there is a large collection of proteins. Differential splicing (page 362) and post-translational modifications (page 56) results in more proteins than there are genes. To refer to these collectively, the term '**proteome**' was coined. This is analogous to the term '**genome**', which refers to the entire collection of genes. The two terms are not exactly comparable because the genome of all cells in an organism is the same (gametes excepted) whereas the proteome varies from cell to cell and within a given cell from time to time even though their genomes are the same. At any one time, the proteome represents the functional portion of the genome. For example, a liver cell makes liver-specific proteins but not brain-specific or muscle-specific proteins and *vice versa*. Thus the proteomes of liver, brain and muscle cells overlap, but differ to a considerable extent. The proteins of a given cell may change during differentiation or in response to physiological needs. In a liver cell, some enzymes are needed in greater amounts after eating than during fasting so that the proteome may vary from hour to hour. Of medical relevance, a comparison of the proteomes of normal and diseased cells may reveal differences in individual proteins, correlated with the disease state. It may be desirable, therefore, to look at whole collections of proteins at once to see how the proteome varies in development, in response to physiological needs, and in disease such as cancer .

As described, 2-D gel electrophoresis made it possible to separate hundreds or thousands of proteins in a crude cell extract. We have explained that MS methods can be applied to single protein spots on the 2-D gels so that each of the spots can be studied and in many cases identified in databases. The speed of MS makes rapid throughput and automation possible. Modern mass spectrometers can characterize over 1000 proteins per day so that they can be identified from databases by the procedures described above. Another aspect of this approach is the study of the proteins present in organelles and complexes in the cell. Mitochondria and nuclear pore complexes are two examples where this approach has been used. Many or most proteins do not function in isolation but participate in larger assemblies to perform a specific task. If the organelle complexes can be isolated by methods that do not dissociate their proteins, then their components can be identified. The nuclear pore complex protein components were determined in this way.

As briefly touched on in Chapter 1, this study of whole proteomes or large collections of proteins such as those in complexes or organelles is known as '**proteomics**'. Proteomics differs from conventional 'protein chemistry' essentially in being concerned with large numbers of proteins at once and since genes control protein synthesis gives us an indirect view of genome function as a whole. It is assuming a prominent role in molecular biology.

By analogy the large-scale study of genes is called '**genomics**'. This is being fostered by techniques in DNA analysis such as microarrays or 'DNA chips' (Chapter 28), which are as revolutionary as MS is in proteomics.

The combination of genomics and proteomics represents a major new approach to the study of life by examining how entire collections of genes and proteins function together to constitute the living process rather than by examining the constituent parts. Its development in a few years has been spectacular.

Bioinformatics and databases

We mentioned earlier that protein and DNA databases had uses and importance beyond that of specific tasks such as protein sequencing, important though this is.

BOX 5.1 Database of website addresses

A useful introduction for students is the National Centre for Biotechnology Information website (**www.ncbi.nlm.nih.gov/About/ primer/index.html),** which contains educational material on a range of topics.

A few selected database website addresses are given below to give reality to the topic. (It is not expected that many users of this book will find them immediately useful without further formal instruction.) A chapter on web use in general and selected databases is recommended in case it particularly interests you. **Baxevanis, A. D.** (2003). The Molecular Biology Database Collection: 2003 Update. *Nucleic Acids Res.*, **31**, 1–12.

Primary sequence databases containing publicly available DNA sequences

GenBank
www.ncbi.nlm.nih.gov/Genbank/GenbankOverview.html

EMBL – the European Molecular Biology Laboratory
www.ebi.ac.uk/ebi_docs/embl_db/ebi/topembl.html

DDBJ – the DNA DataBank of Japan
www.ddbj.nig.ac.jp/

Databases of protein sequences

SWISS-PROT
www.ebi.ac.uk/swissprot/Information/information.html

PIR – Protein Information Resource
www-nbrf.georgetown.edu/pir/

PDB – Database of 3-D protein structures
Structure data determined by X-ray crystallogaphy and NMR.
www.rcsb.org/pdb/

BLAST – basic local alignment search tool
Tool to probe sequence data banks with sequence.
www.ncbi.nlm.nih.gov/blast/psiblast.cgi

ClustalW
Tool for multiple sequence alignment.
www.ebi.ac.uk/clustalw

ExPASy – Expert Protein Analysis System
Tool for protein analysis.
http://au.expasy.org/tools/

A bioinformatics overview

This account of bioinformatics is only to give a general idea of the importance of the subject and to let you have some idea of what it is about. (Some database website addresses are given in Box 5.1.) Bioinformatics courses are being offered by a number of universities. Specific training would probably be required by most if they are to practice in it as a principal activity, for it requires multiple skills and intersects the fields of molecular biology (including molecular genetics, molecular evolution, and biochemistry), computer science, mathematics, and statistics, depending on the project. However, bioinformatics to some degree is probably used by many, or most, research workers in the area.

The most fundamental aspects in bioinformatics are the efficient storage of data in databases, the rapid retrieval of these data and programs for its analysis in various ways including the alignment of sequence data. DNA databases are intimately involved in protein bioinformatics since, increasingly, amino acid sequences are generated from DNA sequences. The development of automated DNA sequencing, which, as mentioned, has allowed the sequencing of the entire human genome, and more and more plant, animal and microbial genomes each year have provided a vast amount of data. The primary sequence databases are publicly available. These are GenBank in the USA, the European Molecular Biology Laboratory (EMBL), and the DataBank of Japan (DDBJ).

In addition to these, there are numerous databases that specialize in specific molecules, specific organisms, or specific methodologies. Of these, databases of protein sequences, derived from translations of the DNA, are probably the most widely used. These include SWISS-PROT and the Protein Information Resource (PIR).

The most widely used tool for the retrieval of these data is the Basic Local Alignment Search Tool (BLAST). With the BLAST family of programs you can use a sequence of interest of either DNA or amino acids to search a DNA (BLASTN) or protein (BLASTP) sequence database for similar sequences. The purpose behind retrieving sequence data in this manner is to discover sequences that are potentially homologous, to then make comparisons among sequences to infer the evolutionary history, function or structure of the sequence of interest (see below). However, to accomplish such analyses, the sequences, which usually vary in length, must first be aligned to ensure that comparisons are made among homologous sites along each sequence. The most commonly used publicly available tool for multiple sequence alignment is ClustalW. These are usually the first steps in any research that employs bioinformatics tools.

What can one do with such data?

- As proteins have evolved from a few ancestral species, their amino acid sequences diverged. Some changes may have little effect on the mechanism of action of a protein but those which are essential to activity are usually strongly conserved since any alteration could inactivate it.

Using databases that hold 3-D protein structures, comparisons of homologous proteins may suggest which motifs in the structures are essential for activity and thus point to the mechanism of action of the protein.

- The tertiary structure patterns may reveal evolutionary relationships since these are usually preserved more than sequence.

- It is often found that a particular activity or role of a protein is associated with a particular domain or motif. Families of different enzymes for example are often found to include the same structural characteristic (e.g. often a domain (page 55)). To give a few examples, one large class of membrane proteins has a helix of hydrophobic residues that just spans the hydrophobic section of the lipid bilayer. It is possible, therefore, to scan possible databases for all proteins with this characteristic. In the important field of cell signalling (Chapter 27) a large family of enzymes all have a domain to transfer a phosphoryl group from ATP to proteins. Consequently it is possible to ask the question of how many such protein kinases exist in a genome by just looking for proteins with domains with close homology to this. The list of possibilities in the protein field is almost endless.

- Many of the types of uses mentioned above can be performed at the DNA level. One application is to identify DNA sequences that code for the protein sequences of interest. This method could also locate genes in DNA sequences. It is relatively easy to identify a prokaryotic gene by searching for an open reading frame. That is a nucleotide sequence long enough to code for a protein which is not interrupted by a 'nonsense triplet', which does not represent any amino acid. By contrast, in eukaryotes, genes are interrupted by introns (page 360) that may contain nonsense triplets, so this approach is not applicable. However, the ends of introns have specific base sequences and these can be searched for to locate genes in new sequences.

- Bioinformatics is an exciting, rapidly growing field because of the advent of rapid sequencing tools, leading to the production of voluminous amounts of data.

Appendix: Introduction to obtaining molecular structures of proteins starting with data from the Protein Data Bank

The Protein Data Bank (PDB) is the worldwide depository of information about the 3-D structures of proteins and nucleic acids. Protein databases derived from the PDB are indispensable resources in all aspects of the molecular biology of proteins. For those who have had no previous contact with the PBD a preliminary general statement may be useful. When research workers determine the 3-D structure of a protein, their experimental data, typically obtained by X-ray crystallography or NMR spectroscopy, is deposited in the PBD together with other information about the protein. The information in databases is freely available. To make use of the data it is often translated into a visual display of the protein's molecular structure on the computer screen. To do this it is necessary to use a molecular graphics viewing program (see below). The PDB supports a number of such programs.

Once you are familiar with the methodology it is relatively simple to obtain a molecular structure in a few minutes, but for those who have no experience in PDP use it can seem somewhat mysterious. To break the ice, a simple exercise in getting the structure of human haemoglobin is given below.

Steps in displaying the human haemoglobin molecule

Each set of atomic coordinates deposited in the PDB is assigned a code ID (1A3N in this case). After opening the PDB website (**http://www.rcsb.org/pdb/home/home.do**), the search can be made either for the PDB ID (the 4 character identification code given to it by the database) or the name of the molecule. (Use the North American spelling of 'hemoglobin', not the British 'haemoglobin'.) Fig. 5.6 shows the page when 1A3N is entered. The web page that appears on the screen gives information

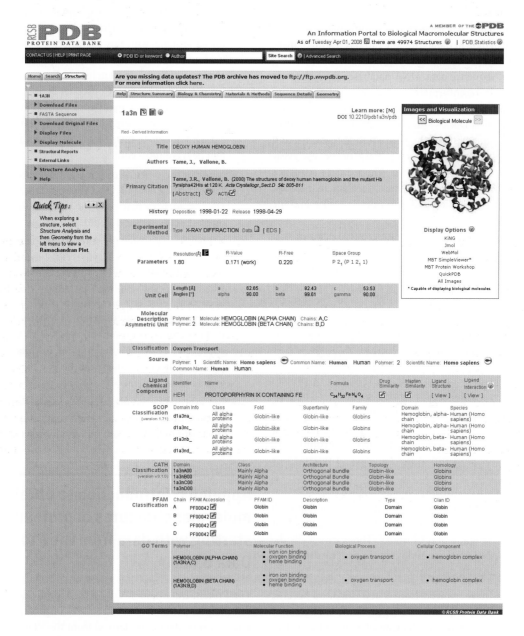

Fig. 5.6 Web page from the Protein Data Bank for ID 1A3N.

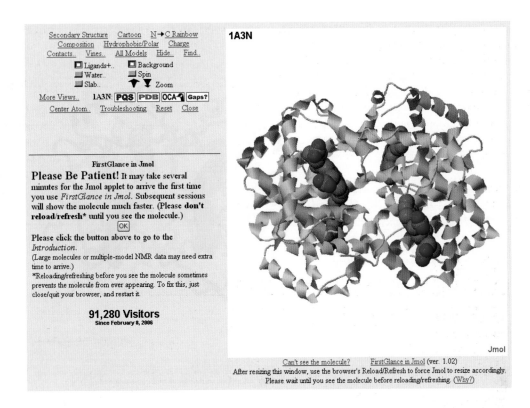

Fig. 5.7 The opening page for the FirstGlance display option.

about the structure determination, reference to papers and the atomic coordinates. (These details are not needed for this exercise.)

In the left-hand column under 'Display molecule' a pull-down menu will give a list of molecular graphics programs for interactive display options. Click on the 'FirstGlance' display option to continue. The screen that now appears is shown in Fig. 5.7. For a beginner this is probably one of the easiest ways to look at the 3-D structures of proteins. The viewing platform works with all popular web browsers and computer platforms, without installing further plug-in software, and is designed to enable the readers of scientific journals to see the main features of published 3-D models in a few clicks. (This viewer is also used with DNA and RNA databases.)

On opening 'FirstGlance', read the Introduction by clicking on the OK button. This will give you instructions on how to manipulate the various images you will be able to access.

One click options at the top left of the page can display major structural features of the molecule, such as secondary structure, amino and carboxy (or 3′ and 5′) termini, composition (protein, DNA, RNA, ligands, and solvent), and the distributions of hydrophobic, polar, and charged amino acids. In each case further options allow the molecule to be displayed with or without ligands, such as haem or water molecules. In the lower left panel accompanying each display, appropriate explanatory notes describe the colours and/or symbols used in each display.

The image that comes up at first (Fig. 5.7) is for a cartoon display, but to get all the information about this display, click on the cartoon button. In the box lower left, explanatory notes on the colours and shapes for this particular display are now given. Similarly, choosing the N→C Rainbow (Fig. 5.8) option reveals a different set of explanatory notes.

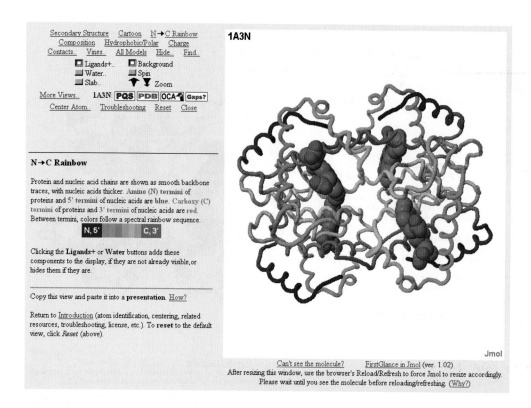

Fig. 5.8 The page that shows the four polypeptide chains of the haemoglobin molecule, each coloured in a rainbow sequence from the amino terminal (blue) to the carboxy terminal (red).

Summary

The classical methods of protein purification are often relatively simple procedures which can be applied on a large scale and are therefore used as preliminary purification steps. The simplest is differential centrifugation which sediments organelles on the basis of mass; these can be discarded or collected depending on the location of the protein of interest. Purification methods for soluble proteins include selective precipitation by high concentrations of ammonium sulphate, or ethanol. The salts are removed by dialysis in a cellophane bag.

More specific separations can be achieved by column chromatography. Molecules are separated as they run through column packings. Separations are variously based on molecular sieving, ion exchange, specific molecular affinities, or hydrophobicity. Column chromatography is used on a preparative scale but also is important analytically on a small scale.

In isoelectric focusing, a pH gradient is established using polymers called ampholines. Proteins migrate to the pH at which they have zero charge (isoelectric point). Two-dimensional gel electrophoresis separates proteins by isoelectric focusing in the first dimension and sodium dodecylsulphate (SDS) elec-

trophoresis in the second. It can separate many proteins in a single run.

In the past decade or so a revolution in protein technology has superseded many of the earlier methodologies. The newer methods work on minute amounts of protein and are rapid. The protein databases, which record vast amounts of information on all aspects of proteins, now complement these technologies.

Amino acid sequencing of a protein was a laborious task. It is often now easier to determine the base sequence of the gene for a given protein (Chapter 28) and translate this into amino acid sequence using the genetic code. DNA databases give gene nucleotide sequences (among much else). The completion of the human genome project means that the sequence of every human protein for which the gene can be identified is obtainable in this way. Emulsion sequencing greatly speeds up the sequencing of complete genomes (Chapter 28, page 462).

Mass spectrometry (MS) has revolutionized much of protein methodology. It permits rapid identification of proteins in databases. Using the minute amount of protein from a spot in a two-dimensional gel, a protein can be identified in a database by its molecular mass and its peptide 'fingerprint'. These can be obtained rapidly by MS. It can also perform amino acid sequencing more quickly than by any other method.

Methods for determining the three-dimensional structures of proteins are X-ray diffraction, now commonly performed with synchrotron radiation, and nuclear magnetic resonance spectroscopy.

The ability rapidly to identify minute amounts of proteins has lead to proteomics in which large numbers of proteins are studied at once. Bioinformatics, which involves the computer-assisted use and analysis of database information is itself a major science and 'mining the databases' is a fruitful research activity.

 Further reading

X-ray crystallography

Hendrickson, W. A. (2000). Synchrotron crystallography. *Trends Biochem. Sci.*, **25**, 637–43.

Mass spectrometry

Papayannopoulos, I. E. (1995). The interpretation of collision-induced dissociation tandem mass spectra of peptides. *Mass Spect. Rev.*, **14**, 49–73.

Yates, J. R. 3rd. (2000). Mass spectrometry. From genomics to proteomics. *Trends Genet.*, **16**, 5–8.
Covers large-scale DNA sequencing to protein-expression studies and the identification of proteins in complexes.

Wilm, M. (2000). Mass spectrometric analysis of proteins. *Adv. Protein Chem.*, **53**, 1–30.
Advanced comprehensive review.

Griffiths, W. J., *et al.* (2001). Electrospray and tandem mass spectrometry in biochemistry. *Biochem. J.*, **355**, 545–61.

Jonsson, A. P. (2001). Mass spectrometry for protein and peptide characterisation. *Cell Mol. Life Sci.*, **58**, 868–84.

Mann, M., Hendrickson, R. C., and Pandey, A. (2001). Analysis of proteins and proteomes by mass spectrometry. *Annu. Rev. Biochem.*, **70**, 437–73.
Advanced comprehensive review.

Protein microarrays

Weinberger, S. R., Dalmasso, E. A., and Fung, E. T. (2001). Current achievements using ProteinChip® array technology. *Curr. Opin. Chem. Biol.*, **6**, 86–91.

Schweitzer, B. and Kingsmore, S. F. (2002). Measuring proteins on microarrays. *Curr. Opin. Biotechnol.*, **13**, 14–19.

Proteomics

Blobel, G. and Wozniak, R. W. (2000). Proteomics for the pore. *Nature*, **403**, 835–6.
News and Views article describing work by Rout and colleagues which gives a complete structure of the nuclear pore.

Pandey, A. and Mann, M. (2000). Proteomics to study genes and genomes. *Nature*, **405**, 837–46.

Maccoss, M. J. and Yates, J. R. (2001). Proteomics: analytical tools and techniques. *Curr. Opin. Clin. Nutr. Metab. Care*, **4**, 369–75.

Martin, D. B. and Nelson, P. S. (2001). From genomics to proteomics: techniques and applications in cancer research. *Trends Cell Biol.*, **11**, S60–5.

Moore, D. W. (2001). *Bioinformatics: sequence and genome analysis.* Cold Spring Harbor Laboratory Press.

Graves, P. R. and Haystead, T. A. (2002). Molecular biologist's guide to proteomics. *Microbiol. Mol. Biol. Rev.*, **66**, 39–63.

Petricoin, E. F., *et al.* (2002). Clinical proteomics: translating benchside promise into bedside reality. *Nat. Rev. Drug Disc.*, **1**, 683–95.

Rappsilber, J. and Mann, M. (2002). What does it mean to identify a protein in proteomics? *Trends Biochem. Sci.*, **27**, 74–8.

Zhu, H., Bilgin, M., and Snyder, M. (2003). Proteomics. *Annu. Rev. Biochem.*, **72**, 783–812.
An advanced review.

Andersen, J. S., and Mann, M. (2006). Organellar proteomics: turning inventories into insights. *EMBO Reports*, **7**, 874–9.

How mass spectrometry-based proteomics methods are being applied to organellar proteomics.

Problems

1 Describe, without chemical detail, four methods for determining the primary structure of a protein.

2 What is meant by the term proteome?

3 What is meant by the term proteomics? It has come into research prominence only relatively recently. What has been a major factor in this?

4 When proteins are separated by polyacrylamide gel electrophoresis, it is common to include sodium dodecylsulphate (SDS) in the gel and reagents. What is the reason for this?

5 List the various types of column-chromatographic separation of proteins.

6 Suppose you have a minute amount of an unidentified protein as a spot on a gel. How could it be sufficiently characterized rapidly to identify a corresponding protein in a protein database?

7 Mass spectrometry has been known for a long time but only relatively recently has it been applied to proteins. What caused this change?

8 Protein databases have assumed great importance. Briefly explain their relevance and use.

9 Briefly explain the basis of protein sequencing by mass spectrometry.

10 List the methods by which the three-dimensional structures of proteins are determined.

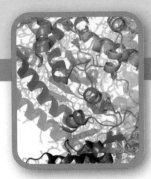

Enzymes

Dalton?

Given that a reaction has a negative ΔG value (Chapter 3, page 28), what determines whether it actually takes place at a perceptible rate in the cell? This question follows on from Chapter 3, where we explained that on the question of whether a reaction is theoretically possible (*may* occur), energy considerations have absolute authority. If the free energy change involved in the reaction under the prevailing conditions is negative, it may occur; if it is zero or positive it cannot occur and nothing in the universe can alter that. Note however that as mentioned in Chapter 3, chemical *conversions* (such as the synthesis of compound X—Y from reactants XOH and YH), involving increase in energy levels, do occur in the cell by incorporating ATP breakdown into the process, resulting in a negative ΔG for the overall process. The sole qualification of the rule that a reaction must have a negative ΔG is that it refers to *net* chemical change in the reaction. At equilibrium, where the free energy change of the reaction is zero there cannot be any net chemical change but it can be shown that this is because the reaction in one direction is exactly balanced by that in the reverse direction. The qualification has no relevance to the practical business of life which is based on net chemical changes.

We have touched briefly on what determines whether a reaction occurs in Chapter 3, but we will elaborate on this here, for it is of basic importance. Energy considerations determine whether a reaction *may* occur, not whether it *does* occur. This is just as well, because energy considerations say that everything combustible around you may burst into flame or would have done so long ago. But petrol does not ignite by itself, and sugar in the bowl on the table does not burn. Nonetheless, petrol in a car cylinder burns and sugar inside the body is oxidized to CO_2 and H_2O.

From these considerations you can see that there is a barrier to the occurrence of chemical reactions and, in the case of biochemical reactions, it is sufficiently large to prevent them occurring at a finite rate. Even though a reaction has a strong negative ΔG something restrains the reaction from happening. In the case of petrol in a car cylinder, a spark from the spark plug overcomes that barrier. This brings us to one of the most fundamental problems that had to be solved before life could

exist. How can chemical reactions be caused to occur in the cell? The answer lies in **enzyme catalysis**. It is worth reflecting on what a formidable obstacle this problem posed for the development of life, for it leads to an appreciation of what an astonishing phenomenon enzyme catalysis is. It was necessary to develop a means whereby at low temperatures, at almost neutral pH, in aqueous solutions, and at low reactant concentrations, otherwise stable molecules undergo the rapid chemical conversions needed for life.

Enzyme catalysis

An enzyme is a catalyst which brings about (usually) one particular chemical reaction but itself remains unchanged at the end of the reaction (as required in the definition of a catalyst). There are thousands of biochemical reactions, each catalysed by a separate enzyme. Multifunctional enzymes with several catalytic activities on the same molecule and multienzyme complexes exist for special situations but the general principle remains. It automatically follows from what we have said about energy considerations that an enzyme cannot affect the equilibrium or direction of a reaction; it can only promote a reaction subject to the limitations of energy considerations. This is little different from saying that you cannot get energy for nothing.

An enzyme is a protein molecule. Proteins are built up of 20 different species of amino acids linked together to form one or more long chains. Enzymes are therefore large molecules – a molecular weight of 10,000 Da corresponds to a small enzyme, and they range in size up to hundreds of thousands of daltons in molecular weight. One of the reasons why they are so large is that the long chain(s) of which they are constituted must be folded such that there is an **active site** on the surface so shaped that it is a three-dimensional pocket or cleft into which the compounds attacked by the enzyme (known as **substrates**) fit with exquisite precision. This site occupies a very small part of the protein in most cases. As with all such specific protein–ligand binding (a ligand is any combining molecule) the substrate attaches reversibly by noncovalent, or weak, bonds. As explained earlier (page 36), the specificity of all such attachments arises from the fact that several weak bonds are needed. It follows that unless there is a precise fit between the interacting groups on the substrate and enzyme, the attachment will not occur. The active site is also called an **active centre** or **catalytic site**. Almost everything in life depends on this simple principle by which specific interactions of proteins with other molecules (which may also be proteins) are achieved.

The nature of enzyme catalysis

To explain enzyme catalysis, we must look at the nature of chemical reactions. A reaction occurs in two stages. Consider a reaction converting substrate to product (S→P). In this reaction, S must first be converted to the **transition state**, $S^{\ddagger}$, which might be thought of as a 'halfway house' in which the molecule is distorted to an electronic configuration that readily converts to P. The transition state has an exceedingly brief existence of 10^{-14}–10^{-13} of a second. The overall reaction S→P must have a negative free energy change; otherwise it could not occur. An important thermodynamic principle is that the free energy change of a reaction is determined solely by the free energy difference between the starting and final products. The 'energy pathway' or energy profile, the route by which the reaction takes place, cannot affect the overall free energy change of the reaction. Hence it is in no way a contradiction that the transition state ($S^{\ddagger}$) is at a higher free energy level than S. The free energy change for S→$S^{\ddagger}$ is positive. It is called the **energy of activation** for that reaction (Fig. 6.1).

This energy hump constitutes a barrier to chemical reactions occurring. If it were not present and the energy profile S→P were a straight downward slope then, as stated earlier, everything that could react would do so.

It follows that energy of activation must be supplied in some way to permit a reaction to occur. In a car cylinder, the spark causes a few molecules of petrol to be activated to the transition state which then react, and in oxidizing, produce enough heat to activate further molecules and so on which results in the explosion. In a noncatalysed reaction in solution, the energy to surmount the hump is supplied by collisions between molecules. Provided that colliding molecules are appropriately oriented, and of sufficient kinetic energy, reactant molecule(s) can be distorted into the appropriate higher energy transition state which enables the reaction to occur. The rate of formation of the transition state therefore determines the overall reaction rate. High temperatures, which increase molecular motion and increase collision frequency between molecules, facilitate this. Hence the organic chemist usually employs high temperatures to promote reactions. At physiological temperatures, however, around 37°C, most biochemical reactions, uncatalysed, proceed at imperceptible rates. (to the earn n this).

As stated, each enzyme has an active site to which the substrate(s) bind. The binding has several effects.

- It positions substrate molecules in the most favourable relative orientations for the reaction to occur.
- The active site is perfectly complementary, not to the substrate in its ground state (as the unactivated substrate molecule is referred to), but to the **transition state**, which is intermediate between the reactant molecules and products.

A substrate in its transition state, for this reason, binds to the enzyme more tightly than does the substrate in its ground state. This is due to the structure of the active centre, which has evolved to do this. A transition state tightly bound to the enzyme is at a lower energy level than the same transition state in free solution as occurs in a noncatalysed reaction. (The binding liberates energy). The transition state has too ephemeral an existence to measure just how tightly it binds to the enzyme, but **transition state analogues** have been synthesized. These are stable molecules similar in structure to the transition state. It has been found that these bind to enzymes with remarkably high affinity – in one case, binding to the enzyme thousands of times more tightly than the substrate.

Another way of looking at this is that formation of the transition state involves a partial redistribution of electrons. The amino acid groups in the structure of the active centre of the enzyme are of such a nature, and so positioned, that they stabilize the electron distribution of the transition state. (You will find this explained in greater detail when you come to the mechanism of the enzyme chymotrypsin, later in this chapter.) The fact that the active centre is a less perfect fit to the substrate than it is to the transition state results in the substrate being strained on binding to the active centre, and this favours

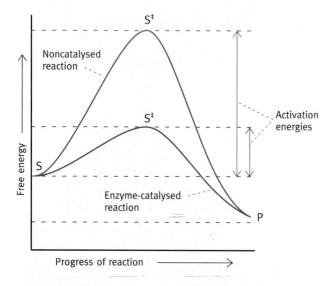

Fig. 6.1 Energy profiles of noncatalysed and enzyme-catalysed reactions. The inverse relationship between the rate constant and the activation energy of the reaction is exponential so that the rate of the reaction is extremely sensitive to changes in the activation energy. S, substrate; $S^{\ddagger}$, transition state; P, products.

transition state formation. The net effect is to lower the activation energy of the reaction; the energy hump barrier to the reaction is reduced (Fig. 6.1) and the reaction rate is increased. This is the central 'secret' of enzyme catalysis on which life depends. Once formed, the transition state rapidly converts to products. The products bind less tightly to the enzyme and diffuse away. The catalysis rate is very sensitive to changes in the activation energy, there being an inverse exponential relationship. A very small reduction in the activation energy, equivalent to the small amount of energy released on formation of a single average hydrogen bond (page 37), can increase the rate of the reaction by a factor of 10^6. Enzymes increase the rate of chemical reactions sometimes by a factor of 10^7–10^{14}. The enzyme urease, which destroys urea, reduces the energy of activation by 84 kJ mol^{-1} and increases the reaction rate 10^{14} times. Most molecules are capable of participating in many different chemical reactions, each with its own transition state. In an uncatalysed reaction, promoted by high temperatures, molecules collide unpredictably and different transition states are formed, resulting in a variety of side reactions. By contrast, an enzyme catalyses a specific reaction for which there are only a limited number of well-defined products.

From what has been said, it might be deduced that if one could produce a protein with a high affinity for the transition state of a given chemical reaction, it would catalyse that reaction. Such has been found to be the case. Antibodies (described in Chapter 30) are proteins which bind tightly to specific structures in molecules, and moreover they can be produced to bind to selected molecules. It was found that an antibody against the transition state involved in the hydrolysis of a synthetic ester behaved as a hydrolytic enzyme towards that ester. The term **abzyme** has been coined for such proteins, *ab* denoting 'antibody'. (The antibody was raised against a stable analogue of the transition state.)

The induced-fit mechanism of enzyme catalysis

The binding of an enzyme to its substrate implies that the relationship between an enzyme and its substrate(s) is simply a 'lock-and-key' model in which the active site is envisaged to be a rigid structure in an unchanging form and the substrate (the key) fits to it (Fig. 6.2(a)). A more recent concept, however, is the '**induced-fit**' mechanism, which is based on the view that the enzyme is not a rigid structure analogous to a lock, but rather is a flexible structure capable of changing its conformation slightly in an interactive way when its substrate binds. ('Conformation' refers to the particular arrangement of the protein chain(s) in three-dimensional space.) In other words, it changes its shape slightly, which has the important effect of altering the spatial arrangements of groups on the molecule. Conformational changes in proteins, as will become clear later in the book, are of central importance to the existence of life in many ways. This induced-fit mechanism was first established for the enzyme **hexokinase**. The enzyme catalyses the transfer of

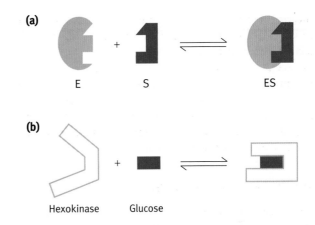

Fig. 6.2 (a) Lock-and-key model of enzyme mechanism. E, enzyme; S, substrate. **(b)** Induced-fit model of the hexokinase mechanism.

a phosphoryl group from ATP to glucose (a hexose). The enzyme has two 'wings' to its structure. In the absence of glucose, these have an 'open' conformation, but on binding of glucose the wings close in a jaw-like movement that results in the creation of the catalytic site (Fig. 6.2(b)). The postulated conformational change has been proved to occur by X-ray crystallographic studies (Fig. 6.3).

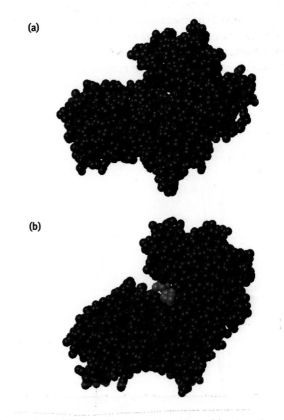

Fig. 6.3 Space-filling models of yeast hexokinase **(a)** unbound (Protein Data Bank Code 1HKG) and **(b)** with a glucose analogue (red) bound (2YHX).

Enzyme kinetics

Enzyme kinetics refers to the study of enzymes by determining their reaction rates. Quite apart from the insight they give into enzyme action, a basic knowledge of the topic is of importance in medicine and pharmacology. Measurement of enzyme activities is a basic requirement also in much of biochemistry and other biological sciences, as well as clinical medicine. Such measurements require a knowledge of what factors influence the reaction or the reaction rate values may be meaningless. Many drugs work by inhibiting enzymes and it is essential in developing these to understand the different types of inhibition by which compounds may work. Substrate concentration effects on enzyme activities are of significance in devising drug therapies, where the rate of enzyme destruction of the drug assumes critical importance. The subject is also important in considering aspects of metabolic control discussed in Chapter 16.

In a typical case, an enzyme rate is measured by incubating it with substrate(s) and any required activators at a defined temperature and pH and following the production of a reaction product with time, or alternatively following the disappearance of substrate. As seen in Fig. 6.4, the reaction, with a fixed amount of enzyme, gradually diminishes in rate with time; this is due to accumulation of product (P), which inhibits the enzyme by binding to the enzyme (E), forming EP, and/or depletion of substrate. In the case of a delicate enzyme, its denaturation could also be a factor. The initial rate of reaction is denoted by V_0. For meaningful quantitative assays of enzyme activity, it is necessary to ensure that initial velocities are measured, which typically means using short time periods before there is a significant amount of product formed. In this situation the reaction velocity is linearly proportional to the amount of enzyme added.

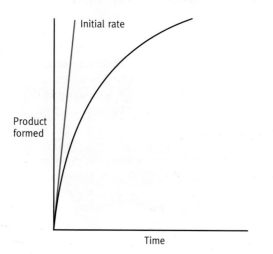

Fig. 6.4 Time course of a typical enzyme reaction in which the amount of enzyme is constant.

Hyperbolic kinetics of a 'classical' enzyme

Almost a century ago, Leonor Michaelis and Maud Menten proposed a model of the way in which enzymes act to fit the observed kinetics of enzyme catalysis for a single substrate enzyme:

$$E + S \rightleftharpoons ES \rightarrow E + P$$

What the equation says is that substrate binds reversibly at the active site of the enzyme to form ES (enzyme-substrate complex), the reaction occurs, and the products (P) diffuse away into the solution. At low [S] the rate of an enzyme-catalysed reaction is determined by the rate of formation of ES. The binding of S to form ES is a reversible equilibrium so that at low [S], formation of ES is relatively low and will vary with changes in [S] so that the rate of reaction varies accordingly. As [S] increases, so will the proportion of enzyme in the form of ES increase and therefore the rate of reaction increases. However, as [S] further increases, the point is reached at which the rate of formation of ES ceases to be rate limiting so that the catalytic activity of the enzyme or the rate of product release then becomes limiting. The catalytic rate of the enzyme is an inherent property and varies from enzyme to enzyme. Some enzymes work extremely rapidly, and others more slowly. At high levels of [S], the enzyme is said to be **saturated** with substrate. The rate of the reaction is then called V_{max} (for maximum velocity), and further increases in the substrate have no effect on the rate of catalysis. V_{max} is a function of the amount of enzyme present in a given experiment and the rate at which each molecule of enzyme catalyses the reaction. If the molar concentration of the enzyme in an experiment is known, the turnover number or K_{cat} can be calculated. This is the number of molecules of substrate converted to product by a molecule of enzyme at saturating levels of substrate per second, values ranging up to 4×10^7/s for catalase. In the case of many or most enzymes, when the velocity of enzyme activity is plotted against substrate concentration a **hyperbolic curve** is obtained, as shown in Fig. 6.5. An enzyme displaying these kinetics is referred to as a **Michaelis–Menten** enzyme and the kinetics as Michaelis–Menten kinetics or **hyperbolic kinetics**. Enzymes of this type are sometimes referred to as 'classical' enzymes because for a long time they were the only ones known. (Other types are referred to below.)

In the derivation of the equation which describes the relationship between the velocity of an enzyme reaction and the substrate concentration certain simplifying assumptions are made, namely that only a single substrate is involved, and that initial velocities (V_0) are measured so that the concentration of product is negligible compared with that of the substrate. This eliminates the complication of product binding significantly to the enzyme. It is also assumed that the system is in a **steady state** in which the rate of formation of ES is exactly balanced by the rate of its removal and that the substrate is in vast molar excess

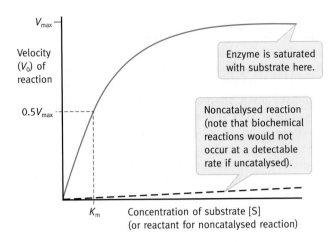

Fig. 6.5 Effect of substrate concentration on the reaction velocity catalysed by a classic Michaelis–Menten type of enzyme. K_m, Michaelis constant. The dashed line shows, for comparison, the effect of reactant concentration on a noncatalysed chemical reaction. Note that the two lines are drawn to illustrate their shapes, not their relative rates.

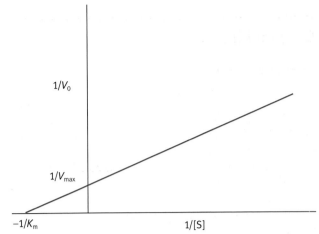

Fig. 6.6 Double reciprocal plot of an enzyme reaction.

over the enzyme. The latter two conditions are usually met because steady state kinetics are established almost instantly and the molar amount of enzyme is usually negligible. The equation which describes the relationship between the velocity of an enzyme reaction and the substrate concentration which results in the hyperbolic curve shown in Fig. 6.5 is known as the **Michaelis–Menten equation**:

$$V = \frac{[S]V_{max}}{[S] + K_m}.$$

The K_m value, expressed in units of molar concentration, is a very useful constant. What might, in molar terms, be a low concentration of S for one enzyme, may be a saturating concentration of substrate for another. It depends on how tightly the substrate binds to the enzyme, or as it is usually expressed, what the **affinity** of the enzyme is for its substrate. This depends on the precise relationship of the substrate to its enzyme – the nature and number of the weak bonds established between them. This affinity is given a numerical value as the **Michaelis constant (K_m)** of the enzyme. It is defined as that concentration of S at which the enzyme is working at half maximal velocity, as shown in Fig. 6.5, and is therefore independent of the total amount of E. At the K_m value of substrate concentration, half the total number of enzyme active sites are occupied and half vacant. Other methods of plotting graphs are used to determine K_m values more precisely; if instead of plotting reaction velocity (V_0) against [S] as is done in Fig. 6.5, $1/V$ is plotted against $1/S$ this gives a straight line which, if extrapolated back (Fig. 6.6) intercepts the horizontal axis at the reciprocal of the K_m. This is known as a **double reciprocal plot**, or a **Lineweaver–Burke** plot after its authors (Fig. 6.6). The intercept of the line with the

vertical axis also gives a value for $1/V_{max}$. *The higher the K_m, the lower is the affinity of the enzyme for its substrate.* K_m values can be used to compare the affinities of different enzymes for their substrates. They are of interest from the viewpoint of metabolism for they tell us how an enzyme will respond to changes in the concentration of substrate in the cell. The affinity of a protein for a ligand is a function of the equilibrium between the protein–ligand complex and the free components. In the case of many enzymes, the K_m represents a true affinity constant (the reciprocal of the dissociation constant) but only for those in which the rate of dissociation of ES back to E + S is much faster than the catalytic step of ES to E + P. The reason is that since an affinity constant represents the position of the equilibrium E + S⇌ES, if ES is rapidly removed by conversion to E + P, a true equilibrium is not established. The K_m is a true affinity constant in those cases where this requirement is met which is true of many enzymes. However, even where it is not, the *apparent* affinity of the enzyme for its substrate, which is in all cases reflected in the K_m value, is a useful way of comparing how different enzymes will respond to substrate concentration changes. As a general statement, most enzymes have K_m values such that they are working in the cell at sub-saturating substrate levels. For many enzymes the K_m is in the range of 10^{-4}–10^{-6} M.

Allosteric enzymes

There is another very important class of enzymes which do not have hyperbolic kinetics. These are regulatory enzymes whose activities are controlled by chemical signals in the cell and which in turn control metabolism. Metabolic control is a major subject described in Chapter 16 and a description of the 'special' kinetics of these enzymes with special reference to aspartate transcarbamylase (page 247) is deferred until then.

For the relationship of K_m to the rate of reaction in allosteric enzymes, see Chapter 16, page 246.

General properties of enzymes

Nomenclature of enzymes

Usually enzyme names end in '-ase', preceded by a term which most often indicates the nature of the reaction it catalyses and/or indicates the substrate. Thus *amylase* attacks *amylose*. A *dehydrogenase* removes hydrogen atoms from a substrate: lactate dehydrogenase is an example. This is not always the case, for proteolytic enzymes often end in '-in'; *pepsin*, *chymotrypsin*, *plasmin*, and *thrombin* are examples. We will explain names more fully when the enzymes are considered in biochemical systems. An international committee has systematized all enzyme names, but as the systematic names are necessarily rather long, there are also shorter recommended names for everyday use. The systematic names and reference numbers are usually given once in published research papers and reference works to avoid any ambiguity.

Isozymes

It is quite common to find that the same reaction is catalysed by a number of distinguishable different enzymes, called isozymes or isoenzymes. There is considerable similarity in the amino acid composition of the different isozymes catalysing a given reaction, suggesting that they all have the same evolutionary ancestor (page 54) but the genes coding for them have diverged somewhat to suit particular roles of the isozymes in the body. Isozymes are usually found in different tissues or in different locations in cells. The reason for multiple versions of the same enzyme is to tailor them to the specific needs of the cell. Thus in some cases the isozymes differ in substrate affinities, or have different regulatory mechanisms or other properties. This is not too surprising because different tissues have quite different roles. A classic example is that of hexokinase and glucokinase (see page 167) – same reaction (but different range of specificities), different K_m and different physiological role.

Isoenzymes often have a number of different subunits, or protein molecules which join together to form the complete enzyme. Thus an enzyme which you will meet later (page 180), **lactate dehydrogenase**, has four subunits of two different types, one known as H because it is the main one in the heart enzyme and the other M because of its association with muscle. Various combinations of H and M subunits produce the different isozymes in different tissues. The isozymes can often be separated because of their different electrical charge. Placed in a gel with a voltage gradient across it they migrate at different rates, a process known as **electrophoresis** (see Chapter 5).

Enzyme cofactors and activators

In the simplest case, the enzyme protein combines with a single substrate, reaction occurs, and the products leave the active site to make way for another molecule. Frequently the enzyme requires a **cofactor** for activity. This may be a metal ion such as Mg^{2+} or Zn^{2+}, which participates in the reaction mechanism (e.g. carboxypeptidase, page 98), or it may be an organic molecule attached to the enzyme in which case it is known as a **prosthetic group**. The protein part is then called an **apoenzyme** (*apo* = detached or separate) and the complete enzyme with the prosthetic group attached, a **holoenzyme**. The prosthetic group is sometimes a vitamin derivative, and some vitamins activate several enzyme species. Since each vitamin molecule in this case activates an enzyme molecule which can bring about reactions in vast numbers of substrate molecules, it explains why small amounts of vitamins can have a huge effect on the body. **Coenzymes** also have vitamins as components and behave somewhat similarly but are not firmly attached to the enzyme (details will be given at appropriate places later in the book). They are very much like substrates in that in most cases they react and leave the enzyme. Their role is to couple enzymic reactions together. For example the coenzyme nicotinamide adenine dinucleotide (NAD^+; page 178) is reduced to $NADH + H^+$ (equivalent to NAD^+ with two hydrogen atoms attached) in dehydrogenation reactions.

$$AH_2 + NAD^+ \rightarrow NADH + H^+$$

The reduced coenzyme leaves the enzyme and becomes the substrate for a second one.

$$B + NADH + H^+ \rightarrow BH_2 + NAD^+$$

The overall result is that the coenzyme acts as the intermediary in transferring hydrogen atoms from one substrate to another. This type of activity plays a prominent role in the release of energy from foodstuffs as described in later chapters.

Effect of pH on enzymes

The activity of an enzyme is influenced by pH in several ways. The protein is influenced by the state of its ionizable groups and the function of the active centre may be likewise dependent on this. The ionization of the substrate itself may also be affected. The rate of catalysis is therefore dependent on the pH. Enzyme pH activity profiles vary from one to another but the optimum is often around neutral pH; a typical plot is shown in Fig. 6.7(a). Exceptions occur such as the case of the digestive enzyme, pepsin, which functions in the acidic stomach contents; its pH optimum is near 2.0.

Effect of temperature on enzymes

Temperature also affects enzyme activity rates. As the temperature increases, the rate of most chemical reactions increases (approximately twofold for each 10°C) but, because of the inherent instability of most protein molecules, the enzyme is

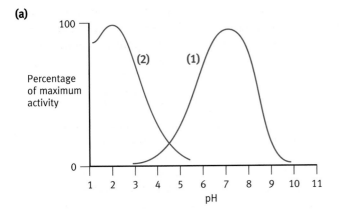

(a)

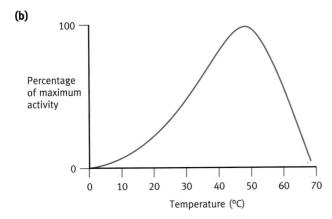

(b)

Fig. 6.7 Effect of pH and temperature on enzyme activity. **(a)** Effect of pH on enzyme activity. Curve 1 is typical of the majority of enzymes with maximal activity near physiological pH. Curve 2 represents pepsin, an exceptional case, since the enzyme functions in the acid stomach contents. **(b)** Effect of temperature on a typical enzyme. The precipitous drop at high temperatures is due to enzyme destruction (though a number of heat-stable enzymes are known). Note that, as described in the text, the temperature 'optimum' of an enzyme has little significance, since the shape of the curve will depend on the length of time the enzyme is maintained at a given temperature before measuring.

inactivated at higher temperatures. Thus a typical enzyme optimum temperature plot would appear as shown in Fig. 6.7(b) but, although perhaps useful in a practical sense, this has little absolute significance since the optimum will depend on the experimental time period used in measuring the rates; the shorter the time the less will be the destructive effect of higher temperatures. A few enzymes are stable to high temperatures but, in general, temperatures over 50°C are destructive and some enzymes are more labile still.

Effect of inhibitors on enzymes

The activity of an enzyme can be affected by inhibitory compounds. Inhibitors may be reversible or irreversible. In the former case the enzyme and inhibitor exist in a reversible

equilibrium ($E + I \rightleftarrows EI$). Irreversible inhibitors bind to the enzyme and do not dissociate from it to an appreciable extent; the extreme case of this is where the inhibitor becomes covalently attached. The effect of aspirin on an enzyme, described below, is one such case. Penicillin is another.

Competitive and noncompetitive inhibitors

Reversible inhibition of an enzyme may be of different types. One class, called **competitive inhibitors**, simply mimic the substrate and compete with the latter for binding to the active site. The degree to which the latter is occupied by the inhibitor will determine the degree of inhibition. The inhibition will be a function of the relative affinities for the enzyme of substrate and inhibitor and their relative concentrations. It is possible to distinguish between competitive and noncompetitive inhibition using the double reciprocal plot already described and which is shown in Fig. 6.8. A competitive inhibitor will have no effect at infinite substrate concentration since the substrate will completely win in the competition to bind to the active site. The intersection of the reciprocal plot with the vertical axis (which represents infinite [S]) will be the same whether inhibitor is present or not (i.e. V_{max} is unchanged). It will, however, change the K_m so that the intersection with the horizontal axis which gives the reciprocal of the K_m value is changed by the inhibitor (K_m is increased).

Inhibitors of enzymes play an important part in the treatment of diseases. For example, physostigmine is a competitive inhibitor of acetylcholinesterase and is used for patients with myasthenia gravis (page 494). The so-called 'statin' drugs, which are used to treat people with high cholesterol levels, are competitive inhibitors. The drug simvastatin, for example, is a competitive inhibitor of an enzyme that controls the synthesis of cholesterol (page 174). These compounds originated from fungi.

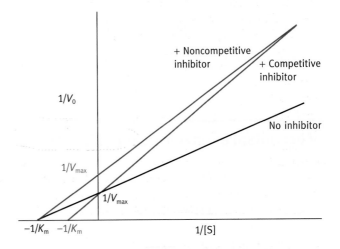

Fig. 6.8 Double reciprocal plot of enzyme reactions in the presence of competitive and noncompetitive inhibitors respectively. See text for explanation of inhibitors.

A **noncompetitive inhibitor** binds to the enzyme at a position separate from the active site so that there is no competition with the substrate; this is readily seen in a double reciprocal plot, as shown in Fig. 6.8, in which it is seen that the V_{max} at infinite substrate concentration is reduced while the K_m is unchanged.

Noncompetitive inhibitors work in various ways. For example, a heavy metal such as mercury may react with a thiol group essential for catalytic activity. Removing the metal with a thiol compound or a metal-chelating agent may reverse this.

In other cases, an inhibitor may covalently acylate the active site of the enzyme in an irreversible manner. Thus aspirin inactivates an enzyme (cyclooxygenase) involved in prostaglandin synthesis (page 232) as follows:

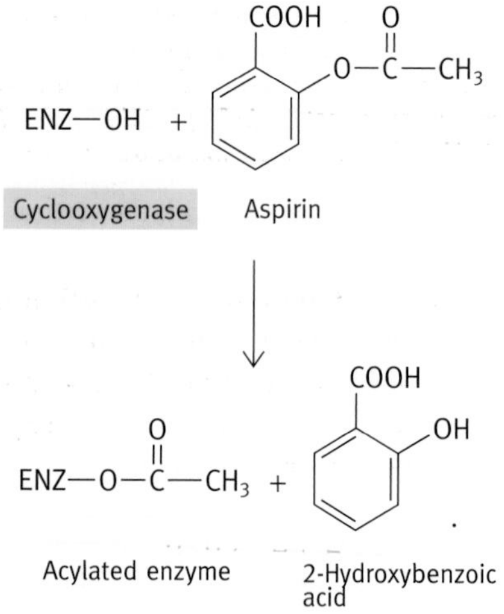

Prostaglandins are involved in pain and inflammation, so inhibition of their synthesis by aspirin ameliorates these symptoms (page 232, Box 14.2).

A third type of inhibition is known as *uncompetitive* (subtly different from noncompetitive). In this, the inhibitor binds to the enzyme in a noncompetitive way but binds only to the ES form of the enzyme.

Allosteric inhibitors are a special class, discussed on page 246.

Mechanism of enzyme catalysis

We can illustrate how one class of enzymes works in structural terms, using the enzyme **chymotrypsin** as an example. Enzymes may increase reaction rates by 10^{14} times. A 10^{14}-fold increase means that an enzyme catalyses in 1 second an amount of chemical reaction that without the enzyme would require hundreds or thousands of years.

The chemistry of the chymotrypsin reaction that follows involves little more than a proton jumping from one group to another and back again. It is one of the most satisfying illustrations of the remarkable abilities of proteins to perform specific chemical tasks at incredible speeds and by mechanisms, which are, in concept, very simple.

Mechanism of the chymotrypsin reaction

Chymotrypsin is a digestive enzyme produced by the pancreas; it hydrolyses specific peptide bonds of proteins in the diet. Its digestive role is described on page 347.

The **active centre** of an enzyme is usually in a cleft of the protein, into which fits the substrate to be attacked, as already stated. In the case of chymotrypsin, the natural substrate is a polypeptide and it hydrolyses those peptide bonds whose carbonyl group is donated by a large hydrophobic amino acid residue (mainly the aromatic ones phenylalanine, tyrosine, and tryptophan but also methionine). The active centre has a large hydrophobic pocket to accommodate the bulky hydrophobic group. The enzyme is an **endopeptidase** – it hydrolyses internal peptide bonds of proteins, in contrast to an **exopeptidase** which hydrolyses terminal peptide bonds (*endo* = within; *exo* = without). The reaction given here in nonionized structures for clarity is as follows:

$$R\text{-}CONH\text{-}R' + H_2O \longrightarrow RCOOH + R'NH_2$$

R represents a large hydrophobic residue of a polypeptide substrate that provides the carbonyl group of the peptide bond. R′ represents the rest of the peptide.

Chymotrypsin is one of a group of proteases known as **serine proteases** because the active centre contains a serine residue. Hydrolysis of the peptide bond takes place in two stages. In the first, the serine-OH becomes acylated and the first product (R′NH$_2$) is released. In the second stage the acyl enzyme is hydrolysed and the second product (RCOOH) released. (Nonionized structures are used for clarity; R and R′ are as defined in the equation already given above.)

Stage 1:

$$RCONHR' + Enz\text{-}OH \longrightarrow RCOO\text{-}Enz + R'NH_2$$

Peptide substrate | Enzyme with serine–OH group | Intermediate acyl enzyme | First product

Stage 2:

The ester bond in the intermediate acyl enzyme is hydrolysed by water, releasing the second product (RCOOH) and restoring the enzyme to its original state.

$$RCOO\text{-}Enz \; + \; HOH \; \longrightarrow \; RCOOH \; + \; Enz\text{-}OH \; .$$

                                    Second   Enzyme in
                                    product   original state

There are two main questions to be considered.

(1) Why is the —OH of the serine residue so reactive in stage 1 of the enzymic reaction? Serine itself is a very stable molecule and its hydroxyl group unreactive at neutral pH; other serine residues in the chymotrypsin molecule are inert, so why is that one reactive? The peptide bond of the substrate to be attacked is likewise quite stable at neutral pH and on its own does not react at a perceptible rate.

(2) Why does water so readily hydrolyse the ester bond in stage 2 of the reaction? A carboxylic ester is relatively stable at neutral pH in aqueous solution. To answer these questions we must look at the structure of the catalytic centre of the enzyme.

The catalytic triad of the active centre

Projecting into the active site of the enzyme are the side chains of three amino acid residues that are part of the polypeptide chain comprising the enzyme – aspartate, histidine, and serine, known as the **catalytic triad**. Although quite widely separated in the polypeptide chain of the enzyme, folding of the chain brings them together in the active centre as diagrammatically shown in Fig. 6.9. This is an important role of enzyme proteins – to bring together and fix in optimal relationships the reactive groups involved in catalysis. This is a reason why enzymes are large.

A few points about these groups first.

- The side-chain carboxyl group of aspartate (the ionized form of aspartic acid) has a pK_a of about 4 and is therefore dissociated at physiological pH.
- The serine —OH group with a pK_a of about 14 is not significantly dissociated.
- The imidazole side chain of histidine is the interesting one with a pK_a in its protonated form of about 6 (see below). This means that at physiological pH there is a rapid equilibrium between the protonated and unprotonated states. On the Brønsted–Lowry definition of an acid being a proton donor and a base being a proton acceptor, in its protonated form histidine is an acid and in its unprotonated form it is a base.

Thus the histidine side chain in its protonated form can readily function as a proton donor in which case it is acting as a **general acid**. In its unprotonated form, it can accept a proton and function as a general base. It can thus promote **general acid–base catalysis**, as will shortly be explained

It might be helpful to remember some chemical principles. A covalent bond involves two atoms sharing a pair of electrons

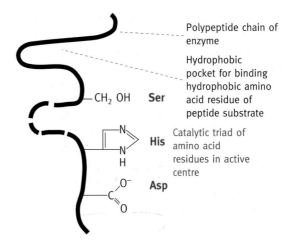

Fig. 6.9 The active site of chymotrypsin. Folding of the polypeptide brings together the three amino acid residues of serine, histidine, and aspartate. In the unfolded chain they are quite widely separated. The substrate specificity of chymotrypsin for peptides whose carbonyl group is donated by a hydrophobic amino acid derives from the specific hydrophobic binding site of the active centre.

between them. Each electron is attracted to both atoms and this mutually holds the two atoms together. Formation of the bond releases energy, which is why it forms. A simple example is the formation of a covalent bond between two hydrogen atoms to form a hydrogen molecule (H_2):

$$H\cdot + H\cdot \rightarrow H:H$$

Each hydrogen atom has a single electron so that the two atoms contribute equally to formation of the bond. In some molecules or ions, an atom has an unshared pair of electrons that can be donated to another atom to form a covalent bond. The chemical entities containing such donor atoms are called **nucleophiles**. The acceptor of the pair of electrons is called an **electrophile**. The process of forming such bonds is known as a **nucleophilic attack**. (For nomenclature reasons in chemistry, in the one specific case of a proton (H^+), which has no electron, the chemical entity supplying the electrons is called a base and not a nucleophile, but the principle is unchanged.)

$$X: \quad + \quad Y \quad \rightarrow \quad X^+ : Y^-$$
    Nucleophile   Electrophile

One of the nitrogen atoms of the imidazole side chain of histidine has such an unshared pair of electrons that can interact with a proton. The dissociation of the imidazole group of histidine occurs as follows:

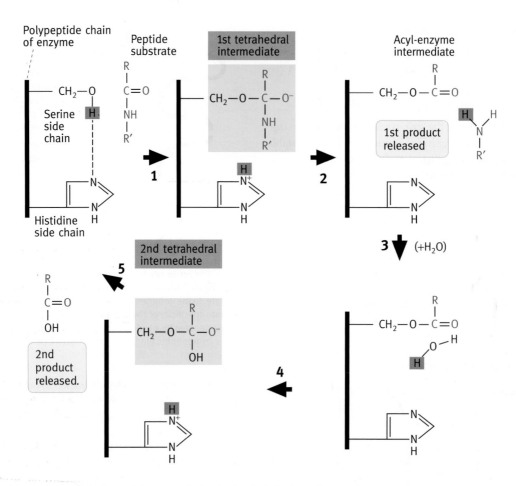

Fig. 6.10 Reactions at the catalytic centre of chymotrypsin involved in the hydrolysis of a peptide substrate. For ease of presentation the enzyme polypeptide chain is given as a straight line with only two of the catalytic triad residues shown. The role of the third residue (aspartate) is described later. The broken line represents a hydrogen bond. The steps are described in the text.

In donating two electrons to H^+ to form the covalent bond in the reverse reaction above, the nitrogen atom acquires a positive charge. Note that the imidazole ring of histidine can exist in two tautomeric forms. In the unprotonated form, the single hydrogen atom can be on either of the two nitrogen atoms. With that background we can proceed to the mechanism of chymotrypsin catalysis.

The reactions at the catalytic centre of chymotrypsin

All three amino acid residues of the catalytic triad – aspartate, histidine, and serine – are essential for the enzyme to function properly but actual chemical changes (reactions) occur only between the histidine and serine so we will deal with these first to keep it simple. The role of aspartate will be explained later.

A peptide substrate molecule attaches to the catalytic site of chymotrypsin by the binding of its hydrophobic group to a specific nonpolar pocket such that the carbonyl carbon (C=O) atom of the peptide bond to be attacked is close to the serine —OH. Simultaneously the hydrogen atom of the serine —OH

transfers to the histidine nitrogen atom and the oxygen atom forms a bond with the carbonyl carbon atom of the substrate as shown in step 1 of Fig. 6.10. This produces the first tetrahedral intermediate shown in the yellow box in the figure. (The term tetrahedral refers to the organization of bonds of the carbon atom of the bond to be broken; a tetrahedral carbon has its four bonds pointing to the vertices of a tetrahedron.) In step 1 the formation of the tetrahedral intermediate requires the C=O of the carboxyl group to become single-bonded thus forming an oxyanion as shown. This is stabilized by interacting with an oxyanion 'hole' in the enzyme.

That is what happens, but how can serine react in this way? The serine —OH group is normally unreactive at neutral pH. For the oxygen atom to form the bond with the carbon atom of the substrate, it has to lose its hydrogen atom as a proton, but, with a pK_a of 14, the —OH group does not significantly dissociate except in strongly alkaline solutions whose pH is near the pK_a of the group. The answer lies in the ability of the histidine N atom to abstract the hydrogen – it is acting as a **general base** (defined, we remind you, as a group that accepts a proton). It

Fig. 6.11 The function of aspartate in the catalytic triad. In situation **(a)** the histidine is held in the form shown, by a strong hydrogen bond with aspartate. This results in the nitrogen with the unshared pair of electrons facing the serine proton which it can abstract. If the histidine is not held by being hydrogen bonded to aspartate the situation shown in **(b)** could result with the protonated nitrogen facing the serine and therefore the serine proton would be less efficiently abstracted.

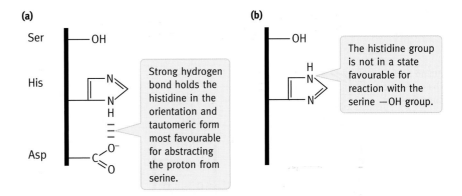

does so because the serine and histidine of the catalytic triad are oriented so that the hydrogen of the —OH group is perfectly positioned to interact with the nitrogen atom of the histidine.

The histidine with its acquired proton is now a **general acid** – it can donate a proton. (You can now see why the process is referred to as **general acid–base catalysis**.) The proton is transferred to the tetrahedral intermediate causing it to break down as shown in step 2 of Fig. 6.10, to liberate the first product ($R'NH_2$) and form the acylated enzyme.

We are halfway there; the next step is to hydrolyse the ester bond of the acyl-enzyme intermediate and thus liberate the second product of peptide hydrolysis (RCOOH) and restore the enzyme to its original state ready for reaction with the next substrate molecule. It is basically a repeat of the strategy used in the first stage. What is required is for the oxygen atom of water to make a nucleophilic attack on the carbonyl carbon atom of the ester bond of the acyl-enzyme intermediate. When a molecule of water enters the active site (Fig. 6.10, step 3) the histidine group, acting as a general base abstracts a proton (just as it did from serine in the previous stage of the reaction). The water oxygen atom makes a nucleophilic attack forming the second tetrahedral intermediate (Fig. 6.10, step 4). Exactly as before, the protonated histidine is now a general acid; it donates its acquired proton back to the tetrahedral intermediate (Fig. 6.10, step 5) resulting in completion of the reaction and liberation of the second product (RCOOH). The serine —OH is restored to its original state ready for reaction with the next substrate molecule.

The actual mechanisms by which the chemical changes are brought about are thus remarkably simple, involving little more than histidine acquiring a proton and giving it back at each of the two stages.

What is the function of the aspartate residue of the catalytic triad?

If, by genetic engineering, the aspartate residue is deliberately converted to an asparagine residue in which the carboxyl group now becomes an amide group that does not significantly dissociate, the catalytic activity of chymotrypsin falls by a factor of 10,000. The aspartate carboxylate anion is thus essential but nevertheless it does not undergo any chemical reaction during the catalytic process. Why then is it needed? The aspartate carboxylate anion forms a strong hydrogen bond with the histidine side chain as shown in Fig. 6.11. Its main function is to hold the histidine residue in the orientation and tautomeric form shown in the figure, so that the nitrogen atom which accepts the proton from the serine residue is always facing the latter, optimally positioned to abstract the proton from the —OH group. If the aspartate is converted to asparagine, the hydrogen-bonding potentiality is very much weaker and this immobilizing effect on the histidine residue is missing. It is historically interesting that initially it was believed that this residue in the catalytic triad was in fact asparagine. When it was appreciated that aspartate would make more sense mechanistically, the structure determination was re-checked revealing that this residue is indeed aspartate. The reference by **D. M. Blow** in the Further reading section at the end of this chapter describes the fascinating discovery of the mechanism of chymotrypsin action.

Other serine proteases

The catalytic triad mechanism has been adopted by a variety of hydrolytic enzymes. The serine proteases chymotrypsin, trypsin, and elastase have the same mechanism, all with the aspartate, histidine, and serine residues. They all hydrolyse peptides but have different specificities for the component amino acid residues forming the peptide bond. The active centres differ in the pockets in the active sites required for the binding of the substrates; they will accept only the particular amino acid side chains of their specific substrates. The pocket in chymotrypsin is hydrophobic (Fig. 6.12(a)), whereas that of trypsin has a negatively charged aspartate residue (different from the one in the catalytic triad) to which binds the partially charged basic side chains of trypsin-specific substrates (Fig. 6.12(b)). That of elastase is smaller and access to it restricted by threonine and valine residues so that the enzyme is specific for peptide bonds whose carbonyl group is contributed by amino acid residues with small side chains (Fig. 6.12(c)). The three enzymes described above are structurally closely related to one

Amino acid side chain of typical substrate that binds to pocket

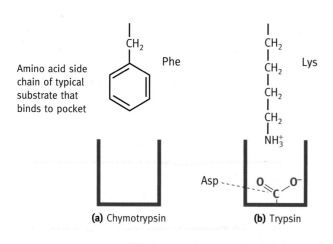

Phe

Lys

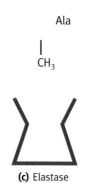

Ala

(a) Chymotrypsin

(b) Trypsin

(c) Elastase

Fig. 6.12 Simplified diagram of the pockets in the active centres of chymotrypsin, trypsin, and elastase into which fit the amino acid side chains of their respective substrates. That of chymotrypsin accommodates a bulky hydrophobic group such as the side chains of phenylalanine or tryptophan, that of trypsin accommodates the positively charged side chain of lysine or arginine that bind to the negatively charged aspartic acid residue present in the binding site. The elastase pocket accepts only smaller amino acid side chains of substrate molecules, the entrance to the binding site being restricted by the side chains of valine and threonine residues.

another and clearly have an evolutionary relationship. The bacterial proteolytic enzyme, subtilisin (from *B. subtilis*) is a totally different protein but still has the identical catalytic triad; the independent convergent evolution of this emphasizes its basic importance. Several enzymes, unrelated in function to the proteinases, also have the catalytic triad. Acetylcholinesterase (page 116) is an example.

A brief description of other types of protease

As well as the serine proteases there are three other classes of protease in terms of the structures of their catalytic sites. These are the thiol, aspartic, and zinc proteases. The **thiol proteases** are very similar to the serine types but instead of an activated serine hydroxyl group as in chymotrypsin, there is an activated thiol group of cysteine. An intermediate thioester (RCO—S—Enz) is formed instead of the carboxylic ester RCO—O—Enz) as in chymotrypsin. The plant proteolytic enzyme papain (found in the latex juice of the papaya) is an example.

Pepsin, whose digestive role is described on page 147, belongs to the **aspartic protease** class. It has a catalytic diad of two aspartate residues: one is unprotonated and can accept a proton, the other is protonated and can donate one. The two act in turn as a general acid and a general base respectively and reverse their roles with each round of reaction. Several enzymes of this class are known, such as the kidney enzyme **renin** which is involved in blood-pressure control. The finding that HIV (the AIDS virus) has an aspartic protease required for its replication has heightened interest in the group. Inhibition of this enzyme is a site for therapeutic attack on the disease.

A third class of proteases is the metalloproteases which depend on a metal ion, usually zinc, in the active site. An example of the class of **zinc proteases** is carboxypeptidase A, a digestive enzyme, with a preference for hydrophobic amino acid residues, which hydrolyses off the terminal carboxyl residue from peptides. It exhibits a remarkably large conformational change on substrate binding (see page 89 for the induced-fit theory). Mechanistically the catalytic process has strong resemblances to that of chymotrypsin in that general acid–base catalysis is involved – a proton is removed from water by transfer, in this case to a glutamate residue. The activated water molecule then attacks the carbonyl carbon of the peptide, forming a tetrahedral transition state, as shown in Fig. 6.10, reaction 4). Activation of the water molecule in this case is promoted by its binding to a zinc atom, which is bound to the enzyme active site.

Summary

An enzyme is a catalyst that brings about a specific reaction but is unchanged in the process. It cannot affect the equilibrium of a reaction. The free energy change of a reaction determines whether it may proceed but says nothing about whether it will occur.

For a chemical reaction to happen, an energy barrier has to be overcome. The reactants must be activated to a higher energy transition state, which rapidly converts to products. The energy for this is the activation energy, supplied in uncatalysed reactions by molecular collisions.

An enzyme is a protein that binds substrates and lowers the activation energy of the reaction it catalyses. It does so because the active centre binding the substrate is perfectly complementary not to the substrate itself but to the transition state. It also aligns substrates and provides a catalytic site. This is a small area of the protein that contains chemical groupings essential for the catalysis.

On binding of substrates to enzymes the protein often changes shape, a process known as the induced-fit mechanism.

Enzyme kinetics explores enzymes by measuring reaction rates. The Michaelis–Menten equation derived from these explains the hyperbolic course of a reaction in terms of the substrate binding to the enzyme and the enzyme-substrate complex then breaking down to liberate products. An important constant is the K_m or Michaelis constant, which is the substrate concentration at which the enzyme has half its maximal activity. It is useful in comparing enzyme affinities for substrates and relevant to drug therapies.

Other types of allosteric enzyme exist in which the reaction progress is sigmoidal, rather than hyperbolic as in 'classical' enzymes. Allosteric enzymes are controlled. They undergo conformational changes that affect their activities (discussed in Chapter 16).

Enzymes are affected by temperature, pH, and inhibitors. Inhibitors may compete with the substrate for the active site or be noncompetitive and inhibit by binding elsewhere. A graphical procedure, the double reciprocal plot, distinguishes between the two. Enzymes often need activators, prosthetic groups, or coenzymes attached to them. Activators may be metal ions; prosthetic groups are permanently attached small organic molecules; coenzymes differ from these in that they attach transitorily.

Different enzymes catalysing the same reaction are known as isoenzymes. They have characteristics tailored to their particular roles, usually in different tissues.

Chymotrypsin is a digestive enzyme that hydrolyses proteins and the mechanism of which is understood in detail. Its active site consists of a hydrophobic substrate-binding site and a catalytic triad of amino acid residues (serine, histidine, and aspartate). The serine —OH group is reactive because the histidine group is perfectly positioned to abstract the serine —OH proton. During the hydrolysis of the peptide bond an intermediate is formed that is decomposed into an acylated serine group by the donation of the abstracted proton from the histidine group. The function of the aspartate residue is to hydrogen bond the histidine residue and fix it in the favourable orientation.

The second stage of the reaction is to hydrolyse the acylated serine complex liberating the bound peptide-acyl group. A water molecule is activated by the abstraction of a proton by the histidine group forming an intermediate that is also decomposed by the donation of the abstracted proton from the histidine residue.

The mechanism of this catalysis is astonishingly simple, involving essentially a histidine residue activating the serine —OH group by abstracting a proton in the first half of the reaction and then activating a water molecule in a similar manner.

Other proteases use different but related mechanisms including the aspartic class of which an essential enzyme in the AIDS virus is one. Inhibitors have been developed to specifically attack this enzyme on which the virus replication depends.

 Further reading

Steitz, T. A., Shoham, M., and Bennett, Jr., W. S. (1981). Structural dynamics of yeast hexokinase during catalysis. *Phil. Trans. Roy. Soc. SeriesB, London*, **293**, 43–52.
Discusses the conformational change of the enzyme on binding of glucose.

Srere, P. A. (1984). Why are enzymes so big? *Trends Biochem. Sci.*, **9**, 387–90.
A necessarily speculative, but interesting article.

Koshland, Jr., D. E. (1987). Evolution of catalytic function. *Cold Spring Harbor Symp. Quant. Biol.*, **LII**, 1–7.
A concise clear summary of the basic roles of active sites and of enzyme conformational change in catalysis.

Kraut, J. (1988). How do enzymes work? *Science*, **242**, 533–40.
Enzyme catalysis explained by the principle of transition state stabilization. Also describes catalytic antibodies specific for transition state analogues.

Schramm, V. L. (1998). Enzymatic transition state analog design. *Annu. Rev. Biochem.*, **67**, 693–720.

Catalytic mechanisms

Blow, D. M. (1997). The tortuous story of Asp . . . His . . . Ser: structural analysis of a-chymotrypsin. *Trends Biochem. Sci.*, **22**, 405–8.
A personal story of how the classical elucidation of the mechanism of this enzyme took place.

Dodson, G. and Wlodawer, A. (1998). Catalytic triads and their relatives. *Trends Biochem. Sci.*, **23**, 347–52.
A review of the mechanism of chymotrypsin catalysis together with a discussion of other enzymes using the 'serine' catalytic triad and their evolutionary relationships.

Hooper, N. M. (ed.) (2002). Proteases in biology and medicine. *Essays Biochem.*, **36**, 1–167.
A complete issue with 12 reviews on many aspects of proteases, including caspases, cancer, proteasomes, and blood clotting.

Benkovic, S. J. and Hammes-Schiffer, S. (2003). A perspective on enzyme catalysis. *Science*, **301**, 1196–201.

Problems

1   What information can be drawn from a plot of the rate of activity of an enzyme against temperature?

2   The oxidation of glucose to CO_2 and water has a very large negative $\Delta G^{0'}$ value, and yet glucose is quite stable in the presence of oxygen. Why is this so?

3   Explain how an enzyme catalyses reactions.

4   Describe how you would determine whether an inhibitor of an enzyme reaction is a competitive or noncompetitive one. Draw a graph to illustrate your answer.

5   In the case of some enzymes a K_m value is a true measure of the affinity of the enzyme for the substrate. In other cases it is not. Explain this.

6   A competitive inhibitor for a specific enzyme works by combining with its active site and blocking access to it by its substrate. Transition state analogues have been found to be very effective in some cases. Why would you expect such a molecule to be more effective than a competitive analogue of the substrate of the same enzyme?

7   If you want to compare the amount of an enzyme in different preparations by measuring reaction rates what precaution must you take if the measurements are to be meaningful?

8   In the active centre of chymotrypsin there is a serine residue that makes a nucleophilic attack on the carbonyl carbon atom of the peptide substrate forming a covalent bond to it. For this, the proton of the serine —OH has to be removed at the same time. Other serine residues in the protein are totally inert in this respect, as is free serine. Explain how this reaction is caused to occur in the catalytic centre of the enzyme.

9   The active centre of chymotrypsin contains an aspartate residue; although this does not participate in the chemical reaction involved in catalysis, its conversion to asparagine lowers the activity of the enzyme 10,000 fold. Why is this so?

10 Chymotrypsin, trypsin, and elastase all have the same catalytic mechanism but have different specificities. Explain the reason for this.

11 Explain the meaning of the terms thiol protease and aspartyl protease.

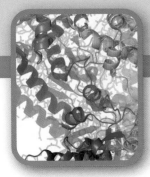

Chapter 7

The cell membrane and membrane proteins

In Chapter 3 we discussed the importance of three types of weak, noncovalent bonds as well as the hydrophobic force resulting from water molecules rejecting nonpolar molecules. Formation of weak bonds (as with all bonds) involves a decrease in free energy so that only when formation of such bonds is maximized do you have the preferred most stable structure. This forms a link to the topic now to be dealt with – the cell membrane, which is an assembly of molecules held together by noncovalent bonds.

Other links are to Chapter 1 where the need for the first self-replicating system to be contained by a membrane was discussed and to Chapter 2 where the variety of membranous structures inside most eukaryotic cells was described.

Basic lipid architecture of membranes

The basic structure of all biological membranes is the lipid bilayer, the constituents of which are **polar lipids**. A polar lipid (Fig. 7.1) is an **amphipathic** or **amphiphilic** molecule (*amphi* = both kinds of) because it has a polar part, the so-called head

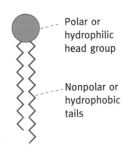

Polar or
hydrophilic
head group

Nonpolar or
hydrophobic
tails

Fig. 7.1 An amphipathic molecule of the type found in cell membranes.

group, and a pair of nonpolar hydrophobic tails. In the presence of water, such molecules are forced into structures in which the hydrophilic head groups have maximum contact with water, to maximize noncovalent bond formation. Conversely the hydrophobic tails are forced into minimum contact with water because such contact forces water into a higher-energy arrangement (page 38) and therefore is resisted. Also, the contact between hydrophobic tails maximizes van der Waals interactions (page 38), again minimizing the energy level of the structure.

When polar lipids of the type in membranes are agitated in an aqueous medium, one of the structures formed is called a **liposome** – a hollow spherical lipid bilayer. As mentioned in Chapter 1 the formation of liposomes may be relevant to the origin of cells. A lipid bilayer has two layers of polar lipids with their hydrophobic tails pointing inwards and their hydrophilic heads outwards in contact with water and each other as shown in Fig. 7.2. The presence of two hydrophobic tails favours this arrangement rather than that of a micelle, which is a solid sphere with all the tails pointing to its centre. Single tails favour micelle formation.

If a synthetic liposome is sectioned and stained with a heavy metal that attaches to the polar heads, the lipid bilayer appears in the electron microscope as a pair of dark parallel 'railway lines' owing to absorption of electrons by the metal stain. A living cell membrane treated in the same way has an identical appearance.

The polar lipid constituents of cell membranes

There are a number of membrane polar lipids, structures that look quite different from one another on paper, but in space-filling models they all have the same basic shape as the molecule shown in Fig. 7.1, with a polar head and two hydrophobic tails. Before dealing with the structures of the various polar lipids, a brief discussion of the nature of lipids in general may be useful.

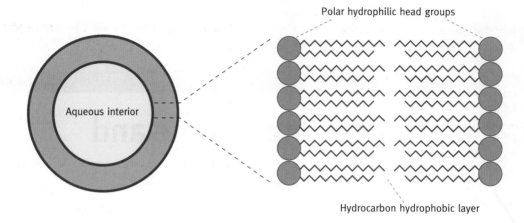

Polar hydrophilic head groups

Aqueous interior

Hydrocarbon hydrophobic layer

Fig. 7.2 A synthetic liposome made of a lipid bilayer structure.

A lipid is a fat. **Neutral fat** is derived from fatty acids; *neutral fats never occur in membranes*. A fatty acid has the structure RCOOH where R is a long hydrocarbon chain. The most common lengths are C_{16} and C_{18}. These may be called hexadecanoic and octadecanoic acids, respectively, if fully saturated or, by their common names, palmitic and stearic acids. They can also be represented as 16:0 and 18:0, indicating the number of carbon atoms and the number of double (unsaturated) bonds.

Stearic acid (C_{18}) has the structure:

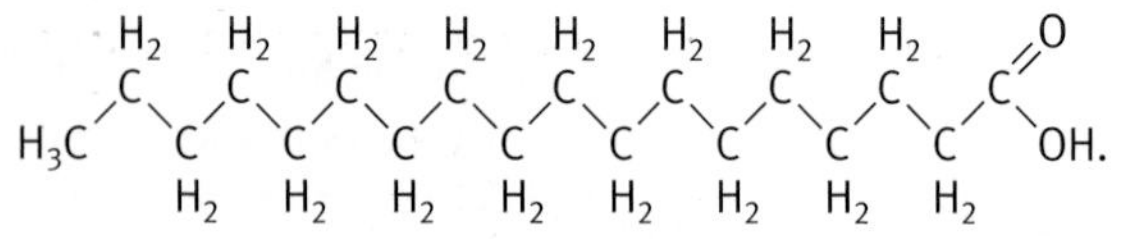

For convenience, saturated fatty acids are often represented simply as:

or $CH_3(CH_2)_{16}COOH$.

At neutral pH as, say, the sodium salt, a fatty acid is a soap – what you wash yourself with:

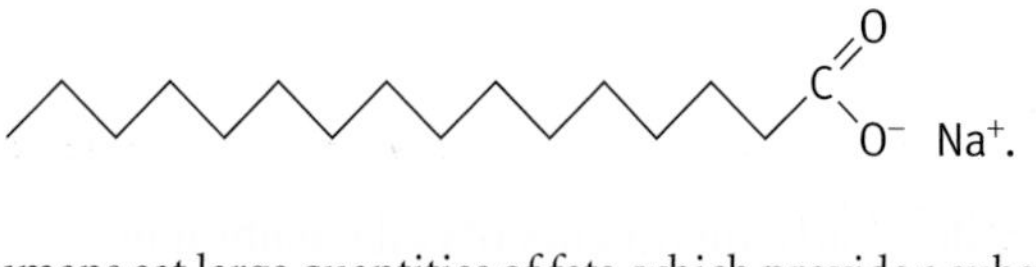

Humans eat large quantities of fats, which provide a substantial proportion of their food, but soaps are not good to eat. They do not taste nice and their detergent action could disrupt cell membranes. Instead we eat mainly neutral fat, more properly called **triacyglycerols (TAGs)** in which three molecules of fatty acids are esterified to the three hydroxyl groups of glycerol as shown in Fig. 7.3. A carboxylic acid attached in ester linkage is an acyl group, and there are three acyl groups attached to one

Ester bond

Glycerol — Three molecules of fatty acid — One molecule of neutral fat (or triacylglycerol (TAG))

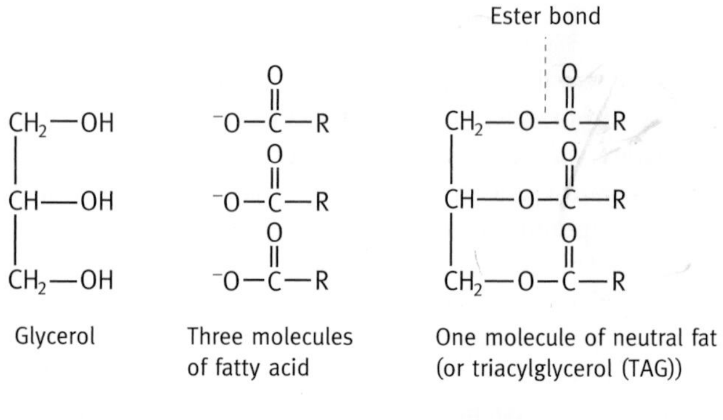

Fig. 7.3 A triacylglycerol and its component parts.

glycerol molecule, hence the name triacylglycerol. The term triglyceride is sometimes used but is not chemically correct.

If neutral fat is boiled with NaOH or KOH, the ester bonds are hydrolysed, forming soaps and glycerol, which is how soap is made. Neutral fat is suitable as a food; it is not a detergent. As stated, *neutral fat does not occur in membranes*. We have discussed it to help in understanding what a polar lipid is by contrast. A TAG molecule has no polar groups and therefore forms oily droplets or particles in water – it could never produce a lipid bilayer structure. The membrane polar lipids can be classified according to their structures.

Glycerophospholipids

Membrane lipids containing glycerol are the most abundant form and are known as **glycerophospholipids**. They are based on glycerol-3-phosphate.

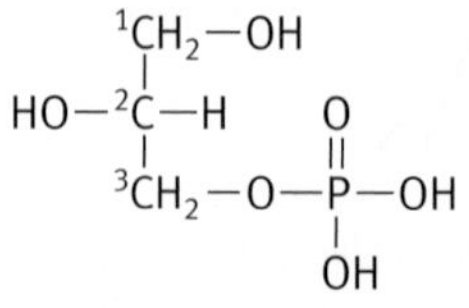

sn-Glycerol-3-phosphate

The prefix sn, for stereospecific numbering, refers to the nomenclature system used here. The central carbon atom of this compound is asymmetric; in glycerol-3-phosphate as found in the cell, the secondary hydroxyl is represented in this projection to the left and the carbon atoms numbered from the top. If two fatty acids are esterified to the primary (C-1) and secondary (C-2) hydroxyl groups of glycerol-3-phosphate, the product is **phosphatidic acid**.

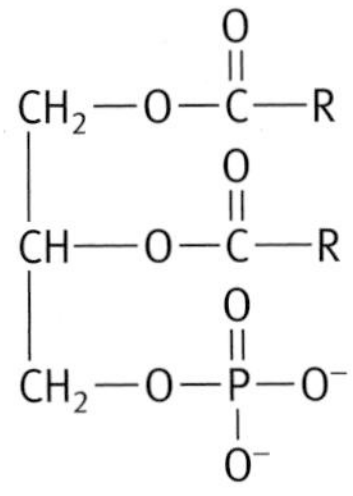

Phosphatidic acid

We can attach other polar molecules to the phosphoryl group. If phosphatidic acid is attached to something else (e.g. 'X') it becomes a **phosphatidyl group** (e.g. phosphatidyl-X) with the structure:

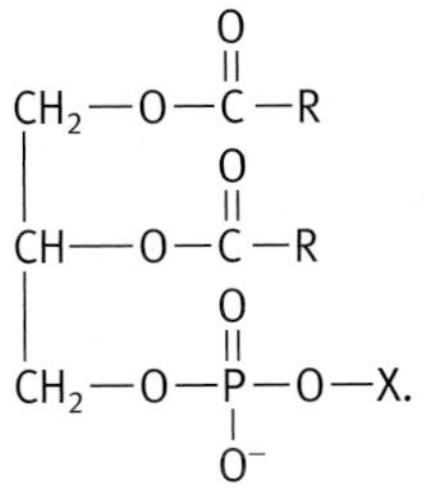

If X is also a highly polar molecule we now have an extremely polar head group.

The structure above gives a misleading impression of the shape of the molecule. A space-filling model would look somewhat like this:

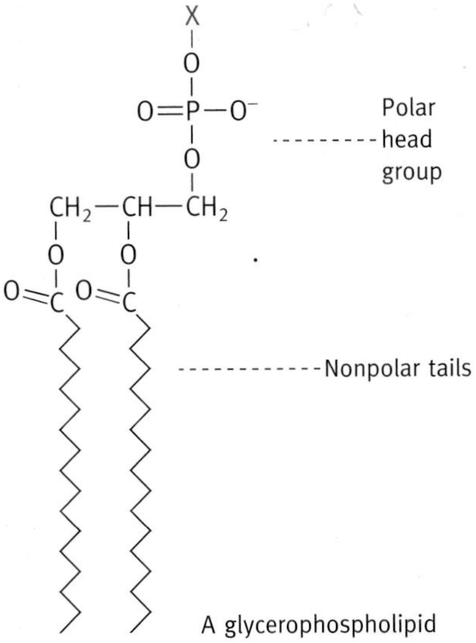

A glycerophospholipid

What are the polar groups attached to the phosphatidic acid?

Living cells use a variety of structures to form their membranes, some quite complex, but all structures used have *the same overall shape and amphipathic properties*. Several different polar groups are attached to different phosphatidic acid molecules.

One polar substituent is **ethanolamine** ($HOCH_2CH_2NH_3^+$) giving the phospholipid **phosphatidylethanolamine** (PE for short, or the trivial name is **cephalin**). Its structure is:

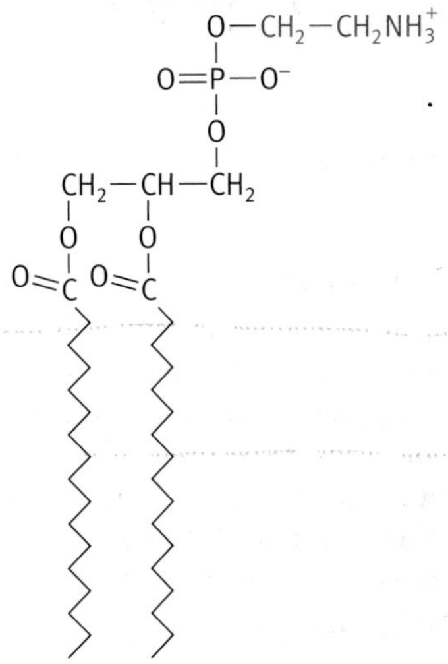

If the nitrogen of ethanolamine is trimethylated, it is **choline**:

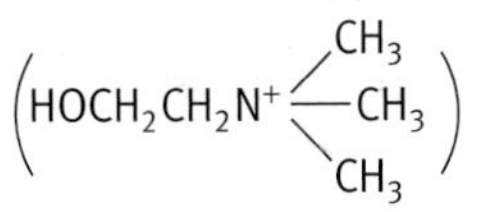

and the phospholipid derived from it is **phosphatidylcholine** (PC) or **lecithin**. Its structure is the same as that given for PE above, apart from the three methyl groups on the nitrogen atom.

If the ethanolamine is carboxylated we have a **serine** substituent giving **phosphatidylserine** or PS (serine is $HOCH_2CHNH_2COOH$). The attachment is like this:

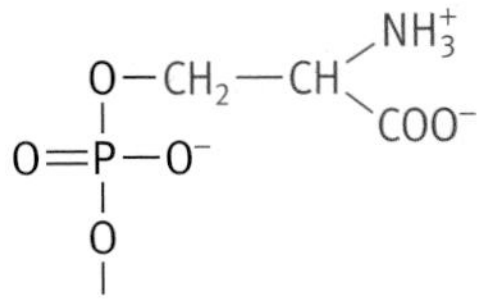

Another quite different polar substituent is the hexahydric alcohol, **inositol**:

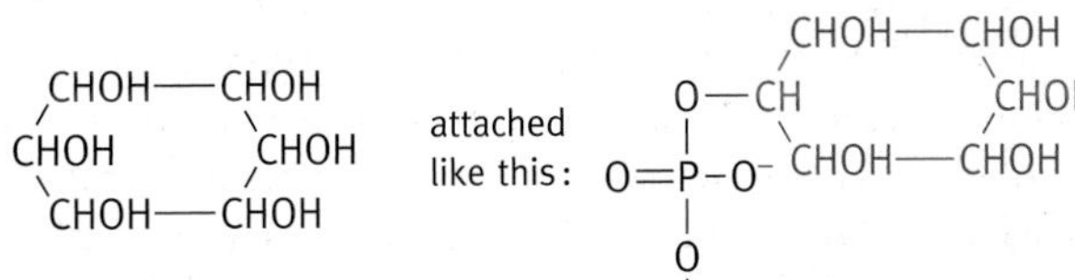

Inositol

giving **phosphatidylinositol**, usually abbreviated to PI.

Different ways have been used to ring the changes on glycerol-based polar lipids to form components of lipid bilayers. **Cardiolipin** or **diphosphatidylglycerol** has two phosphatidic acids linked by a third glycerol unit.

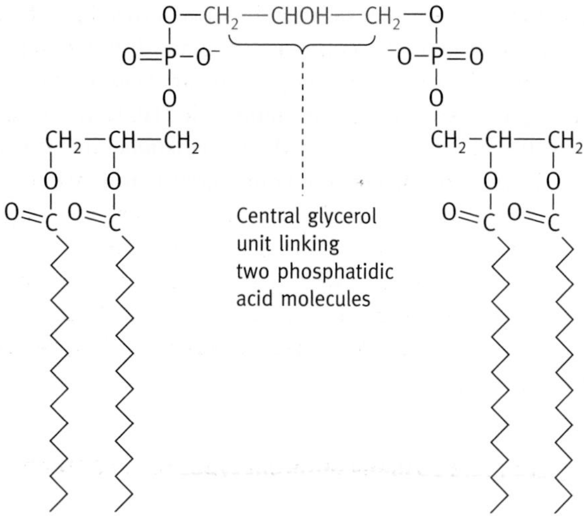

Central glycerol unit linking two phosphatidic acid molecules

The resultant structure still has the amphipathic shape to fit into lipid bilayers. It occurs in the inner mitochondrial membrane and in bacterial cell membranes.

All of the above polar lipids are based on glycerol. However, the process of evolution has produced molecules of almost identical overall shape derived from a different structure called **sphingosine**.

Sphingolipids

The basic molecule of **sphingosine** is not all that different from glycerol; the middle (C-2) hydroxyl group of glycerol is replaced by an —NH_2 group and a hydrogen on carbon atom 1 is replaced by a C_{15} hydrocarbon group – more or less a permanently 'built-in' hydrocarbon tail. One tail is not enough for a bilayer constituent. A second tail, a fatty acid, is attached to the central —NH_2 group by means of a —CO—NH— linkage forming a **ceramide**, similar in shape to a diacylglycerol.

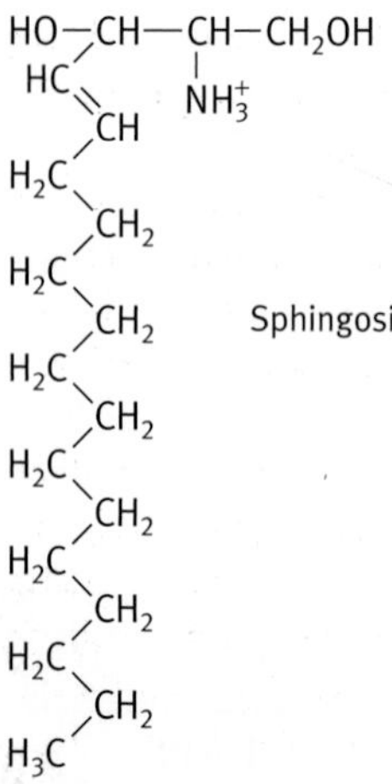

Sphingosine

A ceramide

If now we add a phosphorylcholine group, as in lecithin, the product is **sphingomyelin**:

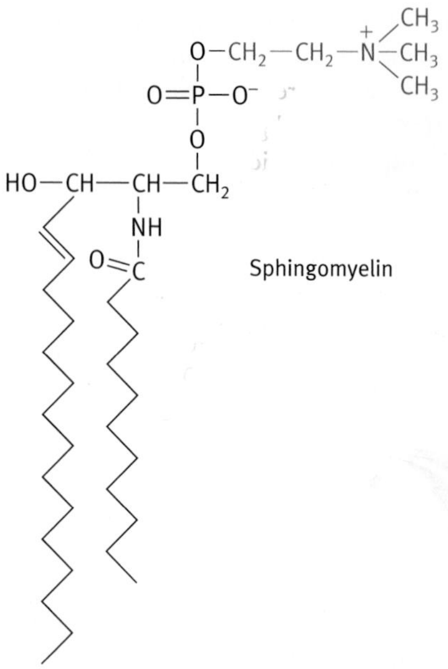

Sphingomyelin

which is similar in shape to lecithin. It is prevalent in the myelin sheath of nerve axons.

Evolution has been described as a tinkerer – it goes on modifying things and if the change is beneficial it is preserved. It has tinkered with the sphingosine-based polar molecules by adding different polar groups to the free —OH of the ceramide molecule. Sugars are highly polar and are used for this purpose producing **glycosphingolipids**.

Using a single sugar as the polar group we get a **cerebroside**, important in brain cell membranes. Cerebrosides contain either glucose or galactose. The latter is a stereoisomer of glucose in which the C-4 hydroxyl is inverted.

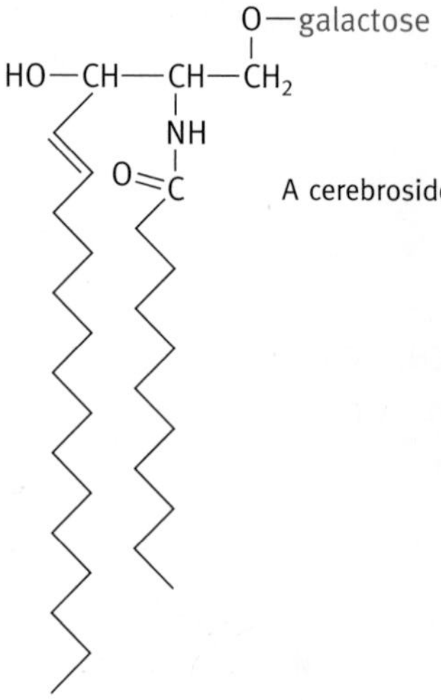

A cerebroside

There is a wide variety of sugars in nature. Some are **amino sugars** such as glucosamine (whose structure is given below) or derivatives of them.

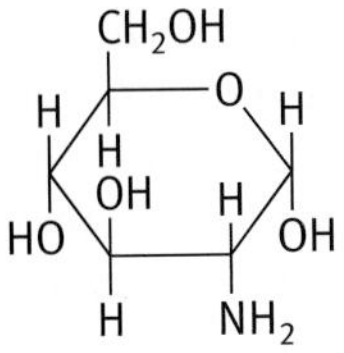

Combinations of the different sugars (**oligosaccharides**) can produce a variety of molecules in which a small number of different sugars are linked together in a branched oligosaccharide structure. (*Oligo* = few, so an oligosaccharide is a small polymer of sugars – perhaps 3–20 sugars rather than the hundreds or thousands found in polysaccharides.) Such an oligosaccharide is highly polar. When one of these is attached to a ceramide we have a **ganglioside**.

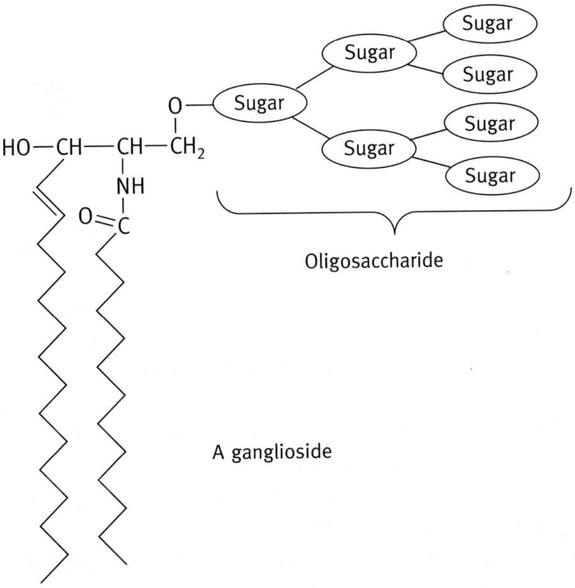

Oligosaccharide

A ganglioside

N-Acetylglucosamine and **sialic acid** are shown below because they have special interest as components of gangliosides.

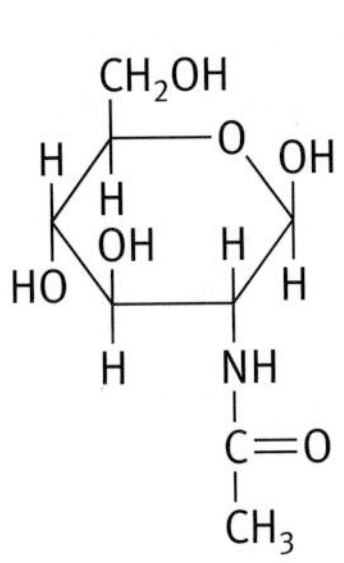

N-Acetylglucosamine 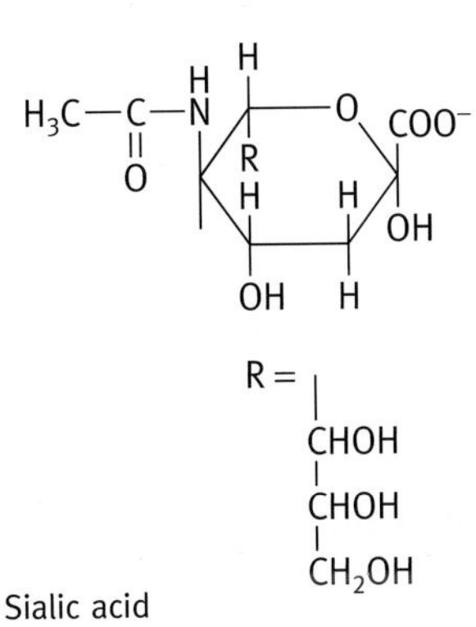   Sialic acid
(also called *N*-acetylneuraminic acid)

Sialic acid is involved in infections by the influenza virus. Gangliosides, based on sphingosine, are what distinguishes the human blood groups O, A, and B.

Membrane lipid nomenclature

The names of individual polar lipids have been given above, but alternative collective terms are sometimes used. The terms **membrane lipids** and **polar lipids** include any lipid found in cell membranes. A **phospholipid** is any lipid containing phosphorus. Those based on glycerol are **glycerophospholipids**, as opposed to sphingomyelin which is a single specific sphingosine-based phospholipid. The ceramide-based membrane lipids with carbohydrate polar groups and lacking any phosphoryl group are called **glycolipids** or **glycosphingolipids** to indicate that they are based on sphingosine. A plasmalogen is a glycerophospholipid in which one of the hydrophobic tails is linked to glycerol by an ether bond (not shown). (Cholesterol, another membrane component described below, is also classified as a lipid.)

Why are there so many different types of membrane lipid?

It is not fully understood why there are so many different types of membrane lipid. The different membrane lipids confer different properties on the membrane surface. The choline substituent of lecithin (PC) has a positive charge, the serine of PS is a zwitterion (page 45), and the carbohydrate of a cerebroside has no charge. Different cells have quite different membrane lipid compositions. Cerebrosides and gangliosides are common in brain cell membranes, and the cell membranes of the myelin sheath of nerve axons are rich in glycosphingolipids. As well as different cells having different lipid compositions, the outer and inner halves of the bilayer of the one membrane are different from one another and different membranes within the cell have different compositions. For example, glycolipids are always on the outer side of the cell membrane so that their sugars point outwards from the cell into the external aqueous environment. This asymmetry is preserved by the fact that transverse movement of lipids, known colloquially as 'flip-flop' (from one side of the membrane to another), is severely restricted; such movement would involve transferring the polar heads through the central hydrocarbon layer to get to the other side. This is energetically unfavourable. Proteins catalysing energy-dependent membrane flip-flop are involved in the creation and maintenance of this asymmetry.

The limited flip-flop movement of membrane lipids contrasts with their potential for rapid lateral movement within the plane of the bilayer. The assembly of lipids into the bilayer structure does not involve covalent bonds and, in general, they move around freely.

Some particular membrane lipids (e.g. PI) have a special role as the source of chemical intracellular signals. Also cell signalling in some cases is dependent on specific proteins associating with patches of the cytoplasmic membrane rich in particular lipids (Chapter 27). Specific protein–lipid associations in general appear to be a probable reason for the diversity of membrane lipids. Indeed the association and dissociation of proteins from membranes is increasingly recognized to play a vital part in the regulation of the activities of cells. Given the wide variety of different membrane proteins and their functions, it is not surprising that different types of membrane lipids are needed.

Some membrane lipids are of special medical interest. For example, there is a clinical interest in gangliosides, because of genetic diseases called **glycosphingolipidoses**. One example is Tay–Sachs disease (Box 26.1) in which there is mental retardation and early death. Another is Gaucher's disease. In these, the glycosphingolipids are not broken down properly in lysosomes (page 417) and residues of them accumulate causing severe brain disorders.

The fatty acid components of membrane lipids

The fatty acyl 'tail' components of a membrane lipid such as lecithin vary. In length they may range from C_{14} to C_{24} (almost always an even number, for reasons explained in Chapter 13), but C_{14}–C_{18} are most common. A variety of fatty acids, often unsaturated with one or more double bonds, are present in membrane lipids. The usual ones are C_{18} and C_{16} with one double bond in the middle (oleic and palmitoleic acids, respectively), linoleic (C_{18} with two double bonds), and arachidonic acid (C_{20} with four double bonds). The **nomenclature** of such acids is dealt with in a later chapter (page 227).

The two fatty acyl residues in a single phospholipid molecule may be the same or different, but usually in a glycerophospholipid the fatty acid attached to the C-1 —OH group is saturated and that attached to the C-2 —OH group is unsaturated. The degree of saturation of fatty acid tails is of great importance because the central hydrocarbon core of the bilayer must be a two-dimensional fluid rather than solid. Unsaturated hydrocarbon tails reduce the temperature at which a bilayer loses its fluidity.

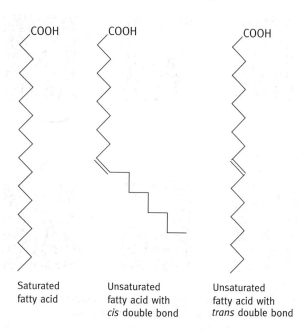

Saturated fatty acid | Unsaturated fatty acid with *cis* double bond | Unsaturated fatty acid with *trans* double bond

Membrane fluidity is essential to allow lateral movement of transmembrane proteins so that they can interact with one another. In addition, such proteins undergo conformational changes needed for ion channels, transporters, and receptors to function. Without a fluid bilayer such changes might be difficult to accommodate. (The proteins referred to are described later.) A saturated fatty acid tail is straight, but a double bond in the *cis* configuration introduces a kink into it as illustrated, and kinks favour fluidity. Natural unsaturated fatty acids are almost always in the *cis* configuration (Box 7.1).

The saturated chains that are straight pack together comfortably and interact, but the kinked ones cannot do so and stay fluid at lower temperatures. This physical effect of unsaturation can be seen by comparing hard mutton fat with olive oil. Both are TAGs but the olive oil has unsaturated fatty acyl tails.

The importance of maintaining membrane fluidity is illustrated by the fact that bacteria adjust the degree of unsaturation of the fatty acid components of their membrane bilayers according to growth temperature. In special situations, such as during hibernation, animals modulate the degree of saturation of cell membrane components to cope with lower body temperature.

What is cholesterol doing in membranes?

From its structure, as conventionally drawn (Fig. 7.4(a)), cholesterol seems an improbable membrane constituent. However, the conformation of the ring system, shown in Fig. 7.4(b), shows that the molecule is elongated, the steroid nucleus being rigid and the hydrocarbon chain flexible. It is an amphipathic molecule, the —OH group being weakly polar.

Cholesterol in membranes acts as a 'fluidity buffer'. On the one hand, at temperatures above the melting point of a lipid bilayer, it reduces fluidity because of its rigid structure, but, on the other hand, at lower temperatures it increases fluidity because it prevents close-packing of the flexible fatty acid tails. It thus 'buffers' fluidity. The observed effect is that cholesterol 'blurs' the melting point of a lipid bilayer. Without cholesterol, the transition from solid to liquid is sharper than when cholesterol is present. A red blood cell membrane may be about 25% cholesterol. On the other hand, bacterial membranes have no cholesterol and animal cell mitochondria have very little. Plants contain other sterols, known as **phytosterols**, but no cholesterol. Fungal membranes contain ergosterol, not cholesterol. (See Box 7.5 where amphotericin B is discussed. This has affinity for ergosterol over cholesterol.)

The self-sealing character of the lipid bilayer

The lipid bilayer is effectively a two-dimensional fluid. The bilayer gives cells flexibility and a self-sealing potential, the latter being essential when a cell divides (Fig. 7.5(a)).

This versatility is also essential in **endocytosis** – the process by which cells can engulf material (Fig. 7.5(b)). The reverse can

BOX 7.1 *Trans* fatty acids

As is widely known, there is considerable concern that intake of *trans* fatty unsaturated fatty acids are harmful to health. Natural foods have very little of these present. They originate mainly from ruminant bacteria in cattle and sheep and are associated with the meat. The main source in commercial foods comes from partial hydrogenation of unsaturated oils. This is used to solidify the oils to varying extent for use in the food industry such as manufacture of pastries. Dietary regulations or guidelines are being introduced in various countries to minimize or eliminate partially hydrogenated foods.

The structures of *cis* and *trans* unsaturated fatty acids are given on page 106. It can be seen that the *cis* variety are kinked while the *trans* isomers are straight chain and in this respect resemble saturated fatty acids. There is good evidence that *cis* unsaturated acids are essential to maintain the liquidity of cell membranes presumably by preventing the tight side-by-side packing of the fatty acid tails of membrane polar lipids.

Since *trans* fatty acids have the straight chain structure it might be presumed that they might be similar to saturated fatty acids in this respect, but there is no direct evidence that this is a significant effect.

However, there is evidence that *trans* fatty acids have undesirable physiological effects and may contribute to cardiovascular disease, though the biochemical mechanism(s) by which these occur are not understood. Dietary studies indicated that *trans* fats, compared with the same amount of saturated or *cis* unsaturated fats, raises the level of blood **low-density lipoproteins (LDLs)** and lowers level of **high-density lipoproteins (HDLs)**. LDLs are known as 'bad' cholesterol, and HDLs as 'good' cholesterol. The reduction in the ratio of HDL cholesterol to total cholesterol is well known to be associated with an increased risk of cardiovascular disease. The *trans* fats were also found to raise the level of serum triglycerides as compared with the effects of comparable amounts of saturated or *cis* fats.

There are other effects of *trans* fats reported; these are reviewed in the Mozaffarian reference. The authors of this review concluded that near elimination of manufactured *trans* fats from the diet could prevent in the USA many thousands of cardiovascular events each year.

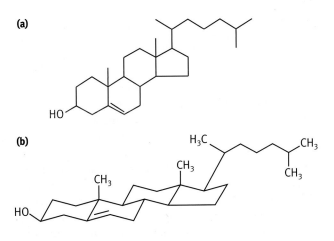

Fig. 7.4 (a) Structure of cholesterol as conventionally drawn. **(b)** Structure drawn to give a better indication of the actual conformation of cholesterol.

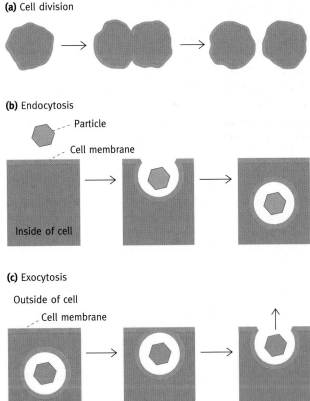

Fig. 7.5 (a) Cell division; **(b)** endocytosis; **(c)** exocytosis for the release of, for example, a digestive enzyme from cells.

also happen. If the cell needs to secrete a substance such as a digestive enzyme it is synthesized inside the cell and enclosed within a membrane-bounded vesicle that migrates to the plasma membrane, fuses with it, and releases its contents to the outside. The process is called **exocytosis**. A good example of this is the secretion of digestive enzymes into the intestine by pancreatic cells (Fig. 7.5(c)).

Permeability characteristics of the lipid bilayer

The ability of small molecules to diffuse through a lipid bilayer is related to their fat solubility. Strongly polar molecules such as ions traverse the bilayer extremely slowly if at all. Large, more weakly polar molecules such as glucose penetrate very slowly, although smaller ones such as ethanol or glycerol diffuse across more readily. Ionized groups of polar molecules and inorganic ions are surrounded by a shell of water molecules which must be stripped off for the solute to pass through the lipid bilayer hydrocarbon centre, and this is energetically unfavourable. The lipid bilayer is therefore almost impermeable to such molecules and to ions, but slow leaks inevitably occur.

Water molecules, despite being polar, apparently pass through the lipid bilayer with sufficient ease for the needs of some cells. Presumably this is due to the small size of the molecule and its lack of charge, but there is some uncertainty as to how it traverses the bilayer. Cells involved in water transport, such as those of renal tubules and secretory epithelia, have a transmembrane protein, **aquaporin**, which allows free movement of water. Many cells have **porins** (see below). The lipid bilayer also readily allows gases in solution such as oxygen to diffuse through it.

A high proportion of molecules of biochemical interest are polar in nature and cannot pass through the lipid bilayer at rates commensurate with cellular needs. This means that although the bilayer structure is ideal for holding in the contents of a cell, special arrangements must be made to permit rapid movement of molecules across the membrane as needed. Membrane proteins are responsible for this.

Membrane proteins and membrane design

A beautiful aspect of cell membrane structure is that it provides a flexible system that can be modified in the evolutionary sense. Different membranes have different functions and hence have different protein molecules in their membranes. If a new protein is evolved to act in a membrane it can be inserted into the membrane. It is the ultimate in flexible design. In the cell, membrane proteins are inserted into the membrane as the proteins are synthesized. Proteins experimentally isolated from membranes have been incorporated into synthetic phospholipid liposomes where they function exactly as in their parent cell membrane.

The membrane structure containing proteins is called a **lipid fluid mosaic** (Fig. 7.6). In this, laterally mobile protein molecules are present in a two-dimensional lipid layer. Such proteins are called **integral proteins** (labelled I in Fig. 7.6). Other proteins are called **peripheral proteins** (labelled P in Fig. 7.6) because they associate with the periphery of the membrane (see below).

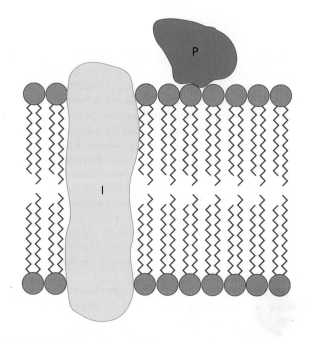

Fig. 7.6 Lipid fluid mosaic model for membrane structure. I, integral protein; P, peripheral protein.

Structures of integral membrane proteins

Most integral membrane proteins are so structured that they are held in the lipid bilayer by noncovalent forces. Integral membrane proteins with α helical structures are amphipathic with ends made of hydrophilic amino acids in contact with water and the middle section of hydrophobic ones as shown diagrammatically in Fig. 7.7(a). The essential feature of such integral membrane proteins is that the hydrophobic section corresponds in length with the hydrocarbon middle zone of the membrane lipid bilayer. The parts of the protein projecting from the membrane that are in contact with water, or the polar head groups of membrane lipids, have hydrophilic residues as illustrated in Fig. 7.7(b). This is a diagram of **glycophorin**, a major component of the erythrocyte membrane. On the external surface of the membrane is a large N-terminal section of polypeptide, rich in hydrophilic amino acids; attached to serine, threonine, and asparagine residues are carbohydrates (page 56), making this part of the protein extremely hydrophilic. On the inside face of the bilayer is a shorter C-terminal section devoid of carbohydrate attachments but also containing hydrophilic amino acids. Connecting the two external sections and spanning the lipid bilayer is a stretch of 19 amino acid residues which, in the form of an α helix, are sufficient to just span the hydrocarbon central layer of the membrane, which is about 30 nm wide. In this section of the chain, isoleucine,

(a) Polypeptide
chain of protein

| Hydrophilic amino acids | Hydrophobic amino acids | Hydrophilic amino acids |

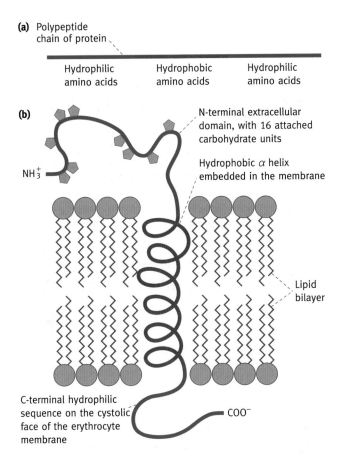

(b)

N-terminal extracellular domain, with 16 attached carbohydrate units

NH_3^+

Hydrophobic α helix embedded in the membrane

Lipid bilayer

C-terminal hydrophilic sequence on the cystolic face of the erythrocyte membrane

COO^-

Fig. 7.7 (a) The structural plan of an integral membrane protein. **(b)** Glycophorin, a protein of the erythrocyte membrane. The α helix, containing about 19 hydrophobic amino acid residues, is approximately 30 Å in length, which is sufficient to span the nonpolar interior of the lipid bilayer.

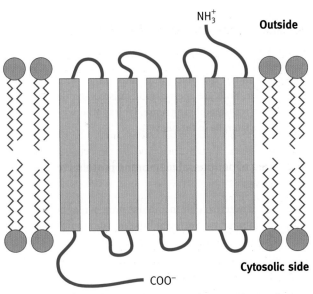

NH_3^+ **Outside**

COO^-

Cytosolic side

Fig. 7.8 Topological representation of bacteriorhodopsin, with seven α helices spanning the lipid bilayer. The α helices are not actually arranged linearly, as shown here for convenience, but are clustered compactly together.

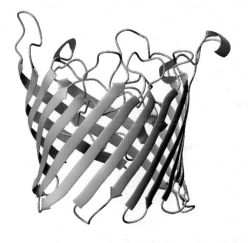

Fig. 7.9 The β-barrel structure of a bacterial porin (Protein Data Bank code 1BH3). β strands are yellow, α helices are red, and everything else is grey.

leucine, valine, methionine, and phenylalanine, all strongly hydrophobic and good α-helix-formers, predominate. There are no strongly hydrophilic residues, except in those cases where the protein is a transmembrane aqueous pore lined with hydrophilic residues (see below).

In some proteins, the polypeptide chain loops back and crosses the bilayer several times – for this, alternating hydrophilic and hydrophobic sections are required, each of the hydrophobic sections being about 19 amino acids long and forming α helices, which nicely span the bilayer. One of the best-studied examples is **bacteriorhodopsin**, a protein present in the membrane of the purple bacterium *Halobactium halobium* found in brine ponds. The protein crisscrosses the membrane seven times, forming a cluster of seven α helices spanning the membrane and connected by hydrophilic loops (Fig. 7.8). The cluster has a light-absorbing pigment at its centre to capture light energy which drives the pumping of protons from the cell to the outside. The energy of the proton gradient so formed is used to drive the synthesis of ATP by a mechanism to be described in Chapter 12.

The transmembrane **porins** are different. These are proteins found in most cell plasma membranes, including the outer membrane of certain bacteria and in the outer mitochondrial and chloroplast membranes. These outer membranes are relatively porous to small molecules due to porins. These are water-filled channels with a β barrel structure (Fig. 7.9) varying in different cases from 8 to 20 antiparallel strands with alternating hydrophilic and hydrophobic residues, arranged so that the hydrophobic residues are outside, in contact with the hydrophobic lipid bilayer components, while the hydrophilic residues point inwards. There is some selectivity in what the porin channels conduct; in bacteria some allow anions to cross and others conduct sugars.

If the protein tended to move out of the membrane, the hydrophobic groups of the protein would come into contact with water and hydrophilic groups into contact with the hydrocarbon layer. Both are energetically resisted and the protein is fixed in the transmembrane sense. Unless otherwise restrained it can move laterally in the bilayer.

Anchoring of peripheral membrane proteins to membranes

A peripheral water-soluble membrane protein may be associated with a membrane by hydrogen bonding and ionic attractions. However, an alternative method exists in the case of certain proteins. In these, the proteins have attached to them a fatty acid whose hydrocarbon chain is inserted into the lipid bilayer, thus anchoring the protein to the membrane (Fig. 7.10).

As so often is the case, evolutionary tinkering has produced variations on the one theme. *N*-**Myristoylation** of the protein is one in which myristic acid (C_{14}) is linked to an N-terminal glycine of the protein:

$$R—CO—NH—CH_2—CO—polypeptide.$$

In addition, C_{14}, C_{16}, and C_{18} fatty acids may anchor proteins in a similar fashion, but attached to serine or threonine groups by ester linkage or to cysteine by thioester linkage. More complex acids are also found in ether linkage for the same purpose.

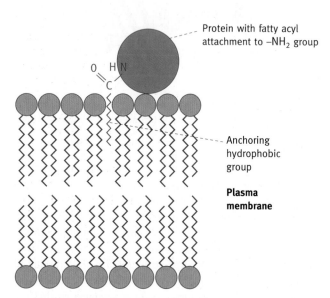

Fig. 7.10 A fatty acid anchoring molecule joined by CO—NH linkage to the amino group of a protein. Alternative linkages such as ester and thiol ester can occur with appropriate side groups of the protein.

Glycoproteins

Many membrane proteins have attached to them branched oligosaccharides, on the external surface. The carbohydrates are covalently attached to the side chains of asparagine or serine (page 56). The sugars which make up the oligosaccharides include various isomers of glucose and amino sugars.

In glycophorin of the red blood cell (Fig. 7.7(b)), half the weight of the protein molecule is carbohydrate. The exact function of these carbohydrate attachments is an area of uncertainty but may, in some cases, have recognition roles on cell surfaces. Glycoproteins are found elsewhere than in membranes such as in the blood, as one example.

Functions of membranes

Membranes have a wide variety of functions; for the moment we will deal with the plasma membrane surrounding the cell. We will come to the internal membranes in later chapters.

The main functions of such membranes, apart from retaining cell contents, are:

- transport of substances in and out of the cell
- ion transport and nerve impulse conductance
- cell signalling (a major topic dealt with in Chapter 27)
- maintaining the shape of the cell
- cell–cell interactions.

Membrane transport

Many, or most of the substances that must be taken up by the cell or have to be removed from the cell, cannot diffuse across the lipid bilayer. The lipid bilayer has to be impermeable to most molecules other than small hydrophobic ones because it must retain the needed cell constituents against leakage. For transport of polar structures such as sugars, amino acids and inorganic ions, specific membrane proteins are needed. Although inorganic ions are very small, they cannot diffuse through a cell membrane because they are surrounded by a shell of water molecules and, as explained earlier, removal of these is an energy barrier. Simple holes in the membrane will not do – they would allow nonspecific leakage in and out and would be lethal. Usually transport systems handle only specific molecules so there are many different transport systems and therefore many different transport proteins.

The problem is a major one for all life forms. Bacteria must take in essential nutrients. So must all cells but the problem for say a mammal is greatly compounded because the chemical environment in the blood is very different from that inside cells. (Possibly blood evolved from sea water pumped through the body.) The sodium and potassium ion concentrations of blood and intracellular space are different and if equilibrated would be lethal. A remarkable situation is that animals use the Ca^{++} ion as one of the most potent regulators of much of cell chemistry, and yet Ca^{++} is present extracellularly at high levels. For this to work calcium levels inside the cell must be kept extremely low. Because there is such a steep gradient of the ion across the membranes, signals can cause instant fluxes of the ion to be transported into the cell. Pumps then immediately withdraw the calcium to terminate the signal as required. The result is that an animal as complex as a mammal is a mass of never-ending membrane pumping. Bearing in mind that nerve activity is also a matter of ion movement across membranes (described shortly) the brain is energetically an expensive organ to maintain. The energy ultimately derives from ATP and perhaps 30% of your total energy expenditure is on membrane pumping in general.

Active membrane transport

We can separate transport systems into active and passive types. An active sytem is transport that requires performance of work, which usually involves the hydrolysis of ATP directly or indirectly. It requires work to be done because the movement of substances occurs against concentration gradients. However, rapid influxes of ions into a cell can occur by simply opening a channel to allow the ion to rush through driven by a concentration gradient. This happens in nerve impulse transport. The maintenance of the gradient in these cases requires work.

The amount of energy required to transport a solute into a cell, against a gradient, can be calculated from the equation for chemical reactions given in Chapter 3 (see Box 7.2).

We will now deal with various types of active transport.

The sodium/potassium pump

A good example of active transport is the Na^+/K^+ pump present in animal cells. This pumps Na^+ out of cells and K^+ into cells using ATP energy. Animal cells have a high internal K^+ concentration (140 mM) and a low Na^+ concentration (12 mM) as compared with those of the blood (4 mM and 145 mM respectively). The ion gradients are necessary for electrical conduction in excitable membranes and, in some cases, for driving solute transport across membranes.

Mechanism of the Na^+/K^+ pump

The Na^+/K^+ pump is also called the Na^+/K^+ ATPase because ATP is hydrolysed to ADP + Pi as Na^+ is pumped out of the cell and K^+ in. The protein is a complex of four polypeptides or subunits. There are two identical α subunits and two β subunits. Some proteins have the property of slightly changing their shape when specific ligands bind to them. This may be due to a change in the conformation of a protein, or to a relative change in position of subunits of a protein complex. The covalent attachment of a phosphoryl group to the Na^+/K^+ ATPase by transference from ATP also causes a **conformational change**. Such a change can alter the ability of a protein to bind a given ligand (if it has been designed to do so by evolution). Thus, in one form, the protein of the pump binds Na^+ but not K^+ and another form K^+ but not Na^+.

In the model shown in Fig. 7.11, the Na^+/K^+ ATPase protein exists in two conformations. Form (a) is open to the interior of the cell and binds Na^+ but not K^+. ATP phosphorylates this, (on an aspartyl residue) yielding ADP + phosphorylated protein. In this form it is effectively open to the outside but no longer binds

---

BOX 7.2 Calculation of energy required for transport

$$\Delta G = \Delta G^{0'} + RT\,2.303\log_{10}\frac{[\text{products}]}{[\text{reactants}]}$$

For transport of a solute whose structure does not change, $\Delta G^{0'}$ is zero and the equation becomes:

$$\Delta G = RT\,2.303\log_{10}\frac{[C_2]}{[C_1]}$$

where C_1 is the concentration outside and C_2 is the concentration inside, taken in this example to be in the ratio of 10/1 so that

$$\Delta G = (8.315\times10^{-3}\text{ kJ mol}^{-1}\text{ K}^{-1})(298\text{K})2.303\log_{10}10/1$$
$$= 5.706\text{ kJ mol}^{-1}.$$

Thus the transport of 1 mol, under these conditions (at 25°C), requires 5.706 kJ. Note that the above calculation applies to the transport of an uncharged solute. With a charged solute, generation of an electrical potential requires an additional term to correct for this.

BOX 7.3 Cardiac glycosides

There is an interesting medical aspect. Cardiac glycosides are a group of compounds found in the foxglove plant (*Digitalis purpurea*). They are steroids or steroid glycosides; steroids resemble cholesterol in structure; glycosides are molecules with sugars attached. Such compounds inhibit the Na^+/K^+ ATPase by preventing the removal of the phosphoryl group from the transport protein. This 'freezes' the pump in one form (Fig. 7.11(b)) and stops the ion transport. The cardiac glycosides have long been used clinically as treatment for congestive heart failure. The compounds are lethal in sufficient amounts, but in appropriate doses, the partial inhibition of the Na^+/K^+ ATPase increases the Na^+ concentration inside heart muscle cells and thus lowers the Na^+ gradient from outside to inside. This has the effect of raising the cytoplasmic Ca^{2+} level because there is another system which transports Na^+ into the cell and Ca^{2+} out of the cell. The ejection of Ca^{2+} is driven by the Na^+ gradient (see below); if the latter is lowered by cardiac glycosides then ejection of Ca^{2+} is reduced and the internal level of Ca^{2+} rises which stimulates heart muscle contraction; the role of Ca^{2+} in contraction is dealt with in Chapter 8. Ouabain, the African arrow-tip poison, has similar effects to cardiac glycosides.

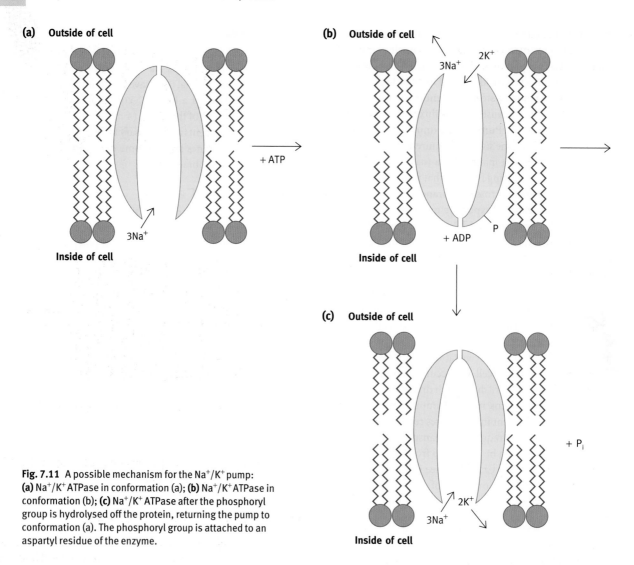

Fig. 7.11 A possible mechanism for the Na^+/K^+ pump: **(a)** Na^+/K^+ ATPase in conformation (a); **(b)** Na^+/K^+ ATPase in conformation (b); **(c)** Na^+/K^+ ATPase after the phosphoryl group is hydrolysed off the protein, returning the pump to conformation (a). The phosphoryl group is attached to an aspartyl residue of the enzyme.

Na^+ (which diffuses away) but does bind K^+ which therefore attaches. This gives form (b) (Fig. 7.11) (See also Box 7.3). The —P group is now hydrolysed from the protein giving Pi and the protein reverts to form (a) to which the K^+ no longer binds. The latter therefore enters the cell and more Na^+ attaches to the pump (Fig. 7.11(c)). Evidence from antibody studies shows that the whole protein does not revolve in the process as might seem an obvious mechanism.

The net result is that hydrolysis of ATP pumps Na^+ out and K^+ in. The ratio is 3 Na^+ out and 2 K^+ in.

The overall equation is:

$$3Na^+ (in) + 2 K^+ (out) + ATP + H_2O \rightarrow$$
$$3Na^+ (out) + 2 K^+ (in) + ADP + Pi$$

A large family of related ATP-dependent pumps transport specific solutes across membranes

These use much the same phosphorylating strategy as the sodium /potassium pump. One very important example is the Ca^{++}-ATPase of muscle. As described in the next chapter (page 131) vertebrate striated muscle is triggered to contract by a neuronal signal that causes liberation of Ca^{++} into the muscle cytoplasm (known as the sarcoplasm). This pump terminates the contraction by the ATPase almost instantly withdrawing the ion into a reservoir. Another example is provided by the ATP-driven proton pump which causes acidification of lysosomal vesicles.

ATP-binding cassette (ABC) proteins

One of the largest groups of transport systems is known as **ATP-binding cassette (ABC) proteins**. These are multidomain transport proteins all of which have two cytoplasmic ATP-binding domains (the unit being called a cassette) and two transmembrane ones. At the expense of ATP breakdown, small molecules like drugs are transported across the membrane. A wide variety of such systems exist and are found in both bacteria and eukaryotes. In eukaryotes a protein known as the multidrug resistance protein (MDR protein) is an ABC transporter. It ejects pharmacological agents used in cancer treatment. The cancer cell may increase its amount of MDR protein and become resistant to the therapy. An increase of such resistance to one drug may result in multiple drug resistance – hence its name. The protein is also referred to as **P (permeability) glycoprotein**.

A number of genetic diseases are associated with deficiencies in ABC transporters. The most important is **cystic fibrosis**, which affects 1 in 2500 Caucasians. The responsible gene called the **CFTR** gene has been isolated. It codes for an ABC transporter protein known as the **cystic fibrosis transmembrane conductance regulator protein**. This is a regulated chloride channel in epithelial cells. A number of different mutations of the CFTR gene are known to cause the disease but most often the regulatory domain of the CFTR is affected. In patients with the disease, chloride secretion from the cells is decreased in the lungs and the ducts of the pancreas and other glands. For reasons not understood this leads to the formation of viscous mucus which blocks the bronchioles of the lungs. Death may result from lung infection or heart failure.

Co-transport systems

The examples of ATP-dependent transport proteins given above all involve a solute simply being transported across a membrane at the expense of ATP hydrolysis. The ATP is *directly* involved with the transport. Such systems are called **uniports**. There is, however, a thermodynamically feasible alternative strategy. When the uniport establishes an ion or other gradient, this gradient has potential energy and ions can flow down the gradient. If appropriately harnessed the flow of ions can be made to perform work. The energy still originally comes from ATP but not directly. There are two basic types of co-transport systems.

The Na^+/K^+ ATPase, described above, produces a steep Na^+ gradient across the cell membrane, which has potential energy and, given a chance, the accumulated exterior Na^+ will flow back into the cell. In Na^+ cotransport the transport protein conducts Na^+ into the cell only so long as another substance hitches a thermodynamic ride on the back of this flow. Such a transport system is called a **symport** (it simultaneously transports two components). The glucose symport protein permits movement of Na^+ and glucose across the membrane *when both are present* but not when only one is present. It can thus transport glucose against a concentration gradient utilizing the potential energy of the Na^+ gradient. The Na^+ entering the cell is then pumped out by the Na^+/K^+ ATPase to maintain the Na^+ gradient so that it is ATP hydrolysis which indirectly supplies the energy for glucose uptake. Absorption from the gut of glucose and amino acids occurs by this mechanism (Fig. 7.12), separate symport proteins being required for the different substances transported.

Antiport systems also exist. As mentioned above (Box 7.3), in connection with cardiac glycoside action on the heart, the cotransport of Na^+ can be used in another way – to pump out Ca^{2+}. This is an **antiport** system (Fig. 7.13). Na^+ is transported into the cell only if at the same time as Ca^{2+} is transported out.

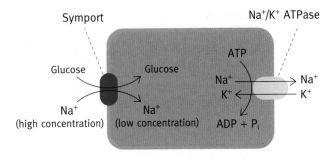

Fig. 7.12 The Na^+/glucose cotransport system – a symport.

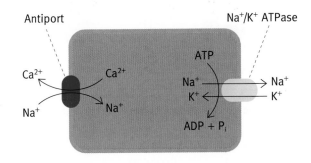

Fig. 7.13 The Na^+/Ca^{2+} cotransport system – an antiport.

Again the energy in the Na^+ gradient established by the Na^+/K^+ ATPase system is the driving force.

Passive transport or facilitated diffusion

In passive transport, a protein permits the movement of a substance across the membrane so that the movement will be down whatever concentration gradient exists across the membrane; no energy is involved in the actual transport process. This is **facilitated diffusion**. We have already described **porins**, the hydrophilic β barrel channels which allow this. Another good example is the anion-transport protein in red blood cells which lets HCO_3^- and Cl^- ions pass through the membrane in either direction (Fig. 7.14). The purpose of this was discussed earlier (page 67). Another important example of facilitated diffusion is the glucose transporters possessed by many animal cells. They allow glucose to diffuse passively across the membrane. When we come to discuss the utilization of blood glucose by cells these become very important.

Gated ion channels

An important class of passive-transport systems is that of the gated ion channels. These are aqueous pores, highly selective for specific ions, that open and close on receipt of a signal. They are found in large numbers and of varying specificities in most membranes, and the different channels respond to different

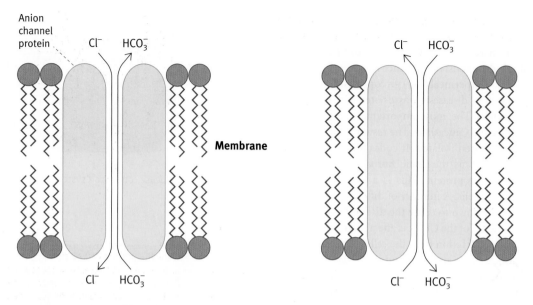

Fig. 7.14 An anion channel of red blood cells. Cl^- and HCO_3^- may move in either direction, according to concentration gradients across the membrane. The counterflow of the ions prevents any electrical potential developing across the membrane.

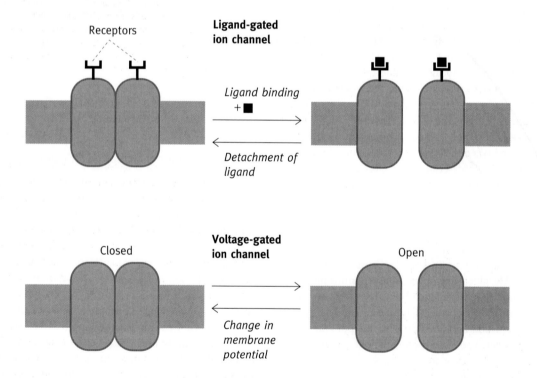

Fig. 7.15 Ligand-gated and voltage-gated ion channels. Examples of ligand-gated channels are the acetylcholine-gated Na^+/K^+ channels found on postsynaptic membranes and on muscle membranes at neuromuscular junctions (also known as the acetylcholine receptors) but the principle applies to other neurotransmitters and to a wide variety of chemical signals in other types of cells. Examples of voltage-gated channels are the Na^+ and K^+ channels found in nerve axons. Movement of ions is passive, driven solely by concentration gradients.

signals. Figure 7.15 illustrates ligand-gated and voltage-gated channels. The most important **gated pores** are those for Na^+, K^+, **and Ca^{2+}** (each selective for one ion) and the Na^+/K^+ **channel**. The passage of ions through these channels, when open, is the result of concentration gradients across the membrane so that the flow proceeds in either direction as determined by these. Since the gradients are steep in respect of ions such as Na^+, K^+, and Ca^{2+} the rate of movement of these ions through open channels is very much faster, perhaps 1000-fold, than the rate of transport of Na$^+$ and K$^+$ ions brought about by the Na$^+$/K$^+$ ATPase pump. This becomes an important factor in nerve impulse generation (see page 120).

Mechanism of the selectivity of the potassium channel

The most intensively studied ion channels are the gated channels of the nerve cells, described below. However, a major advance in understanding how different channels exert their strict selectivity for certain ions has come from the crystallization of a bacterial potassium selectivity channel protein which permitted the determination of its three-dimensional structure by X-ray diffraction. The membrane-spanning units of this protein are homologous with subunits of eukaryotic sodium/ potassium channels.

The selectivity filter of the channel consists of four identical transmembrane protein subunits, creating a cone-shaped channel. Figure 7.16 shows the tetrameric structure in end-on view. The extracellular end of the pore is at the pointed end of the cone (lower end, Fig. 7.17). Here hydrated ions can enter the cavity where a potassium ion (green) is shown surrounded by eight water molecules (red). (Remember that the cavity is completed by two additional monomers, above and below those two shown side-by-side in the figure.) The channel then narrows into the selectivity part of the pore where there are several sites. The narrow section is formed by four connecting polypeptide loops belonging to each of the four protein subunits (two are shown in Fig. 7.17). The polypeptide chains of these loops are oriented so that the peptide carbonyl oxygens (C=O groups shown red in the figure) point into the channel. There are groups of eight of these for each site in the complete channel. The selectivity loops consist of a sequence, Thr-Val-

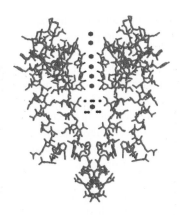

Fig. 7.17 Potassium channel (Protein Data Bank code 1K4C) diagram. Two of the four protein molecules that make up the pore are shown, with potassium ions (green spheres) passing through. The K$^+$ at the bottom, in aqueous solution, is surrounded by eight water molecules (red), and in this form is too large to traverse the pore. The K$^+$ ion has to shed the water molecules in order to pass through the channel, but this is energetically unfavourable. Carbonyl oxygen atoms of the polypeptide chain that line the pore (in red) facilitate this by substituting for the eight water molecules. Once through the pore each K$^+$ ion becomes hydrated again (not shown). The hydrated sodium ion is too large to pass through the selectivity filter, while the filter cannot facilitate the shedding of its water molecules, since unhydrated sodium ions, being smaller than potassium, cannot interact with the oxygen atoms lining the pore. Illustration taken from PDB Molecule of the Month series.

Gly-Tyr-Gly, which is highly conserved and found in all K$^+$ channels.

The channel allows the free movement of potassium ions through the pore but almost completely blocks the passage of sodium ions. For each 10,000 K$^+$ ions only one Na$^+$ ion is allowed to pass despite the fact that the sodium ion is smaller than the potassium ion. The selectivity is based on thermodynamic principles. The potassium ion in solution is comfortably surrounded by eight water molecules which must be removed for the ion to traverse the pore. Removal of these water molecules is energetically unfavourable since it means breaking their noncovalent attractions to the ion. The peptide carbonyl groups lining the selectivity filter channel are spaced so that they exactly mimic the arrangement of eight water molecules around the ion. The potassium ion can therefore easily slip from the water molecules into the thermodynamically similar situation within the filter channel so that there is no energy barrier. In Fig. 7.17 before the selectivity filter, in the large cavity, as mentioned, you see one potassium ion (green) surrounded by the eight water molecules (red). The ions passing through the narrow selectivity filter can be seen to be surrounded by four oxygen atoms of the peptide-bond carbonyl oxygens (eight for the four subunit chains) thus mimicking the structure of a hydrated ion. The ions move through the filter from one site to the next. Why can sodium ions not do the same? Being smaller,

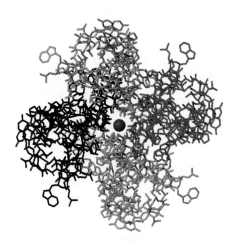

Fig. 7.16 Potassium channel (Protein Data Bank code 1BL8) stick diagram, showing an end view of the four proteins making up the channel, looking from the inside of the cell, with K$^+$ ions making their way in the pore.

a dehydrated ion cannot attach to the oxygens in the channel for the latter are too far apart for this. The hydrated sodium ion is too large to pass through the selectivity filter, while the filter cannot facilitate the shedding of its water molecules. The barrier, the energy requirement for shedding the hydrating molecules, is not removed in the case of sodium ions as it is for potassium. The high selectivity of the channel for K^+ is coupled with a very rapid flow. It is believed that this results from the K^+ bound to the selectivity sites being sequentially displaced from one site to the next by incoming K^+ ions. The simplicity of the mechanism for achieving such selectivity is astonishing.

It is likely that the selectivity filters of other K^+ channels are similar to the bacterial channel described above, but have different gating mechanisms attached to them. K^+ channels in nerve cells are voltage-gated – they open and close as described below.

Nerve-impulse transmission

Nerve impulse is traditionally part of physiological teaching, but the basic biochemical principals constitute one of the most beautiful and ingenious applications of molecular biology. Mechanistically it is an astonishingly simple and intellectually satisfying system.

The ionic gradients established across cell membranes are made good use of by cells that have gated ion channels in their membranes. When channels open, rapid flows (fluxes) of ions occur through the membrane, as appropriate to their function. The most spectacular use of this principle is in nerve conduction, which we will use as an important and best-studied example. The gated ion channels involved in nerve conduction are those for Na^+/K^+, Na^+ alone, K^+ alone, and Ca^{2+}.

A nerve impulse is transmitted by a series of neurons or nerve cells that have a central cell body and a thin axon; axons may be very long, even metres in length, and terminate in branches. The gap between one neuron and another is called a **nerve synapse** (Fig. 7.18); the signal is transmitted chemically across synapses by **acetylcholine** (or other neurotransmitter), which is released from the neuron ending to stimulate the next neuron. The signal molecules combine with receptors on the

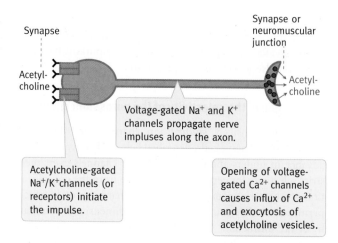

Fig. 7.19 Simplified diagram of the events in the transmission of a nerve impulse along a neuron, starting with the acetylcholine stimulation of the postsynaptic membrane and ending with the release of acetylcholine into a nerve synapse or into a neuromuscular junction. The acetylcholine-gated channel conducts both Na^+ and K^+ ions; there are two separate voltage-gated channels for Na^+ and K^+ respectively for the propagation of the impulse along the axon.

postsynaptic membrane – the starting membrane of the next neuron. We confine our discussion here to acetylcholine as the transmitter substance. The acetylcholine released into the synapse is rapidly hydrolysed by **acetylcholinesterase** bound to the synaptic membrane and the neuron becomes ready to accept a new signal. The anticholinesterase **organophosphate nerve gases** and insecticides inhibit the hydrolysis. Snake venoms such as cobra toxin attach to the channels in the postsynaptic membrane and inactivate them.

The same type of receptor occurs at neuromuscular junctions; acetylcholine liberated by motor nerve endings binds to the receptors and triggers muscle contraction. **Curare**, the arrow poison, blocks the action of acetylcholine here and paralyses muscles. To relax voluntary muscles during surgery, anaesthetists use modern derivatives of curare-like nondepolarizing (see below) muscle relaxants such as **vecuronium**, which block the signal from motor nerves to skeletal muscles. All act essentially like curare, but are much shorter acting and have less propensity for causing histamine release and hypotension than curare. Fig. 7.19 gives a simplified overview of how a neuron conducts a signal. See also Box 7.4.

The acetylcholine-gated Na^+/K^+ channel or acetylcholine receptor

The channel consists of a pore created by five protein subunits arranged in a circle. Two of the subunits have a binding site for acetylcholine. Each subunit contains an α helix kinked in the middle thus restricting the channel size (Fig. 7.20). In the closed form, hydrophobic groups project into the channel but when two molecules of acetylcholine bind to the receptor the helices tilt slightly and cause the groups to swing out of the way leaving

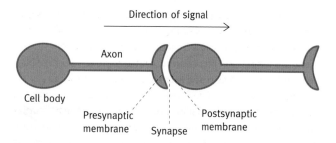

Fig. 7.18 A synapse between two neurons. A nerve impulse in the first one causes the liberation of a neurotransmitter such as acetylcholine into the cleft. This stimulates the succeeding neuron.

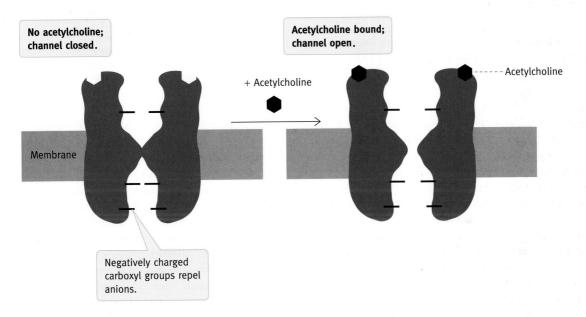

Fig. 7.20 Diagram of the acetylcholine-gated Na^+/K^+ channel of postsynaptic membrane and the receptor at neuromuscular junctions. The channel is composed of five subunits arranged to form a pore but only two are shown for simplicity; two are identical and each has an acetylcholine-binding site which induces the allosteric change to open the channel on ligand binding. It opens for only a fleeting period.

the channel open to Na^+ and K^+ ions. The gate remains open only for about a millisecond, even if acetylcholine is still bound, because the gate rapidly becomes desensitized and closes. The channel is highly selective for Na^+ and K^+ ions; anions are repelled by strategically placed negatively charged carboxyl groups of amino acid side chains.

How does acetylcholine binding to a membrane receptor result in a nerve impulse?

Consider a nerve synapse at which acetylcholine is liberated by the presynaptic neuron (the membrane at the end of the neuron carrying the impulse to the synapse). It diffuses the short distance to the postsynaptic membrane of the next neuron where it triggers a nerve impulse in that neuron. As with other cells, neurons have a high level of K^+ and a low level of Na^+ inside relative to the levels outside, the result of the Na^+/K^+ pump.

In the resting stage, K^+, high in concentration inside, leaks out because of special K^+ 'leak' channels creating a negative charge inside and a positive one outside, since the membrane is impermeable to anions. The K^+ leakage is self-limiting, because the internal negative charge so created holds K^+ ions back and an equilibrium is established in which the resting cell potential across the membrane is about −60 mV (more negative inside). This results in a separation of electric charges at the membrane

(a)

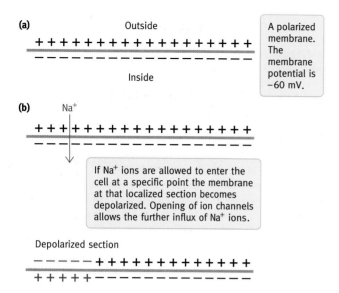

(b)

If Na$^+$ ions are allowed to enter the cell at a specific point the membrane at that localized section becomes depolarized. Opening of ion channels allows the further influx of Na$^+$ ions.

Fig. 7.21 Diagram to illustrate the meaning of the term 'polarized membrane' and its depolarization. In talking about depolarization in the context of neuronal function it is important to remember that only localized sections of membrane are referred to, not the whole cell membrane, and also that the depolarization is transitory as explained in the text. **(a)** A polarized membrane. **(b)** Depolarization by Na$^+$ entry.

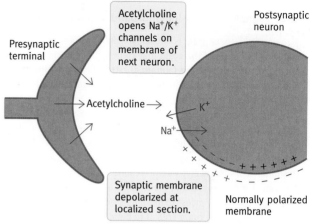

Fig. 7.22 Transmission of a nerve impulse across a nerve synapse. The presynaptic membrane liberates acetylcholine (or other neurotransmitter) on arrival of a nerve impulse. The acetylcholine binds to the receptors (the Na$^+$/K$^+$-gated channels) on the postsynaptic membrane of the next neuron. The resultant channel opening causes an influx of Na$^+$ ions and a smaller efflux of K$^+$ ions. This causes a local depolarization of the postsynaptic membrane. This depolarization triggers the propagation of the nerve impulse along the axon of the neuron as described in the text.

(which is an electrical insulator) with a surplus of negative charges inside and positive outside. They attract one another other across the lipid bilayer as shown in Fig. 7.21(a), creating an electrical or **membrane potential**. The membrane is then said to be **polarized**. If you placed an electrode inside the cell and another outside, a voltage would be recorded by a meter placed in the circuit connecting the two.

When Na$^+$/K$^+$ channels in the postsynaptic membrane open in response to acetylcholine binding, the resulting Na$^+$ influx (Fig. 7.21(b)) is greater than the efflux of K$^+$ because the negative charge inside the membrane opposes the latter. While the ion fluxes are large, they only occur for about a millisecond and there is negligible effect on the overall Na$^+$ and K$^+$ gradients between the inside and the outside of the cell. The membrane polarization in the postsynaptic membrane in the vicinity of the channels is reversed – it changes from about −60 mV to +65 mV and the section is then said to be **depolarized** (Fig. 7.21(b)). The channels now close. The synaptic transmission events are summarized in Fig. 7.22.

Nerve impulse propagation is driven by ion gradients

Nerve impulse propagation uses ion gradients that exist across the cell membranes to supply the required energy, the propagation mechanism involving nothing more than the opening and closing of voltage-gated channels in the correct sequence. Nothing physically moves along the length of the axon in the sense of a flow of molecules or ions. All that is needed is that, at the end of the neuron, the presynaptic membrane becomes locally depolarized.

The mechanism of nerve-impulse propagation along the axon was elucidated by the classic work of Hodgkin and Huxley in Cambridge, England, in 1952. The acetylcholine-gated channel, as described, creates a small patch of depolarized membrane at the postsynaptic membrane (Fig. 7.23(a)). This, in turn, partially depolarizes the adjacent section of membrane due to the diffusion of ions along the axon for a short distance (Fig. 7.23(b)). In the axon membrane there are separate **voltage-gated Na$^+$ channels** and **K$^+$ channels**. When the membrane in which they are located is partially depolarized by this, more Na$^+$ channels open. The resultant influx of Na$^+$ fully depolarizes the local section of membrane. The Na$^+$ channels are inactivated after about a millisecond. (This will be returned to shortly.) At this point, the depolarization peaks and the membrane potential reverts to that of the resting stage – it is repolarized. The repolarization results from opening of voltage-gated K$^+$ channels, which open slightly later than the Na$^+$ channels and allow efflux of K$^+$ ions. The K$^+$ channels also are rapidly inactivated so that the resting potential is restored. The whole thing takes about 3 ms.

The depolarization of a small section of membrane followed by its repolarization can be measured experimentally by having electrodes placed across the membrane linked to a cathode ray oscillograph. The result is shown in Fig. 7.24 (black curve). The electrical 'spike' of voltage change is known as an **action potential**. At the risk of overexplaining what may be obvious, this spike is the experimental measurement of the changes in membrane polarization. It is the changes in membrane polarization that conduct the impulse; the action potential spike is not

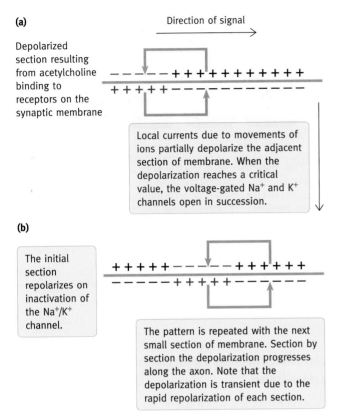

(a)

Direction of signal →

Depolarized section resulting from acetylcholine binding to receptors on the synaptic membrane

Local currents due to movements of ions partially depolarize the adjacent section of membrane. When the depolarization reaches a critical value, the voltage-gated Na^+ and K^+ channels open in succession.

(b)

The initial section repolarizes on inactivation of the Na^+/K^+ channel.

The pattern is repeated with the next small section of membrane. Section by section the depolarization progresses along the axon. Note that the depolarization is transient due to the rapid repolarization of each section.

Fig. 7.23 How a nerve impulse is propagated along the axon. **(a)** Starting with the small depolarized section caused by acetylcholine at the synaptic membrane as shown in the previous figure, **(b)** when a small section of membrane is depolarized, local currents due to ion movements (brown arrows) cause partial depolarization of the adjacent section of membrane. When the membrane potential in the latter has fallen to the threshold value of −40 mV, opening of voltage-gated Na^+ channels causes rapid further depolarization of that section. Closing of the Na^+ channels and opening of the K^+ channels restores the polarization but meanwhile the next section has become sufficiently depolarized to trigger channel opening in that section. The same cycle is repeated until the end of the neuron is reached. The impulse is prevented from going backwards by a mechanism described in the text.

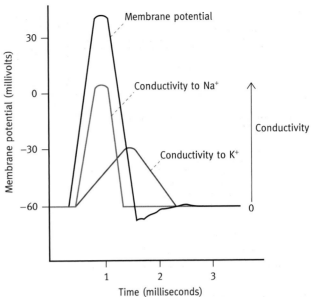

Fig. 7.24 Events in an action potential. The membrane potential (black line) is a measurement of the depolarization and repolarization of the membrane section. The conductivity is a measurement of the degree of opening of the voltage-gated Na^+ channels (red line) and of the K^+ channels (blue line) in the axon membrane which cause the polarization changes. These curves represent experimental measurements of the depolarization and repolarization events at the membrane. It is the latter which propagate the nerve impulse along the axon.

something in addition to this but simply a demonstration of it. The conductivity of the patch of membrane to Na^+ ions and to K^+ ions during an action potential indicates the opening of the relevant channels (red and blue curves in Fig. 7.24). What we have achieved so far is the depolarization of a short section of membrane adjacent to the synapse. This depolarization propagates itself along the axon, one small section at a time, until finally the distal synaptic membrane is depolarized causing the release of the transmitter substance to carry the impulse across the synaptic cleft. How does this depolarization travel along the axon occur? At each section exactly the same is repeated so that the next small patch of membrane is depolarized by spread of the Na^+ ions along the inside face of the membrane so there is another action potential generated; this spreads to the next and

so on producing an action potential at one small section of membrane after another very rapidly until at the end of the neuron the synaptic membrane is likewise depolarized.

Mechanism for ensuring the nerve impulse only goes forward

The Na^+ ions from the depolarized section can spread in both directions. What is to stop the nerve impulse travelling in the backwards direction as well? The Na^+ and K^+ channels have a special property which copes with this. There is a **refractory period** during which the channels which have opened and closed cannot be re-opened until after a slight delay. By the time they are capable of re-opening, the depolarization has moved too far along the axon for it to affect the channels; they will open only when the next impulse arrives. The impulse can only go forwards.

Mechanism of control of the voltage-gated Na^+ and K^+ channels

The Na^+ and K^+ channels show considerable structural homology, suggesting that they have a common ancestor. The Na^+ channel is a protein with four transmembrane domains linked on the cytoplasmic side by extensive hydrophilic peptide loops and smaller loops on the outside. The four domains are arranged in the membrane to form the channel. Each of the four is comprised of six transmembrane *a* helices so it is a large protein. In each of the transmembrane domains, one of the transmembrane helices is a voltage sensor rich in basic residues which slightly changes its position in response to membrane-

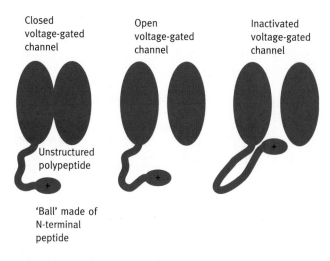

Closed voltage-gated channel

Open voltage-gated channel

Inactivated voltage-gated channel

Unstructured polypeptide

'Ball' made of N-terminal peptide

Fig. 7.25 The general principle of the 'ball-and-chain' inactivation of the voltage-gated K$^+$ channels. Only two of the subunits forming the pore are shown. Restoration of the original closed state occurs after the inactivated state. The Na$^+$ channels are believed to be inactivated by a mechanism similar in principle but the blocking peptide is a cytoplasmic loop of the protein connecting two transmembrane domains rather than an N-terminal 'ball'.

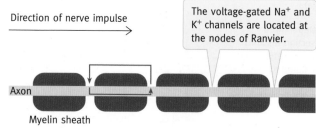

Direction of nerve impulse

The voltage-gated Na$^+$ and K$^+$ channels are located at the nodes of Ranvier.

Axon

Myelin sheath

Fig. 7.26 Myelinated nerve found in motor and many other neurons. Breakdown of the insulating myelin sheath occurs in multiple sclerosis, and the resulting defect in transmission of action potentials causes the symptoms of this disease. The red arrows represent the local currents caused by depolarization at the nodes from one node to the next so that depolarization and triggering of action potentials leaps along the axon in a saltatory (jumping) fashion: this greatly speeds up the rate of transmission.

potential changes and this is what causes channel opening and closing. However, we have pointed out above that the channels, opened by a depolarization of the membrane, are inactivated after about 1 ms. Although in this state they do not allow passage of ions, this does not involve restoration of the original closed state until after the refractory period mentioned above. How does this inactivation occur?

In the case of the K$^+$ channel, a 'ball-and-chain' mechanism illustrated in Fig. 7.25 involves an N-terminal peptide (the ball) tethered to the channel by a section of polypeptide such that the ball is free to move. When the channel opens, the positively charged ball, attracted by a negative charge, fits into and blocks the channel. The Na$^+$ channel may have a similar inactivation mechanism, but in this case it is one of the cytoplasmic loops connecting two of the domains that is believed to be involved rather than a terminal peptide 'ball'.

At the next synapse (or at neuromuscular junctions) the membrane depolarization activates voltage-gated Ca^{2+} channels. The resultant inflow of Ca^{2+} causes acetylcholine-filled vesicles to discharge the neurotransmitter from the synaptic membrane (see Fig. 7.19).

The central feature of nerve-impulse propagation mechanism is that the signal strength is maintained right along the lengths of axons. There is no modulation of the strength of a nerve impulse: it is all or nothing. If the initial signal is sufficient to trigger the initial depolarization, the impulse will travel. It is the frequency of the impulses that is controlled. The signal strength is maintained by the electrochemical gradient across the neuronal membrane, generated by the Na$^+$/K$^+$ ATPase; it is boosted every time an action potential is generated.

Myelinated neurons permit more rapid nerve-impulse transmission

In motor neurons, which trigger muscle contraction, there is another refinement. The axon, instead of being bare, is insulated by a myelin sheath (Fig. 7.26), which prevents ion movement (signal leakage) across it. The myelin sheath is interrupted every couple of millimetres at places called the **nodes of Ranvier** and it is here that the voltage-gated Na$^+$ and K$^+$ channels are located so that active depolarization of the membrane occurs at the nodes. The axon between the nodes is heavily insulated by the myelin sheath. This enables the depolarization to rapidly spread passively to the next node. The result is that the action potentials leap from node to node in what is called saltatory conduction. In **multiple sclerosis**, breakdown of the myelin sheath insulation impairs this, causing nerve impulses to travel along the nerve much more slowly. (See page 104 for the structure of a main myelin component.)

Why doesn't the Na$^+$/K$^+$ pump conflict with the propagation of action potentials?

Since the action potential propagation depends on migration of Na$^+$ ions across the neuron membrane, it might be thought that the Na$^+$/K$^+$ pump in continually restoring the balance would be confusing the mechanism. This does not occur because the local ion movements in nerve conduction are much faster than those caused by the Na$^+$/K$^+$ pump which acts as a slow 'trickle charger', keeping the ion gradient topped up. The membrane potential derives from the relatively small number of ions near the membrane surfaces and the movements of ions in generating action potentials involve only a minute proportion of the ions in the bulk of the cells and extracellular medium. There is little concentration change involved.

BOX 7.5 Membrane-targeted antibiotics

In the unceasing battle for survival, microorganisms throw chemical missiles at one another and natural selection ensures that any such weapons are targeted to wherever they will do a lot of harm. One of these targets is the ionic gradients across cell membranes. Membrane antibiotics destroy these gradients by allowing their equilibration across the membrane. Two classes of antibiotic are used to achieve this, mobile ion carriers (**ionophores**) and channel formers.

Valinomycin is the best-known ionophore, produced by a *Streptomyces* species, and is active against *Mycobacterium tuberculosis*. It is an unusual 12-membered ring containing D-valine and L-valine, and two hydroxy acids linked by peptide and ester linkages. The outside is hydrophobic while the carbonyl oxygens point inwards so that they can chelate monovalent ions, especially K^+. K^+ in solution is attached to water molecules by weak bonds, breaking of which requires energy so there is a thermodynamic barrier to it losing its attached water molecules. Valinomycin offers the K^+ an equally comfortable thermodynamic environment chelated to its carbonyl groups so that the ion can slip from its water cage into the valinomycin cage easily, the free energies of the two states being similar. (The same strategy is used in the selectivity filter of potassium channels; page 115.) The folded antibiotic enclosing the ion with its lipid-soluble exterior carries the ion across the membrane. Since there is equilibrium between a hydrated ion and a valinomycin-enclosed ion, the effect is to equilibrate the ions across the membrane. It picks up K^+ inside the cell where its level is high and releases it outside where

it is low. Several such ionophores are known. Nonactin, produced by *Streptomyces* species, is selective for K^+. The antibiotic A23187 exchanges Ca^{2+} for H^+; calcium ions are extremely important in cell regulation so disruption of normal transport is an effective weapon. It is a useful biochemical research tool.

The second type of membrane antibiotic is the channel formers, of which the gramicidins, produced by *Bacillus brevis*, are the best known. Gramicidins A, B, C, and D are dimers of linear peptides, 15 residues long, consisting of alternating L- and D-amino acids. They insert themselves into lipid bilayers, each dimer spanning only one half of the layer. Two of these temporarily associate end to end to form a contiuous channel across the membrane, and allow the free passage of H^+, K^+, and Na^+ ions. This destroys the ion gradient essential to the life of the cell. Gramicidin S is quite different. It is a cyclic 10-residue peptide. Polymixins are of this type and active against Gram-negative bacteria. Clinical use of these agents is usually only in the form of ointments for superficial applications since they may affect animal cell membranes.

A different type of action is shown by amphotericin B. This is a polyene antibiotic acting against many fungal infections. It allows loss of low-molecular-weight substances from cells by creating pores formed by several molecules of antibiotic complexed with membrane cholesterol. It lyses red blood cells. Mitochondria are not affected since the antibiotic is active only against sterol-containing membranes. Candidin, nystatin, and fungichromin have similar activities.

Role of the cell membrane in maintaining the shape of the cell

Eukaryotic cells have an internal scaffolding which maintains the shape of the cell and is involved in amoeboid motility. It is called the **cytoskeleton** (Chapter 8). This is made of protein microfilaments which pervade the cytoplasm and, at various places, are attached to integral proteins in the cell membrane. Such membrane proteins are not then free to move laterally in the lipid bilayer, as are other proteins.

An extreme example of the attachment of membrane proteins to the cytoskeleton is in the red blood cell which has special 'cell-shape' proteins. The cell is a biconcave disc (Fig. 7.27), which has the advantage of presenting a large surface area for gaseous exchange, but it is always on the move and therefore subject to shearing forces as it squeezes through capillaries, demanding a robust but flexible cell membrane. Underneath the membrane is a scaffolding of fibres of the protein, **spectrin**, anchored to the anion-channel protein (page 122) by a protein appropriately called **ankyrin**. The name spectrin comes from the fact that you can release the red cell contents but retain the empty membrane with its cytoskeleton, producing a red cell ghost (spectre). The anion-channel protein is a large protein which projects into the cytoplasm providing a cytoskeleton attachment point (Fig. 7.28). Spectrin is also linked to glycophorin by other specific linking proteins.

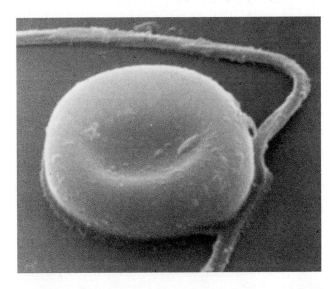

Fig. 7.27 Scanning electron micrograph of an erythrocyte. Courtesy Professor W. G. Breed, Department of Anatomy, University of Adelaide.

In a small number of people the cytoskeleton of red blood cells is deficient because of faulty spectrin or ankyrin due to a genetic defect. The cells are abnormally shaped and tend to be destroyed by the spleen. The diseases are called **hereditary spherocytosis** and **hereditary elliptocytosis**.

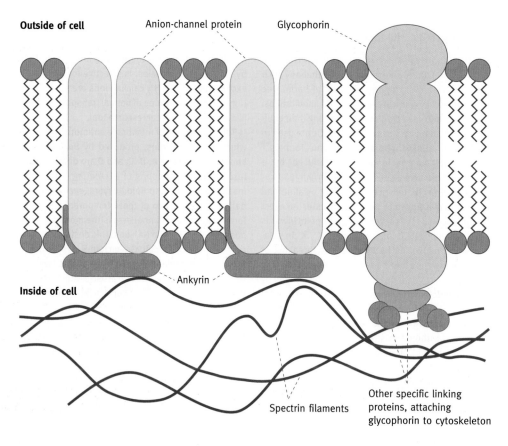

Fig. 7.28 Diagram to illustrate attachment of anion-channel protein and glycophorin to the cytoskeleton.

Cell–cell interactions – tight junctions, gap junctions, and cellular adhesive proteins

In an epithelial tissue such as the lining of the intestine, the products of food digestion are selectively taken up by the cells and then transported from the cell by appropriate systems in the membrane of the opposite side of the cell facing the blood vessels. Special bands of membrane proteins encircle the cells and these bind to corresponding proteins of the neighbouring cells, creating '**tight**' **junctions** or nonleaky cell–cell contacts.

Adjacent cells in some cases exchange molecules so as to coordinate the chemical activities throughout a tissue. This is achieved by proteins forming a tunnel between cells, known as **gap junctions** which will allow smaller molecules such ATP to pass, but not proteins. The coordination of heart cell contractions depends on gap junctions.

Another function of membrane proteins is to promote adhesion between cells of a tissue to form that tissue. If different types of embryo cells are mixed together they will re-associate with cells of their own type – kidney cells to kidney cells and so on. This is a function of tissue-specific cell–cell adhesion proteins called **cadherins** (see also page 61). Another family of this class of proteins, the **N-CAMS** (for nerve cell adhesion molecules), is important in nervous tissue formation.

Summary

Biological membranes have a lipid bilayer structure made up of a variety of different amphipathic lipids held together by non-covalent bonds. They are arranged with their hydrophobic tails pointing to the middle of the bilayer and their hydrophilic sections to the outside. Lipid bilayers are two-dimensional fluids which can self-seal. This permits endocytosis, a process by which cells take in material, and enables cells to eject molecules in a reverse process called exocytosis.

The fatty acid components may be saturated or unsaturated; the latter, in a *cis configuration*, are essential in maintaining the bilayer in a fluid condition. *Trans* unsaturated fatty acids resemble saturated fatty acids in that they are straight chain, not kinked. Cholesterol also plays a moderating influence on membrane fluidity.

Proteins are embedded in the bilayer as required for the functions of various membranes and often retain their lateral mobility. The membrane proteins have structures in keeping with their location in the hydrophobic membrane interior with hydrophobic amino acid residues in the central region and hydrophilic residues in domains that will be on either side of the membrane.

Porin proteins form hydrophilic channels across the membrane and these have hydrophobic amino acids on the outside face of the protein that is in contact with the hydrophobic layer of the membrane and hydrophilic ones pointing into the channel. Most membrane proteins are glycosylated on their exterior domains.

Lipid bilayers are largely impermeable to hydrophilic molecules and transport systems are needed for molecular traffic. These may be active, driven by ATP breakdown, or passive, driven by concentration gradients only. Nearly all cells have the Na^+/K^+ ATPase which pumps sodium out and potassium in. The ion gradient so formed is used to drive transport of other molecules in symport and antiport mechanisms.

Many channels are selective gated pores controlled by ligand attachments or membrane potential. The structure and mode of action of the potassium channel have been elucidated. It is based on an ingenious device basically thermodynamic in concept that allows the K^+ ions to shed their attached water molecules and pass through the pore. Na^+ ions, although smaller, cannot do so.

Nerve-impulse conduction is the most spectacular membrane activity involving gated ion channels opening and closing at specific times. The energy for the nerve impulse derives from the ion gradient established by the Na^+/K^+ pump.

Membranes have other functions; one of the most important of these in eukaryotes is cell signalling (Chapter 26). The cell membrane has a role in maintaining the shape of the cell by interactions between the cytoskeleton (Chapter 8) and integral membrane proteins. Cell-cell interactions such as tight junctions, gap junctions, and cell-cell adhesions cadherins, Chapter 4, page 61) all involve special membrane proteins.

 ## Further reading

Membrane lipids and movement

Higgins, C. F. (1994). Flip-flop: the transmembrane translocation of lipids. *Cell*, **79**, 393–5.
Discusses how lipids get to where they should be and how asymmetry of the lipid bilayer is achieved.

Rooney, S. A., Young, S. L., and Mendelson, C. R. (1994). Molecular and cellular processing of lung surfactant. *FASEB J.*, **8**, 957–67.
Reviews a more physiological role of phospholipids – that of lining the alveoli.

Shin, J.-S. and Abraham, S. N. (2001). Caveolae – not just craters in the cellular landscape. *Science*, **293**, 1447–8.
Caveolae are small pits in mammalian cell membranes, now attracting much attention.

Deurs, B., et al. (2003). Caveolae: anchored, multifunctional platforms in the lipid ocean. *Trends Cell Biol.*, **13**, 92–100.

Mozaffarian, D., Katan, M. B., Ascherio, A., Stampfer, M. J., and Willett, W. C. (2006). Trans fatty acids and cardiovascular disease. *New Engl. J. Med.*, **354**, 1601–13.

Membrane proteins

Macdonald, C. (1985). Gap junctions and cell-cell communication. *Essays Biochem.*, **21**, 86–118.

Von Heijne, G. (1994). Membrane proteins: from sequence to structure. *Annu. Rev. Biophys. Biomol. Struct.*, **23**, 167–92.
A review tracing the steps from sequence to structure of integral membrane proteins.

Von Heijne, G. (1995). Membrane protein assembly: rules of the game. *BioEssays*, **17**, 25–30.
Discussion of how proteins can be 'stitched' into membranes with multiple transmembrane helices.

Avery, J. (1999). Synaptic vesicle proteins. *Curr. Biol.*, **9**, R624.
A single-page quick guide.

Borst, P. and Elferink, R. O. (2002). Mammalian ABC transporters in health and disease. *Annu. Rev. Biochem.*, **71**, 537–92.

Kaplan, J. H. (2002). Biochemistry of Na, K-ATPase. *Annu. Rev. Biochem.*, **71**, 511–35.

Ion channels and porins

Neher, E. and Sakmann, B. (1992). The patch clamp technique. *Sci. Am.*, **266**(3), 44–51.
Enables studies to be made on ligand-gated and voltage-gated ion channels.

Changeux, J.-P. (1993). Chemical signalling in the brain. *Sci. Am.*, **269**(5), 30–7.
The acetylcholine receptor and how it works.

Catterall, W. A. (1995). Structure and function of voltage-gated ion channels. *Annu. Rev. Biochem.*, **64**, 493–531.
Covers sodium, calcium, and potassium channels.

Doyle, D. A. (1998). The structure of the potassium channel: molecular basis of K^+ conduction and selectivity. *Science*, **280**, 69–76.

Choe, S., Kreusch, A., and Pfoffinger, P. J. P. (1999). Towards the three-dimensional structure of voltage-gated potassium channels. *Trends Biochem. Sci.*, **24**, 345–9.

Kozono, D., *et al.* (2002). Aquaporin water channels: atomic structure and molecular dynamics meet clinical medicine. *J. Clin. Invest.*, **109**, 1395–9.

Mitter, G. (2003). The puzzling portrait of a pore. *Science*, **300**, 2020–2.
Reviews new light on the mechanism of voltage-gated potassium channels in neurons.

Blaustein, R. O. and Miller, C. (2004). Ion channels: Shake, rattle or roll? *Nature*, **427**, 499–500.
Short News and Views article about the voltage sensor in voltage-gated ion channels.

Roosild, T. P., Lê, K.-T., and Choe, S. (2004). Cytoplasmic gatekeepers of K$^+$-channel flux: a structural perspective. *Trends Biochem. Sci.*, **29**, 39–45.
Structural knowledge of K$^+$-selective channels has started to provide a basis for understanding the biophysical machinery underlying their electrophysiological properties.

Problems

1 Describe the lipid bilayer structure of membranes.

2 What is the role of cholesterol in eukaryotic membranes?

3 Give the structure of phosphatidic acid.

4 Name three glycerophospholipids based on phosphatidic acid. Name the substituent in each case attached to the latter.

5 What is the significance of *cis*-unsaturated fatty acyl tails in membrane phospholipids?

6 Why don't inorganic ions readily pass through membranes?

7 What is meant by facilitated diffusion through a membrane? Give an example.

8 Compare the structure of a triacylglycerol and a polar lipid. Why can the former never be used in lipid bilayers?

9 Explain the rationale behind the structure of sphingosine.

10 Explain how an ion gradient across a cell membrane can be harnessed to provide the energy for the active transport of an unrelated molecule or ion.

11 What is a gated ion channel? Name two different types.

12 Digitalis is used to strengthen the heartbeat in patients with congestive heart failure. What is the mechanism of this?

13 Discuss the principle of the mechanism of the selectivity filter of a potassium ion channel.

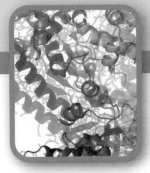

Chapter 8

Muscle contraction, the cytoskeleton, and molecular motors

Contrary to one's initial impressions, there is a lot of mechanical work occurring in almost all eukaryotic cells. This is driven by adenosine triphosphate (ATP)-fuelled molecular motors, of several different makes, causing contractions or carrying loads along tracks according to their designated roles. Muscle contraction is a specialized case in which the molecular motors are collected together in vast numbers and organized so that together they can produce the remarkable contractile forces that muscles develop. In this case the motors are fixed in position and act on ratchet-like 'tracks' that are forced to slide. Molecular motors present in nonmuscle cells such as liver and most other cells are in motion running along tracks and pulling loads of macromolecules, vesicles, and organelles between various parts of the cell. The intracellular motor activity resembles that of a busy city. In this second category the tracks against which the motors act are fixed so that the motors move. The tracks in this case constitute the **cytoskeleton**, which pervades almost all multicellular eukaryotic cells. As described later, the cytoskeleton has several important roles in cellular function in addition to providing transport tracks.

Prokaryotes, because of their small size, do not need an internal skeleton. Some do, however, have external flagellae equipped with molecular motors to propel cells through liquid.

In this chapter we will first deal with the mechanism of muscle contraction, which provides a good basis for understanding other molecular motors. Following this we will describe the cytoskeleton and the molecular motors that operate on its tracks.

Muscle contraction

A reminder of conformational changes in proteins

The most astonishing thing about muscle contraction and all molecular motors is that movement occurs at all. The only basis for biological processes are molecules and, for contraction or movement, individual molecules must move in a directional fashion. What type of molecular activity is available for this? The only type we know of is conformational change in proteins, minute changes in shape of individual protein molecules that occur as a result of different ligands binding to them. The conformational changes are each minute, but collectively they can add up to the gross movement of muscles.

As indicated, for movement to occur, whatever exerts force must have something to exert it against. A 'molecular motor' exerting force by conformational change must always have a partner structure to react against – if the motor is fixed, the partner molecule will move; if the partner molecule is fixed, the motor molecule will move. To anticipate the mechanism of muscle contraction, the motor is **myosin**, which is fixed, and the structure it acts against, and slides, is the **actin filament**.

Types of muscle cell and their energy supply

The two main classes of muscles are **striated** and **smooth**. Striated muscle is found in skeletal muscle, which is under voluntary nerve control and contracts rapidly. It derives its name from its appearance under the microscope (Fig. 8.1). Heart muscle is striated but not identical in structure to skeletal muscle, and is under involuntary control. Smooth muscle is found in the intestine and blood vessels, which are typically under involuntary nervous control and also frequently under control by hormones. It contracts slowly and can maintain the contraction for extended periods. It is not striated.

In all muscles the reserve of ATP, on which contraction depends, is only enough for a brief period of intensive contraction. This explains a role for a fast-acting reserve in skeletal muscles of creatine phosphate, a high-energy phosphate compound. When ATP is converted to adenosine diphosphate (ADP) and inorganic phosphate (P_i) by contraction, ATP is

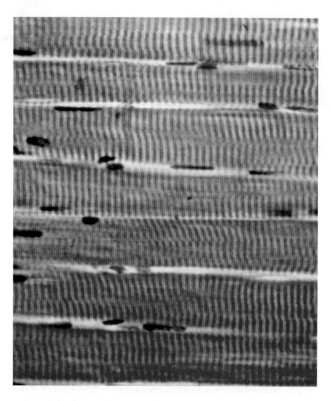

Fig. 8.1 Photomicrograph of myofibrils of striated skeletal muscle. © Robert Harding.

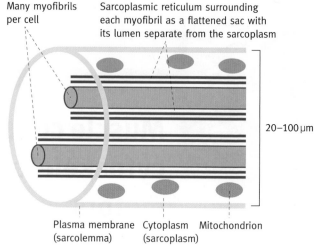

Fig. 8.2 A myofibre (muscle fibre) or muscle cell from striated muscle. Bundles of myofibres make up a muscle.

rapidly regenerated by a reaction catalysed by the enzyme, creatine kinase, using the phosphoryl group of creatine phosphate.

> Contraction event: $ATP + H_2O \rightarrow ADP + P_i$
> ATP-regeneration event: $ADP + creatine\text{—}\circled{P} \rightarrow ATP + creatine$

During muscle recovery, ATP formed by oxidative metabolism replenishes the pool of creatine phosphate by the creatine kinase reaction, which is reversible in the presence of high levels of ATP and low levels of ADP.

Creatine phosphate has the structure shown below. Because of the reactivity of the phosphoryl group, there is a spontaneous slow (noncatalysed) formation of **creatinine**; this has no function and is excreted in the urine at a daily rate proportional to muscle mass. Since creatinine is formed at a relatively constant rate in the body and its only fate is to be excreted in the urine, plasma (or serum) creatinine is used as an indicator of kidney function adequacy.

Creatine phosphate

Creatinine

Structure of skeletal striated muscle

A muscle is composed of long multinucleated cells called myofibres (Fig. 8.2). The plasma membrane (the **sarcolemma**) has nerve endings associated with it, at neuromuscular junctions, that deliver the nervous signal that triggers contraction. The cell contains many mitochondria, in keeping with the high demand for ATP to drive contraction. Running lengthwise within it are multiple **myofibrils**, each surrounded by a membranous sac, the **sarcoplasmic reticulum.**

Structure of the myofibril

The myofibril is the structure that does the contracting. Each is divided into segments (**sarcomeres** (Fig. 8.3(a)) bounded by **Z discs** of proteins (Z for *zwischen* or *between*). On contraction, the Z discs are pulled closer together, thus shortening the individual sarcomeres and hence also the myofibrils (Fig. 8.3(b)). This in turn shortens the myofibre, causing muscle contraction.

How does the sarcomere shorten?

The Z discs of protein at each end of the sarcomeres have attached to them thin **actin filament 'rods'** pointing to the center of the sarcomere, made of the protein, actin. The filaments are attached to the Z discs only at one of their ends. In vertebrate muscle in cross section, the **thin filaments** are in multiple hexagonal arrays on the two discs (Fig. 8.3(c)). Inside each array is a **thick filament** which has finger-like projections that do the actual work of contraction. By a ratchet-like mechanism the finger-like projections 'claw' the thin filaments towards the centre and in doing so pull the Z discs closer together (Fig. 8.3(b)). The process is known as the **sliding-filament model.**

That, then, is the overall picture of muscle contraction. To understand how contraction happens we must look at the molecular structures involved.

(a)

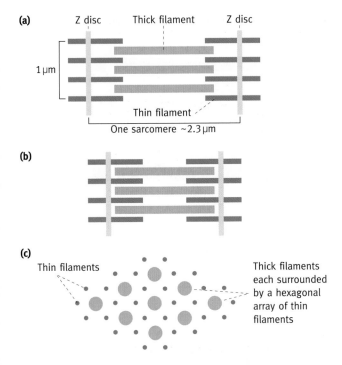

(b)

(c)

Fig. 8.3 Arrangement of thick and thin filaments in a sarcomere. **(a)** Relaxed. The striated appearance of myofibril sections in electron microscopy is caused by the amount of protein the beam traverses; **(b)** contracted; **(c)** arrangement in cross-section. The diagram shows only a few filaments but each sarcomere has a large number, all in lateral register.

Structure and action of thick and thin filaments

The thin filaments are made of the protein actin. The **G actin** protein molecule (G for globular) (Fig. 8.4(a)) has two dissimilar parts connected by a narrow waist, giving it a polarity. It polymerizes into long fibres, in a head-to-tail polarity (Fig. 8.4(b)). **A thin filament** consists of two such fibres coiled around each other with a long pitch, the ends being referred to as (+) and (−).

(a)

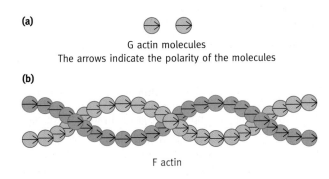

G actin molecules
The arrows indicate the polarity of the molecules

(b)

F actin

Fig. 8.4 (a) Globular (G) actin. **(b)** Fibrous (F) actin made from the polymerization of G actin. The two strands are actually closely apposed forming a rod-like structure, but are presented here and in Fig. 8.10 in more open form for clarity.

(a)

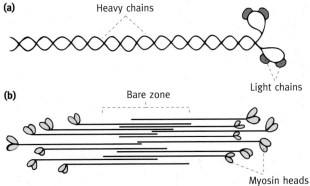

(b)

Fig. 8.5 (a) A myosin molecule consisting of a dimer structure of two heavy chains, each terminating in a myosin head. The latter has two dissimilar light chains attached to it. **(b)** A thick filament made of about 300 myosin molecules arranged in a bipolarfashion.

Myosin, the protein of the **thick filament** is a dimer made up of a straight rod section formed from two polypeptides in coiled-coil configuration (Fig. 8.5(a)) – that is, two α helices (page 50) coiled around each other in a long helical form. Each α helix has regularly spaced residues that form hydrophobic attachments to its partner, the two chains forming a rigid rod structure. Each chain terminates in a globular head; two other small polypeptide chains, called light myosin chains, are attached to each myosin head.

Thick filaments are formed from several hundred myosin molecules arranged in a bipolar fashion as shown in Fig. 8.5(b) and are held in a central position relative to the thin filaments by other proteins involved in sarcomere structure. One is **titin**, a huge protein. The thin filaments are anchored to the Z discs at their (+) ends so that the myosin heads, at both ends of the thick filament, are oriented the same way, relative to the thin-filament polarity (Fig. 8.6).

How does the myosin head convert the energy of ATP hydrolysis into mechanical force on the actin filament?

The myosin head is an ATPase enzyme. It hydrolyses ATP to ADP and P_i. It undergoes conformational changes at the expense of this, which results in it exerting a force on the actin filament. The actual 'power stroke' in contraction does not occur on ATP hydrolysis, as you might expect. Instead it causes the myosin head–ADP/P_i complex to take up a different conformational state; you might think of it as adopting a 'high-energy' conformation. It is when the P_i + ADP leave the protein that the liberation of free energy occurs, so that the 'power stroke' in muscle contraction correlates, not with ATP hydrolysis *per se*, but with this release from the myosin head. The energy transfer is mediated by conformational changes in the myosin head.

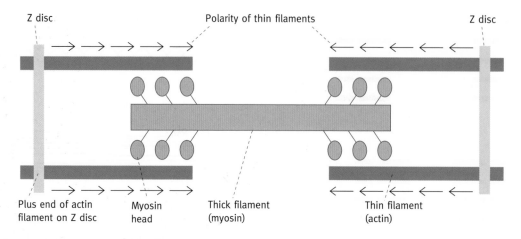

Fig. 8.6 Diagram showing the arrangement of thick and thin filaments in a sarcomere. The arrows in the latter are to show the polarity of the actin filaments. In a contraction event, in effect, the myosin heads track along the actin filaments towards their (+) ends and, in doing so, claw the two discs together. The heads are shown as simple shapes for clarity but their actual structures are described below.

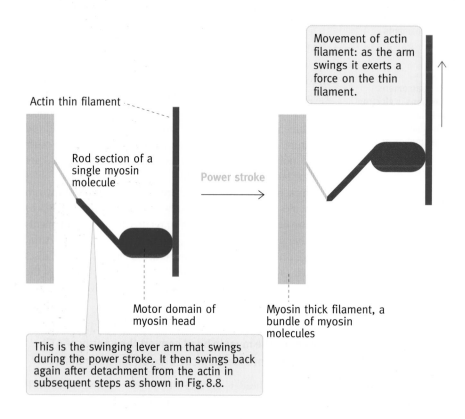

Fig. 8.7 Diagrammatic representation of the swinging-lever-arm mechanism of muscle contraction. The myosin head has a motor domain which binds to the actin filament without changing its angle of attachment. The motor domain is the site of nucleotide (ATP/ADP) attachment. It is connected to the myosin rod by an α helix which forms the lever arm; this is surrounded by the myosin light chains (not shown). When the power stroke is induced by P_i + ADP dissociation from the motor domain (see text) the small resultant conformational change is amplified by the lever arm which swings through an arc of 70°, sufficient to displace the actin filament by about 10 Å. Note that for clarity only a single myosin head is shown; each myosin molecule has two heads.

Mechanism of the conformational changes in the myosin head

A myosin head has a compact region which, by secondary forces, attaches to the actin filament (the right-hand side in Fig. 8.7). Its orientation to the actin filament is always perpendicular. There also is an extended α helix (shown to the left-hand side of the diagram) which, at its distal end, joins to the rod part of the myosin molecule. During contraction, when the ATP is hydrolysed on the myosin head, with the ADP + P_i still bound, a small conformational change occurs in the head. This causes the α helix to adopt the primed, 'high-energy' conformation ready for the power stroke which occurs on the release of ADP + P_i. On this happening, the lever arm swings relative to the actin-binding part and this forces the head to move. Since this is attached to the actin filament the latter is forced to slide as illustrated in Fig. 8.7. Figure 8.8 shows the three-dimensional structures of the myosin head before and after the power stroke. On the right of each structure is the site which attaches to the actin. You will notice that the lever is surrounded by the two myosin light chains (not represented in Figs 8.7 and 8.9). These stabilize the lever arm.

Fig. 8.8 The three-dimensional structure of the myosin head: (upper) before the power stroke with the swinging lever arm in the upright position; (lower) the head immediately after the powerstroke.

Let us now look more closely at the steps involved in this cycle, starting with Fig. 8.9(a) in which the head has just finished its power stroke in a contraction. (Only one of the two heads of myosin is represented in the diagram in the interests of clarity.) Vast numbers of individual heads are involved in any contraction.

The following sequence of events then occur as depicted in Fig. 8.9.

- In (a) the myosin head is attached to the actin filament having just completed its previous power stroke in which a force was applied to the actin filament. The head cannot detach from the actin until a molecule of ATP binds. This resembles the state of muscles in rigor mortis in which muscles are contracted, but head detachment does not occur after ATP depletion.

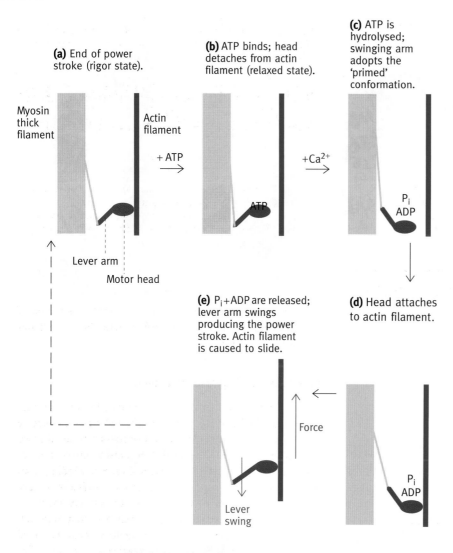

(a) End of power stroke (rigor state).

Myosin thick filament

Actin filament

+ ATP

Lever arm

Motor head

(b) ATP binds; head detaches from actin filament (relaxed state).

ATP

+Ca²⁺

(c) ATP is hydrolysed; swinging arm adopts the 'primed' conformation.

P_i
ADP

(e) P_i+ADP are released; lever arm swings producing the power stroke. Actin filament is caused to slide.

(d) Head attaches to actin filament.

Force

Lever swing

P_i
ADP

Fig. 8.9 Swinging-lever-arm model of muscle contraction. The pink bar is the thick filament made of multiple myosin molecules; the thin pink line represents an individual myosin molecule; only one of two heads is shown (red) for clarity. The green line is the thin actin filament. ATP, P_i, and ADP represent molecules bound to the myosin head. The essence of the model is that the power stroke is associated with P_i and ADP release and the force is exerted by the proximal subunit of the myosin head (the lever arm) swinging in a lever-like fashion. See text for individual steps.

BOX 8.1 Muscular dystrophy

An incompletely understood aspect of muscle contraction relates to muscular dystrophy. This term covers a group of related inherited diseases in which there is a progressive muscular weakness and muscle wasting. There are many different forms of muscular dystrophy, with age of onset varying from birth (congenital muscular dystrophy) through to adulthood (some types of limb girdle muscular dystrophy). The clinical course may be static or progressive. In some patients there is involvement of the respiratory and cardiac muscles, which can result in respiratory failure (the major cause of death) or cardiomyopathy, respectively.

The best-known and most common form is Duchenne muscular dystrophy. This is an X-linked disorder that predominantly affects boys. Weakness becomes clinically obvious in the pre-school age group and is progressive; affected boys usually lose the ability to walk by their teenage years and there is progressive weakness of respiratory muscles, resulting in death in the late second or third decade of life.

The affected gene in this disease has been cloned; it encodes a very large protein called dystrophin. This protein is attached at one end to structural elements of the muscle cell cytoskeleton (page 133) and at the other to a transmembrane complex of proteins, which are, in turn, linked to proteins of the extracellular matrix that surrounds muscle cells. The linkage of intracellular structural components to the membrane and thence to proteins of the extracellular matrix (through the dystrophin-associated-protein complex) is thought to play a part in stabilizing, or strengthening the muscle cell membrane to withstand the contractile forces involved in muscle contraction. In Duchenne muscular dystrophy, dystrophin is absent so that the connection of the intracellular cytoskeleton is broken. Some other forms of muscular dystrophy (inherited in an autosomal fashion) result from mutations in the genes encoding other members of the dystrophin-associated-protein complex.

- In (b) a molecule of ATP has bound, causing detachment of the myosin head from the actin filament. This is the state existing in relaxed muscle.

- In (c) ATP hydrolysis occurs but the products, P_i and ADP, are still attached to the myosin head. A conformational change occurs in the head causing the lever arm to adopt the 'primed' conformation.

- In (d) the head attaches to the actin filament resulting in state (e) (the ADP and P_i are still attached).

- In (e) the power stroke occurs in which P_i and ADP are released and the actin filament is forced to slide, thus causing a contractile force. This returns the head to our starting point shown in (a).

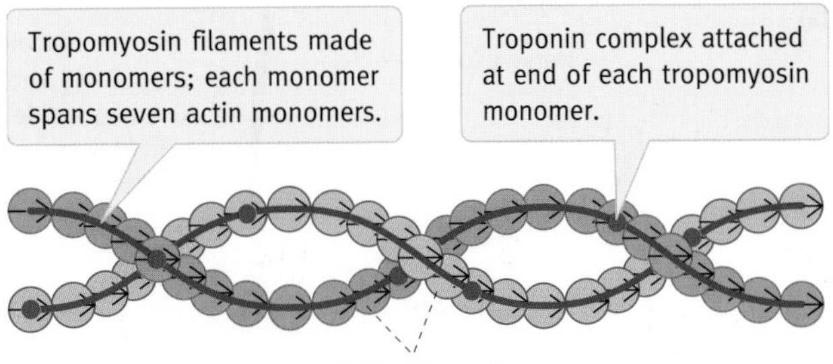

Tropomyosin filaments made of monomers; each monomer spans seven actin monomers.

Troponin complex attached at end of each tropomyosin monomer.

Actin filaments

Fig. 8.10 The relationship of the actin and tropomyosin molecules to each other. Each tropomyosin molecule binds to seven actin monomers forming a continuous filament in each of the actin filament grooves. Each tropomyosin molecule also has a troponin complex bound at one end.

Control of voluntary striated muscle

In skeletal muscles, contraction is initiated by a nerve impulse causing Ca^{2+} ions to be liberated into the myofibril from the sarcoplasmic reticulum which surrounds each myofibril within the muscle cell. The reticulum membrane has **voltage-gated Ca^{2+} channels**, normally closed. On receipt of a nerve impulse to the muscle cell, they open and release Ca^{2+} from the lumen of the reticulum into the myofibril, causing contraction. The reticulum membrane is rich in a Ca^{2+} ATPase (an ATP-dependent pump) that pumps Ca^{2+} from the cytosol (known as the sarcoplasm) surrounding the myofibril into the lumen of the reticulum. This depletes the myofibril of Ca^{2+} and terminates muscle contraction.

How does Ca^{2+} trigger contraction?

Thin filaments have associated with them additional protein molecules called **tropomyosin**. This is an elongated molecule that lies along each of the two helical grooves between the two actin fibres of a thin filament (Fig 8.10). A tropomyosin molecule has on it seven actin attachment sites, each binding to an actin monomer within the groove and successive molecules overlap to form continuous threads along the thin filament.

Each tropomyosin molecule has attached to it, at one end, an additional complex of three more globular proteins called **troponin** in which Ca^{2+} causes a conformational change. This allows the myosin head to attach to the actin filament and starts the **myosin-actin power cycle**. Withdrawal of Ca^{2+} terminates the contraction event.

For a long muscle to contract, all of the sarcomeres in a myofibril and all the myofibres in the muscle need to respond to a motor nerve impulse essentially simultaneously, for otherwise the contraction will be uncoordinated. The nerve impulse causes liberation of acetylcholine from the nerve ending at the neuromuscular junction; this causes a local depolarization of the plasma membrane that rapidly propagates throughout the latter (see page 131). As described, the depolarization causes the voltage-gated Ca^{2+} channels of the sarcoplasmic reticulum to open and release the ion on to the myofibril. In order to ensure that the signal reaches all of the sarcoplasmic reticulum within a cell very rapidly, the plasma membrane is invaginated into transverse (T) tubules that enter the cell at Z discs and make direct contact with the sarcoplasmic reticulum membrane. This permits the electrical signal to reach, virtually simultaneously, all of the contractile units controlled by that nerve impulse (Fig. 8.11).

BOX 8.2 Malignant hyperthermia

An inherited condition known as **malignant hyperthermia** exists in some humans, pigs, dogs and poultry. Following the administration of so called triggering agents, a depolarizing muscle relaxant (suxamethonium) or one of the volatile anaesthetics such as halothane, a hypermetabolic process is initiated within the skeletal muscle. The pathological process is thought to be an increase in the opening time and conductance of the calcium channel of the sarcoplasmic reticulum, the **ryanodine receptor protein**. In approximately 50–80% of humans, the faulty calcium channel is associated with a mutation in the gene for this protein. The increased myoplasmic calcium levels catalyse increased muscle contraction, increased mobilization of energy through glycogen breakdown and gylcolysis, resulting in increased aerobic and anaerobic metabolism. If unrecognized or left untreated, respiratory and metabolic acidosis, breakdown of muscle fibres (rhabdomyolysis), renal failure, and death may occur. Treatment includes withdrawal of the triggering agents, hyperventilation with 100% oxygen, administration of a specific antidote, dantrolene and active cooling. The clinical incidence is 1:15,000 children and 1:50,000 adults where triggering agents are used. The faulty gene prevalence is unknown, but estimated to be about 1:5000.

(a)

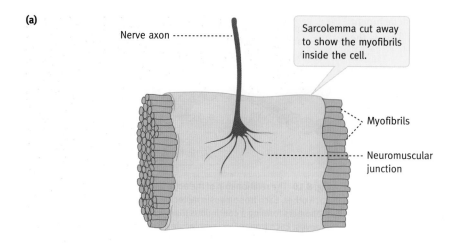

Nerve axon

Sarcolemma cut away to show the myofibrils inside the cell.

Myofibrils

Neuromuscular junction

(b)

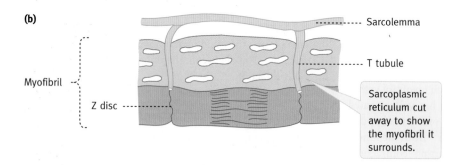

Sarcolemma

T tubule

Myofibril

Z disc

Sarcoplasmic reticulum cut away to show the myofibril it surrounds.

Fig. 8.11 (a) Plasma membrane (sarcolemma) of the myofibre, showing the neuromuscular junction. **(b)** The transverse (T) tubules that carry the plasma membrane depolarization signal to the sarcoplasmic reticulum (SR). It is postulated that the plasma membrane depolarization is directly transmitted to the SR Ca^{2+} channels causing rapid Ca^{2+} release from the SR.

Smooth muscle differs in structure and control from striated muscle

Smooth muscle is found in the walls of blood vessels, in the intestine, and in the urinary and reproductive tracts. The long spindle-shaped cells have a single nucleus and associate together to form a muscle in patterns appropriate to their function, such as an annular arrangement in blood vessels and a criss-cross network in the bladder.

The basic principles of contraction are the same as in striated muscle but the contractile components are not so highly organized. There are no myofibrils or sarcomeres, the latter being the reason for the absence of a striated appearance under the microscope. Instead, actin filaments run the length of the cell (which is small, compared with a striated muscle cell) and are anchored into the cell membrane.

Control of smooth muscle contractions

Although Ca^{2+} controls this also, the mechanism is different from that in striated muscle. A smooth muscle contracts typically about 50 times more slowly than a striated muscle. There is no requirement for contraction to be almost instantaneous

throughout the structure. The contraction signal to cells can spread at a more leisurely pace. A nerve impulse from the autonomic system causes Ca^{2+} gates in the **plasma membrane** to open and allow an inrush of the ion into the cells from the outside. There is no sarcoplasmic reticulum. The relatively slow diffusion of Ca^{2+} throughout the cell can be tolerated because of the small distances involved and the slow response requirements. Special junctions between cells allow a neurological signal to spread throughout the muscle.

In the head of each myosin molecule, as already described, are two small polypeptides known as **myosin light chains**. In smooth muscle, one of these (designated the **p-light chain**) inhibits the binding of the myosin head to the actin fibre, and thus prevents contraction. Ca^{2+} activates a myosin kinase that catalyses phosphorylation of the p-light chain by ATP, and abolishes its inhibitory effect, thus triggering contraction.

The Ca^{2+} does not directly activate the kinase; but does so via the protein, **calmodulin**. When the Ca^{2+} levels fall due to pumping out from the cell by a Ca^{2+} ATPase in the cell membrane, the process reverses; a phosphatase enzyme catalyses dephosphorylation of the myosin light chain abolishing the binding between myosin and actin and causing muscle relaxation. The scheme is summarized in Fig. 8.12.

As well as neurological control of smooth muscle contractions, several hormones also exert control; some prostaglandins

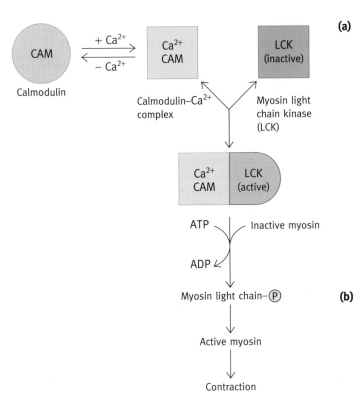

Fig. 8.12 Mechanism of activation of smooth muscle contraction by Ca^{2+}.

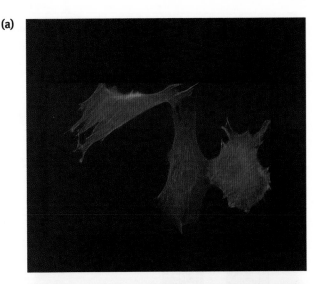

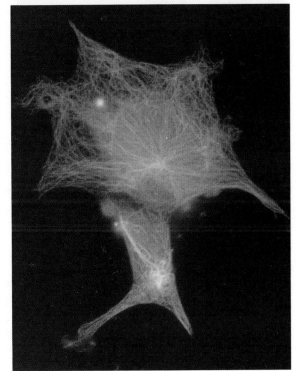

cause muscle contraction (page 232). Norepinephrine (noradrenaline) causes contraction of certain blood vessel muscles.

The cytoskeleton

An overview

The eukaryotic cell cytoskeleton is a complex scaffolding of thin protein filaments which pervades all parts of the cell. There are three classes of the cytoskeletal components. The first two are the **actin filaments** and **microtubules**. They are not visible in the microscope unless appropriately stained. Fig. 8.13 (a) and (b) show examples of these. The third group, **intermediate filaments**, are quite different structurally and in function. They are intermediate in size between the other two and are treated as part of the cytoskeleton.

The cytoskeleton has several roles. It connects the interior of cells to the extracellular matrix; it confers shape on animal cells; it is involved in transport within cells; it is involved in cytokinesis and in chromosome separation at cell division; it confers motility on cells.

Motility is not something we associate with most cells in the body, but, quite apart from cells such as macrophages and other white cells of the body which migrate through tissues by amoeboid-like action, motility is a widespread property of

Fig. 8.13 (a) Actin filaments in mouse myoblasts visualized with an actin antibody. The filaments pervade the cell and bundles of them (called stress fibres) are attached at focal points to the cell membrane, often at points at which the membrane makes contact with a solid surface. The fibres are rapidly disassembled and assembled. Photograph kindly provided by Dr P. Gunning, Children's Medical Research Institute, Sydney, Australia. **(b)** Microtubules in cytoplasm of a cell radiating from the microtubule-organizing centre (MTOC) or centrosome (see text).

animal cells. In embryonic development for example, they migrate over surfaces by a crawling action. This also occurs in normal wound healing. Yet another form of movement dependent on microfilaments is due to cilia, whose beating causes movement

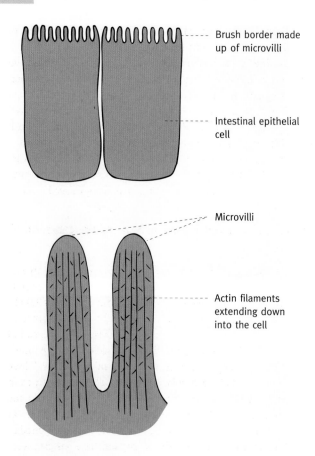

Fig. 8.14 Diagram of actin filaments in microvilli. Note that the filaments have a complex cross-linking and anchoring arrangement with other proteins to form a robust cytoskeletal structure.

Labels in figure:
- Brush border made up of microvilli
- Intestinal epithelial cell
- Microvilli
- Actin filaments extending down into the cell

of surface mucous layers in the respiratory passages, or of flagellae which sperm use for swimming.

Perhaps the most remarkable role of the cytoskeleton is that, as already stated, it provides transport tracks within cells for molecular motors to run along pulling loads to various parts of the cell. There is constant internal transport in eukaryotic cells for which the cytoskeleton fibres provide the tracks. There has to be active transport of materials within eukaryotic cells. The distances are too big for diffusion to be adequate as it is in prokaryotes. A nerve axon may be half a metre or more in length (several metres, in giraffes and whales); synthesis of proteins and vesicles occurs in the nerve cell body and may have to be transported to the tip of the axon. Membrane-enclosed transport vesicles carry proteins to their cellular destinations (page 136) pulled along cytoskeletal tracks by ATP-powered protein motors. These motors have other roles in causing movement. For example they are needed to separate chromosomes at the time of cell division so that at mitosis, chromosomes on the spindle move apart into daughter cells. The intermediate filaments confer toughness on structures.

The cytoskeleton is in a constant dynamic state

In one sense the cytoskeleton is a bewildering structure because actin fibres and microtubules are subject to continual assembly and collapse. We are not used to the concept that a highly organized transport system can be based on such a flickering, apparently random, ephemeral set up. Rather than a nicely laid road system it is more like a complex of microscopic tracks that may disappear at any moment and reappear pointing in a different direction. It is hardly possible to envisage how such a system can handle such complex traffic with such an organized outcome.

There are a few exceptions to the dynamic instability of actin fibres and microtubules. Muscle thin filaments are permanent; another example is brush border 'fingers', the microvilli (Fig. 8.14) of the intestinal cells which an actin framework stabilizes and increases the absorptive area of the cells. Microtubules running the length of cilia and flagellae are also permanent and the centrosomes (page 324) seem to be made of fused microtubules; their mechanism of function is not understood.

We will now deal in turn with the actin filaments, then the microtubules and finally with the intermediate filaments.

The role of actin and myosin in nonmuscle cells

Actin is an abundant protein in most eukaryotic cells. Myosin homologues also are an almost universal constituent of such cells but are present in smaller amounts than myosin is in muscles.

Actin filaments are particularly dense near the plasma membrane where bundles of them are anchored into the membrane to form stress fibres, involved in maintaining cell shape (Fig. 8.13(a)) and in causing amoeboid movement.

Assembly and collapse of actin filaments

Actin filaments of the cytoskeleton reversibly assemble and disassemble rapidly from their monomers that bind to each other by noncovalent bonds only. As described for muscle actin (page 127), cytoskeletal actin has a head to tail asymmetry, and they polymerize in a head to tail orientation by a self-assembly process so that the filaments have a polarity with ends referred to as plus and minus.

Free actin monomers have a molecule of ATP attached to them; this is hydrolysed to ADP shortly after incorporation into the growing end of a growing actin polymer. The monomer is an ATPase but only after it has been incorporated into a growing actin filament. Actin-ATP polymerizes much more readily than does actin-ADP so that a terminal monomer in the ADP form is likely to dissociate. Continued elongation of the filament requires new actin-ATP monomers to be added before the ATP on the previously added unit is hydrolysed. Unless this happens the filament will tend to depolymerize. Specific capping proteins which attach to the ends of completed actin filaments presumably stabilize them if and when they reach a suitable target. If not they will have a terminal ADP monomer and collapse. A variety of associated proteins will bind their origin sites. The constantly changing arrangement of actin filaments is associated with cell shape changes, cell movement and internal transport. When a phagocyte engulfs a particle, actin filaments are seen to extend into the pseudopodia and push it out (these are the footlike extensions the cell puts out in the direction of movement). It is a very dynamic situation.

Recently, there have been major advances in this field. A new family of proteins called **formins** have been found in all classes of muilti-cellular eukaryotic organisms. They are large proteins with several distinct proteins domains. As well as initiating actin filament assembly they move with the growing ends, and protect the ends from termination by capping proteins. It seems that formins may also play a role in microtubule assembly. They are known to initiate new unbranched actin filaments such as stress fibres and the rings of actin filaments which are responsible for cell separation in cytokinesis. Another complex of proteins (called the Arp 2/3 complex), distinct from formins, causes branch actin filaments from existing actin filaments. (See Goode review in Further reading for an authoritative account of the entire field.)

Mechanism of contraction in nonmuscle cells

Myosin II, which is the 'conventional' muscle-like myosin molecule (with two heads) found in skeletal muscle, is also present in nonmuscle cells. When contracting mechanisms are needed in these cells, myosin II molecules aggregate into bipolar filaments of about 16 myosin molecules analogous to but much smaller than striated muscle thick filaments. In an action very similar to that described for muscle (page 126) these then exert a contracting force on adjacent actin filaments (Fig. 8.15) and exert a pull on the cell membrane to which the filaments are anchored. Control of contraction is by phosphorylation of myosin as in smooth muscle.

Cytokinesis, or the constriction of a dividing cell into two daughter cells, provides a good example of this type of contractile system. An annulus of actin fibres assembles at the equatorial plane where constriction occurs. Actin filaments are anchored at the plasma membrane. Overlapping of these filaments (with opposite polarities) allows cytoplasmic myosins to exert a contraction force, rather like that in muscle. The set-up is disassembled as the contraction progresses. The action constricts the cell and ultimately separation into two daughter cells occurs.

The role of actin and myosin in intracellular transport of vesicles

The rod-like 'tail' of the muscle-type 'conventional' myosin molecule is designed so that, as described, myosin can assemble itself into bipolar filaments both in muscle and nonmuscle cells and cause contractions to occur. A family of other types of myosin exist, numbered in order of their discovery. Instead of the long rod-like tail of myosin II, the tail attached to the motor head is small. They are sometimes called **minimyosins.** These molecules cannot form bipolar bundles needed for contraction but single motor molecules run along actin filaments, at the expense of ATP hydrolysis, by the same basic mechanism as described for muscle. The tail of the minimyosin is designed to attach to another structure (the cargo) such as a vesicle membrane which is moved along the actin filament. The members of the family have different tails for binding to specific vesicle membranes or other molecular cargoes. They have been found in yeast and vertebrates and in almost all tissues including brain. With one exception they run towards the (+) end of the actin fibre.

We now turn to a different class of transport systems in which both the tracks and the motors are different from the actin-myosin ones.

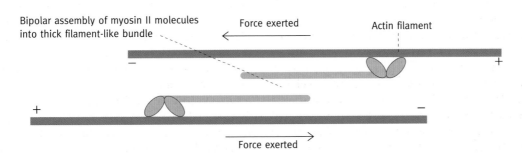

Fig. 8.15 Diagram of the pulling action of myosin II on actin filaments. Note that ATP hydrolysis supplies the energy. The achievement of useful work by this system will depend on appropriate anchoring of the actin filaments to the cell membrane so as to change cell shape. See page 127 for an explanation of the (+) and (−) polarities of the actin filaments. Each myosin molecule has a region capable of interacting with other molecules and thus allowing self-assembly of bipolar filaments. The diagram is a cross-sectional representation – the bipolar assemblies are cylindrical. In the arrangement of actin fibres and myosin bundle shown, a contraction results from the relative sliding of the actin filaments.

Microtubules, cell movement, and intracellular transport

Cytoplasmic microtubules are made by polymerization of **tubulin** protein subunits, which are dimers of α and β tubulin molecules that give the dimer a polarity. Tubulin reversibly polymerizes to form a hollow tube bounded by 13 longitudinal rows of subunits. The polymerization of the dimers occurs in a head-to-tail fashion so that a microtubule has a polarity, with the ends referred to as plus (+) and minus (−). Microtubules are basically unstable in the cell. They can assemble and disassemble very rapidly. The instability is associated with unprotected ends – where these occur, the microtubules undergo collapse by depolymerizing into free tubulin subunits. In the cell, the (−) ends are associated with the **microtubule-organizing centre** (**MTOC**), a structure near the nucleus and containing a pair of small bodies, the centrioles, made of fused microtubules (Fig. 8.16). The cell is pervaded by microtubules radiating out from the MTOC (Fig.8.13(b)).

The MTOC protects the (−) end. Microtubules grow (and collapse) in the cell from the (+) ends. It is thought that when the microtubule reaches an appropriate destination, target proteins 'cap' and protect the microtubule. Microtubules grow out of the MTOC in random directions. Those which make contact with an appropriate component of the cell will, on this model, be capped and stabilized while the remainder collapse.

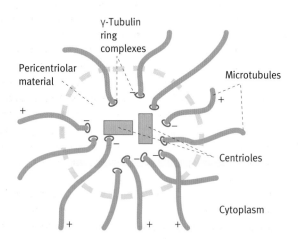

Fig. 8.16 Diagram of microtubules radiating out from the microtubule-organizing centre (MTOC). The (+) and (−) indicate the polarity of the microtubules. The centrioles are a pair of tube-like structures made of fused microtubules; the radiating microtubules originate in the material surrounding the centrioles. (There is no precise boundary to this.) Each microtubule is initiated at a γ-tubulin complex, which includes several other proteins.

But what protects a growing microtubule *before* it reaches a target protein? Free tubulin subunits have GTP attached to them; a GTP-tubulin subunit added to a growing microtubule protects the (+) end. Tubulin is a latent GTPase; its GTP is hydrolysed to GDP and P_i after addition of a subunit to a growing tubule, but this does not happen until after a slight delay. Thus newly added subunits will be in the GTP form, so temporarily capping the microtubule, which protects the end from collapse. A tubulin-GDP subunit is not protective so that unless a new tubulin-GTP monomer is added before hydrolysis of the GTP on the last added monomer is hydrolysed, the microtubule depolymerizes. The GTP hydrolysis in this model is acting as a clock. Note that this situation parallels that in actin-filament polymerization, described earlier, except that there it is ATP rather than GTP. In cilia and flagella, where permanent microtubules occur (see below), covalent modification of the protein with cross-links between them occurs after assembly.

It may be that formins, already mentioned above, are involved in microtubule assembly, as well as in actin fibre assembly.

Molecular motors: kinesins and dyneins

Two types of microtubule-associated motors have been identified, **kinesin** and **dynein**. both of which are dependent on ATP for their movement. The motor domains of kinesins and myosin show structural similarities which suggest that they evolved from a common ancestor. Kinesin travels along a microtubule in the (−) → (+) direction, and dynein in the opposite direction. There are families of kinesin and dynein molecules specialized for different functions. The heads which perform the actual movement along the microtubule are probably a constant motif within each family while different tail structures attach to specific cargoes such as vesicles (Fig. 8.17(b)). Kinesin and dynein molecules (supplied with ATP) will walk along microtubule fibres immobilized on a solid, just as myosin heads will walk along immobilized actin fibres.

The best-known kinesin, known as **conventional kinesin**, is illustrated in Fig. 8.17(a). There are two heads which power the movement along microtubules. The motor heads are about half the size of those of the myosins. The movement differs from that of the myosins in that the two heads indulge in a walking-like action along the microtubule, one attached at a time with the heads swivelling past one another at each step. The neck of the molecule is a flexible section which allows this. The movement is driven by ATP hydrolysis. The other end of the molecule binds to the cargo, such as a vesicle.

As stated, the kinesin travels towards the (+) end of a micro-tubule, which means in general towards the periphery of the cell. In a nerve axon they pull vesicles from the cell body outwards along the axon. A fairly common genetic defect exists in which the kinesins are deficient, resulting in a **peripheral neuropathy** causing pains in hands and feet.

(a)

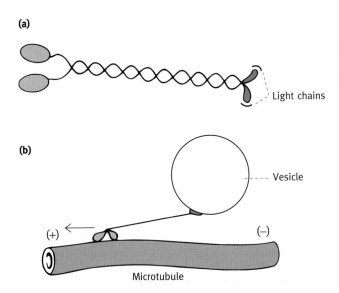

Light chains

(b)

Vesicle

(+)   (−)

Microtubule

Fig. 8.17 **(a)** Diagram of the structure of a kinesin molecule. **(b)** Diagram illustrating how a kinesin molecule can transport vesicles along a microtubule. Also see Fig. 8.18 for an electron micrograph of vesicles being transported within a cell.

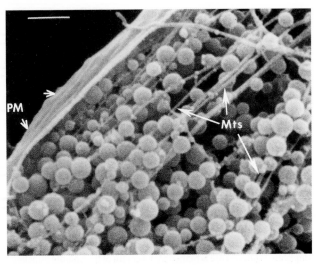

Fig. 8.18 Vesicle transport by microtubules. Scanning electron micrograph of pigment-containing vesicles being transported along microtubules in a chromatophore of a squirrel fish. The change of colour of the cell is effected by the movement of the vesicles to and from the cell centre. Mts, microtubules; PM, plasma membrane. Scale bar = 0.5 μm.

Dynein is a much larger molecule, again with two motor heads. It travels along microtubules in the direction opposite to that of kinesin. Dynein motors are found in almost all eukaryotic cells and also have the task of transporting vesicles.

A striking example of vesicle transport along microtubules is shown in the beautiful scanning electron micrograph in Fig. 8.18. Some fish and amphibia have a camouflage mechanism which rapidly changes the colour of the skin. This is achieved by vesicles containing pigment either moving along microtubules to the periphery of the cells or becoming evenly distributed giving the background appearance.

Role of microtubules in cell movement

This is confined to cilia and flagella in eukaryotes. Cells lining the respiratory passages of lungs have large numbers of cilia whose beating motion sweep along mucus and its entrapped foreign particles. Sperm propel themselves with flagellae. Cilia are smaller than flagella, but other than that the organelles are basically similar in structure. Microtubules, in this case permanent structures, run down the length of the organelle, originating in a basal body closely resembling a centriole. Associated with the microtubules are dynein molecules. Using ATP energy, they 'walk' along microtubules (towards the (−) end) with their tails attached to a partner microtubule; the microtubule pairs are cross-linked and cannot slide relative to one another, so causing a bending wave motion in the cilium or flagellum, instead of a sliding one. Bacterial flagella are different structures and they rotate.

Role of microtubules and molecular motors in mitosis

Chromatids, as the pairs of new chromosomes are called in mitosis (page 323), are held together by their centromeric regions and have to be separated (Fig. 8.19). The nuclear membrane disappears and the MTOC divides, one copy of each migrating to opposite ends of the cell. From these, microtubules grow out to form the mitotic spindle. The duplicated chromosomes become arranged in the central plane of the spindle. There are two types of microtubule in the spindle; the first type, called kinetochore fibres, attach to the centromeres of each chromatid (see Fig. 8.19). The second type are polar fibres which overlap at their positive ends (explained below).

Two actions are involved in chromosome segregation. First the chromosomes are pulled towards the poles. This is caused by shortening of the kinetochore fibres attached to the centromeres. Microtubules cannot contract; the shortening is due to disassembly of the microtubule at the attachment point to the chromosome. It is possible that a motor is involved but the process is not fully understood in detail.

In the second action, kinesins, which operate between the overlapping polar fibres, pull them past each other and drive the centrosomes to opposite ends of the cell. Kinesin tails are bound to the one fibre and their motor heads to the overlapping one of opposite directionality (Fig. 8.19). In moving towards the (+) end of the latter they reduce the overlap of the two fibres and in doing so drive the centrosomes to the opposite sides of the cell. Also dynein molecular motors associated with the astral microtubules pull them apart. These move to the negative end of the astral microtubules.

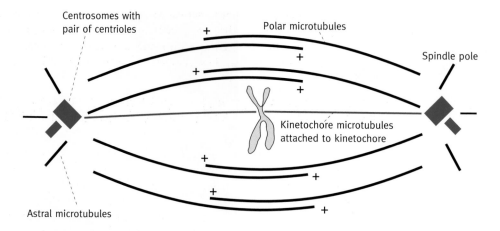

Fig. 8.19 Mitotic spindle at the metaphase state. The microtubules originate at the centrosomes. The duplicated chromosomes (only one duplicate is shown) are arranged at the centre of the spindle attached to the kinetochore fibres at the complex of proteins known as the kinetochore (see Fig. 21.10). The overlapping polar fibres propel the centrosomes apart; it is believed that kinesins in the overlapping regions of the polar fibres cause the separation. The astral microtubules assist in the separation of the spindle poles, (+)-directed kinesin motors being involved. The kinetochore fibres are progressively shortened at the kinetochore attachment site by disassembly. This pulls the duplicated chromosomes apart to opposite ends of the cell in preparation for cell division.

BOX 8.3 Effects of drugs on the cytoskeleton

Actin and microtubule filaments are assembled and disassembled with great speed. The correct organization of these processes is essential to the development, multiplication, and life of the cell. This is illustrated dramatically by the effects of a number of plant and sponge-derived drugs that are produced as a defensive measure against predators. The drugs work by binding either the monomer forms of the filaments or the polymerized form, which perturbs the equilibrium to one side or the other. Some work on actin and some on microtubules.

The poisonous mushroom *Amanita phalloides*, or death cap, produces the deadly toxin **phalloidin**; this binds to F actin and stabilizes actin filaments so preventing their turnover. **Cytocholasin** is a toxin of fungal origin that binds to the growing (+) ends of actin filaments;

it prevents the assembly and disassembly of the filaments and so inhibits for example the formation of the contractile ring needed to separate daughter cells after mitosis. It acts on mammalian cells.

Turning to microtubules, **colchicine** is an alkaloid produced by the autumn crocus; it binds tightly to tubulin monomers, preventing microtubule assembly and therefore spindle formation. It promotes depolymerization of microtubules and freezes mitosis at metaphase. **Vinblastine** and **vincristine** are alkaloids from *Vinca rose*, which also freeze mitosis at metaphase by binding to spindle fibres. These drugs may have some anticancer potential. **Paclitaxel**, first isolated from the bark of the Pacific yew tree, but since synthesized chemically, interferes with normal microtubule growth during cell division. Sold as **Taxol**®, it is used in cancer chemotherapy.

Intermediate filaments

The actin network has microfilaments 6 nm in diameter; the microtubule filaments are 20 nm in diameter. The third type of filaments in eukaryotic cells average 10 nm in diameter – hence their name of **intermediate filaments** (IFs). These have a different role from actin filaments and microtubules. In short, the role of IFs is to confer mechanical strength to cells. They are found in vertebrates, but not in all cells, and only in a few nonvertebrates.

IFs comprise a diverse group of homologous proteins. Typically there is a core filament about 350 amino acid residues in length, with the ends varying in the different types of IF. They are elongated molecules whose central α helical sections form

coiled coil dimers. These in turn laterally pack together forming robust structures.

Various types of IF occur in different eukaryotic cells and are expressed at specific stages of development and differentiation, suggesting that they play important roles. These include **keratin** IFs in epidermal cells, which confer skin toughness; a number of **lamin** proteins form a network associated with the inner surface of the inner nuclear membrane. Intermediate filaments in general are not ephemeral, as most actin and microtubules are, but lamins are the exception in that at mitosis the nuclear membrane is disassembled after phosphorylation. **Neurofilaments** exist in nerve cells to give mechanical support to the long axons; **desmin** filaments are located in the Z discs of sarcomeres. At the extreme end of the spectrum the **keratins** of hair, fingernails, and horses' hooves are made from IFs.

IFs, however, are not essential for cell growth and division; cells in culture, in which IF formation does not occur due to mutations, still grow and divide. This is possibly because they are not subject to the mechanical stresses experienced by cells in functional tissues. It seems, however, that IFs are not inert. They are subject to multiple phosphorylations, which may determine their interactions. (For a comprehensive account, see the **Bishr Omary** *et al.* review in Further reading).

Summary

Muscle contraction and all molecular motors depend on conformational changes in proteins. Skeletal muscle cells have multiple myofibrils running through their length, each divided into sarcomeres which are the contractile units. The sarcomeres have at each end a strong Z disc and projecting from these are thin filaments made of the fibrous protein actin pointing to the centre of the sarcomere. They form a hexagonal 'cage' inside of which a thick filament is suspended. The thick filament is a bundle of hundreds of myosin molecules arranged in a bipolar fashion. These are rod like molecules, each a dimer arranged as a coiled coil and terminating in a pair of globular heads.

Contraction is explained by the sliding-filament model in which sarcomeres are caused to shorten by thick filaments. The myosin heads contact the actin filaments. A cycle of events takes place driven by ATP hydrolysis to ADP and P_i, in which the heads exert a force on the thin filaments pulling the Z discs towards the centre of the sarcomere, thus shortening it. The ATP hydrolysis however does not coincide with the power stroke. The power stroke occurs when P_i and ADP are released and the actin filament is forced to slide. The myosin heads exert their force by a swinging lever arm mechanism in which a lever-like arm of the myosin swings from one position to another.

Contraction is triggered by nerve impulses, which cause release of calcium ions from the sarcoplasmic reticulum, a sac surrounding the myofibrils. The ions initiate the contractile cycle of the myosin heads by causing a conformational change in a tropomyosin complex attached to the actin filaments, which somehow controls the triggering of the contraction cycle. The contraction is terminated by removal of the Ca^{2+} ions by an ATP-driven pump, which returns the ions to the sarcoplasmic reticulum. Smooth muscle has a different structure and control system.

Muscle is a specialized grouping of molecular motors but almost all eukaryotic cells have individual molecular motors that run along tracks pulling loads. A cell is like a busy city in its transporting activities. Their function is to transport membrane vesicles containing newly synthesized macromolecules to parts of the cell that they would not reach by diffusion. The tracks are of two types comprising what is known as the cytoskeleton.

Actin filaments are made of the same protein as the thin filaments of muscle. Assembly of the filaments involves hydrolysis of ATP. They are assembled and disassembled as needed and this occurs with disconcerting rapidity. (This does not occur in muscle thin filaments.) The molecular motors using these belong to the myosin family. They have a pair of globular heads as in muscle myosin but the long rod-like tail is replaced by a short one that attaches to the vesicles to be transported.

Microtubules are hollow tubes made of polymerized tubulin protein monomers, the process involving GTP. They also develop and collapse as needed.

ATP-driven molecular motors known as kinesins and dyneins move along the microtubules (which have polarity) in opposite directions pulling cargoes. Kinesins play an important role in chromosome movements during cell division as well as in the transport of organelles.

Further reading

Muscle contraction

Holmes, K. C. (1997). The swinging lever arm hypothesis of muscle contraction. *Curr. Biol.*, 7, R1 12–18.

Clarke, M. (1998). Sighting of the swinging lever arm of muscle. *Nature*, 395, 443.
A brief News and Views account of the three-dimensional structure of the myosin motor head.

Weir, A. (2000). Muscular dystrophy. *Curr. Biol.*, 10, R92.
A one-page quick guide giving essentials of the molecular aspects of the disease.

Spence, H. J., Chen, Y.-J., and Winder, S. J. (2002). Muscular dystrophies, the cytoskeleton and cell adhesion. *BioEssays*, 24, 542–52.

Smooth muscle contraction

Small, J. V. (1995). Structure–function relationships in smooth muscle: the missing links. *BioEssays*, 17, 785–92.
Discusses the structural organization of the smooth muscle cell and control of contraction.

Green, N. M. and MacLennan, D. H. (2002). Calcium calisthenics. *Nature*, 418, 598–9.
A detailed look at sarcoplasmic Ca^{2+} ATPase.

Halsall, P. J. and Hopkins, P. M. (2002). Malignant hyperthermia. *Br. J. Anaesth.*, 3, 5–9.

Actin filament assembly

Higgs, H. N. (2005). Formin proteins: a domain based approach. *Trends Biochem. Sci.*, **30**, 342–53.

Popowicz, G. M. *et al.* (2006). Filamins: promiscuous organiser of the cytoskeleton. *Trends Biochem. Sci.*, **31**, 411–9.

Filamins – a group of proteins involved in the assembly of the actin network.

Goode, B. L., and Eck. M. J. (2007). Mechanism and function of formins in the control of actin assembly. *Annu. Rev. Biochem.*, **76**, 593–627.

A comprehensive review probably more suitable for instructors.

Intermediate filaments

Helfand, B. T., *et al.* (2004). Intermediate filaments are dynamic and mobile elements of cellular architecture. *J. Cell Sci.*, **117**, 133–41.
Commentary on motility associated with IF assembly, disassembly, and subcellular organization.

Bishr Omary, M., Ku, Nam-On, Guo-Zhong Tao, and Liao, J. (2006). 'Heads and tails' of intermediate filament phosphorylation: multiple sites and functional insights. *Trends Biochem. Sci.*, **31**, 383–94.

Cell movement

Banting, G. and Higgins, S. J. (eds.) (2000). Molecular motors. *Essays Biochem.*, **36**, 1–29.
A complete issue devoted to 13 reviews on molecular motors.

Cross, R. A. and Carter, N. J. (2000). Molecular motors. *Curr. Biol.*, **10**, R177–9.
One of the 'primer' series which gives essential information in a couple of pages.

Sablin, E. P. (2000). Kinesins and microtubules: their structures and motor mechanisms. *Curr. Opin. Cell Biol.*, **12**, 35–41.
Reviews mainly structural aspects.

Schliwa, M. and Woehlke, G. (2001). Switching on kinesin. *Nature*, **411**, 424–5.
A News and Views article on the structure and mechanism of kinesin.

Burgess, S. A. (2003). Dynein structure and power stroke. *Nature*, **421**, 715–18.

Various authors (2003). A special issue on molecular machines. *BioEssays*, **25**, 1145–235.
A thematic issue on molecules or molecular complexes that perform functions involving the conversion of chemical energy into molecular or chemical work.

Microtubules in cell movement

Byard, E. H. and Lange, B. M. H. (1991). Tubulin and microtubules. *Essays Biochem.*, **26**, 13–25.
Excellent general review.

Goldstein, L. S. B. (1993). With apologies to Scheherazade: tails of 1001 kinesin motors. *Ann. Review Genetics*, **27**, 319–51.
An in-depth review of these minimotors with useful overviews.

Walker, R. A. and Sheetz, M. P. (1993). Cytoplasmic microtubule-associated motors. *Annu. Rev. Biochem.*, **62**, 429–51.
Detailed review of the minimotors and their roles.

Wallee, R. B. and Sheetz, M. P. (1996). Targeting of motor proteins. *Science*, **271**, 1539–44.
Discusses the movements of kinesins and dyneins on microtubules.

Microtubules in vesicle transport

McNiven, M. A. and Ward, J. B. (1988). Calcium regulation of pigment transport *in vitro. J. CellBiol.*, **106**, 111–25.
A research paper but readable.

Allkan, V. J. and Schroer, T. A. (1999). Membrane motors. *Curr. Opin. CellBiol.*, **11**, 476–82.
Reviews membrane vesicle traffic along actin and microtubule tracks.

Endow, S. A. (2003) Kinesin motors as molecular machines. *BioEssays*, **25**, 1212–19.
Transport processes within the cytoplasm which move vesicles and other organelles along actin filaments and microtubules.

Microtubules in mitosis

Cande, W. Z. and Hogan, C. J. (1989). The mechanism of anaphase spindle formation. *BioEssays*, **11**, 5–9.

Glover, D. M., Gonzalez, C., and Raff, J. W. (1993). The centrosome. *Sci. Am.*, **268**, 32–9.
Useful review of this relatively little-known structure.

Desai, A. (2000). Kinetochores. *Curr. Biol.*, **10**, R508.
A one-page guide.

Sharp, D. J., Rogers, G. C., and Scholey, J. M. (2000). Microtubule motors in mitosis. *Nature*, **407**, 41–7.

Wittman, T., Hyman, A., and Desai, A. (2001). The spindle: a dynamic assembly of microtubules and motors. *Nat. Cell Biol.*, **3**, E28–E34.

Problems

1   How is contraction of a voluntary striated muscle sarcomere controlled?

2   How is smooth muscle contraction controlled?

3   Can you see any similarities in principle between the mechanisms of ATP synthesis by ATP synthase and its utilization by myosin for contraction?

4   Explain with simple diagrams the mechanism by which myosin causes movement of the actin filament.

5   Actin is found in nonmuscle cells. What are its roles there?

6   What are microtubules? What controls their assembly and collapse?

7   Microtubules with unprotected ends undergo collapse. What protects the ends as they form?

8   What are kinesin and dynein? How does their movement differ from that of myosins?

9   At cell division, chromosomes on the metaphase equatorial plate move apart. Microtubules are attached to the kinetochores and shorten as the chromosomes move apart. Does this mean that microtubules contract? Explain your answer.

10 What are intermediate filaments? What are their functions?

11 What do the proteins G actin and tubulin have in common?

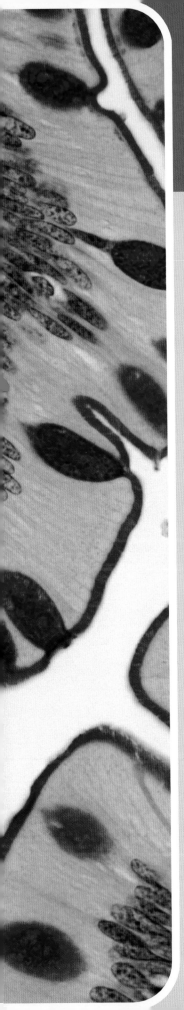

Metabolism

Small intestine
Light micrograph of a section through the small intestine (jejunum). The jejunum is the site of absorption of the
products of digestion. It is deeply folded and lined with finger-like projections (villi). These are lined with microvilli
(purple) and together they greatly increase the surface area and absorptive capacity of the jejunum. The villi are also
covered in goblet cells (purple circles), whose secretions protect and lubricate the gut and help in the formation of
faeces. Beneath the goblet cells are enterocytes (blue), which absorb the nutrients from digestion.
INNERSPACE IMAGING/SCIENCE PHOTO LIBRARY

Chapter 9

Food digestion, absorption, distribution to the tissues, and appetite control

First, where does this topic fit in the overall arrangement of the book? In Chapter 3 we looked at what is necessary for chemical reactions to occur and at the chemistry of weak bonds. Then in Chapter 4, the structure of protein molecules laid the basis for the key reactions and processes on which life depends, and in Chapter 5 came methods for protein investigation. Next, we dealt with the properties of enzymes, which bring about the chemical reactions (Chapter 6). Chapter 7 dealt with the cell membranes because this information is essential for understanding digestion and absorption. We now progress to the major area of metabolism.

The purposes of metabolism can be summed up as follows.

- A major role of metabolism is to oxidize food to provide energy in the form of adenosine triphosphate (ATP).
- Food molecules are converted to new cellular material and essential components.
- Waste products are processed to facilitate their excretion in the urine.
- Some specialized cells in human babies and hibernating animals oxidize food to generate heat. This is exceptional since heat is, in general, a byproduct of metabolism.

If you have already heard of metabolic pathways such as glycolysis and the citric acid cycle, we suggest you put them out of your mind for the time being. We do not, in this chapter, give any details of the metabolic pathways or the metabolites involved since we first want to give the broad picture.

Food digestion, absorption, and distribution to the tissues is a logical place at which to start, for that is where metabolism begins. At a gross level, the subject may seem mundane; at the molecular level it is as elegant and interesting as the topics with a more exciting image.

In this chapter we will discuss:

- what are the chemical components of food

- how it is prepared for absorption into the bloodstream – what are the bonds in these compounds that need to be broken
- how it reaches the bloodstream
- the traffic of food molecules between the blood and tissues, and between different tissues
- how this is regulated to satisfy different physiological needs – in essence, the broad logistics of fuel movements in the body.
- finally there is what has been described as almost the major medical problem facing the developed world – that of appetite and regulation of food intake.

Chemistry of foodstuffs

There are three main classes of food – proteins, carbohydrates, and fats.

- **Proteins**, as described in Chapter 4, are large polymers of amino acids linked together to form polypeptide chains, which, in turn, may assemble into dimers or larger aggregates.
- **Carbohydrates** are the sugars and their derivatives; the name comes from their empirical formulae with carbon atoms and the elements of water in the ratio of 1:1 (CH_2O). Although simple monosaccharide sugars such as glucose occur in food, most of the carbohydrate in food is in the form of disaccharides, such as sucrose and lactose, or polysaccharides such as starch.
- **Fats** in the diet are mainly in the form of triacylglycerols (TAGs), sometimes called neutral fats (page 102). Butter, olive oil, and the visible fat of meat are typical examples. Polar lipids (page 103) are also present.

Digestion and absorption

With the exception of monosaccharides such as glucose, all of the foodstuffs mentioned above are digested by hydrolysis into their constituent parts in the small intestine (a minor exception is the absorption of dipeptides and tripeptides). To be absorbed, substances must cross membranes to enter the mucosal cells lining the intestine. TAGs cannot do this, nor can proteins, and among the carbohydrates only monosaccharides are absorbed. Thus, digestion consists mainly of the conversions:

- Proteins, held together by peptide bonds → amino acids
- Carbohydrates, held together by various glycosidic bonds → sugar monomers (monosaccharides)
- TAGs, held together by ester bonds → fatty acids and monoacylglycerol.

Anatomy of the digestive tract

The following regions (of non-ruminants) are involved.

- In the **mouth** food is masticated and lubricated for swallowing; limited starch digestion occurs.
- The **stomach** contains hydrogen chloride (HCl), which 'sterilizes' food and denatures proteins; partial digestion of proteins occurs.
- The **small intestine** is the major site of digestion and absorption of all classes of food. It is lined by fine finger-like processes, the villi, which are covered by epithelial cells – these are known as brush border cells because the microvilli of the epithelial cells resemble bristles on a brush (Fig. 8.14). The microvilli are on the external membrane of the epithelial cells of the villi, giving the large surface area needed for absorption.
- The **large intestine** is involved in the removal of water. It is also the site of bacterial fermentation of some fibre and other components resistant to normal digestion.

What are the energy considerations in digestion and absorption?

So far as digestion goes, there are no thermodynamic problems. Hydrolytic reactions such as hydrolysis of proteins to amino acids, of disaccharides and polysaccharides to monosaccharides, and of fats to fatty acids and monoacylglycerols are all exergonic processes – they have negative ΔG values sufficient to push the equilibrium entirely to the side of hydrolysis. Hydrolytic reactions (splitting or lysis by water) in biochemistry are invariably of this type. Absorption is a different matter since it is often an active process in which molecules are absorbed against a concentration gradient, and energy is needed.

A major problem in digestion – why doesn't the body digest itself?

Food is, chemically, little different from the tissues of the animal that eats it. A fearsome array of enzymes is produced by the digestive system to completely digest the food into its components, and those enzymes have to be produced inside living cells, which, if exposed to their action, would be destroyed. There are two major types of defence against this.

Zymogen or proenzyme production

The enzymes are produced as inactive **proenzymes** or **zymogens** that are activated only when they reach the stomach and small intestine. Thus, glands producing digestive enzymes secrete most of them as inactive proteins and are never themselves exposed to the destructive processes. In the case of enzymes that pose no threat (amylase, the enzyme that hydrolyses starch, is one), proenzymes are not involved. The question also arises as to how cells selectively secrete digestive enzymes or other proteins, but this is best left until we deal with protein targeting (Chapter 26) for the mechanism is complex and not directly related to digestion. We will deal with the mechanisms of zymogen activation when we come to particular enzymes.

Protection of intestinal epithelial cells by mucus

The cells lining the intestinal tract are protected from the action of activated digestive enzymes, the main line of defence being the layer of mucus that covers the epithelial lining of the gut. The essential components of mucus are the **mucins**. These are large glycoprotein molecules – proteins with large amounts of carbohydrates attached to their polypeptide chains in the form of oligosaccharides that contain a mixture of sugars you have already met in membrane glycoproteins (page 56): glucosamine, fucose, and sialic acid to name a few. The mucins form a network of fibres, interacting by noncovalent bonds and resulting in a gel containing more than 90% water due to the hydrophilic carbohydrates that protects intestinal cells. The carbohydrates may protect the mucin proteins themselves from digestion. The mucin gel is quite permeable to low molecular weight digestion products, but much less permeable to digestive enzymes. The mucins are synthesized and secreted by special goblet cells in the epithelial lining of the gut. The amount of mucin secreted is controlled.

Digestion of proteins

In a normal or 'native' protein, the polypeptide chain is folded up, the shape being largely determined by weak bond interactions. In this compact folded form, many of the peptide bonds are hidden away inside the molecule where they are not accessible to hydrolytic enzymes. An important early step in digestion

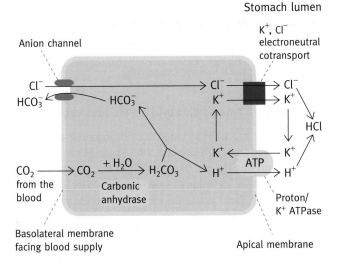

Fig. 9.1 Mechanism of gastric hydrochloric acid (HCl) secretion.

is to denature the native proteins. This is done by the acid in the stomach where, due to HCl secretion, the pH is about 2.0. This partially disrupts the polypeptide folding, making the polypeptide chain susceptible to proteolysis.

HCl production in the stomach

The stomach epithelial lining contains **parietal** or **oxyntic** cells that secrete acid. In essence, the process consists of secreting H^+ – this is similar to the ejection of Na^+ from cells against a concentration gradient (page 113), which is achieved by a Na^+/K^+ ATPase in the membrane. In a similar manner, the acid-secreting cells have a H^+/K^+ ATPase. Using energy from ATP hydrolysis, they eject H^+ and import K^+, the latter recycling back to the exterior of the cell. The process is shown in Fig. 9.1. Where do the protons come from? An enzyme, **carbonic anhydrase**, converts CO_2 inside parietal cells to carbonic acid, which dissociates as shown.

$$CO_2 + H_2O \rightleftharpoons H_2CO_3 \rightleftharpoons H^+ + HCO_3^-$$
Carbonic anhydrase

The resultant protons are pumped into the stomach lumen and the bicarbonate ions then exchange with Cl^- in the blood via an anion-transport protein (Fig. 9.1). We have already met this type of exchange in the anion channel of the red blood cell. The Cl^- ions then exit to the stomach lumen (cavity) to form HCl with the secreted protons, as shown in the figure.

Pepsin, the proteolytic enzyme of the stomach

To indicate an inactive precursor of an enzyme, the suffix 'ogen' is used; for example, pepsinogen is the inactive form of pepsin. Alternatively, the term proenzyme is used in some cases.

Cells of the stomach epithelium secrete pepsinogen. The secretion is stimulated by the hormone gastrin released by stomach cells into the blood in response to food. Pepsinogen is pepsin with an extra stretch of 44 amino acids on the polypeptide chain. This additional segment covers, and blocks, the active site of the enzyme. When pepsinogen meets HCl in the stomach, a change in conformation exposes the catalytic site which self-cleaves – it cuts off the extra peptide from itself to form active pepsin. As soon as a small amount of active pepsin is produced in this way, it converts the rest of the secreted pepsinogen to pepsin.

Pepsin is unusual for an enzyme in that it works optimally at acid pH (Fig. 6.7(a)), most enzymes requiring a pH near neutrality. It hydrolyses peptide bonds of proteins within the molecule, producing a mixture of peptides. It is therefore an *endo*enzyme or *endo*peptidase; that is, it does not attack terminal peptide bonds at the end of molecules, but only those within the molecule. The derivation is from the Greek *endo*, meaning within.

Proteolytic enzymes (proteases) are usually specific in their action – they hydrolyse only peptide bonds adjacent to certain amino acid residues. Pepsin has this characteristic so that only partial digestion of proteins can occur in the stomach. It hydrolyses only those peptide bonds in which one of the aromatic amino acids – tyrosine, phenylalanine, or tryptophan – supplies the NH of the peptide bond.

In ruminants, another stomach enzyme, **rennin**, acts on the casein of milk, causing clotting, so that it does not pass too rapidly through the stomach and escape gastric digestion.

Completion of protein digestion in the small intestine

The chyme, as the partially digested stomach contents are called, enters the duodenum at the start of the small intestine. The acid stimulates the duodenum to release hormones (**secretin** and **cholecystokinin**) into the blood, which stimulate the pancreas to release pancreatic juice. This is alkaline and (together with bile juice) neutralizes the HCl giving a slightly alkaline pH suitable for pancreatic enzyme action and terminating pepsin activity.

A battery of pancreatic proteases is produced from clusters of cells in the pancreas and secreted into the intestine as pancreatic juice via the main pancreatic duct. There are three endopeptidases – **trypsin**, **chymotrypsin**, and **elastase** – all entering the small intestine in the form of the inactive proenzymes, trypsinogen, chymotrypsinogen, and proelastase, respectively. An exopeptidase, **carboxypeptidase**, secreted as an inactive proenzyme, chops off terminal amino acids from the C-terminal end of peptides.

$$H_2N-CH-CO-NH-CH-CO-NH-----NH-CH-COOH$$
$$\qquad\quad | \qquad\qquad\qquad | \qquad\qquad\qquad\qquad\quad |$$
$$\qquad\quad R \qquad\qquad\qquad R' \qquad\qquad\qquad\qquad R''$$

Amino terminal end   Carboxy terminal end
or N-terminal end   or C-terminal end

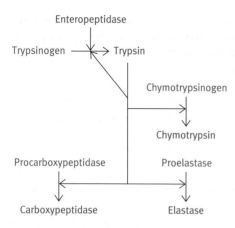

Fig. 9.2 Activation of the pancreatic proteolytic enzymes. Proenzymes (zymogens) are shown in red; activated proteolytic enzymes are shown in green.

The mechanism by which these enzymes work has been given in Chapter 6, page 94.

Activation of the pancreatic proenzymes

As with pepsinogen activation, pancreatic proenzymes are activated by proteolytic cleavage of the proenzymes. The activation process is triggered by a specialized enzyme secreted by the cells of the small intestine, an active enzyme (not a proenzyme in this case) called **enteropeptidase**, which hydrolyses a single specific peptide bond of trypsinogen, activating it to trypsin. The initial amount of active trypsin so produced now activates all proenzymes (including trypsinogen itself) so that all are rapidly activated in a proteolytic cascade (Fig. 9.2). The activation details differ for different enzymes. This is an elegant mechanism for ensuring that the array of *active* enzymes is present only in the intestinal lumen. Their premature activation in the pancreas is deleterious – if it occurs, the disease pancreatitis ensues. Blockage of the pancreatic duct, or damage to the gland, can trigger this disease. After synthesis, the proenzymes are stored in the cells producing them in membrane-bound secretory vesicles. On hormonal or neurological stimulation these fuse with the cell membrane and release their contents by exocytosis (page 107). As a safeguard against accidental leakage of the vesicles before secretion, the cells contain a **trypsin-inhibitor protein** capable of inactivating any trypsin that might accidentally come in contact with the cytoplasm. The inhibitor protein fits the trypsin active site so perfectly that a sufficient number of weak bonds are formed to make the combination of the two almost completely irreversible. The mechanism by which enzymes to be secreted are produced and enveloped in the secretory vesicles is more suitably dealt with later (page 416).

An additional enzyme, called **aminopeptidase** (an **exopeptidase**), attached to the outside (lumen side) of intestinal cells progressively hydrolyses off terminal amino acids from the amino ends of peptides. Thus the three endopeptidases (trypsin, chymotrypsin, and elastase), each with a different preference for the peptide bonds they attack, chop in the middle of polypeptides, while carboxypeptidase and aminopeptidase chop at each end, and proteins are finally converted to free amino acids in the lumen of the intestine.

Absorption of amino acids into the bloodstream

Amino acids are transported from the intestine across the cell membrane of epithelial cells (the brush border cells; see Fig. 9.6) and into the cell. The brush border cells actively concentrate amino acids inside themselves from where they diffuse into blood capillaries inside the villi. Because there are so many different amino acids, there are different transport pumps in the membrane. The cotransport mechanism (page 113) is one in which in which the Na^+ gradient is used to drive the uptake of some of the amino acids. Figure 7.12 shows glucose being transported by a cotransport mechanism but it applies equally well to amino acids using different transport proteins. In **Hartnup disease**, transport deficiencies impair absorption from the intestine and reabsorption in kidney tubules. This causes deficiencies in essential amino acids (page 283). Absorption of di- and tripeptides is not affected.

We will take up the subject of what happens to the amino acids in the blood on page 169.

Digestion of carbohydrates

Structure of carbohydrates

The main carbohydrates in the diet are starch and other polysaccharides, and the disaccharides **sucrose** and **lactose** (the latter from milk). Free **glucose** and **fructose** also occur. The aim of digestion is to hydrolyse the dietary components to monosaccharides.

In polysaccharides and disaccharides, monosaccharides are linked together by **glycosidic bonds** to form glycosides. This is an important bond from several viewpoints and needs to be described. Glucose has the following structure.

α-D-Glucose   β-D-Glucose

In α-D-glucose (in the pyranose six-membered ring configuration), the —OH on carbon atom 1 points below the plane of the ring (imagine it is below the plane of the paper),

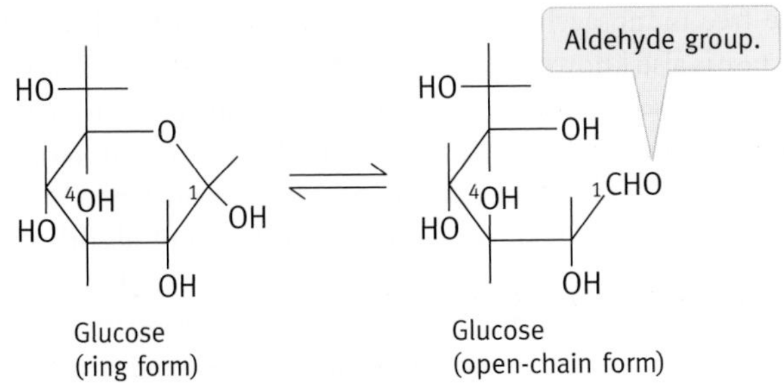

Fig. 9.3 Mutarotation of glucose.

and in β-D-glucose it points above it. In free monosaccharides, the two are in free equilibrium in solution by **mutarotation** (via an open-chain structure). The mutarotation occurs as shown in Fig. 9.3.

The glycosidic bond

Suppose we have two glucose molecules.

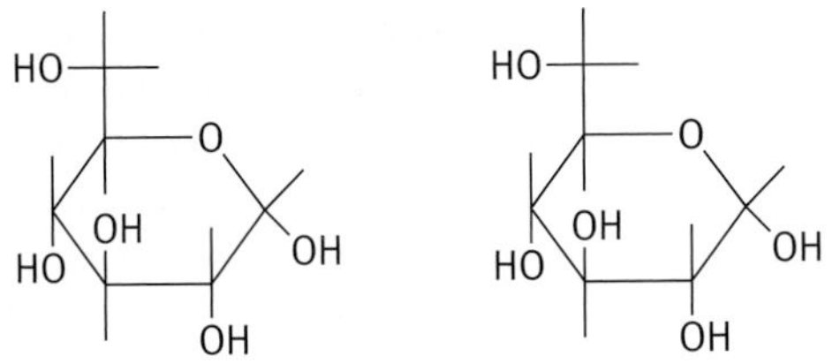

They can (by removal of the elements of water) be joined together by a glycosidic bond.

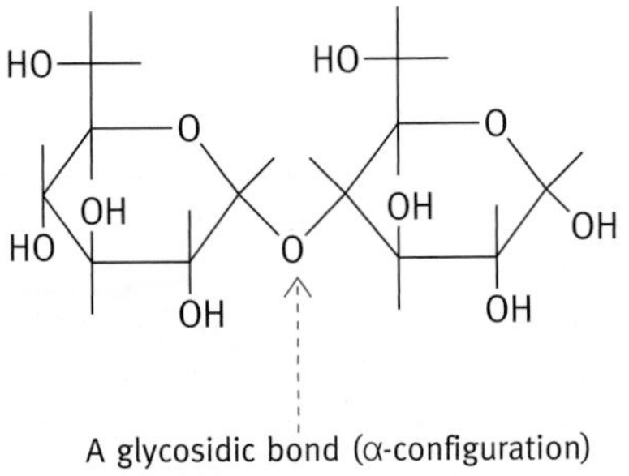

A glycosidic bond (α-configuration)

The glycosidic bond fixes the configuration on carbon atom 1 of the first monosaccharide into one form – it no longer mutarotates. In the example shown, the glycosidic bond is between carbon atoms 1 and 4 of the two units and is in α-configuration. The compound is therefore glucose-$(1 \rightarrow 4)$-α-glucose. It is a disaccharide plentiful in malted barley and has the trivial name **maltose**. As will be described shortly, disaccharides with β-glycosidic bonds also exist.

Digestion of starch

From the structure of maltose above it can be seen that glucose units can be joined together indefinitely to form a huge

polysaccharide molecule. The amylose component of starch, a molecule hundreds of glucose units (glucosyl units) long, is precisely this. If we represent an α-glucosyl unit as ⬭ then amylose is:

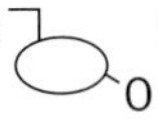

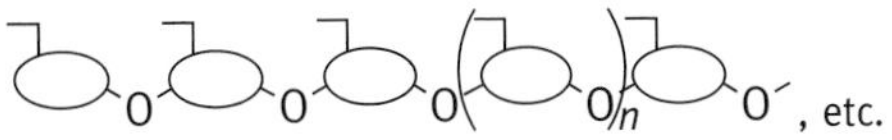

where n is a large number.

The second component of starch is **amylopectin**. This molecule also is a huge polymer of glucose but instead of it being one long chain it consists of many short chains (each about 30 glucosyl units in length) cross-linked together. The glucosyl units of the chains are $(1 \rightarrow 4)$-α-glycosidic bonds as in amylose but the cross-linking between chains is by $(1 \rightarrow 6)$-α-glycosidic bonds, as shown below. (We omit bonds and groups not relevant to the topic in hand.)

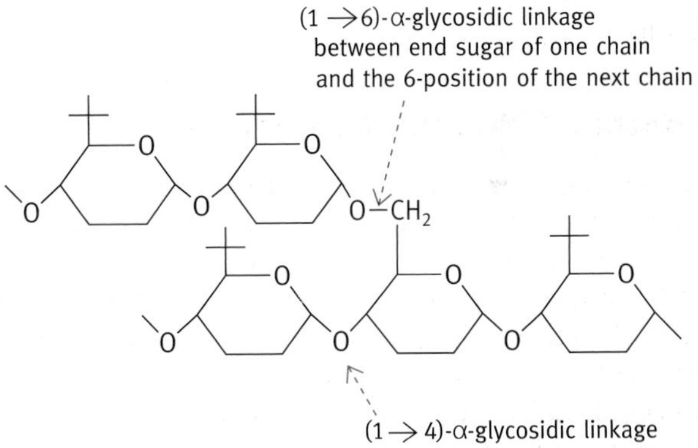

Many chains are linked together in this way. Amylopectin is therefore like this:

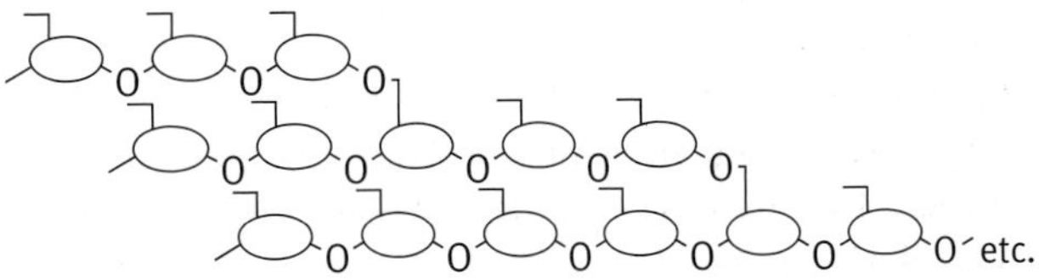

Digestion of starch is initiated by α-**amylase**, which is present in the saliva and pancreatic juice. It is an **endoenzyme**, hydrolysing glycosidic bonds anywhere inside the molecule except that, in amylopectin, it can't attack a bond near the $(1 \rightarrow 6)$ linkage. The amylase thus trims off the molecule but leaves cores containing the $(1 \rightarrow 6)$ bonds plus the nearby glucosyl units. This limit dextrin, as it is called (the limit of hydrolysis), is hydrolysed by an intestinal enzyme (an amylo $(1 \rightarrow 6)$-α-glucosidase) that hydrolyses the $(1 \rightarrow 6)$ linkages. The two-unit and three-unit sugars produced by amylase digestion are attacked by intestinal α-**glucosidase** (**maltase**) which produces free glucose. Salivary amylase has only a brief time to attack starch while food is in the mouth, for stomach acid

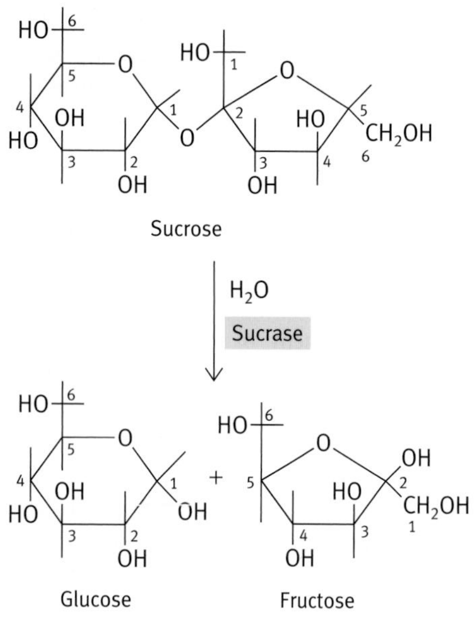

Fig. 9.4 Sucrose conversion to glucose and fructose by sucrase.

the free monosaccharides. Because lactose is a β-galactoside, lactase is also known as β**-galactosidase**. Most adults, to varying degrees, lose the ability to produce lactase as they leave childhood, and Asians particularly may develop **lactose intolerance**, though it also occurs in Caucasians. They are unable to hydrolyse the sugar and, since the disaccharide is not absorbed, it passes to the large intestine where it is fermented by bacteria giving rise to severe discomfort (due to gas production), and diarrhoea. If products containing lactose are avoided, the disease symptoms disappear. Yoghurt, which has less lactose than milk, may be tolerated and cheese, having little lactose, usually presents no problem.

Absorption of monosaccharides

Absorption of glucose into epithelial cells occurs by the Na^+ cotransport mechanism as shown in Fig. 9.6. The Na^+ is continually pumped to the outside by the Na^+/K^+ ATPase (page 111) so this maintains the Na^+ concentration difference and drives the cotransport. The glucose has to exit the cells on the side opposite that of the lumen so as to enter the blood capillaries for transport to the liver by the hepatic portal vein. The glucose transporter for this movement is the facilitated-diffusion type (page 113); the monosaccharide moves down the cell/blood concentration gradient created by active uptake from the intestine (Fig. 9.6). Fructose is absorbed from the gut by a Na^+-independent passive-transport system.

destroys the enzyme. Most of the starch digestion therefore occurs in the small intestine. The digestion of the other two main dietary carbohydrates, sucrose and lactose, also occurs in the small intestine.

Digestion of sucrose

Sucrose is a dimer of glucose and fructose. It is hydrolysed in the small intestine by the enzyme, **sucrase**, as shown in Fig. 9.4. Very large amounts of sucrose can occur in the diet. The absorption of large amounts of fructose presents some special metabolic problems discussed on page 258.

Digestion of lactose

The other major disaccharide of food is lactose. Lactose is galactose-$(1{\rightarrow}4)$-β-glucose, the principal sugar in milk. Galactose has the carbon 4-OH of glucose inverted.

The 'curly' bond shown in Fig. 9.5 is used to simplify the presentation of the structure; it avoids having to invert the second glucosyl unit. An intestinal enzyme attached to the external membrane of the epithelial cells, **lactase**, hydrolyses lactose to

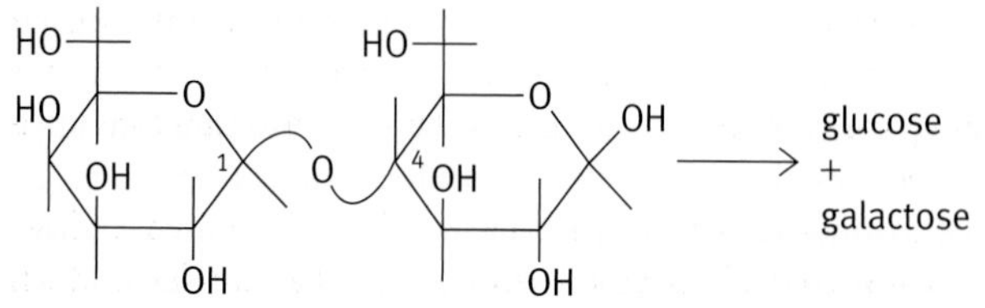

Fig. 9.5 Lactose conversion to glucose and galactose by lactase.

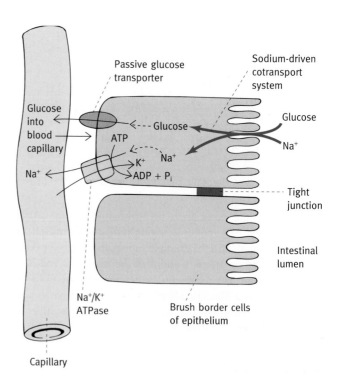

Fig. 9.6 Absorption of glucose from the intestinal lumen by cotransport with Na^+.

Amino acids and sugars and other non-lipid molecules absorbed from the intestine are collected by the portal blood system which delivers digestion products directly to the liver. This arrangement means that these digestion products transit the liver before being released into general circulation – an advantageous arrangement since the liver is responsible for removing most toxic foreign compounds entering the body via the intestine, and for processing some of the absorbed nutrients.

Digestion and absorption of fat

In Chapter 7, page 102, we described what fat is – TAG or neutral fat. Since there are no polar groups, liquid fat in water forms insoluble droplets with minimum contact between the lipid and water and cannot be absorbed as such.

The main digestion of fat occurs in the small intestine by the action of the pancreatic enzyme, **lipase**, which mainly attacks *primary* ester bonds (red arrows in Fig. 9.7); the middle ester bond is a secondary ester bond and is not significantly attacked. Lipase is secreted as a proenzyme, and in the small intestine the first step in activation is by trypsin which hydrolyses off a specific small peptide. But then another pancreatic protein, **colipase**, is also needed to activate lipase; it binds to the enzyme in a 1:1 ratio.

Simple as this digestion reaction is, there are problems. Fat is physically unwieldy in an aqueous medium, and as lipase can

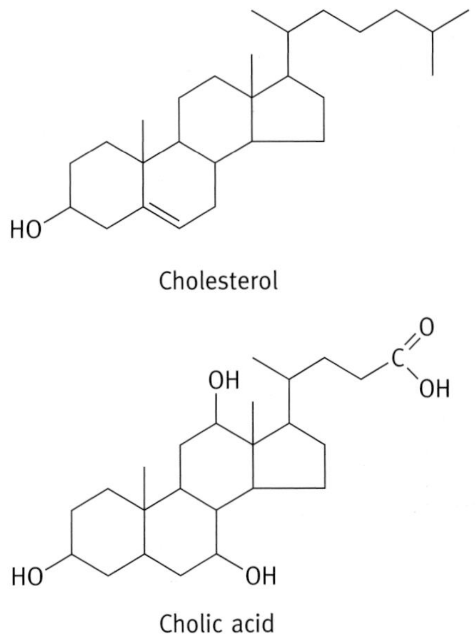

Fig. 9.8 Structures of cholesterol and a bile acid, cholic acid.

attack only at the oil/water interface, the area of this in a simple fat/water mixture is insufficient for the required rate of digestion. The available surface area is increased by emulsifying the fat. The monoacylglycerol and free fatty acids produced by lipase action together with bile salts acting as biological detergents help to emulsify the oily liquid into droplets.

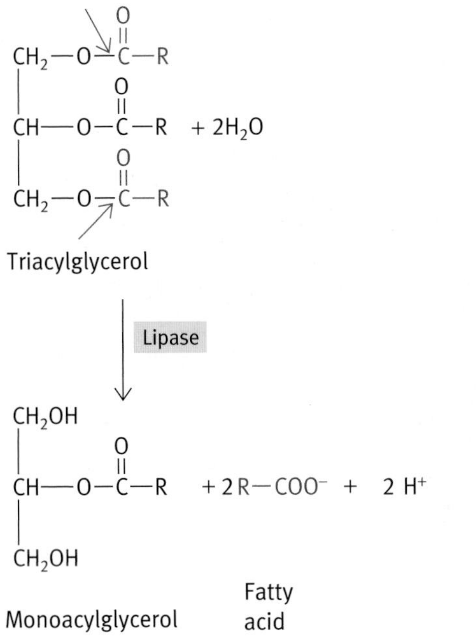

Bile acids from the liver are stored in the gall bladder until discharged into the duodenum. They are produced from cholesterol, whose structure they resemble (Fig. 9.8). The main change is that the hydrophobic side chain of the cholesterol molecule is converted to a carboxyl group, and —OH groups may be added.

The main bile acid is cholic acid; others varying in the number and position of hydroxyl groups exist. The cholic acid mostly has attached to it either glycine or the sulphonic acid, taurine. Glycine is $NH_3^+CH_2COO^-$ Taurine is $NH_3^+CH_2SO_3^-$ (it does *not* occur in proteins). If cholic acid is represented as $RCOO^-$ then glycocholic acid is $RCONHCH_2COO^-$ and taurocholic acid $RCONHCH_2SO_3^-$. These conjugated acids have lower pK_a values (approximately 3.7 and 1.5, respectively) than the parent cholic acid (approximately 5.0). Possibly the reason for the conjugation is to ensure full ionization in the intestinal contents to make them better detergents – the ionized forms are called bile salts.

The products of lipase action are still relatively insoluble in water but they must be moved from the emulsion to the cells lining the intestine to be absorbed. This movement is facilitated by bile salts.

The monoacylglycerols and fatty acids, in the presence of bile salts, form **mixed micelles**. These are disc-like particles in which bile salts are arranged around the edge of the disc surrounding a more hydrophobic core containing digestion products of

Fig. 9.7 Digestion of triacylglycerol to monoacylglycerol and fatty acid by lipase.

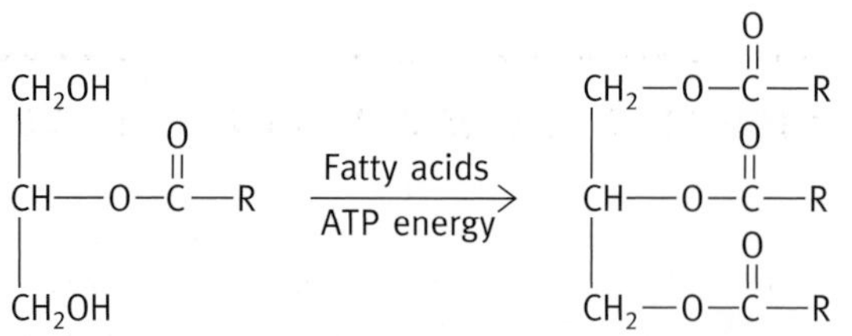

$$CH_2OH$$
$$CH—O—\overset{\overset{O}{\|}}{C}—R \quad \xrightarrow[\text{ATP energy}]{\text{Fatty acids}} \quad$$

Fig. 9.9 Summary of resynthesis of fat.

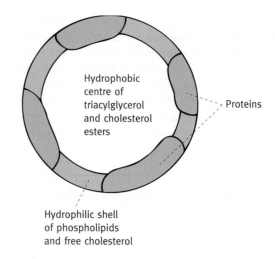

Hydrophobic centre of triacylglycerol and cholesterol esters

Proteins

Hydrophilic shell of phospholipids and free cholesterol

Fig. 9.10 Cross-section of a chylomicron.

fat, together with cholesterol and phospholipids. The particles are smaller than emulsion droplets, giving a clear suspension. They carry higher concentrations of lipid digestion products than is possible in true solution. In this form, the lipid digestion products diffuse to enter the epithelial cells; probably the micelle breaks down at the cell surface and the free lipids diffuse in. Bile salts are also partially reabsorbed and transported back to the liver.

Resynthesis of TAG in intestinal cells

The products of digestion of fat absorbed into the intestinal cells are resynthesized into fat, the fatty acids being re-esterified producing TAGs (Fig. 9.9).

The mechanism of this resynthesis is not given here because it would divert too much from the topic in hand – it is dealt with later (page 228). The TAG, together with absorbed cholesterol in esterified form (see below), must now be transported from the intestinal cell to the tissues of the body. TAG cannot diffuse out of the cell through membranes and, in any case, it would be insoluble in the blood – it cannot simply be ejected into the circulation.

The solution is that TAG and cholesterol are arranged inside the epithelial cells into fine particles, called **chylomicrons**. They are released from the cell by exocytosis (page 107), since they are too large to traverse the membrane in any other way.

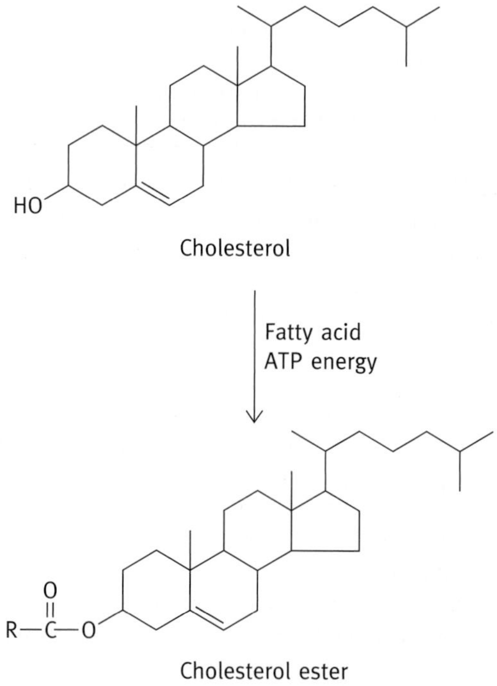

Chylomicrons

Chylomicrons (Fig. 9.10) are spherical particles, in the middle of which are hydrophobic molecules – TAG and cholesterol esters. Cholesterol has a hydrophilic —OH group and a hydrophobic part. To form chylomicrons, cholesterol is converted to **cholesterol ester** by attachment of a fatty acid to its —OH group as shown in the simplified diagram of Fig. 9.11. The enzyme which does the esterification is called **acyl-CoA:cholesterol acyltransferase (ACAT)**. It requires coenzyme A (page 181, but details are not needed here).

In making a cholesterol ester from cholesterol, the polar —OH group, which would interfere with packing cholesterol into the hydrophobic centre of chylomicrons, is eliminated. It illustrates the importance that the polarity/hydrophobicity characteristics of biological molecules play in life. Hydrophobic

Cholesterol

Fatty acid
ATP energy

Cholesterol ester

Fig. 9.11 Simplified diagram of the esterification of cholesterol.

particles of TAG and cholesterol ester would coalesce into an insoluble mass unless they were stabilized. For this, there is a 'shell' containing phospholipids, some free cholesterol which is weakly amphipathic and, importantly, some special proteins.

There are several different proteins involved, a major one being **apolipoprotein B (apoB)**, necessary for chylomicron synthesis. ApoB is a glycoprotein, the carbohydrate attachment providing a highly polar group. The prefix *apo* means 'detached' or 'separate'. Thus an apolipoprotein is a protein normally found in lipoproteins but now is detached or separate. The whole chylomicron structure is called a **lipoprotein**. The stabilization by the hydrophilic shell allows it to remain as a suspended particle in the lymph chyle (see below) and blood. The overall composition of chylomicrons is about 90% or more TAG, giving them a low density.

The chylomicrons are not released directly into the blood (unlike absorbed amino acids and monosaccharides) but into the lymph vessels. The suspension of chylomicrons in **lymph** fluid is called chyle. A few words on lymph might be useful here. As blood circulates through the capillaries, a clear lymph fluid containing protein, electrolytes, and other solutes filters out to bathe cells in an interstitial fluid. All tissues have a fine network of lymph capillaries closed at the fine ends into which lymph drains. The lymph capillaries join into lymphatic ducts and discharge the lymph back into the major veins of the neck via the thoracic duct. It is not actively pumped but movements of the body propel it along. After a fatty meal, chylomicrons, entering the bloodstream via lymph, give the blood a milky appearance. They circulate in the blood and their contents are used by tissues.

Digestion of other components of food

There are components in food other than the ones dealt with so far and digestive enzymes exist to hydrolyse them into their constituent parts. Thus phospholipids in food are hydrolysed by phospholipases; nucleic acids are hydrolysed by ribonuclease (RNase) and deoxyribonuclease (DNase). These enzymes are in the pancreatic juice. Plant material contains fibre – carbohydrate molecules such as cellulose, which in humans constitute the fibre that is not hydrolysed and is important in the diet. Its components are cellulose, lignin, hemicelluloses, pectin, and gum. Fibre has several important beneficial effects. It provides bulk and speeds up movement through the intestine, and may provide protection against carcinogens by speeding up the rate of their elimination, and also by binding some of them. Fibre can lower blood cholesterol levels in some cases by binding the molecule and reducing its absorption. Bacterial fermentation metabolizes part of the fibre in the large intestine.

Herbivores, for which cellulose is the main food component, have microorganisms in the rumen digestive tract which produce cellulases which hydrolyse the fibre. The microorganisms convert the liberated carbohydrate to short fatty acids such as acetate and propionate which are the main source of energy for the animals.

We now have the various products of digestion just having reached the blood – fat and cholesterol go into the general circulation via the lymph system as chylomicrons, and everything else is headed for the liver in the portal vein and thence into the general circulation.

Storage of food components in the body

Animals take in food at periodic intervals; they do not eat continuously. The body is, in effect, continually exposed to cyclical feast and famine conditions. The periods between meals can for humans vary from short intervals during the day to longer periods during sleep and very long periods during starvation. The biochemical machinery in the body has to cope with these situations.

After a meal, the blood is loaded with digestion products absorbed from the intestine. To state the obvious, this material is not held in the blood until it is all used up by metabolism, but rather it is rapidly cleared from the blood by uptake into the tissues so that blood levels quickly return to normal. After a fatty meal, the lipaemia (presence of milky blood plasma) is cleared within a few hours. Similarly, blood glucose may increase after a meal from its normal 90 mg dl^{-1} to 140 mg dl^{-1} in an hour (5 mM and 7.8 mM, respectively) but it reverts back to normal, in a nondiabetic person, in 2 hours. The tissues are not using all of this food at once; most of it is stored.

How are the different food components stored in cells?

Glucose storage as glycogen

It would not be practicable for cells to store glucose as the free monosaccharide – the osmotic pressure of the glucose at high concentrations would be excessive. The osmotic pressure of a solution is proportional to the number of particles of solute in the solution. If, therefore, large numbers of glucose molecules are joined together to form a single macromolecule, the osmotic pressure, exerted by the store of glucose residues, is accordingly reduced. The resultant polymerized molecule may fall out of solution as a granule. In animals, glucose is polymerized into a highly branched so-called 'animal starch', or glycogen, with the same chemical bondings as in amylopectin (page 149), but more highly branched. When needed, glycogen is broken down again. A very important fact is that *glycogen storage in animals is limited*. In humans, the glycogen reserves of liver, which are used to supply glucose to the blood for utilization by other tissues, especially the brain, are exhausted after about 24 hours of starvation. This has a profound effect on the biochemistry of an animal, as will become apparent later in the chapter. Muscle also has large glycogen reserves: these are used to provide energy for ATP production, needed in muscle contraction, but it only serves the muscle – free glucose is not produced from glycogen in muscle and cannot be released from it to the rest of the body, as happens in the liver. Other tissues do not store glycogen (kidney is a minor exception).

In all of the above, we have talked about glucose. What about other monosaccharides such as galactose and fructose, present in lactose and sucrose respectively? Glucose is the monosaccharide of central metabolic importance; other monosaccharides

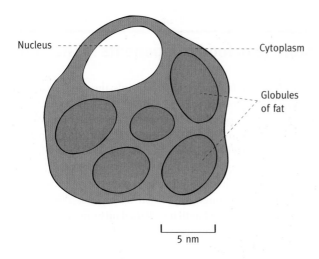

Fig. 9.12 A fat cell.

are converted to glucose (or glycogen), or else to compounds on the main glucose-metabolizing pathways.

Storage of fat in the body

Fat is stored by cells as TAG. The bulk of it is stored in the fat cells of adipose tissue, which is distributed in many areas of the body. A loaded fat cell under the microscope looks like droplets of oil surrounded by a thin layer of cytoplasm and membrane as depicted in Fig. 9.12.

Unlike glycogen storage, *that of fat is essentially unlimited* and large reserves of energy are stored in the body as triacylglycerol in the adipose cells. In a typical human the reserves of fat, in terms of stored energy, are about 50 times those of the glycogen reserves, and can amount to 30 kg or more in total.

Since this limited glycogen storage causes the metabolic problems described later, it maybe asked why the body stores so much fat, and so little glucose, when the latter is so essential to life in animals? Fat is more highly reduced (less oxidized) than is carbohydrate and hence contains more energy per unit mass, and, additionally, glycogen in the cell is hydrated while fat is not. This means that the latter occupies much less weight and volume per unit of stored potential energy than does glycogen. If the energy equivalent of fat stored in our bodies was in the form of glycogen we would need to be much larger, perhaps twice the size, than we are (migrating birds might then be too heavy to take off). Since glucose storage is so limited and starch and sucrose intake in the diet potentially almost unlimited in humans of modern societies, it follows that glucose in excess of that used for glycogen synthesis must be stored in another way – in fact as fat. The position then is as shown in Fig. 9.13.

An important fact is that, while glucose is readily converted to fat, *in the human body there is no net conversion of fatty acids to glucose.* (The glycerol moiety of TAG *can* be converted to glucose but this represents a small fraction of the energy available in the three long-chain fatty acids.)

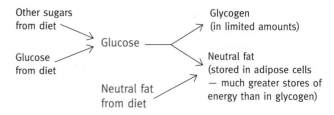

Fig. 9.13 Storage of sugars and fat from the diet in the post-absorptive period.

Are amino acids stored by the body?

Digestion of the third main class of food, protein, results in amino acids being absorbed and taken into the blood, but there is no dedicated storage form of amino acids in animals. Plant seeds have storage proteins whose only function appears to be storage of amino acids in a convenient form to supply the developing embryo. In animals, dietary amino acids in the blood are taken up by tissues as needed for the synthesis of cellular proteins, neurotransmitters, and other nitrogenous components, and then those in excess of immediate needs are degraded, the amino group ($-NH_2$) being removed and, in mammals, converted into urea and excreted in the urine. Urea is highly water soluble, neutral, and nontoxic. This leaves the carbon-hydrogen 'skeletons' of the amino acids, which contain chemical energy. These are converted to glycogen or fat, depending on the particular amino acid, or they can be oxidized to release energy or used to provide other metabolites (Fig. 9.14), according to the physiological needs at the time.

In one limited sense, there is a storage of amino acids in the body – in the proteins of all cells. Muscle proteins are quantitatively of greatest importance. However, they are all functional proteins mainly involved in contraction and are not dedicated to storage. When amino acids must be made available to the body in general, these proteins are broken down and the result

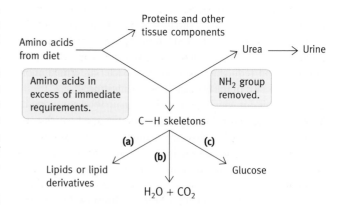

Fig. 9.14 Summary of the fate of amino acids in the diet. Which of the metabolic routes **(a)**, **(b)**, or **(c)** is followed depends on the particular amino acid, the physiological state, and biochemical control mechanisms.

is muscle wasting. You will shortly see in what situation this becomes essential.

Characteristics of different tissues in terms of energy metabolism

The different tissues in the body have special biochemical characteristics. However, in our present context of looking at overall food traffic in the body, several organs are of overriding importance. These are the liver, the skeletal muscles, the brain, the adipose tissue (fat) cells, and the red blood cells. (We are excluding regulatory organs such as the pancreas and adrenal glands from the list though their roles are crucial.)

- The liver has a central role in maintaining the blood glucose level; it is the glucostat of the body. When blood glucose is high, such as after a meal, the liver takes it up and stores it as glycogen. When blood glucose is low, it breaks down the glycogen and releases glucose into the blood.

In starvation, after about 24 hours, the reserves of glycogen in the liver are exhausted by this activity; the concentration of glucose in the blood would decline to lethal levels if nothing were done about it for the brain cannot function without an adequate supply. During starvation, there will, for a prolonged period, be reserves of fat that are released into the blood by the fat cells of adipose tissue. This will keep the muscles and other tissues happy, but not the brain and red blood cells, which cannot use fatty acids. The liver saves the situation by converting amino acids to glucose, a process called **gluconeogenesis** (Chapter 15). The main sources of amino acids in this situation are muscle proteins, which are broken down to provide them. This means destruction of functional muscle proteins, causing muscle wasting, but this is preferable to death resulting from low blood glucose.

The liver also plays an important role in fat metabolism. In starvation, the fat cells release fatty acids into the blood that can be directly used by most tissues – brain and red blood cells excepted, as stated. However, in such conditions, where massive fat utilization is occurring, the liver converts some of the fatty acids in the blood to small compounds called **ketone bodies** (page 157) and releases them into the blood for use by other tissues (red blood cells again excepted). Of special importance, the brain can adapt to utilize ketone bodies which, during starvation, can supply perhaps half of that organ's energy needs. The rest must come from glucose. For the moment, ketone bodies might be regarded as fatty acids partially metabolized by the liver. The liver is also the major site of fatty acid synthesis from glucose and other foods that are in excess of those required to replenish glycogen stores. The fat is not stored but is exported to adipose cells and other tissues as TAG in the form of very-low-density lipoproteins (VLDLs described on page 171). Accumulation of fat in liver is pathological.

So, energy-wise, the liver stores excess glucose as glycogen until the stores have been filled and releases glucose when

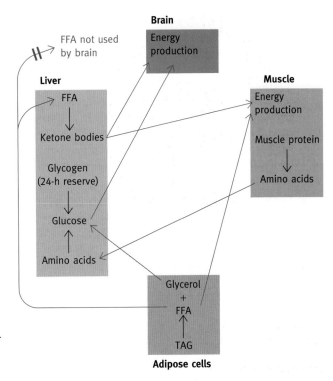

Fig. 9.15 Main fuel movements in the body during starvation. TAG, triacylglycerols; FFA, free fattyacids.

needed, as long as there are glycogen reserves left. In starvation it produces glucose mainly from amino acids released by muscle protein breakdown and ketone bodies from fat, which are turned out into the blood for use by the brain and, with respect to glucose, by red blood cells (Fig. 9.15). When food intake is normal, it synthesizes fatty acids from glucose and other dietary components and exports them. It preferentially oxidizes fatty acids for energy supplies. The liver has many other functions but for now we are concentrating on energy supplies in the body.

It might be useful, at this point, to summarise the fairly complex metabolic characteristics of tissues.

- The brain must, as stated, have a continuous supply of glucose in the blood. It has no significant fuel reserves. If the blood becomes hypoglycaemic (low glucose level), the rate of glucose entry into the brain is reduced since it involves passive transport and nerve cell function is impaired. Convulsions and coma result. However, as already emphasized, in starvation the brain can adapt to use ketone bodies produced by the liver from fat for perhaps half of its energy needs. This helps to economize on scarce glucose.

- Skeletal muscles require large-scale ATP production for muscle contraction. They can utilize most types of energy source. They take up glucose from the blood and store it as glycogen but, *in contrast to the liver, they do not release any glucose back into the blood*. They oxidize fatty acids, ketone bodies, and amino acids as well as glucose. In the presence of a high level of free fatty acids and ketone bodies in the blood, these are preferentially oxidized. (In starvation, muscles sacrifice their protein to supply the liver with amino acids for glucose synthesis.) In vigorous contraction when the energy needs may outstrip the oxygen supply, muscle can metabolize glucose (or glycogen) anaerobically causing lactate accumulation. This is very inefficient in ATP generation but in emergency can occur on a large scale and may make the difference between survival or not.

- The role of the fat cells of adipose tissue can be stated simply. After a meal, they take up fatty acids from lipoproteins supplied in the diet or synthesized by the liver. These are stored as TAG. Glucose is needed to supply the glycerol moiety. As soon as the blood glucose level falls as in mild fasting (or in starvation) they release reserves of fat into the blood as fatty acids for the rest of the body to use.

- Red blood cells can utilize only glucose, producing lactate, which is released from the cell. They are terminally differentiated cells (that is, fully formed and do not divide) without mitochondria and cannot oxidize foodstuffs as can other cells. But their Na^+/K^+ ATPase pumps must run and they have other energy-requiring processes.

Overall control of fuel distribution in the body by hormones

With that general description of the major organs involved in food logistics in the body, let us now see in overview how food distribution is directed – how the various organs are coordinated in their activities to cope with different physiological situations in humans. The main controls are the hormones **insulin**, **glucagon**, and **epinephrine** (**adrenaline**) as will be described below. These situations in humans are as follows.

- Excess food available – the postprandial condition after a good meal.
- Mild fasting – this refers to the condition between meals; for example, before breakfast after the night's fasting.
- Starvation – fasting that goes on for longer than about 1 or 2 days.
- Emergency responses – this relates to conditions, for example, where vigorous muscular action is needed to avoid a threat.

The overall picture of food traffic in the body in these nutritional situations is summarized below.

Postprandial condition

In the post-absorptive stage, dietary components are at a high level in blood. Glucose is taken up by the liver and muscle and used to replenish glycogen stores. Beyond this, excess glucose is taken up mainly by the liver and converted into TAG. The healthy liver does not retain fats in large amounts as do the adipose cells but transfers it to other tissues *via* very-low-density lipoprotein (VLDL) (page 155).

Amino acids are taken up by all tissues and used for the synthesis of protein or other components. Any in excess of immediate needs are converted to fat or glycogen and the nitrogen excreted.

Cells help themselves to the fatty acids of the TAG in chylomicrons as needed (see page 170 for the mechanism). Cells of lactating mammary glands are active in this to supply milk fats, and adipose cells take them up to convert to stored TAG.

It is clear that with all of this uptake and release of foodstuffs going on there has to be a system to control it all, otherwise there would be metabolic chaos. This is the role of hormones released by endocrine glands, which release their chemical signals into the blood where they reach all cells and instruct target cells (those designed to receive the signals) on what they should be doing in terms of food logistics.

The main 'signal' for storage of food components to take place is the pancreatic hormone, **insulin**, whose release from the pancreas into the blood occurs in response to high blood glucose levels. The high insulin level is the signal to the tissues that times are good and that food should be stored both as glycogen and fat. Correlated with this is the low level of glucagon, another pancreatic hormone, whose release is *inhibited* by high blood glucose levels. Glucagon has the opposite effect to insulin. It signals that the tissues are short of fuel and that storage organs should release some into the blood. Note that the brain and red blood cells are not affected by insulin; they go on using glucose from the blood all the time.

Fasting condition

The insulin-stimulated storage activity in the post-absorptive period lowers the blood glucose and amino acids to normal levels and clears the chylomicrons. With time, the blood glucose level begins to fall and, with it, insulin secretion. The hormone is destroyed after release, so its blood level falls in a few minutes when secretion stops. In concert with these events, the pancreas releases glucagon, whose level rises in response to low blood glucose. This stimulates the liver to break down glycogen and release glucose into the blood. Glucagon also stimulates adipose tissue to hydrolyse triacylglycerol and release free fatty acids and glycerol into the blood. Muscles and liver, and other tissues, use the fatty acids to provide energy and this conserves glucose for the brain. Glycogen synthesis and fat synthesis from glucose cease due to the low insulin/high glucagon ratio. There is no point in synthesizing and breaking them down at the same time.

Prolonged fasting or starvation

After about 24 hours the liver has no glycogen left to provide blood glucose and no other tissue, other than the quantitatively unimportant kidney, is capable of releasing glucose. The insulin level is low and that of glucagon high. In this situation the adipose cells pour out free fatty acids into the blood, for use by muscle and other tissues; the fat reserves in adipose cells are sufficient perhaps for weeks. Muscles break down their proteins to amino acids. The liver uses these to synthesize glucose, a process activated by glucagons. (Gluconeogenesis, as the process is called, also occurs in other circumstances; see Chapter 15.) In times of stress, the steroid **glucocorticord hormones** are released from the adrenal cortex: the main one in humans is **cortisol**. It stimulates gluconeogenesis and promotes muscle protein breakdown.

As starvation proceeds, the high glucagon causes the adipose tissue to turn out more and more free fatty acids into the blood and the liver metabolizes some of these into ketone bodies which it turns out into the blood. As mentioned, **ketone bodies** can be regarded as 'predigested' or partly metabolized fatty acids. There are two components, acetoacetate and β-hydroxybutyrate.

$$CH_3COCH_2COO^- \text{ Acetoacetate}$$
$$CH_3CHOHCH_2COO^- \text{ } \beta\text{-Hydroxybutyrate}$$

The name 'ketone bodies' is a misnomer; they are not 'bodies', nor is β-hydroxybutyrate a ketone (it has no C=O group), but the term is an old one and still often used. As mentioned, the brain in starvation adapts to obtain part of its energy from them, which economizes on glucose.

The emergency situation – fight or flight

When an animal such as a human is presented with a dangerous situation, the adrenal glands release **epinephrine (adrenaline)** from the adrenal medulla into the blood in response to a neurological signal from the brain. Adipose tissue is innervated and can also be stimulated by epinephrine and norepinephrine (noradrenaline) released from nerve endings. Epinephrine, as it were, presses the biochemical panic button. It overrides normal control and stimulates the liver to pour out glucose into the blood and the adipose cells to release free fatty acids so that muscles have no shortage of fuel. It also stimulates glycogen breakdown in skeletal muscle enabling the cells to produce ATP, at the maximal rate. The pattern has an overall logic of ensuring that muscles can react maximally and instantly to escape the threat.

Regulation of food intake: appetite control

The epidemic of **obesity** in many developed countries with its association with type 2 diabetes and other health problems has greatly stimulated research interest into what are the normal controls on food intake, weight homoeostasis, and energy balance.

If the intake of energy by a human exceeds the expenditure of energy then the excess will end up stored in the form of fat. Even a small imbalance between food intake and energy output would, over long periods, cause excessive weight changes. The chance of arriving at a correct balance between the two, without specific controls to achieve it, is remote. It is now known that a number of controls exist to achieve this balance but they are very complex and involve their signals being integrated in the brain as well as by more direct effects on metabolism in peripheral tissues. It seems that the known control systems are not sufficient to prevent the current epidemic of obesity. This may be due to the temptations of readily available food present in many areas of modern living. However a possibility is that the appetite and weight controls were not primarily intended to guard against obesity but rather the reverse to guard against death by starvation, which was a more likely danger in the times when these controls were being evolved. In that situation it may be that a physiology that was very 'thrifty' and readily stored excess energy as fat would be a positive survival mechanism in times of food shortage (see **Marx** reference in Further reading).

Despite considerable advances having been made in the past decade to our understanding, there is still no widely applicable therapeutic remedy to the problem of obesity, which some have described as the world's greatest medical problem. We will now present the main known aspects of appetite and weight control.

Hormones that control appetite

There are several hormones that control appetite produced by the digestive tract in response to the absence or presence of food.

Ghrelin increases appetite; it is a peptide hormone produced by the stomach when it is empty of food; its level in the blood rises precipitously and falls just as quickly after a meal. It is short acting and functions to stimulate apetite. The other known relevant digestive tract hormones inhibit appetite.

Cholecystokinin is liberated by the intestine in the presence of food and helps to give a feeling of satiety. Its role is to terminate meals. There are other neural inputs from the gastrointestinal tract to the brain which play a role in signalling satiety.

PYY-3-36 is produced, in the presence of food, by endocrine epithelial cells of the small intestine and the start of the colon and liberated into the blood which carries it to the brain. This neuropeptide inhibits appetite. Injection of PYY-3-36 into experimental animals or humans, in amounts sufficient to give normally achieved blood levels, suppresses appetite for about 12 hours.

The digestive tract is not the only source of appetite control hormones. The cells of adipose tissue produce some important ones.

Leptin is produced by adipocytes, the fat storage cells. When they have heavy reserves of fat, leptin production increases and conversely decreases as the reserves are depleted during starvation. It suggests that leptin would cause the brain to inhibit appetite. Leptin is involved in longer-term control of eating and thus of maintaining constancy of body weight. It is carried to the brain in the blood and combines with receptors. Leptin also has been found to have more direct effects on fatty acid metabolism in muscles (see below).

Leptin was a discovered in 1994 when its gene was pinpointed in a mutant strain of extremely obese mice, which lacked the hormone. Injection of the latter caused weight loss in the mutant mice. In very rare cases, in a few children lacking leptin production, injection of the hormone corrected the obesity, giving rise to hopes of it being used to control obesity in general. Unfortunately it appears that obese humans often have normal or even high levels of leptin suggesting, as one possibility, resistance to the hormone, such as lack of receptors (see recent work on amylin below).

Adiponectin is also produced by adipocytes. Low levels are associated with obesity. Its is known to have effects on muscle fatty acid oxidation as is the case also with leptin (see below). Adiponectin and leptin are two of several **adipokines** or **adipocytokines** (see page 432 for cytokines) secreted by adipose tissue.

Insulin is produced by the pancreas when blood glucose is high as occurs after meals. Its glucose-homeostatic role is best known but it also may effects on appetite, resembling those of leptin, though less pronounced.

Amylin is produced by the pancreatic β-cells, along with insulin, in response to food intake. The hormone is a peptide, 37 amino acids in length. It exerts a short-term satiety effect *via* receptors in the posterior part of the brain. Administration of amylin to rats caused decrease in food intake, and some weight loss.

Recent work (see **Roth** *et al.* in Further reading) has shown that injection of both leptin and amylin into diet-induced obese rats caused weight loss greater than either hormone alone. The loss was mainly in the fat content, and the effect on leptin responsiveness occurred only with amylin. Further clinical studies suggested that this applied also to human obese or overweight subjects, in which 12.7% greater weight loss occurred with leptin and an amylin analogue, than with either alone. It was suggested that amylin may restore sensitivity to leptin. The authors suggest that a multifactorial approach with different hormones may be advantageous in obesity control studies.

The hypothalamus integrates appetite control hormone effects

An important appetite control centre is located in the **hypothalamus**, a small area at the base of the brain. In a region of this, known as the **arcuate nucleus**, there are two subsets, or groups, of neurons with opposing effects on appetite. They are controlled by some of the circulating hormones described above. One of these groups (the **NPY/AgRP** producing set) produce two neuropeptides (**Neuropeptide Y** and **Agouti-related peptide**, the latter discovered in agouti mice) which *increase* appetite. The other group, known as the **pro-opiomelanocortin** (**POMC**) set, produce neuropeptides which *inhibit* appetite.

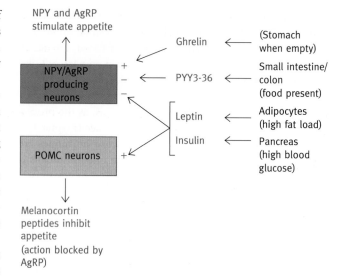

Fig. 9.16 Simplified diagram of appetite control *via* hypothalamic neurons. See text for explanation. The green plus sign represents stimulation of the target neurons and the red minus sign, inhibition. NPY, neuropeptideY; AgRP, agouti-related peptide; POMC, pro-opiomelanocortin.

They work in opposition for the AgRP neuropeptide blocks the action of the POMC neuropeptides. It is another example of push–pull control so common in control mechanisms. Fig. 9.16 shows the way in which the various circulating hormones interact with the arcuate nucleus to control appetite.

- At times of the day when the stomach is empty, ghrelin is secreted and stimulates the NPY/AgRP subset to produce their peptides and stimulate eating.

- After intake of food, ghrelin secretion stops and the appetite inhibiting hormone, PYY-3-36 is produced due to the presence of food in the intestine. This inhibits the NPY/AgRP group in the brain from producing its appetite stimulating peptides.

- Leptin, produced when fat reserves are high, inhibits production of the appetite stimulating NPY/AgRP neurones and stimulates production of the appetite inhibiting POMC neuropeptides.

- Insulin has similar, probably less pronounced, effects to those of leptin.

- In fasting or starvation the reverse happens. Ghrelin is produced by the stomach, stimulating appetite. In the absence of food in the intestine, PYY-3-36 is not produced so its inhibitory influence is not present. Leptin production falls as fat reserves decline so its inhibitory influence is diminished or absent. Insulin will be minimal. Ghrelin therefore has the field to itself and its stimulation of appetite is unopposed.

Other controls on eating exist, though less is known of the way in which they work. Maintenance of body weight is a balance between food intake and expenditure so this is more complex. The metabolic rate and hence of energy expenditure is affected by considerations such as the level of thyroid hormone, as well the amount of physical activity. It has also been shown that leptin increases energy expenditure and there is evidence that it increases the metabolism of fat rather than allowing it to be deposited in fat cells. It is known to inhibit an enzyme in muscle necessary for fat synthesis. We deal with the major metabolic controls on energy metabolism in Chapter 16, and, with new developments relative to the control of fat synthesis.

Summary

Digestion of food

In digestion, the polysaccharides and proteins of food are hydrolysed into their monomer subunits (simple sugars and amino acids). This is required for them to be absorbed by the intestinal epithelial cells into the bloodstream. TAGs (triacylglycerols or fats) are hydrolysed into fatty acids and monoacylglycerol, the process being aided by bile salts which emulsify the fats to give a large surface area for the enzyme lipase, secreted by the pancreas, to attack.

The body has to guard against self-digestion by the proteases released by the stomach lining and pancreas. Glycoproteins secreted from epithelial cells of the intestine form mucins, which coat and protect cells. The proteolytic enzymes are secreted in an inactive zymogen form and are activated only when they reach the lumen of the intestine.

In the stomach, pepsin partly digests protein but the main digestion and absorption occur in the small intestine. Pepsin is unusual in working optimally at about pH 2. This is maintained in the stomach by secretion of HCl from parietal cells by an ATP-dependent system.

Starch is hydrolysed by pancreatic amylase, which is secreted into the intestine as an active enzyme presumably because it poses no threat to the cells of the pancreatic duct. It is also present in the saliva.

Sugars and amino acids are absorbed into the bloodstream and taken by the hepatic portal vein to the liver and thence into the general circulation.

Digestion products of fat (monoacylglycerol and fatty acids) are resynthesized into TAGs in the epithelial cells and sent into the lymphatics parceled as chylomicrons and thence into the blood, to be distributed around the body. Chylomicrons are lipoproteins, each a complex of phospholipids, cholesterol, and TAG together with a specific collection of protein molecules. The phospholipids keep the particle in suspension fortransport.

In the large intestine water is absorbed and some digestion of dietary fibre occurs due to the presence of bacteria.

Distribution of absorbed digestion products to the tissues

Storage of glucose: the blood after a meal becomes loaded with absorbed food components. These are rapidly cleared from the blood. Glucose is stored in muscle and liver as glycogen, a polymer of glucose. Muscles use it to supply energy but liver has the important function of using it to release free glucose when needed such as in starvation, into the bloodstream. Unless the blood glucose levels are adequate, the brain cannot function normally. Storage of glycogen in the liver is limited to about a 24-h supply of glucose during starvation. When this is exhausted it synthesizes glucose. Muscle does not release glucose into the circulation.

Storage of fat is primarily in the fat cells of adipose tissue where it occurs in large amounts as TAG. Fat storage is virtually unlimited. It has a higher energy content than glycogen and is not hydrated. If the energy stores of fat were replaced by the equivalent calorific value as glycogen we would be much larger in size. Excess glucose is converted to fat by the liver and other sugars are converted to glucose. Glucose can be converted to fat but the reverse cannot happen in animals.

Amino acids are not stored as such.

Characteristics of tissues in terms of energy metabolism

Liver has a central role in maintaining blood glucose levels. It synthesizes fats and distributes it to other tissues. Muscles store glycogen but only for their own use.

Hormonal regulation of food distribution

Insulin, released in response to high glucose levels, stimulates fat and glycogen storage. Glucagon released from the pancreas when blood glucose is low stimulates release of glucose from the liver and of fatty acids from fat cells. Insulin and glucagon have opposite effects.

Regulation of food intake – appetite control

This is now a most important medical area of investigation due to the epidemic of obesity in developed countries. Several hormones are involved in regulating food intake. Ghrelin stimulates appetite and is produced when the stomach is empty. A numbers of other hormones inhibit appetite. These are the peptide hormone PYY-3-36 (produced by the intestinal epithelial cells in the presence of food), leptin and adipoleptin (produced by fat cells with the rate of release increasing as the fat stores increase). They have effects on fat oxidation. Insulin (produced by the pancreas when blood glucose levels are high) has lesser effects on appetite control than leptin. Ghrelin and the inhibitory hormones work in opposing ways on appetite centres in the brain.

There is great interest in the possibility of hormones being used to control obesity though experiments with leptin have not been successful except in a few obese children who lacked the hormone completely.

Further reading

Digestion and absorption of fats

Riddihough, G. (1993). Picture of an enzyme at work. *Nature*, **362**, 793.
News and Views summary of how pancreatic lipase is activated in the digestion of fats.

Cholesterol transport

Attie, A. D. (2007). ABCA1 at the nexus of cholesterol, HDL and artherosclerosis. *Trends Biochem. Sci.*, **32**, 172–9.

Insulin resistance

Alper, J. (2000). New insights into Type 2 diabetes. *Science*, **289**, 37–8.

Arner, P. (2003). The adipocyte in insulin resistance: key molecules and the impact of thiazolidinediones. *Trends Endocrin. Metab.*, **14**, 137–45.
A wide-ranging review relating to obesity and diabetes. Requires knowledge of cell signalling (Chapter 27).

Appetite control

Mechoulam, R. and Fride, E. (2001). A hunger for cannabinoids. *Nature*, **410**, 763–4.
Links between cannabinoids and leptin.

Schwartz, M. W. and Morton, G. J. (2002). Keeping hunger at bay. *Nature*, **418**, 595–7.
News and Views article on the hormones that control eating.

Batterham, R. L., *et al.* (2002). Gut hormone PYY (3-36) physiologically inhibits food intake. *Nature*, **428**, 650–2.

Bjorbaek, C. and Hellenberg, A. N. (2002). Leptin and melanocortin signaling in the hypothalamus. *Vitam. Horm.*, **65**, 281–311.

Flier, J. and Maratos-Flier, E. (2000). Energy homeostasis and body weight. *Curr. Biol.*, **10**, R215–17.

Marx, J. (2003). Cellular warriors at the battle of the bulge. *Science*, **299**, 846–9.

Flier, J. S. (2004). Obesity wars: Molecular progress confronts an expanding epidemic. *Cell*, **116**, 337–50.
A comprehensive review dealing with leptin and other relevant adipocyte signalling molecules.

Muolo, D. M. and Newgard, C. B. (2006). Obesity related derangements in metabolic regulation. *Annu. Rev. Biochem.*, **75**, 367.

Wagenmakers, A. (ed). (2006). The biochemical basis of health aspects of exercise. *Essays in biochemistry*, Portland Press U.K.
A student-oriented wide-ranging text which covers aspects of insulin and nitric oxide signalling and their roles in diabetes type 2, obesity, and vascular problems.

Roth, J. D., *et al.* (2008). Leptin responsiveness restored by amylin agonism in diet-induced obesity: evidence from nonclinical and clinical studies. *Proc. Natl. Acad. Sci. U.S.A.*, **105**, 7257–62.

Problems

1 Which digestive enzymes are produced in an inactive zymogen form? Why should this be? Why do you think amylase is not produced as an inactive zymogen?

2 Explain how the inactive proteases are activated in the intestine.

3 If pancreatic proenzymes are prematurely activated due to physical and chemical damage or pancreatic duct blockage, what is the result?

4 Why is digestion of foods necessary?

5 Many people, especially Asians, suffer severe intestinal distress on consuming milk or milk products. Why is this so?

6 Write down the structure of a neutral fat. What is the alternative name for this? Indicate which groups in the molecule are primary esters.

7 Fat in water forms large globules with little fat/water interface for lipase to attack. Explain how the digestive system copes with the physical intractability of lipid.

8 Amino acids and sugars absorbed into the intestinal cells move into the portal bloodstream and are carried via the liver and thence to the rest of the body. What happens to the absorbed digestion products of fat?

9 Why do cells of the body store glucose as glycogen? Why not save trouble and store it as glucose?

10 Triacylglycerol (TAG) is stored in much greater quantities in the body than is glycogen. Why is this so?

11 Can glucose be converted to fat? Can fatty acids be converted in a net sense to glucose? Do amino acids have a special dedicated long-term storage form in the body? Give brief explanations.

12   In terms of food logistics, describe the chief metabolic characteristics of the liver.

13   Can the brain use fatty acids for energy generation?

14   What are the chief metabolic characteristics of fat cells?

15   What is the main energy source used by red blood cells?

16   What are the main hormonal controls on the logistics of food movement in the body in different nutritive states?

17   At one time ketone bodies were regarded as entirely abnormal. Discuss this briefly.

18   Descibe the role of leptin. Discuss whether this hormone is likely to be of therapeutic use in obese humans.

19   Appetite control involves a competition between enhancing and inhibitory controls. Briefly describe the competing hormones.

Chapter 10

Mechanisms of transport, storage, and mobilization of dietary components

In this chapter we deal with the biochemical mechanisms involved in fuel transport, storage, and release processes described in Chapter 9. Cholesterol is included (though not an energy-yielding fuel) because it is absorbed from the diet and transported by the same lipoproteins as fat.

Glucose traffic in the body

The structures of carbohydrates are given on pages 148–50.

Mechanism of glycogen synthesis

As stated earlier (page 156), after a meal when blood glucose levels are high and the insulin/glucagon ratio is high, glucose is taken up by liver and muscle (and, to a minor extent, by kidney tubules) and used to replenish their glycogen stores. Glycogen does not occur in any other cells of the body.

The synthesis of glycogen from glucose is an **endergonic process** – it requires energy. Hydrolysis of the glycosidic bond joining glucose units in glycogen has a $\Delta G^{0'}$ value of about -16 kJ mol^{-1}, which would give an equilibrium totally to the side of glucose for the simple reaction glycogen + H$_2$O → glucose. Therefore energy must be supplied from high-energy phosphoryl groups to form the glycosidic bond.

Glycogen synthesis occurs by enlarging pre-existing glycogen molecules (called 'primers') by the sequential addition of glucose units. The synthesis of a glycogen granule is initially primed by the protein, **glycogenin**, which transfers eight glucosyl residues (see below) to a tyrosine —OH on itself. The protein remains inside the glycogen granule. The new units are always added to the nonreducing end of the polysaccharide. Glucose is an aldose sugar. This is seen clearly in the open-chain form which is in spontaneous equilibrium in solution with the ring form (Fig. 10.1), the latter predominating.

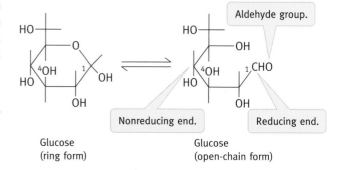

Fig. 10.1 The spontaneous equilibrium reaction of the two forms of glucose in solution.

An aldehyde is a reducing agent, so that the carbon 1 end of the ring sugar is the reducing end and carbon 4 the nonreducing end. The glycogen chain thus always has a nonreducing end.

(The glucose unit is represented as ⬭$_O$.)

The process of glycogen synthesis in essence therefore is as follows.

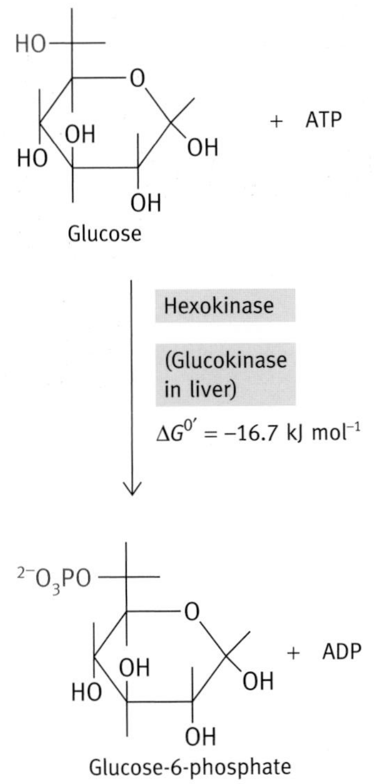

Glucose + ATP

Hexokinase

(Glucokinase in liver)

$\Delta G^{0'} = -16.7$ kJ mol^{-1}

Glucose-6-phosphate + ADP

Fig. 10.2 Phosphorylation of glucose by adenosine triphosphate (ATP).

Glucose-6-phosphate (G-6-P)

Phosphoglucomutase

$\Delta G^{0'} = -7.3$ kJ mol^{-1}

Glucose-1-phosphate (G-1-P)

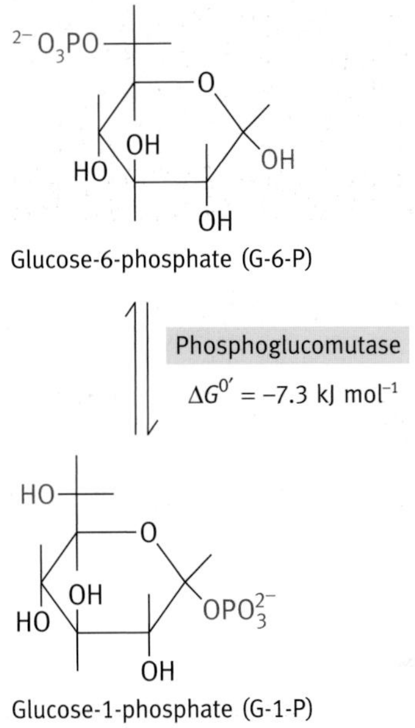

Fig. 10.3 Mutation of phosphoglucose.

The process is repeated over and over again, elongating the polysaccharide chain.

How is glycogen synthesis driven energetically?

When glucose enters the cell it is phosphorylated by adenosine triphosphate (ATP) (Fig. 10.2). The reaction is catalysed in brain and muscle by the enzyme hexokinase. A **kinase** transfers a phosphoryl group from ATP to something else – in this case, glucose, which is a hexose sugar – hence the name, **hexokinase**. In liver, a different enzyme catalysing the same reaction is called glucokinase, essentially occurring only at high glucose concentration, but hexokinase also is present and works when blood glucose levels are low (see page 167 for an explanation of this).

The $\Delta G^{0'}$ of this reaction is strongly negative, making it irreversible. The charged glucose-6-phosphate molecule is unable to traverse the cell membrane so that glucose phosphorylation has the effect of trapping the molecule inside the cell and causing entry of more glucose from the blood, since it enters by facilitated diffusion (page 113). The phosphoryl group is now switched around to the 1 position by a second enzyme, phosphoglucomutase, a freely reversible reaction which mutates (changes) phosphoglucose (Fig. 10.3).

The $\Delta G^{0'}$ of hydrolysis of the phosphoryl ester group of glucose-1-phosphate (G-1-P) is -21.0 kJ mol^{-1}, which is about the same as that of hydrolysis of the glycosidic bond in glycogen. You might expect from these values that the glucose unit would be transferred from G-1-P to a glycogen primer and the

synthesis would be done. It was thought years ago that this was what happened. But it is not so. If glycogen synthesis occurred directly from G-1-P, the process would be freely reversible and hence uncontrollable.

G-1-P is converted to the activated form, UDPG

An extra step is inserted that renders glycogen synthesis thermodynamically irreversible. A compound similar to ATP (Fig. 3.4), uridine triphosphate (UTP), exists in which the adenine of ATP is replaced by uracil (U) whose structure does not matter in this context (but is given on page 299 where it is relevant). (Uridine is uracil-ribose, a structure analogous to adenosine.) The cell synthesizes UTP (the energy coming from ATP). In glycogen synthesis, **uridine diphosphoglucose** (**UDP-glucose** or **UDPG**) is made by a reaction between UTP and G-1-P as shown in Fig. 10.4.

This enzyme, for systematic nomenclature reasons, is named for the reverse reaction (which does not occur in the cell). This reaction (were it to occur) would cleave UDPG with pyrophosphate. The enzyme is therefore a pyrophosphorylase and is called **UDP-glucose pyrophosphorylase**; inorganic pyrophosphate is not available in the cell because it is rapidly destroyed and therefore the reverse reaction is inoperative.

UDPG is the 'activated' or reactive glucose compound that donates its glucosyl group to glycogen. Wherever in animals an 'activated' sugar has to be produced for a chemical synthesis, as a general rule it is a UDP derivative. Whether there is a fundamental reason for this or an accidental quirk that evolution was locked into isn't known. (Plants use the corresponding adenine compound for starch synthesis.) Synthesis of glycogen using UDPG as glucosyl donor occurs as shown in Fig. 10.5 catalysed by the enzyme **glycogen synthase** ('synthase' was used wherever

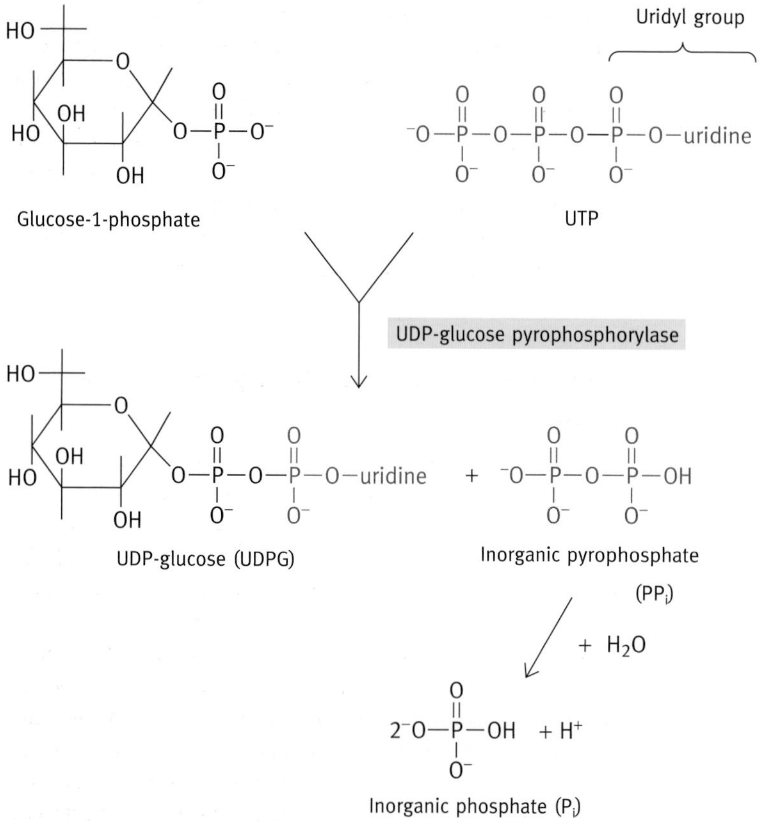

Glucose-1-phosphate

UTP

Uridyl group

UDP-glucose pyrophosphorylase

UDP-glucose (UDPG)

Inorganic pyrophosphate (PP$_i$)

+ H$_2$O

Inorganic phosphate (P$_i$)

Fig. 10.4 Formation of uridine diphosphoglucose (UDPG) by a reaction between uridine triphosphate (UTP) and glucose-1-phosphate (G-1-P) catalysed by UDPG pyrophosphorylase. See text for an explanation of the name of this enzyme.

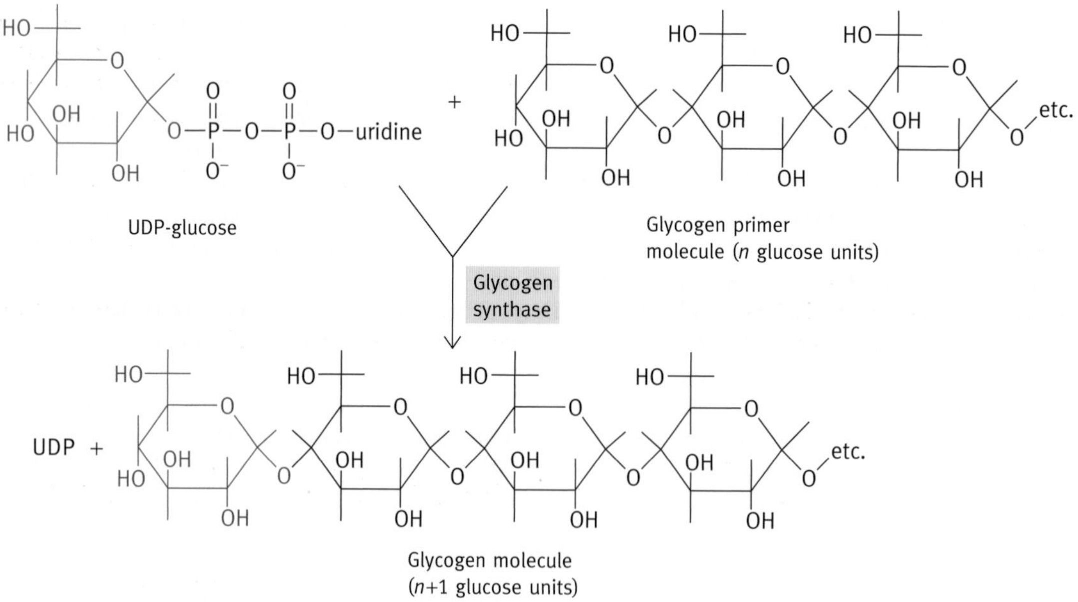

UDP-glucose

Glycogen primer molecule (*n* glucose units)

Glycogen synthase

UDP +

Glycogen molecule (*n*+1 glucose units)

Fig. 10.5 Synthesis of glycogen by the elongation of a glycogen primer molecule using uridine diphosphate (UDP)-glucose (UDPG) as glucosyl donor.

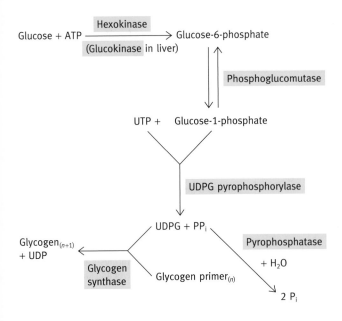

Fig. 10.6 Summary of glycogen synthesis from glucose.

a synthesizing enzyme does *not* directly use ATP and 'synthetase' where ATP is involved, but now either can be used). The UDPG pyrophosphorylase produces inorganic pyrophosphate. This is hydrolysed by inorganic pyrophosphatase, which, as explained on page 35, has a $\Delta G^{0'}$ value of -33.5 kJ mol^{-1}, and thus pulls the process of UDPG synthesis from G-1-P and UTP completely to the right. Glycogen synthesis in the cell is not therefore directly reversible by the route of its synthesis. The synthesis is summarized in the scheme shown in Fig. 10.6.

Adding branches to glycogen

There is one more point about glycogen synthesis. If more and more glucosyl groups were added to primer molecules, the result would be very long polysaccharide chains – which is precisely what plants synthesize as the amylose starch component. Glycogen is different – instead of consisting of long chains of glucosyl units, it is a highly branched molecule. This structure is achieved by another enzyme, called the branching enzyme. When the glycogen synthase has made a 'straight-chain' exten-

sion more than 11 units in length, the branching enzyme transfers a short block of terminal glucosyl units from the end of an $(1 \rightarrow 4)$-α-linked chain to the C(6)—OH of a glucose unit on the same or another chain (Fig. 10.7). The energy in the $(1 \rightarrow 6)$-a link is about the same as in the $(1 \rightarrow 4)$ link so the reaction is a simple transfer one. It creates more and more ends both for glycogen synthesis and breakdown.

Thus, to return to the starting point, blood glucose, after a meal, is taken up by tissues and converted into glycogen by the reactions outlined.

Breakdown of glycogen to release glucose into the blood

The liver, as mentioned, stores glycogen, not for itself (as is the case in muscle) but for other tissues and especially the brain and red blood cells. In fasting, such as between meals, it breaks down glycogen (high glucagon/low insulin are the signals for this) and releases glucose to keep up the blood level, for otherwise the brain would cease to function normally.

As with synthesis, **glycogen breakdown** occurs at the non-reducing end. Instead of splitting off (lysing) glucose by water *(hydrolysis)* it splits it off with phosphate *(phosphorolysis)*. The enzyme is called **glycogen phosphorylase** (Fig. 10.8). The reaction involves the coenzyme, pyridoxal phosphate, which we meet later in transamination reactions (page 285). It acts as a general acid-base catalyst (see page 97).

The G-1-P so formed is converted by **phosphoglucomutase**, working in the reverse direction, to glucose-6-phosphate and this is hydrolysed (in liver and kidney only) by **glucose-6-phosphatase** to give free glucose which is released into the blood. Muscle cells break down glycogen by the phosphorylase reaction for metabolism, not for glucose release.

$$\text{Glucose-1-phosphate} \rightleftharpoons \text{glucose-6-phosphate}$$

$$\text{Glucose-6-phosphate} + H_2O \rightarrow \text{glucose} + P_i$$

A summary of the production of glucose from glycogen by liver and kidney is shown in Fig. 10.9.

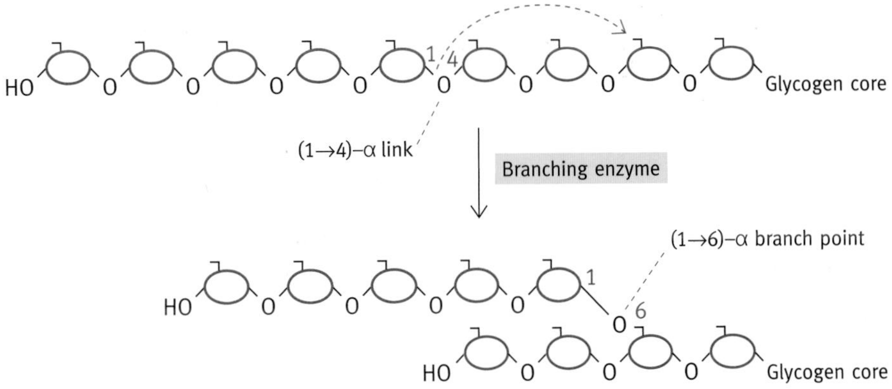

Fig. 10.7 Action of the branching enzyme in glycogen synthesis.

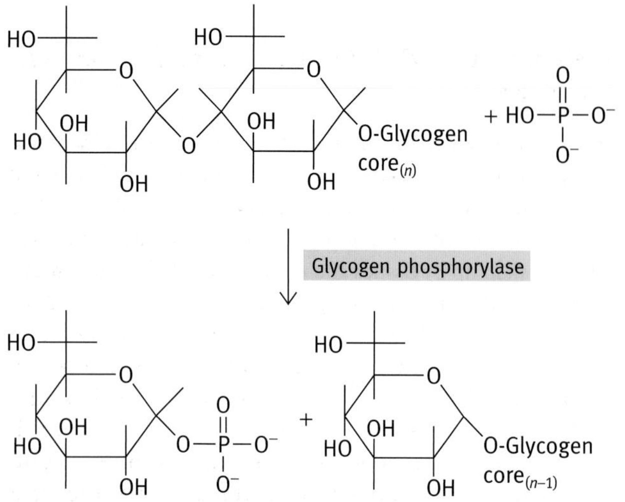

Fig. 10.8 Action of glycogen phosphorylase.

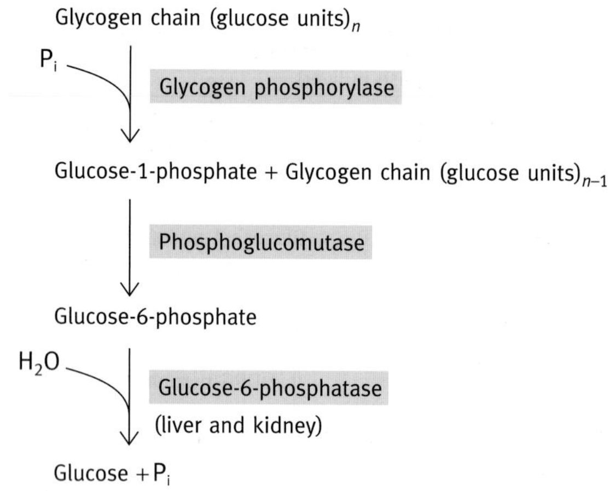

Fig. 10.9 Degradation of glycogen by phosphorolysis and the ultimate release of free glucose into the blood by the liver and kidney.

Glucose-6-phosphatase is located in the membrane of the endoplasmic reticulum (ER; page 19). It is a remarkable enzyme in that its active site is exposed to the lumen of the ER, not the cytoplasm. Glucose-6-phosphate is transported through the ER membrane (by a transport protein) where it is hydrolysed. The products, glucose and P_i, are transported back into cytoplasm by separate transport systems.

Removing branches from glycogen

The branch points of glycogen create a problem, for phosphorylase cannot function within four glucose units of such points –

the enzyme is big and presumably the branch gets in the way of it attaching to the site of the glucosyl bond to be attacked. A pair of enzymes takes care of this problem, so that the breakdown can go on. The first of these, the **debranching enzyme**, transfers three of the glycosidic units of the branch to the 4-OH of another chain. This makes them part of a chain long enough for phosphorylase to work on. The last unit, in (1→6) linkage, is hydrolysed off by an (1→6)-a-glucosidase activity of *the same enzyme* (which has two activities) and this opens up the chain for further attack by phosphorylase (Fig. 10.10). Glucose uptake, storage, and release in the liver are summarized in Fig. 10.11.

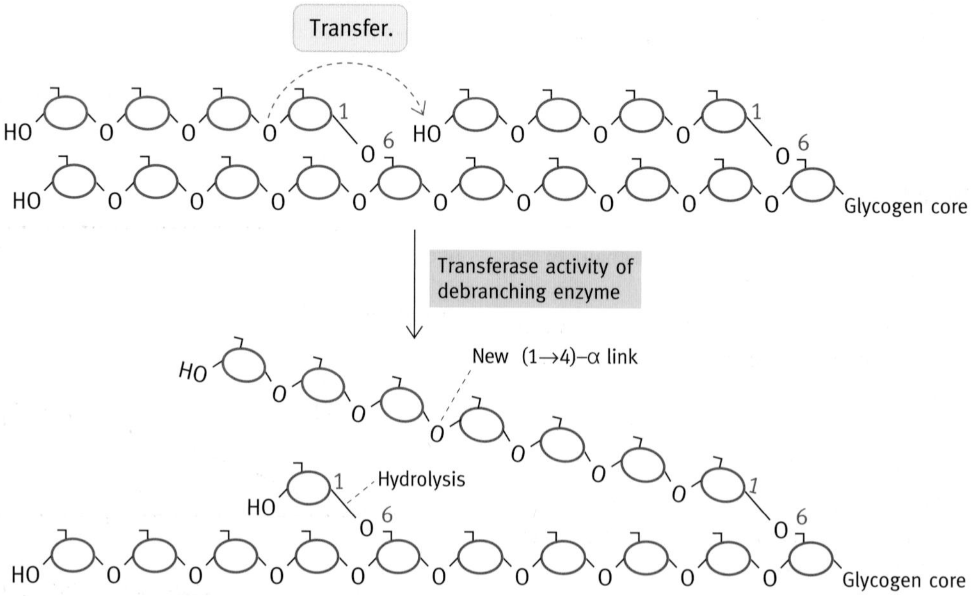

Fig. 10.10 The debranching process. Before debranching, the structure above cannot be attacked by glycogen phosphorylase. After the transferase and hydrolysis actions of the debranching enzyme, both chains are now open to phosphorylase attack.

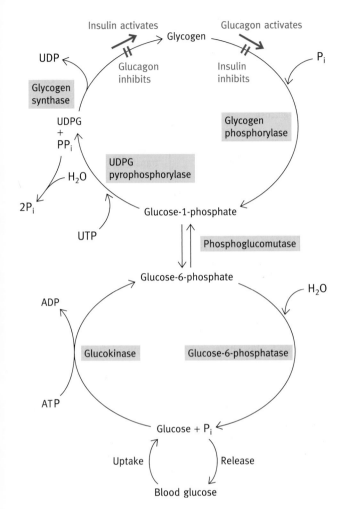

Fig. 10.11 Summary of glucose uptake, storage, and release in the liver. Note that glucagon and insulin do not act directly on glycogen metabolism enzymes (see Chapter 16).

Why does liver have glucokinase and the other tissues, hexokinase?

Once glucose enters a cell it is phosphorylated to glucose-6-phosphate. Glucokinase in liver and hexokinase in brain and other tissues catalyse the same reaction. Why are there two types? In starvation, the entry of glucose into muscle and adipose cells is restricted because of the lack of insulin (the mechanism of control of glucose uptake is described on page 251), while the entry of glucose into brain, liver and red blood cells is *not insulin dependent*. In starvation, the liver synthesizes glucose from amino acids supplied by muscle protein breakdown so that it can keep the blood glucose level up to permit normal brain function. It would not make sense for the liver to take up this glucose in competition with the brain. The main uptake by liver is due to glucokinase, which has a much lower affinity for glucose than has hexokinase; the K_m for glucose of glucokinase is 10 mM, and that of hexokinase is 0.05 mM (Fig. 10.12). This means that the liver takes it up less readily when its concentration in blood is low since its passive entry depends on phosphorylation within the cell. After a meal, when blood glucose levels are high and when insulin signals the liver to make glycogen, glucokinase works maximally. Also hexokinase is inhibited by glucose-6-phosphate while glucokinase is less inhibited. Thus uptake of glucose can proceed in liver even at high glucose-6-phosphate levels in the cells. In addition to this automatic control, the glucose transporter of liver has a lower affinity than that of most other cells which has the same effect. The glucose transporter of the brain is fully saturated, that is to say, it is working at its maximal rate, at normal blood glucose levels.

Key issues in the interconversion of glucose and glycogen

There are some important points to note, especially the following.

- The glycogen synthesis and degradation pathways are different and therefore independently controllable. (Control is a major topic: see page 251.)
- Glucose-6-phosphatase is found in liver and kidney only so that these can release glucose into the blood.
- Most importantly, insulin activates glycogen synthesis, appropriate when glucose is plentiful in the blood. The effect of glucagon is the reverse.

Thus, insulin promotes glycogen synthesis in muscle and liver; glucagon promotes release of glucose from liver. How these controls are achieved is the subject of Chapter 16.

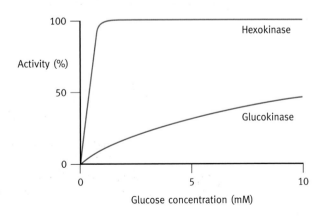

Fig. 10.12 The response of hexokinase and glucokinase to glucose concentration.

What happens to other sugars absorbed from the intestine?

The diet of animals usually results in large amounts of other sugars being absorbed into the portal bloodstream. The sugar in milk is lactose (page 150), which on hydrolysis in the small intestine yields galactose and glucose. Sucrose yields fructose and glucose. Fructose has a separate metabolic route (page 258). So far as energy metabolism goes, the policy is to convert these other sugars either to glucose or to compounds on the main glucose metabolic pathways in the liver. What happens to galactose has special medical interest.

Galactose metabolism

To convert galactose to glucose, the conformation of the H and OH on carbon atom 4 are inverted or epimerized.

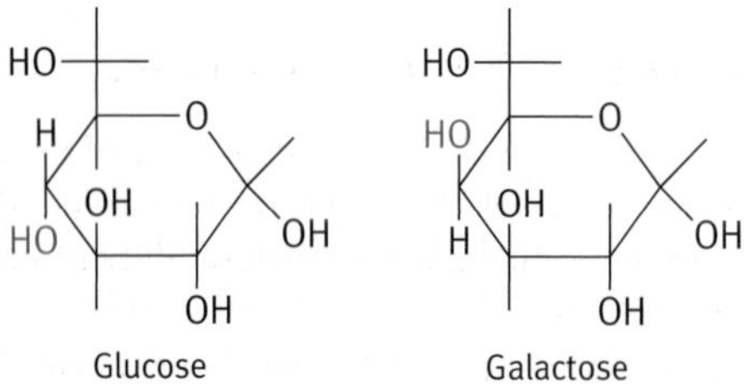

Glucose   Galactose

This is not done directly but by an epimerase that attacks not galactose itself, but UDP-galactose to form UDPG. (UDPG is also involved in glycogen synthesis as described above.)

First, galactose is phosphorylated by galactokinase to galactose-1-phosphate. (This contrasts with hexokinase and glucokinase which phosphorylate glucose on the 6-OH.)

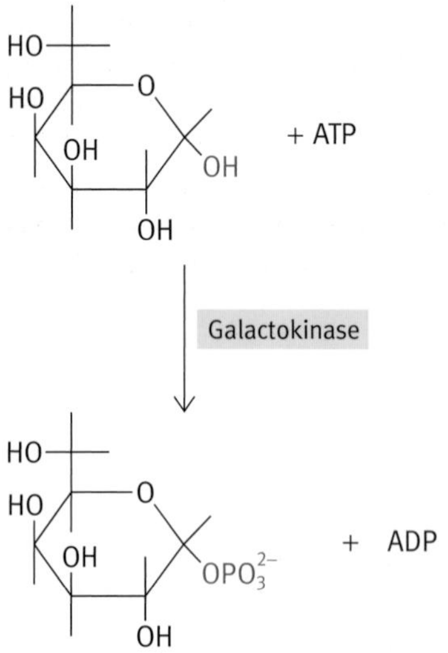

You might have expected this to react with UTP by analogy with UDPG formation but that is not what happens. Instead, galactose-1-phosphate displaces G-1-P from UDPG forming UDP-galactose and G-1-P as follows.

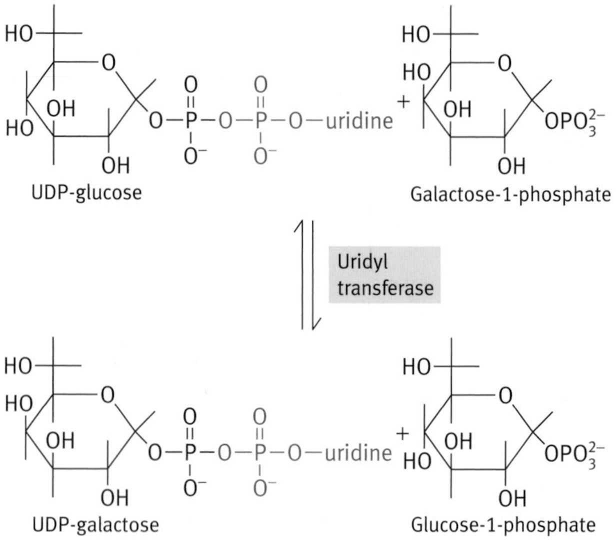

UDP-glucose   Galactose-1-phosphate

UDP-galactose   Glucose-1-phosphate

You will see that what happens in this reaction is that the —P— uridine group (uridyl) is transferred from UDPG to galactose-1-phosphate and the enzyme is therefore a **uridyl transferase**. Now epimerization of UDP-galactose occurs.

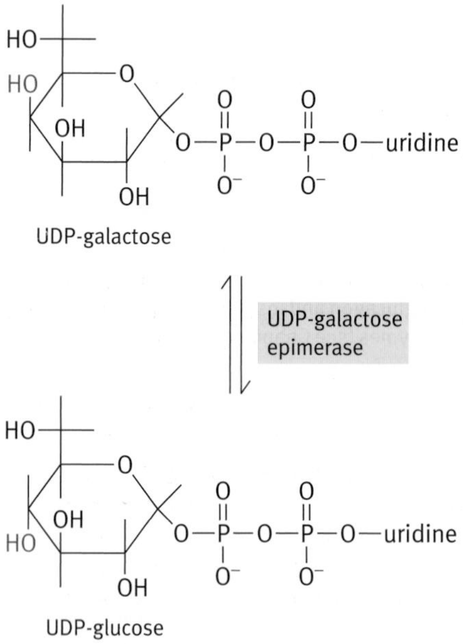

UDP-galactose

UDP-glucose

Putting the three reactions together, the following sequence occurs.

$$\text{Galactose} + \text{ATP} \longrightarrow \text{Galactose-1-P} + \text{ADP}$$
Galactokinase

$$\text{Galactose-1-P} + \text{UDP-glucose} \rightleftharpoons \text{Glucose-1-P} + \text{UDP-galactose}$$
Uridyl transferase

$$\text{UDP-galactose} \rightleftharpoons \text{UDP-glucose}$$
UDP-galactose epimerase

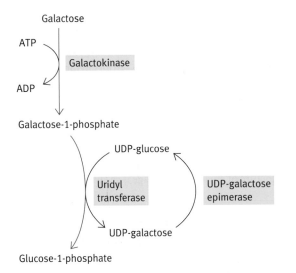

Fig. 10.13 How galactose enters the main metabolic pathways by conversion of galactose to glucose-1-phosphate. The green arrows represent the net effect of the reactions.

The net effect is to convert galactose to G-1-P (as summarized in Fig. 10.13).

BOX 10.1 Uridyl transferase deficiency and galactosaemia

Deficiencies in enzymes involved in galactose metabolism give rise to galactosaemias. The best known one is the infant genetic disease in which the uridyl transferase is missing and galactose cannot be converted to glucose. Accumulation of galactose and its metabolic products lead to impairment of brain development and blindness. Elimination of galactose from the diet during the 2 months after birth avoids this. UDP-galactose is needed for glycolipid and glycoprotein synthesis (page 229), but there is no problem since, in this galactosaemia, the epimerase is normal. This enzyme is reversible so that UDPG, synthesized from UTP and G-1-P, can be converted to UDP-galactose as needed. The patient can therefore survive without any intake of galactose in the diet. The galactosaemic children so treated develop normally.

Another hereditary defect is caused by a deficiency of galactokinase. This galactosaemia is a mild disorder; it causes early cataracts.

Amino acid traffic in the body (in terms of fuel logistics)

Since there is no dedicated storage of amino acids, there is no whole-body story comparable with that of glycogen and fat traffic, but amino acid traffic does occur. The movement of greatest importance is that occurring during muscle wasting in starvation. As explained, muscle proteins break down to produce amino acids some of which are transported to the liver to provide substrates for glucose synthesis. It will be more convenient, however, to deal with this in detail later (page 237). We will therefore move on to the traffic of fat in the body – a very important subject.

Fat and cholesterol traffic in the body

Fat and cholesterol are completely different both in structure and biochemistry. Medically they are quite distinct. However, as mentioned in the introduction, cholesterol is absorbed from the diet and transported by the same lipoproteins as is fat, probably because they both present the same problem of water insolubility. It is therefore convenient to discuss the two together.

Uptake of fat from chylomicrons into cells

After a fatty meal, the blood is loaded with **chylomicrons**. The chylomicron (page 152) has a shell of phospholipids, cholesterol, and proteins surrounding a core of triacylglycerol (TAG) and cholesterol esters.

Let us consider the fate of TAG first. It is taken up (but not as such) by the adipose cells for storage and by the lactating mammary gland for secretion in milk, by muscle and other tissues as an energy source and also by liver.

Free fatty acids (FFA), not esterified with glycerol, readily pass through cell membranes; TAG cannot do so. The TAG of chylomicrons is hydrolysed in the capillaries by a **lipoprotein lipase** attached to the *outside* of endothelial cells lining the blood capillaries to produce glycerol (which is transported to the liver) and FFA which enter the adjacent cells (Fig. 10.14). Utilization of glycerol in liver is described on page 239.

The amount of lipase activity present in the capillaries of a particular tissue determines the amount of FFA release in that tissue and hence the amount of FFA uptake by it. Adipose tissue is rich in the enzyme, as is lactating mammary gland, while tissues with less need for fat have less of it in the capillaries. The

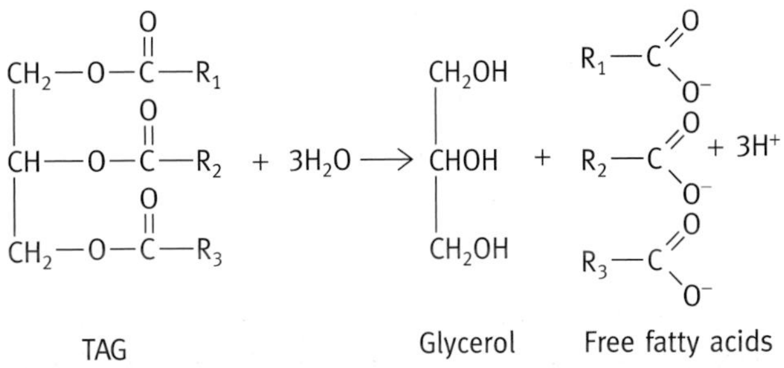

Fig. 10.14 Breakdown of triacylglycerol (TAG) by lipoprotein lipase.

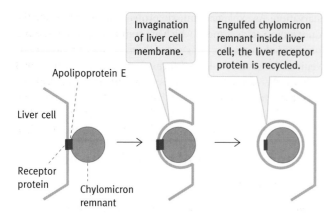

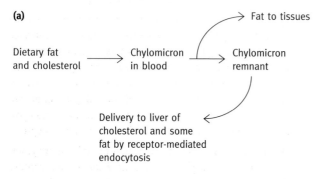

Fig. 10.15 Uptake of chylomicron remnants into a liver cell by receptor-mediated endocytosis. The receptor on the liver cell is specific for apolipoprotein E present on the chylomicron remnant. See Fig. 26.7 for further details of the fate of the engulfed particle and of the mechanism of endocytosis.

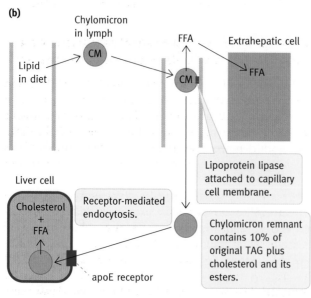

Fig. 10.16 (a) Transport of fat and cholesterol by chylomicron route. **(b)** Mechanism for the transport of fat and cholesterol from the intestine to the tissues by chylomicrons. CM, chylomicron; FFA, free fatty acids; HDL, high-density lipoprotein; apoE, apolipoprotein E; TAG, triacylglycerol.

amount of lipoprotein lipase is varied according to physiological need; for example, a high-insulin, low-glucagon ratio after a meal causes increases in the synthesis of the enzyme.

The progressive removal of TAG from chylomicrons reduces them in size to **chylomicron remnants**, containing all the cholesterol and its esters and about 10% of the TAG of the original chylomicron. The remnants are taken up by the liver by receptor-mediated endocytosis (Fig. 10.15), thus delivering fat and cholesterol from the intestine to the liver. Cholesterol esters are hydrolysed in lysosomes (see page 174 for LDL endocytosis).

A picture of chylomicron usage is shown in Fig. 10.16.

Logistics of fat and cholesterol movement in the body

An overview

The movement of lipid and cholesterol around the body is a subject of great importance, particularly from a medical viewpoint. Cholesterol is essential in the body in membranes, and for hormone synthesis, but it is dangerous in excess because of its insolubility, and because it is not broken down.

The liver is the major source of lipid and cholesterol circulating in the blood in the form of **lipoproteins** (chylomicrons, as described, carry absorbed fat and cholesterol from the intestine, and are present only after a meal). The liver synthesizes TAG from glucose and other metabolites and it receives about 10% of the absorbed fat due to uptake of chylomicron remnants. Nonetheless, it is not a storage organ for lipid – as mentioned, a 'fatty liver' in which there are extensive depositions of fat is pathological. The liver exports its TAG to peripheral tissues in the form of **very-low-density lipoprotein (VLDL)**, a lipoprotein similar in *structure* to a chylomicron. The rationale appears to be that liver, as a major source of fat, supplies this to major fat users such as muscle, which oxidize it as a source of energy, and to adipose cells for storage.

The liver is the major site of cholesterol synthesis in the body, and also receives dietary cholesterol via the chylomicron remnants. This also is exported to peripheral tissues in the VLDL resulting in an outward flow, from liver to peripheral tissues, of lipid and cholesterol.

There is also a reverse flow of cholesterol from peripheral tissues to the liver

This flow is summarized in (Fig. 10.17). The flow (described later in Fig. 10.20) is from high-density lipoprotein (HDL), which picks up cholesterol from peripheral cells, and returns it to the liver, either directly or by cholesterol ester transport to VLDL or low-density lipoprotein (LDL). The physiological rationale for this apparent 'equilibration' of cholesterol between the liver and the rest of the body is not self-evident. It might be a way of ensuring that all cells have adequate

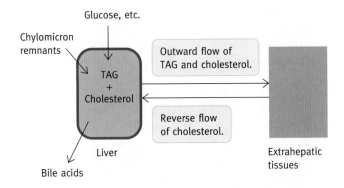

Fig. 10.17 Overview of the major movements of fat (triacyglycerol, TAG) and cholesterol to and from the liver. The liver synthesizes both components from metabolites and receives them from chylomicron remnants. It also converts cholesterol to bile acids.

cholesterol supplies with any excess returned to the liver, which excretes part of it into the intestine in the form of bile acids.

With that introduction we shall proceed to an account of the mechanisms involved, but it should be kept in mind that, in detail, it is an extraordinarily complex business, is incompletely understood, and is the subject of a massive continuing research effort.

Utilization of cholesterol in the body

Cholesterol is an important constituent of animal cell membranes. Because of the effects of its oversupply in causing cardiovascular disease, mechanisms for its removal from the body are of great interest.

The main route for disposing of cholesterol is bile acid formation by the liver. (The structures of cholesterol and of a bile acid are shown on page 152.) About 0.5 g per day is disposed of in humans by this route. However bile acids are partly reabsorbed from the gut and reused. One therapeutic approach to lowering blood cholesterol levels in patients is to administer a compound that complexes bile acids in the intestine and prevents this reabsorption; another, more recent approach, is to inhibit its production, as described on page 174. Cholesterol is also used in the adrenal glands and gonads for the synthesis of steroid hormones. (Examples of the structures of the latter are given page 434.)

Cholesterol ester, whose structure and synthesis are given on page 173, is the storage form of cholesterol in cells and is the form in which much of the cholesterol in lipoproteins is carried.

Lipoproteins involved in fat and cholesterol movement in the body

There are two lipoproteins produced by the liver, **very-low-density lipoprotein (VLDL)** and **high-density lipoprotein**

(HDL). The production of lipoproteins involves the ER and the Golgi apparatus, which packages them into membrane-bound vesicles for release by exocytosis. In the rat, about 80% of lipoprotein is made by the liver and most of the rest by intestinal cells as chylomicrons. The VLDL is converted by TAG removal by cells to **intermediate-density lipoprotein (IDL)** and then **low-density lipoprotein (LDL)** so that, including chylomicrons, five different lipoproteins are found in circulation. Figure 10.18 shows electron micrographs of different lipoproteins.

Apolipoproteins

Each type of lipoprotein has its own particular associated set of **apolipoproteins** (the name given to proteins of lipoproteins when detached). A dozen or more are already known. Their functions are not established in every case but they have several roles.

- Some are required as components for the production of lipoproteins. In the case of chylomicrons the main one is apoB$_{48}$ and, in that of VLDL, it is apoB$_{100}$.

- Some are required as destination-targeting signals – apoproteins designed to bind to specific receptors on the surface of cells. Binding leads to uptake of the bound lipoprotein by receptor-mediated endocytosis (see Fig. 10.15). In this way individual lipoproteins are taken up only by designated cells. Examples of this targeting function are apoB$_{100}$ on LDL, which binds to LDL receptors, and apoE on chylomicron remnants, which binds to liver receptors.

- Some apolipoproteins are required to activate enzymes. Thus apoCII on chylomicrons is necessary for lipoprotein lipase activity, which removes fatty acids from TAG.

Mechanism of TAG and cholesterol transport from the liver and the reverse cholesterol transport in the body

Let us start with VLDL and the outward flow of TAG and cholesterol from the liver. VLDL secretion is inhibited during the high-insulin post-absorptive phase because the hormone reduces the amount of the required apoB$_{100}$ available for VLDL synthesis. In people who are deficient in the synthesis of this apolipoprotein, accumulation of TAG in the liver occurs due to the failure of its export via VLDL.

Once VLDL is released, the TAG is progressively removed by lipoprotein lipase action in the way already described for chylomicrons. As the amount of fat diminishes, the percentage of cholesterol and its esters rises (Table 10.1), causing an increase in density, and the lipoprotein structures shrink in size. The end product of the process is LDL, formed via the intermediate state, IDL. LDL and IDL are taken up by peripheral tissue cells via receptor-mediated endocytosis so that the remaining TAG and cholesterol are delivered to extrahepatic tissues. The number of LDL receptors on cells is regulated to control LDL uptake. This outward flow of cholesterol from the liver is counterbalanced by a reverse flow as described below.

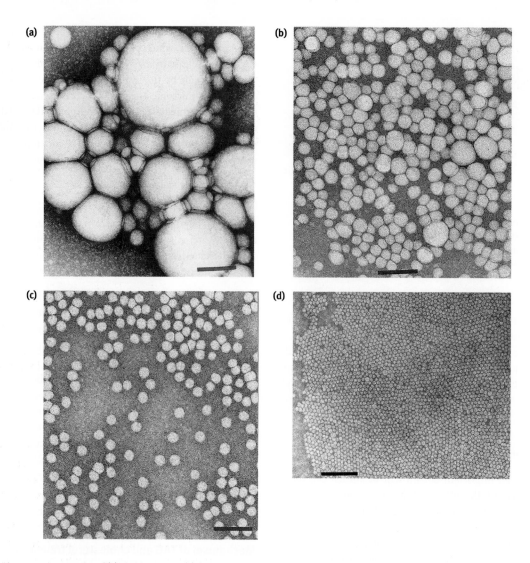

Fig. 10.18 Electron micrographs of **(a)** chylomicrons; **(b)** very-low-density lipoproteins (VLDL); **(c)** low-density lipoproteins (LDL); and **(d)** high-density lipoproteins (HDL). Scale bars, 1000 Å. Kindly provided by Dr Trudy Forte of the Lawrence Berkeley Laboratory, University of California, USA.

Table 10.1 Approximate content of fat and cholesterol as percentage dry weight of different lipoproteins

	Percentage dry weight of				
	Chylomicron	VLDL	IDL	LDL	HDL
Neutral fat (TAG)	80	50	30	10	8
Cholesterol + cholesterol ester	8	20	30	50	30

VLDL, very-low-density lipoprotein; IDL, intermediate-density lipoprotein; LDL, low-density lipoprotein; HDL, high-density lipoprotein; TAG, triacylglycerol.

The role of HDL in cholesterol transport

HDL is produced mainly by the liver in an 'immature' disc-like form, known as **nascent HDL**, made of phospholipid and apoproteins but containing little TAG or cholesterol. The nascent HDL picks up cholesterol from extrahepatic cells. Free cholesterol picked up by HDL as such would remain in the peripheral layer of the lipoprotein with its —OH group in contact with water, but, once in the HDL, it is converted to cholesterol ester. Hydrophobic force (page 38) causes the latter to migrate to the centre of the HDL, away from contact with water. The accumulation of the ester causes the HDL to fatten out into a spherical particle. Cholesterol esterification *in cells* requires ATP but HDL has none of this available and an alternative strategy is used. A fatty acyl group of lecithin, which is present in the hydrophilic shell, is

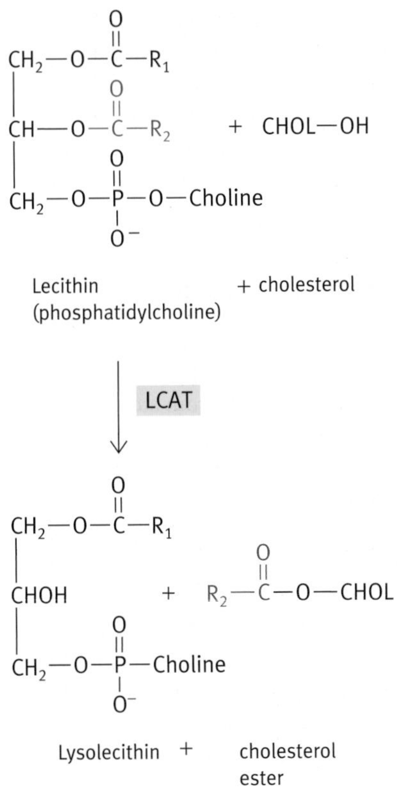

$$CH_2-O-\overset{\displaystyle O}{\overset{\displaystyle \|}{C}}-R_1$$

$$CH-O-\overset{\displaystyle O}{\overset{\displaystyle \|}{C}}-R_2 \quad + \quad CHOL-OH$$

$$CH_2-O-\overset{\displaystyle O}{\overset{\displaystyle \|}{P}}-O-Choline$$
$$\underset{\displaystyle O^-}{|}$$

Lecithin   + cholesterol
(phosphatidylcholine)

↓ **LCAT**

$$CH_2-O-\overset{\displaystyle O}{\overset{\displaystyle \|}{C}}-R_1$$

$$CHOH \quad + \quad R_2-\overset{\displaystyle O}{\overset{\displaystyle \|}{C}}-O-CHOL$$

$$CH_2-O-\overset{\displaystyle O}{\overset{\displaystyle \|}{P}}-Choline$$
$$\underset{\displaystyle O^-}{|}$$

Lysolecithin +   cholesterol
                      ester

Fig. 10.19 Reaction catalysed by lecithin: cholesterol acyltransferase (LCAT), where R_1 and R_2 are fatty acyl groups; CHOL—OH is cholesterol.

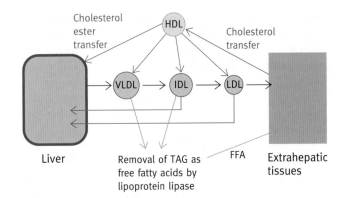

Fig. 10.20 The role of lipoproteins in the outward and reverse flow of cholesterol from, and to, the liver. The red lines represent transfer of cholesterol esters. The transfer of cholesterol esters between lipoproteins is reversible and more complex than shown. In addition, triacylglycerol (TAG) interchange occurs. The situation shown is that in the fasting condition in which chylomicrons are not present. After a meal containing fat, chylomicrons participate in the cholesterol ester transfer from high-density lipoprotein (HDL) and, in delivering cholesterol to the liver *via* chylomicron remnants, also contribute to the reverse cholesterol flow. VLDL, very-low-density lipoproteins; IDL, intermediate-density lipoproteins; LDL, low-density lipoproteins; FFA, free fatty acids.

transferred to cholesterol in an energy-neutral reaction by an enzyme associated with HDL, called **lecithin: cholesterol acyltransferase (LCAT)** (Fig. 10.19). (The ATP-requiring synthesis which occurs in cells is catalysed by an enzyme called acyl-CoA: cholesterol acyltransferase (ACAT).

The HDL transfers its cholesterol ester to VLDL, IDL and LDL and also to chylomicrons (present after a fat-containing meal) by means of a **cholesterol ester transfer protein (CETP)**. Although IDL and LDL deliver their contents to extrahepatic tissues as described above, a proportion return to the liver (they both use the LDL receptor). This is the explanation of the seemingly pointless recycling back to the liver of IDL and LDL now enriched by cholesterol from HDL; it is the reverse flow of cholesterol. This is illustrated in Fig. 10.20.

The transfer of cholesterol ester between lipoproteins occurs reversibly and a more complex traffic occurs between all of them; in addition, transfer of TAG takes place between the various components, catalysed by CETP which actually exchanges cholesterol ester for TAG. The precise reasons for this traffic complexity are not known.

How does cholesterol exit cells to be picked up by HDL?

How does cholesterol get out of the peripheral donor cell to be picked up? The mechanism has been elucidated. In an isolated island in Chesapeake Bay (Tangier Island) some of the inhabitants exhibited a condition known as **Tangier disease** in which high concentrations of cholesterol accumulate in lymphatic organs such as tonsils and spleen, the spleen becoming enlarged. Genetic studies pinpointed the deficient gene responsible for the condition. It codes for the enzyme ABCA1 transporter (ATP-binding cassette A1 transporter). This protein belongs to a large super family of membrane glycoproteins which transport molecules through membranes (described on page 112), but its main function was in doubt. It now appears that an important role is to transport cholesterol out of cells to be picked up by HDL. The principle apolipoprotein of HDL is known as apoA-I. The ABCA1 transporter transports cholesterol and phospholipid to HDL particles. As well as accepting cholesterol from intestine and other peripheral tissues, it also enables macrophages to unload excess cholesterol, so that ABCA1 is anti-artherogenic (see **Attie** reference in Further reading). This results in cholesterol being sent back to the liver where, as mentioned, some of it can be disposed of as bile salts.

It has long been known that high levels of HDL cholesterol (colloquially known as 'good cholesterol') correlate with low risk of coronary disease and clinical syndromes are known in which low levels of HDL in the blood lead to increased risk of coronary occlusions.

The control of cholesterol levels is very important, and several mechanisms have been identified. Cholesterol levels in liver increase with high levels of saturated fat in the diet. However, **unsaturated fats** especially the **omega-3 (ω-3)** and **omega-6** (ω-6) **fatty acids** (see Box 14.1; page 227) cause decreased levels of cholesterol and also of TAG. The biochemical mechanism is unknown. In liver the cells respond to high cholesterol levels by down regulation (decrease in the number) of the LDL receptors, which reduces the uptake of LDL and IDL from the blood. As a result, blood LDL cholesterol rises since uptake of LDL and IDL is the route by which cholesterol in the reverse flow re-enters the liver.

LDL cholesterol is colloquially known as 'bad cholesterol' because high levels are associated with increased **atherosclerosis** risk. A high **LDL cholesterol/HDL cholesterol ratio** correlates with the incidence of **coronary heart disease**. The high LDL levels lead to the development of plaques in blood vessels, a complex process in which cholesterol is involved. These, if present in coronary arteries, may burst causing blockages and result in heart attacks. An extreme case of this is found in genetic disease **familial hypercholesterolaemia**, in which people have defective LDL receptors leading to very high levels of LDL cholesterol and early vascular disease.

In addition to control of intake, the rate of synthesis of cholesterol is controlled. If the cellular level is low, synthesis and intake are increased; if the level is high, synthesis and intake are inhibited so that cholesterol homeostasis is achieved. How do these control mechanisms work? HMG-CoA (3-hydroxy-3-methylglutaryl-coenzyme A) is a metabolite you have not met yet in this book; it is described on page 233 where we briefly deal with cholesterol synthesis, but at this point it is sufficient that it is needed for cholesterol synthesis. The first metabolite committed to cholesterol synthesis is formed from HMG-CoA by **HMG-CoA reductase** and this is the main control point (Box 10.2). At low cholesterol levels in the cell, the genes coding for this enzyme and for LDL receptor protein are activated so that cholesterol synthesis and intake of cholesterol into cells are activated. At high cellular cholesterol levels the genes are not activated and, since there is a turnover of the two proteins, the level of reductase falls and cholesterol synthesis diminishes, as

also does intake via LDL receptors. The novel mechanism of this control is given in the Osborne reference at the end of this chapter. There are further controls. The HMG-CoA reductase is inhibited at high cholesterol levels; the enzyme becomes phosphorylated, in which form it is inactive. Also at high cholesterol levels the enzyme is more rapidly destroyed, so that three separate controls operate.

Release of FFA from adipose cells

After a meal, adipose cells store fat. In fasting they release free fatty acids into the blood to supply the body with a source of energy. A lipase releases FFA from stored TAG but there is an important difference here. The lipase inside the adipose cell is *not* the same (lipoprotein) lipase as in the capillary walls. It is a *hormone-sensitive lipase, activated by a glucagon signal to the cells and inhibited by an insulin signal to the cells.* The mechanism of this is described in Chapter 16. You can see how it all fits in. After a meal, the insulin level is high and that of glucagon low. This instructs adipose cells to take up glucose from the blood and FFA from chylomicrons and VLDL and convert them into TAG; release of FFA is inhibited. In fasting, the reverse occurs; glucagon is high and insulin low, so the lipoprotein lipase *inside the cell* is activated and tissues are supplied with FFA released in the blood.

The adipose cell hormone-sensitive lipase is also activated by epinephrine (adrenaline). Thus, in response to an epinephrine-liberating alarm, the adipose cells pour out FFA to supply the muscles with FFA to support any required muscular activity. How epinephrine does this is an important topic, but it is more suitable to deal with it in Chapter 16. Norepinephrine (noradrenaline) liberated from the sympathetic nervous system that innervates adipose tissue has the same effect.

How are FFA carried in the blood?

It would be unsatisfactory for relatively high levels of FFA to be carried in the blood since, at neutral pH, they are detergents. Instead they are carried adsorbed to the surface of the blood protein, **serum albumin**. This protein has hydrophobic patches on its surface and is the carrier for many molecules with hydrophobic sections. The adsorbed FFA are in equilibrium with FFA in free solution so that, as cells take them up, in order to maintain the equilibrium, more dissociate from the protein carrier into the serum where they are available to cells. The serum albumin cannot traverse the blood–brain barrier so the brain cannot be supplied with FFA. However, ketone bodies, derived from the partial metabolism of fats, are carried in simple solution. These can enter the brain to provide, in starvation, part of the energy, as already described. The rest must be supplied by glucose.

BOX 10.2 Inhibitors of cholesterol synthesis

Therapeutical drugs, known as statins, have been developed to reduce cholesterol levels. These resemble mevalonic acid (page 233) in structure, an intermediate in cholesterol synthesis formed by HMG-CoA reductase. They competitively inhibit this very effectively by combining with the active site of the enzyme. Simvastatin and lovastatin are examples, sold under proprietary names.

Other approaches for reducing total and LDL cholesterol levels are inhibitors of cholesterol absorption and bile acid-sequestering agents, which prevent their reabsorption from the intestine.

Summary

After a meal, glucose is stored as glycogen. The glucose is 'activated' as uridine diphosphoglucose (UDPG) and then polymerized into glycogen by glycogen synthase. Glycogen is broken down in times of need by glycogen phosphorylase to supply the cells with an energy source.

In liver, glycogen phosphorylase has another important function; it is used to liberate free glucose into the blood stream to maintain sufficient levels for the use, primarily, of brain and red blood cells. Other sugars absorbed from the intestine are converted to glucose or its derivatives and also stored as glycogen.

Glycogen metabolism is controlled by two hormones; insulin, which stimulates synthesis, and glucagon, which stimulates breakdown (described in Chapter 16). Epinephrine also promotes breakdown

Galactose conversion to glucose involves a pathway involving UDP-galactose. A genetic disease, galactosaemia, in infants affects brain development. The disease is due to the absence of an enzyme, uridyl transferase, needed to convert UDPG-galactose to UDPG, a process involving UDPG. Early detection of the disease and dietary restriction to avoid galactose intake avoids the tragic repercussions and normal development occurs.

Amino acids from the diet are not stored but are utilized at once for protein synthesis and other tissue components, the excess being disposed of at once. The carbon skeletons are converted to fat or glycogen or oxidized depending on the particular amino acid and the metabolic controls operating at the time. In terms of energy, the main 'traffic' in amino acids is in the synthesis of glucose during starvation. The topic of amino acid metabolism is dealt with in Chapter 19.

Absorbed fat present in the blood as chylomicrons is hydrolysed by a surface enzyme on cells lining the capillaries, to release fatty acids which are taken up by cells of adjacent tissues. Massive storage of fat as triacylglycerol occurs in the cells of adipose tissue after feeding. This is released into the blood during fasting as free fatty acids for use by other tissues, a hormone-controlled process (Chapter 16). The chylomicron remnants remaining, after tissues have taken up much of the fat, are taken up by the liver by receptor-mediated endocytosis. In the liver the fat and cholesterol are liberated from these, the latter being stored as cholesterol ester.

The liver also synthesizes cholesterol and fat from glucose and other food molecules and exports these to other tissues. This occurs in the form of chylomicron-like structures called very-low-density lipoproteins (VLDLs). The fat is removed by tissues from VLDL by the same process as described for chylomicrons. As the fat removal continues, the VLDL become low-density lipoproteins (LDLs), which now have a high proportion of cholesterol. These are taken up by the tissues by receptor-mediated endocytosis. This outward delivery of cholesterol from liver to extrahepatic tissues is counterbalanced by a reverse flow back to the liver. High-density lipoproteins (HDLs) pick up cholesterol from cells and transfer it to LDL, part of which is taken up by the liver. The cholesterol removal as HDL is brought about by ABCA1 transporter.

The pathway of cholesterol movement is complex and the reasons for the two-way traffic not completely understood. It is however of great medical importance because the liver converts some of the cholesterol to bile acids and is a route for cholesterol removal to some extent. Excess LDL in the blood is associated with atherosclerosis, causing heart attacks. The 'statin' drugs inhibit cholesterol synthesis and are widely used therapeutically to reduce cholesterol levels in blood.

Further reading

Glycogen synthesis

Alonso, M. D., Lomako, J., Lomako, W. M., and Whelan, W. J. (1995). A new look at the biogenesis of glycogen. *FASEB J.*, 9, 1126–37.

The role of glycogenin

Van Schaftingen, E., Detheux, M., and Da Cunha, M. V. (1994). Short-term control of glucokinase activity: role of a regulatory protein. *FASEB J.*, 8, 414–19.
Describes the occurrence, properties, and control of this enzyme.

Lipid transport and lipoproteins

Nilsson-Ehle, P., Garfinkel, A. S., and Schotz, M. C. (1980). Lipolytic enzymes and plasma lipoprotein metabolism. *Annu. Rev. Biochem.*, 49, 667–93.
General summary of lipoproteins, including useful information at the physiological level.

Gannon, B. and Nedergaard, J. (1985). The biochemistry of an inefficient tissue. *Essays Biochem.*, 20, 110–64.
Deals with heat production by brown fat cells.

Angelin, B. and Gibbons, G. (1993). Lipid metabolism. *Curr. Opin. Lipidol.*, 4, 171–6.
Overview of lipoproteins.

Tall, A. (1995). Plasma lipid transfer proteins. *Annu. Rev. Biochem.*, **64**, 235–57.
Comprehensive review of cholesterol ester transport protein (CETP), lipid transport, and forward and reverse cholesterol transport between liver and extrahepatic tissues.

Osborne, T. E. (1997). Cholesterol homeostasis: Clipping out a slippery regulator. *Curr. Biol.*, **7**, R172–4.
Describes an unusual control of cholesterol synthesis

Gura, T. (1999). Gene linked to faulty cholesterol transport. *Science*, **285**, 814–15.
A short informative report on the discovery of the way in which cholesterol is removed from cells. The work relates to cholesterol transport out of cells and Tangier disease.

Redgrave, T. G. (1999). Chylomicrons. Chapter 3 in 'Lipoproteins in Health and Disease', edited by Betteridge, D., Illingworth, R., and Shepherd, J., ISBN: 0340552697, Oxford University Press.

Ballantyne, C. M. (2003). Current and future aims of lipid-lowering therapy: changing paradigms and lessons from the Heart Protection Study on standards of efficacy and safety. *Am. J. Cardiol.*, **92**, 3K–9K.

Brewer, H. B. and Santamarina-Fogo, S. (2003). Clinical significance of high-density lipoproteins and the development of artherosclerosis: focus on the role of adenosine triphosphate-binding cassette protein A1 transporter. *Am. J. Cardiol.*, **92**, 10K–16K.

Huff, M. W. (2003). Dietary cholesterol, cholesterol absorption, postprandial lipemia and artherosclerosis. *Can. J. Clin. Pharmacol.*, **10**, (Suppl. A), 26A–32A.

Borst, P. and Elferink, R. O. (2002). Mammalian ABC transporters in health and disease. *Annu. Rev. Biochem.*, **71**, 537–92.

Rader, D. J. (2003). Regulation of reverse cholesterol transport and clinical implications. *Am. J. Cardiol.*, **92**, 42–9.

Attie, A. D. (2007). ABCA1: at the nexus of cholesterol, HDL and atherosclerosis. *Trends Biochem. Sci.*, **32**, 172–9.
Reviews removal of cholesterol from cells and formation of HDL particles.

Problems

1   Starting with glucose-1-phosphate, explain how the synthesis of glycogen is made thermodynamically irreversible.

2   Have you any comments on the name of enzyme UDP-glucose pyrophosphorylase?

3   Which tissues release glucose into the blood as a result of glycogen breakdown? Why is this so?

4   What reaction do the two enzymes glucokinase and hexokinase catalyse? Why should the liver have glucokinase while brain and other tissues have hexokinase?

5   In the genetic disease galactosaemia, infants are unable to metabolize galactose correctly. Removal of galactose from the diet can prevent the deleterious effects of the disease. UDP-galactose epimerase is reversible; why is this important in connection with the above statement?

6   How is triacylglycerol (TAG) removed from chylomicrons and utilized by tissues?

7   Explain what very-low-density lipoprotein (VLDL) is and its likely role.

8   What is the main route of cholesterol removal from the body? Why is this of medical interest?

9   Cholesterol is esterified in high-density lipoprotein (HDL). Conversion of cholesterol to its ester is an energy-requiring process but HDL has no access to ATP. How is the esterification achieved?

10   How is fat released from adipose cells and under what conditions? How is it transported in the blood for transportation to other tissues? Compare this to the transport of fat from the liver to other tissues.

11   A genetic disease exists in which cholesterol levels in the blood are very high. Explain the cause of this.

12   How is fat from the adipose tissue transported and utilized by cells? How is fat from the liver transported and utilized by cells?

13   What is the reverse flow of cholesterol and its significance.

Chapter 11

Principles of energy release from food

The release of energy in the form of adenosine triphosphate (ATP) from glucose, fats, and amino acids involves fairly long and somewhat involved metabolic pathways. There is the possibility that, if we go at once into these, the overall strategy of energy generation by cells may be lost among the detail. We will, therefore, at this stage, deal with major stages of the processes, picking out landmark metabolites only, and view the strategies on a broad basis. When this has been done, subsequent chapters will deal in more detail with the pathways; in short, we want you to see the wood before the trees.

Overview of glucose metabolism

Before we go into the stages of glucose oxidation we need to say a little about biological oxidation in general.

Biological oxidation and hydrogen-transfer systems

Oxidation does not necessarily involve oxygen; it involves the removal of electrons, and chemically this is the definition of oxidation. It may involve *only* electron removal such as in the ferrous/ferric system:

$$Fe^{2+} \rightarrow Fe^{3+} + e^-$$

or it may be the removal of electrons accompanied by protons from a hydrogenated molecule. In biochemical systems it commonly involves enzymic removal of two hydrogen atoms from a metabolite molecule such as:

$$-CH_2-CH_2- \rightarrow -CH{=}CH- + 2H^+ + 2e^- \text{ or}$$

$$-CHOH-CH_2- \rightarrow -CO-CH_2- + 2H^+ + 2e^-$$

In such chemical oxidation systems, the electrons must be transferred from the electron donor to an electron acceptor. Depending on the particular electron acceptor, the electron transferred may be accompanied by the proton, in which case it is a hydrogen atom being transferred, or alternatively the proton may be liberated into solution, and only the electron transfered.

The ultimate electron acceptor in the aerobic cell is oxygen. Oxygen is **electrophilic** – it has an avidity for electrons. When it accepts four electrons, protons from solution join up with it to produce water molecules:

$$O_2 + 4e^- + 4H^+ \rightarrow 2H_2O$$

However, oxygen is only the *ultimate* electron acceptor in the cell – there are other electron acceptors, which form a bucket brigade, carrying electrons from metabolites to oxygen by a chain of intermediate carriers handing electrons from one to the next, until finally they are handed to oxygen. They accept electrons and hand them on to the next acceptor, and thus are electron carriers. This is the **electron transport chain**, which plays a predominant part in ATP generation. The essential concept is given in Fig. 11.1. Two of the electron carriers involved in energy production are of such central importance in metabolism that we must now describe them. It is essential for you to be completely familiar with these electron carriers.

NAD⁺—an important electron/hydrogen carrier

The first carrier involved in the oxidation of many metabolites is NAD⁺, which stands for **nicotinamide adenine dinucleotide.** You have already met a nucleotide in the form of adenine monophosphate (AMP) (Fig. 3.3); a **nucleotide** has the general structure, base—sugar—phosphate, the base in AMP being adenine. (The structures of adenine and other bases are dealt with later in Chapter 21 and need not concern us now.) In the case of NAD⁺, a dinucleotide is formed by linking the two phosphate

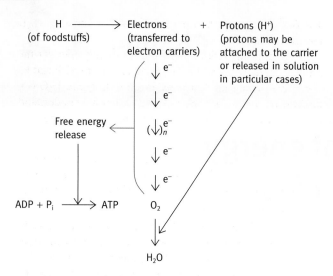

Fig. 11.1 The concept of electron transfer to oxygen and ATP generation.

groups of two nucleotides (which is unusual and not how nucleotides are linked together in nucleic acids, as will be evident in later chapters). The general structure of NAD^+ is

Base—sugar—phosphate—phosphate—sugar—base

The two bases are adenine (as in ATP) and nicotinamide, respectively – giving the structure of NAD below.

Adenine   Nicotinamide

| |

Ribose   Ribose

| |

Phosphate—Phosphate

NAD^+ is a **coenzyme** – a small organic molecule that participates in enzymic reactions. It differs from an ordinary enzyme substrate only in that its reduced form leaves the enzyme and attaches to a second enzyme where it donates its reducing equivalents to a second substrate. NAD^+ thus acts catalytically by being continually reduced and reoxidized, and in doing so transfers electrons from one molecule to another. It also transfers a proton, and so transfers one hydrogen atom and one electron (see below).

The 'business end' of the molecule is the nicotinamide group. It is derived from the vitamin, nicotinic acid or niacin. Niacin is an essential vitamin for some animals, but humans can make it from the tryptophan, which is an essential amino acid. The rest of the molecule fits the coenzyme to appropriate enzymes but does not undergo any chemical change.

Nicotinamide has the structure:

When linked in NAD^+ the structure is:

where R is the rest of the NAD^+ molecule.

NAD^+ can be reduced by accepting two electrons from two hydrogens on a metabolite. One proton is transferred as a hydride ion ($H:^-$), to give the following structure (the second proton being liberated into solution):

Reduced NADH $+ H^+$

NAD^+ is the coenzyme for several dehydrogenases that catalyse this type of reaction:

$$AH_2 + NAD^+ \rightleftharpoons A + NADH + H^+.$$

The reduced NAD^+ can then diffuse to a second enzyme and participate in a reaction such as:

$$B + NADH + H^+ \rightleftharpoons BH_2 + NAD^+.$$

In this way NAD^+ acts, effectively, as the carrier for the transfer of a pair of hydrogen atoms from A to B even though it carries only one proton:

$$AH_2 + B \rightarrow A + BH_2.$$

In biochemistry such reactions are often presented in the form of whirligigs.

In case it requires extra emphasis, *NADH + H^+ can add two H atoms to a substrate* since the second electron it transfers to an acceptor molecule is joined by a proton from solution as shown above. In equations, reduced NAD^+ is written as $NADH + H^+$. In the text, the term NADH is used, but it always implies $NADH + H^+$.

FAD and FMN are also electron carriers

Another (hydrogen) carrier is **flavin adenine dinucleotide (FAD)**. In this case the electrons are transferred as hydrogen atoms. FAD is synthesized from vitamin B_2 or riboflavin. The important feature of this molecule is that it can (in combination with appropriate proteins) accept two H atoms to become $FADH_2$. FAD is a prosthetic group – it is a permanent attachment to its apoenzyme, unlike NAD^+, which moves from one dehydrogenase to another. Its role will become clear; for the present it is sufficient that FAD can be reduced. The chemistry of this reduction is given below, where a related carrier, flavin mononucleotide (FMN), is also described.

FAD has the structure: isoalloxazine ring structure—ribitol—phosphate—phosphate—ribose—adenine. It is very unusual in having the linear ribitol molecule instead of ribose.

Oxidation–reduction reactions occur in the isoalloxazine structure:

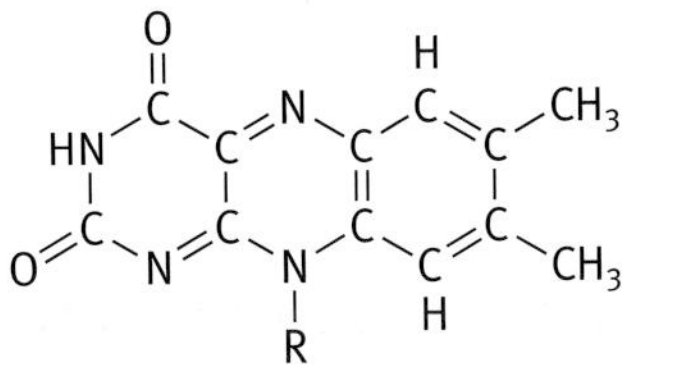

FAD (oxidized form) 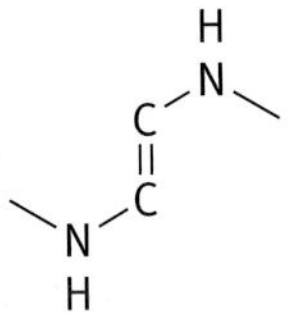 $FADH_2$ (reduced form)

FMN has the structure: isoalloxazine – ribitol – phosphate. It is reduced in the same way as FAD.

Energy release from glucose

Often glucose is first stored in liver and muscle as glycogen in animals and then broken down to glucose-1-phosphate, which is oxidized. This does not affect the overall account given below except that, as described in later chapters, the initial steps of metabolism of glycogen and glucose are slightly different and for these the ATP yield from glycogen is three per glucosyl unit, not two as from glucose. (In the case of glucose an ATP molecule is used to form glucose-1-phosphate.) For convenience, in this section, we will refer to glucose oxidation.

The main stages of glucose oxidation

Overall, glucose is oxidized as follows:

$$C_6H_{12}O_6 + 6O_2 \rightarrow 6CO_2 + 6H_2O$$

The $\Delta G^{0'}$ for this reaction is $-2820 \text{ kJ mol}^{-1}$, liberating large amounts of energy.

In the cell, this oxidation process is accompanied by the synthesis of more than 30 molecules of ATP from ADP and P_i. The entire process of glucose oxidation to CO_2 and H_2O can be divided up into three stages. If you cannot fully understand the summaries below do not be concerned – they will become clear shortly.

- Stage 1: **glycolysis.** This results in the lysis or splitting of glucose into two C_3 fragments to ultimately yield pyruvate, accompanied by reduction of NAD^+. *This occurs in the cytoplasm of cells.*

- Stage 2: the **citric acid cycle.** This is also known as the **Krebs** (after its discoverer) or **tricarboxylic acid (TCA) cycle** (because citric acid is a tricarboxylic acid). In mitochondria the carbon atoms of pyruvate are converted to an acetyl group and CO_2, and in the cycle, electrons from the acetyl groups are transferred to electron carriers. No oxygen is involved in this stage. Carbon atoms are released as CO_2, the oxygen largely from water. The cycle is located inside mitochondria in eukaryotes and in the cytoplasm of bacteria.

- Stage 3: the **electron transport system.** Electrons are transported from the electron carriers to oxygen where, with protons from solution, water is formed. It is in stage 3 that most of the ATP is generated. This occurs in the inner mitochondrial membrane in eukaryotes and in the cell membrane in prokaryotes.

Stage 1 in the release of energy from glucose: glycolysis

Glycolysis does not involve oxygen, and only two ATP molecules per molecule of glucose lysed are produced from ADP (three if you start from glycogen). The end products are pyruvate and NADH as shown in Fig. 11.2. In **aerobic glycolysis,** when the oxygen supply is plentiful, the NADH is reoxidized to NAD^+ *via* mitochondria in eukaryotic cells (page 210); the pyruvate is taken up by the mitochondria where it is metabolized to CO_2 and H_2O *via* the citric acid cycle.

Anaerobic glycolysis

However, oxygen is not necessarily always plentiful since the delivery in blood may not keep up with its usage, especially in muscle during vigorous activity. Since NAD^+ acts catalytically and is present in cells in small amounts, for glycolysis to proceed the NADH must be recycled back to NAD^+; no NAD^+, no glycolysis. If the capacity of mitochondria to do this is inadequate at high glycolytic rates and/or there is insufficient oxygen availability, glycolysis needed to generate ATP for contraction would be impaired unless another way were found to regenerate NAD^+ from NADH. An 'emergency' system comes into play in this situation. The NADH is re-oxidized by reducing pyruvate to produce lactate. The reaction is:

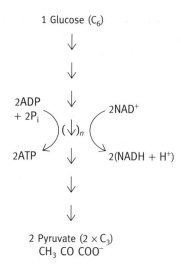

1 Glucose (C$_6$)

2ADP + 2P$_i$ → 2ATP

2NAD$^+$ → 2(NADH + H$^+$)

$(\downarrow)_n$

2 Pyruvate (2 × C$_3$)
CH$_3$ CO COO$^-$

Fig. 11.2 The net result of aerobic glycolysis of glucose. NADH is oxidized by mitochondria – see page 210.

$$CH_3COCOO^- + NADH + H^+ \rightleftharpoons CH_3CHOHCOO^- + NAD^+.$$
Pyruvate   Lactate dehydrogenase   Lactate

The enzyme catalysing this reaction is **lactate dehydrogenase.** The production of lactate from glucose is known as **anaerobic glycolysis**, as opposed to aerobic glycolysis which forms pyruvate. The object of anaerobic glycolysis is not to produce lactate but to reoxidize NADH and thus permit continued ATP production from glycolysis (Fig. 11.3). Since there is a lot of lactate dehydrogenase in muscle, the NADH can be rapidly reoxidized in this way and this in turn permits glycolysis to proceed at a very fast rate. The advantage of this is that, although only two ATP molecules are generated per molecule of glucose,

relatively vast amounts of glucose can be broken down which could be important when muscles are very active. If you are being chased by a tiger, the extra ATP could be very important for survival. The lactate formed enters the blood but is not wasted; it is used mainly by the liver to resynthesize glycogen, as described later.

As a parenthetic note, yeast can live entirely on anaerobic glycolysis in the absence of oxygen using an analogous trick to reoxidize NADH. The pyruvate is converted to acetaldehyde + CO$_2$; a second enzyme, alcohol dehydrogenase, reduces the acetaldehyde to ethanol.

$$CH_3COCOO^- + H^+ \longrightarrow CH_3CHO + CO_2$$
Pyruvate decarboxylase

$$CH_3CHO + NADH + H^+ \rightleftharpoons CH_3CH_2OH + NAD^+$$
Alcohol dehydrogenase

Because of the low yield of ATP, vast quantities of glucose are broken down, producing alcohol and CO$_2$. To illustrate the scale of this by a homely example, the latter causes the explosion of bottles if, during beer-making, it is bottled too early while glycolysis is still vigorous. Pyruvate decarboxylase does not occur in animals, which is just as well perhaps, since vigorous exercise might then have been literally intoxicating.

As already mentioned in the introduction, in the above account, for simplicity, we have talked about glucose being the carbohydrate that is glycolysed. It can indeed be free glucose, but in muscle it is mainly glycogen, the storage form of glucose, that is broken down. We also again remind you that in the next chapter the detailed mechanisms of glycolysis, the citric acid cycle, and the electron transport system will be given. At this stage the main thing is to get an overview.

Stage 2 of glucose oxidation: the citric acid cycle

The **mitochondria**, as described in Chapter 2, are small organelles located in the cytoplasm of the cell. They are where most of the ATP is produced. The inner membrane is the site of ATP generation. Its area is increased by invagination to form **cristae** whose extent reflect the level of ATP demand (Fig. 11.4). The mitochondrion is filled with a concentrated solution of enzymes – it is called the **matrix**. It is here that stage two of glucose metabolism mainly occurs, only one reaction being located in the inner membrane.

Aerobic glycolysis, as stated, produces pyruvate and NADH in the cytoplasm. To be further oxidized, the pyruvate must enter the mitochondria. For NADH oxidation, the electrons are transported in and oxidized, leaving the NAD$^+$ in the cytoplasm (page 210) to participate in glycolysis. A transport system in the inner mitochondrial membrane takes the pyruvate from the cytoplasm into the mitochondrion.

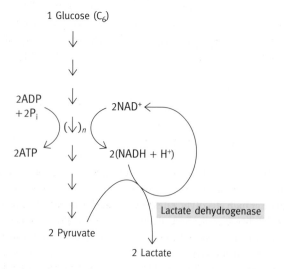

1 Glucose (C$_6$)

2ADP + 2P$_i$ → 2ATP

2NAD$^+$

2(NADH + H$^+$)

$(\downarrow)_n$

2 Pyruvate

Lactate dehydrogenase

2 Lactate

Fig. 11.3 The net result of anaerobic glycolysis of glucose.

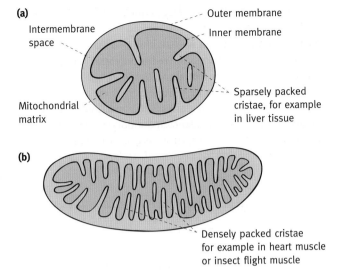

(a)
Outer membrane
Intermembrane space
Inner membrane
Mitochondrial matrix
Sparsely packed cristae, for example in liver tissue

(b)
Densely packed cristae for example in heart muscle or insect flight muscle

Fig. 11.4 Mitochondria from **(a)** liver and **(b)** heart tissue. The number of cristae reflects the ATP requirements of the cell. See Fig. 2.5 for an electron micrograph of a mitochondrion.

Some cells such as mature erythrocytes have no mitochondria and so must generate ATP by glycolysis only; they are glucose-dependent.

How is pyruvate fed into the citric acid cycle?

We now come to an enzyme reaction of major importance in which pyruvate, transported into mitochondria, is converted to a compound which is at the crossroads of energy metabolism. This is acetyl-coenzyme A (acetyl-CoA), a compound not previously mentioned in this book.

What is coenzyme A?

Coenzyme A is usually referred to as CoA for short, but is written in equations as CoA—SH because its thiol group is the reactive part of the molecule.

Unlike NAD^+ and FAD, CoA is *not an* electron carrier, but an acyl group carrier (A for acyl). Like NAD^+ and FAD it is a dinucleotide and, as so often happens with cofactors, incorporates a water-soluble vitamin in its structure, in this case **pantothenic acid**. Pantothenic acid has an odd sort of structure.

$$HO—CH_2—\underset{\underset{CH_3}{|}}{\overset{\overset{CH_3}{|}}{C}}—\underset{\underset{OH}{|}}{\overset{\overset{H}{|}}{C}}—\overset{\overset{O}{||}}{C}—NH—CH_2—CH_2—C\overset{O}{\underset{OH}{\diagdown}}$$

Pantothenic acid

You would normally expect the vitamin moiety of a coenzyme to play a part in the reaction for which the coenzyme is required (as with, for example, riboflavin in FAD and nicotinamide in NAD^+), but the pantothenic acid moiety is just 'there' and apparently quite inert. It presumably provides a recognition group to help bind the CoA to appropriate enzymes

but why this particular structure is used is not obvious. It may be a quirk that happened by chance early in evolution and the cell became locked into its use.

The structure of CoA is:

Adenine   β-Mercaptoethylamine
Phosphate—Ribose   Pantothenic acid
Phosphate——Phosphate

The **β-mercaptoethylamine** part is the business end of the molecule.

$$RCO—\underbrace{NHCH_2CH_2—SH}$$

β-Mercaptoethylamine moiety

The CoA molecule carries acyl groups as thiol esters. For example, acetyl-CoA can be written as $CH_3CO—S—CoA$. *The thiol ester is a high-energy compound* (unlike a carboxylic ester). It has a $\Delta G^{0'}$ of hydrolysis of approximately -31 kJ mol^{-1} as compared with about -20 kJ mol^{-1} for a carboxylic ester. The difference is due to the fact that the latter is resonance stabilized (page 32) and so has a lower free-energy content than a thiol ester, which is not resonance stabilized. With that description of CoA we can now return to the question of what happens to pyruvate transported into mitochondria.

Oxidative decarboxylation of pyruvate

The pyruvate is subjected to an *irreversible* oxidative decarboxylation in which CO_2 is released (decarboxylation), a pair of electrons is transferred to NAD^+ (oxidation), and an acetyl group is transferred to CoA. (Note that this is different from the **nonoxidative** decarboxylation carried out by yeast pyruvate decarboxylase, described earlier. In the latter there is, as implied, no oxidation, NAD^+ is not involved, and the product is acetaldehyde, not an acetyl group.) The large negative free-energy change means that the oxidative decarboxylation reaction is irreversible. This has important metabolic repercussions (page 193). The reaction, catalysed by **pyruvate dehydrogenase**, is as follows:

$$\text{Pyruvate} + \text{CoA}—\text{SH} + NAD^+ \rightarrow \text{Acetyl}—\text{S}—\text{CoA} + NADH + H^+ + CO_2$$

$$\Delta G^{0'} = -33.5 \text{ kJ mol}^{-1}$$

The acetyl group of acetyl-CoA is now fed into the citric acid cycle. At this stage we will not be concerned with the reactions in the cycle. The essential point is that the carbon atoms of the acetyl group of acetyl-CoA produced from pyruvate are converted to CO_2 while three molecules of NAD^+ are reduced to NADH. In addition, a molecule of FAD is reduced to $FADH_2$, the electrons coming indirectly, in part, from water (page 194). Almost as a sideline, the cycle generates one 'high-energy'

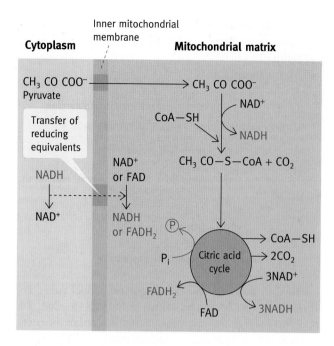

Fig. 11.5 This shows what happens to pyruvate and NADH generated in glycolysis and the generation of further NADH and FADH$_2$ inside mitochondria. The FAD is always attached to an enzyme. The source of the reducing equivalents is dealt with in the next chapter. The NADH and FADH$_2$ are oxidized by the electron transport pathway of the inner membrane, described shortly.

phosphoryl group from P$_i$ for each acetyl group fed in. Stage 2 of glucose oxidation is illustrated in Fig. 11.5.

In summary, a molecule of pyruvate from the cytoplasm is converted in the mitochondria by pyruvate dehydrogenase and the citric acid cycle to three molecules of CO_2 and, in the process, three molecules of NAD^+ and one molecule of FAD are reduced. We have hardly started to make ATP – only two molecules in glycolysis and two in the cycle (per starting glucose molecule) but there are almost 30 still to be made.

So far, it's been mainly preparation of fuel – the big return in the form of ATP generation comes in stage 3.

Stage 3 of glucose oxidation: electron transport to oxygen

The oxidation of NADH and FADH$_2$ takes place in the inner mitochondrial membrane, which contains the chain of electron carriers.

The electron transport chain – a hierarchy of electron carriers

Redox potentials

The problem we are now concerned with, we remind you, is the transference of electrons from NADH and FADH$_2$ to oxygen with the formation of water:

$$NADH + H^+ + \tfrac{1}{2}O_2 \rightarrow NAD^+ + H_2O$$

The $\Delta G^{0'}$ of this reaction is -220 kJ mol^{-1}. To understand the process, we need to discuss the **redox potential** (also known as the **oxido-reduction potential**) of compounds.

Compounds capable of being oxidized are, by definition, electron donors; in any oxido-reduction reaction there must be an electron acceptor (oxidant) and a donor (reductant). The reaction:

$$X^- + Y \rightleftharpoons X + Y^-$$

can be considered to occur by two theoretical half-reactions, as follows:

(a) $X^- \rightleftharpoons X + e^-$;
(b) $Y + e^- \rightleftharpoons Y^-$.

Each of these involves the reduced and oxidized forms of each reactant which are called **conjugate pairs** or reduction-oxidation couples or, the most convenient term, **redox couples**. X and X^- are such a couple and Y and Y^-, another couple. Clearly, any real-life oxidation-reduction reaction must involve a pair of redox couples, because one must donate electrons and one must accept them.

Different redox couples have different affinities for electrons. One with a lesser affinity will tend to donate electrons to another of higher affinity. The redox potential value (E'_0) is a measurement of the electron affinity or electron-donating potential of a redox couple. This is of importance in biochemistry for it is an indicator of the direction in which electrons will tend to flow between reactants. Equally important E'_0 values are directly related to free-energy changes (see below).

E'_0 values are expressed in volts – the more negative, the lower the electron affinity, the greater the tendency to hand on electrons, the greater the reducing potential, and the higher the energy of the electrons.

Determination of redox potentials

The reason why this chemical property is expressed in volts is due to the method of redox potential determination. As stated above, two redox couples must be involved in an oxido-reduction reaction but, since electron transfer is involved, they can be in two separate vessels (half-cells) if they are connected by a copper wire to conduct electron flow. The method is to use the $2H^+ + 2e^- \rightleftharpoons H_2$ equilibrium (catalysed by the platinum black electrode) in one half-cell as the reference redox couple, and compare the unknown sample in the second half-cell with it. Electrons will flow through the wire according to the relative electron affinities of the two systems. The positive ions in each half-cell (H^+ in the hydrogen electrode half-cell and, for example, ferrous and ferric ions in the sample half-cell) are accompanied by anions. Change in the positive ions due to loss or gain of electrons in the half-cells necessitates a compensating

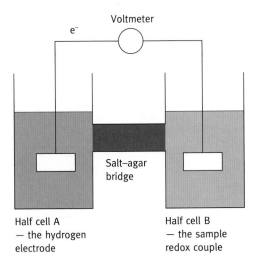

Voltmeter

Half cell A
— the hydrogen
electrode

Half cell B
— the sample
redox couple

Salt–agar
bridge

Fig. 11.6 Apparatus for measurement of redox potentials. The reference hydrogen electrode in A contains the redox couple $H_2/2H^+$ catalysed by platinum black on the electrode. The sample half-cell B contains equimolar amounts of the two components of the redox couple whose redox potential is being determined. If B is more reducing than A, electrons will flow from B to A and reduce $2H^+$ to H_2 if the half-cells are connected by a wire. The E_0' value is therefore more negative than that of the hydrogen electrode whose E_0' is assigned the value of -0.42 V. As protons in A are reduced to hydrogen atoms, anions will flow from A to B via the agar-salt bridge to preserve charge neutrality. If the sample in B is less reducing than the reference hydrogen electrode, a reverse series of events will occur and the sample redox couple will have an E_0' value more positive than -0.42 V.

BOX 11.1 Calculation of the relationship between the $\Delta G^{0'}$ value and the E_0' value

There is a direct relationship between the $\Delta G^{0'}$ value and the E_0' value of an oxido-reduction reaction, quantified by the Nernst equation:

$$\Delta G^{0'} = -nF\Delta E_0',$$

where n equals the number of electrons transferred in the reaction, F is the Faraday constant (96.5 kJ V^{-1} mol^{-1}), and E_0' is the difference in redox potential between the electron donor and electron acceptor.

If we consider the oxidation of NADH, the half-reactions are:

(1) $NAD^+ + 2H^+ + 2e^- \rightarrow NADH + H^+$ $E_0' = -0.32$ V,

(2) $\frac{1}{2}O_2 + 2H^+ + 2e^- \rightarrow H_2O$ $E_0' = 0.816$ V.

For the overall reaction, subtracting (1) from (2), we get:

$$NADH + H^+ + \tfrac{1}{2}O_2 \rightarrow NAD^+ + H_2O,$$

$$\Delta E_0' = +0.816 - (-0.32) = +1.136 \text{ V}.$$

Therefore:

$$\Delta G^{0'} = -2(96.5 \text{ kJV}^{-1} \text{ mol}^{-1})(+1.136 \text{ V}) = -219.25 \text{ kJ mol}^{-1}.$$

---

anion migration from one half-cell to the other. The agar-salt bridge shown in Fig. 11.6 provides the route for this. The electrical potential between the half-cells is measured by a voltmeter inserted into the copper wire connecting electrodes in each half-cell. The reference (hydrogen electrode) half-cell is arbitrarily assigned the value of zero and the relative value of the sample cell is the redox potential of the redox pair in it. In physics, the convention is that electrical current flows in the opposite direction to electron flow and, therefore, the half-cell that is donating electrons has the more negative voltage. Thus, if the sample half-cell is more reducing than the reference half-cell, its redox potential value is more negative.

The E_0 values (written without a prime) are standard values measured at 1.0 M concentrations of components or H_2 gas at 1 atmosphere pressure and 1.0 M HCl (pH 0). In biochemistry, values are adjusted to pH 7.0 instead of pH 0 and this brings the redox potential of the reference half-cell to a value of -0.42 V. Redox potentials so corrected are written as E_0' values. The half-reaction $NAD^+ + 2H^+ + 2e^- \rightarrow NADH + H^+$ has an E_0' value of -0.32 V, and that of the half-reaction $\frac{1}{2}O_2 + 2H^+ + 2e^- \rightarrow H_2O$, an E_0' value of $+0.82$ V. The very large difference indicates that NADH has the potential to reduce oxygen to water but the

reverse will not occur. The relationship between the $\Delta G^{0'}$ value and the E_0' value is shown in Box 11.1.

Electrons are transported in a stepwise fashion

In the cell, the transport of electrons to oxygen does not happen in a single step. In the electron transport system of the mitochondria there is a chain of electron carriers of ever-increasing redox values (decreasing reducing potentials) terminating in the ultimate electron acceptor, oxygen. In effect, electrons from NADH and $FADH_2$ bump down a staircase, each step being a carrier of the appropriate redox potential and each fall releasing free energy (Fig. 11.7). The free energy is thus liberated in manageable parcels – manageable in the sense that it can be harnessed into mechanisms that (indirectly) result in ATP generation from ADP and P_i rather than being wasted as heat, as occurs in simple burning of glucose. The oxidation of NADH and $FADH_2$ thus drives the conversion of ADP and P_i to ATP. Hence the complete process is called **oxidative phosphorylation**. Thirty or more ATP molecules are synthesized from the oxidation of one molecule or glucose (the exact number is discussed in the next chapter).

In Fig. 11.8, all three stages are put together in one scheme.

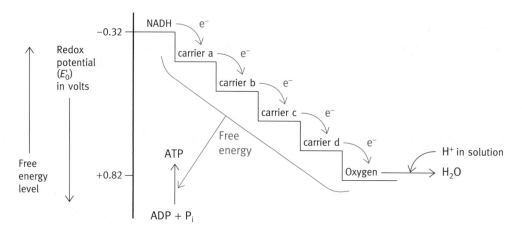

Fig. 11.7 Principle of the electron transport chain. (The number of carriers shown is arbitrary; each is a different carrier.) With some carriers only electrons are accepted, protons being liberated into solution, while in the case of other carriers protons accompany the electrons. The final reaction with oxygen involves protons from solution.

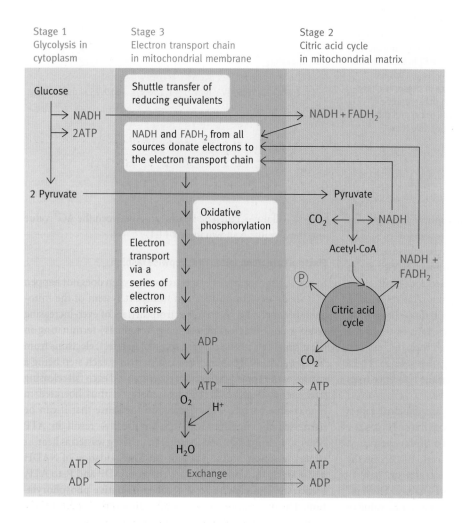

Fig. 11.8 Summary of the oxidation of glucose. For ease of presentation only products are shown; obviously, reduced NAD is formed from NAD^+, etc. The same scheme applies to glycogen except that three ATP molecules are produced per glucose unit. We indicate the production of one—Ⓟ from the citric acid cycle rather than as one ATP because, in this case, GDP rather than ADP is the acceptor. Energetically, it amounts to the same thing and is explained in the next chapter. The ATP generated is transported to the cytoplasm by a mechanism involving ADP intake. An important point to note is that NAD^+ and FAD collect electrons from different metabolic systems including fatty acid oxidation and not just from glucose (see Fig. 11.10). NADH and $FADH_2$ donate electrons to different points in the electron transport chain. $FADH_2$ donates electrons to the chain at a point lower down than does NADH. See Figs 12.28 and 12.29 for details of shuttles.

Energy release from oxidation of fat

In addition to glucose and glycogen breakdown to supply the energy for ATP generation, the body oxidizes fat for the same purpose. In terms of energy production, it is the fatty acid components of neutral fat (triacylglycerol, TAG) that are quantitatively important, the glycerol portion being less significant. Chemically, fatty acids are quite different from glucose and you might well expect their oxidation to be correspondingly different. It is at this point that you might get your first glimpse of how, despite complexity in detail, the metabolism of different foodstuffs dovetails together with majestic simplicity. Glucose, as described, is oxidized so that acetyl-CoA is formed and fed into the citric acid cycle. Fatty acids are also manipulated so that carbon atoms are detached two at a time as acetyl-CoA which is also fed into the citric acid cycle. In the preliminary manipulations, NAD^+ and FAD are reduced in both systems, and the electrons they carry are fed into the electron transport system as in glucose oxidation. Figure 11.9 gives the relationship between glucose and fat oxidation.

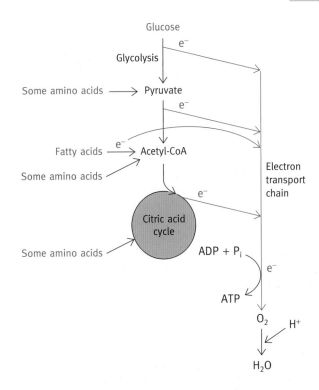

Fig. 11.10 Relationship of glucose, fat, and amino acid oxidation for energy generation.

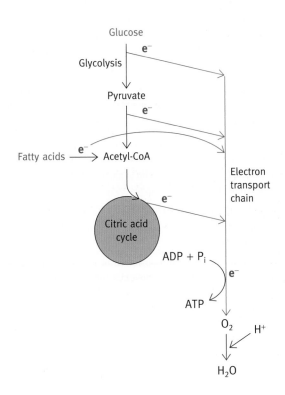

Fig. 11.9 Relationship of fat and glucose oxidation. This figure emphasizes the collection of electrons for the electron transport pathway.

Energy release from oxidation of amino acids

The situation with regard to amino acid oxidation is more complex in detail but similar in concept. There are 20 different amino acids and, as explained earlier, if these are present in excess of immediate requirements their amino groups are removed and carbon-hydrogen skeletons used as fuel. Once again, the metabolism of the latter shows a simplicity of concept. The carbon-hydrogen skeletons are converted to pyruvate, or to acetyl-CoA, or to intermediates in the citric acid cycle, so that they also join the same metabolic path (as shown in Fig. 11.10). The citric acid cycle thus plays a central role in metabolism.

The interconvertibility of fuels

Although this chapter is concerned primarily with energy generation by food oxidation, a brief note here can throw light on the fuel logistics already described in earlier chapters. We described how glucose, in excess, can be converted to fat. This is because fatty acids can be synthesized from acetyl-CoA:

Glucose → pyruvate → acetyl-CoA → fatty acids

However, in animals, fatty acids *cannot be* converted, in a net sense, to glucose, because to synthesize the latter (the topic of Chapter 15) pyruvate is needed. In animals, acetyl-CoA cannot be converted to pyruvate because the pyruvate dehydrogenase reaction is irreversible and, therefore, fatty acids cannot be converted to glucose but glucose can be converted to fatty acids and thence to TAG (Fig. 11.11).

Those amino acids that give rise to pyruvate or a citric acid cycle acid can be converted to glucose (glucogenic amino acids). This is why, in starvation, muscle proteins are destroyed and the amino acids so produced transported to the liver, which converts them to glucose. There are organisms such as bacteria and plants that do have the capacity to convert fat and C_2 compounds such as acetate to glucose, but this involves a special cycle known as the **glyoxylate cycle** (a modified citric acid cycle) that is *not* present in animals. This is also described later (page 241).

The directions of metabolism, whether fat and glucose are oxidized or synthesized, are the result of control mechanisms to be described in Chapter 16.

The association of vitamins with metabolism affords us a convenient point to give a brief survey of vitamins (Box 11.2).

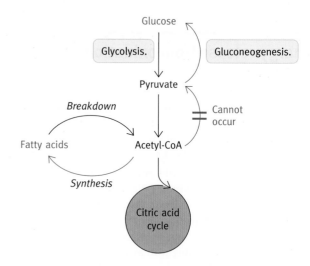

Fig. 11.11 Why glucose can be converted into fats but fats cannot be converted into glucose in animals. The reverse pathways in red are not completely the same as the forward pathways and are described in later chapters. In plants and bacteria, fat can be converted into glucose but not by the reversal of the pyruvate dehydrogenase reaction.

BOX 11.2 A survey of vitamins

A range of vitamins is essential in the diet and they have widely different structures and functions. We have already described in this chapter the roles of nicotinic acid, B_2 or riboflavin, and pantothenic acid. Rather than compiling the structures and functions of all the vitamins together in one section we will give a brief summary of them here together with page references to where they are dealt with in the appropriate places. Their functions are, we believe, more understandable if presented in the context of what they do.

The vitamins are traditionally divided into fat-soluble and water-soluble ones.

Fat-soluble vitamins

- Vitamin A or retinol is involved in the visual process (page 450) and the control of development.
- Vitamin D is converted to calcitriol; this is a signalling molecule, which activates genes involved in the absorption of calcium from the intestine.
- Vitamin E (α-tocopherol) is an antioxidant (page 490).
- Vitamin K is involved in blood clotting (page 486).

Water-soluble vitamins

- Ascorbic acid (vitamin C) is involved in hydroxylation reactions, of especial importance in collagen formation (page 58). It is an important antioxidant.
- Biotin is involved in carboxylation reactions (page 199).
- Folic acid is involved in 1-carbon transfer reactions (page 301), which occur in nucleotide metabolism.
- Niacin, or nicotinic acid (B_3), is a component of NAD^+ (page 178).
- Pantothenic acid (B_5) is a component of coenzyme A (page 181).
- Pyridoxin group (B_6) is involved in amino acid metabolism and in glycogen phosphorylase (page 165).
- Riboflavin (B_2) is a component of FAD and FMN (page 179).
- Thiamin (B_1) is involved in pyruvate metabolism (page 193) and the citric acid cycle.
- Vitamin B_{12} (cobalamin) is involved in a number of unusual carbon transfer reactions (pages 219, 308).

Summary

Biological oxidation involves removal of electrons and transfer to another acceptor molecule which need not be oxygen. In biochemical systems it commonly involves enzymic removal of two hydrogen atoms from a metabolite molecule. In the cell a variety of electron/hydrogen carriers exist in the transfer of electrons to oxygen. Nicotinamide adenine dinucleotide (NAD$^+$) is an important one. It is a dinucleotide containing the vitamin nicotinamide which can accept two electrons and a hydrogen atom, forming NADH. It functions as a coenzyme for dehydrogenases. Flavin adenine dinucleotide (FAD) is of similar structure but the accepting group is the vitamin riboflavin. Flavin mononucleotide (FMN) is a single-nucleotide form. They are hydrogen carriers being reduced respectively to FADH$_2$ and FMNH$_2$.

Glucose oxidation occurs in three stages. Stage 1, glycolysis, occurs in the cytoplasm, stage 2, the citric acid cycle, occurs in the mitochondrial matrix, and stage 3, the electron transport system, is in the inner mitochondrial membrane, the latter being the main site of ATP generation.

Stage 1 Glycolysis produces pyruvate from glucose or glycogen or, in the absence of sufficient capacity to reoxidize NADH by mitochondria, lactic acid.

Stage 2 Pyruvate enters the mitochondria and is converted to acetyl-CoA. Coenzyme A (CoA) is of central importance; it is a dinucleotide containing the vitamin pantothenic acid. The citric acid cycle extracts high-energy electrons from the acetyl groups in the form of reduced electron carriers, NAD$^+$ and FAD, while the carbon atoms of the acetyl group are converted to CO$_2$.

Stage 3 The electron transport system is a hierarchy of carriers of different energy potentials and the electrons move from one to the other down the energy gradient to oxygen. The energy potential of the transfer of electrons from one carrier to the next is quantified by the redox value, which can be directly measured in a simple apparatus. The energy released during the electron transport is ultimately trapped as ATP.

The release of energy from the oxidation of fatty acids differs only in their preliminary conversion to acetyl-CoA, which is oxidized in mitochondria, as is the case for pyruvate. The conversion of pyruvate to acetyl-CoA is irreversible so that, in animals, fat cannot be converted to glucose. Pyruvate is needed for this.

Energy release from oxidation of amino acids is more complicated since there are 20 different amino acids, each with its own pathway of metabolism. However they are all converted to pyruvate, acetyl-CoA or to citric acid cycle intermediates.

The detailed mechanisms by which these processes occur are given in the next chapter.

Problems

1 What are the three major phases involved in the oxidation of glucose and where do they occur?

2 Write down the overall structure of NAD$^+$ in words; show the structure of the electron-accepting group in the oxidized and reduced form. Explain how NAD$^+$ acts as a hydrogen carrier between substrates.

3 What is FAD and what is its role?

4 Explain the difference between aerobic and anaerobic glycolysis in muscle and the circumstances in which they occur. What is the point of anaerobic glycolysis?

5 Write down the structure of coenzyme A in words; also give the structure of its acyl-accepting group. What is the $\Delta G^{0'}$ of hydrolysis of a thiol ester? How does this compare with that of a carboxylic ester?

6 The pyruvate dehydrogenase reaction is of central importance. Write down the reaction.

7 What normally happens to the acetyl-CoA generated in the pyruvate dehydrogenase reaction?

8 Glycolysis and the citric acid cycle produce NADH and FADH$_2$. What happens to these?

9 The redox couple FAD + 2H$^+$ + 2e$^-$ → FADH$_2$ has an E_0' value of −0.219 V. That of $\frac{1}{2}$O$_2$ + 2H$^+$ + 2e$^-$ = 0.816 V. Calculate the $\Delta G^{0'}$ value for the oxidation of FADH$_2$ by oxygen to water. The Nernst equation is $\Delta G^{0'} = nF\Delta E_0'$ where $F = 96.5$ kJ V^{-1} mol^{-1}.

10 What is the major source of acetyl-CoA other than the pyruvate dehydrogenase reaction?

11 Can glucose be converted to fat? Explain your answer.

12 Can fatty acids be converted to glucose in animals? Explain your answer.

Chapter 12

Glycolysis, the citric acid cycle, and the electron transport system

In Chapters 10 and 11 you saw the overall pattern of the way in which various foodstuffs are metabolized and we outlined the strategies of the three stages of the oxidation of food components absorbed from the diet.

We now want to fill in the mechanisms of the metabolic pathways, starting with carbohydrate oxidation. In subsequent chapters we will deal with how fatty acids are metabolized to join up to the glucose oxidation pathway at acetyl-CoA and then do the same with amino acid metabolism. A potential problem in studying these pathways is to forget their purposes – to get lost in the detail. If necessary, keep on going back to the previous chapter to refresh your memory on where pathways are heading.

Regulation of these pathways is dealt with in Chapter 16, but if it is preferred to deal with them now the appropriate sections can be found in that chapter (page references are given below). We have chosen to have a separate chapter on regulation, because it enables us to deal with the general strategies of control that apply to all aspects of metabolism. It also allows a more integrated approach.

Stage 1 – glycolysis

This, we remind you, results in the splitting (lysis) of glucose or a glucosyl unit of glycogen (glyco) into two molecules of pyruvate and the reduction of NAD^+.

Glucose or glycogen?

So far, for simplicity, we have talked mainly of glucose catabolism. However, in liver, skeletal muscle, and parts of the kidney, glucose is stored as glycogen, and glycolysis may be proceeding from this rather than from free glucose. There is a difference between the two.

When glycogen is broken down, glucose-6-phosphate is produced via glucose-1-phosphate. In the liver this can be hydrolysed to release free glucose into the blood. However, glucose-6-phosphate is on the glycolytic pathway also and in all tissues can be broken down to pyruvate.

Free glucose, obtained from the blood, is also converted to glucose-6-phosphate by phosphorylation using adenosine triphosphate (ATP). You have met this reaction already (page 163) because it is the same as that involved in glycogen synthesis.

Whether glucose-6-phosphate goes to glycogen or to pyruvate or to blood glucose in the case of liver depends on how the metabolic control switches are set, according to physiological needs, and this is a major later topic (Chapter 16). The relationship between glycolysis, glucose, and glycogen is shown in Fig. 12.1.

Why use ATP here at the beginning of glycolysis?

It may seem odd that a pathway designed to produce ATP should start by using up ATP. Why use it here? The reason is that glycolysis involves phosphorylated compounds and ATP must be used to phosphorylate glucose – it has the necessary energy potential. Glucose-6-phosphate is a low-energy phosphoryl compound so certainly we have lost a high-energy phosphoryl group in using ATP. Think of it as an investment for, as you'll see, a 100% profit is made on the ATP used in glycolysis of glucose

Why is glucose-6-phosphate converted to fructose-6-phosphate?

The next step is to convert glucose-6-phosphate, an aldose sugar, to fructose-6-phosphate, its ketose isomer.

To (apparently) digress for a moment, organic chemistry textbooks describe a test-tube reaction called the aldol condensation. In this, an aldehyde and a ketone (or another aldehyde) condense together as shown in Fig. 12.2. The reverse of the reaction can be used to split an aldol into two parts – an aldol being the β-hydroxycarbonyl compound shown. Glucose-6-

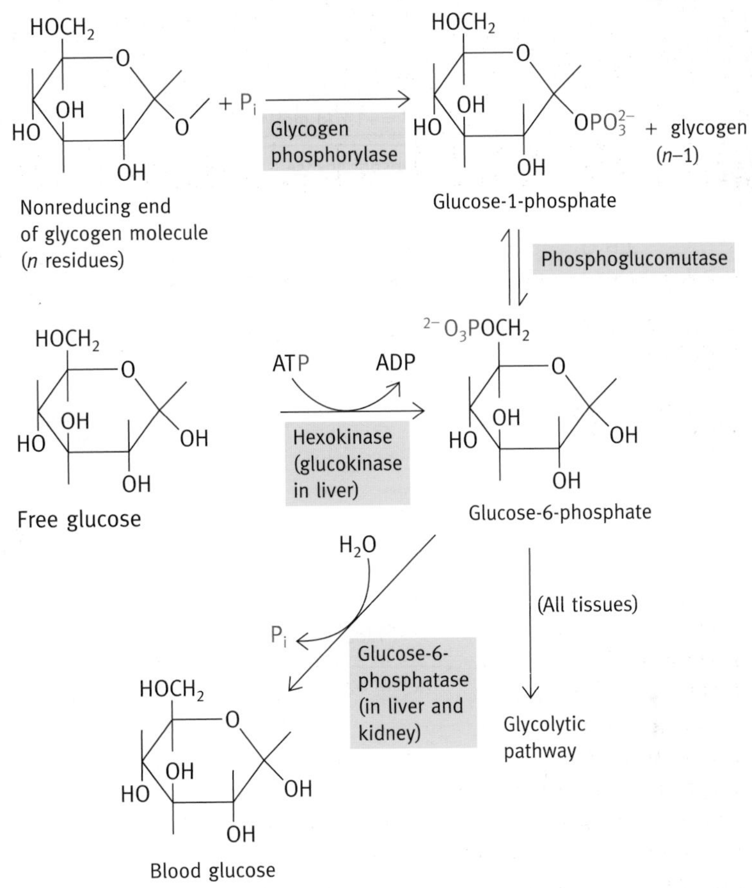

Fig. 12.1 Production of glucose-6-phosphate from glycogen or free glucose and its fate. Which routes are operative depends on control mechanisms described in Chapter 16.

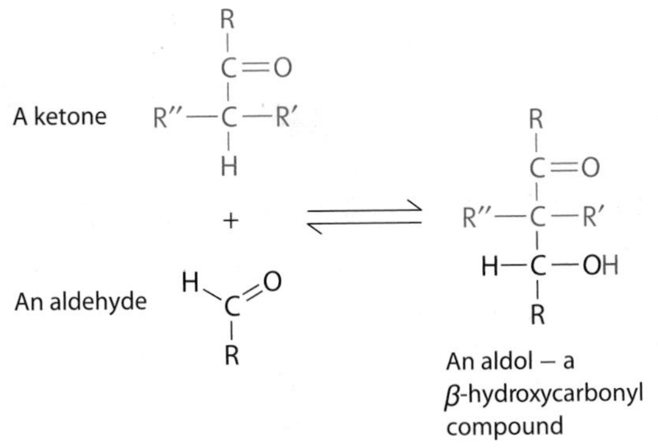

Fig. 12.2 The aldol condensation. A chemical reaction between an aldehyde and a ketone (or aldehyde).

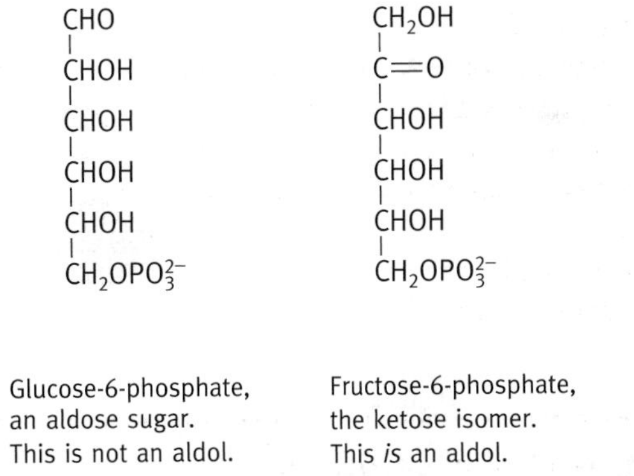

Glucose-6-phosphate, an aldose sugar. This is not an aldol.

Fructose-6-phosphate, the ketose isomer. This *is* an aldol.

Fig. 12.3 The straight-chain formulae of an aldose sugar and its ketose isomer. The glucose-6-phosphate is in equilibrium with the six-membered ring (pyranose) form and the fructose-6-phosphate with the five-membered ring (furanose) form. The ring structures are shown in Fig. 12.5.

phosphate is not an aldol, but fructose-6-phosphate is, as seen in the straight-chain formulae in Fig. 12.3. By forming the fructose isomer, the sugar phosphate can be split into two by the aldol reaction. The glucose-6-phosphate is isomerized into fructose-6-phosphate by the enzyme **phosphohexose isomerase**. Before splitting, another phosphoryl group from ATP is transferred to the fructose-6-phosphate by the enzyme phosphofructokinase (PFK), yielding fructose-1:6-bisphosphate.

Splitting fructose bisphosphate to two C$_3$ compounds

The fructose-1:6-bisphosphate is now split by the enzyme **aldolase** catalysing the aldol reaction; the second phosphate means that each of the two C$_3$ products has a phosphoryl group giving glyceraldehyde-3-phosphate and dihydroxyacetone

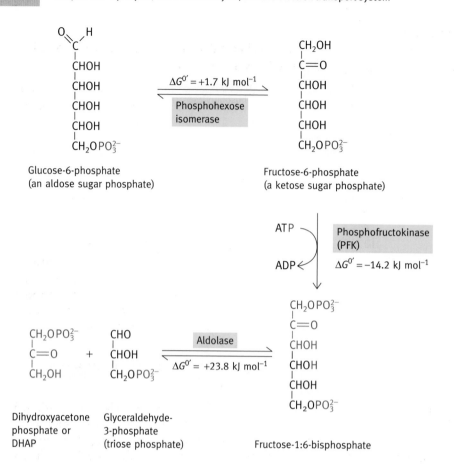

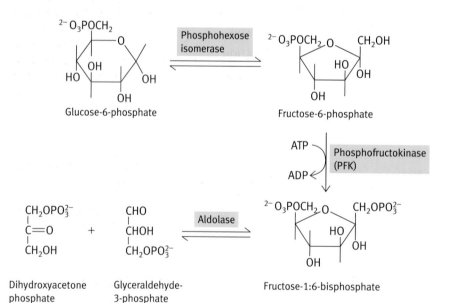

Fig. 12.4 Conversion of glucose-6-phosphate into two C_3 compounds. The $\Delta G^{0'}$ value for the aldolase reaction in the forward direction would appear to preclude its occurrence, but see text for explanation of this. Straight-chain formulae for the sugars are used here for clarity; the reactions are commonly presented as in Fig. 12.5.

Fig. 12.5 This is the same scheme as in Fig. 12.4 but presented in the more usual form with sugars as ring structures.

phosphate (Fig. 12.4). In Fig. 12.5, the same reactions are presented with the more commonly used ring structures for the sugars; the fructose-6-phosphate is in the five-membered ring (furanose) configuration. The $\Delta G^{0'}$ for the aldolase reaction is +24.3 kJ mol^{-1}, which would seem to preclude its ready occurrence. There are, however, special considerations applying to this reaction (see below). In cellular conditions, the ΔG is small and the reaction freely reversible.

A note on the $\Delta G^{0'}$ and ΔG values for the aldolase reaction

The reaction catalysed by aldolase has a $\Delta G^{0'}$ value of $+24.3\ \text{kJ mol}^{-1}$ and is freely reversible while reactions with smaller $\Delta G^{0'}$ values paradoxically are irreversible. The explanation is that $\Delta G^{0'}$ values are determined at 1 M concentrations of reactants and products; since concentrations in the cell are more likely to be at 10^{-3}–10^{-4} M actual ΔG values are always different from $\Delta G^{0'}$ values. Nonetheless, the latter are usually useful guides to metabolic events. This is not true, however, in the case of aldolase, where the correlation between $\Delta G^{0'}$ values and ΔG values in the cell is very poor. The reason is that, in the aldolase reaction, one molecule of the reactant, fructose-1:6-bisphosphate (F-1:6-BP), gives rise to two molecules of product, glyceraldehyde-3-phosphate (GAP) and dihydroxyacetone phosphate (DHAP). The relationship between $\Delta G^{0'}$ and ΔG values is given in the equation:

$$\Delta G = \Delta G^{0'} + RT \ln \frac{[\text{products}]}{[\text{reactants}]},$$

that is,

$$\Delta G = \Delta G^{0'} + RT \ln \frac{[\text{GAP}] \times [\text{DHAP}]}{[\text{F-1:6-BP}]}.$$

Because there are two products of low concentration, the $RT\ln([\text{products}]/[\text{reactants}])$ moiety of the equation has a large negative value, giving a ΔG compatible with ready reversibility in the cell. To illustrate this, from the actual intracellular concentrations of fructose-1:6-bisphosphate, glyceraldehyde-3-phosphate, and dihydroxyacetone phosphate, as determined in rabbit skeletal muscle, a small ΔG value of $-1.3\ \text{KJ mol}^{-1}$ can be calculated.

Interconversion of dihydroxyacetone phosphate and glyceraldehyde-3-phosphate

Glyceraldehyde-3-phosphate and dihydroxyacetone phosphate are isomeric molecules. An enzyme, **triose phosphate isomerase,** interconverts these two compounds.

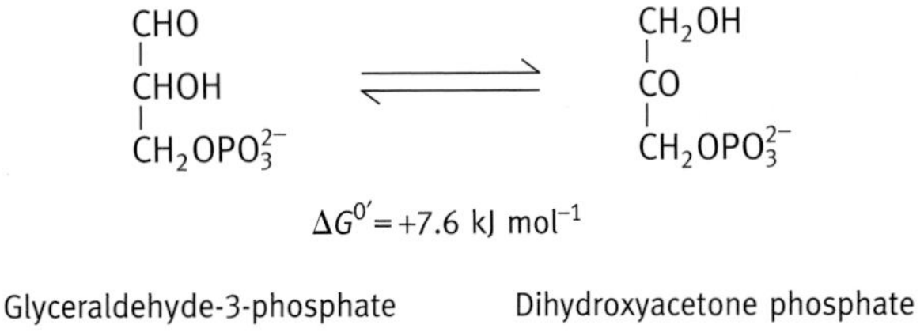

$$\Delta G^{0'} = +7.6\ \text{kJ mol}^{-1}$$

Glyceraldehyde-3-phosphate   Dihydroxyacetone phosphate

The two compounds are in equilibrium but, since glyceraldehyde-3-phosphate is continually removed by the next step in glycolysis, all of the dihydroxyacetone phosphate is progressively converted to glyceraldehyde-3-phosphate.

Glyceraldehyde-3-phosphate dehydrogenase – an oxidation linked to ATP synthesis

The aldehyde group of glyceraldehyde-3-phosphate is oxidized by **glyceraldehyde-3-phosphate dehydrogenase,** using NAD^+ (page 178) as electron acceptor. You would expect this to produce a carboxyl group (and so it does, ultimately) but oxidation of a —CHO group to —COO^- has a large negative ΔG value, sufficient in fact to generate a high-energy phosphate compound from inorganic phosphate (P_i) on the way (Fig. 12.6).

The mechanism by which this is achieved is as follows. At the active site of the enzyme there is the amino acid cysteine which has a thiol or sulphydryl group (—SH) on its side chain. The

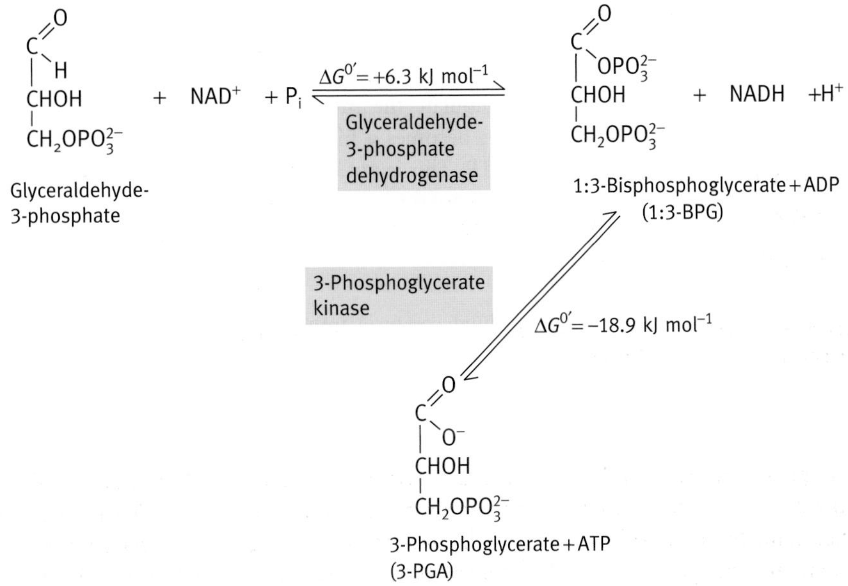

Fig. 12.6 Conversion of glyceraldehyde-3-phosphate to 3-phosphoglycerate.

aldehyde glyceraldehyde-3-phosphate condenses with the thiol to form a thiohemiacetal.

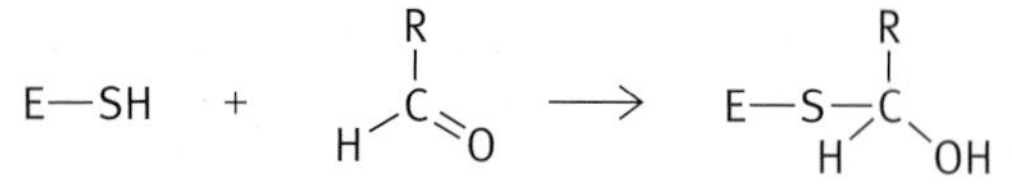

| Enzyme with thiol group | Glyceraldehyde-3-phosphate | Enzyme–thiohemiacetal complex |

The complex is now oxidized on the enzyme active site, the electrons being accepted by NAD^+, and a thiol ester is formed with the enzyme thiol group.

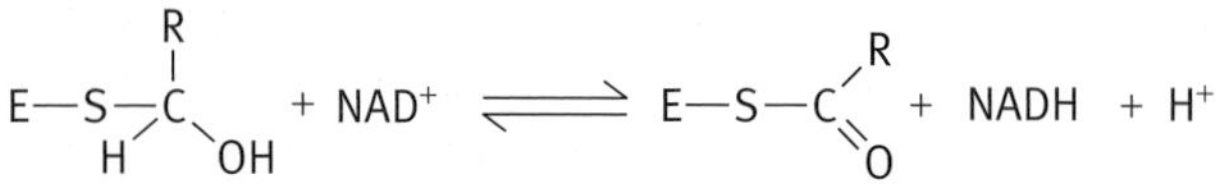

A thiol ester (R—CO—S—), as explained earlier (page 191), is a high-energy compound – of the same order as that of a high-energy phosphate compound. It is thermodynamically feasible, therefore, for P_i to react as follows.

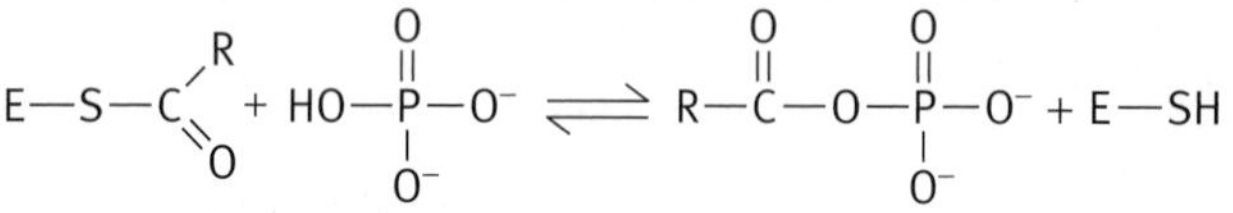

The $RCO—O—PO_3^{2-}$ group is also high energy and so its phosphoryl group can be transferred to ADP forming ATP.

The responsible enzyme is **phosphoglycerate kinase** because, in the reverse direction, it transfers a phosphoryl group from ATP to **3-phosphoglycerate (3-PGA)**. By convention, kinases are always named from the ATP side of the reaction.

The generated phosphoryl group in this process is attached to the actual substrate (1:3-bisphosphoglycerate) of an enzyme. For this reason it is called **substrate-level phosphorylation**, a point to which we will refer later. 3-Phosphoglycerate is a low-energy phosphate compound and cannot phosphorylate ADP.

The next steps in glycolysis manipulate the molecule so that this low-energy phosphate ester becomes a high-energy phosphoryl group, transferable to ATP. This is not energy for nothing – you will see how it is done within the law.

The final steps in glycolysis

The phosphoryl group of 3-phosphoglycerate is transferred from the 3 to the 2 position as shown.

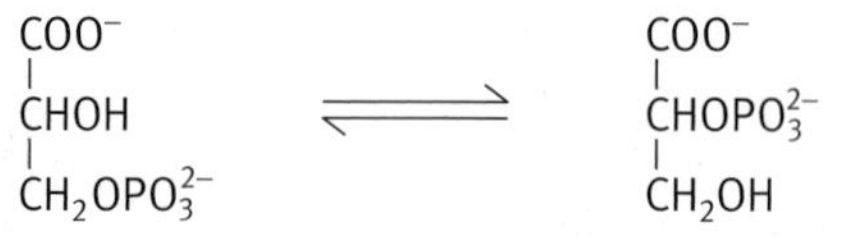

$$\Delta G^{0'} = +4.4 \text{ kJ mol}^{-1}$$

| 3-Phosphoglycerate | 2-Phosphoglycerate |

This is called the **phosphoglycerate mutase** reaction. The reaction is not really an *intramolecular* transfer of the phosphoryl group (though it is in the enzyme from plants). The enzyme from rabbit muscle contains a phosphoryl group that it donates to the 2-OH group of 3-phosphoglycerate forming 2:3-bisphosphoglycerate. The 3-phosphoryl group is now transferred to the enzyme to replace the donated phosphate, so that the net effect is the reaction shown above.

The next step in glycolysis is that a water molecule is removed from the 2-phosphoglycerate. Enzymes catalysing such reactions are usually called **dehydratases** but in this particular case in glycolysis, the old established name is **enolase** (because it forms a substituted enol).

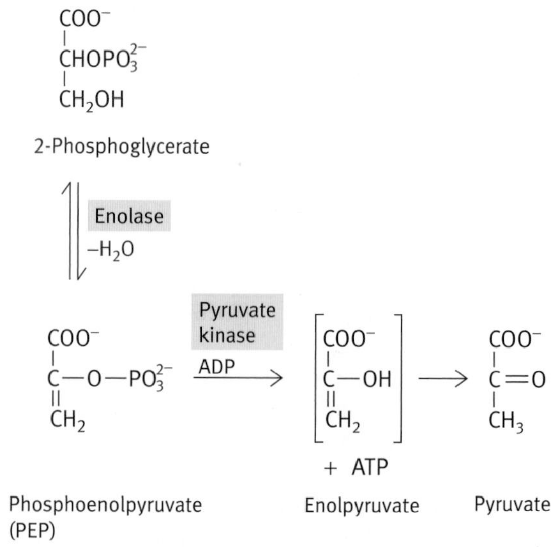

| Phosphoenolpyruvate (PEP) | Enolpyruvate | Pyruvate |

The enolase reaction has a $\Delta G^{0'}$ of only $+1.8$ kJ mol^{-1}, but the enolphosphate compound is of the 'high-energy' type, with a $\Delta G^{0'}$ of hydrolysis of -62.2 kJ mol^{-1}; a reason for this is that the immediate product of the reaction, the enol form of pyruvate, spontaneously converts to the keto form, a reaction with a large negative $\Delta G^{0'}$ value.

The phosphoryl group is transferred to ADP by the enzyme **pyruvate kinase**; this name might misleadingly imply that pyruvate can be phosphorylated by ATP by reversal of the reaction; the name of the enzyme derives from the convention, mentioned earlier, that a kinase is named from the reaction involving ATP even though, in this case, that reaction never occurs because the substrate, enol-pyruvate, does not occur, except fleetingly. It spontaneously changes to pyruvate. The irreversibility of the conversion of phosphoenolpyruvate to pyruvate has important metabolic repercussions as you will see when we come later to gluconeogenesis. (There is a potential source of confusion arising from the fact that, in certain plants and microorganisms, pyruvate *is* directly converted to phosphoenolpyruvate by a quite different enzyme that utilizes two phosphoryl groups from ATP. However, this reaction does *not occur* in animals.)

The complete glycolytic pathway is shown in Fig. 12.7.

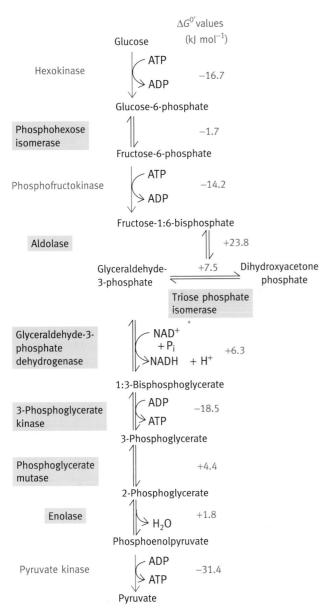

Fig. 12.7 The glycolytic pathway. Irreversible reactions are indicated in red. The free reversibility of the aldolase reaction would appear to be inconsistent with such a large $\Delta G^{0'}$ value (see text for explanation).

Anaerobic glycolysis

In vigorously exercising muscle, and in red blood cells without mitochondria, glycolysis produces lactic acid rather than pyruvate. This has already been explained in Chapter 11, page 180.

The ATP balance sheet from glycolysis

Starting with glucose, two molecules of ATP were used to form fructose-1:6-bisphosphate. The phosphoglycerate kinase produced two ATP molecules per original glucose and the pyruvate kinase two – a total of four and a net gain of two.

In muscle, glycolysis may start with glycogen. In this case, the energy in the glycosidic bonds of glycogen are preserved, because the initial reaction is to split off the units by inorganic phosphate (phosphorolysis) giving glucose-1-phosphate. This saves one ATP so that the yield of ATP per glycogen unit is three. The same is true of liver, but the glucose-6-phosphate formed from glucose-1-phosphate is largely converted to free glucose which is released into the blood rather than glycolysed (see page 165).

Transport of pyruvate into the mitochondria

The other product of glycolysis besides NADH is pyruvate. Unless it is reduced to lactate in the cytoplasm (page 180) the pyruvate is transported into the mitochondrial matrix by an antiport type of membrane transport protein (page 113), which exchanges it for OH^- inside the matrix.

Conversion of pyruvate to acetyl-CoA – a preliminary step before the citric acid cycle

Before we get to the cycle itself we must deal with the preparation of pyruvate to enter the cycle, by which we mean its conversion to acetyl-CoA.

As outlined earlier, pyruvate in the mitochondrial matrix is converted to acetyl-CoA, which feeds the acetyl group into the citric acid cycle (the structure of CoA is described on page 181).

The overall reaction catalysed by pyruvate dehydrogenase is:

$$\text{Pyruvate} + NAD^+ + CoA{-}SH \rightarrow \text{Acetyl}{-}S{-}CoA + NADH + H^+ + CO_2.$$

Pyruvate dehydrogenase, the responsible enzyme, is a very large complex composed of many polypeptides. It essentially consists of three different enzyme activities aggregated together for efficiency, each catalysing one of the intermediate steps in the process (Fig. 12.8). The first step is decarboxylation of pyruvate to produce CO_2 and a hydroxyethyl group $CH_3CHOH{-}$ attached to the cofactor **thiamin pyrophosphate** (TPP). TPP is derived from thiamine, or vitamin B_1, deficiency of which impairs the ability to metabolize pyruvate, among other effects. The hydroxyethyl group, in a series of steps, is converted to the acetyl group of acetyl-CoA with the reduction of NAD^+ (see Fig. 12.8). The process is known as an **oxidative decarboxylation** for obvious reasons. (Detailed structures of cofactors are given below for reference purposes.)

This conversion of pyruvate to acetyl-CoA is irreversible; the $\Delta G^{0'}$ of the reaction is -33.5 kJ mol^{-1}. As you will see later, this is of profound significance for irreversibility means that fatty acids can never be converted in a net sense to glucose in the animal body though, as mentioned, bacteria and plants have a

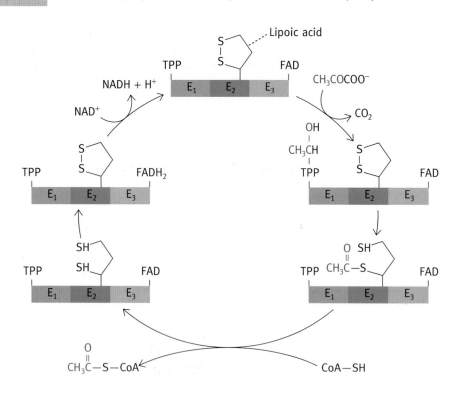

Fig. 12.8 Mechanism of the pyruvate dehydrogenase reaction. TPP, thiamin pyrophosphate; E1–E3, enzyme regions of complex. (The $FADH_2$ is at an unusually low redox potential in this enzyme and can reduce NAD^+.)

special mechanism for achieving this. The acetyl-CoA now enters the citric acid cycle.

Components involved in the pyruvate dehydrogenase reaction

TPP has the structure:

TPP-hydroxyethyl has the structure:

Lipoic acid in its reduced form has the structure:

and in its oxidized form the structure is:

Lipoic acid is attached to a lysine side chain of the enzyme by —CO—NH— linkage. In Fig. 12.8 it is represented by the disulphide structure. The three enzymes represented by E_1, E_2, and E_3 are part of a very large protein complex. Its complex regulation is given on page 259.

Stage 2 – the citric acid cycle

A preliminary overview of the cycle is given in Chapter 11, page 182.

The citric acid cycle produces a combustible fuel in the form of reducing equivalents of NADH and $FADH_2$, part of which, in effect, comes from splitting water. This fuel is burned in the next stage, the electron transport system, to produce ATP from ADP and P_i. Unlike schemes to run cars on water, the cycle is thermodynamically sound because it uses the free energy made available from the destruction of the acetyl group of acetyl-CoA to drive the process.

It must be noted that the splitting of water is not a direct 'head on' process as occurs in photosynthesis and oxygen is not

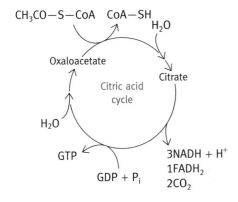

Fig. 12.9 The inputs and outputs of the citric acid cycle. Individual cycle reactions are not indicated.

liberated as such, but as CO_2. Rather it is an indirect splitting that can easily be overlooked and is seldom referred to in texts. We need to explain this aspect more clearly for it is central to appreciating what the cycle is all about and much of ATP production requires it. It also may remove potential confusion about how eight high-energy electrons (plus a ninth hydrogen atom on CoA—SH) can arise from the destruction of an acetyl group with only three hydrogen atoms (see below).

Acetyl-CoA enters the cycle by a reaction with oxaloacetate to produce citrate. Do not bother at this stage with how this occurs – it is described later. The reaction involves the input of a molecule of water. With one complete 'turn' of the cycle (again explained later) oxaloacetate is re-formed and the acetyl group of acetyl-CoA has disappeared. As a result of the cycle reactions, the products from one acetyl group are as follows: two molecules of CO_2; the reduction of three molecules of NAD^+ to NADH; and that of one molecule of FAD to $FADH_2$. In addition, the CoA—S— of acetyl-CoA becomes CoA—SH (Fig. 12.9). (A single high-energy phosphoryl group, as GTP, is produced from P_i.) If you add up all the reducing equivalents of the three NADH and one $FADH_2$, there are eight. (Remember that NAD^+ accepts two electrons.) The formation of the thiol group of CoA—SH (from CoA—S—) requires one more – a total of nine reducing equivalents (effectively equivalent to nine H atoms).

The CH_3CO—S—CoA supplies three of these so that there is a shortfall of six. There is no involvement of oxygen in the cycle so there is also a shortfall of three oxygen atoms to produce the two molecules of CO_2. The source of these 'missing' atoms includes two molecules of H_2O. You will see below that, as well as the input of H_2O into citrate synthesis, a second molecule of water enters the cycle. But, this still leaves a shortfall of two hydrogen and one oxygen atoms. It will be more convenient to explain the source of these later (page 198).

Thus, we see the remarkable feat – electrons of H_2O are raised up the energy scale to reduce NADH and $FADH_2$ and, of course, electrons from the acetyl group also are utilized for this purpose. Again, it is emphasized that this does not mean that

the components of H_2O go *directly* to these products. The reactions involved in converting the acetyl group to its products provide the free energy to convert H_2O into reducing equivalents. In fact, the whole cycle has a negative ΔG value and so proceeds thermodynamically 'downhill'.

With that preamble we can now turn to the reactions of the cycle.

A simplified version of the citric acid cycle

Possibly one obstacle to learning the cycle is that the progression of reactions doesn't make much sense until you have completed it, so we will first look at a simplified version devoid of detail.

Be sure you know the structure of oxaloacetic acid, for that is where it all starts and finishes. Acetate is CH_3COO^-, the oxalo group is $^-OOC-C\overset{O}{\underset{}{\Vert}}-$, so oxaloacetate is:

$$\overset{O}{\underset{}{\Vert}}C-COO^-$$
$$H_2C-COO^-.$$

The acetyl group of acetyl-CoA is joined to oxaloacetate to form citric acid; it is easy to see how citrate can be derived from oxaloacetate.

$$\begin{array}{c}CH_2COO^-\\ |\\ HO-C-COO^-\\ |\\ H_2C-COO^-\end{array}$$

It is helpful to remember that citrate is a C_6 *symmetric* tricarboxylic compound. This is converted to its asymmetric isomer isocitrate, which in the cycle is then progressively converted to α-ketoglutarate (C_5), succinate (C_4), fumarate (C_4), malate (C_4), and oxaloacetate (Fig. 12.10). This means that one turn of the cycle eliminates the acetyl group fed into it.

We suggest that you make yourself completely familiar with these acids of the cycle. With that preparation, a more detailed consideration of this superhighway of carbon-compound metabolism can be given.

Mechanisms of the citric acid cycle reactions

We might, for convenience, divide the reactions into three groups.

1 The synthesis of citrate is the reaction feeding acetyl groups into the cycle.

2 The 'top part' of the cycle involves conversion of C_6 citrate to C_5 α-ketoglutarate.

3 The 'lower part' of the cycle involves the conversion of succinate to oxaloacetate.

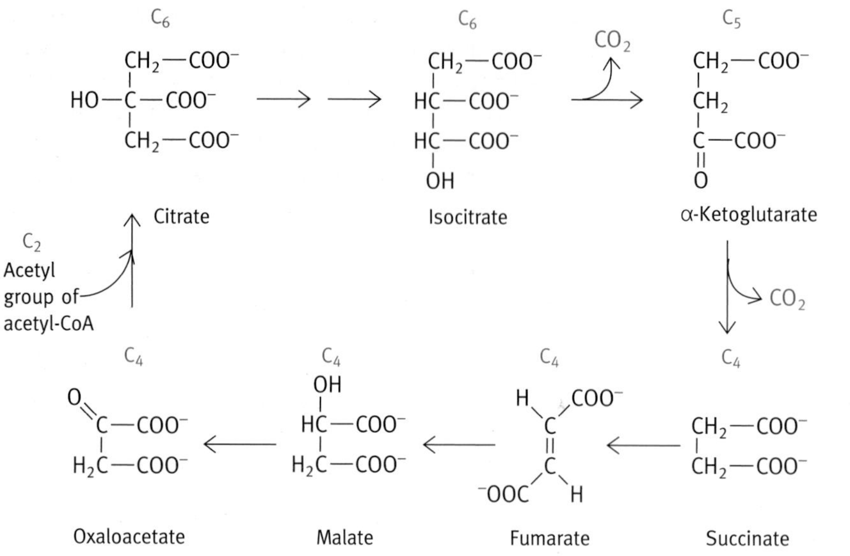

Fig. 12.10 Simplified citric acid cycle showing the component acids and their sequence. The purpose of this diagram is for you to learn the structures of the main cycle acids and their interrelationships without complicating details. When you are familiar with these it will be easier to appreciate the full cycle.

The synthesis of citrate

The name of the enzyme involved here is **citrate synthase**. The enzyme catalyses the condensation of acetyl-CoA with oxaloacetate by an aldol reaction to give citryl-CoA. This is hydrolysed to citrate. Citrate formation has a large negative $\Delta G^{0'}$ value (-32.3 kJ mol^{-1}), due to the hydrolysis of a thiol ester, and hence the reaction is irreversible.

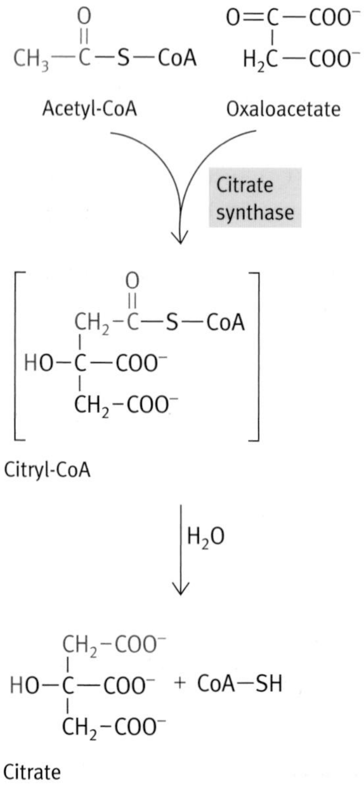

Conversion of citrate to α-ketoglutarate

Citrate → isocitrate

The strategy of this reaction is to switch the hydroxyl group of citrate (a symmetric molecule) from the 3 position to the 2 position, giving isocitrate (an asymmetric molecule). The logic of this will become apparent. This isomerization of citrate is catalysed by a single enzyme that reversibly removes water and adds it back across the double bond in either direction.

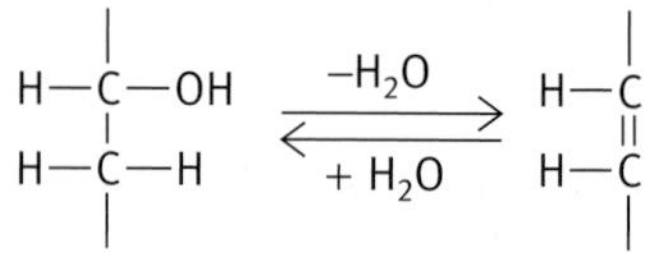

Enzymes catalysing such reactions are, as mentioned, usually called dehydratases but in this case, due to long tradition, its name is aconitase because the unsaturated intermediate product is *cis*-aconitate (found first in the plant genus *Aconitum*).

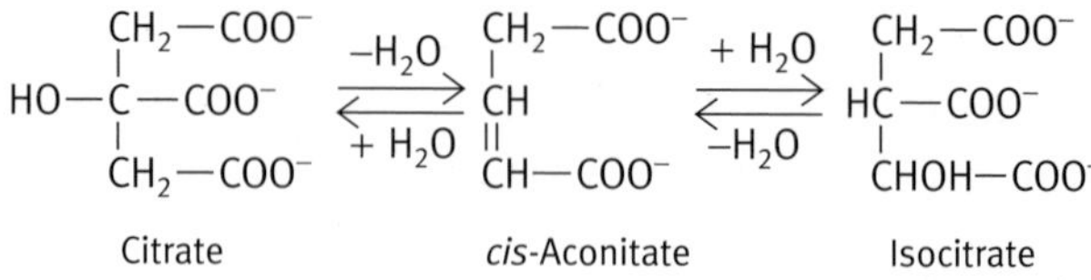

Since isocitrate is now metabolized further in the cycle the net effect is that aconitase catalyses the reaction sequence:

$$\text{Citrate} \rightarrow cis\text{-aconitate} \rightarrow \text{isocitrate}$$

Isocitrate dehydrogenase

You have already met one example of NAD$^+$-requiring dehydrogenases in lactate dehydrogenase (page 180). This again is a common reaction of the type:

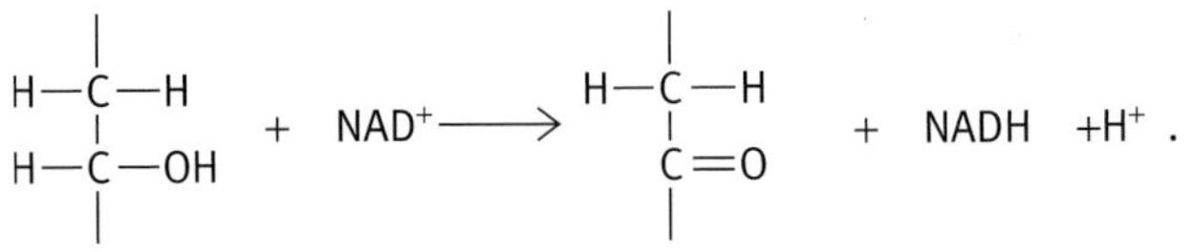

$$H-\underset{|}{\overset{|}{C}}-H \quad + \quad NAD^+ \longrightarrow \quad H-\underset{|}{\overset{|}{C}}-H \quad + \quad NADH \;+H^+ \; .$$
$$H-\underset{|}{\overset{|}{C}}-OH \qquad\qquad\qquad \underset{|}{C}=O$$

In the cycle isocitrate dehydrogenase catalyses the reaction:

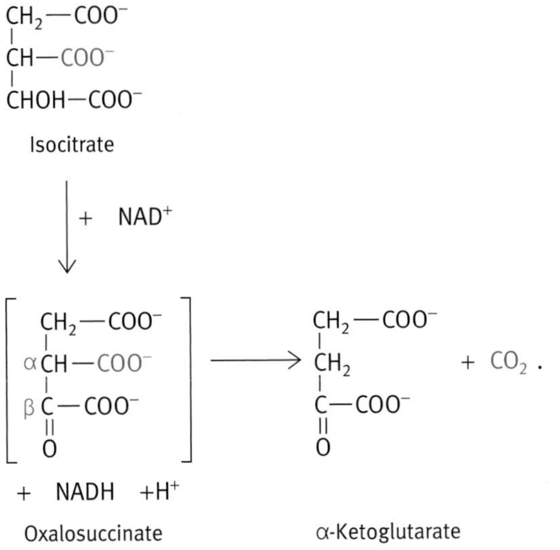

$$CH_2-COO^-$$
$$\underset{|}{C}H-COO^-$$
$$CHOH-COO^-$$

Isocitrate

$\downarrow \quad + \;\; NAD^+$

$$\begin{bmatrix} CH_2-COO^- \\ \alpha\,CH-COO^- \\ \beta\,\underset{\parallel}{C}-COO^- \\ O \end{bmatrix} \longrightarrow \begin{array}{l} CH_2-COO^- \\ CH_2 \\ \underset{\parallel}{C}-COO^- \\ O \end{array} \quad + \; CO_2 \; .$$

$+ \;\; NADH \;+H^+$

Oxalosuccinate   α-Ketoglutarate

The immediate product is oxalosuccinate which is a β-keto acid (the keto group is β to the centre COOH group). Such acids are unstable and readily lose the carboxyl group as CO_2. This happens on the surface of the isocitrate dehydrogenase so the product is the C_5 acid α-ketoglutarate as shown.

The C_4 part of the cycle

α-Ketoglutarate resembles pyruvate in that both are α-keto acids. We can write both as:

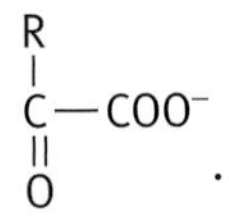

$$\underset{\parallel}{\overset{R}{\underset{O}{C}}}-COO^- \; .$$

For pyruvate R = $-CH_3$; for α-ketoglutarate R = $-CH_2CH_2COO^-$. We have already seen that pyruvate dehydrogenase converts pyruvate to acetyl-CoA and CO_2. This reaction format is repeated here. The enzyme requires **thiaminpyrophosphate**.

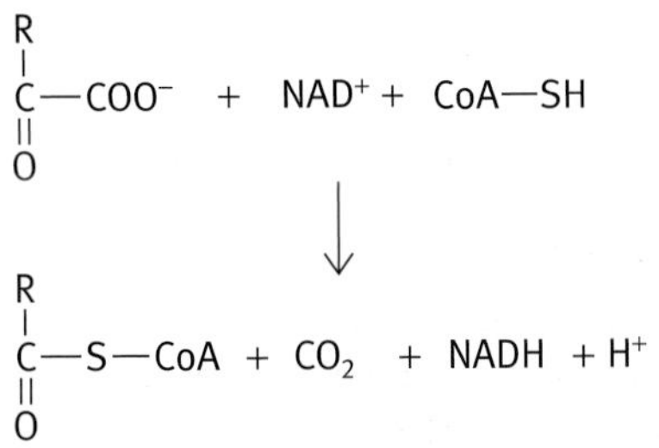

$$\underset{\parallel}{\overset{R}{\underset{O}{C}}}-COO^- \; + \;\; NAD^+ + \;\; CoA-SH$$

$\downarrow$

$$\underset{\parallel}{\overset{R}{\underset{O}{C}}}-S-CoA \; + \;\; CO_2 \; + \;\; NADH \;+ H^+$$

An equivalent enzyme complex exists, using the same set of cofactors as in pyruvate dehydrogenase, for α-ketoglutarate and precisely the same equation applies as above except that R = $-CH_2CH_2COO^-$ the enzyme complex attacks α-ketoglutarate rather than pyruvate. The product of **α-ketoglutarate dehydrogenase** is therefore succinyl-CoA, analogous to acetyl-CoA. Thiamine deficiency in the diet greatly reduces energy release from foodstuffs.

However, in the cycle, whereas acetyl-CoA is used to form citrate, succinyl-CoA is broken down to free succinate plus CoA—SH. In principle, the simplest way to do this would be to hydrolyse succinyl-CoA. However, the $\Delta G^{0'}$ of this hydrolysis of the thiol ester is −35.5 kJ mol⁻, enough energy to raise P_i to a high-energy phosphate compound. So, why waste this energy? Why not trap it instead? This is precisely what happens.

Generation of GTP coupled to splitting of succinyl-CoA

The reaction is:

$$\text{Succinyl-CoA} + GDP + P_i \rightleftharpoons \text{succinate} + GTP + CoASH;$$
$$\Delta G^{0'} = -2.9 \; KJ \; mol^{-1}.$$

The enzyme is named for the back reaction: hence **succinyl-CoA synthetase**. It synthesizes succinyl-CoA from succinate and GTP but, in the citric acid cycle, it works in the other direction, of course. GDP is used in liver and kidney because the GTP is used in gluconeogenesis. Other tissues that do not synthesize glucose for the rest of the body use ADP, as do plants. They use a different isoenzyme.

The mechanism of the reaction is as follows: P_i displaces CoA producing succinyl phosphate:

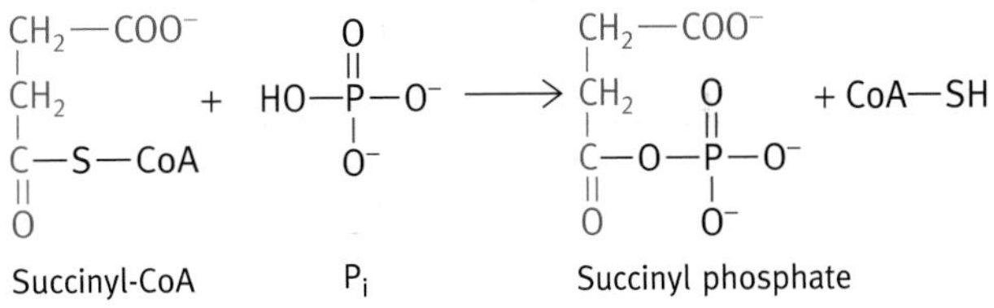

$$\begin{array}{l} CH_2-COO^- \\ CH_2 \\ \underset{\parallel}{C}-S-CoA \\ O \end{array} + \; HO-\underset{\underset{O^-}{|}}{\overset{\overset{O}{\parallel}}{P}}-O^- \longrightarrow \begin{array}{l} CH_2-COO^- \\ CH_2 \quad\quad O \\ \underset{\parallel}{C}-O-\underset{\underset{O^-}{|}}{\overset{\overset{\parallel}{}}{P}}-O^- \\ O \end{array} + \; CoA-SH$$

Succinyl-CoA P_i Succinyl phosphate

The phosphoryl group is now transferred to GDP, giving succinate and GTP:

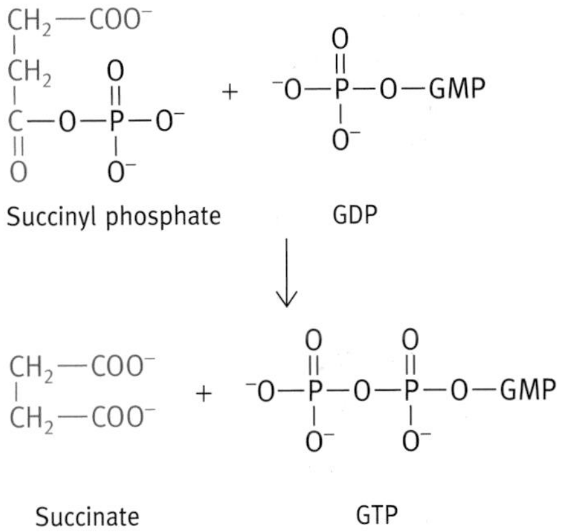

$$\begin{array}{l} CH_2-COO^- \\ CH_2 \quad\quad O \\ \underset{\parallel}{C}-O-\underset{\underset{O^-}{|}}{\overset{\overset{\parallel}{}}{P}}-O^- \\ O \end{array} + \; ^-O-\underset{\underset{O^-}{|}}{\overset{\overset{O}{\parallel}}{P}}-O-GMP$$

Succinyl phosphate   GDP

$\downarrow$

$$\begin{array}{l} CH_2-COO^- \\ CH_2-COO^- \end{array} + \; ^-O-\underset{\underset{O^-}{|}}{\overset{\overset{O}{\parallel}}{P}}-O-\underset{\underset{O^-}{|}}{\overset{\overset{O}{\parallel}}{P}}-O-GMP$$

Succinate   GTP

Earlier we pointed out that, to balance the output of CO_2 and reducing equivalents from the cycle with the input components, in addition to the two H_2O molecules entering the cycle we need two more hydrogen atoms and one oxygen. These arise from the involvement of inorganic phosphate in the breakdown of succinyl-CoA as described above. In case this is not clear, the *actual* reactions given above can be notionally regarded for balance-sheet purposes as being *equivalent* to the two reactions:

$$GDP + P_i \rightarrow GTP + H_2O;$$

$$\text{Succinyl-CoA} + H_2O \rightarrow \text{succinate} + \text{CoASH}.$$

We emphasize that the reaction *does* not proceed in that way but it illustrates how the 'missing' elements of H_2O are supplied to the cycle to put the balance sheet in order.

Conversion of succinate to oxaloacetate

First, succinate is dehydrogenated by **succinate dehydrogenase** whose electron acceptor is FAD (page 179), firmly bound to the enzyme and which can reversibly accept a pair of hydrogen atoms. Why is NAD^+ used for the other dehydrogenation reactions of the cycle and FAD here? It's a question of redox potentials. On page 182, we described how electrons will flow from a lower redox potential (higher reducing potential or energy level) to electron acceptors of higher redox potential (lower reducing potential or energy level). In the case of succinate dehydrogenase the reaction is of the *desaturation* type:

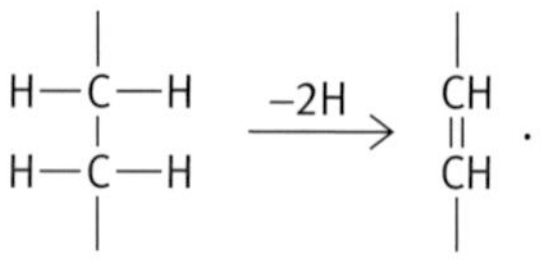

The redox potential or reducing potential of this system is such that it cannot reduce NAD^+ but can reduce FAD (which is more strongly oxidizing than NAD^+). The reaction is therefore a *dehydrogenation*:

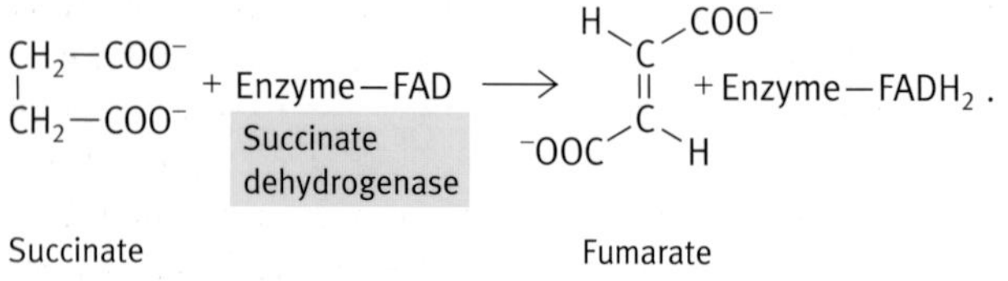

The rest of the cycle, conversion of fumarate to oxaloacetate, is plain sailing because you have already met the reaction types involved. A molecule of water is added to fumarate (*cf.* aconitase, above). The enzyme should logically be called fumarate hydratase but, by long usage, is called **fumarase**. The *hydration* reaction is:

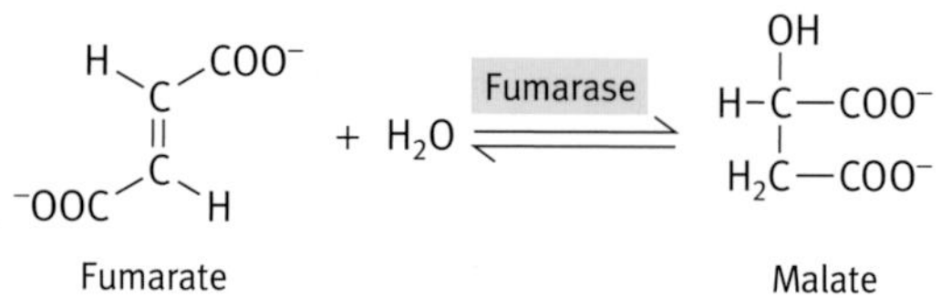

The malate so produced is *dehydrogenated* by malate dehydrogenase, an NAD^+ enzyme, and so the cycle is back to the starting point, oxaloacetate.

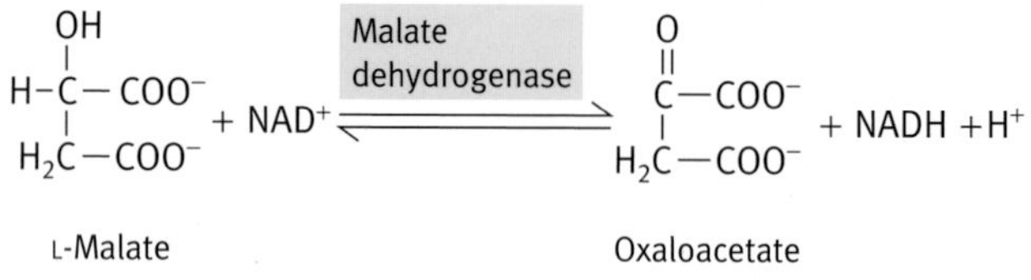

The $\Delta G^{0\prime}$ of this reaction is $+29.7$ kJ mol^{-1}, which is very unfavourable; the reaction proceeds because the next reaction, the conversion of oxaloacetate to citrate, is strongly exergonic and pulls the reaction over.

The complete cycle is shown in Fig. 12.11.

What determines the direction of the citric acid cycle?

The cycle operates in a unidirectional way, as shown in Fig. 12.13. This is due to three of the reactions having a sufficiently large negative ΔG value to be irreversible. These are the synthesis of citrate from acetyl-CoA and oxaloacetate ($\Delta G^{0\prime} = -32.3$ kJ mol^{-1}), the decarboxylation of isocitrate to α-ketoglutarate ($\Delta G^{0\prime} = -20.9$ kJ mol^{-1}), and the α-ketoglutarate dehydrogenase reaction ($\Delta G^{0\prime} = -33.5$ kJ mol^{-1}). This results in the cycle operating in one direction even though the equilibrium of the malate dehydrogenase reaction is in favour of the reverse direction ($\Delta G^{0\prime} = +29.7$ kJ mol^{-1}). The overall operation of the cycle reactions has a negative ΔG value. Control of the cycle is given on page 259.

Stoichiometry of the cycle

The daunting overall equation for the process is (for reference purposes only):

$$CH_3CO\text{--}S\text{--}CoA + 2H_2O + 3NAD^+ + FAD + GDP + P_i \rightarrow$$
$$2CO_2 + 3NADH + 3H^+ + FADH_2 + CoA\text{--}SH + GTP;$$

$$\Delta G^{0\prime} = -40 \text{ kJ mol}^{-1}.$$

Topping up the citric acid cycle

The cycle starts with oxaloacetate condensing with acetyl-CoA and ends with oxaloacetate so that the latter component is not used up. The cycle acids occupy a special place in metabolism in that they are not necessarily available from the diet in large

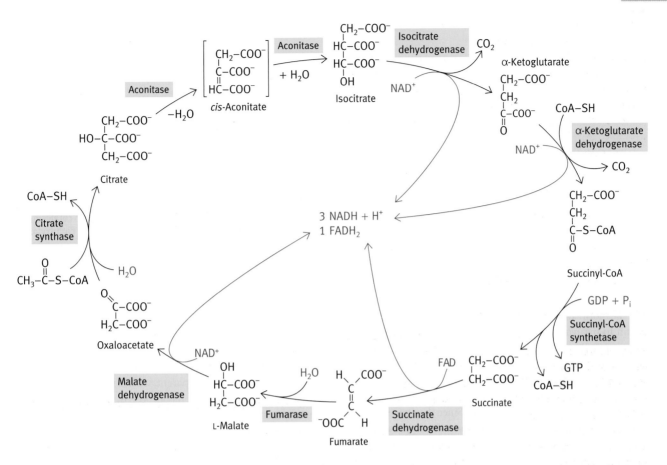

Fig. 12.11 Complete citric acid cycle. Red highlights the production of reducing equivalents from the cycle. Blue highlights the supply of the elements of H_2O to the cycle. (The conversion of citrate to isocitrate involves removal and addition of H_2O but there is no net gain.) Note that the FAD is not free but is attached to the succinate dehydrogenase protein. The involvement of water in the synthesis of citric acid is explained on page 194.

amounts. Carbohydrates in the diet give rise to large quantities of C_3 acid in the form of pyruvate production but cycle acids (C_4, C_5, and C_6) are not available in such quantities. Fats provide large amounts of the C_2 (acetyl groups), but since these are completely destroyed in the cycle, they do not make any net contribution to cycle acids. Certain amino acids can provide cycle acids but, by the same token, cycle acids are withdrawn to synthesize some amino acids (described in Chapter 19) and other metabolites. A deliberate method of topping up cycle acids to keep the energy-generating mitochondrial reactions running properly is essential and there is such a provision in cells. An important reaction for this, called an **anaplerotic** or 'filling-up' reaction, is that of pyruvate plus CO_2 being converted to oxaloacetate, using energy from ATP hydrolysis.

The enzyme is called **pyruvate carboxylase** (and quite different from pyruvate decarboxylase of yeast, please note). This is the crucial point at which C_3 acids can be converted to C_4 acids, so pyruvate carboxylase is an enzyme of central importance.

Pyruvate carboxylase: biotin, the cofactor for CO_2 activation

It requires the vitamin, biotin, for its function. Wherever 'activated' CO_2 is needed for synthetic reactions catalysed by a group of carboxylase enzymes, biotin is the cofactor. Biotin is a water-soluble B group vitamin. It becomes covalently bound to its enzyme where it accepts a carboxy group from bicarbonate to form carboxybiotin, the reaction being thermodynamically driven by the conversion of ATP to ADP and P_i (see reaction below). Carboxybiotin is a reactive, but stable, form of CO_2 that can be transferred to another molecule that is to be carboxylated. The $\Delta G^{0'}$ for the cleavage of CO_2 from carboxybiotin is -19.7 kJ mol^{-1}. In this case pyruvate is the substrate, but other carboxylation enzyme systems are also biotin-dependent.

$$ATP + \begin{array}{c} COO^- \\ | \\ C=O \\ | \\ CH_3 \end{array} + HCO_3^- \longrightarrow \begin{array}{c} O \\ \| \\ C-COO^- \\ | \\ H_2C-COO^- \end{array} + ADP + P_i + H^+$$

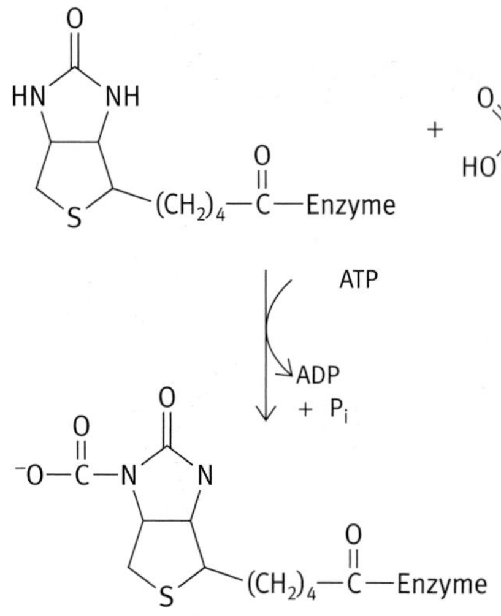

Pyruvate carboxylase has two catalytic sites – one to carboxylate the biotin and the other to transfer the carboxy group from biotin to pyruvate. (In some bacteria, the two activities reside on separate enzymes.) The attachment of the biotin to the long lysyl side chain of the protein provides a flexible arm to permit the biotin to oscillate between the two sites (Fig. 12.12).

Despite the stated importance of pyruvate carboxylase, *Escherichia coli* cells do not possess it. Since they have the citric acid cycle, how can they manage without the topping-up reaction? The answer is that *E. coli*, in addition to the normal citric acid cycle, has a modified form (the glyoxylate cycle) described on page 240, which obviates the need for it.

We will return to pyruvate carboxylase later, for it has metabolic importance other than its anaplerotic role for the citric acid cycle.

Stage 3 – the electron transport chain that conveys electrons from NADH and FADH$_2$ to oxygen

A preliminary overview of this stage is given in Chapter 11, page 182.

Remember that what we are doing is looking at the three major stages involved in glucose (or glycogen) oxidation. Stage 1 was glycolysis, stage 2 was the citric acid cycle, and now we come to the final stage. Energy-wise we have not achieved much yet: per starting molecule of glucose only a trivial yield of four ATP molecules – two from glycolysis and two from the cycle (via GTP), with one extra if a glucosyl unit of glycogen is the starting compound. The other products per mole of starting glucose are 10 NADH (two from glycolysis, two from pyruvate dehydrogenase, six from the cycle), and two FADH$_2$ from the cycle. (Don't forget that one molecule of glucose produces two pyruvate molecules and hence supports two turns of the cycle.)

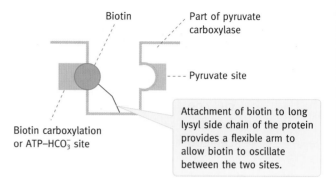

Fig. 12.12 The role of biotin in the active site of enzymes catalysing carboxylation reactions. The example shown is pyruvate carboxylase.

The oxidation of the NADH and FADH$_2$ will produce most of the ATP (from ADP and P$_i$) generated by the oxidation of glucose.

The electron transport chain

The basic principles of the electron transport chain are given on page 183. It would probably be useful to read this again. For reasons that will become apparent, we will now discuss electron transport, pure and simple. Its purpose *is*, definitely, ATP production from ADP and P$_i$, but just for the time being forget all about this and concentrate on electron movement to oxygen.

Where does it take place?

Electron transport carriers exist in or on the inner mitochondrial membrane. As we have already seen (Fig. 11.4), the inner membrane is folded into cristae. This increases the amount of inner membrane present, the density of cristae in a mitochondrion being related to the energy requirements of the cell.

Nature of the electron carriers in the chain

Haem is the prosthetic group of several electron carriers – called **cytochromes** because of their colour (red). The different cytochromes are called c_1, c, a, and a_3 (in order of their participation in the chain; the role of two b cytochromes is given later). The essentials of the haem structure are shown in Fig. 12.13, and its full structure in Fig. 4.19.

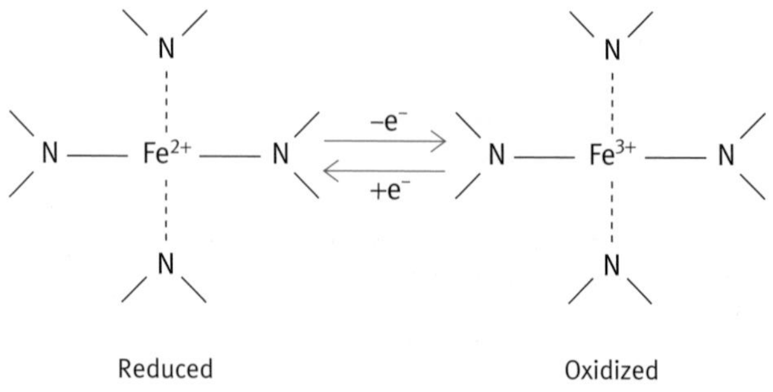

Fig. 12.13 Diagrammatic representation of haem (but take a look at the actual structure in Fig. 4.19).

The important thing about the haem molecule is that, as the prosthetic group of the cytochrome electron carriers, the Fe atom oscillates between the Fe^{2+} and Fe^{3+} states as it accepts an electron from the preceding carrier and donates it to the next carrier in the chain. Note the difference from haem in haemoglobin which remains in the Fe^{2+} form. The characteristics of the haem molecule are modified by the specific protein to which it is attached, and variations in the haem side groups occur in different cytochromes and in their precise attachment to their apoproteins. Thus, it is not a contradiction that different cytochromes have different redox potentials (electron affinities) and yet have haem as their prosthetic group.

Another type of electron carrier, based on iron, are the so-called non-haem iron proteins. In these, the iron is bound to the thiol side group of the amino acid cysteine of the protein and also to inorganic sulphide ions forming **iron-sulphur complexes**, or iron-sulphur centres. The simplest type is shown in Fig. 12.14. As with the cytochromes, the iron atom in these can accept and donate electrons in a cyclical fashion, oscillating between the ferrous and ferric state. Such iron-sulphur centres are associated with flavin enzymes. They accept electrons from FAD – enzymes such as succinate dehydrogenase and the dehydrogenase involved in fat oxidation (described in Chapter 13). Another type of carrier is an FMN-protein. FMN (flavin adenine mononucleotide) consists of the flavin half of FAD (see page 179). It carries electrons from NADH to an iron-sulphur centre. All of the iron-sulphur centres transfer electrons to ubiquinone (see below).

As well as these protein-bound electron carriers, there is one carrier not bound to a protein. This is a molecule, illustrated in Fig. 12.15, called **ubiquinone**, because it can exist as a quinone and is found ubiquitously. It is often referred to as coenzyme Q (CoQ), UQ, or Q. Q is an electron carrier because it can accept protons as shown in Fig. 12.15; it can exist as the free radical, semiquinone intermediate, thus permitting the molecule to

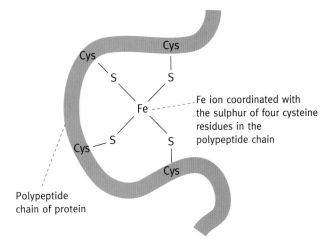

Fig. 12.14 An iron-sulphur centre. There are several types of these, increasing in complexity and numbers of Fe and S atoms. The simplest form is shown here.

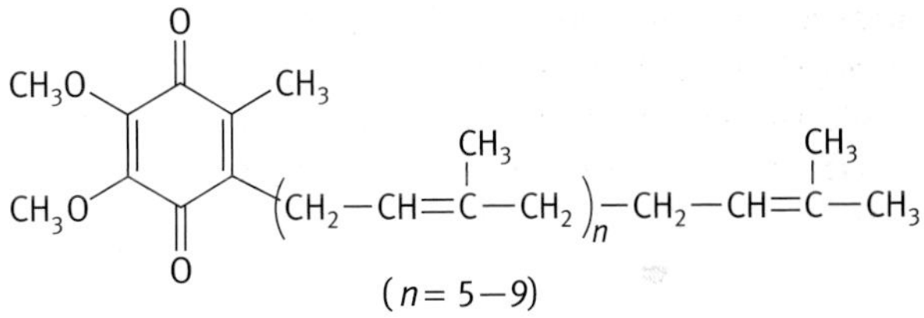

hand over a single electron to the next carrier rather than a pair of electrons. The very long hydrophobic tail on the molecule (as many as 40 carbon atoms long in 10 isoprenoid groups) makes it freely soluble *and mobile* in the nonpolar interior of the inner mitochondrial membrane.

Structure of ubiquinone (coenzyme Q)

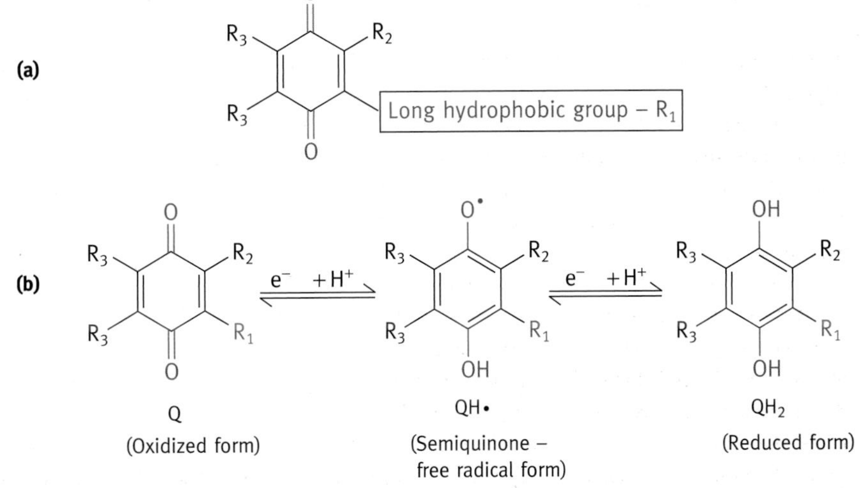

$(n = 5-9)$

Fig. 12.15 (a) Ubiquinone or coenzyme Q structure (in the oxidized form), **(b)** Oxidized, semiquinone, and reduced forms of ubiquinone (Q). The semiquinone, QH· can exist as the anion, $Q^{·-}$. R_1, long hydrophobic group; R_2, —CH_3; R_3, —O—CH_3. The full structure of ubiquinone is given above.

(a)

Long hydrophobic group – R_1

(b)

Q
(Oxidized form)

QH·
(Semiquinone –
free radical form)

QH_2
(Reduced form)

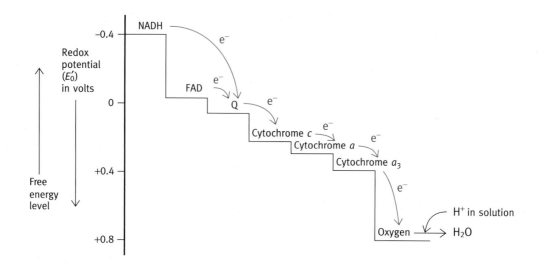

Fig. 12.16 The approximate relative redox potentials of some of the main components of the electron transport system in mitochondria. The arrows indicate electron movements. The role of cytochrome *b* components is shown in Fig. 12.21.

To summarize, we have in the electron transport chain an FMN-protein, non-haem iron-sulphur proteins, Q not bound to a protein and freely mobile in the membrane, and haem proteins known as cytochromes. An important point is that one of the latter, **cytochrome *c***, is a small water-soluble protein molecule (molecular weight ~12.5 kDa, just over 100 amino acids) that is loosely attached to the outside face of the inner mitochondrial membrane so that it also is free to move. All the other proteins of the respiratory complexes are built into the membrane structure as integral proteins in fixed positions.

Arrangement of the electron carriers

In Chapter 11 we discussed the redox potentials of electron acceptors and explained that electrons flow from a carrier of higher reducing potential (low redox potential or lower electron affinities) to one of lower reducing potential (more oxidizing, or higher redox potential or higher electron affinities). The electron carriers in the chain are of different redox potentials. Their electron affinities increase progressively down the chain.

The redox potentials are directly related to $\Delta G^{0'}$ values, as already discussed on page 182. The electron carriers are arranged in the electron transport chain such that there is a continuous progression down the free-energy gradient (increasing redox potentials) with the corresponding release of free energy as the electrons move from one carrier to the next (Fig. 12.16). They form, as it were, a bucket brigade carrying electrons down the hill. In considering glucose oxidation, the task in this stage of metabolism is to transfer electrons from NADH and FADH$_2$ to oxygen. The whole scheme involves a somewhat formidable list of steps but (fortunately) the carriers are grouped into the four complexes shown in Fig. 12.17 so that we will not give the detailed carrier list. These complexes are built into the structure of the inner mitochondrial membrane, interconnected by the mobile electron carriers, ubiquinone and cytochrome *c*. Q takes

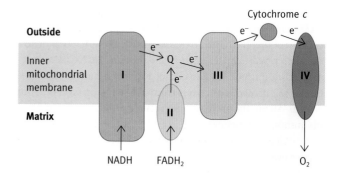

Fig. 12.17 The electron transport chain with electron carriers grouped into four main complexes. Complex I, NADH:Q oxidoreductase; complex II, succinate:Q oxidoreductase; complex III, QH$_2$:cytochrome *c* reductase; complex IV, cytochrome oxidase. Q, ubiquinone or coenzyme Q. FADH$_2$ is generated in the cycle from succinate by succinate dehydrogenase. Note that the complexes are located in the inner mitochondrial membrane. Q and cytochrome are mobile carriers capable of physically transporting electrons from one site in the membrane to another. Cytochrome *c* is surface-located. Note that FADH$_2$ exists attached to flavoprotein enzymes. The major ones are succinate dehydrogenase and fatty acyl-CoA dehydrogenases involved in fat oxidation; the latter is described are Chapter 13.

electrons from complexes I and II and delivers them to complex III. Cytochrome c is the intermediary between complexes III and IV. Complex I carries electrons from NADH to Q. Complex II carries electrons from succinate and other substrates (fatty acids, glycerol phosphate) via FADH$_2$ to Q; complex III uses QH$_2$ to reduce cytochrome c. Complex IV transfers electrons from cytochrome c to oxygen. Complexes I, III, and IV are, for convenience, referred to as NADH:Q oxidoreductase, QH$_2$:cytochrome c oxidoreductase, and cytochrome oxidase, respectively. Complex IV, cytochrome oxidase, is also a multi-subunit structure; electrons are donated to it by cytochrome c on the outer face of the inner mitochondrial membrane. Cytochrome oxidase contains the haem proteins, cytochromes a and a_3, and copper centres which participate in the final transference of electrons to oxygen. The final reduction catalysed by cytochrome oxidase is:

$$O_2 + 4e^- + 4H^+ \rightarrow 2H_2O.$$

We now come to the point of all this electron transport.

Oxidative phosphorylation – the generation of ATP coupled to electron transport

If you remember how ATP is generated in glycolysis, you may be puzzled. In glycolysis there is substrate-level phosphorylation. In this, the phosphorylation (generation of ATP from ADP and P_i) is inseparably linked to the reactions in glycolysis. Either the reactions occur involving ATP synthesis, or they do not occur. You cannot have the relevant reactions without ATP generation since this is an intrinsic component of the reactions. The same is true of ATP generation in the citric acid cycle via GTP.

By contrast, we have described the electron transport reactions without even mentioning ATP production. The purpose of electron transport is to generate ATP, but electron transport to oxygen proceeds very readily in broken-up mitochondria without any ATP generation at all. This situation perplexed and frustrated biochemists for decades. If electron transport generates ATP in the cell, why can it readily occur without this in broken-up mitochondria? In cells, or in intact 'healthy' mitochondria, electron transport results in ATP generation but, in broken mitochondria, electron transport occurs very happily

without it. It conflicted with the type of substrate-level phosphorylation that was known from glycolysis.

The chemiosmotic theory of oxidative phosphorylation

The solution was discovered by the English biochemist, Peter Mitchell, who in a private laboratory in 1961 produced a concept of how electron transport causes ATP synthesis that was so novel that it was at first hardly taken seriously by most and strongly opposed by many. It took a long time for it to be accepted (and for a Nobel Prize to be awarded to Mitchell in 1978). The concept is based on the simple notion that gradients have the ability to do work. A gradient of water pressure can be used to generate electricity, a gradient of air pressure to drive a windmill, etc. A chemical gradient is no different. Molecules or ions will migrate from a high concentration to a low concentration and, if a suitable energy-harnessing device is interposed, useful work can be done.

Applying this concept, two things are needed for ATP generation coupled to electron transport: firstly, electron transport must create a gradient of some sort; and, secondly, the gradient must be allowed to flow back through a device that uses the energy of the gradient to synthesize ATP from ADP and P_i. Mitchell's concept thirdly required the existence of intact vesicles across whose membrane a gradient could be established. It has to have an inside and an outside, as separate compartments, explaining why broken mitochondria do not make ATP. The membrane must therefore be impermeable to the solute of which the gradient consists.

Mitchell discovered that electron flow caused protons to be ejected from inside the mitochondrion (or $E.$ $coli$ cell) to the outside, thus creating a proton gradient across the membrane (Fig. 12.18); in other words, the pH of the external solution

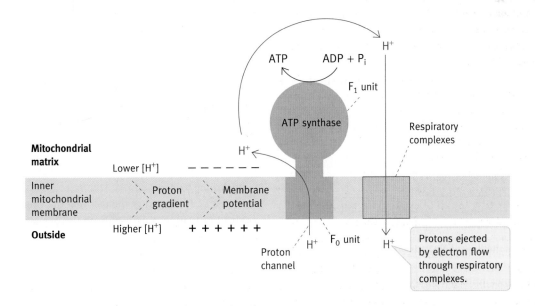

Fig. 12.18 The ATP-generating system of the inner mitochondrial membrane. The entry of H$^+$ into the matrix via the F$_0$ unit is discussed on page 205.

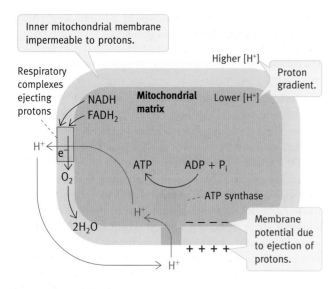

Fig. 12.19 Generation of ATP in mitochondria by the chemiosmotic mechanism. Note that FADH$_2$, produced by the dehydrogenation of fatty acids (described in Chapter 13), enters the same pathway as that utilized by the oxidation of succinate.

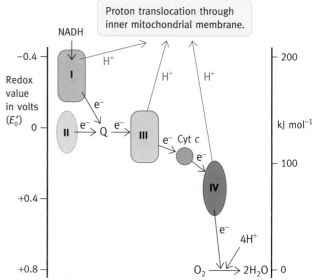

Fig. 12.20 Approximate positions on the redox potential scale of the electron transport complexes. The scale on the right gives the approximate free energy released when a pair of electrons is transferred from components to oxygen. Q, ubiquinone.

dropped, an exciting moment in the history of biochemistry. A membrane (or charge) potential, negative inside and positive outside, is also generated by the proton expulsion and also contributes to the total energy gradient or **proton-motive force** available for ATP synthesis. The inner mitochondrial membrane itself is virtually impermeable to protons. This is a prerequisite for the system, but inserted into the membrane are special proton-conducting channels. Protons flow from the outside through these channels back into the mitochondrial matrix and the energy of this flow is harnessed to the formation of ATP from ADP and P$_i$. Mitchell's theory has a majestic simplicity. All aerobic life on this planet is driven by the creation of a pH gradient across a membrane. Dinosaurs roamed on it; from whales to aerobic bacteria, from plants to humans – all are driven by it. It is one of the great concepts in biochemistry.

The proton-conducting channels are knob-like structures that completely cover the inner surface of the cristae. These are, in fact, **ATP synthase** complexes that convert ADP and P$_i$ to ATP, the process being energetically driven by the proton flow. The process will be described shortly.

The overall **chemiosmotic mechanism** in mitochondria is shown in Fig. 12.19. Two main questions can be asked,

- How does the flow of electrons from NADH and FADH$_2$ to O$_2$ cause protons to be pumped from the matrix side of the inner mitochondrial membrane to outside the membrane?
- How does the flow of protons into the mitochondrion drive the synthesis of ATP from ADP and P$_i$?

How are protons ejected?

Protons are transferred from the matrix to the cytosolic side of the membrane by three separate complexes. These are I (NADH: Q oxidoreductase), III (QH$_2$: cytochrome coxidoreductase), and IV (cytochrome oxidase). Complex I is a huge complex with a large domain extending into the matrix where the NADH binds. The mechanism by which complex III ejects protons was the first to be established. The proton-pumping mechanism of complex IV is not fully understood, though the three-dimensional structure of this complex has been obtained.

Complex II, which reduces Q to QH$_2$ using FADH$_2$ as the reductant (Fig. 12.17) does not pump protons because the free-energy drop in the process is insufficient. FAD reduction to FADH$_2$ occurs mainly by succinate dehydrogenase and in fatty acid metabolism. The FADH$_2$ has a higher redox potential (lower free energy) than NADH. Proton pumping by complex III is described below, followed by that of complex IV (cytochrome oxidase).

Figure 12.20 illustrates the fact that the energy for proton-gradient formation is derived from the free energy released as electrons are transported down the electron carrier chain.

The Q cycle in complex III ejects protons from mitochondria

The mechanism by which protons are translocated as a result of electron flow is established in the case of complex III. The *principle* of Mitchell's original idea is as simple as it is ingenious. In essence, hydrogen *atoms* are assembled on the matrix side of the

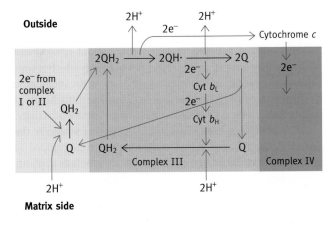

Outside

$2H^+$ $2H^+$

$2e^-$ → Cytochrome c

$2QH_2$ → $2QH\cdot$ → $2Q$

$2e^-$

Cyt b_L

$2e^-$

Cyt b_H

$2e^-$ from complex I or II

QH_2

Q QH_2 ← Q

Complex III   Complex IV

$2H^+$ $2H^+$

Matrix side

Fig. 12.21 The mechanism of proton translocation by complex III as a result of electron transport in mitochondria. The red arrows represent physical diffusion of components rather than chemical transformations; the latter are indicated by black arrows and electron transport by blue arrows. (However, note that the blue arrow representing electron transport by cytochrome c is effected by the physical diffusion of the latter from complex III to complex IV and its return after oxidation.) Cytochrome b protein spans the membrane so that electrons are transferred from the outer face to the inner face. The molecules of Q and QH_2 are in equilibrium with a membrane pool of these, but this is not shown in order to simplify the diagram arrangement. The electrons from complex I (as QH_2) arise mainly from NADH generated in glycolysis and the citric acid cycle; those from complex II come from the succinate → fumarate step of the cycle via an FAD-protein. As will be described in the next chapter, electrons from fatty acid oxidation also enter complex III via complexes I and II. Q, ubiquinone; QH·, semiquinone; Cyt, cytochrome. Cyt b_L and b_H refer to haem sites on cytochrome b of low and high redox potentials respectively.

inner mitochondrial membrane, using protons from the matrix and electrons from the transport chain. The H atoms are assembled on Q, forming the reduced form, QH_2, which now diffuses to the opposite face of the membrane where the reverse happens – electrons are stripped off the hydrogen atoms by the electron transport system and the resultant protons escape to the outside. The essentials of the process are the positioning of the responsible catalytic proteins on opposite sides of the membrane and a mobile carrier to transport the hydrogen atoms across the membrane from one face to another. It is called the **Q cycle.**

That is the principle for proton transport in complex III. The actual mechanism is shown in Fig. 12.21; it looks complicated but is in fact very simple with few reactions involved, and achieves what Mitchell's original idea proposed: assemblage of H atoms on one side attached to Q and release on the other side. Note first of all that the red arrows simply represent physical movement of ubiquinone and its derivatives; the second point is that, in effect, two distinct processes are going on, both of which eject a pair of protons. In the membrane there are pools of Q and QH_2.

Now in more detail, let us start with QH_2 formation by complexes I and II, and whose electrons originate from NADH and

FADH$_2$, respectively. In this, two protons are taken up from the matrix as shown to the left of the diagram and two electrons from NADH (in complex I) or FADH$_2$ (in complex II) form hydrogen atoms on Q. The QH_2 now migrates to a site, in complex III, *on the external face of the inner membrane*, where one electron is removed and passed on to reduce cytochrome c, which transports it to complex IV Remember that cytochrome c is also mobile. The site on complex IV that accepts electrons from cytochrome c is exposed on the external face of the inner membrane and cytochrome c is also located on this face of the membrane. A proton is ejected, leaving the half-oxidized quinone, QH- (page 201). A further electron is now removed from the latter to form Q, but in this case the electron is handed on, not to cytochrome c but to cytochrome b, to which we will return shortly. A proton is ejected. That is the end of that half of the story – a molecule of QH_2 from complexes I/II has been oxidized, two electrons passed (one to cytochrome c and one to cytochrome b), and two protons ejected from the matrix to the outside. The Q so formed now returns to the general pool.

There is another part to the story. A *second* molecule of QH_2 is oxidized in the same way as the first, resulting in the ejection of two more protons. From each of the two molecules of QH_2 we thus have two electrons passing on to complex IV via cytochrome c and two to cytochrome b. The latter transfers the electrons to a different haem site on the cytochrome b whose redox potential is greater (lower energy) and in this way transports the two electrons to the matrix side of the membrane. They are donated here to a molecule of Q and, with a pair of protons from the matrix, hydrogen atoms on Q are assembled as QH_2. The QH_2 now migrates back to the site of the external face and the merry-go-round starts again. (The diagram in Fig. 12.21 shows the same molecule of Q going round the cycle, for clarity, but molecules of Q and QH_2 will enter and exit the pool in a dynamic equilibrium – it amounts to the same thing.) This somewhat convoluted double process actually oxidizes only one molecule of QH_2 (since a molecule of Q is reduced in the whole process), but achieves the ejection of four protons. The *net* effect of the reactions *within complex* III can be summarized in the equation

$$QH_2 + 2H^+ \text{ (matrix)} + 2 \text{ cyt } c \text{ (Fe}^{3+}) \rightarrow Q + 4H^+ \text{ (outside)} + 2 \text{ cyt } c \text{ (Fe}^{2+}).$$

Complex IV also contributes to the proton gradient

Complex IV oxidizes reduced cytochrome c, forming water, and it also contributes to the formation of the proton gradient across the mitochondrial membrane. Water is formed by the reaction:

$$4 \text{ cyt } c \text{ (Fe}^{2+}) + 4H^+ + O_2 \rightarrow 4 \text{ cyt } c \text{ (Fe}^{3+}) + 2H_2O.$$

During the oxidation process, protons are actively pumped out of the mitochondrion by a mechanism not yet fully understood but possibly involving protein conformational changes.

For the oxidation of four reduced cytochrome *c* molecules, four protons are transported out of the mitochondrial matrix (two per electron pair). The protons used to form water are taken from the matrix and this increases the proton gradient. The result of the oxidation of four reduced cytochrome *c* molecules is to remove a total of four protons as water, and four ejected into the cytosolic compartment.

To briefly digress on a separate note of importance, in the process of water formation, electrons are added to oxygen. Although oxygen is a safe molecule, addition of a single electron to an oxygen atom yields a dangerous, highly reactive free radical, superoxide. Addition of the two electrons forms peroxide which is also potentially dangerous. Cytochrome oxidase adds four electrons to oxygen, forming water from H^+ but without releasing intermediate species. However, oxygen free radicals are generated in other ways and this constitutes a topic discussed in Chapter 29.

ATP synthesis by ATP synthase is driven by the proton gradient

Paul Boyer of UCLA, who won a Nobel prize for his work on ATP synthase, referred to it as a 'splendid molecular machine' and added 'All enzymes are beautiful, but ATP synthase is one of the most beautiful as well as one of the most unusual and important'; all of which is true.

ATP synthase is the name of the structure with one part visible as a knob (the F_1 unit) projecting into the matrix on the inside surface of the inner mitochondrial membrane and the other anchored in the membrane itself (the F_0 unit; o = oligomycin; see Box 12.1). The knobs are diagrammatically represented in Figs 12.18 and 12.21. A simplified diagram, showing the major components of the ATP synthase is shown in Fig. 12.22. The reaction catalysed by the synthase is:

$$ADP^{3-} + P_i^{2-} + H^+ \rightarrow ATP^{4-} + H_2O.$$

The standard free energy change of this is about +29.3 kJ and so cannot proceed without a large input of energy which is supplied by the proton gradient established across the membrane by electron transport. Protons flow back into the mitochondrial matrix through the ATP synthase. It is a major metabolic activity.

ATP synthase is found in aerobic organisms wherever the energy derived from electron transport has to be trapped as ATP. It is not confined to mitochondria – chloroplasts in plants use the same device. So do *E. coli* cells which are roughly the size of a mitochondrion, the cell membrane in this context being equivalent to the inner mitochondrial membrane. Its ATP synthase units project from the cell membrane into the cytoplasm and the electron transport system in the membrane creates a proton gradient from outside to inside by pumping protons from the inside to the outside of the bacterial cell. The structures from all sources are essentially the same.

Structure of ATP synthase

The complete structure of ATP synthase is shown in Fig. 12.23 (in this case from *E. coli*, but it applies to mitochondria). It may look formidable, but we can look at the two parts separately, F_0 in the membrane and F_1 the knob projecting into the mitochondrial matrix, and how they function.

The F_1 unit and its role in the conversion of ADP + P_i to ATP

The F_1 unit is a ring formed by six protein subunits arranged in a barrel-like structure in external appearance more or less like the segments of an orange (Fig. 12.23). All F_1 subunit proteins are designated by Greek letters. The 'knob' consists of a hexamer of three α protein subunits and three β subunits, the two alternating. Each β subunit has a catalytic (enzymic) site which synthesizes ATP from ADP + P_i, so there are three such sites per F_1 unit located at interfaces with the α subunits. The narrow cavity of the barrel is occupied by an elongated asymmetric shaft, the γ subunit, projecting beyond the barrel to form a short 'stalk' which connects the F_1 to the F_0 unit in the membrane. (In Fig. 12.23, the visible part of the γ subunit is shown in yellow and its extension inside the hexamer in a darker shade.) Another subunit called ε forms part of the stalk structure.

Figure 12.24 shows two sections of F_1 as ribbon diagrams, determined by X-ray diffraction, with the asymmetric γ subunit shaft in the centre. We will come to the other subunits in due course, so do not be concerned with them now.

Activities of the enzyme catalytic centres on the F_1 subunit

If we consider for the moment an enzymic site on a *single β subunit*, the sequence of events in the synthesis of a molecule of ATP as proposed in Boyer's model is as follows (Fig. 12.25(a)).

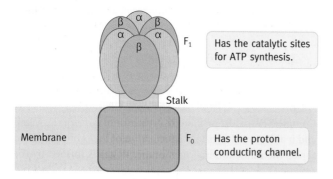

Fig. 12.22 Simplified diagram showing the major components of ATP synthase. The F_0 has multiple subunits and is integral with the lipid bilayer membrane. It is the proton-conducting channel. The F_1 is composed of a hexamer ring of alternating α and β subunits enclosing two central subunits γ and ε which project downwards and contact the F_0 unit.

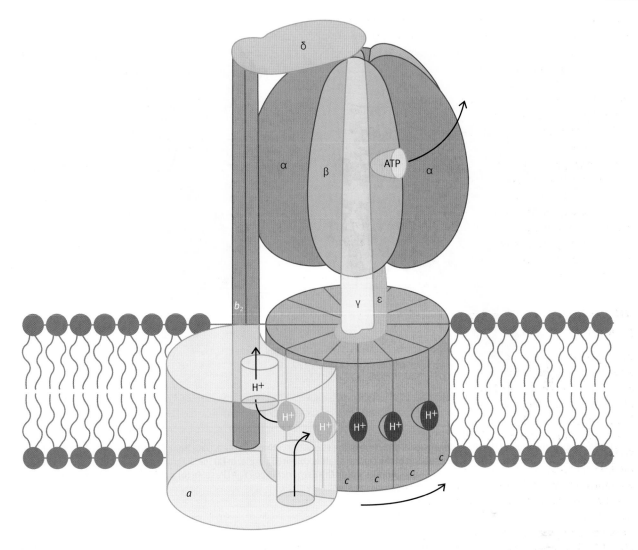

Fig. 12.23 Model of the *E. coli* ATP synthase. All ATP synthases are essentially the same except for a small variation in the number of *c* subunits. We will describe the model in terms of the mitochondrial location since this is most relevant to the text. The F_0 consists of a ring of *c* subunits integrated into the membrane lipid bilayer. Adjacent to it is the *a* subunit also integrated into the bilayer. This has two nonconnecting proton-conducting half channels, one open to the outside of the membrane and the other to the inside of the mitochondrion. Flow of protons from the outside through the F_0 into the mitochondrion matrix causes rotation of the ring of *c* subunits. This drives the rotation of the 'stalk' formed by the γ and ε subunits which project as a central asymmetric 'shaft' into the barrel-like hexamer of subunits constituting the F_1 in the mitochondrial matrix. The F_1 hexamer has three α and three β subunits surrounding the central shaft. Each β subunit has an active site near the interface with the adjacent α subunit in which synthesis of ATP from ADP and phosphate occurs. As the central shaft rotates, it contacts in succession the surrounding subunits and causes conformational changes in the active sites involved in ATP synthesis (see Fig. 12.25). The actual energy-requiring step is the release of ATP from the sites. This is supplied via the conformational changes. The δ and b_2 subunits have the role of preventing the rotation of the hexamer of the F_1 as the central shaft rotates; it is equivalent to bolting down the casing of an electric motor to stop it rotating as the shaft rotates. A more detailed description of the ATP synthase and the mechanism of its action are given in the text.

- The site is open and nothing is bound to it (the **O state**).
- A conformational change in the protein converts the site to a low-affinity state; ADP and P_i now bind to it loosely but there is no catalytic activity (the **L state**).
- A further conformational change produces a tight-binding state – the ADP and P_i become tightly bound. This is now catalytically active and ATP is formed (the **T state**).
- A conformational change opens up the site, ATP escapes and the site is back to the original open state.

The model postulates that each site progresses sequentially through the three conformations. (It is easy to fall into the error of imagining that the sites rotate – they do not.) ATP synthase is a most unusual enzyme in that catalytic activity is dependent on cooperation between the subunits. At any point in time one of

(a) **(b)**

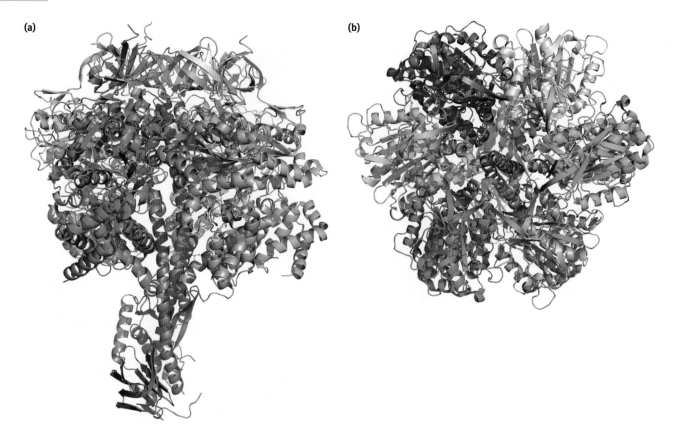

Fig. 12.24 Ribbon diagrams of the three-dimensional structure of the F_1 of ATP synthase (Protein Data Bank Code 1JNV), with the γ subunit in the central cavity (coloured yellow-brown). The ε subunit of the central shaft is coloured purple. In **(a)** the diagram is a longitudinal section of the entire head, showing the conformation of the ε and γ subunits within the *Escherichia coli* F_1 ATPase. In **(b)** a cross-section of the head is shown, giving the relative arrangement of the α and β subunits.

the three β subunit sites is in the O state, one in the L state, and one in the T state (see Fig. 12.25(b)). A site in the T state with a molecule of ATP bound to it will convert to the open O state and release the ATP when the preceding site on the bsubunit in the L state becomes occupied by ADP and P_i, which at the same time converts to the T state itself.

The ADP and P_i now are tightly bound and ATP synthesis proceeds.

You may be puzzled that in the third step above, we simply state that ATP is formed from ADP and P_i, which seems energetically wrong. However, the researchers working on this problem found that *when ADP and P_i are firmly bound to the active site of the enzyme* there is little free energy change in the formation of ATP. The equilibrium constant is about 1 as compared with 10^{-5} for ADP and P_i in free solution (see page 30 if you need to be reminded about equilibrium constants and free energy). This does not conflict with what you have learned about the energetics of ATP, for it applies only to reactants tightly bound to the enzyme. *Energy is needed to release the ATP* so that the conversion of ADP + P_i *in solution* to ATP *in solution* requires the expected energy input. This is supplied by the conformational changes which the enzyme undergoes during the

catalytic cycle. These are caused by the rotating asymmetric γ subunit sequentially contacting the F_1 subunits. It is not known how the energy transference occurs, but each β subunit in turn undergoes a conformational change which puts it into a 'high energy state' allowing ATP release.

Structure of the F_0 unit and its role

F_0 built into the inner mitochondrial membrane is the motor which is driven to rotate by a flow of protons from outside the inner membrane into the inside of the mitochondrion. It rotates the γ subunit inside the F_1 to which it is connected. The F_0 consists of a ring of c subunits (Fig. 12.23; F_0 proteins are denoted by italic letters rather than the Greek ones used for F_1 protein), varying in number from 10 to 14 in the various ATP synthases. Do not worry about the H^+ depicted on the ring of subunits of the F_0 – we will come to these shortly.

Each c subunit is a single α helical polypeptide in the shape of a hairpin so that each has two 'arms' spanning the lipid bilayer. The crucially important feature to note is that in the middle of the α helix of one of these arms of each c subunit is an aspartate residue which is thereby placed in the centre of the hydrophobic

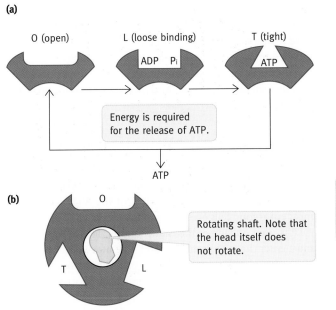

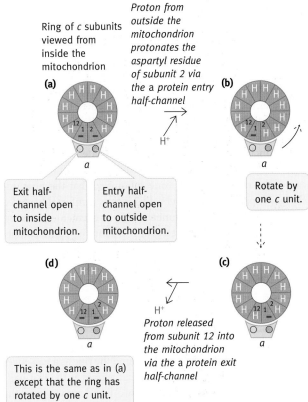

Fig. 12.25 The catalytic sites of ATP synthase as proposed in the Boyer model: **(a)** the changes that occur in a single site of one β subunit of F_1 during the synthesis of ATP. **(b)** The three β subunits work in a cooperative manner and the conversions in one site are coordinated with the other two sites. This means that at any one time an F_1 unit has one subunit in the O state, one in the L state, and one in the T state. The rotating shaft is shown as a notional asymmetric shape to convey that it is believed that it successively interacts with the subunits as it rotates. The actual structure of the shaft within the F_1 barrel is given in Fig. 12.24.

Fig. 12.26 Diagram to illustrate the principle of F_0 rotation. In **(a)** the central aspartyl group in c units 1 and 2 are unprotonated and in contact with the hydrophilic environment provided by the two half-channels of subunit a. If now the residue in subunit 2 is protonated from the *outside* of the mitochondrion *via* the entry half-channel of a as in **(b)**, the ring of c units will rotate to bring the uncharged residue into hydrophobic contact with the lipid bilayer. At the same time, subunit 12 moves into the hydrophilic region provided by the exit half-channel of a as shown in **(c)** and the proton is lost to the in side of the mitochondrion. This restores the situation to that in (a) except that the ring has moved by one subunit as shown in **(d)**. Repetition of this cycle causes stepwise F_0 rotation. The net result is that protons flow from the outside of the membrane to the inside driven by the concentration and charge gradient and in doing so rotate F_0. A molecular model of this diagrammatic figure has been used as the cover picture for this edition (Protein Data Bank code 1C17).

lipid bilayer. Adjacent to the ring of c subunits is the large a protein.

Mechanism by which proton flow causes rotation of F_0

To explain this we will use Fig. 12.26 which gives a different view of the ring of 12 c subunits seen in plan from the F_1 side.

It is essential to note that the c ring is surrounded by the hydrophobic lipid bilayer except for the two c subunits which interface with the a protein. Ten of the subunits will be in the hydrophobic environment of the surrounding lipid bilayer. This energetically requires the central aspartyl residue of each of these to be in the protonated uncharged —COOH state rather than the unprotonated charged —COO$^-$ state. In the case of the two c subunits adjacent to the a protein the situation is different because, it is postulated, there are two half-channels in the a protein, as shown in Fig. 12.23 as a transparent shape. These expose their aspartyl residues to a hydrophilic environment. They are therefore in the unprotonated —COO$^-$ state. Figure 12.23 shows that two half-channels do not make a direct connection between the two faces of the membrane, so that protons cannot flow via them directly across a protein. One half-channel is open to the matrix *inside* the mitochondrial

membrane (the left-hand one in Fig. 12.26) while the other (on the right) is open to the *outside*.

What causes the ring to rotate? The answer is basically extremely simple but you will need to follow Fig. 12.26 closely. The state of the aspartyl residue in the centre of each sub unit is shown as a white H for the uncharged state (—COOH) and a green minus sign for the ionized state (—COO$^-$). In Fig. 12.26(a) the aspartyl residues of subunits 1 and 2 are charged since each is exposed to one of the hydrophilic half channels of protein a. Those of the other ten of the c ring subunits, in contact with the hydrophobic lipid bilayer, are, as stated, uncharged.

The ring cannot move in this state since it would bring the charged group of subunit 2 into the hydrophobic environment (which is thermodynamically 'forbidden'). The aspartyl group of subunit 2 is open to the half-channel which connects to the *outside* of the mitochondrial membrane where there is a high concentration of protons. This causes its central aspartyl group to become protonated from the *outsidepool* thus converting it to the uncharged, protonated —COOH state (Fig. 12.26(b)), which is 'uncomfortable' in a hydrophilic environment. This causes the ring to move by one subunit so that the now un-charged aspartyl group of subunit 2 is thus comfortably placed in a hydrophobic environment. This movement, however, brings the uncharged aspartyl group of subunit 12 into contact with the hydrophilic half-channel which is open to the *inside* of the mitochondrion *where the proton concentration is low* (Fig. 12.26(c)) causing it to lose its proton to the matrix. This produces the state in Fig. 12.26(d) which is the same as in Fig. 12.26 (a) except that the ring has moved by one *c* subunit. Repetition of the same cycle of events causes stepwise rotation of the ring. Each proton joining the aspartyl group from outside the mitochondrion is thus carried round the ring as a passenger on its *c* subunit until after 11 moves it arrives at the exit half-channel of the aprotein (Fig. 12.26(c)) and moves into the mitochondrion (Fig. 12.26(d)). Rotation of the *c* ring relative to the *a* protein has been demonstrated by experiments in which the *a* and *c* proteins were chemically cross-linked.

Thus the proton flow, reinforced by the membrane potential, by this complex pathway from the outside to the inside of the mitochondrial inner membrane generates the force which rotates the γ subunit inside F_1 The rotation of F_0 is unidirectional. This is the result of much higher proton concentration outside the membrane than inside. This means that protona-tion of aspartyl residues occurs preferentially from the outside and deprotonation to the inside.

If the F_1 unit is detached from the F_0 its reactions are reversible in the presence of ATP, which is hydrolysed. It is an ATPase. Yoshida's group in Japan has, in an ingenious experi-ment (described in the **Noji *et al.*** reference at the end of this chapter), directly visualised the reverse-direction rotation of the γ subunit in such detached F_1 units as ATP is hydrolysed.

To summarize, the shaft is asymmetric and sequentially con-tacts the F_1 hexamer subunits. In some way, not understood, this transmits rotational energy into conformational changes in the F_1 subunits. This supplies the energy to allow ATP release.

It is estimated that for each molecule of ATP synthesized, three protons flow through the F_0, though this is not certain to be the actual figure. The value need not necessarily be a whole number. It is the world's smallest rotary motor and one of the most remarkable enzymes known.

What is the role of the elongated subunit *b* dimer and the *d* subunit on the left of the structure in Fig. 12.23? This is rather interesting. To digress briefly, an electric motor needs to have

its outer casing bolted down to a bench or whatever to stop it rotating as the shaft inside it turns, The two protein subunits have the same job in ATP synthesis. They bolt down the F_1 to the membrane to restrain it as the γ subunit rotates.

An interesting movie showing the ATP synthase in action may be seen on the website: http://www.mrc-dunn.cam.ac.uk/research/atp_synthase/

Transport of ADP into mitochondria and ATP out

The inner mitochondrial membrane is impermeable to most compounds, and to electrons. Special transport systems (trans-locases) have been developed. Most of the ATP synthesis in most eukaryotic cells occurs in mitochondria while most of the ATP is used outside of the mitochondria. Hence ADP and P_i must enter the mitochondrion and ATP move out. The highly charged molecules cannot diffuse passively across the inner mitochondrial membrane and a special transport mechanism exists. ATP–ADP translocase exchanges ATP inside the mito-chondrion for an ADP outside the mitochondrion (Fig. 12.27).

Where does the energy for this ATP–ADP exchange come from? As already explained, electron transport generates not only a pH gradient across the inner mitochondrial membrane but also a membrane potential across the inner mitochondrial membrane, positive outside and negative inside, due to the ejection of H^+ ions (see page 118 if you are not clear what is meant by a membrane potential). ATP carries four negative charges out, while ADP carries only three in. Thus the ATP-ADP exchange tends to neutralize this membrane potential. Therefore the exchange of ATP for ADP costs the equivalent of one proton. The transport of the P_i needed along with ADP for ATP generation is catalysed by a phosphate translocase in the mitochondrial membrane (Fig. 12.29). It carries $H^2PO_4^-$ into the matrix driven by the proton gradient.

Another transport problem occurs in the oxidation of cytoplasmic NADH generated in glycolysis. Two different 'shuttle' mechanisms exist to cope with this. These will now be described.

Reoxidation of cytoplasmic NADH from glycolysis by electron shuttle systems

In the aerobic situation, the NADH generated in glycolysis is reoxidized by transferring its electrons into mitochondria. This is the 'normal' route of reoxidation of NADH. NADH itself cannot enter the mitochondrion; there are two systems for transferring its electrons into mitochondria. In these, protons from NADH are transported into the mitochondrion, leaving NAD^+ in the cytoplasm.

The glycerol phosphate shuttle

The first, the glycerol phosphate shuttle, involves dihydroxy-acetone phosphate (generated by the aldolase reaction). An enzyme in the cytoplasm transfers electrons from NADH to

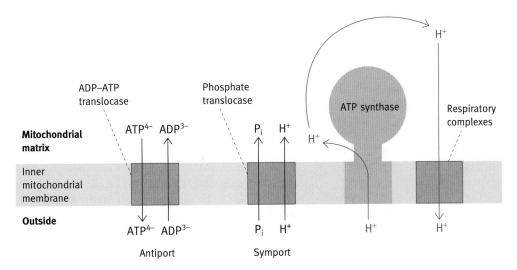

Fig. 12.27 Diagram of transmembrane traffic in mitochondria involved in ATP generation. All of the traffic is *via* specific transport proteins. Other transport systems for metabolites exist.

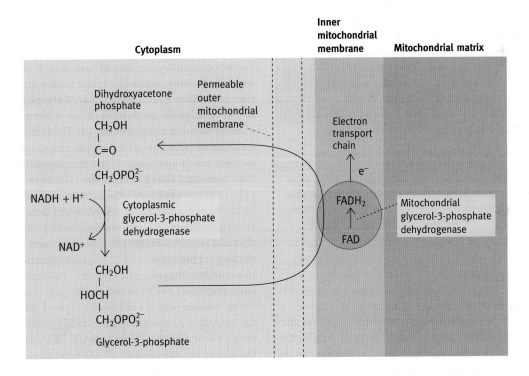

Fig. 12.28 The glycerol phosphate shuttle that transfers electrons from cytoplasmic NADH to the electron transport chain of mitochondria.

dihydroxyacetone phosphate giving glycerol-3-phosphate (Fig. 12.28). The enzyme is called **glycerol-3-phosphate dehydrogenase**, working in reverse in the above reaction.

Glycerol-3-phosphate can reach the inner mitochondrial membrane (the outer one being highly permeable) where a different type of glycerol-3-phosphate dehydrogenase, *built into the membrane* – with an FAD prosthetic group transfers electrons from glycerol-3-phosphate to the mitochondrial electron transport chain. The dihydroxyacetone phosphate so produced cycles (shuttles) back into the cytoplasm to pick up more electrons (Fig. 12.28). Note that glycerol-3-phosphate does not have to enter the mitochondrial matrix but its pair of electrons gain access to the electron chain carrying electrons to oxygen, located in the inner mitochondrial membrane. The net effect is to transfer electrons from cytoplasmic NADH to the mitochondrial electron transport chain.

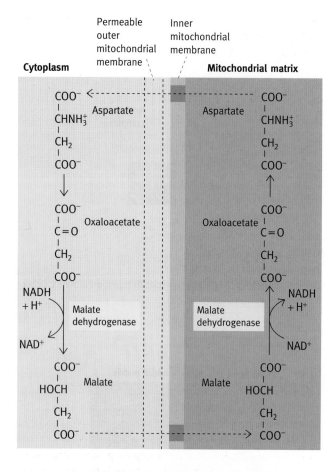

Fig. 12.29 The malate–aspartate shuttle for transferring electrons from cytoplasmic NADH to mitochondrial NAD⁺. The mechanism of the interconversion of oxaloacetate and aspartate is dealt with on page 285. This shuttle, unlike the glycerophosphate shuttle, is reversible, and can operate as shown, bringing NAD⁺ inside the mitochondrion, only if the NADH/NAD⁺ ratio is higher in the cytoplasm than in the mitochondrial matrix.

The malate–aspartate shuttle

Another shuttle, the **malate–aspartate shuttle**, transfers electrons from cytoplasmic NADH to mitochondrial NAD⁺. The mitochondrial NADH thus generated is then oxidized by the electron transport chain. This shuttle system involves transfer of electrons from NADH to oxaloacetate to form malate in the cytoplasm which is transported into the mitochondrion, by a specific carrier, where it is reoxidized to oxaloacetate, mitochondrial NAD⁺ being reduced (Fig. 12.29). The net effect is that NADH outside reduces NAD⁺ inside. This shuttle is a little more complicated in that the oxaloacetate, so formed, cannot traverse the mitochondrial membrane to get back to the cytoplasm. It is converted to aspartate, which is transported, again by a specific carrier, to the cytoplasm and reconverted to oxaloacetate there; hence the name, the malate–aspartate shuttle. At this stage we will not give the mechanism of aspartate $\rightleftharpoons$

oxaloacetate inter-conversions, since it will be more convenient to do this later when we deal with amino acid metabolism.

Different tissues probably use the two shuttles to different extents. The two differ in a significant way – the glycerol phosphate shuttle results in cytoplasmic NADH reducing the FAD of the prosthetic group of glycerol-3-phosphate dehydrogenase. The FADH₂ has a higher redox potential than NADH (lower energy); it hands on its electrons to the electron transport chain at a point that is further along the chain from that at which NADH hands on its electrons (page 204). The net ATP generation from the oxidation of a *cytoplasmic molecule* of NADH by this glycerol phosphate route is 1.5 molecules. The aspartate-malate shuttle starts with one molecule of cytoplasmic NADH and ends up with one molecule of mitochondrial NADH whose oxidation generates 2.5 molecules of ATP.

The balance sheet of ATP production by electron transport

It requires an estimated three protons to flow through the ATP synthase to generate one molecule of ATP from ADP and P_i, assuming that the latter two are already inside the mitochondrion. As described, the transport of a molecule of ADP into the mitochondrion and that of one of ATP to the cytoplasm requires the energy equivalent of one proton entering the mitochondrion. Hence four protons have to be pumped out of the matrix to drive the production of one molecule of ATP made available in the cytoplasm of the cell. For each pair of electrons transported from NADH to oxygen, the consensus is that 10 protons are pumped out of the mitochondrial matrix (four from complex I, four from complex III, and two from complex IV). Thus the oxidation of one molecule of NADH (located inside the mitochondrion) will produce 2.5 molecules of ATP. For the oxidation of one molecule of FADH₂ (that is, a pair of electrons from succinate or fatty acid), six protons are pumped yielding 1.5 molecules of ATP. (Earlier estimates were 3 and 2, respectively.) The values are known as P/O ratios since a pair of electrons reduce one atom of oxygen.

The molecules of NADH produced in glycolysis, located in the cytoplasm, require separate consideration; these donate their pairs of electrons either to NAD⁺ located inside the mitochondrion or to mitochondrial FAD, depending on which shuttle mechanism the cell uses. Thus, a *cytoplasmic molecule* of NADH may give rise to either 2.5 or 1.5 molecules of ATP.

Yield of ATP from the oxidation of a molecule of glucose to CO_2 and H_2O

Starting from free glucose rather than glycogen, the net yield of ATP from the complete oxidation of the molecule is either 30 or 32 depending on which shuttle is used for the cytoplasmic NADH. To summarize: two from glycolysis at the substrate level (remember that, although four ATP molecules are generated in glycolysis, two were used at the start so the net gain is two). In the citric acid cycle, two molecules of ATP (*via* GTP in liver and kidney) are produced per molecule of glucose at the succinyl-CoA stage (one per turn of the cycle but two acetyl-CoA molecules are produced per glucose). Thus we have four molecules of ATP produced at the substrate level; all the rest come from electron transport.

Glycolysis produces, per molecule of glucose, two molecules of NADH, which are located in the cytoplasm. These will give rise either to a total of five or three molecules of ATP depending on the shuttle used. Per molecule of glucose, pyruvate dehydrogenase produces two molecules of NADH and the citric acid cycle, six. Oxidation of these will produce 20 molecules of ATP. Oxidation of the $FADH_2$ generated from the succinate → fumarate step produces a further three ATP molecules.

The total is therefore 2 + 5 (or 3) + 2 + 20 + 3 = 32 (with the malate-aspartate shuttle used) or 30 (with the glycerol-3-phosphate shuttle used). These values are estimates – with substrate-level phosphorylation, ATP generation is always a whole number, but there are not necessarily whole-number relationships between electron transport, proton ejection, and ATP generation.

An *E. coli* cell is equivalent in this context to a mitochondrion, the cell membrane equating to the inner mitochondrial membrane and the bacterial cytoplasm to the mitochondrial matrix. In such cells there is no need for shuttle mechanisms to transport NADH electrons to the respiratory pathway. In *E. coli* there is no transport of ATP and ADP needed into the cytoplasm. The ATP yield from the oxidation of a molecule of glucose in *E. coli* is therefore greater.

Is ATP production the only use that is made of the potential energy in the proton-motive force?

The answer is almost, but not quite. In newborn babies, heat production to maintain body temperature is helped by brown fat cells – brown because they are rich in mitochondria that contain the coloured cytochromes. The generation of ATP in mitochondria is dependent on the inner mitochondrial membranes being impermeable to protons, thus forcing the latter to enter the mitochondrial matrix only via the ATP-generating channels. If you made a hole in the membrane the protons

would simply flood through it, effectively acting as a short circuit, no ATP would be generated, and the energy would be liberated as heat. In brown fat cell mitochondria, this is essentially what happens, channels permitting non-productive (that is, no ATP synthesis) proton flow being made by a special protein, **thermogenin**. Chemicals that transport protons unproductively through membranes (**dinitrophenol** is the classical one) also 'uncouple' oxidation from ATP generation (see Box 12.1).

Bacteria also harness energy by pumping protons across the membrane to the outside of the cell and generating ATP as described by reversed proton flow. However, the proton gradient is also used in uptake of solutes into the cell, the H^+ gradient being used for cotransport (lactose uptake is an example) just as the Na^+ gradient is used in animal cells (page 113). Remarkably also, the cilia of bacteria are rotated by a flow of protons through the protein machinery that rotates the cilium; it runs on 'proticity' rather than the electricity used by an electric motor. The proton-driven motor of cilia is reminiscent of the F_0 'motor' of ATP synthase.

 Summary

Glycolysis is stage 1 in the complete oxidation of glucose or glucosyl units of glycogen. It causes the lysis of the C_6 glucose molecule into the two C_3 molecules of pyruvate (hence the name glycolysis). It occurs in the cytosol. It produces a net gain of only two ATP molecules (three if we start with glycogen) but prepares the glucose for the next stage, the citric acid cycle.

In glycolysis there is one oxidation step which reduces NAD^+ to $NADH + H^+$. Since NAD^+ is limited in amount it must be reoxidized via the mitochondrion or glycolysis would halt. Under normal conditions the NADH is reoxidized *via* mitochondria. NADH cannot itself enter the mitochondrion but shuttle mechanisms transfer the electrons from NADH either to NAD^+ or to FAD inside the mitochondrion. During vigorous exercise, NADH is formed too rapidly for the normal oxidation route to cope with it. It is rapidly reoxidized by lactate dehydrogenase reducing pyruvate to lactate.

The citric acid cycle is stage 2 of the oxidation of glucose. Pyruvate is transported into the mitochondrial matrix where it is converted by pyruvate dehydrogenase to acetyl-CoA, which enters the citric acid cycle. The first reaction is the conversion of the acetyl group to citrate by condensation with oxaloacetate catalysed by citrate synthase.

In a single turn of the citric acid cycle, the electrons from the acetyl group (plus extra ones originating indirectly from water) are transferred to NAD^+ and FAD, regenerating oxaloacetate. The carbon atoms are removed as CO_2. During the cycle only two ATP molecules are produced per molecule of glucose (in the equivalent form of GTP in liver) but it has generated fuel in the form of NADH and $FADH_2$ for stage 3, which is the transfer of electrons from these carriers to oxygen, forming water. This is associated with a large release of free energy.

The **electron transport system** is stage 3 of the oxidation of glucose. It generates most of the ATP. As the electrons move along the hierarchy of electron carriers from NADH and $FADH_2$, the released free energy is used to generate a proton gradient (augmented by a membrane charge potential) across the inner mitochondrial membrane by proton pumping. Protons are pumped out of the mitochondia. This is the celebrated Mitchell chemiosmotic theory. The electron carriers are grouped into four complexes. Proton pumping occurs in complexes I, III, and IV but not II. The mechanism of pumping by complex I is not known. In complex III it is achieved by the Q cycle, Q being ubiquinone, which is a mobile electron carrier. Cytochrome c is also mobile and connects complexes III and IV, the latter being cytochrome oxidase which transfers electrons to oxygen, forming water. Proton pumping here probably involves conformational changes in the protein subunits.

The proton gradient is used to drive ATP synthesis by the molecular machines known as ATP synthase of the inner mitochondrial membrane. They are minute rotating motors driven by proton flow. The rotation causes conformational changes in the ATP synthase subunits, the energy of which drives the condensation of ADP + P_i to ATP. The mechanism of these remarkable rotary machines is now almost fully established. The ATP is transported out into the cytoplasm by exchange with ADP, the process being driven by the membrane potential. The yield of ATP per molecule of glucose cannot be calculated with absolute precision but approximately 30 molecules are produced from ADP + phosphate, which is lower than previous estimates.

In prokaryotes there are no mitochondria but the cell membrane is equivalent to the inner mitochondrial membrane in the present context.

Further reading

Schatz, G. (2007). The Magic Garden. *Annu.* Rev. Biochem., **76**, 673–678.
A highly readable personal review of mitochondrial function and assembly and how the field has developed.

Electron transport chain

Slater, E. C. (1983). The Q cycle, an ubiquitous mechanism of electron transport. *Trends Biochem. Sci.*, **8**, 239–42.
Describes the proton-pumping system present in both mitochondria and chloroplast.

Cecchini, G. (2003). Function and structure of Complex II of the respiratory chain. *Annu. Rev. Biochem.*, **72**, 77–109.
A research-level review.

ATP synthase

Abrahams, J. P., Leslie, A. G. W., Lutter, R., and Walker, J. E. (1994). Structure at 2.8 Å resolution of F_1ATPase from bovine heart mitochondria. *Nature*, **370**, 621–8.
The classical paper giving the three-dimensional structure determination.

Noji, H., Yasuda, R., Yoshida, M., and Kinosita, Jr., K. (1997). Direct observation of the rotation of F_1ATPase. *Nature*, **386**, 299–302.

A seminal article demonstrating physical rotation of the central subunit of the enzyme.

Boyer, P. D. (1999). What makes ATP synthase spin? *Nature*, **402**, 247–9.
A News and Views article summarizing in a succinct way the rotary mechanism of the enzyme. It relates to the research article below.

Rastogi, V. K. and Girvin M. E. (1999). Structural changes linked to proton translocation by subunit c of the ATP synthase. *Nature*, **402**, 263–8.
Proposes a mechanism based on structural work by which proton flow through the F_0 may produce rotation.

Hutcheon, M. L., Duncan, T. M., Ngai, H., and Cross, R. L. (2001). Energy-driven subunit rotation at the interface between subunit α and the c oligomer in the F_0 sector of *E. coli* ATP synthase. *Proc. Natl.Acad. Sci. U.S.A.*, **98**, 8519–24.

Capaldi, R. A. and Aggeler, R. (2002). Mechanism of the F_1F_0-type ATP synthase, a biological rotary motor. *Trends Biochem. Sci.*, **27**, 154–60.
Review on the function of the rotary motor from structural, genetic, and biophysical studies.

Boyer, P. D. (2002). Reflections; A research journey with ATP synthase. J. *Biol. Chem.*, **277**, 39045–61.
A long, but easily readable account of the Nobel Prize winner's research career, leading to elucidation of the ATP synthase mechanism.

Mitochondrial genetics and ageing

Nagley, P. and Wei, Y.-H. (1998). Ageing and mammalian mitochondrial genetics. *Trends Genet.*, **14**, 513–17.
An account of mitochondrial mutations and their possible relationship to ageing.

Problems

1 Explain the chemical rationale for the phosphohexose isomerase reaction.

2 What is meant by substrate-level phosphorylation? Give an example of such a system. How does this differ in basic terms from oxidative phosphorylation?

3 The reaction for which the enzyme pyruvate kinase is named never occurs in the cell. Discuss this.

4 How many ATP molecules are generated in glycolysis, from:
(a) glucose
(b) glycosidic unit of glycogen?

5 In calculating how many molecules of ATP are produced as a result of the oxidation of cytoplasmic NADH, we cannot be sure of the exact answer in eukaryotes. Why is this so?

6 What is the chemical rationale for isocitrate being oxidized before loss of CO_2 occurs?

7 Explain how the citric acid cycle acids can be 'topped up' by an anaplerotic reaction. Can acetyl-CoA participate in this?

8 What is the cofactor involved in carboxylation reactions? Explain how it works.

9 Outline the arrangement of respiratory complexes in the electron transport chain.

10 What characteristic do ubiquinone and cytochrome *c* have in common? What are their physical locations in the cell?

11 What is the immediate role of electron transfer in the respiratory chain?

12 It is stated in the text that the yield of ATP from the complete oxidation of a molecule of glucose in eukaryote cells is either 30 or 32 molecules; in the case of *E. coli* it is stated that the yield is greater than this. Why does this difference in statements exist?

13 Explain what is meant by the statement that the citric acid cycle is a water-splitting device.

14 By means of diagrams and brief notes, explain in outline how the proton gradient generated by electron transport is harnessed into ATP production by ATP synthase.

15 The active sites in the F_1 subunits of ATP synthase are said to be cooperatively interdependent, a most unusual situation. Explain what this means.

16 Give a brief account of the complexes that constitute the electron transport system of the inner mitochondrial membrane with particular reference to the creation of the proton gradient across the membrane.

17 Briefly explain the basic physicochemical principle by which the F_o unit of ATP synthase is caused to rotate.

18 Which of the following is out of place? Oxaloacetate, malate, GDP, acetyl-CoA, H_2O, NAD^+?

Chapter 13

Energy release from fat

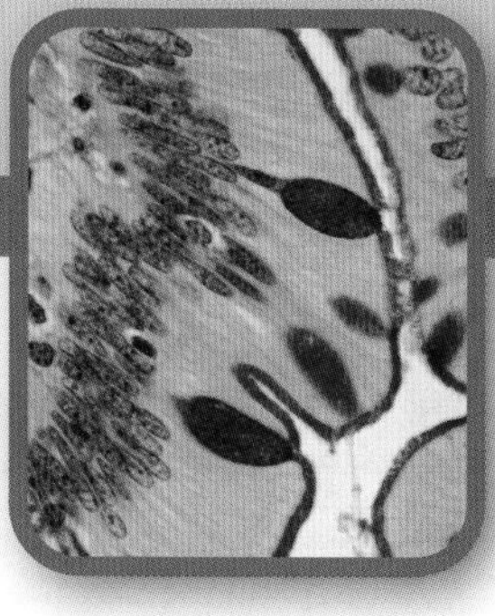

In the previous sections you have seen how energy in the form of adenosine triphosphate (ATP) is obtained in the cell starting with the oxidation of glucose or the breakdown of glycogen.

Fat is the other major source of energy for ATP production. It provides perhaps half of the total energy needs of heart and resting skeletal muscles. By far the largest amount of stored energy occurs as fat for, as mentioned earlier, there appears to be no limit to the amount of neutral fat that can be stored in the body in the adipose cells.

The subject of fat oxidation and concomitant ATP production is greatly simplified because fat oxidation involves the citric acid cycle and electron transport system that we have discussed in glucose oxidation. As already described (page 185), the two systems (glucose oxidation and fat oxidation) converge at acetyl-CoA so all we are mainly concerned with in fat oxidation is the relatively simple task of chopping up fatty acids by removing two carbon atoms at a time as acetyl-CoA.

The processing of fatty acid molecules involves NAD^+ and FAD reduction the reduced forms of which are oxidized by the same pathways we have already discussed. It is efficient to use the same machinery for obtaining energy from all classes of dietary components. Regulation of the pathways in this chapter are dealt with in Chapter 16 on metabolic control, page 259.

A few simple points first.

- Fatty acid oxidation occurs inside mitochondria (Fig. 11.4, page 181).

- Before oxidation can occur, free fatty acids are released from triacylglycerol (TAG) stores by hydrolysis by the hormone-sensitive lipase (Chapter 16, page 260). This also produces a molecule of glycerol per TAG hydrolysed and this is metabolized separately. It is manipulated to enter metabolism in the glycolysis pathway – it is phosphorylated and oxidized to give dihydroxyacetone phosphate of glycolysis fame. Thus, in starvation, the glycerol moiety *can* give rise in the liver to glucose by the gluconeogenesis pathway (Chapter 15).

- Free fatty acids for oxidation are obtained by peripheral tissues from that released into the blood by adipose cells when the glucagon level is high. They also enter the cells as a result of lipoprotein lipase attack on chylomicrons or very-low-density lipoprotein produced by the liver (VLDL, page 170). Chylomicrons exist after feeding, while fatty acids are released from adipose cells in starvation (Chapter 9, page 157).

- Free fatty acids from adipose cells are carried as ionized molecules attached in a freely reversible manner to serum albumin. They readily diffuse into cells so that the amount entering cells increases as their blood level rises as a result of release from adipose cells. As the fatty acids are taken up by cells, more will dissociate from the serum albumin carrier protein to maintain the equilibrium between free and bound fatty acids.

- During conversion to acetyl-CoA, the fatty acid is always in the form of an acyl-CoA. The first stage in oxidation of fatty acids is always to convert the carboxylic acids to the fatty acyl-CoA compounds, a reaction known as fatty acid activation.

- Fatty acids are broken down by removing two carbon atoms at a time as acetyl-CoA, a process known as *β*-**oxidation**.

Mechanism of acetyl-CoA formation from fatty acids

'Activation' of fatty acids by formation of fatty acyl-CoA derivatives

The term 'activation' of a carboxylic acid refers to the fact that the thiol ester is a high-energy (or reactive) compound. The activation reaction is:

$$R\,COO^- + ATP + CoA{-}SH \rightleftharpoons R\,CO{-}S{-}CoA + AMP + PP_i;$$

$$\Delta G^{0'} = -0.9 \text{ kJ mol}^{-1}.$$

The free-energy change of this reaction is small (because of the high energy of the thiol ester) but hydrolysis of the inorganic pyrophosphate (PP_i) by the ubiquitous enzyme, inorganic pyrophosphatase, makes the overall process strongly exergonic and irreversible ($\Delta G^{0'} = 32.5 \text{ kJ mol}^{-1}$). (See page 35 if you have forgotten this point.)

There are three different fatty acid-activating enzymes for short-chain, medium-chain, and long-chain acids, respectively – called **fatty acyl-CoA synthetases** (sometimes called **thiokinases**).

Transport of fatty acyl-CoA derivatives into mitochondria

Activation of fatty acids occurs in the outer mitochondrial membrane, which is effectively the same compartment as is the cytoplasm because of its permeability to most metabolites, but their conversion to acetyl-CoA occurs only in the mitochondrial matrix. The acyl group of fatty acyl-CoA is carried through the inner mitochondrial mitochondrial membrane, without the CoA, by a special transport mechanism and then handed over to CoASH *inside* the mitochondrion to become fatty acyl-CoA again. The high-energy nature of the acyl bond is preserved during the transport – otherwise it could not reform fatty acyl-CoA inside the mitochondrion without further energy expenditure. To achieve this, on the external face of the inner membrane of the mitochondrion, the acyl group is transferred to a rather odd hydroxylated molecule, **carnitine**.

Carnitine

Although a carboxylic ester is usually of the low-energy type, the structure of carnitine is such that the fatty acyl—carnitine bond is of the high-energy type – the acyl group has a high group-transfer potential. Presumably this a reason for the evolution of carnitine as the carrier molecule in this transport system. The fatty acyl-carnitine so formed is transported into the mitochondrial matrix where the reverse reaction occurs – carnitine is exchanged for CoASH and the free carnitine is taken back to the cytoplasm to collect another fatty acyl group.

Fatty acyl-carnitine

The scheme is shown in Fig. 13.1.

The acyl transfer from fatty acyl-CoA to carnitine is catalysed on the cytoplasmic side of the inner membrane by an enzyme called **carnitine acyltransferase I** and that on the matrix side by **carnitine acyltransferase II**. Genetic defects are known in which

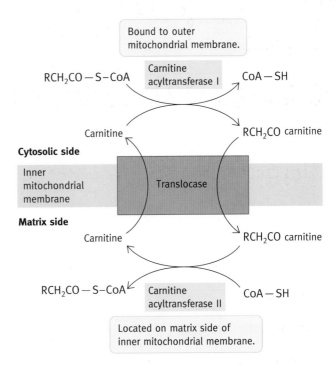

Fig. 13.1 Mechanism of transport of long-chain fatty acyl groups into mitochondria where they are oxidized in the mitochondrial matrix. The acyl—carnitine bond is an unusual ester bond in that it has a high group-transfer potential – the compound is of the high-energy type so that exchange of carnitine for CoASH inside the mitochondrion occurs without need for energy input. See text for structures.

there is a carnitine deficiency or deficiency in the carnitine acyl-transferase. In some cases this can manifest itself as muscle pain and abnormal fat accumulation in muscles.

Conversion of fatty acyl-CoA to acetyl-CoA molecules inside the mitochondrion by β-oxidation

There are four separate reactions in this process, three of which are analogous in reaction types to those converting succinate to oxaloacetate in the citric acid cycle namely, **dehydrogenation** by an FAD enzyme, **hydration**, and an **NAD⁺-dependent dehydrogenation**. We suggest that you refresh your memory on these by having a quick look on page 198 at the reaction sequence: succinate $\rightarrow$ fumarate $\rightarrow$ malate $\rightarrow$ oxaloacetate. The corresponding reactions on fatty acyl-CoA derivatives are shown in Fig. 13.2. In these the β carbon atom (C_3) is converted to a C=O group, forming a β-ketoacyl-CoA. The ketoacyl-CoA is cleaved by CoASH, splitting off two carbon atoms as acetyl-CoA and forming a shorter fatty acetyl-CoA derivative. Because the molecule is split by the —SH group of CoASH, the enzyme is called a **thiolase**. The thiolase reaction preserves the free energy as a thiol ester of fatty acyl-CoA. When the fatty acyl group has been shortened, by successive rounds of acetyl-CoA production, to the C_4 stage (butyryl-CoA) the next round of reactions produces acetoacetyl-CoA, which is finally split by CoASH into

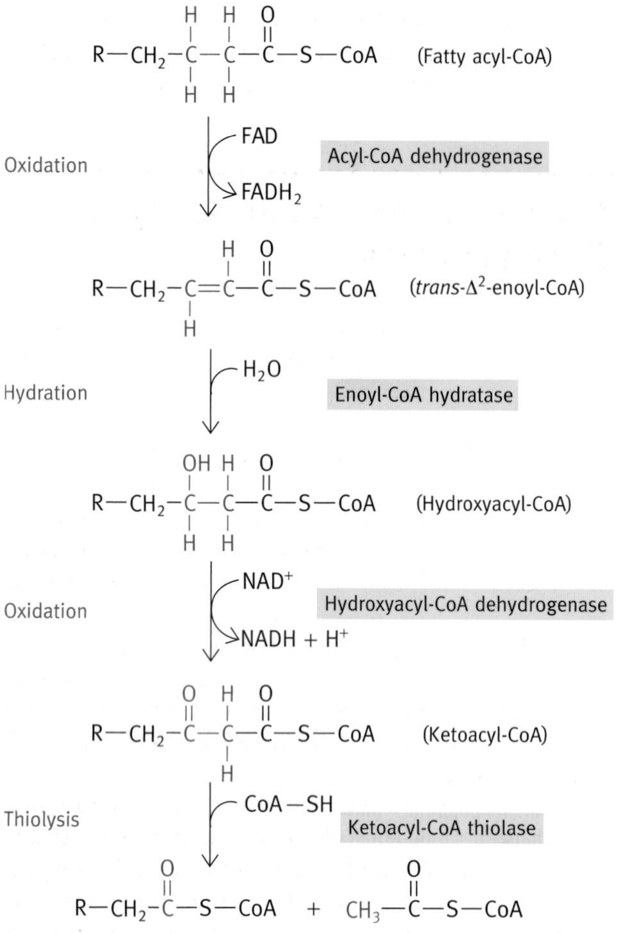

$$R-CH_2-\overset{\overset{\displaystyle H}{|}}{C}-\overset{\overset{\displaystyle H}{|}}{\underset{\underset{\displaystyle H}{|}}{C}}-\overset{\overset{\displaystyle O}{||}}{C}-S-CoA \quad \text{(Fatty acyl-CoA)}$$

Oxidation ⟍ FAD → FADH$_2$ | Acyl-CoA dehydrogenase

$$R-CH_2-\overset{\overset{\displaystyle H}{|}}{C}=\overset{\underset{\underset{\displaystyle H}{|}}{}}{C}-\overset{\overset{\displaystyle O}{||}}{C}-S-CoA \quad \text{(trans-}\Delta^2\text{-enoyl-CoA)}$$

Hydration ⟍ H$_2$O | Enoyl-CoA hydratase

$$R-CH_2-\overset{\overset{\displaystyle OH}{|}}{C}-\overset{\overset{\displaystyle H}{|}}{\underset{\underset{\displaystyle H}{|}}{C}}-\overset{\overset{\displaystyle O}{||}}{C}-S-CoA \quad \text{(Hydroxyacyl-CoA)}$$

Oxidation ⟍ NAD$^+$ → NADH + H$^+$ | Hydroxyacyl-CoA dehydrogenase

$$R-CH_2-\overset{\overset{\displaystyle O}{||}}{C}-\overset{\overset{\displaystyle H}{|}}{\underset{\underset{\displaystyle H}{|}}{C}}-\overset{\overset{\displaystyle O}{||}}{C}-S-CoA \quad \text{(Ketoacyl-CoA)}$$

Thiolysis ⟍ CoA—SH | Ketoacyl-CoA thiolase

$$R-CH_2-\overset{\overset{\displaystyle O}{||}}{C}-S-CoA \quad + \quad CH_3-\overset{\overset{\displaystyle O}{||}}{C}-S-CoA$$

Fig. 13.2 One round of four reactions by which a fatty acyl-CoA is shortened by two carbon atoms with the production of a molecule of acetyl-CoA. Note the similarity of reaction types in the desaturation, hydration, and ketoacyl formation with the succinate → fumarate → malate → oxaloacetate steps of the citric acid cycle.

two molecules of acetyl-CoA. A specific thiolase in mitochondria performs this reaction:

$$CH_3COCH_2CO-S-CoA + CoA-SH \rightarrow 2CH_3CO-S-CoA$$
$$\text{Acetoacetyl-CoA} \qquad\qquad\qquad \text{Acetyl-CoA}$$

The reaction sequence in each round involves conversion of saturated (only single bonds in the chain) fatty acyl-CoAs to β-ketoacyl-CoAs. The process is therefore referred to as **β-oxidation of fatty acids**. The NADH and the FADH$_2$ (the latter on the acyl-CoA dehydrogenase) feed electrons into the electron transport chain exactly as already described (see Fig. 12.17 for a summary of this).

Energy yield from fatty acid oxidation

A molecule of palmitic acid (C$_{16}$), called palmitate at pH 7 in its ionized form, is converted to eight molecules of acetyl-CoA and

in this process we generate seven FADH$_2$ molecules and seven NADH molecules. The acetyl-CoA is metabolized by the citric acid cycle and the NADH and FADH$_2$ oxidized by the electron transport chain (Chapter 12). NADH oxidation generates, per molecule, 2.5 ATP molecules and from FADH$_2$, 1.5 ATP molecules. (Since the NADH is generated in the mitochondrial matrix, it does not have to be carried in by a shuttle mechanism.)

If you count it all up (not forgetting the one GTP per acetyl-CoA from the citric acid cycle) the oxidation of one mole of palmitic acid generates 106 moles of ATP from ADP and P$_i$, allowing for the two ATP consumed in formation of palmitoyl-CoA. Effectively two ATP molecules are consumed at the beginning though only one directly participates. Due to PP$_i$ release, two high-energy phosphate groups are released as P$_i$, forming AMP. A kinase phosphorylates this using ATP and thus formation of the acyl-CoA costs the two molecules of ATP converted to ADP. Overall, this represents about a 33% efficiency in trapping the free energy in usable form, based on $\Delta G^{0'}$ values.

Oxidation of unsaturated fat

Olive oil and other TAGs have a high content of **monounsaturated fatty acids** with a *cis*-configured double bond (see page 106 for explanation of this). For example, palmitoleic acid has a *cis*-configured double bond between carbon atoms 9 and 10. So far as oxidation goes, this fatty acid is treated by the cell in exactly the same way as palmitic acid for three rounds of β-oxidation. At this point the product is *cis*-Δ^3-enoyl-CoA:

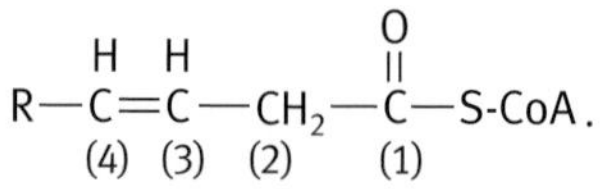

$$R-\overset{\overset{\displaystyle H}{|}}{\underset{\underset{\displaystyle (4)}{}}{C}}=\overset{\overset{\displaystyle H}{|}}{\underset{\underset{\displaystyle (3)}{}}{C}}-\underset{(2)}{CH_2}-\overset{\overset{\displaystyle O}{||}}{\underset{\underset{\displaystyle (1)}{}}{C}}-S\text{-}CoA.$$

The double bond in this position prevents the acyl-CoA dehydrogenase from forming a double bond between carbon atoms 2 and 3, as is required in β-oxidation of a saturated acyl-CoA.

An extra isomerase enzyme takes care of this by shifting the existing double bond into the required 2–3 position; it generates the *trans*-isomer in doing so:

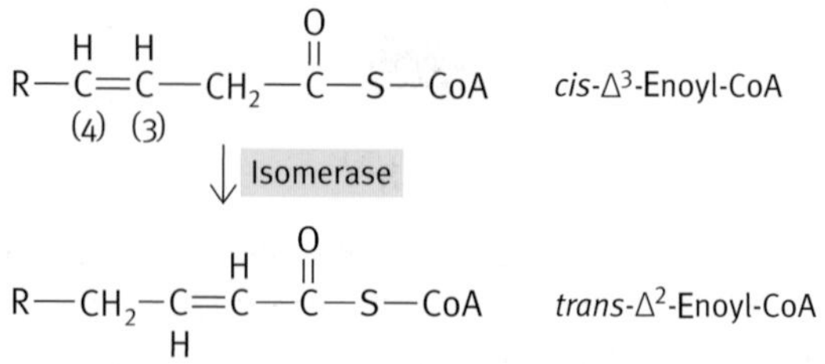

$$R-\overset{\overset{\displaystyle H}{|}}{\underset{\underset{\displaystyle (4)}{}}{C}}=\overset{\overset{\displaystyle H}{|}}{\underset{\underset{\displaystyle (3)}{}}{C}}-CH_2-\overset{\overset{\displaystyle O}{||}}{C}-S-CoA \qquad \textit{cis-}\Delta^3\text{-Enoyl-CoA}$$

↓ Isomerase

$$R-CH_2-\overset{\overset{\displaystyle H}{|}}{\underset{\underset{\displaystyle H}{|}}{C}}=C-\overset{\overset{\displaystyle O}{||}}{C}-S-CoA \qquad \textit{trans-}\Delta^2\text{-Enoyl-CoA}$$

This is now on the main pathway of fat breakdown and so the problem of monounsaturated fat oxidation is solved. (If you look at Fig. 13.2, you will see that, in oxidation of saturated acyl-CoAs, it is the transisomer that is generated.)

Polyunsaturated fatty acids pose additional problems – for example, linoleic acid has two double bonds (Δ^9 and Δ^{12}). In effect, one (Δ^9) is dealt with as described above while the second (Δ^{12}) is handled by using an additional enzyme at the appropriate stage, again putting the molecule on the normal metabolic path of β-oxidation (the steps are not given here).

Oxidation of odd-numbered carbon-chain fatty acids

A small proportion of fatty acids in the diet (for example those from plants) have odd-numbered carbon chains, β-oxidation of which produces, as the penultimate product, a five-carbon β-ketoacyl-CoA instead of acetoacetyl-CoA. Cleavage of this by thiolase produces acetyl-CoA and the three-carbon propionyl-CoA

$$CH_3—CH_2—CO—CH_2—CO—S—CoA + CoA—SH \rightarrow$$

$$CH_3—CH_2—CO—S—CoA + CH_3—CO—S—CoA$$
$$\text{Propionyl-CoA} \qquad \text{Acetyl-CoA}$$

Propionyl-CoA is carboxylated to succinyl-CoA and hence on to the citric acid cycle mainstream by the following reactions. The epimerase catalyses the conversion of D- to L-methylmalonyl-CoA.

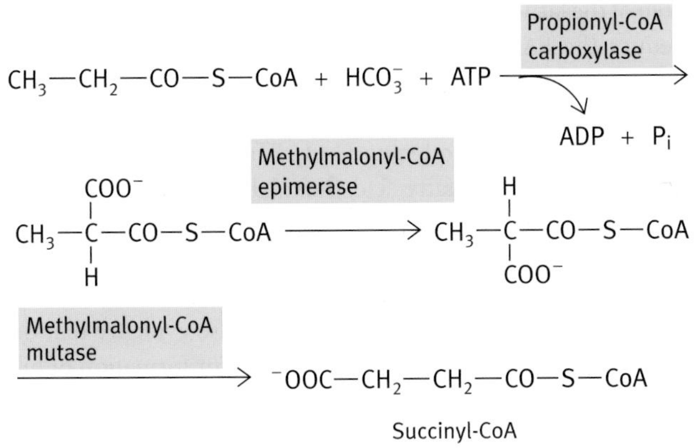

$$CH_3—CH_2—CO—S—CoA + HCO_3^- + ATP \xrightarrow{\text{Propionyl-CoA carboxylase}} ADP + P_i$$

$$\text{Methylmalonyl-CoA epimerase}$$

$$CH_3—\overset{COO^-}{\underset{H}{C}}—CO—S—CoA \longrightarrow CH_3—\overset{H}{\underset{COO^-}{C}}—CO—S—CoA$$

$$\text{Methylmalonyl-CoA mutase} \longrightarrow {}^-OOC—CH_2—CH_2—CO—S—CoA$$
$$\text{Succinyl-CoA}$$

It is the last reaction that is of great interest, for it involves the most complex coenzyme of all, **deoxyadenosylcobalamin**, a derivative of **vitamin B_{12}**.

Propionate is also formed in the degradation of three amino acids (valine, isoleucine, and methionine) and from the cholesterol side chain. Propionate is a major product of ruminant digestion so the pathway is of particular importance in these animals. It comes from bacterial digestion of plant material in ruminants.

Deficiency of the methylmalonyl-CoA mutase or the inability to synthesize the required cofactor from vitamin B_{12} leads to **methylmalonic acidosis**, usually fatal early in life.

Ketogenesis in starvation and type 1 diabetes mellitis

So far we have explained that fatty acids are converted to acetyl-CoA, which then joins the citric acid cycle and is oxidized. This is true for tissues in 'normal' circumstances but it has to be qualified for the liver in a particular circumstance.

The body can be in a physiological situation where fat metabolism is the main source of energy. This occurs in starvation after exhaustion of glycogen stores; the same can occur in untreated type 1 diabetics where the inability to metabolize carbohydrate effectively results in an almost analogous glucose 'starvation' irrespective of glucose availability. In this situation, the fat cells are pouring out free fatty acids due to high levels of glucagon (page 156) and the activation of hormone-sensitive lipase (page 260) and the liver may produce excessive amounts of acetyl-CoA; 'excessive' in the sense that there is too much produced to be fed into the citric acid cycle.

The liver cell, *in effect*, joins two acetyl groups together, by a mechanism to be described shortly, to form **acetoacetate** ($CH_3COCH_2 COO^-$), which is partly reduced to β-hydroxybutyrate ($CH_3 CHOH CH_2 COO^-$). The two, known as **ketone bodies** by tradition, are released into the blood. They are water soluble and transported in the blood to extrahepatic tissues. (As explained, the term 'ketone bodies' is an historical misnomer – they are molecules, not bodies, and β-hydroxybutyrate has no keto group.)

How is acetoacetate made from acetyl-CoA?

Acetoacetyl-CoA is formed from acetyl-CoA by reversal of the ketoacyl-CoA thiolase reaction:

$$2CH_3CO—S—CoA \rightleftharpoons CH_3COCH_2CO—S—CoA + CoA—S$$

One would have imagined that free acetoacetate would be formed by simple hydrolysis of the acetoacetyl-CoA (thus pulling the equilibrium over). However, it is not so. Instead, a third molecule of acetyl-CoA is used to form **3-hydroxy-3-methylglutaryl-CoA (HMG-CoA)** by an aldol condensation, followed by hydrolysis of the thiolester bond to give acetoacetate. The scheme is shown in Fig. 13.3.

One cannot be sure why the synthesis of acetoacetate proceeds in this way. HMG-CoA is synthesized by many animal cells. It is a precursor of cholesterol, which is an essential constituent for their membranes. Ketone body formation occurs in mitochondria (where fat conversion to acetyl-CoA occurs).

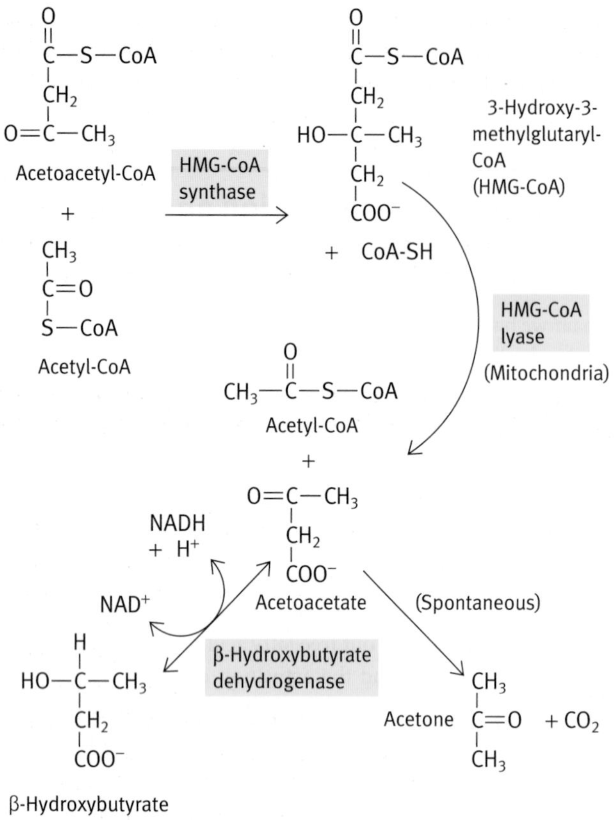

Fig. 13.3 Ketone body production in the liver during excessive oxidation of fat in starvation or type 1 diabetes. The process occurs in mitochondria. HMG-CoA is also the precursor of cholesterol but this occurs in the cytosol of rat liver where HMG-CoA synthase also occurs.

Formation of HMG-CoA for cholesterol synthesis takes place in the cytoplasm where a separate HMG-CoA synthase isomer occurs attached to the endoplasmic reticulum membrane.

Utilization of acetoacetate

Acetoacetate can be used by peripheral tissues to generate energy. (See page 154 for a discussion of the role of ketone bodies in the body.) In mitochondria acetoacetate is converted to acetoacetyl-CoA by an acyl-exchange reaction in which succinyl-CoA is converted to succinate:

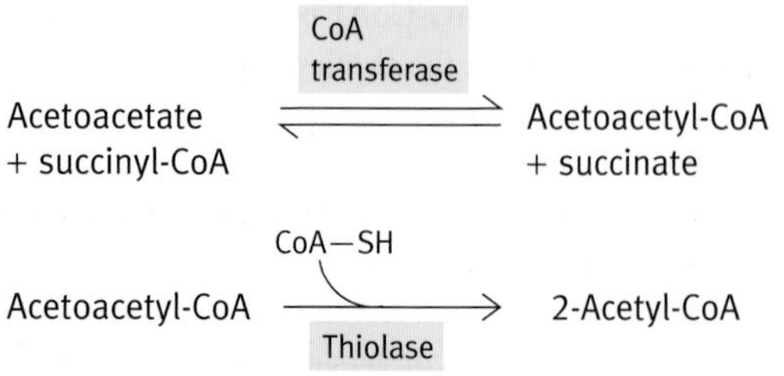

Acetoacetyl-CoA is cleaved by a thiolase using CoASH to form two molecules of acetyl-CoA. β-Hydroxybutyrate is also utilized by being dehydrogenated first to acetoacetate.

Acetoacetate, being a β-keto acid, tends to decarboxylate spontaneously, and produce acetone (CH_3COCH_3), a volatile solvent which is exhaled. In untreated type I diabetics with high concentrations of ketone bodies, acetone gives rise to a characteristic fruity smell in the breath.

Excessive production of ketone bodies leads to **ketoacidosis**, a potentially dangerous condition (page 262).

Peroxisomal oxidation of fatty acids

The bulk of fatty acid oxidation occurs in mitochondria, but some are oxidized in **peroxisomes**. These are vesicles bounded by a single membrane present in mammalian cells; they cannot synthesize proteins and receive their enzymes from the cytoplasm by a special transport mechanism. They contain flavoprotein oxidase enzymes, which attack a number of substrates using molecular oxygen generating not water, but hydrogen peroxide.

$$RH_2 + O_2 \rightarrow R + H_2O_2$$

The substrates include some phenols and D-amino acids and very-long-chain fatty acids ($>C_{18}$) which are not oxidized by mitochondria. The latter are shortened to C_8-acyl-CoA which is presumably released into the cytoplasm and transported into mitochondria for conventional oxidation, probably along with acetyl-CoA as well. The fatty acid oxidation is by β-oxidation (page 218), producing acetyl-CoA. The electrons from the first step of the fatty acid oxidation chain (acyl-CoA dehydrogenase) reduce the FAD prosthetic group of the enzyme to $FADH_2$. In mitochondria this is reoxidized by the cytochrome chain to produce ATP. In peroxisomes, the $FADH_2$ is reoxidized by molecular oxygen producing H_2O_2. The NADH produced by the later step in the fatty acid β-oxidation (hydroxyacyl-CoA dehydrogenase, Fig. 13.2) is possibly reoxidized by export of the reducing equivalents to the cytoplasm since there is no cytochrome system in peroxisomes. Peroxisomes do not generate ATP by fatty acid oxidation. There is also evidence that the cholesterol side-chain oxidation required for formation of bile acids occurs in peroxisomes and that synthesis of some complex lipids requires peroxisomes. Biogenesis of these organelles is not fully elucidated. A number of lethal genetic diseases are known in which peroxisome biogenesis is not normal. An example is the **Zellweger syndrome**, often fatal by the age of 6 months; in this there are abnormally high levels of C_{24} and C_{26} long-chain fatty acids and of bile acid precursors.

The metabolic roles of peroxisomes are rather a mixed bag, and they are the least well understood of the organelles, but their essential roles are underlined by the existence of these diseases.

The H_2O_2 generated in peroxisomes is potentially a dangerous oxidant. It is destroyed by the enzyme catalase present in peroxisomes. This is a haem-protein enzyme, which catalyses the reaction:

$$2H_2O_2 \rightarrow 2H_2O + O_2$$

Where to now?

We have so far dealt with energy production from glucose and from fat. The remaining third major food component is in the form of the amino acids. It would be logical to deal with energy production from these next but it is more convenient at this stage to remain with fat and carbohydrate metabolism and deal with fat and glucose synthesis and then with the control of metabolism. The reason for this is that energy production from individual amino acids essentially consists of converting them to compounds on the main glycolytic and citric acid cycle pathways, so that there is very little information to be given in terms of energy production *per se*. The main biochemical interests of amino acid metabolism lie elsewhere, as will be seen when we deal with this in Chapter 19.

Summary

Energy release from fat involves the oxidation of fatty acids released from triacylglycerol. They are first converted to fatty acyl-CoAs, and the acyl groups transported into the mitochondria. Fatty acyl-CoA cannot enter the mitochondrion but an enzyme of the outer membrane transfers it to carnitine, a curious small molecule, and a transport system transfers the acylcarnitine into the matrix where another enzyme transfers the acyl group back to CoA. In the matrix the fatty acyl-CoA is converted to acetyl-CoA by a process called β-oxidation in which carbon atoms are released two at a time in the form of acetyl-CoA. The fatty acid chain of acyl-CoAs is dehydrogenated by a series of enzymes, which produces β-ketoacyl-CoAs. From these, acetyl units are split off by the enzyme ketoacyl-CoA thiolase which attaches each to CoASH and releases acetyl-CoA. The fatty acid chain is sequentially shortened until it is completely converted to acetyl-CoA. This is oxidized in the citric acid cycle precisely as described for that derived from glucose in Chapter 12.

If fatty acid metabolism is proceeding very rapidly such as occurs in diabetes or starvation, acetoacetate and β-hydroxybutyrate are formed from the excess acetyl-CoA; these are water soluble and circulate in the blood to be used by other tissues. They are collectively known as ketone bodies (even though the hydroxybutyrate is not a ketone nor are they 'bodies'). The brain can use them for about half its energy needs, thus reducing the glucose requirement of the brain, which, in starvation, is precious. It has to be synthesized by the liver using amino acids from muscle-protein breakdown (Chapter 15). Excessive production of the acids can be a serious complication in untreated type 1 diabetes.

Peroxisomal oxidation of fatty acids occurs to some extent. It may be a way of oxidizing fatty acids longer than C_{18}, which are not oxidized by mitochondria. This oxidation does not feed into the electron transport chain but generates hydrogen peroxide.

Further reading

McGarry, J. D. and Foster, D. W. (1980). Regulation of hepatic fatty acid oxidation and ketone body production. *Annu. Rev. Biochem.*, **49**, 395–420.
Despite the regulatory title, the article provides a general review of ketone body formation.

Lardy, H. and Shrago, E. (1990). Biochemical aspects of obesity. *Annu. Rev. Biochem.*, **59**, 689–710.
Very readable account relating obesity, hormones, and metabolism.

Dansen, T. B. and Wirtz, K. W. A. (2001). The peroxisome in oxidative stress. *IUBMB Life*, **51**, 223–30.

Problems

1 Peripheral tissues obtain their free fatty acids for oxidation from the blood. Explain three ways in which free fatty acids become available to cells.

2 Which cells of the body do not use free fatty acids for energy supply?

3 In breaking down fatty acids to acetyl-CoA:
 (a) What is always the first step?
 (b) Where does this occur?
 (c) Where does fatty acid breakdown to acetyl-CoA occur in eukaryotes?

(d) How do fatty acid groups reach this site of breakdown?

4   Illustrate similarities in the oxidation of fatty acids to acetyl-CoA with a section of the citric acid cycle.

5   What is the yield of ATP from the complete oxidation of a molecule of palmitic acid? Explain your answer.

6   Explain how a monounsaturated fat (Δ^9) is broken down to acetyl-CoA.

7   Is acetyl-CoA, derived from fatty acid breakdown, always fed into the citric acid cycle? Explain your answer.

8   HMG-CoA is an intermediate in both acetoacetate synthesis and cholesterol synthesis. Where do these two processes occur?

9   What are peroxisomes and what is their function?

Chapter 14

Synthesis of fat and related compounds

In the previous chapters on metabolism we have been dealing with catabolic systems that break down fats and carbohydrates to yield energy. We are now switching to the other side of metabolism – the building up of molecules or **anabolism**. Note that regulation of the pathways in this can be found in Chapter 16.

Fat synthesis occurs mainly to convert carbohydrate in excess of that needed to replenish glycogen stores to a more suitable storage form, triacylglycerol (TAG). Alcohol and certain amino acids also give rise to fat. As already explained, if this were not done, to store the amount of energy contained in our storage fat as glycogen we would be much larger in size. Fat synthesis occurs in times of dietary plenty.

Mechanism of fat synthesis

General principles of the process

If you refer to the overall diagram of metabolism (Fig. 11.11), you will see that, as well as fatty acids being converted to acetyl-CoA, acetyl-CoA can be converted into fatty acids. The fact that fatty acids are synthesized from acetyl-CoA, two carbon atoms at a time, explains why most natural fatty acids have even numbers of carbon atoms. Carbohydrate intake in excess can lead to fat deposition – glucose goes to pyruvate, which goes to acetyl-CoA, which goes to fat. However, since the step catalysed by pyruvate dehydrogenase is irreversible, acetyl-CoA cannot be converted in a net sense to pyruvate and hence fat cannot be converted into glucose in animals. (Bacteria and plants have, as you will see, a special trick for doing so – page 240.)

You will by now be familiar with the concept that, in metabolic pathways, at least some reactions are different in the forward and reverse directions. This is true for fat breakdown and synthesis. For some steps the two pathways use the same reactions (in opposite directions), but there are others that are different in the two directions. These make both pathway directions thermodynamically favourable, irreversible, and separately controllable.

Synthesis of malonyl-CoA is the first step

To render synthesis of fatty acids from acetyl-CoA thermodynamically favourable, we must have a reaction that is irreversible. Refresh your memory on this, if necessary, by reading page 29 again, since it is essential to appreciate this for understanding of the very first reaction in fat synthesis. In this, the acetyl-CoA molecule is carboxylated by CO_2, using ATP breakdown as energy source, to form malonyl-CoA, but, in the next reaction in fat synthesis, the CO_2 is lost. This seems pointless unless you remember that the process is thus made thermodynamically irreversible. It is to do with energy, rather than chemical change. Malonic acid has the structure HOOC—CH_2—COOH so malonyl-CoA is HOOC—CH_2—CO—S—CoA. The enzymic reaction below is catalysed by **acetyl-CoA carboxylase** – it adds a carboxyl group to acetyl-CoA, forming malonyl-CoA.

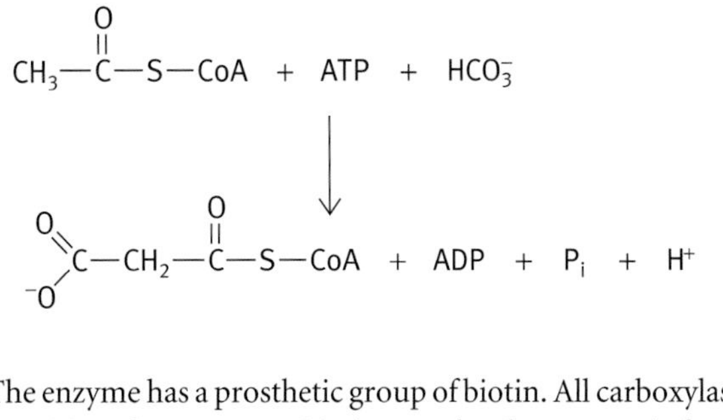

$$CH_3-\overset{\overset{\displaystyle O}{\|}}{C}-S-CoA \ + \ ATP \ + \ HCO_3^-$$

$$\downarrow$$

$$\overset{O}{\underset{-O}{\diagdown}}C-CH_2-\overset{\overset{\displaystyle O}{\|}}{C}-S-CoA \ + \ ADP \ + \ P_i \ + \ H^+$$

The enzyme has a prosthetic group of biotin. All carboxylases using ATP to incorporate CO_2 into molecules, we remind you, have this feature. As an intermediate in the reaction, an activated CO_2–biotin complex is formed (page 199).

To synthesize fatty acids we have to add two carbon units at a time, starting with acetyl-CoA. *Malonyl-CoA is the active donor of two carbon atoms in fatty acid synthesis*, despite the fact that the malonyl group has three carbons. But before this, a few words of explanation about the acyl carrier protein or ACP.

The acyl carrier protein (ACP) and the β-ketoacyl synthase

When fatty acids are broken down to acetyl-CoA, all of the reactions occur, not as free fatty acids, but as thiol esters with CoA. When fatty acids are synthesized, all of the reactions also occur on fatty acyl groups bound as thiol esters, but instead of CoA being used to esterify the reactants, *half* of the CoA molecule is used. We remind you of the structure of CoA.

Phosphate —pantothenate —NHCH$_2$CH$_2$ —SH
Phosphate —ribose —adenine
phosphate

Phosphopantotheine
moiety in box

The half in the box is 4-phosphopantotheine and this is the 'carrier' thiol used in fatty acid synthesis. It is not free, as is CoA, but is bound to a protein called the acyl carrier protein (ACP). You can think of ACP as a protein with built-in CoA or, alternatively, as a giant CoA molecule with the AMP part of CoA replaced by a protein.

The enzyme, **β-ketoacyl synthase**, also known as **condensing enzyme**, has a reactive thiol. This also is needed for acyl thiol ester formation. In this case the reactive thiol group is that of the amino acid cysteine. The ACP and β-ketoacyl synthase in mammals are part of a large multifunctional complex.

Mechanism of fatty acyl-CoA synthesis

We will start with the situation shown in Fig. 14.1 (a) in which the two thiol groups on the fatty acyl-CoA-ACP synthase complex are vacant.

CH$_3$CO— is transferred from acetyl-CoA to the ACP thiol by a specific transferase (Fig. 14.1 (b)). This is then further transferred to the β-ketoacyl synthase thiol group (Fig. 14.1 (c)), leaving the ACP site vacant. Another transferase transfers the malonyl group of malonyl-CoA to the latter, forming malonyl-ACP (Fig. 14.1(d)). The synthase now transfers the acetyl group to the malonyl group, displacing CO$_2$ and forming a β-ketoacyl-ACP; in this initial case it is β-ketobutyryl-ACP (Fig. 14.1(e)). The latter is reduced by activities on the same protein complex (described below) to form butyryl-ACP (Fig. 14.1 (f)). The butyryl group is transferred to the β-ketoacyl synthase (Fig. 14.1(g)). The resultant situation is precisely analogous to that shown in Fig. 14.1(c) since both have a saturated acyl synthase complex (acetyl and butyryl, respectively). A series of reactions

now ensues, identical to those in Fig. 14.1 (starting with the reaction c → d), except that acetyl- is now butyryl- and the end product is hexanoyl-ACP. Five more rounds produces a palmitoyl-ACP. When C$_{16}$ is reached, a hydrolase releases palmitate (Fig. 14.1 (h)),

$$CH_3(CH_2)_{14}CO—S—ACP + H_2O \rightarrow CH_3(CH_2)_{14}COO^- + ACP—SH$$

Note that the β-ketoacyl-ACP synthase reaction leading to stage (e) is irreversible because of the energy considerations of the decarboxylation involved.

If longer-chain acids are required to be formed such as stearic acid (C$_{18}$), a separate system elongates the palmitate.

Organization of the fatty acid synthesis process

The sequence of reactions given above for the conversion of acetyl-CoA to palmitate is relatively simple or complex according to your viewpoint but, in animal tissues, the organization of the enzyme activities involved is remarkable from any viewpoint. In *Escherichia coli*, the situation is straightforward, in that there is a series of separate enzymes, as is usual in metabolic pathways, and, after each reaction, the products diffuse to the next enzyme. In animals, the situation is different; all of the catalytic activities of the steps following malonyl-CoA formation reside on one giant protein known as palmitate synthase, the activities of which have not been separated. (A pair of such proteins collaborate as the functional dimer complex.) The growing fatty acid chain oscillates from the ACP-thiol group to the β-ketoacyl synthase thiol group but it is never released from the enzyme complex until palmitate synthesis is completed. The elongation step and reductive reactions all occur with the substrate attached to the ACP. The long flexible arm of the 4-phosphopantetheine is presumably needed to permit the different intermediates to interact with the appropriate catalytic centres of the complex. The advantage of a single large complex is that the process of synthesis can be more rapid since each intermediate is positioned to interact with the next catalytic centre rather than having to diffuse away and find the next enzyme.

The reductive steps in fatty acid synthesis

In the above sequence of events involved in saturated fatty acyl-ACP synthesis, the β-ketoacyl group attached to the ACP is reduced by a succession of three steps shown in Fig. 14.2. The reductant in these steps is **NADPH** (not NADH), the reduced form of nicotinamide adenine dinucleotide phosphate (NADP$^+$), an electron carrier, so far not mentioned in this book, and which will now be described. (See page 178 if you want to be reminded of the chemistry of NAD$^+$ reduction. That of NADP$^+$ is the same.)

What is NADP$^+$?

NAD$^+$, we remind you, is:

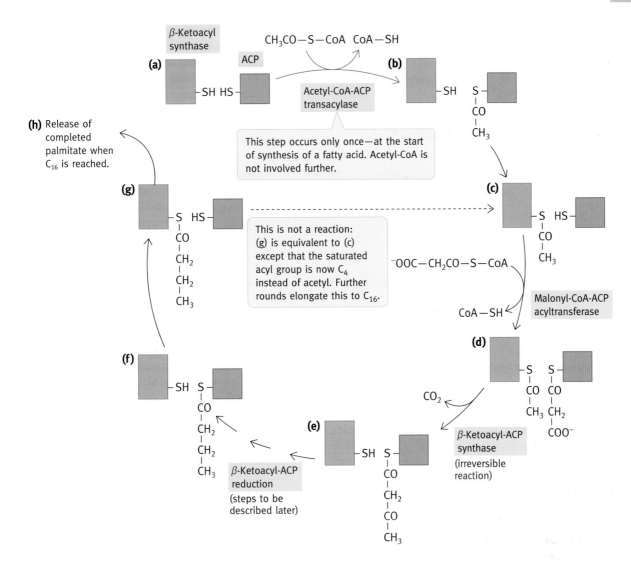

Fig. 14.1 The steps involved in the synthesis of fatty acids. Although the β-ketoacyl synthase of animals is a domain of a single protein molecule, it is shown here as being separate (orange). This is because the functional enzyme unit is a dimer arranged head to tail and the transfers between the two —SH groups occur between the domains on the two dimer-constituent protein molecules. Note that the thiol of the ACP (blue) is on the phosphopantetheine moiety and that of the synthase is on a cysteine residue of the protein. After seven successive rounds of reactions, the resultant palmityl-ACP is hydrolysed to release free palmitate. ACP, acyl carrier protein.

P—ribose—nicotinamide
|
P—ribose—adenine.

NADP$^+$ is:

P—ribose—nicotinamide
|
P—ribose—adenine
|
P.

The extra phosphoryl group is attached to the 2′ hydroxyl group of the ribose which is attached to adenine.

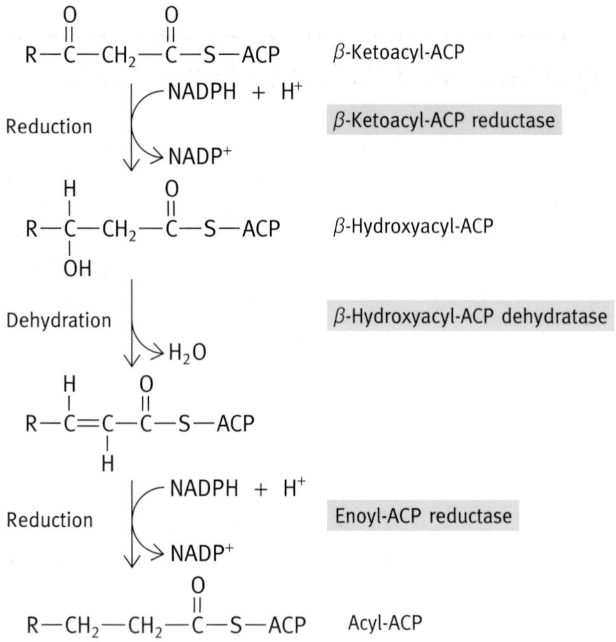

$$R—\overset{\overset{\displaystyle O}{||}}{C}—CH_2—\overset{\overset{\displaystyle O}{||}}{C}—S—ACP \qquad \textit{β-Ketoacyl-ACP}$$

Reduction — NADPH + H⁺ → NADP⁺   β-Ketoacyl-ACP reductase

$$R—\overset{\overset{\displaystyle H}{|}}{\underset{\underset{\displaystyle OH}{|}}{C}}—CH_2—\overset{\overset{\displaystyle O}{||}}{C}—S—ACP \qquad \textit{β-Hydroxyacyl-ACP}$$

Dehydration → H₂O   β-Hydroxyacyl-ACP dehydratase

$$R—\overset{\overset{\displaystyle H}{|}}{C}=\overset{\overset{\displaystyle}{}}{\underset{\underset{\displaystyle H}{|}}{C}}—\overset{\overset{\displaystyle O}{||}}{C}—S—ACP$$

Reduction — NADPH + H⁺ → NADP⁺   Enoyl-ACP reductase

$$R—CH_2—CH_2—\overset{\overset{\displaystyle O}{||}}{C}—S—ACP \qquad \text{Acyl-ACP}$$

Fig. 14.2 Reductive steps in fatty acid synthesis.

The extra phosphoryl group has no effect on the redox characteristics of the molecule – it is, in fact, purely an identification or recognition signal. An NAD⁺ enzyme will not react with NADP⁺ and *vice versa* (with the odd exception of little significance).

For a long time it was thought that this was just an odd quirk of nature, no more than a curiosity. However, eventually it was realized that the use of the two coenzymes constitutes an important form of **metabolic compartmentation**. To understand this, remember that in energy release the aim is to oxidize reducing equivalents, while in, for example, fat synthesis the aim is to use reducing equivalents for reductive synthesis. The aims are diametrically opposite. The cell keeps these metabolic activities separate by metabolic compartmentation – one, the oxidative part, uses NADH as electron carrier; the other, the reductive part, uses NADPH.

Separate mechanisms exist to reduce NAD⁺ and NADP⁺. You already know how NAD⁺ is reduced in glycolysis, the pyruvate dehydrogenase reaction, the citric acid cycle, and fat oxidation. How is NADP⁺ reduced? This is explained in the next section.

Fatty acid synthesis takes place in the cytoplasm

In terms of tissues, the main sites of fatty acid synthesis in the body are the liver and adipose cells but many other tissues also produce fat. Adipose cells are the main site of storage.

Within a cell, **palmitate synthesis** from acetyl-CoA occurs in the cytosol. This is in contrast to **fatty acid oxidation** which occurs in the **mitochondria**. The major source of acetyl-CoA

for fat synthesis is the pyruvate dehydrogenase reaction (page 193), which is located in the mitochondrial matrix. Acetyl-CoA cannot cross the mitochondrial membrane to the site of fatty acid synthesis in the cytosol so acetyl residues must be transported from mitochondria to the cytosol. What is done is part of an ingenious scheme. The acetyl-CoA in the mitochondrion is converted to citrate by the citric acid synthase reaction of the citric acid cycle. The citrate is transported by a mitochondrial membrane system into the cytosol where it is cleaved back to acetyl-CoA and oxaloacetate by a separate enzyme, called **ATP-citrate lyase** or the **citrate cleavage enzyme**. This reaction is coupled to hydrolysis of ATP to ADP and inorganic phosphate (P_i), thus ensuring it goes to irreversible completion. (Remember that the citrate-synthesizing reaction in mitochondria is irreversible, so a different enzyme is needed for its cleavage.)

$$Citrate + ATP + CoA—SH + H_2O \rightarrow acetyl—CoA$$
$$+ oxaloacetate + ADP + P_i$$

That is not the end of the story. The oxaloacetate cannot get back into the mitochondrion, the membrane of which is impervious to it, and for which no transporter exists. It is reduced in the cytoplasm to malate by NADH (note, not NADPH); the malate so formed is oxidatively decarboxylated to pyruvate and CO_2 by an enzyme (the '**malic**' enzyme) that uses NADP⁺ (note, *not* NAD⁺), thus producing NADPH, which is needed for fat synthesis (Fig. 14.3). The pyruvate is now transported back into the mitochondrion (Fig. 14.4). The pyruvate transported into the mitochondria can be converted back to oxaloacetate by the **pyruvate carboxylase reaction**.

$$Pyruvate + HCO_3^- + ATP$$

(Pyruvate carboxylase)

$$Oxaloacetate + ADP + P_i + H^+$$

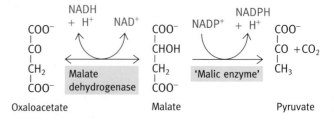

Fig. 14.3 Reduction of NADP⁺ for fatty acid synthesis. The net effect of the two reactions is to transfer reducing equivalents from NADH to NADP⁺. The pyruvate so produced in the cytoplasm enters the mitochondrion. The source of the oxaloacetate is shown in Fig. 14.4.

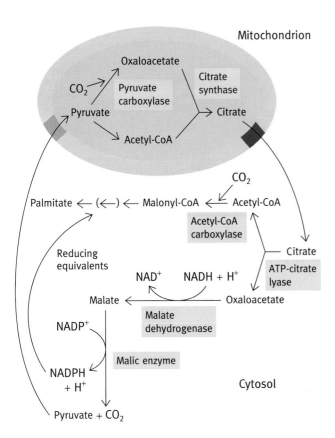

Fig. 14.4 Source of acetyl groups (acetyl-CoA) and reducing equivalents (NADPH + H$^+$) for fatty acid synthesis. The ATP-citrate lyase reaction involves the breakdown of ATP to ADP and P$_i$. This ensures complete breakdown of citrate.

Synthesis of unsaturated fatty acids

As explained, the body requires unsaturated fatty acids for the production of polar lipids for membrane synthesis to achieve lipid bilayer fluidity, and they also have other roles. A separate enzyme system in liver can introduce a single double bond into the middle of stearic acid, generating oleic acid. Δ^9 indicates that the double bond is between carbon atoms 9 and 10 of the fatty acid, with the carboxyl carbon being number one.

$$CH_3(CH_2)_7CH=CH(CH_2)_7COOH$$
Oleic acid, $18:1(\Delta^9)$

However, the system cannot make double bonds between this central double bond and the methyl end of the molecule. Linoleic acid, which has two double bonds, and linolenic acid with three, thus cannot be synthesized by the body.

$$CH_3(CH_2)_4CH=CH\ CH_2\ CH_2=CH(CH_2)_7COOH$$
Linoleic acid, $18:2(\Delta^{9,12})$

$$CH_3\ CH_2\ CH=CH\ CH_2\ CH=CH\ CH_2\ CH=CH(CH_2)_7\ COOH$$
α-Linolenic acid, $18:3(\Delta^{9,12,15})$

Since they are needed for membrane components and for synthesis of eicosanoid regulatory molecules (page 232) these two acids must be obtained from the diet (they are known as omega fatty acids – see Box 14.1); plants have enzymes that can desaturate in the terminal half of the fatty acids. The liver can elongate linoleic acid and introduce an extra double bond to give the C$_{20}$ arachidonic acid (20:4, $\Delta^{5,8,11,14}$; see below) with

Citrate leaves the mitochondrion only when it is at a high concentration; this occurs when carbohydrate is plentiful. Citrate does not occur in the cytosol at other times.

Thus, the citrate mechanism not only transports acetyl groups out of the mitochondrion, it also generates NADPH for fat synthesis. The reduction in the cytoplasm of oxaloacetate to malate by NADH and the oxidation of malate to pyruvate by NADP$^+$ constitutes a neat mechanism for transferring electrons from the NADH metabolic 'compartment' into the NADPH 'compartment' used for reductive synthesis reactions. In addition, citrate activates the first reaction committed to fat synthesis – the acetyl-CoA carboxylase, producing malonyl-CoA. Control of this enzyme is discussed in Chapter 16, page 259.

For every acetyl-CoA molecule produced in the cytosol from citrate, we generate *one* NADPH molecule. However, the formation of each —CH$_2$—CH$_2$— group from —CH$_3$CO— by palmitate synthase requires *two* NADPH molecules (see Fig. 14.2). Yet another mechanism for producing the extra NADPH is required – this will be explained in Chapter 17 for it belongs to a different metabolic topic, the pentose phosphate pathway.

> **BOX 14.1 Omega fatty acids and diet**
>
> Another system of nomenclature for unsaturated fatty acids designates the carbon of the end methyl group as ω (omega), and numbers the unsaturated bonds of the carbon chain from there. So linolenic acid is an ω-3 fatty acid, indicating a double bond on the third carbon counting from the ω carbon atom, and linoleic is an ω-6 fatty acid.
>
> Since mammals cannot make double bonds between Δ^9 and the methyl end of the molecule, fatty acids such as linolenic and linoleic are referred to as essential fatty acids and need to be supplied in the diet. ω-3 and ω-6 acids such as these two are found in certain plant seeds, such as linseed and canola, while other members of the ω-3 and ω-6 families with longer carbon chains are plentiful in fish oils such as those of cod and salmon.
>
> The potential for the reduction of serum cholesterol levels by increasing the ratio of polyunsaturated fatty acids to saturated fatty acids in the diet is now well known, as is the correlation between high levels of cholesterol and coronary heart disease.

four double bonds. It can also elongate acids to the C_{22} and C_{24} acids involved in the lipids of nerve tissues. All of these transformations occur as the CoA derivatives.

Synthesis of TAG and membrane lipids from fatty acids

TAG is synthesized for storage purposes. For attachment of a fatty acid in ester bond to glycerol, the acid must be activated to the form of acyl-CoA, a reaction catalysed by acyl-CoA synthetase, as already described.

$$RCOOH + CoA—SH + ATP \rightarrow RCO—S—CoA + AMP + PP_i$$

A carboxylic acid ester has a lower free energy of hydrolysis than has a thiol ester and hence activation of the acid makes TAG synthesis from the fatty acyl-CoA exergonic. Surprisingly perhaps, the acceptor of acyl groups is not glycerol but glycerol-3-phosphate, which arises mainly by reduction of the glycolytic intermediate, dihydroxyacetone phosphate (DHAP; see Fig. 12.28). Adipocytes require glucose therefore to provide this for TAG synthesis. Phosphorylation of glycerol can occur in the liver but not in adipose cells, a point of significance. The reaction is:

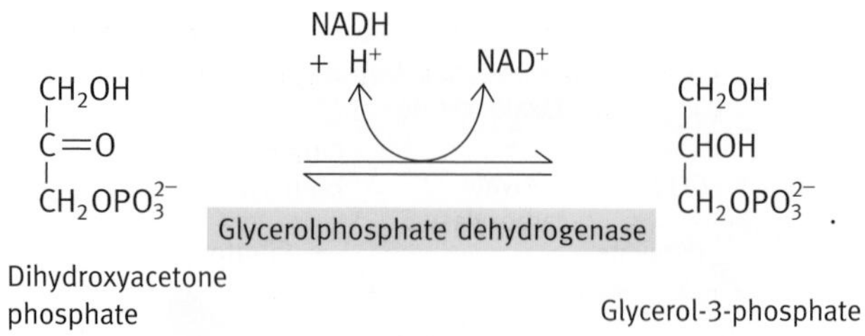

The steps in TAG synthesis are shown in Fig. 14.5.

Synthesis of new membrane lipid bilayer

There are two aspects in the synthesis of glycerophospholipids, the major components of membranes. First, there are the metabolic reactions by which their synthesis is achieved, which are largely understood. However, there is a different, perhaps more interesting, problem of how these result in the formation of new membrane. In a sense membrane synthesis poses a unique problem. There are many different types of membrane in a eukaryotic cell that have to be increased along with cell growth and division. Membrane synthesis occurs by producing new phospholipids *in situ* in pre-existing membranes. In this

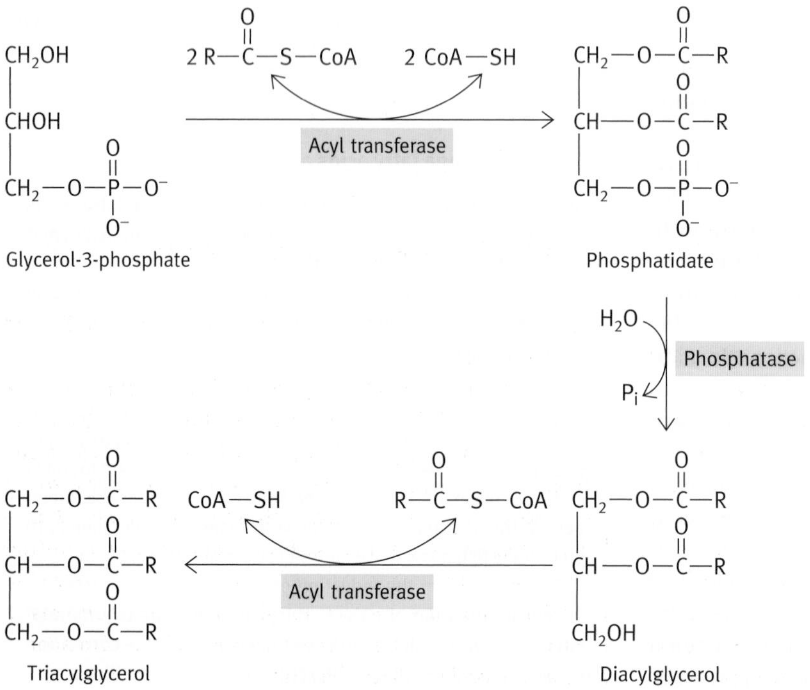

Fig. 14.5 Reactions involved in the synthesis of triacylglycerol (neutral fat) from glycerol-3-phosphate.

section we will first describe the metabolic routes of synthesis that are necessary before we can then discuss the second problem of membrane synthesis.

Synthesis of glycerophospholipids

If you turn back to page 103 you will see that membranes contain glycerol-based phospholipids in which a polar alcohol is esterified to the phosphoryl group of phosphatidic acid.

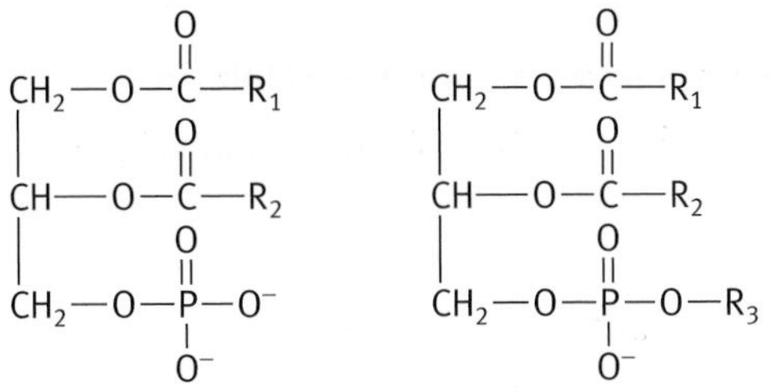

Phosphatidate   Phospholipid

R_3 may be serine, ethanolamine, choline, inositol, or diacylglycerol. The most prevalent membrane lipids in eukaryotes are phosphatidylethanolamine and phosphatidylcholine so we will consider their synthesis first; ethanolamine is $NH_2CH_2CH_2OH$ and choline is:

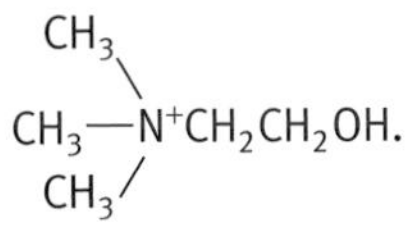

Both can be written as the alcohols 'R—OH'. The alcohols are 'activated' to participate in the synthesis; this occurs in two steps. First, a phosphorylation by ATP:

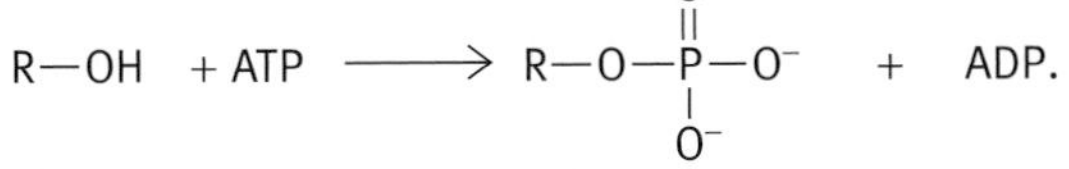

Secondly, a reaction with **cytidine triphosphate (CTP)**. CTP is the same as ATP except that it has the base cytosine (C) in place of adenine (cytosine-ribose is called cytidine). We will come to its precise structure later in the book but it is not needed here. All cells have CTP for it is needed for nucleic acid synthesis (Chapters 22 and 23). Whether CTP involvement in membrane lipid synthesis rather than ATP is due to an accidental quirk of evolution or whether there is a good reason for it is unknown, but the fact is that CDP compounds are always the 'high-energy' donors of groups in this area (just as for equally unknown reasons it is always UDP-glucose for processes such

as glycogen synthesis in animals). The reaction is, in fact, very reminiscent of UDP-glucose formation from glucose-1-phosphate and UTP (page 163).

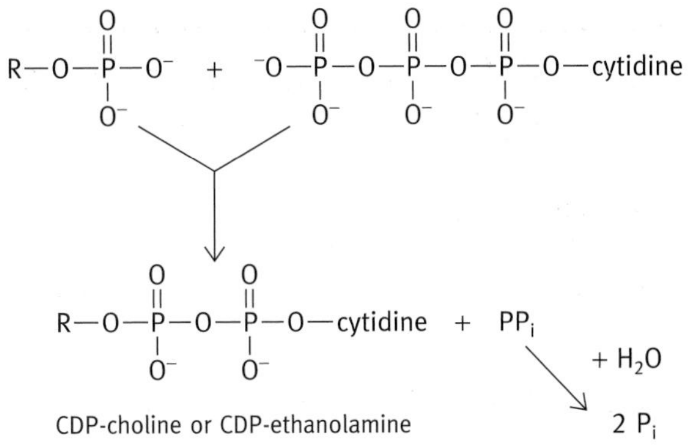

CDP-choline or CDP-ethanolamine   2 P_i

Hydrolysis of inorganic pyrophosphate (PP_i) to P_i makes the reaction strongly exergonic.

The final reaction in phosphatidylethanolamine and phosphatidylcholine synthesis is as follows:

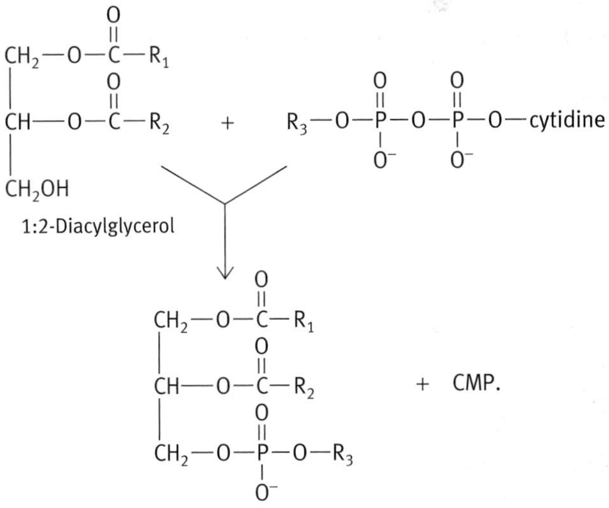

Phosphatidylcholine or phosphatidylethanolamine

The diacylglycerol comes from hydrolysis of phosphatidic acid by a phosphatase.

In the reactions above we join an alcohol (ethanolamine or choline) to CDP and then transfer phosphoryl alcohol to another alcohol (diacylglycerol); the head group is 'activated'. Energetically, it would be the same to join diacylglycerol to CDP and transfer a diacylglycerol phosphoryl group to ethanolamine or choline – that is, to activate the diacylglycerol instead of the polar alcohol or head group. This is what happens in phosphatidylinositol and cardiolipin (page 103) synthesis in eukaryotes. In the latter, instead of inositol, the reactant is another molecule of diacylglycerol:

$$CH_2-O-\overset{\overset{O}{\parallel}}{C}-R_1$$
$$CH-O-\overset{\overset{O}{\parallel}}{C}-R_2 + CTP \rightarrow$$
$$CH_2-O-\overset{\overset{O^-}{\mid}}{\underset{O^-}{P}}-O^-$$

Phosphatidate

$$CH_2-O-\overset{\overset{O}{\parallel}}{C}-R_1$$
$$CH-O-\overset{\overset{O}{\parallel}}{C}-R_2 \qquad + PP_i$$
$$CH_2-O-\overset{\overset{O}{\parallel}}{\underset{O^-}{P}}-O-\overset{\overset{O}{\parallel}}{\underset{O^-}{P}}-O-cytidine$$

+ Inositol

$$CH_2-O-\overset{\overset{O}{\parallel}}{C}-R_1$$
$$CH-O-\overset{\overset{O}{\parallel}}{C}-R_2 \quad + \quad CMP$$
$$CH_2-O-\overset{\overset{O}{\parallel}}{\underset{O^-}{P}}-O-inositol$$

Phosphatidylinositol

The situation is summarized in the scheme in Fig. 14.6. This scheme illustrates the principle of the two ways in which glycerophospholids can be synthesized. It should be emphasized that the field is a complex one in which extensive

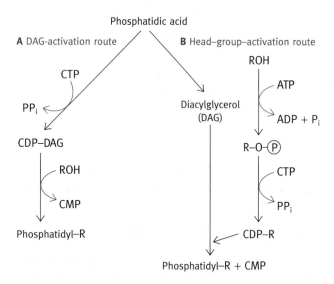

Fig. 14.6 The two pathways (A and B) for synthesis of glycerophospholipid. (ROH, the polar head group to be attached to the phospholipid.) Different cells may use different routes for synthesizing a given phospholipid. In mammals, phosphatidylcholine (PC), phosphatidylserine (PS), and phosphatidylethanolamine (PE) are synthesized by route B; the three are interconvertible by decarboxylation of PS to PE, methylation of PE to PC, and head-group exchange between PE and PS and free ethanolamine or serine. Synthesis of cardiolipin and phosphatidylinositol in mammals follows route A. In *Escherichia coli* phospholipid synthesis occurs by route A.

interconversion of the phospholipids occurs. For example, serine and ethanolamine can be exchanged, phosphatidylethanolamine can be methylated to phosphatidylcholine, phosphatidylserine can be decarboxylated to phosphatidylethanolamine. Phosphatidylcholine can be converted to any of the others by head-group exchange. The particular route of synthesis of a given phospholipid may vary in different organisms.

Sphingolipids (page 104) are produced from palmitoyl-CoA and serine by a complex set of reactions, which will not be given here.

Synthesis of new membrane lipid bilayer

In eukaryotic cells, the main site of phospholipid synthesis is the membrane of the smooth endoplasmic reticulum (ER). It also takes place in the outer mitochondrial membrane. Fatty acyl-CoAs and glycerophosphate in the cytoplasm are converted to phosphatidic acid by enzymes located on the cytoplasmic surface of the ER. The phosphatidate is converted to phosphatidyl derivatives of serine, choline, ethanolamine, and inositol as described earlier, again by enzymes on the cytoplasmic surface of the ER. The newly synthesized phospholipids are inserted into the outer leaflet of the lipid bilayer of the ER membrane (possibly at the fatty acyl-CoA stage) and the synthesis completed *in situ*. As stated above, extensive interconversion of the phospholipid head groups results in the formation of the various membrane components. Cells cannot produce new membranes *de novo*. They can only increase existing membranes which enlarge their structures. At cell division, each daughter cell receives half.

In this way, new lipid membrane is synthesized. The inner and outer layers of a bilayer are different in phospholipid composition (page 105). The newly synthesized phospholipids are in the outer leaflet of the ER lipid bilayer, and there has to be a transfer of phospholipids to the inner leaflet to maintain the balance (Fig. 14.7). Simple flipping of phospholipids from the outer to the inner layer would involve the polar head group traversing the hydrophobic centre of the bilayer, which is energetically unfavourable. A family of phospholipid-transporting enzymes, or so-called flippases, are known. They are driven by ATP hydrolysis. It is not fully clear how the asymmetries in different membranes are achieved and maintained.

There is another problem too. The new membrane synthesized in the ER has to be transported to other sites such as the plasma membrane for cell growth and division to occur. There are two possible mechanisms for this; first, membrane vesicles are budded off the ER and these migrate to their target membrane and fuse with it thus adding new membrane to say the plasma membrane. Phospholipid-transfer proteins are known which pick up phospholipids from one membrane and transport them to another membrane, and this is the second possible way in which newly synthesized phospholipids may be delivered to target membranes. Membrane synthesis is not completely understood.

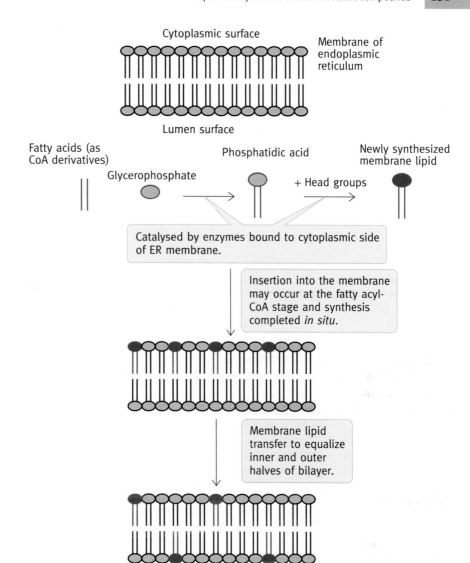

Fig. 14.7 Synthesis of new membrane lipid bilayer. Enzymes which synthesize new membrane lipids are located at the cytoplasmic surface of the endoplasmic reticulum (ER); the newly synthesized lipids are located in the cytoplasmic half of the bilayer. Transfer to the inner half to maintain bilayer equality is by a 'flippase'. The new membrane is transported from the ER to other membranes in the cell in vesicle form or *via* phospholipid transport proteins.

Synthesis of prostaglandins and related compounds

The Greek word *eikosi* means 20; this is the basis for the name of a group of compounds called **eicosanoids**, all of which contain 20 carbon atoms and are related to polyunsaturated fatty acids. Although present in the body at very low levels, they have a wide range of physiological functions.

There are three main groups named after the cells in which they were first discovered. These are the **prostaglandins**, the **thromboxanes**, and the **leukotrienes**. Prostaglandins were first discovered in semen and so named because it was thought that they originated from the prostate gland (actually it is from the seminal vesicles). However, it is now known that very many tissues produce prostaglandins. Thromboxanes were discovered in blood platelets or thrombocytes and leukotrienes in leucocytes.

The prostaglandins and thromboxanes

A number of different prostaglandins exist, varying in their detailed structures, with subclasses such as PGE, and PGF. A subscript number indicates the number of double bonds in the side chains attached to the cyclopentane ring structure. Fig. 14.8(b) shows the structure of PGE_2 as an example. In humans the compounds with two double bonds are most important and are synthesized from arachidonic acid, the structure of which is shown in Fig. 14.8(a). Other prostaglandins are derived from related unsaturated fatty acids. Arachidonic acid is present in cells, mainly as the fatty acid component of a phospholipid. For prostaglandin synthesis it is released by a phospholipase.

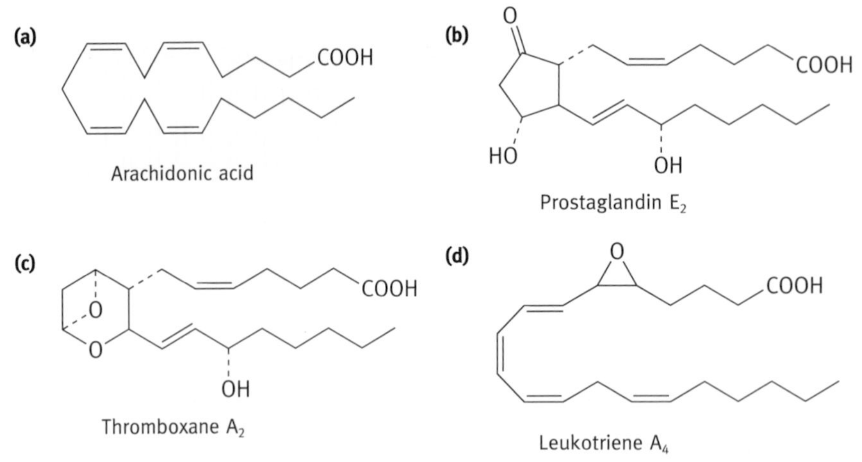

(a)

Arachidonic acid

(b)

Prostaglandin E$_2$

(c)

Thromboxane A$_2$

(d)

Leukotriene A$_4$

Fig. 14.8 Structure of prostaglandins and related compounds: **(a)** Arachidonic acid; **(b)** prostaglandin E$_2$ (PGE$_2$); **(c)** thromboxane A$_2$ (TXA$_2$); **(d)** leukotriene A$_4$ (LTA$_4$). Probably only those specially interested in this area would need to memorize these structures.

BOX 14.2 **Nonsteroidal anti-inflammatory drugs (NSAIDs)**

Pain relief is an area of major unmet need and drugs to achieve pain control are of extreme interest. One of the oldest drugs and still much used is aspirin; as stated in the text, this inhibits the cyclooxygenase (COX) reaction necessary for the conversion of arachidonic acid to the family of prostaglandins.

Prostaglandins have a wide range of protective physiological effects including protection of the mucosal membrane lining of the upper gastrointestinal tract and for maintenance of kidney function.

Prostaglandins also cause pain, inflammation, and fever. As well as their normal protective roles, they are formed and liberated at sites of tissue and cell damage. In the arthritic diseases where cell damage at joints result in their production, prostaglandins contribute to the joint pains from which millions of people suffer. There is great interest, therefore, in drugs that inhibit prostaglandin synthesis.

Besides aspirin, many similarly acting drugs, such as ibuprofen, have been produced as anti-inflammatory pain-killing agents. They are about the most commonly used group of drugs and are collectively known as nonsteroidal anti-inflammatory drugs or NSAIDs. These are in contrast to the steroidal drugs such as cortisone and synthetic glucocorticoids (mentioned in the Box 27.1 on page 434).

Unfortunately the NSAIDs can have adverse effects because, as well as suppressing the synthesis of the pain-producing prostaglandins, they also inhibit the normal production of protective prostaglandins. One of the best-known adverse effects is irritation of the lining of the upper gastrointestinal tract leading to bleeding and formation of ulcers. This is especially important in the treatment of rheumatoid arthritis, where high NSAID doses for a prolonged period may be necessary.

There was therefore tremendous interest in the development of a new class of NSAIDs, known as **COX-2 inhibitors**, which selectively inhibit the production of the inflammatory pain-producing prostaglandins at the site of cell damage. We will now explain the basis of this development, summarized in Fig. 14.9.

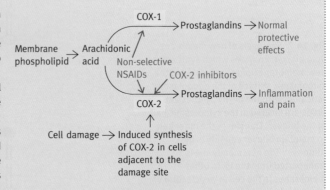

Fig. 14.9 Action of cyclooxygenase isoenzymes and effects of selective inhibition.

The COX that is involved in prostaglandin synthesis occurs in two isoforms (isoenzymes). **COX-1** is present in almost all cells and is needed for the normal production of prostaglandins. The other form, **COX-2** is normally present in cells in small or zero amounts. At the sites of cell damage, adjacent cells are induced to synthesize COX-2 and this results in increased synthesis of prostaglandins causing inflammation and pain.

The 'older' nonselective NSAIDs, such as aspirin, inhibit both COX-1 and COX-2 effectively but, as stated above, may cause the damage to the gastrointestinal mucosal membrane already referred to. COX-2 inhibitors selectively inhibit COX-2 with lesser effects on COX-1 and so while they are effective painkillers, they have reduced undesirable gastrointestinal side effects. However, recent data show that COX-2 selective drugs may cause cardiovascular side effects such as heart attacks or strokes, possibly through inhibition of the synthesis of protective prostacyclins (PGI, a prostaglandin subclass) within the lining of blood vessels. As a result, the use of some of the COX-2 inhibtors has now been severely curtailed.

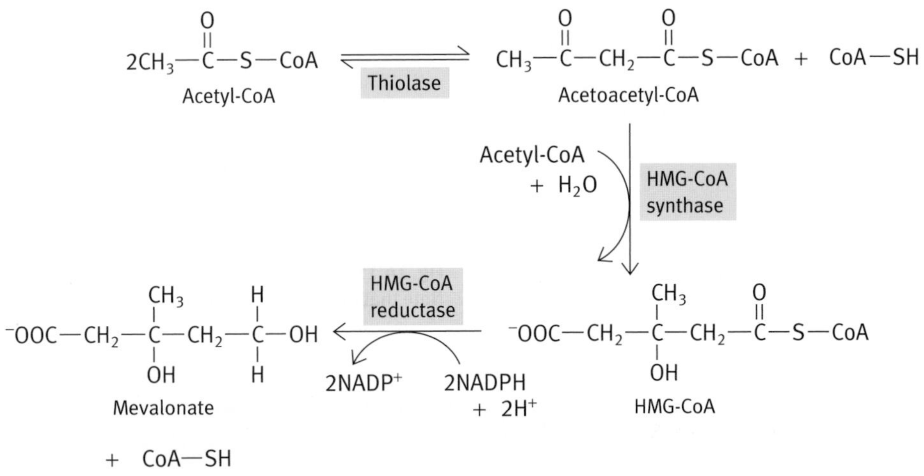

Fig. 14.10 The synthesis of mevalonate from acetyl-CoA. HMG-CoA, 3-hydroxy-3-methylglutaryl-CoA.

The first step in synthesizing prostaglandins is carried out by a **cyclooxygenase**, which forms the ring structure from arachidonic acid (Fig. 14.8(a)). This enzyme is inhibited by **aspirin** (acetylsalicylic acid), which covalently modifies the enzyme by acetylating an essential serine group near the active site of the enzyme.

The prostaglandins have a wide variety of physiological effects. They are released immediately after synthesis and act as local hormones on adjacent cells by combining with receptors. They cause pain, inflammation, and fever; they cause smooth muscle contraction and are involved in labour; they are involved in blood pressure control; they suppress acid secretion in the stomach. Aspirin, by its inhibition of cyclooxygenase (Box 14.2), suppresses many of these effects.

Thromboxanes (Fig. 14.8(c)) affect platelet aggregation and thus blood clotting. They are formed by further conversion of certain of the prostaglandins, so that aspirin inhibits their formation also; in particular, a low dose (100 mg per day) inhibits platelet thromboxane formation resulting in decreased platelet aggregation. This therapy has been found to reduce the risk of heart attacks by reducing the danger of blood-clot formation blocking the coronary arteries.

Leukotrienes

The structure of one of the leukotrienes is given in Fig. 14.8(d). Leukotrienes arise from arachidonic acid by a different route involving a lipooxygenase reaction. They cause protracted contraction of smooth muscle, are a factor in asthma symptoms by constricting airways, and play a role in white cell function.

Synthesis of cholesterol

Cholesterol is an essential component of cell membranes and the precursor of bile salts and steroid hormones. The liver

and, to a lesser extent, the intestine are the most active in its synthesis and from the liver a two-way flux 'equilibrates' the cholesterol of the liver and rest of the body, as described in Chapter 10, page 171.

Remarkably, the sole starting material for synthesis of this large molecule is acetyl-CoA from which, by a long metabolic pathway, cholesterol is produced. The first few stages of the synthesis are of special interest for this is where control of cholesterol occurs and we will confine this account to the production of mevalonic acid, the first metabolite committed solely to cholesterol synthesis. This involves the reactions shown in Fig. 14.10, which occur in the cytosol.

HMG-CoA (3-hydroxy-3-methylglutaryl-CoA) is also the precursor of acetoacetate, one of the ketone bodies, but for this purpose it is synthesized inside the mitochondrion.

The regulation of cholesterol levels in cells and the 'statin' drugs have already been described (see page 174).

Conversion of cholesterol to steroid hormones

The steroid hormones are, as well known, important signalling molecules. The sex hormones such as testosterone and oestradiol are involved in determining secondary sex characteristics. They are produced in the gonads. The adrenal cortex produces two main classes of steroid:

- The **glucocorticoids** have several effects, among which are control on carbohydrate metabolism including gluconeogenesis.
- The **mineralocorticoids** are responsible for maintaining ion balance in the body.

The mode of action of hormones is a major topic dealt with in Chapter 27. All steroid hormones are derived from cholesterol. The pathways of production of some of the main ones are shown in Fig. 14.11. The conversion of cholesterol into steroids involves cleavage of the side chain, a reaction in which cytochrome P450 participates, producing pregnenolone, which is the common precursor of all steroids. The structures of testosterone and progesterone are shown in Fig. 27.3.

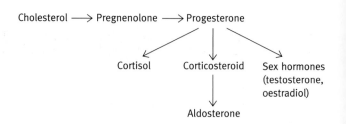

Fig. 14.11 Outline of cholesterol conversion to steroid hormones. Note that individual conversions involve more than single steps.

Summary

Fatty acid synthesis occurs in the cytosol using acetyl-CoA as the starting substrate. A multienzyme complex called fatty acid synthase (or palmitate synthase) produces palmitate from this. The reductant in the process is NADPH (not NADH). Further lengthening of the palmitate is catalysed by a separate enzyme system on the cytosolic face of the endoplasmic reticulum.

The acetyl-CoA from which fatty acids are synthesized is produced from pyruvate inside the mitochondrial matrix. It is not directly transported out to the cytosolic site of synthesis but by an ingenious mechanism. In the matrix, acetyl-CoA is converted to citrate by the (irreversible) enzyme citrate synthase (the first step in the citric acid cycle). For fatty acid synthesis, citrate is transported out into the cytosol. There, an ATP-citrate lyase releases acetyl-CoA and oxaloacetate. This reaction depends on the hydrolysis of ATP making it irreversible also. NADH is used to reduce the oxaloacetate to malate. Malate is converted to pyruvate by the malic enzyme, which reduces $NADP^+$ to NADPH in the process. NADPH is involved in reductive synthesis in the cell whereas NADH is involved in oxidative processes. Thus the system supplies both acetyl-CoA and the reductant.

Fatty acids are synthesized by the fatty acid synthase cycle. Oddly enough, the donor of C2 units in the process is not acetyl CoA but the C3 molecule malonyl-CoA. This is formed by carboxylation of acetyl-CoA. The CO_2 is released in the donor reaction;

its purpose is to render fatty acid synthesis irreversible. TAGs are synthesized from glycerophosphate and fatty acyl-CoAs.

Membrane lipids are synthesized on the endoplasmic reticulum membrane. Unsaturated fatty acids are produced by separate enzyme systems and used to synthesize prostaglandins and related compounds. Prostaglandins are involved in pain, inflammation and fever. They cause contraction of smooth muscle and have other physiological actions. Prostaglandins and thromboxanes are synthesised from arachidonic acid, with the first step being the formation of a ring structure by cyclooxygenase. Aspirin (acetylsalicylic acid) is a potent inhibitor. It does this by acetylating an essential serine in the enzyme. Thromboxanes are involved in platelet aggregation and blood clotting. Aspirin has a protective action in reducing blood clotting and the blocking of coronary arteries. Leukotrienes involved in smooth muscle contraction in airways and in white cell function are made from arachidonic acid by a different route.

Cholesterol is also synthesized from acetyl-CoA; the first committed step is the production of mevalonate by the enzyme HMG-CoA reductase. This is a crucial control point and is the target for the cholesterol-lowering statin drugs. Mevalonate is converted to cholesterol by an involved pathway (not described here). Cholesterol is the precursor of steroid hormones.

Further reading

Fatty acid synthase

Smith, S. (1994). The animal fatty acid synthase: one gene, one polypeptide, seven enzymes. *FASEB J.*, **8**, 1248–59.
Illustrated article on the structure and function of this remarkable protein.

Membrane lipid synthesis

Dawidowicz, E. A. (1987). Dynamics of membrane lipid metabolism and turnover. *Annu. Rev. Biochem.*, **56**, 43–61.
Review on membrane lipids, their site of synthesis, transport, and assembly into membranes.

Rooney, S. A., Young, S. L., and Mendelson, C. R. (1994). Molecular and cellular processing of lung surfactant. *FASEB J.*, **8**, 957–67.
Reviews a more physiological role of phospholipid – that of lining the alveoli.

Kent, C. (1995). Eukaryotic phospholipid biosynthesis. *Annu. Rev. Biochem.*, **64**, 315–43.
An account of the various pathways for the synthesis of the membrane phospholipids.

Kent, C. and Carman, G. M. (1999). Interactions among pathways for phosphatidylcholine metabolism, CTP synthesis and secretion through the Golgi apparatus. *Trends Biochem. Sci.*, **24**, 146–50.
Prostaglandins, thromboxanes, and cyclooxygenases.

Evans, J. F. and Kargman, S. L. (2004). Cancer and cyclooxygenase-2 (COX-2) inhibition. *Curr. Pharm. Des.*, **10**, 627–34.

Foral, P. A., *et al.* (2003). Gastrointestinal-related adverse effects of COX-2 inhibitors. *Drugs Today*, **39**, 939–48.

Fitzpatrick, F. A. (2004). Cyclooxygenase enzymes: regulation and function. *Curr. Pharm. Des.*, **10**, 577–88.

Garaj, N. M. (2003). Cyclooxygenase-2 inhibitors. *Anesth. Analg.*, **96**, 1720–38.

Ramalingam, S. and Belani, C. P. (2004). Cyclooxygenase-2 inhibitors in lung cancer. *Clin. Lung Cancer*, **5**, 245–33.

Wong, D., Wang, M., Cheng, Y., and FitzGerald, G. A. (2005). Cardiovascular hazard and non-steroidal anti-inflammatory drugs. *Current Opinion in Pharmacol.*, **5**, 204–10.

Smith, W. L. (2008). Nutritionally essential fatty acids and biologically indispensable cyclooxygenases. *Trends Biochem. Sci.*, **33**, 27–37.

Problems

1. An early step in fatty acid synthesis is that acetyl-CoA is carboxylated, to be followed immediately by decarboxylation of the product. What is the point of this?

2. By means of a diagram, outline the steps in a cycle of elongation during fatty acid synthesis, omitting the reductive steps.

3. Discuss briefly the physical organization of the fatty acyl synthase complex in eukaryotes. How does this differ from that in *Escherichia coli*? What is the advantage of the eukaryotic situation?

4. Write down the structures of NAD^+ and $NADP^+$. Explain the reason for the existence of both NAD^+ and $NADP^+$.

5. What are the main tissues in which fatty acid synthesis occurs in animals?

6. Palmitate synthesis from acetyl-CoA occurs in the cytoplasm. However, pyruvate dehydrogenase, the main producer of acetyl-CoA used in fat synthesis, occurs inside the mitochondrion. Acetyl-CoA cannot traverse the mitochondrial membrane. How does the fatty acid-synthesizing system in the cytoplasm obtain its supply of acetyl-CoA?

7. The synthesis of fatty acid involves reductive steps requiring NADPH. What are the sources of the latter?

8. Describe the synthesis of triacylglycerol (TAG) from fatty acids.

9. What is the role of CTP in lipid metabolism?

10. (a) What are eicosanoids?
    (b) What are they made from?
    (c) Briefly describe their physiological significance.
    (d) What is the relevance of aspirin in this area of metabolism?

11. Describe a type of drug that is used to reduce the rate of cholesterol synthesis.

Chapter **15**

Synthesis of glucose (gluconeogenesis)

The control of gluconeogenesis is given on page 255 in the next chapter.

The body is able to synthesize glucose from any compound capable of being converted to pyruvate (or oxaloacetate), by the process of gluconeogenesis. This excludes fatty acids and acetyl-CoA but includes lactate (which is released into the blood by red blood cells and by muscle during vigorous exercise) as well as most of the amino acids, so that a variety of compounds can be converted to glucose and stored as glycogen. However, in addition to this 'routine' metabolic role, gluconeogenesis can literally make the difference between life and death, despite the fact the glucose is such a common molecule prevalent in most diets. In case you have forgotten its significance, in starvation it assumes an overriding importance (page 155).

The brain, as stated several times earlier, requires a constant supply of glucose, which means that the blood glucose level must be kept within normal limits or coma and death will ensue. However, the liver's total store of glucose in the form of glycogen is limited and after about 24 hours without food the store is exhausted. Nevertheless, people do not die as a result of a day's starvation. Supply of blood sugar in this situation is dependent on the liver – no other organ, other than the quantitatively unimportant kidney, can fill this need. If fasting is prolonged beyond about 24 hours, the liver must therefore synthesize glucose. A minimum of 100 g per day is needed in humans.

Fat mobilization results in ketone body production by the liver, which can supply part of the brain's energy needs and therefore has a glucose-sparing effect (page 155) as the blood ketone level rises, but it cannot replace synthesis of glucose. Although the effects of glucose deprivation on brain are the most dramatic, red blood cells also are dependent on glycolysis of glucose for energy supply, since they have no mitochondria. The body normally has sufficient fat to supply energy for weeks of starvation, but fatty acids do not pass through the blood–brain barrier and cannot be used by the brain.

The main starting point for the gluconeogenesis pathway in liver is pyruvate although the glycerol moiety of triacylglycerol (TAG) is also used (page 239). To re-emphasize a crucial point, whereas pyruvate can give rise to acetyl-CoA and therefore lead to fat synthesis, the reverse cannot happen. Acetyl-CoA cannot be converted in animals to pyruvate and therefore fatty acids cannot be converted into glucose in a net sense. (It can in *Escherichia coli* and in plants, page 241.)

We will first describe the pathway by which pyruvate is converted into glucose and after that discuss the broader biochemical implications.

Mechanism of glucose synthesis from pyruvate

If you examine the glycolytic pathway (Fig. 12.7), there are three reactions that are irreversible because of thermodynamic considerations:

- phosphorylation of glucose by hexokinase (or glucokinase) using ATP
- phosphorylation of fructose-6-phosphate by phosphofructokinase
- conversion of phosphoenolpyruvate (PEP) to pyruvate.

Glucose is synthesized *via* the intermediate metabolites and reversal of reactions found in glycolysis, but a way has to be found to bypass the irreversible reactions.

The first thermodynamic barrier in the process is the conversion of pyruvate to PEP. Because the spontaneous conversion of the enol form of pyruvate to the keto form has a very large negative $\Delta G^{0'}$ value, the PEP $\rightarrow$ pyruvate reaction in glycolysis is irreversible in animal cells (there is no enol pyruvate substrate) but, as mentioned earlier, because of the convention of naming kinases after the reaction involving ATP, it is called pyruvate kinase nonetheless. In animals, a roundabout route involving two reactions is used for the conversion of pyruvate to PEP.

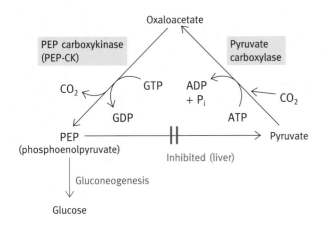

Fig. 15.1 The generation of phosphoenolpyruvate (PEP) for gluconeogenesis from pyruvate in the liver. Note that this scheme makes sense only if the PEP → pyruvate reaction is prevented. How this is done is described in Chapter 16 on metabolic control (page 258).

Two —$\textcircled{P}$ groups are used in the process, making the process thermodynamically favourable. The route is as follows:

(1) Pyruvate + ATP + HCO_3^- → Oxaloacetate + ADP + P_i + H^+

   (catalysed by pyruvate carboxylase)

(2) Oxaloacetate + GTP → PEP + GDP + CO_2

   (catalysed by PEP carboxykinase , or PEP-CK)

   Sum. Pyruvate + ATP + GTP + H_2O → PEP + ADP + GDP $+ P_i + 2H^+$

The scheme is shown diagrammatically in Fig. 15.1.

It is not known why GTP is used in the second reaction rather than ATP. Energetically it is just the same. Reaction 1, catalysed by pyruvate carboxylase, which synthesizes oxaloacetate, has an important role in topping up the citric acid cycle (page 198) quite separate from its role in gluconeogenesis. The second enzyme is usually referred to as **PEP-CK**. Its full name is **phospho-enolpyruvate carboxykinase** – because in the reverse direction it carboxylates PEP and transfers a phosphoryl group.

Pyruvate carboxylase occurs only in mitochondria, whereas PEP-CK occurs in both mitochondria and the cytosol. Since oxaloacetate cannot be made in the cytosol there is clearly a problem in converting pyruvate to glucose. There are two solutions, both of which operate. Either PEP is formed in mitochondria and transported out or oxaloacetate is reduced to malate and transported out. The malate is dehydrogenated to oxaloacetate and this has the advantage of supplying NADH for gluconeogenesis.

Once PEP is formed, the reactions of glycolysis are all reversible until fructose-1:6-bisphosphate is reached. The formation of this in glycolysis from fructose-1-phosphate is irreversible, but the step is bypassed by the simple device of hydrolysing off the phosphoryl group from the 1:6 compound as shown:

Fructose-1:6-bisphosphate

+ H_2O

Fructose-1:6-bisphosphatase

Fructose-6-phosphate

+ P_i

Similarly, when glucose-6-phosphate is reached, the glucokinase (or hexokinase) reaction of glycolysis is irreversible but, in liver, **glucose-6-phosphatase** produces glucose which is secreted from the cell. Muscle and adipose cells do not have this enzyme and cannot release glucose into the blood. The complete gluconeogenesis reactions are shown in Fig. 15.2. Note that whether at any given time glycolysis or gluconeogenesis is taking place depends on control mechanisms described in Chapter 16.

There are thus four enzymes involved in gluconeogenesis that do not participate in glycolysis: pyruvate carboxylase, PEP-CK, fructose-1:6-bisphosphatase, and glucose-6-phosphatase. The levels of these enzymes are about 20–50 times greater in rat liver than in rat skeletal muscle, in keeping with the importance of gluconeogenesis in liver.

What are the sources of pyruvate used by the liver for gluconeogenesis?

In starvation, after the glycogen reserves have been exhausted, the main source of pyruvate originates from the breakdown of muscle proteins. Muscle protein breakdown is promoted by the stress hormone **cortisol**, produced during starvation (page 157). Hydrolysis of muscle protein produces 20 amino acids. Although the amino acid **alanine** represents only a small proportion of proteins, of the total amino acids leaving the muscle to be transported to the liver, more than 30% is in the form of this amino acid (see below). Alanine is a **glucogenic amino acid** – its carbon skeleton is capable of being converted into glucose by the liver and the same is true of most of the other amino acids. (The metabolism of amino acids is separately discussed in Chapter 19, so do not be concerned with the mechanism of the reactions mentioned in this section.)

How is alanine formed from the protein-breakdown products in muscle? Several of the amino acids give rise to citric acid

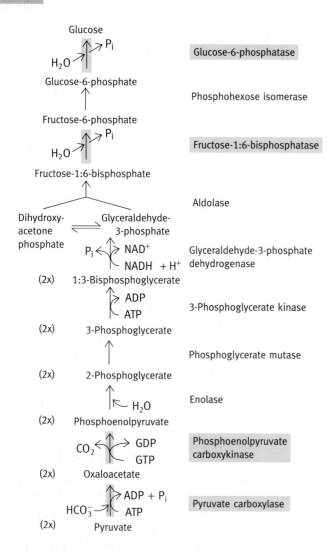

Fig. 15.2 The complete gluconeogenesis pathway from pyruvate to glucose. Reactions in yellow are different from those occurring in glycolysis.

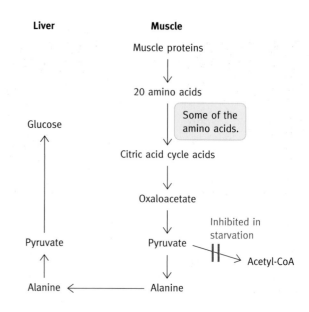

Fig. 15.3 Mechanism by which breakdown of muscle proteins supplies the liver with a source of pyruvate for gluconeogenesis during starvation. Note that the scheme is dependent on the pyruvate dehydrogenase not using the pyruvate (see text). Glutamine is also produced in muscle as a gluconeogenic substrate for the liver in addition to alanine.

cycle acids which are converted to oxaloacetate via reactions of the cycle. The oxaloacetate can be converted into pyruvate by the reactions shown in Fig. 15.1 and converted into alanine for release into the blood. The amino group of alanine comes from the other amino acids. Note, however, that this formation of alanine in muscle depends on the muscle not oxidizing the pyruvate to acetyl-CoA which, in starvation, would defeat the whole object of the exercise which is glucose synthesis. In the situation of starvation the muscle will be preferentially utilizing fatty acids and ketone bodies for energy generation so that acetyl-CoA will be plentiful. As will be described in Chapter 16, a high acetyl-CoA/CoA ratio leads to the inactivation of pyruvate dehydrogenase. The pyruvate produced from the amino acids is therefore reserved for alanine synthesis.

In summary, the net effect is that many of the amino acids derived from muscle protein breakdown are converted to alanine, which is carried in the blood to the liver. (The amino acid

glutamine is also synthesized in muscle for transport to the liver for glucose synthesis but the principle is similar.) In the liver, the alanine is converted back to pyruvate and thence to glucose. The overall process is summarized in Fig. 15.3.

As the concentration of blood ketone bodies rises during starvation, the brain progressively uses more of these for energy generation and so reduces its utilization of glucose and lessens the demand for gluconeogenesis (which, however, always remains essential). This is important for it requires about 2 g of muscle protein to be broken down for each gram of glucose made, a rate of loss that, if continued, would reduce the period that could be survived in starvation.

A second source of pyruvate for gluconeogenesis, of less importance in starvation but important in more normal situations, is **lactate**, produced by the anaerobic glycolysis of glucose or glycogen. As stated, mature red blood cells have no mitochondria and rely on glycolysis for ATP generation. In normal nutritional situations, a major source of lactate is **muscle glycolysis** – in strenuous muscle activity, the glycolysis rate within muscle may exceed the capacity of mitochondria to reoxidize NADH and lactate is produced (page 180). This travels via the blood to the liver where it is converted to pyruvate and thence to glucose (or glycogen). This constitutes a physiological cycle, called the Cori cycle after its discoverers (Fig. 15.4). The cycle has two main effects; it 'rescues' lactate for further use and, secondly, it counteracts **lactic acidosis**. The introduction of large amounts of lactic acid into the blood may exceed its buffering power and cause a deleterious fall in pH of the blood. The synthesis of glucose from lactate involves the uptake of

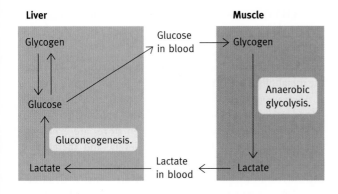

Liver   **Muscle**

Fig. 15.4 The Cori cycle. Note that this is a physiological cycle involving muscle and liver. The muscle has very low levels of three enzymes essential for gluconeogenesis. Excess lactate is produced in muscle during vigorous action during which glycolysis outstrips the capacity of mitochondria to reoxidize reduced NAD^+ – that is, during anaerobic glycolysis.

two protons (when $NADH + H^+$ is used for reduction of 1:3-bisphosphoglycerate).

Synthesis of glucose from glycerol

Another metabolite used for gluconeogenesis in the liver is **glycerol** released from **TAG hydrolysis** mainly in adipose tissue. This is taken up by the liver and converted to glucose by the route shown in Fig. 15.5. The enzyme **glycerol kinase**, which phosphorylates glycerol as the first step in the conversion, is very low in amount in adipose tissue and therefore glycerol is not converted to glucose (or TAG) in that tissue, but the enzyme is present in liver. This system is logical in that, in a situation where blood glucose supply is of vital importance, the glycerol produced in fat cells is transported to the liver and

converted to blood glucose. Adipose cells therefore do not use the glycerol themselves; supplying the brain with glucose takes priority over that.

During prolonged starvation, after a small initial drop in blood glucose level, the latter remains constant for several weeks and fatty acids supplied by adipose cells likewise remain at a constant level so that the mechanisms are extremely effective. It is, however, perhaps surprising that the whole business of muscle wastage during starvation could, on the face of it, have been avoided by animals using a simple metabolic trick present in bacteria and plants that permits fat to be converted to glucose – the glyoxylate cycle (page 240). One wonders if there are probably good evolutionary reasons for animals not using it.

Synthesis of glucose from propionate

The synthesis of glucose from propionate is of minor importance except in ruminants, where propionate is a major digestion product. It is converted into succinyl-CoA by the metabolic route for oxidizing odd-numbered fatty acids, present in all animals (page 219). Succinyl-CoA, being a component of the citric acid cycle, is on the normal metabolic highway, and is glucogenic.

We will digress briefly to discuss the effect of ethanol metabolism, which has special relevance to gluconeogenesis in the liver.

Effects of ethanol metabolism on gluconeogenesis

Ethanol is oxidized to acetaldehyde which is further oxidized to acetate, part of which enters the blood to be used by other tissues. Acetate is converted to acetyl-CoA by the acetate-activating enzyme found in most tissues and the acetyl group is then oxidized to carbon dioxide and water by the citric acid cycle or diverted into fat synthesis.

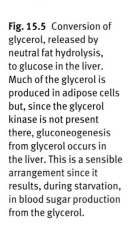

Fig. 15.5 Conversion of glycerol, released by neutral fat hydrolysis, to glucose in the liver. Much of the glycerol is produced in adipose cells but, since the glycerol kinase is not present there, gluconeogenesis from glycerol occurs in the liver. This is a sensible arrangement since it results, during starvation, in blood sugar production from the glycerol.

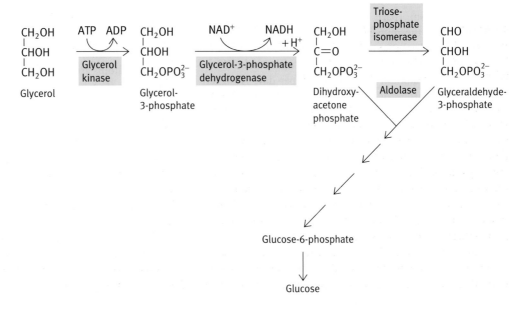

The oxidation of ethanol to acetaldehyde occurs in the liver and is mainly due to alcohol dehydrogenase in the cytosol:

$$CH_3CH_2OH + NAD^+ \xrightarrow[\text{dehydrogenase}]{\text{Alcohol}} CH_3CHO + NADH + H^+.$$

Oxidation of the acetaldehyde occurs in the mitochondrial matrix:

$$CH_3CHO + NAD^+ + H_2O \xrightarrow[\text{dehydrogenase}]{\text{Aldehyde}} CH_3COO^- + NADH + 2H^+.$$

A relatively minor route of ethanol catabolism exists. This is the microsomal ethanol-oxidizing system. Microsomes are small fragments, or vesicles, produced when the endoplasmic reticulum (ER) is disrupted during cell breakage. They do not exist as such naturally but are convenient experimental particles. They contain a cytochrome P450 which uses molecular oxygen and NADPH to produce acetaldehyde from alcohol:

$$CH_3CH_2OH + NADPH + H^+ + O_2 \rightarrow CH_3CHO + NADP^+ + 2H_2O$$

(It might seem odd to use NADPH in an oxidation reaction, but one atom of oxygen oxidizes the alcohol; the other is reduced to water by the NADPH.) The enzyme is inducible (increases in amount) by prolonged heavy intake of alcohol. Many medications are destroyed by P450s and alcohol may alter the rate of this by competing for P450 – something of medical interest.

Effect of ethanol metabolism on the NADH/NAD$^+$ ratio in the liver cell

Because the alcohol dehydrogenase occurs in the cytosol of liver cells, the NADH has to be reoxidized *via* the malate–aspartate shuttle mechanism (page 210), which transport reducing equivalents into the mitochondria. All of this sounds quite harmless, and indeed ethanol produced by microorganisms in the large intestine is absorbed and is a normal metabolite; in humans this can amount to a few grams per day. However, much larger quantities of alcohol may be consumed and the rate of shuttle transfer of reducing equivalents into mitochondria may not keep up with the rate of NAD$^+$ reduction.

The ratio of NADH to NAD$^+$ in a liver cell is normally low. With even a moderate amount of alcohol the level of NADH is increased. Many of the dehydrogenase reactions in which NAD$^+$ participates are close to equilibrium in the cell, which means that the normal cytosolic ratios of oxidized and reduced substrates are affected by the changed NADH/NAD$^+$ ratio. In particular, in liver, the lactate dehydrogenase reaction is affected so that the oxidation of lactate arriving at the liver from extrahepatic tissues to form pyruvate is impaired. We remind you of the lactate dehydrogenase reaction:

$$CH_3COCOO^- + NADH + H^+ \rightleftharpoons CH_3CHOHCOO^- + NAD^+$$
Pyruvate   Lactate

The distortion of the normal reduction pattern of NAD$^+$ affects the metabolism of liver cells.

A serious situation can arise in very heavy drinkers, especially if food intake is restricted during drinking bouts. After 24 hours of food deprivation, the liver glycogen stores are exhausted, and maintenance of blood glucose levels, and therefore of brain function, depends on gluconeogenesis. Gluconeogenesis depends on an adequate supply of pyruvate, which as described, is mainly formed from lactate produced by red blood cells and from alanine derived from muscle protein breakdown. However, in the presence of alcohol-induced abnormally high NADH levels in the liver, pyruvate availability is diminished; that formed from alanine may be reduced to lactate, and lactate, arriving from outside, is less efficiently oxidized to pyruvate. Lactate leaks into the blood causing lactic acidosis, and the reduced pyruvate availability impairs gluconeogenesis. The situation is exacerbated because gluconeogenesis itself is one of the protections against lactic acidosis, two protons being absorbed per molecule of glucose synthesized.

Gluconeogenesis from glycerol could likewise be affected by high NADH/NAD$^+$ ratios by impairing dehydrogenation of glycerol-3-phosphate.

Synthesis of glucose *via* the glyoxylate cycle in bacteria and plants

E. coli can survive quite well with acetate as its sole carbon source. Acetate is converted to acetyl-CoA by an ATP-driven reaction. A net conversion of acetyl-CoA to C$_4$ acids of the citric acid cycle and thence to carbohydrate or any other component of the cell can occur, unlike the situation in animals. In plant seeds, energy stored as TAG is converted to glucose on germination. What allows bacteria and plants to do this?

These organisms possess the normal citric acid cycle but, in addition, can bypass some of its reactions by other reactions not present in animals. If you analyse the citric acid cycle, two carbon atoms are added to oxaloacetate from acetyl-CoA, giving citrate (C$_6$), but then two carbon atoms are lost as two CO$_2$ molecules in forming succinate (C$_4$), so that there is no net conversion of C$_2$ to citric acid cycle intermediates.

The **glyoxylate route** bypasses these losses. Isocitrate is directly split into succinate and glyoxylate – a sort of cycle shortcut:

$$
\begin{array}{l}
COO^- \\
| \\
CHOH \\
| \\
CHCOO^- \\
| \\
CH_2COO^-
\end{array}
\xrightarrow[\text{lyase}]{\text{Isocitrate}}
\begin{array}{l}
CH_2COO^- \\
| \\
CH_2COO^-
\end{array}
+
\begin{array}{l}
COO^- \\
| \\
CHO
\end{array}
$$

Isocitrate   Succinate   Glyoxylate

The C$_2$ glyoxylate now reacts with acetyl-CoA to produce malate – back on the citric acid cycle:

$$CH-COO^- \quad + \quad CH_3-\overset{\overset{\displaystyle O}{\|}}{C}-S-CoA$$
$$\underset{\displaystyle O}{\|}$$

Glyoxylate   Acetyl-CoA

$$+ \quad H_2O$$

Malate synthase

OH
$$CH-COO^- \quad + \quad CoA-SH$$
$$\underset{\displaystyle CH_2-COO^-}{|}$$

Malate

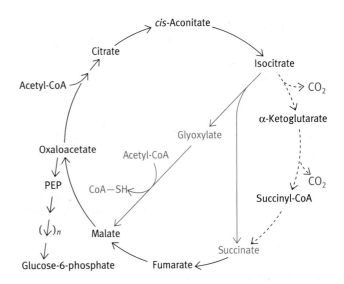

Fig. 15.6 The glyoxylate cycle of plants and bacteria, which permits carbohydrate synthesis from acetyl-CoA. This does not occur in animals. The broken line represents the bypassed section of the citric acid cycle; the red lines are reactions peculiar to the glyoxylate cycle. The two CO$_2$ molecules are highlighted to emphasize that it is these two decarboxylation reactions that must be bypassed. It will be appreciated that, as a result of the glyoxylate reactions, two molecules of oxaloacetate are produced from citrate, one of which is used to form citrate again and one to produce PEP.

The scheme is shown in Fig. 15.6. The net effect is that acetyl-CoA plus oxaloacetate convert to malate plus succinate. Succinate and malate can both be converted to oxaloacetate, one molecule then being available for conversion to glucose, the other to continue the cycle. In plants, this process occurs in membrane-bounded organelles, called glyoxysomes.

On page 198 we described the enzyme pyruvate carboxylase, which is necessary for topping up the citric acid cycle (the anaplerotic reaction). *E. coli* does not possess this enzyme. This is presumably related to the fact that the glyoxylate cycle can generate net increase in cycle acids from acetyl-CoA rendering pyruvate carboxylase unnecessary. The same applies to the need to generate oxaloacetate for carbohydrate synthesis.

It should finally be mentioned that by far the greatest amount of carbohydrate synthesis on Earth occurs during photosynthesis in plants. This involves the fixation of CO$_2$ into another glycolytic intermediate, bisphosphoglycerate, from which glucose synthesis occurs. The mechanism of this is best left until we deal with photosynthesis in Chapter 18.

We have, in the last few chapters, dealt with energy release from glucose and fat, with fat and glucose synthesis. Now is the time to remind you that fat metabolism and carbohydrate metabolism do not go on independently. They form an integrated metabolic activity, each affecting the other, and the whole lot requires regulation. This important topic will be dealt with in the next chapter.

Summary

It is essential, at certain times, for the liver to synthesize glucose by the gluconeogenesis pathway and turn it out into the blood. Unless the brain has adequate supplies of glucose it will cease to function normally.

During starvation, after 24 hours, the liver's glycogen stores are exhausted and to keep up essential supplies of glucose to the brain the liver synthesizes glucose. The substrate for this is pyruvate but any compound convertible to pyruvate can be gluconeogenic. The main source of this in starvation is alanine derived from muscle wasting and transported in the blood to the liver. There it is deaminated to pyruvate. The muscle breakdown is caused by the stress hormone cortisol, which is liberated during starvation.

Gluconeogenesis occurs largely by reversal of glycolytic reactions but there are three irreversible steps in the latter pathway that must be circumvented. Pyruvate is converted to phosphoenolpyruvate *via* oxaloacetate with an input of energy from GTP. The phosphofructokinase and hexokinase reactions are bypassed by phosphatases. The reactions forming phosphoenolpyruvate and the fructose bisphosphatase reaction are important control points in gluconeogenesis (discussed in Chapter 16). Gluconeogenesis also recycles lactate, produced by muscle (during vigorous exercise) and red blood cells, by converting it into glucose or glycogen. The sequence is known as the Cori cycle.

The synthesis of glucose from glycerol arising from triacylglycerol hydrolysis starts with glycerol kinase. It cannot be metabolized by adipose cells, where glycerol kinase is very low. It is released and taken up by the liver. The liver phosphorylates it to glycerol-3-phosphate, then converts it to dihydroxyacetone phosphate which is on the standard gluconeogenic pathway.

One of the dangers of excessive drinking is when it is allied with fasting; the perturbation of the NADH/NAD$^+$ ratio by alcohol metabolism can impair necessary synthesis of glucose for the brain. Muscle does not release free glucose and has no capacity for gluconeogenesis.

In animals, fats cannot be converted to glucose or C3 metabolites. However, the glyoxylate cycle in plants and bacteria, which is a modified citric acid cycle, makes this conversion possible.

Further reading

Felig, P. (1975). Amino acid metabolism in man. *Annu. Rev. Biochem.*, **44**, 933–55.
Discusses amino acid metabolism at the organ level, together with the effects on it of starvation, diabetes, obesity, and exercise.

Snell, K. (1979). Alanine as a gluconeogenic carrier. *Trends Biochem. Sci.*, **4**, 124–8.
Reviews the role of muscle in supplying the liver with metabolites for glucose synthesis.

Problems

1   After 24 hours of starvation, the glycogen reserves of the liver are exhausted but there are still relatively large stores of fat. What is the point of the body going to such trouble to make glucose by gluconeogenesis when virtually unlimited supplies of acetyl-CoA from fatty acids are available to supply energy?

2   Gluconeogenesis requires production of phosphoenolpyruvate (PEP), but pyruvate kinase cannot form this from pyruvate. Why is this and how is the problem overcome?

3   From PEP, gluconeogenesis in the liver produces blood glucose mainly by reversal of glycolytic enzymes. However, two enzymes are involved that do not participate in glycolysis. What are these?

4   Does muscle have glucose-6-phosphatase? Explain your answer.

5   What is the Cori cycle and its physiological role?

6   Adipose cells do not have glycerol kinase, the enzyme that converts glycerol to glycerol-3-phosphate, even though adipose cells produce glycerol from triacylglycerol (TAG) hydrolysis. Liver does have the enzyme. Explain why this appears to be a logical situation.

7   Animals cannot convert acetyl-CoA to carbohydrate. Bacteria and plants can. Explain how they do this.

8   Pyruvate kinase is needed to top up the acids of the citric acid cycle (the anaplerotic reaction). Depletion of the cycle of its acids would impair operation of the cycle. However, *Escherichia coli* does not have this enzyme. Comment on this.

9   In starvation, muscle wasting occurs. What is the relevance of this to gluconeogenesis in the body?

10 Excessive alcohol intake can tend to deprive the brain of glucose if food intake is restricted, as may occur in binge drinking. Explain this.

Strategies for metabolic control and their application to carbohydrate and fat metabolism

This chapter should be read in conjunction with the preceding chapters on metabolism. What we will do in this chapter is to put together all the preceding metabolism to see how the various pathways work and interact.

We have collected together metabolic control mechanisms into a separate chapter, rather than dealing with the topic when the pathways were discussed. There may be advantages in this partly because the pathways themselves are complicated enough on their own but also because it provides the opportunity to give preliminary information on control strategies in general, which have widespread application. Finally it permits an integrated approach to metabolic control.

Although control extends to all facets of metabolism, the control and integration of carbohydrate and fat metabolism is especially important because the flow of chemical change (flux) through these pathways is large. In the body, the direction of the fluxes is changed as meals are followed by periods of fasting. Control of carbohydrate and fat metabolism illustrates all of the principles of metabolic control. The subject is a fairly complex one, so let us be clear on how this chapter deals with it. Enzymes are responsible for metabolism and therefore metabolic control involves control of enzyme activities. We therefore first deal with the strategies by which enzyme activities are regulated. After this we describe how the pathways are kept in balance by regulatory enzymes that sense the levels of metabolites in other pathways, and automatically adjust their rates of activities to the information.

This is not the end, because the metabolic activities of cells are controlled in a wider sense by circulating hormones, which operate in the body as a whole according to current needs. We therefore describe how signals external to the cells – extracellular signals produced by other cells of the body – regulate the metabolic pathways, on top of the automatic controls first described.

But before we deal with any of this the first question to consider is the following one.

Why are controls necessary?

It is obvious when you think of the major pathways that we have dealt with – glycogen synthesis, glycogen breakdown, glycolysis, gluconeogenesis, fat breakdown, fat synthesis, the citric acid cycle, electron transport – that they cannot all be running at full speed (or even necessarily at all) at the same time. If a metabolic pathway is proceeding in one direction, the reactions in the reverse direction must be switched off, otherwise nothing but heat generation would be achieved. In a single pathway such as glycolysis, its required rate will vary according to the energy needs at the time. The metabolic rate of someone playing squash is about six times greater than that of the person at rest. In short, the metabolism of the main energy-yielding materials must be regulated for at least three reasons:

- to avoid the potential problem of futile cycles, or substrate cycles as they are now more properly called (explained below)

- to respond to energy-production needs as energy expenditure varies

- to respond to physiological needs – metabolic pathways need to work in different directions after a meal when metabolites are being stored (Chapter 9) as compared with intervals between meals when stored energy reserves are being utilized. The needs are different again in prolonged starvation and in pathological situations such as diabetes where carbohydrate metabolism is abnormal.

The potential danger of futile cycles in metabolism

The process of gluconeogenesis in the preceding chapter provides a good example for illustrating the potential danger of futile cycles in metabolic systems. One wonders how organisms coped

Gluconeogenesis   **Glycolysis**

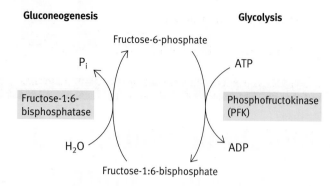

Fig. 16.1 Potential futile cycle at the phosphofructokinase (PFK) step in glycolysis if the reactions were uncontrolled.

with the problem during the evolution of the pathways. Consider the phosphofructokinase (PFK) reaction in glycolysis and its reversal in gluconeogenesis by fructose-1:6-bisphosphatase (Fig. 16.1). In one direction, fructose-6-phosphate is phosphorylated by PFK to yield fructose-1:6-bisphosphate, while in the reverse direction the fructose-1:6-bisphosphatase enzyme hydrolyses the product back to fructose-6-phosphate. This cycle of events would, unchecked, destroy ATP with the generation of heat – more or less equivalent to an electrical short circuit.

Extending this to the whole of glycolysis and gluconeogenesis pathways, these, if uncontrolled, could constitute a giant futile cycle again doing nothing but uselessly destroying ATP. The same applies to glycogen synthesis and breakdown, and fat synthesis and breakdown (Fig. 16.2) or, for that matter, to any synthesis and breakdown pathways.

It is clear from this that the breakdown and synthesis directions of a metabolic pathway must be controlled in a reciprocal manner – activation of one and inhibition of the other. This independent control of the two directions can occur only, as mentioned earlier, at irreversible metabolic steps, for it is here that there are separate reactions in the two directions catalysed

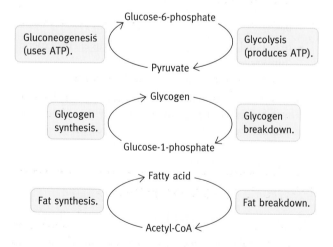

Fig. 16.2 Potential large-scale futile cycles if metabolism were not controlled.

by distinct enzymes which can be separately controlled. A freely reversible reaction is catalysed by the same enzyme in both directions, which cannot be separately controlled. As explained earlier, this is why breakdown and synthesis pathways of a given constituent are usually not exact reversals of one another.

With the realization that controls are operative, the term 'futile cycle' has been replaced by 'substrate cycle' to describe situations such as occurs at the PFK step in glycolysis. Substrate cycling, if allowed to occur at all, may appear to be wasteful. But a limited amount of cycling may be advantageous in making controls on the forward and backward metabolic pathways more sensitive. Suppose you have a pathway in which A is converted to C via B and the reaction A → B occurs at a substrate cycle:

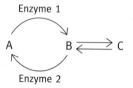

The rate of conversion of A to C can be reduced by inhibiting enzyme 1 or activating enzyme 2. If you do both, the control is more effective; complete shutdown could be achieved without excessively high levels of the inhibitor for enzyme 1. Inhibition of the flux from C to A could similarly be achieved by inhibiting enzyme 2 and activating enzyme 1. The fructose-6-phosphate/fructose-1:6-bisphosphate cycle has dual controls of the type described.

How are enzyme activities controlled?

Metabolic regulation of a pathway means regulation of the rate of one or more reactions in that pathway. There are basically two ways of reversibly modulating the rate of an enzyme activity in a cell. These are:

- to change the amount of the enzyme
- to change the rate of catalysis by a given amount of the enzyme.

(Compartmentation adds a further possibility that availability of substrate may play a role so that control of transport proteins may be relevant.)

There are ways of *irreversibly* activating enzymes, such as the proteolytic conversion of trypsinogen to active trypsin, but in this chapter we are dealing with control mechanisms that are reversible, since this is essential to meaningful metabolic control.

Metabolic control by varying the amounts of enzymes is relatively slow

The level of a protein in a cell can be changed by altering the rate of its production and/or the rate of its destruction. Proteins are relatively short lived – the half-life of enzymes in, say, the liver, might range from about 30 minutes to several days.

This type of control, which is at the gene-activation level, is a relatively long-term affair, with effects being seen in hours or days rather than seconds. It is at the level of adaptation to physiological needs. A few examples are given here.

- Lipoprotein lipase levels in capillaries (page 170) are adjusted to the fat demands of the tissues, the levels increasing in lactating mammary glands.
- The liver changes its enzyme content within hours in response to dietary changes – whether it has to cope with high fat or high carbohydrate intake.
- In the well-fed state, hepatic enzymes involved in fat synthesis increase in amount; a few hours of starvation reverses this.
- Intake of foreign chemicals such as drugs results in a rapid increase in hepatic drug-metabolizing enzymes (page 488).
- A particularly good example is that of enzymes of the urea cycle. Excess nitrogen is converted into urea for excretion into the urine. All of the enzymes of this pathway change in the rat in proportion to the nitrogen content of the diet (page 294).

Control at the level of enzyme synthesis is spectacular in bacteria. For example, if *Escherichia coli* is exposed to lactose as its sole carbon source, synthesis of the enzyme β-galactosidase, needed to hydrolyse the sugar, starts up immediately; an increase in enzyme activity is detectable in minutes and a 1000-fold increase can be measured within hours. When the stimulus for enzyme synthesis is no longer there, reversal of increases in enzyme concentrations occurs relatively slowly since they are returned to lower values only by destruction of the proteins (or dilution by cell growth in bacteria since they multiply so rapidly). The half-lives of regulatory enzymes in mammals tend to be short – perhaps 30 minutes to an hour or so. We will return to this subject later when we deal with control of gene expression (Chapter 23).

Metabolic control by regulation of the activities of enzymes in the cell can be effectively instantaneous

For regulation of *activities*, the amounts of given enzymes are not altered, only the rate at which they work. This method of control is virtually instantaneous or very rapid, depending on the type of control mechanism.

Which enzymes in metabolic pathways are regulated?

In many or most metabolic pathways, there are certain enzymes with special regulatory properties. Often, but not always, these are strategically placed at the first step, which commits the metabolic pathway to the formation of a specific end product.

Consider a metabolic pathway such as:

$$A \rightarrow B \rightarrow C \rightarrow D \rightarrow E \rightarrow \rightarrow \text{utilization for cell synthesis}$$

in which E is an end product needed by the cell. An automatic control mechanism is for the first reaction committed to the synthesis of E to be controlled by the level of E as shown below. This is known as feedback inhibition.

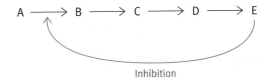

When the level of E is reduced, the inhibition is relieved and E is synthesized. The flux (flow of metabolites through the pathway) is in this sense controlled by the utilization of E, since if more E is produced than can be immediately utilized, its production is automatically diminished.

There are numerous metabolic pathways (such as for the synthesis of amino acids) in which such controls exist. Where the pathway branches, the first enzyme of each branch is often subject to regulation so that formation of each end product is separately controlled. Each end product also partially inhibits the first enzyme of the joint pathway. When it comes to fat and carbohydrate metabolism, the whole system is so complex that the term 'end products' of pathways or 'first' enzymes cannot be defined in quite the same way, so that 'end-product inhibition' is not such a relevant term. Nonetheless the same principle applies, that key intermediates (products) can control metabolically distant enzymic reactions. The control may be feedback, as described, or 'feed-forward' in which metabolites activate downstream enzymes which will be needed to cope with the flow of metabolites coming their way.

The nature of control enzymes

There are essentially two main ways in which the catalytic activity of enzymes is modulated (as distinct from a change in amount).

- **Allosteric control** is effectively instantaneous.
- **Covalent modification** of the protein – mainly by **phosphorylation** and **dephosphorylation** – is rapid but not instantaneous.

Allosteric control of enzymes

Allosteric control of enzymes is of central importance in metabolic regulation. The prefix *allo* means 'other'; it refers to the existence on an enzyme of one or more binding sites other than for substrate. The ligands which bind to the allosteric sites are called **allosteric activators or inhibitors** and collectively allosteric modulators. They do not have to have any structural relationship to the substrates of the enzymes. *At a given substrate concentration*, an allosteric modulator increases or inhibits the activity of the enzyme when it combines at the allosteric site.

Allosteric modulators typically alter the K_m or apparent affinity of the enzyme for its substrates (S). (We have explained earlier (page 91) that the K_m of an enzyme may not be a true affinity constant.) Most enzymes work in the cell at subsaturating levels of [S], so that an *increase in their affinity will increase their activity and a decrease in their affinity has the reverse effect.* At saturating levels of [S], the activity is unchanged, even if the affinity is changed, but this situation does not (usually) occur in the cell.

The mechanism of allosteric control of enzymes

Allosteric enzymes have a multisubunit structure; they are made up of more than one catalytic protein molecule assembled into a single enzyme complex by noncovalent bonds. In some cases the enzyme is a collection of only catalytic subunits but in others there is a complex of catalytic and inhibitory subunits. The latter have no catalytic activity but play a role in the response of the enzyme to allosteric effectors.

A plot of reaction velocity versus substrate concentration of allosteric enzyme is shown in Fig. 16.3. It differs from the Michaelis–Menten enzyme kinetics described earlier (page 90) in that the response of the enzyme velocity to changes in substrate concentration is sigmoidal, rather than hyperbolic.

When an allosteric activator binds to the enzyme, the sigmoid curve is moved to the left, and with an allosteric inhibitor, to the right, as shown in Fig. 16.4. The allosteric activator increases the binding affinity of the enzyme for its substrate; the

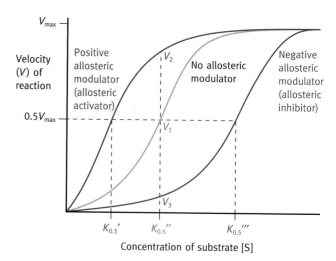

Fig. 16.4 Effect of substrate concentration on a typical allosterically regulated enzyme. V_1, V_2, and V_3 are, respectively, the reaction velocities observed with no allosteric modulator, with an allosteric activator, and with an allosteric inhibitor, at a fixed concentration of S ($K_{0.5}''$). The curves are drawn arbitrarily; actual shapes and positions will be a function of the particular system.

allosteric inhibitor reduces it. The sigmoidal shape of the response to substrate concentration means that over a range of substrate concentrations (centred at that giving half maximum velocity), the rate of enzyme catalysis is more sensitive to substrate concentration change than with a Michaelis–Menten type of enzyme. By the same token, it means that increasing or decreasing the substrate affinity of an enzyme displaying sigmoidal kinetics has a greater effect on reaction velocity at a given substrate concentration than with hyperbolic kinetics. Since, as stated, enzymes in the cell usually are working in the sensitive range of substrate concentration, the sigmoidal response to substrate concentration maximizes the effect of an allosteric modulator.

What causes the sigmoidal response of reaction velocity to substrate concentration?

Binding of a substrate molecule to *one* of the catalytic subunits in an allosteric enzyme has the result that binding of subsequent substrate molecules to the other subunits of the enzyme occurs more readily. This is known as **positive homotropic cooperative binding** of substrate: 'positive' because affinity increases; 'homo' because only one type of ligand, the substrate, is involved; and cooperative because the different catalytic subunits are interacting. There are two models to explain the sigmoid kinetics. These were described when we dealt with haemoglobin, the classical allosteric protein. We will here use the concerted model (page 64) for illustrative purposes.

In this model, either *all* of the subunits in an enzyme bind substrate with low affinity or *all* with high affinity, the two forms being in spontaneous equilibrium (Fig. 16.5), but with the equilibrium strongly to the low-affinity side. The model

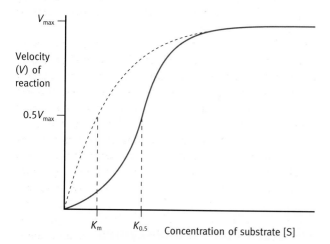

Fig. 16.3 Effect of substrate concentration on the rate of reaction catalysed by a typical allosteric enzyme. The dashed line shows, for comparison, the corresponding curve for a classical Michaelis–Menten enzyme. The term K_m is not strictly applicable to a non-Michaelis–Menten enzyme, and the term $K_{0.5}$ is used instead.

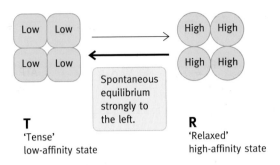

Fig. 16.5 Concerted model of cooperative binding of substrate (S). In this model the enzyme exists in two forms, T and R, the two being in spontaneous equilibrium. Binding of a substrate molecule to the R state swings the equilibrium to the right, thus increasing the affinity of all of the subunits in that molecule. 'High' and 'low' refer to affinities of the enzyme for its substrate. The model assumes that allosteric modulators binding preferentially to the R form will activate. If a modulator binds preferentially to the T form, it will inhibit.

assumes that allosteric modulators alter the position of the equilibrium between the low-affinity (tense or T) state and the high-affinity (relaxed or R) state. If the allosteric modulator binds more tightly to the R state it displaces the T $\rightleftharpoons$ R to the right and results in a higher proportion of molecules being in the R state; it would, at a given substrate concentration, activate the enzyme. As the concentration of the positive allosteric modulator is increased and more of the enzyme converts to the high-affinity state, the substrate-reaction velocity curve tends towards the hyperbolic type. If, by contrast, the allosteric effector is of the negative type, it is assumed that it binds more tightly to the low-affinity T state. This results in more enzyme molecules being in this state and a decreased reaction rate in the collection of enzyme molecules as a whole – it inhibits.

Aspartate transcarbamylase is the classical model of an allosteric enzyme

Aspartate transcarbamylase (ATCase) was one of the earliest allosterically controlled enzymes to be studied in very great structural detail. It is useful to describe here for it illustrates the principle of feedback control of metabolism and the concerted mechanism of allosteric control. It also illustrates the principle that allosteric effectors do not have to have any structural resemblance to the substrate(s) of the enzyme. ATCase is the first metabolic step committed to pyrimidine nucleotide synthesis (described on page 307). It leads to the synthesis of cytidine triphosphate (CTP) (and other pyrimidine nucleotides), which is the allosteric feedback inhibitor of ATCase. It shuts down the pathway when sufficient amounts of pyrimidine nucleotides are present. The enzyme is a large one consisting of six catalytic subunits and six regulatory subunits. These are arranged in the quaternary structure as two catalytic trimers linked by three regulatory dimers. The regulatory dimers do not interact with the catalytic active sites but are the sites for the CTP binding.

When the enzyme is combined with substrate, a massive quaternary structural change occurs. In the absence of substrate, the enzyme is in a compact T (tense) state with a low affinity for substrate but when combined with substrate it expands into a relaxed state with a higher affinity for substrate. The two states are in equilibrium amounts, dependent on the amount of substrate bound – the higher this is, the greater the proportion of enzyme in the high affinity relaxed state. Binding of CTP, the allosteric inhibitor, to the regulatory subunits shifts the equilibrium to the tense state, lowers the affinity for substrate and thus at subsaturating substrate levels, inhibits the enzyme.

Reversibility of allosteric control

Allosteric control is virtually instantaneous in both its application and reversibility. The allosteric modulator attaches to its site by noncovalent bonds and when the concentration of ligand is reduced, it dissociates from combination with the enzyme and everything goes into reverse.

Allosteric control is a tremendously powerful metabolic concept

Allosteric modulators, as mentioned, need have no structural relationship to the substrate of the enzyme regulated. They are often from a separate metabolic pathway. This means that any metabolic pathway or area of metabolism can be connected in a regulatory manner to any other metabolic area, the metabolite(s) of one pathway being regulator(s) of another. The systems controlled are complex; glycogen breakdown feeds glycolysis, which feeds pyruvate into the citric acid cycle and this, in turn, feeds electrons into the electron transport system. This feeds ATP into the energy-utilizing machinery of the cell. Fatty acid breakdown also feeds the citric acid cycle, supplying acetyl-CoA; in reverse, acetyl-CoA formed from pyruvate feeds fatty acid synthesis, and so on.

In such a complex system, each section has to 'know' what others are doing by sensing the levels of key metabolites. Is there enough or too little ATP? Is the citric acid cycle being supplied with too little or too much acetyl-CoA? Is glycolysis too fast or too slow? You can see the chemical chaos that would result unless there was constant second-to-second adjustment of pathways in the light of information reaching each pathway about all other pathways and parts of its own pathway. This is why allosteric control is such an important concept – regulatory enzymes can be designed, in the evolutionary sense, to receive signals from anywhere. Thus glycolysis can be 'informed' on how the citric acid cycle is doing and how the electron transport system is keeping up ATP supply, the information automatically adjusting activities to make a harmonious chemical machine. As Jacques Monod, the originator of the allosteric concept, pointed out, you simply could not have anything as complex as a cell without it. He described it as the 'second secret of life' (DNA being the first).

Allosteric enzymes often have multiple allosteric modulators

The control network *is* complex. A given enzyme may not be just affected by one allosteric signal. It may receive multiple signals from all over the metabolic map, each partially inhibiting or stimulating its rate of reaction. It is a classic case of evolution being a tinkerer; one little adjustment here and another there, by yet another allosteric signal. This gives fine tuning of enzyme activity, and natural selection of the control mechanisms ensures efficient performance of the whole complicated mass of reactions.

Control of enzyme activity by phosphorylation

The second method of enzyme control is by phosphorylation. To be strictly accurate this section ought to refer to covalent modification of enzymes rather than phosphorylation, because other chemical modifications occur, but phosphorylation is of such overwhelming importance in eukaryotic cells that we will confine ourselves to it. Unlike allosteric control, which is of major importance in both prokaryotes and eukaryotes, phosphorylation is less important in prokaryotes but in eukaryotes it is of paramount importance in many vital areas quite apart from metabolic control.

Protein kinases and phosphatases are key players in control mechanisms

The principle is simple. Enzymes called **protein kinases** transfer phosphoryl groups from ATP to specific enzymes. When this happens, the target enzyme undergoes a conformational change such that the enzyme (or an enzyme-inhibitor protein) changes its activity (or inhibitory effect). You can imagine that the addition of such a strongly charged group could have an effect on the conformation of a protein molecule. To reverse the process there are **phosphoprotein phosphatases** (often abbreviated to **protein phosphatases**) which hydrolyse the phosphate from the protein (Fig. 16.6).

The phosphorylation occurs on the hydroxyl group of a serine or threonine of the polypeptide chain of the enzyme, and is identified by the protein kinase by the neighbouring sequence of amino acids around the target —OH group. (In Chapter 27 we describe tyrosine group phosphorylation which has great importance in gene control.) We show the serine phosphorylation process below.

This control of enzymes by phosphorylation is a second general mechanism by which enzymes are controlled – allosteric control was the first, phosphorylation and dephosphorylation of proteins the second. We will shortly describe how the two methods of enzyme control apply to specific metabolic systems.

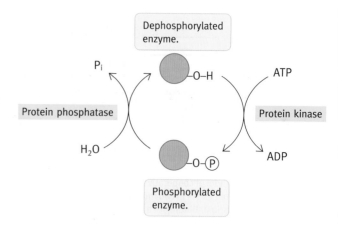

Fig. 16.6 Control of an enzyme by phosphorylation. The phosphorylated enzyme may, in specific cases, be more active or less active than the unphosphorylated enzyme. Note that variations of this may occur in that the activity of an enzyme may be controlled by an inhibitor protein whose inhibitory activity is controlled by phosphorylation.

Control by phosphorylation usually depends on chemical signals from other cells

Allosteric control gives an immediate regulatory mechanism, but control by phosphorylation requires that the phosphorylation and dephosphorylation processes themselves are controlled. The balance between phosphorylation and dephosphorylation determines the activity of the target enzyme. What, therefore, controls the protein kinases and phosphatases? The answer lies for the most part *not* in intracellular automatic controls but in controlling agents such as hormones, external to the cell. (There are a few individual exceptions to this in which phosphorylation is part of the internal metabolic controls – pyruvate dehydrogenase is one (page 193) – but the general point applies.)

To put things in perspective, the internal, allosteric controls coordinate the different pathways and parts of pathways so that metabolic pile-ups and shortages do not occur. Imagine (again) the chaos there would be if glycolysis produced pyruvate irrespective of how much of it the citric acid cycle could handle. A completely isolated cell could carry out this intrinsic control. But, as already implied, in the body there is a second group of controls involving hormones that instruct cells on what their metabolic direction should be – such as whether to store fuel, or release it. This is essential if cells are to do what is beneficial for the body as a whole. Individual cells cannot make these decisions – they must receive signals from other cells. After being so instructed, the internal allosteric controls are still essential to keep the pathways coordinated to avoid bottlenecks and metabolic snarl-ups. It may be likened to the organization of a navy. Each individual ship (cell) has its own internal (allosteric) system of discipline which, in all circumstances, maintains it as an organized unit, but what it does as a unit – where it sails and what it does – depends on external signals from higher naval authorities (endocrine glands and the brain, the brain often controlling hormone release from the endocrine glands).

General aspects of the hormonal control of metabolism

In control of carbohydrate and fat metabolism, the hormones of special importance are **glucagon**, **insulin**, and **epinephrine** (also known as **adrenaline**) and these are what we will deal with here. These are the ones which have immediate and rather dramatic effects and which are invoked on a daily basis as we oscillate between eating and fasting periods between meals with occasional epinephrine-releasing panics thrown in.

Glucagon we remind you is produced by the pancreas when blood glucose is low. It causes release of stored food components. Insulin is produced when blood glucose is high and signals cells to store fuel. Epinephrine, the hormone liberated by the adrenal gland as a result of stressful situations, has a mobilizing effect on food-storage reserves to prepare the body for action.

How do glucagon, epinephrine, and insulin work?

The mechanism of hormone control is a major topic dealt with in Chapter 27, but an introduction to them is needed here because they control metabolism as well as genes. They are chemical signals released into the blood so that all tissues are exposed to them. However, only certain cells, called target cells, respond to a given hormone. Glucagon, epinephrine, and insulin do not enter their target cells; they combine with membrane receptor proteins, specific for each hormone. These are transmembrane proteins, each with an external receptor part displayed like an aerial on the outside of cells ready to combine with its specific hormone or other signalling molecule. Only target cells have the receptors to specific hormones – the rest have none and are 'blind' to the hormone.

The hormones are quickly eliminated from the blood. Typically their half-life is a few minutes, so unless the source

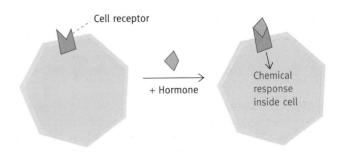

Fig. 16.7 Hormone binding to surface receptor causes chemical events inside the cell.

gland of a given hormone is releasing more, the concentration of the hormone in the blood falls, terminating the signal. However, while the hormone is bound to the receptor, changes occur inside the cell (Fig. 16.7). This leads us to the **second messenger** concept.

What is a second messenger?

If you regard the hormone as the first messenger, then binding to its external receptor on a cell causes a change in the level of a second messenger – a small intracellular regulatory molecule which causes cell responses. Not all hormone signalling involves second messengers, but the ones we are dealing with in this chapter do.

The second messenger for glucagon and epinephrine is cyclic AMP

Cyclic AMP (cAMP) (not 5′ AMP) is adenosine-3′, 5′-cyclic monophosphate, a molecule you have not yet met in this book. It is synthesized from ATP by the enzyme **adenylate cyclase** (Fig. 16.8).

Now things start to come together because inside the cell, cAMP is the allosteric activator of a protein kinase (**PKA**) that phosphorylates serine or threonine —OH groups of specific enzymes, thus modulating their activities. Thus the general mechanism is as in Fig. 16.9.

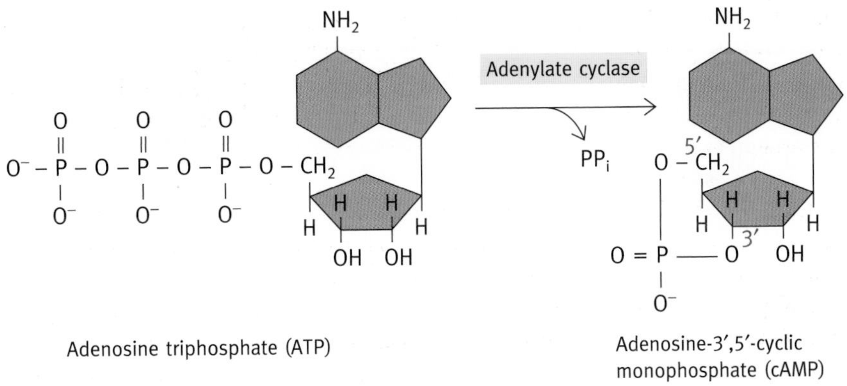

Fig. 16.8 The synthesis of cyclic AMP from ATP by adenylate cyclase.

Adenosine triphosphate (ATP)

Adenosine-3′,5′-cyclic monophosphate (cAMP)

Hormone
(glucagon, epinephrine)

↓

Binds to receptor
of target cell

↓

Increases cAMP level
inside cell

↓

cAMP activates
a protein kinase (PKA)

↓

Protein kinase phosphorylates
specific proteins

↓ ↓ ↓ ↓

This phosphorylation leads to
changes in enzyme activites and
metabolic responses

Fig. 16.9 Steps in the hormonal control of metabolism. Not shown are: (1) that cyclic AMP (cAMP) is continually destroyed; (2) that phosphorylation of proteins is reversed by phosphatases. The metabolic response shown occurs only as long as a hormone is bound to the receptor.

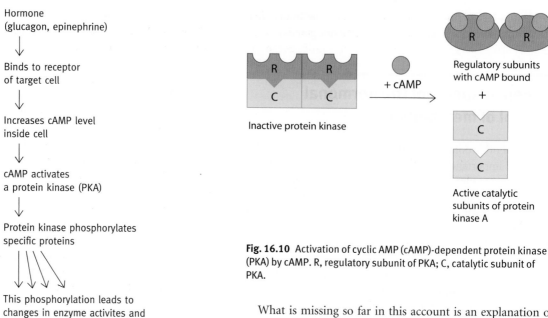

Fig. 16.10 Activation of cyclic AMP (cAMP)-dependent protein kinase (PKA) by cAMP. R, regulatory subunit of PKA; C, catalytic subunit of PKA.

What is missing so far in this account is an explanation of *how* the hormone binding switches on cAMP production and how the latter is switched off. We are deferring description of this until Chapter 27 because it is part of the more general major topic of signal transduction across cell membranes. What we have done so far in this chapter is to give general strategies used by the cell to control metabolism. We can now turn to the application of these to specific pathways.

Control of carbohydrate metabolism

Control of glucose uptake into cells

Glucose cannot readily penetrate lipid bilayers, and therefore membrane transport proteins are needed for its entry into cells. Absorption of glucose from the intestine is active, as we have mentioned (Fig. 7.12). However, transport into other cells of the body is passive – the facilitated-diffusion type in which a specific transporter protein is provided for glucose molecules to traverse the membrane, driven only by the glucose gradient across the membrane.

The way in which cAMP activates PKA is shown in Fig. 16.10. In the absence of cAMP, PKA is a tetramer with two catalytic subunits (C) and two regulatory subunits (R). When cAMP binds to the regulatory subunits, the enzymatically active C monomers are released.

An enzyme, **cAMP phosphodiesterase**, hydrolyses cAMP to AMP (Fig. 16.11) inside the cell so that the activation of cellular PKA depends on continual production of cAMP, which occurs only as long as the hormone is bound to the cell receptor. Thus everything has the required reversibility. If the hormone levels fall, cAMP production ceases, existing cAMP is hydrolysed, and everything goes back to square one, phosphorylated proteins being dephosphorylated by protein phosphatases. Protein phosphatases are usually themselves controlled.

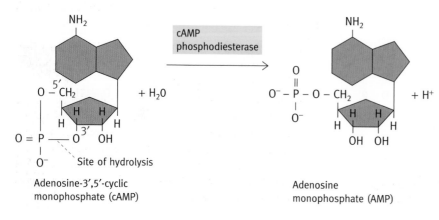

Adenosine-3′,5′-cyclic
monophosphate (cAMP)

Adenosine
monophosphate (AMP)

Fig. 16.11 Reaction catalysed by cyclic AMP (cAMP) phosphodiesterase. The name arises from the fact that cAMP contains a doubly esterified phosphoryl group – that is, a phosphodiester.

There is a family of such glucose transporters, GLUT 1 to GLUT 5, all the isoforms having a common structure characterized by 12 transmembrane sequences. GLUT 4 is insulin-dependent, as in skeletal and heart muscle and adipose tissue, and others are noninsulin sensitive, as in brain (GLUT 3), liver (GLUT 2) and red blood cells (GLUT 1). They have different affinities for glucose, and occur in different tissues to meet the metabolic needs, and hormone situations under different circumstances.

In muscle and adipose cells, where glucose uptake is insulin-dependent, there is a reserve of nonfunctional glucose-transport proteins in the form of membrane vesicles. Insulin causes these to fuse with the cytoplasmic membrane where GLUT 4 is functional, thus increasing the rate of glucose transport (Fig. 16.12). In the absence of insulin the GLUT 4 slowly cycles between the cell membrane and the cytoplasm, with most being in the cytoplasm. Insulin triggers a complex signalling cascade, causing a preponderance of GLUT 4 to be in the membrane. It has been shown in adipose cells that the effect of insulin in recruiting the reserve glucose-transporter protein GLUT 4 into the cell membrane is complete in about 7 minutes. If the insulin is removed, the process reverses, the transporters being returned to the nonfunctional intracellular reserve in about 20–30 minutes.

Since the transport of glucose is by facilitated diffusion not requiring energy, by itself this would simply equilibrate cellular and blood levels, but in the cell the glucose is trapped by phosphorylation and removed by glycogen synthesis or other metabolism and this constitutes a driving force for uptake of the sugar. The first step is the phosphorylation of glucose:

$$\text{Glucose} + \text{ATP} \rightarrow \text{glucose-6-phosphate} + \text{ADP}$$

The enzyme doing this is **hexokinase**, but in liver there is an **isoenzyme** known as **glucokinase**, which carries out the same reaction as hexokinase. Isoenzymes are different enzyme forms carrying out the identical reactions but with different characteristics, such as affinity for substrate or its susceptibility to product inhibition, tailored to the requirements of the tissue in which they are found. Glucokinase has a much lower affinity for glucose than has hexokinase (see Fig. 10.12). This is an important regulatory device. The liver mainly takes up glucose when its blood level is high. When it is low, the liver is turning out glucose so to take it up again rapidly would not be logical. The low affinity of liver glucokinase minimizes this; only when the blood glucose is high does glucokinase work rapidly. When it is low, the brain and other cells dependent on glucose are given priority because they use the high-affinity hexokinase.

There is another reason; hexokinase is inhibited by its product, glucose-6-phosphate, but at physiological concentrations of the latter glucokinase is not. This is important. When the blood glucose is high the liver takes it up and produces concentrations of glucose-6-phosphate for glycogen synthesis, which would inhibit hexokinase. In keeping with its role, the synthesis of this enzyme is increased by insulin.

Insulin also inhibits gluconeogenesis in the liver – it will be present only when blood glucose is high, so why synthesize more?

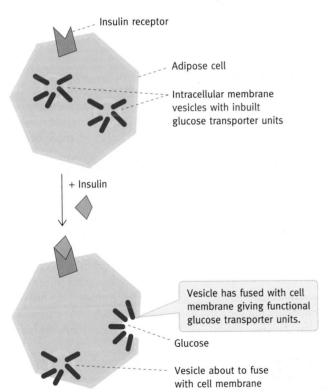

+ Insulin

Insulin receptor

Adipose cell

Intracellular membrane vesicles with inbuilt glucose transporter units

Vesicle has fused with cell membrane giving functional glucose transporter units.

Glucose

Vesicle about to fuse with cell membrane

Fig. 16.12 The effect of insulin in mobilizing glucose transporter units (GLUT 4) in adipose cells, and heart and skeletal muscle. The glucose transport is of the passive facilitated diffusion type. Note that liver and brain are not insulin responsive in this respect.

Control of glycogen metabolism

Glycogen is broken down by glycogen phosphorylase (page 165), and synthesized by glycogen synthase (page 162). Glycogen metabolism control has a vital role in animals, liver and muscle (kidney to a small extent) being the organs involved. It is a complex control system, and it will help if you keep in mind that it has three different levels.

- There is the 'routine' ticking-over level, where the controls are of the automatic allosteric type within cells. It keeps the ATP level topped up.

- There is the physiological level depending on the feeding state. After a meal, the blood glucose is high – **insulin** levels will be high, telling the cells to store food; **glucagon** levels will be very low. As the blood sugar levels fall, insulin secretion will cease and rapidly disappear from the blood, and glucagon levels in the blood rise, telling the liver to release glucose.

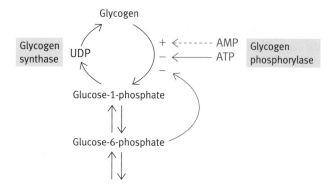

Fig. 16.13 The nonhormonal 'routine' allosteric controls on glycogen metabolism. See text for explanation. Green lines, allosteric control having positive effects; red lines, negative ones.

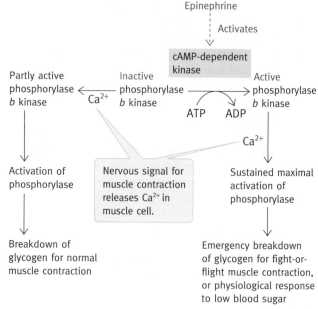

Fig. 16.14 Control of muscle phosphorylase. The enzyme has complex controls. First, it is partially activated allosterically by AMP. Second, it is activated by phosphorylase kinase. Phosphorylase kinase is itself activated in two ways. **(a)** It is partially activated by Ca^{2+} allosterically and **(b)** by a cyclic AMP (cAMP)-dependent protein kinase (PKA) plus Ca^{2+}. The mechanism is described in the text.

- There is the more stressful or emergency level, in which a stressful situation results in the brain signaling the adrenal glands to produce **epinephrine** (**adrenaline**). This presses the panic button for rapid glycogen breakdown. It does the same as **glucagon** in the liver. Muscle, lacking receptors, does not respond to glucagon, but does to epinephrine. Liver responds to both.

In plants, it is interesting that starch metabolism is much the same as glycogen metabolism in animals, with equivalent enzymes, but they have only the 'routine' level of controls. Plants don't feed at intervals, and don't run from predators, so don't need the other controls.

Control of glycogen breakdown in muscle

When a muscle contracts, it hydrolyses ATP to ADP; to regenerate the ATP, glycogen is broken down to provide the energy source. The signal for this in normal 'routine' (nonstressful) situations is AMP (note, not cAMP), which allosterically activates muscle glycogen phosphorylase (Fig. 16.13). AMP is the signal rather than ADP because the enzyme adenylate kinase catalyses the equilibrium:

$$2\,ADP \rightleftharpoons AMP + ATP$$

Because of the position of the equilibrium, AMP is present in greater amounts than ADP and so is a more sensitive indicator that the glycogen phosphorylase needs to be activated to increase ATP synthesis. Most control systems are of the 'push-pull' type. ATP and glucose-6-phosphate allosterically inhibit glycogen phosphorylase (Fig. 16.13). If these are plentiful then breakdown of glycogen is not required.

There is an additional refinement to the muscle control situation. In *normal* muscle contraction (not in the stressful

situation involving cAMP described below), glycogen breakdown is also *partially* activated by the Ca^{2+} release into the cytoplasm (page 131), which triggers the contraction following a motor nerve signal (Fig. 16.14). This mechanism additionally ensures that in *normal* 'routine' contraction, glycogen breakdown keeps pace with energy demand.

There is, however, another situation in which this 'routine' control of glycogen breakdown is overridden. In more stressful situations what you need is to generate ATP at the maximum possible speed. The first minute or so may make the difference between being eaten and not being eaten, and although glycolysis gives only a low yield of ATP, it can occur very rapidly and does not require oxygen. It is therefore logical in an emergency to break down glycogen at the maximum speed to provide glucose-1-phosphate for feeding into glycolysis. This may be important, for the main increased production of ATP by oxidative phosphorylation may take a minute or so while the heart speeds up to supply the extra oxygen needed.

This is known as the **flight-or-fight response**. Binding of epinephrine to cells of muscle and liver causes the production of the second messenger, cAMP (note, not AMP). This overrides other controls, and signals maximum glycogen breakdown. It also stimulates adipose cells to release fatty acids – another energy source.

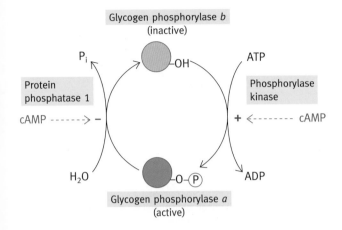

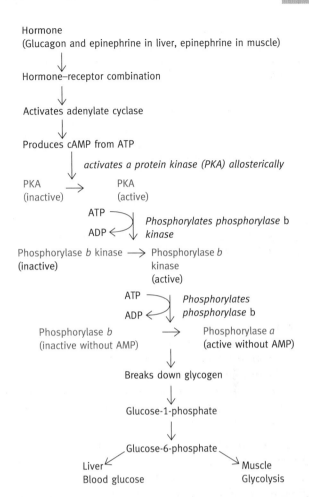

Fig. 16.15 Conversion of glycogen phosphorylase *b* to glycogen phosphorylase *a* by phosphorylase kinase and the reverse by protein phosphatase. It is important to note that the effects of cyclic AMP (cAMP) are not exerted *directly on* the reactions shown – see text. The —OH group is that of a serine residue in the protein.

Fig. 16.16 The amplifying cascade mechanism by which hormones activate glycogen phosphorylase. See page 249 for adenylate cyclase.

Mechanism of muscle phosphorylase activation by cAMP

In muscle, in normal situations, phosphorylase exists in a nonphosphorylated form known as **phosphorylase *b***. This is inactive in the absence of the allosteric activator, AMP, and as explained this *partial* activation by the latter is sufficient for routine needs.

In more demanding situations, **phosphorylase *b*** is converted to **phosphorylase *a***, *which is maximally active even without the presence of AMP*. Phosphorylase *b* is converted to the phosphorylated form, phosphorylase *a*, by a cAMP-activated protein kinase (**phosphorylase kinase *a* or PKA**, Fig. 16.15). Phosphorylase kinase can thus be activated in two ways, one (partially) by allosteric Ca^{2+} activation not involving phosphorylation, as described in Fig. 16.14, and the other by phosphorylation due to hormonal activation of PKA.

The cAMP activation of the phosphorylase kinase, which phosphorylates glycogen phosphorylase *b* is not direct; it activates the protein kinase, **PKA** (A for cAMP), which, in turn, activates phosphorylase kinase (also by phosphorylation), which then activates phosphorylase *b*, converting it to the '*a*' form. The whole scheme from the hormone onwards is shown in Fig. 16.16.

Why so complicated a mechanism? A cell will have only a relatively small number of hormone molecules attached to its receptors. In an emergency, the body requires a big response to the binding of epinephrine to cells. The attachment of a relatively few molecules of hormone to cell receptors must result in a massive breakdown of glycogen. What is needed is a rapid, vastly amplified, response, in which the minute signal

(hormone binding) cascades into a flood of reaction (glycogen breakdown). The multiple steps in phosphorylase activation constitute an amplifying or **regulatory cascade**. Suppose that one molecule of hormone activates one molecule of adenylate cyclase (the enzyme producing cAMP – see page 249), and that the latter produces 100 molecules of cAMP per minute. In this time period, the amplification is a 100-fold. If the cAMP activates a second enzyme which produces an activator at the same rate, the amplification becomes more than 100×100 and so on. In fact, glycogen phosphorylase activation involves four such amplification steps. Regulatory or amplifying cascades are used whenever a massive chemical response is needed from a minute signal. Phosphorylase attacks the end of glycogen chains and it would not be much use producing large numbers of active phosphorylase molecules if they could not find a glycogen end to work on. This is possibly one reason why glycogen is so highly branched, that is to say has many termini, compared with starch – plants do not go into metabolic panic mode.

Control of glycogen breakdown in the liver

The purpose of glycogen breakdown in the liver is mainly for release into the blood rather than for glycolysis as in muscle. Liver phosphorylase *b* is not allosterically partially activated by AMP as is the case in muscle. The glucose-1-phosphate produced by the enzyme is converted to glucose-6-phosphate. Glucose-6-phosphate is hydrolysed to free glucose, which enters the bloodstream to supply the brain and red blood cells. This occurs during fasting periods. The signal for this in liver is the hormone glucagon, secreted by the pancreas in response to low blood glucose. Since insulin is not produced in this situation, the glucagon/insulin ratio is high and hepatic glycogen breakdown is the predominant event. Glucagon activates the liver phosphorylase by the same mechanism as that triggered by epinephrine in muscle; the second messenger for glucagon is cAMP as for epinephrine.

Liver also participates in the fight-or-flight reaction. It responds to epinephrine by pouring out glucose into the blood. The rationale is again to give muscles the maximum supply of fuel to generate ATP in the emergency. An important control is that liver (but not muscle) phosphorylase is allosterically inhibited by free glucose. If there is plenty, then it is logical to inhibit production of more.

Reversal of phosphorylase activation in muscle and liver

Metabolic controls must be reversible – there has to be a switch off mechanism. In the case of glycogen phosphorylase in both liver and muscle the *a* form is converted back to the *b* form by dephosphorylation when the cAMP signal is no longer there. The phosphorylase *b* kinase is likewise inactivated.

The dephosphorylation of both is catalysed by **protein phosphatase 1**. However this must happen only when the cAMP signal is no longer there. There is a **phosphatase inhibitor 1** which inhibits the phosphatase, but the inhibitor is active only when phosphorylated by PKA. What is rather ingenious about the system is that when cAMP is present, it activates PKA which phosphorylates the inhibitor and activates it. Thus cAMP simultaneously leads to phosphorylase conversion of *b* to the *a* form and prevents its conversion back to the *b* form. When cAMP is no longer present, the inhibitor is dephosphorylated and no longer inhibits the phosphatase which then converts phosphorylase *a* back to the low activity *b* form. In liver glucose plays a part in this. It allosterically induces a conformational change in phosphorylase *a* which makes it susceptible to dephosphorylation by the protein phosphatase 1. PKA also phosphorylates glycogen synthase, *which inhibits it*. This inhibition will occur probably only so long as cAMP is present. However, glycogen synthase is not activated simply by the absence of cAMP, for there is another inhibitory mechanism; it is this, that is removed in the presence of insulin, as described

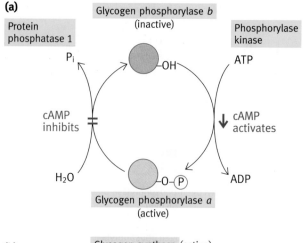

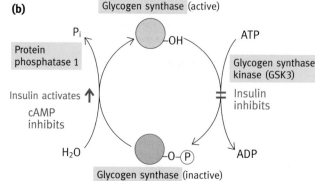

Fig. 16.17 Reciprocal controls on glycogen phosphorylase and glycogen synthase. **(a)** Cyclic AMP (cAMP) causes activation of phosphorylase kinase, which activates phosphorylase by phosphorylating it. The cAMP-stimulated PKA activates the phosphatase-inhibitor protein which prevents inactivation of the phosphorylase by the phosphatase. It also inhibits synthase activation. **(b)** In the presence of insulin, glycogen synthase kinase 3 (GSK3) is inactivated, thus preventing the latter from inhibiting glycogen synthase. In addition insulin causes activation of the phosphatase, thus activating the synthase. Red, inactive; green, active.

below. Possibly the cAMP-dependent inhibition of the synthase is an extra safeguard to avoid a futile cycle of glycogen breakdown and synthesis.

The switchover from glycogen breakdown to glycogen synthesis

Let us look briefly at the physiological context in which these controls are operating. Glycogen metabolism is a balance between glycogen breakdown and glycogen synthesis (Fig. 16.17). Apart from the 'routine' controls where everything is ticking over in a balanced way, glycogen breakdown will predominate in muscle when epinephrine is present to support stressful muscular activity. In liver, it will predominate in two circumstances; one when epinephrine is present and two when glucagon is liberated by the pancreas in response to low blood sugar. When epinephrine is no longer present glycogen breakdown will revert to the routine state in both muscle and liver. In liver, when the blood sugar level has been restored and glucagon is no longer there, the cAMP signal will disappear. The glucose will make phosphorylase *a* susceptible to conversion to *b*.

The switchover from glycogen breakdown to synthesis occurs after feeding and the blood sugar is high. This causes insulin secretion from the pancreas; *insulin is the signal for the activation of glycogen synthase.* It will continue to be secreted until the blood sugar level has been lowered to normal levels. The secretion of insulin and glucagon are reciprocally controlled according to the blood sugar level and, in the absence of insulin, glycogen synthase is inactivated by PKA. *However the inactivation of glycogen synthase which is removed by insulin is different from that produced by PKA in the presence of cAMP.* This will now be explained in the next section.

Mechanism of insulin activation of glycogen synthase

Much of the work on glycogen synthase control has been done on rabbit skeletal muscle but the same may apply to liver (see **Cohen** review at the end of this chapter).

In the situation after feeding, when insulin levels are high and glucagon is low, insulin effects are predominant. There is no cAMP signal in the absence of glucagon so phosphorylase will be in the relatively inactive *b* form.

Glycogen synthase has multiple phosphorylation sites. One group of sites of special interest to us now is a cluster of three serine residues found in the C-terminal end of the enzyme. A key observation was that when insulin activates glycogen synthase in muscle cells, *these three sites are dephosphorylated.* In the absence of insulin, a specific protein kinase called **glycogen synthase kinase 3** (GSK3) phosphorylates the sites. It inactivates glycogen synthase. *In the presence of insulin, GSK3 is inhibited, and thereby the kinase is inactivated.* At the same time insulin activates protein phosphatase 1. This dephosphorylates glycogen synthase which is thereby activated (Fig. 16.18).

How does insulin inactivate GSK3?

The mechanism is illustrated in Fig. 16.18. Insulin combines with the insulin receptors on the external surface of target cells. This activates a signalling pathway inside the cell, which results in the activation of yet another protein kinase, **PKB**. (The mechanism of PKB activation by insulin signalling is described in Chapter 27, page 452.) PKB phosphorylates GSK3, which *inactivates* it. Thus, to summarize, insulin causes inhibition of GSK3 and activation of the protein phosphatase. This latter dephosphorylates and consequently activates glycogen synthase.

The control is reversible. When the insulin level falls, PKB is inactivated by dephosphorylation; this allows GSK3 also to be dephosphorylated, which activates it. The glycogen synthase is now attacked by GSK3 which phosphorylates and inactivates

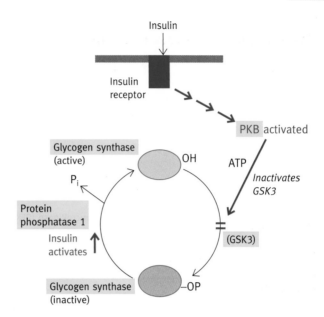

Fig. 16.18 Mechanism of control of glycogen synthase by insulin. The synthase is controlled by phosphorylation which inactivates it and dephosphorylation which activates it. Both PKA and glycogen synthase kinase 3 (GSK3) phosphorylate the enzyme, at different sites, but it is the phosphorylation performed by GSK3 that is reversed by insulin. It does this by activating PKB, a protein kinase that inactivates GSK by phosphorylating it. A protein phosphatase removes the relevant phosphate groups and thus activates the synthase. The mechanism by which insulin activates PKB is dealt with in the chapter on cell signalling (page 442).

it. In the absence of insulin the protein phosphatase needed to activate the synthase is no longer active, and glycogen synthesis ceases.

Control of glycolysis and gluconeogenesis

Allosteric controls

Figure 16.19 shows the main systems. The rationale of controls is as follows: AMP (not cAMP) indicates an increase in the ratio of ADP to ATP (page 252) so signalling that increased glycolysis is wanted to restore ATP levels. As well as activating glycogen phosphorylase, AMP is an allosteric activator of **phosphofructokinase** (PFK) which is the key controlling enzyme in glycolysis subject to multiple controls. At the same time, it inhibits the **fructose-1:6-bisphosphatase.** Activation of glycogen breakdown by AMP increases the level of fructose-6-phosphate which also activates PFK. Activation of PFK will, in turn, increase the level of fructose-1:6-bisphosphate which activates pyruvate kinase, an example of feed-forward control. Evolution is rarely satisfied by one-hit controls and the activating effects of AMP on PFK are balanced by inhibitory effects of ATP. The cell is thus constantly monitoring the ATP/ADP ratio (via AMP), and adjusting the glycolytic speed accordingly. In addition,

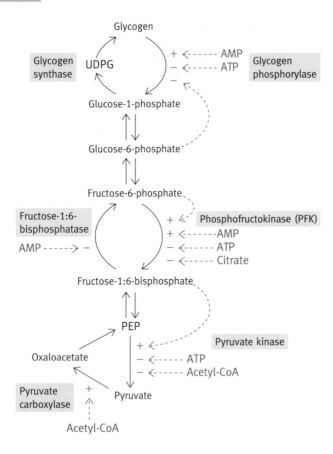

Fig. 16.19 The main intrinsic allosteric controls on glycogen metabolism, glycolysis, and gluconeogenesis. See text for explanation. Green broken lines, allosteric controls having positive effects on activities; red broken lines, allosteric controls having negative effects on activities. UDPG, uridine diphosphoglucose.

latter is transported out of the mitochondria to the cytoplasm where it allosterically inhibits PFK. This again makes sense since if the citric acid cycle is partially shut down; glycolysis, which feeds the cycle, should likewise be inhibited.

The activation of pyruvate carboxylase by acetyl-CoA to produce oxaloacetate needs explanation. Accumulation of acetyl-CoA occurs if the citric acid cycle is deficient in oxaloacetate, since oxaloacetate is needed to accept the acetyl moiety to form citrate. Pyruvate carboxylase catalyses the anaplerotic reaction, which tops up the citric acid cycle, and counteracts this deficiency (page 198). The acetyl-CoA thus automatically activates the synthesis of oxaloacetate. Inhibition of pyruvate dehydrogenase by acetyl-CoA helps to ensure adequate pyruvate for oxaloacetate production. This is also important in gluconeogenesis (see page 237).

However, as with glycogen metabolism, these internal allosteric controls on glycolysis are over-ridden by extracellular signals from hormones.

Hormonal control of glycolysis and gluconeogenesis

The signal for the liver to produce blood glucose is glucagon, *which activates both glycogen breakdown and gluconeogenesis.* Thus, it is logical for cAMP (whose synthesis is increased by glucagon) *in liver* to switch on glycogen breakdown *and* gluconeogenesis. Both produce glucose, but you do not want to activate glycolysis – it needs to be inhibited because you do not want to activate glucose synthesis and its breakdown at the same time (Fig. 16.20(a)). The same applies to the effects of epinephrine-stimulated production of cAMP.

In muscle, the situation is quite different. There, epinephrine stimulation of cAMP production is designed to increase generation of ATP; as mentioned earlier, epinephrine is released in the fight-or-flight reaction in which vigorous muscular contraction maybe called for. It is therefore logical in skeletal muscle for cAMP to switch on glycogen breakdown but, in contrast to liver, it would *not* be logical in muscle to inhibit glycolysis (Fig. 16.20(b)), and thwart the primary aim of

when the ATP level is high, the citric acid cycle flux (the passage of metabolites through the pathway) is diminished (because less ATP synthesis is wanted) and citrate accumulates. The

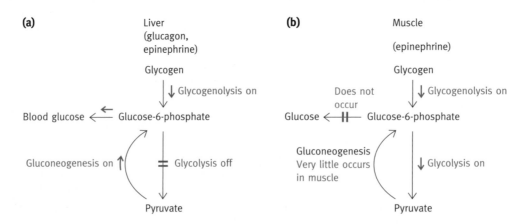

Fig. 16.20 The different control requirements of **(a)** liver and **(b)** muscle in response to glucagon and/or epinephrine. Both hormones have cyclic AMP as second messenger. The term glycogenolysis used here is a synonym for glycogen breakdown.

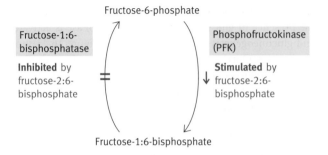

Fig. 16.21 Fructose-2:6-bisphosphate and the control of glycolysis in liver (see text).

producing ATP as rapidly as possible. (Muscle, we remind you, does not produce blood glucose; it has no glucose-6-phosphatase needed for this.) Glycolysis speed-up is required in muscle to prepare for the fight-or-flight reaction.

We thus see that cAMP needs to block glycolysis and activate gluconeogenesis in liver, but activate glycolysis in muscle.

How are these different ends achieved in the two tissues? To answer this we need to turn to an aspect of glycolytic and gluconeogenesis control we have not yet mentioned.

Control of glycolysis and gluconeogenesis pathways by fructose-2:6-bisphosphate

This section may require more than one reading.

Fructose-2:6-biphosphate (F-2:6-BP) is solely a regulatory molecule for PFK. It has not previously been mentioned in this book. The reverse of the PFK reaction, catalysed by fructose-1:6-bisphosphatase (FBPase), has to be reciprocally controlled to avoid a futile cycle (Fig. 16.21). PFK is allosterically inhibited by ATP and is inactive unless this is counteracted by F-2:6-BP. *The rate of glycolysis in muscle and liver parallels the level of F-2:6-BP* so we must consider what controls this level.

F-2:6-BP is synthesized by another enzyme ,which phosphorylates fructose-6-phosphate in the 2 position. It is called **PFK2**, and is bifunctional, with two catalytic sites. One synthesizes F-2:6-BP, the other hydrolyses it back to fructose-6-phosphate. The enzyme does one or the other but not both at the same

time. It is phosphorylated by a kinase, PKA activated by cAMP. *In the phosphorylated form PFK2 hydrolyses F-2:6-BP. When the phosphate group is removed it synthesizes it* (Fig. 16.22). *F-2:6-BP stimulates PFK (note, not PFK2) but inhibits the reverse reaction by FBPase.*

Muscle and liver PFK2 enzymes are different

Epinephrine in muscle and liver and glucagon in liver, as described earlier, cause the production of cAMP as second messenger. The latter activates PKA in both tissues. The liver and muscle have different isoforms of PFK2. The liver one is phosphorylated by PKA, causing it to switch from synthesizing F-2:6-BP to destroying it. It thus inhibits PFK and, therefore glycolysis, and stimulates FBPase which is needed for gluconeogenesis.

Muscle PFK2 has no site for phosphorylation (the serine residue present in liver PFK2 that receives the phosphate group is replaced by an alanine residue in the muscle enzyme). Thus, cAMP does not stimulate destruction of F-2:6-BP in muscle. The level of F-2:6-BP rises in the presence of an epinephrine signal, because it is needed to increase the rate of glycolysis. It is not known, precisely, how this occurs, but the most likely explanation is that, by stimulating glycogen breakdown, the supply of fructose-6-phosphate, the substrate of PFK2, increases and stimulates the synthesis of F-2:6-BP.

Control of pyruvate kinase

Pyruvate kinase (Fig. 16.23) is another enzyme responding to glucagons *via* cAMP. In the liver, but not in muscle, cAMP causes phosphorylation of the enzyme, resulting in its inactivation. The rationale of this is that if blood glucose is low, the liver needs to produce glucose by the gluconeogenesis pathway, not to use it for energy generation. This is therefore a glycolytic switch off, additional to that at the PFK step. Gluconeogenesis is, as stated, a function of the liver mainly in response to glucagon (though other hormones such as cortisol also have a role; see below). We have described how gluconeogenesis from pyruvate requires the following steps, catalysed by pyruvate

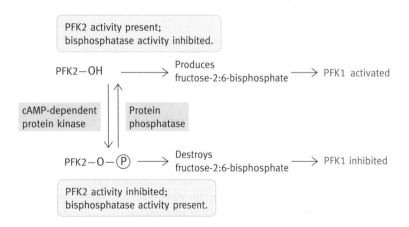

Fig. 16.22 Control of liver PFK2 by phosphorylation by cyclic AMP-dependent protein kinase (PKA). The PFK2 is a double-headed enzyme: (1) it synthesizes fructose-2:6-bisphosphate when unphosphorylated; (2) it destroys the 2:6 compound when phosphorylated.

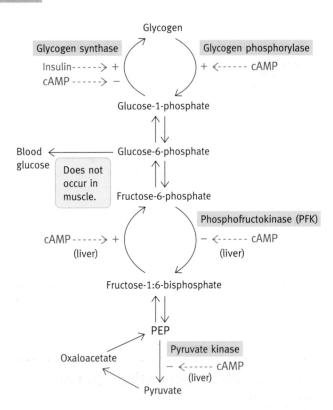

Fig. 16.23 The main external controls in this area of metabolism in liver. Note that 'cAMP' represents the action of glucagon and epinephrine. Its action is not directly on the enzyme being controlled.

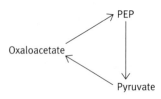

Fig. 16.24 Potential futile cycle. PEP, phosphoenolpyruvate.

carboxylase and phosphoenolpyruvate carboxykinase (PEP-CK) respectively:

$$Pyruvate + ATP + HCO_3^- + H_2O \rightarrow oxaloacetate + ADP + P_i + H^+$$

$$Oxaloacetate + GTP \rightarrow PEP + GDP + CO_2.$$

You can see that there would be a futile cycle, as shown in Fig. 16.24, if pyruvate kinase continues to catalyse the reaction:

$$PEP + ADP \rightarrow pyruvate + ATP$$

For gluconeogenesis, the pyruvate kinase in liver needs to be inactivated so that the PEP is sent up the gluconeogenic path. Inactivation of the liver enzyme by cAMP achieves this.

The net effect of all these controls is that in liver, the PFK in the presence of glucagon is inhibited, the FBPase is activated and gluconeogenesis supplies blood sugar. Figure 16.23 summarizes this. Once again muscle has different needs – it does not synthesize glucose, glycolysis must not be switched off, and its pyruvate kinase is not phosphorylated in response to cAMP elevation by epinephrine.

Glucocorticoid stimulation of gluconeogenesis

In starvation, the main substrate for gluconeogenesis in the liver comes from muscle-protein breakdown to amino acids. During periods of stress, the cortex of the adrenal glands liberate steroid hormones called glucocorticoids, the principal one in humans being **cortisol**. It promotes gluconeogenesis. It has complex effects in the body. It promotes protein breakdown in muscles and other peripheral tissues. This supplies amino acids to the liver, which uses them for gluconeogenesis. It affects the activity of the PEP-CK gene needed for PEP synthesis.

Fructose metabolism and its control differs from that of glucose

In Western societies, large amounts of fructose are consumed, largely in the form of sucrose, but also in fructose drinks and other manufactured foods. Fructose is absorbed from the intestine and is metabolized mainly (or entirely) by the liver. Its metabolism is not insulin controlled, and therefore not directly affected by diabetes, and was therefore thought to be suitable for patients with the disease. It is converted in the liver to fructose-1-phosphate by **fructokinase** and this is split by **aldolase B** to glyceraldehyde and dihydroxyacetone phosphate. (Aldolase B is different from the glycolytic aldolase A.) The glyceraldehyde is phosphorylated to the 3-phosphate, so the products are the same as in glycolysis (glyceraldehyde-3-phosphate and dihydroxyacetone phosphate). Part of this is converted back to glucose.

So, the story so far is not very different from that of glucose metabolism. However, the control situations *are* different and result in fructose being of significance beyond its calorific value in causing increased fat synthesis. The aldolase B reaction by-passes the main glycolytic control of the phosphofructokinase step. The result is that fructose metabolism may swamp the liver cell with NADH reducing equivalents, since the glyceraldehydes-3-phosphate is oxidized by NAD^+. This is similar to what happens with excessive alcohol metabolism (page 239). This, and the rapid formation of pyruvate from fructose leads to increased fat synthesis, and export from the liver as very-low-density lipoproteins (VLDL) (page 170).

Control of pyruvate dehydrogenase, the citric acid cycle, and oxidative phosphorylation

Pyruvate dehydrogenase occupies a strategic position in metabolism as the irreversible reaction producing acetyl-CoA by

Fig. 16.25 The intrinsic control of pyruvate dehydrogenase (PDH) by direct allosteric control and by phosphorylation and dephosphorylation in the mammalian enzyme complex. The multiplicity of these controls reflects the strategic position of PDH whose activity is the gateway for pyruvate to enter the citric acid cycle and the fat-synthesis pathways. Despite their complexity in detail, they largely add up to activation by substrates and inhibition byproducts. (In this context ATP can be regarded as a product of the result of entry of acetyl-CoA into the citric acid cycle.) It is worth noting that control by phosphorylation of proteins is usually associated with extrinsic controls, rather than with the intrinsic ones here.

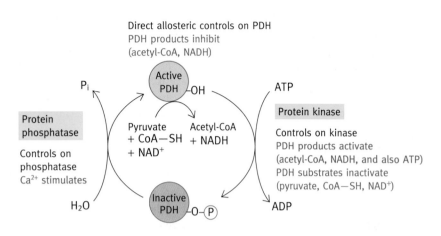

which pyruvate from glycolysis feeds into the citric acid cycle and into fat-synthesis. The pyruvate dehydrogenase complex is regulated in several ways. Acetyl-CoA and NADH, which are products of the enzyme reaction, are allosteric inhibitors. CoA—SH and NAD^+, which are substrates, allosterically activate it. Thus the acetyl-CoA/CoA ratio and the NADH/NAD^+ ratio control pyruvate dehydrogenase. The logic is that if there is a lot of acetyl-CoA and NADH, the production of acetyl-CoA and NADH by pyruvate dehydrogenase should be reduced and vice versa if CoA and NAD^+ are high.

Of major importance is the negative control by high ATP levels. This, in effect, is monitoring the 'energy charge'. If ATP is low, feed in more fuel to the citric acid cycle; if high, cut off the fuel supply. The ATP control of pyruvate dehydrogenase is not a direct one. At high ATP/ADP ratios a **pyruvate dehydrogenase kinase** is activated and this inactivates the dehydrogenase by phosphorylation of the enzyme; this is reversed by a phosphatase (Fig. 16.25). The kinase is actually part of the pyruvate dehydrogenase complex. The inactivating kinase is additionally allosterically activated by NADH and acetyl-CoA. Usually protein kinases are activated by extracellular signals, but this is one of the exceptions.

In the citric acid cycle and electron transport, the intrinsic controls are different in that the major internal controls are the availability of NAD^+ and ADP as substrates. If much of the NAD^+ is present in its reduced form (NADH) then the dehydrogenases in the citric acid cycle are restricted in their activity by lack of substrate. Since NADH accumulates when electron transport to oxygen cannot cope with available NADH produced by the cycle, the cycle is inhibited. Similarly if the ADP/ATP ratio is low, electron transport is inhibited because oxidation and phosphorylation are tightly coupled (called **respiratory control**). This is of major importance. In short, if there is plenty of ATP the furnaces shut down automatically.

In addition to this control by NAD^+ and ADP availability, the cycle is allosterically controlled at the citrate synthase step (ATP inhibits), the isocitrate dehydrogenase step (ATP inhibits and ADP activates), and the α-oxoglutarate dehydrogenase step

(succinyl-CoA and NADH inhibit). It is all very logical and the cycle metabolites are kept in balance one with another – no pile-ups, no shortages.

Controls of fatty acid oxidation and synthesis

Nonhormonal controls

These are illustrated in Fig. 16.26. Acetyl-CoA carboxylase plays a key role here. It converts acetyl-CoA to malonyl-CoA which

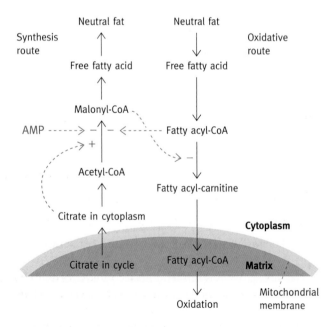

Fig. 16.26 Major intrinsic control points in fat oxidation and synthesis by which the two routes mutually suppress each other. Note that once acetyl-CoA is formed from fat breakdown, its further metabolism is subject to the controls of the citric acid cycle, etc., already described. Dashed lines indicate allosteric effects.

enters the fatty acid synthesis pathway. Fatty acid oxidation must be inhibited to avoid a futile cycle uselessly synthesizing fatty acids and breaking them down again to acetyl-CoA. The two pathways for fatty acid synthesis and their breakdown suppress each other. Fatty acyl-CoAs (the product of the first step in fatty acid oxidation) allosterically inhibit acetyl-CoA carboxylase, the first committed step in fat synthesis while malonyl-CoA inhibits transfer of the fatty acyl groups to carnitine which prevents them being transported to the intra-mitochondrial site of their oxidation. Acetyl-CoA carboxylase therefore is a key controlling enzyme. It is activated by citrate (which also gives rise in the cytoplasm to acetyl-CoA, the substrate of the enzyme). Citrate is transported out of the mitochondria only when its level is high, and this happens only in times of food plenty when it is logical to store fat.

An important control on acetyl-CoA carboxylase is by phosphorylation which inhibits it. This is effected by the AMP-activated kinase to be described shortly.

Breakdown of acetyl-CoA carboxylase is a new type of fat metabolism control

A recent discovery (see **Neels and Olefsky** reference in Further reading) has shown that a protein (called TRB3) mediates the destruction of the carboxylase by proteasomes (page 399). TRB3 is induced by cellular stress; it blocks the action of insulin and in destroying the carboxylase increases fat oxidation (See reference in Further reading). It shifts the balance of control from synthesis to energy production much as AMPK does (see below) but in a different way.

Hormonal controls on fat metabolism

The most important decision in this area is whether fat cells should store fat as triacylglycerol (TAG) or release free fatty acids from stored TAG into the blood. Insulin is the signal for the former, glucagon and epimephrine for the latter. The level of blood glucose is the master controller, for it determines the relative levels of the two hormones.

Fat cells contain a **hormone-sensitive lipase**, glucagon and epinephrine (via cAMP) being the activators. It carries out the reaction:

$$\text{Triacylglycerol} + H_2O \rightarrow \text{diacylglycerol} + \text{free fatty acid}$$

The diacylglycerol is further degraded to glycerol and free fatty acids. The hormone sensitive lipase is controlled by phosphorylation, the phosphorylated enzyme being active. A cAMP-activated protein kinase carries this out and a protein phosphatase reverses it in the absence of cAMP (Fig. 16.27).

Insulin antagonizes these effects; it stimulates fat synthesis in liver and the high insulin/glucagon ratio restricts release of fatty acids from fat cells, and increases glucose uptake.

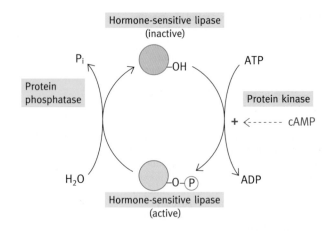

Fig. 16.27 Activation of adipose cell hormone-sensitive lipase by cyclic AMP (cAMP)-dependent phosphorylation. Insulin antagonizes the effects of the catecholamine hormones and glucagon, which increase the cAMP level. Note also that glucagon and epinephrine cause inhibition of fat synthesis in liver and adipose cells, respectively, by preventing dephosphorylation of acetyl-CoA carboxylase.

Fat cells take up fatty acids from VLDL sent out by the liver, and convert them to TAG. For this glycerolphosphate is required (page 239), which is supplied by reduction of the glycolytic intermediate, dihydroxyacetone phosphate. Since dihydroxyacetone phosphate is formed by the glycolytic pathway, uptake of glucose by the adipose cells is needed. Insulin stimulates this. The presence of glucose in the cell, and hence of dihydroxyacetone phosphate facilitates TAG synthesis.

Effects of leptin and adiponectin on fat metabolism

In dealing with obesity and appetite controls (page 157) we mentioned two hormones produced by adipocytes of the fat depots in the body. Leptin is produced when the fat reserves are depleted. It has neurogenic effects (page 158) in the hypothalamus but both it and adiponectin also have a more direct control on fat metabolism. They both activate the AMP-activated kinase described below and inhibit acetyl-CoA carboxylase by phosphorylation.

Responses to metabolic stress

We have already referred to the liberation of the hormone cortisol in times of stress. But also there are situations in which cells become deficient in ATP to a dangerous point. Excessive exercise may cause it in muscle, or oxygen deficiency may do the same. In heart cells ATP shortage could be very serious.

Cells have a general 'emergency' response, essentially shutting down anabolism (synthetic reactions). There is no point in this situation in synthesizing proteins or other cell components, since inhibition of this poses no immediate danger. Two mechanisms are known – one is the AMP-activated kinase, and another is the hypoxia response. These will now be described.

Response to low ATP levels by AMP-activated protein kinase

We have described how the production of ATP by oxidative phosphorylation is controlled. ATP level in the cell is of critical importance and for this reason there is a more general ('global') control in cells which senses their energy state. **AMP-activated kinase (AMPK)** is the central effector of this control. Note that this is not AMP kinase (adenylate kinase), which phosphorylates AMP. The logic is that if AMP is present at increased levels, it indicates a potential deficiency of ATP.

The AMPK is activated by the increased AMP level. It closes down nonessential metabolic systems by phosphorylating key enzymes. It is itself controlled by phosphorylation; another **AMPK kinase** phosphorylates and activates it in the presence of AMP while a protein phosphatase can reverse this. The activated enzyme both shuts down anabolic reactions and increases ATP-generating catabolic processes (Fig. 16.28).

AMPK when activated has a wide range of activities (see **Kemp** *et al.* reference in Further reading).

- It mobilizes the glucose transporting enzyme GLUT 4 and increases glucose uptake by cardiac muscle, skeletal muscle and fat cells.
- It inhibits fat synthesis by phosphorylating acetyl-CoA carboxylase, as described above.
- It activates the glycolysis enzyme PFK2 in cardiac muscle in oxygen deficiency (but not in skeletal muscle).

Cells in the mass of a tumour tend to be deficient in oxygen, so AMPK is also believed to protect tumour cells from oxygen depletion and so has the undesirable effect of assisting cancer development. AMPK has relevance to obesity and type 2 diabetes. By promoting glucose uptake into muscle it may combat

insulin resistance in which glucose transport is deficient so that drug activators of AMPK (which are known) may have antidiabetic value.

Response of cells to oxygen deprivation

Protection against hypoxia

Hypoxia refers to a situation in which the oxygen level in a tissue is low. In almost all mammalian cells (red blood cells excepted since they get their ATP from anaerobic glycolysis) an adequate supply of oxygen is of overriding importance since most of the ATP production depends on it. There are several protective responses to hypoxia, most at the level of gene activation. These include the following:

- increased production of the hormone, **erythropoetin**, causes increased red blood cell production and hence increases the oxygen-carrying capacity of blood
- some glycolytic enzymes and glucose transporters are induced, the rationale of this being that in the absence of oxygen, glycolysis is the only alternative way of generating ATP
- factors are produced that promote **angiogenesis** (the production of new blood vessels).

Mechanism of the hypoxia response

The key to hypoxic gene activation is a family of transcription factors called **hypoxia-inducible factors** (HIFs). A transcription factor is a protein that enters the nucleus and attaches to the control sections of target genes, which it activates to produce mRNAs and hence the synthesis of the proteins coded by those genes. This is the subject of Chapter 23, so here we need only talk in general terms of gene activation. It has a domain (region), which promotes specific gene activation.

In normoxia (normal oxygen levels), HIF protein is destroyed rapidly – its half-life in the cell is about 5 minutes, so that its

Fig. 16.28 Simplified version of the role of AMP kinase in restricting anabolic reactions when increase in the AMP/ATP ratio indicates that ATP levels are suboptimal.

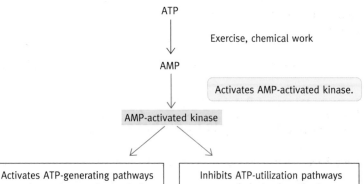

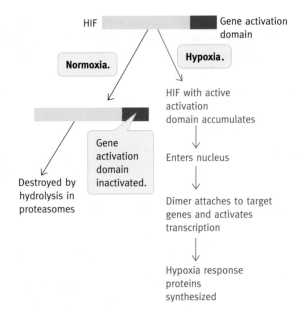

Fig. 16.29 Summary diagram of response to oxygen levels. HIF, hypoxia-inducible factor.

level remains low, therefore HIF has essentially no role to play in normal oxygen conditions. However, in hypoxia its level increases so that it causes the synthesis of protective response proteins. Figure 16.29 gives a general summary of the mechanism.

What is the nature of the switches in hypoxia that prevent proteolytic breakdown of HIF? The control depends on two separate post-translational modifications of the HIF. In normoxia HIF is rapidly destabilized by hydroxylation of two critical proline residues, which leads to proteolytic destruction of the HIF in proteasomes. The proline hydroxylases are rate-limited by oxygen concentration and are believed to be acting as oxygen sensors so that, at low oxygen levels, HIF is not subject to proline hydroxylation and accumulates. An asparagine hydroxylation also occurs.

The hypoxia response mechanism described here has potential medical relevance, because hypoxia in specific tissues is one of the effects of heart attacks, strokes, and other vascular diseases.

It presumably may also have the less desirable effect of helping cells in the relatively oxygen poor centre of tumours to thrive.

Integration of fat and carbohydrate metabolism controls in diabetes

The interrelated nature of metabolic pathways and their control can be illustrated by describing some of the biochemical features seen in diabetes mellitus. There are two forms of the disease.

- **Type 1**, also known as **juvenile-onset diabetes** or **insulin-dependent diabetes mellitus**, is caused by auto-immune attack on insulin-producing cells of the pancreas, so that insulin production is deficient. This type responds to insulin injection. It usually occurs early in life.

- In **type 2** or **maturity-onset diabetes** (**non-insulin-dependent diabetes**), insulin levels may be normal or elevated, but cells are relatively insensitive to the hormone. The onset of this usually occurs after the age of about 35, frequently associated with obesity. Diabetes is one of the most prevalent diseases, affecting more than 100 million people, type 2 being much more prevalent than type 1.

Insulin has widespread effects on mammalian metabolism. In general terms it has the opposite effects of those of glucagon. It promotes glycogen synthesis, increases fatty acid synthesis in the liver and glucose uptake into cells such as muscle and adipose tissue. When the insulin/glucagon ratio is low, glycogen breakdown occurs in liver to supply blood glucose, and the hormone-sensitive lipase of fat cells is activated to release fatty acids into the blood. Insulin also stimulates the synthesis of specific enzymes involved in response to the hormone such as those involved in metabolism of dietary components after feeding. Conversely the synthesis of PEPCK, an enzyme required for gluconeogenesis, is reduced by insulin. The basic mechanism of insulin action is part of the larger cell signalling topic dealt with in Chapter 27.

In uncontrolled diabetes, the blood glucose level may become elevated to the point where it spills over into the urine, carrying with it water, resulting in increased urine volume and thirst. In the type 1 disease, where insulin production is deficient, there is a high glucagon/insulin ratio, the metabolic events that occur resemble some of those found in the starved state. Some of the biochemical features of the type 1 disease are as follows.

- Transport of glucose into muscle and adipose cells is impaired because the glucose transporters are not recruited into their functional sites in the membrane.

- Utilization of glucose is diminished by these cells, leading to high blood glucose levels; the muscles, are especially critical in this respect. Although glucose entry into liver is mainly not insulin-dependent, the high glucagon/insulin ratio promotes gluconeogenesis; glycogen breakdown in the liver is increased, both of these events contributing further to high blood glucose levels.

- The high glucagon/insulin ratio activates lipolysis in adipose cells so that the blood level of free fatty acids is high.

- The liver metabolizes these and produces ketone bodies (page 157) in excess of what can be utilized by peripheral tissues. In uncontrolled type 1 diabetes the blood levels of acetoacetate and β-hydroxybutyrate are elevated. Excretion of these can lead to a long-term depletion of cations such as Na^+, causing **ketoacidosis**, which in uncontrolled diabetes can be fatal. Spontaneous decarboxylation of acetoacetic acid produces acetone, which gives a characteristic fruity smell to the breath of people with untreated diabetes. **Ketosis** is rarely encountered in type 2 diabetes; in this type, insulin is present so the excessive release of fatty acids from the fat cells by hormone-sensitive lipase does not occur. Nonenzymic glycation of proteins increases if the blood glucose level is high. The aldehyde group of glucose reacts with protein amino groups and forms a stable covalent complex. Measurement of the level of glycated hemoglobin is an index of how well the blood glucose level has been controlled for the previous couple of months. (Red blood cells only live for 120 days.)

Summary

Reason for metabolic controls Metabolic pathways must be regulated to avoid futile substrate cycling, and to respond to changing physiological needs. Enzyme activities may be controlled by changing the amount of enzyme, and/or by changing the rate of catalysis of enzymes. The first is slow; the second is almost instantaneous. Control points in metabolic pathways are usually at irreversible steps in which the forward and backward reactions can be separately controlled. Regulation of enzymes at these points is by allosteric mechanisms and/or covalent modification.

Allosteric control is a powerful concept essential for cells to exist. Allosteric enzymes are multisubunit proteins with allosteric sites, to which molecules (modulators) attach and affect the activity. The effect of substrate concentration on their reaction rates is sigmoidal rather than hyperbolic. Attachment of modulators usually alters the affinity of the enzyme for its substrate(s) and this affects its rate of activity at a given suboptimal substrate concentration. Two theoretical models that account for their properties are the concerted model and the sequential model. Both involve conformational changes. Allosteric control coordinates the rates of disparate metabolic pathways and is virtually instantaneous.

Control by phosphorylation involves covalent modification of enzymes by protein kinases. The phosphorylation may activate or inhibit the activity and is reversible by phosphatases. The phosphorylation states of the enzymes are usually regulated by hormones. Glucagon, epinephrine (adrenaline), and insulin are the important ones in our present context. The first two cause the production of a second messenger molecule, cyclic AMP (cAMP), which activates kinases. Insulin control also involves phosphorylation but is more complex.

Control of glucose uptake into cells Glucose does not diffuse across membranes; it must be transported by proteins. The facilitated diffusion occurs by the movement of transport proteins from within the cell to the membrane. These glucose transporters are GLUT isoforms. This process is not insulin-responsive in brain, red blood cells or liver but is controlled by insulin in adipose cells and muscle.

Once glucose enters the cell it is rapidly phosphorylated by hexokinase (most tissues) or glucokinase (liver). The lower affinity of glucokinase means that at times when blood glucose is low and the liver is releasing glucose it does not efficiently take it up again. Hexokinase is inhibited by glucose-6-phosphate at physiological concentrations but glucokinase is not. This allows the liver to take up glucose at high blood sugar levels and to synthesise glycogen-using glucokinase to phosphorylate glucose. Not only does glucokinase only operate efficiently at high blood glucose but it is increased by insulin.

Control of glycogen metabolism Glycogen phosphorylase in the absence of hormonal signals exists in muscle and liver as phosphorylase *b*, which is allosterically activated by AMP and inhibited by ATP. Ca^{2+} released during muscle contraction activates. Epinephrine (and glucagon in the liver) causes cAMP to be formed in muscle and liver. This triggers the conversion of phosphorylase *b* to form *a*, which is fully active without AMP. The conversion is by a catalytic cascade of kinases, which amplifies the response. At the same time cAMP causes inactivation of glycogen synthase so that a futile cycle is avoided. Glycogen synthase is active only when dephosphorylated. The main activating signal is insulin, which causes the dephosphorylation of the synthase at specific sites.

Control of glycolysis and gluconeogenesis The allosteric controls here fit a logical pattern. If ATP levels are suboptimal, AMP, which is a sensitive indicator of this, activates glycogen breakdown and glycolysis. It also inhibits the reverse pathway involved in gluconeogenesis. ATP inhibits glycolysis as does citrate since both signal adequate energy levels.

The hormonal controls differ in muscle and liver. A major control in liver is in response to glucagon. The pancreas liberates glucagon when blood glucose levels are low. The liver's response is to inhibit glycolysis, increase gluconeogenesis and channel glucose-6-phosphate into production of blood glucose by glucose-6-phosphatase. This is achieved by cAMP indirectly regulating the activities of an enzyme PFK2, which controls the level of the fructose-2:6-bisphosphate. The latter compound is a main determinant of the rate of glycolysis. It activates PFK.

PFK2 is an unusual enzyme with two different catalytic sites. In the nonphosphorylated form it synthesizes the 2:6 compound and when phosphorylated if hydrolyses it. cAMP, produced in response to the glucagon signal, activates a kinase which phosphorylates PFK2. This destroys the 2:6-activating molecule and inhibits glycolysis. The final part of this involved control is that the 2:6 compound inhibits fructose-1:6-bisphosphatase. This enzyme is required for gluconeogenesis so the net effect of destruction of the 2:6 compound in the presence of cAMP is to inhibit glycolysis and stimulate gluconeogenesis. Epinephrine has a similar effect to glucagon and is responsible for the flight-or-fight reaction to a threatening situation.

In muscle, epinephrine also produces cAMP but the need here is to maximize glycolysis to produce ATP. In this tissue the PFK2 is a different isoenzyme and is not affected by cAMP. The increased glycogen breakdown caused by cAMP stimulates glycolysis and also, in some indirect way related to this, increases the level of the activating fructose-2:6-bisphosphate. Muscle does not have the gluconeogenesis pathway and does not produce blood glucose.

Gluconeogenesis is also promoted in liver by an additional hormonal control. The starting point of the pathway is phosphoenolpyruvate (PEP). When glucagon raises the level of cAMP it inhibits the pyruvate kinase enzyme, thus channelling PEP into the gluconeogenesis pathway. The stress hormone cortisol liberated in starvation also promotes gluconeogenesis (Chapter 15).

Fructose metabolism and its control differs from that of glucose leading to increased fat synthesis.

Pyruvate dehydrogenase is controlled by a combination of logical allosteric controls in which substrates activate and products inhibit. There is however an inbuilt protein kinase. Most protein kinases are subject to hormonal controls but in this case it is inhibited by products of the enzyme reaction and activated by substrates. This is logical because phosphorylation of the enzyme by the kinase inactivates it.

The citric acid cycle is controlled both allosterically and by availability of NAD^+ and ADP. Electron transport is tightly coupled to ATP production and availability of ADP is of prime importance in control.

Fatty acid oxidation and synthesis are reciprocally controlled allosterically so that either oxidation or synthesis is proceeding and a futile cycle is avoided. An AMP-activated kinase also inhibits fatty acid synthesis by phosphorylating acetyl-CoA carboxylase, the first committed step in the synthetic pathway. This inhibits the enzyme. The malonyl-CoA produced by the carboxylase inhibits the transport of fatty acids into mitochondria, thus inhibiting their oxidation. In fat cells, hormonal control is principally at the level of triacylglycerol (TAG) breakdown and synthesis. Glucagon and epinephrine (*via* cAMP) activate the hormone-sensitive lipase, which liberates fatty acids into the blood while insulin promotes TAG synthesis.

Overall regulation of ATP levels is a safety control mechanism. If the ATP 'charge' is suboptimal, its level is maintained by switching off nonvital synthetic reactions using ATP. This is done by the ubiquitous AMP-activated kinase, which shuts down many processes by specific phosphorylations. This is a response to metabolic stress.

Response of cells to oxygen deprivation is how the body deals with another stress situation. **Hypoxia** is a situation in which the oxygen level in a tissue is abnormally low and protective responses occur. These are to increase erythropoetin production, which increases red blood cell numbers; they increase glycolytic enzymes and glucose transporters. Since glycolysis can produce ATP anaerobically, they increase growth of new blood vessels in the tissue.

These responses require gene activation by a hypoxia-inducible transcription factor (HIF). In normoxia, HIF is inactivated and destroyed due to hydroxylation of proline and asparagine residues. This does not occur in hypoxia in which situation the HIF remains active and induces the protective responses.

The system has medical interest in that heart attacks and other vascular diseases cause hypoxia in specific tissues. It also unfortunately helps hypoxic cells of tumours to survive and to become vascularized thus potentially helping cancers to develop.

Diabetes is one of the most important examples in which metabolic control is abnormal. There are two types of the disease. Type 1 occurs usually in juveniles and is due to a deficiency of insulin. Type 2 usually occurs later in life and is mainly due to cells failing to respond to insulin normally. Insulin may be present in normal amounts. Its onset is often associated with obesity.

Further reading

Substrate cycles

Newsholme, E. A., Challiss, R. A. J., and Crabtree, B. (1984). Substrate cycles: their role in improving sensitivity in metabolic control. *Trends Biochem. Sci.*, 9, 277–80.
Describes the regulatory roles of these cycles.

Control of carbohydrate metabolism

Hue, L. and Rider, M. H. (1987). Role of fructose-2, 6-bisphosphate in the control of glycolysis in mammalian tissues. *Biochem. J.*, 245, 313–24.

Kitamura, K., *et al.* (1989). Purification and characterisation of rat skeletal muscle fructose-6-phosphate, 2-kinase: fructose-2, 6-bisphosphatase. *J. Biol. Chem.*, **264**, 9799–806.

Depre, C., Rider, M. H., and Hue, L. (1998). Mechanisms of control of heart glycolysis. *Eur. J. Biochem.*, **258**, 277–90.
Review on mechanisms of control of heart glycolysis under normal and reduced oxygen supply.

Cornish-Bowden, A. and Cardenas, M. L. (1991). Hexokinase and glucokinase in liver metabolism. *Trends Biochem. Sci.*, **16**, 281–2.
Discusses briefly a common misconception about these enzymes.

Hardie, D. G. (2000). Metabolic control: A new solution to an old problem. *Curr. Biol.*, **10**, R757–9.

Kemp, B. E., *et al.* (2003). AMP-activated protein kinase, super metabolic regulator. *Biochem. Soc. Trans.*, **31**, 162–8.
Review of the metabolite sensing mechanism found in all eukaryotes which responds to ATP depletion by shutting down ATP-utilizing anabolic pathways and increasing catabolic pathways generating ATP, as well as controlling gene expression.

Carling, D. (2004). The AMP-activated protein kinase cascade – a unifying system for energy control. *Trends Biochem. Sci.*, **29**, 18–24.
Review of this downstream component of a protein kinase cascade that acts as an intracellular energy sensor maintaining the energy balance within the cell and whole body.

Jope, R. S. and Johnson, G. V. W. (2004). The glamour and gloom of glycogen synthase kinase-3. *Trends Biochem. Sci.*, **29**, 95–102.
GSK3 is a key component in a number of cellular processes and diseases.

Kahn, B. B. *et al.* (2005). AMP-activated protein kinase: ancient energy gauge provides clues to modern understanding of metabolism. *Cell Metabolism*, **1**, 15–24.
A comprehensive review which deals with the importance of AMPK to metabolism in general including obesity and type 2 diabetes.

Witters, L. A., Kemp, B. E., and Means, A. R. (2006). Chutes and ladders: the search for protein kinases that act on AMPK. *Trends Biochem. Sci.*, **31**, 13–6.
A review of AMP-activated protein kinase written in terms of a game, more familiar to some as Snakes and Ladders.

Watson, R. T. and Pessin, J. E. (2006). Bridging the gap between insulin signalling and GLUT4 translocation. *Trends Biochem. Sci.*, **31**, 215–22.

Control of fat metabolism

McGarry, J. D. and Foster, D. W. (1980). Regulation of hepatic fatty acid oxidation and ketone body production. *Annu. Rev. Biochem.*, **49**, 395–420.
Despite the regulatory title, the article also provides a general review of ketone body formation.

Quant, P. A. (1994). The role of mitochondrial HMG-CoA synthase in regulation of ketogenesis. *Essays Biochem.*, **28**, 13–25.
Summarizes the control of ketogenesis and the physiological repercussions of the process.

Insulin signal transduction

Cohen, P. (1999). The Croonian Lecture 1998. Identification of a protein kinase cascade of major importance in insulin signal transduction. *Phil. Trans. Ser. B, Roy. Soc. Lond.*, **354**, 485–95.
A readable review of this seminal work.

Alper, J. (2000). New insights into Type 2 diabetes. *Science*, **289**, 37–8.

Arner, P. (2003). The adipocyte in insulin resistance: key molecules and the impact of thiazolidinediones. *Trends Endocr. Metab.*, **14**, 137–45.
A wide-ranging review relating to obesity and diabetes.

Brownsey, R. W., Boone, A. N., Elliott, J. E., Kulpa, J. E., and Lee, W. M. (2005). Regulation of acetyl-CoA carboxylase. *Biochem. Soc. Trans.*, **34**, 223–7.

Neels, J. G., and Olefsky, J. M. (2006). A new way to burn fat. *Science*, **312**, 1756–8.
A Science perspective article. Acetyl-CoA carboxylase controls fat storage and utilization in adipose cells. An important regulator of this enzyme modulates its degradation and is a potential therapeutic target for treatment of insulin resistance and obesity.

Hypoxia

Bruick, R. K. and McKnight, S. L. (2001). Oxygen sensing gets a second wind. *Science*, **295**, 807–8.
Briefly reviews hypoxia response.

Kaelin, W. G. (2002). How oxygen makes its presence felt. *Genes Dev.*, **16**, 1441–5.
A 'perspective' on hypoxia response.

Lando, D., *et al.* (2002). FIH-1 is an asparaginyl hydroxylase enzyme that regulates the transcriptional activity of hypoxia-inducible factor. *Genes Dev.*, **16**, 1466–71.

Lando, D., *et al.* (2002). Asparagine hydroxylation of the HIF transactivation domain: a hypoxia switch. *Science*, **295**, 858–61.

Metabolic control analysis

Hofmeyr, J.-H. S. and Cornish-Bowden, A. (2000). Regulating the cellular economy of supply and demand. *FEBS Lett.*, **476**, 47–51.
Discusses the (hitherto somewhat ignored) fact that demand is an important control parameter.

Oliver, S. (2002). Demand management in cells. *Nature*, **418**, 33–4.
A News and Views article on the importance of product removal in metabolic control.

Problems

1 What are the two main ways by which the activities of enzymes may be reversibly modulated?

2 (a) Compare the relationship between substrate concentration and rate of enzyme catalysis in a nonallosteric enzyme and a typical allosteric enzyme.
   (b) What is the advantage of the allosteric enzyme substrate/velocity relationship?

3 If a typical allosterically controlled enzyme is exposed to saturating levels of substrate, what would be the effects of allosteric activators on the reaction velocity?

4 What is the main feature of allosteric control that makes it such a tremendously important concept?

5 What are the salient features of intrinsic regulation and extrinsic regulation by extracellular signals?

6 By means of a diagram, illustrate the main *intrinsic* controls on glycogen metabolism, glycolysis, and gluconeogenesis and explain the rationale.

7 Pyruvate dehydrogenase (PDH) is a key regulatory enzyme. In general, products of the reaction inhibit the reaction. There are three mechanisms of control involved; what are they?

8 What controls the release of insulin and glucagon from the pancreas?

9 What is a second messenger? Name the second messenger for epinephrine and glucagon and explain how it exerts metabolic effects.

10 How does insulin control the rate of glucose entry into fat cells?

11 Explain how cAMP activates glycogen breakdown.

12 Glucagon activates liver phosphorylase via cAMP as its second messenger. Muscle does the same with epinephrine stimulation. However, cAMP has quite different effects on liver and muscle glycolysis. Explain these.

13 Several hormones that elicit different cellular responses nonetheless use cAMP as their second messenger. How can one compound be used for these?

14 Phosphofructokinase is a key control enzyme. What is the major allosteric effector for this enzyme and how is its level controlled in liver?

15 To synthesize glucose in the liver, phosphoenolpyruvate (PEP) is needed. However, production of PEP would achieve little in this regard if it were dephosphorylated to pyruvate by pyruvate kinase. How is this futile cycle avoided? Why would this mechanism not be appropriate in muscle?

16 How does glucagon cause fatty acid release by fatty cells?

17 Metabolic pathways often include at least one reaction with a large ΔG value. What advantages accrue from this?

18 The control of glycogen synthase is to a large extent effected by insulin.
   (a) Describe the physiological rationale for this.
   (b) Explain in outline the nature of this control.

19 What part does glucose play in the regulation of glycogen breakdown in liver?

20 Describe the reciprocal controls that arrange that fatty acids are either synthesized or oxidized but not both at the same time.

21 Why do you think phosphorylation is such a common way of controlling enzyme activity?

22 Discuss the role of the AMP-activated protein kinase (AMPK).

Why should there be an alternative pathway of glucose oxidation? The pentose phosphate pathway

A completely different pathway of glucose oxidation exists called the pentose phosphate pathway, but sometimes called the 'direct oxidation pathway' or the 'hexose monophosphate shunt'. The need for another pathway may puzzle you, since in the preceding chapters we have dealt with a most extensive pathway (glycolysis, the citric acid cycle, and the electron transport pathway) for oxidizing glucose. Hence the question in the chapter title.

Paradoxically, it is not really a pathway to provide for a different method of glucose oxidation nor is it for ATP production, but rather it is to cater for some specialized metabolic needs.

- It supplies ribose-5-phosphate (a pentose sugar) for nucleotide and nucleic acid synthesis (nucleic acid synthesis is dealt with in later chapters). Ribose is also a component of coenzymes NAD and FAD.
- It supplies NADPH for fat and other reductive syntheses.
- It provides a route for excess pentose sugars in the diet to be brought into the mainstream of glucose metabolism.

The pathway is found mainly in the cytoplasmic compartment of cells along with glycolysis. The enzymes are most plentiful in tissues with high demands for NADPH where it is used for reductive syntheses, and in rapidly dividing cells, which require ribose-5-phosphate for DNA synthesis. Quantitatively the main demand is for fatty acid synthesis so that the enzymes of the pathway are plentiful in liver and adipose tissue. Skeletal muscle by contrast has very little but since all cells require ribose for nucleic acid synthesis, all tissues probably have some. (This applies to immature red blood cells; mature ones require NADPH for other reasons; page 490.) Cholesterol and steroid synthesis also require NADPH from the pentose-phosphate-pathway enzymes.

The pentose phosphate pathway has two main parts

In the first part, the oxidative section, glucose-6-phosphate is converted by **glucose-6-phosphate dehydrogenase** to 6-phosphogluconate (*via* 6-phosphogluconolactone), during which $NADP^+$ is reduced to NADPH (Fig. 17.1). Then **6-phosphogluconate**

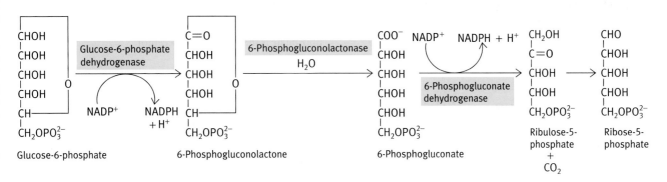

Fig. 17.1 Oxidative reactions of the pentose phosphate pathway. Control is mainly by availability of $NADP^+$.

dehydrogenase reduces $NADP^+$ and generates a β-keto acid, which is decarboxylated to a keto-pentose (**ribulose-5-phosphate**); an isomerase converts the latter to the aldose isomer, ribose-5-phosphate. The oxidative section is irreversible. It produces the two components ribose-5-phosphate and NADPH.

The rate-limiting step is the first reaction, that of oxidation of glucose-6-phosphate to 6-phosphoglucolactone. The rate of this reaction is tightly coupled to the level of $NADP^+$. This governs the allocation of glucose-6-phospate to the pentose phosphate pathway, rather than the glycolytic pathway.

The oxidative section produces equal amounts of ribose-5-phosphate and NADPH

Tissue demands for the two products, ribose-5-phosphate and NADPH, vary greatly. For example, fat synthesis requires large amounts of NADPH, but production of NADPH also produces ribose-5-phosphate by the reactions given in Fig. 17.1, which may be far more than the cell needs for nucleotide synthesis. Conversely, in a rapidly dividing cell that is not synthesizing fat, large amounts of ribose-5-phosphate are needed to synthesize nucleotides, but there is little requirement for NADPH. The requirements for ribose-5-phosphate and NADPH may, in other cells, be anywhere in between this – equal amounts of the two products or more of one than the other. The nonoxidative section takes care of this problem, as described below. As already implied, control of the oxidative section of the pathway is mainly by availability of $NADP^+$.

The nonoxidative section interconverts sugars, according to the cells' needs

The nonoxidative branch is a team act of a pair of enzymes, **transaldolase** and **transketolase**, which together interconvert sugars in accordance with the metabolic needs of the cell. These two enzymes detach, from a ketose sugar phosphate, C_3 and C_2 units, respectively, and transfer them to other aldose sugars (Fig. 17.2). Transketolase is a thiamin pyrophosphate enzyme as is pyruvate dehydrogenase.

Between them, transketolase and transaldolase can work an almost bewildering range of sugar interconversions (which probably few biochemists carry in their heads in detail). If you put together the reactions of the *oxidative* section given above, the following balance sheet emerges.

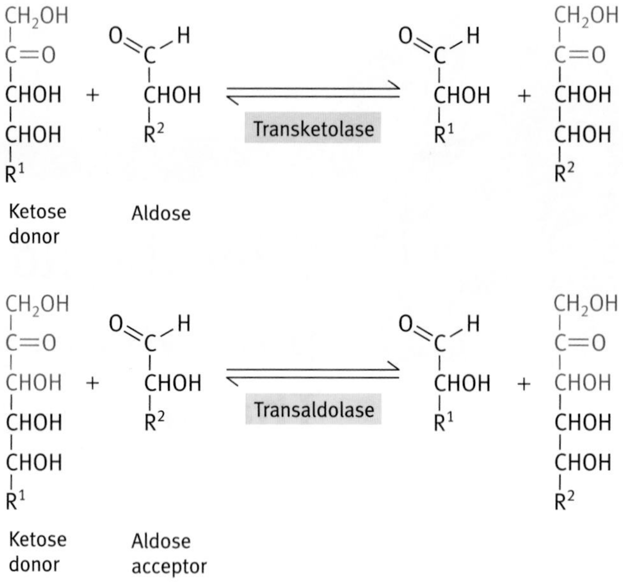

Fig. 17.2 Reactions catalysed by transketolase and transaldolase.

$$\text{Glucose-6-phosphate} + 2NADP^+ + H_2O \rightarrow$$
$$\text{ribose-5-phosphate} + 2NADPH + 2H^+ + CO_2$$

In situations where the needs for ribose-5-phosphate and NADPH are *balanced* the nonoxidative section is not required, because the oxidative part produces both products in appropriate amounts.

Conversion of surplus ribose-5-phosphate to glucose-6-phosphate

However, if, say, a nondividing fat cell requires *more NADPH than ribose-5-phosphate* the nonoxidative section reconverts (recycles) excess ribose-5-phosphate into glucose-6-phosphate according to the stoichiometry:

$$6 \text{ Ribose-5-phosphate} \rightarrow 5 \text{ glucose-6-phosphate} + P_i$$

First, *only part* of the ribose-5-phosphate, an aldose sugar, is converted into **xylulose-5-phosphate**, a ketose sugar, since both transaldolase and transketolase use only ketose sugars as group donors (Fig. 17.3). The remaining part of the ribose-5-phosphate is the aldose sugar acceptor.

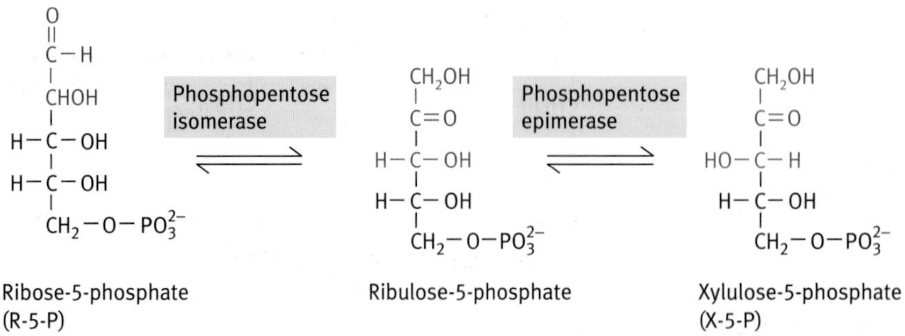

Ribose-5-phosphate (R-5-P)

Ribulose-5-phosphate

Xylulose-5-phosphate (X-5-P)

Fig. 17.3 Conversion of some ribose-5-phosphate to xylulose-5-phosphate. The reason for this is that in reaction 1, Fig. 17.4, the transketolase requires the donor substrate to be a ketose sugar, which is what xylulose-5-phosphate is. The remaining ribose-5-phosphate (an aldose) acts as the receptor in reaction 1.

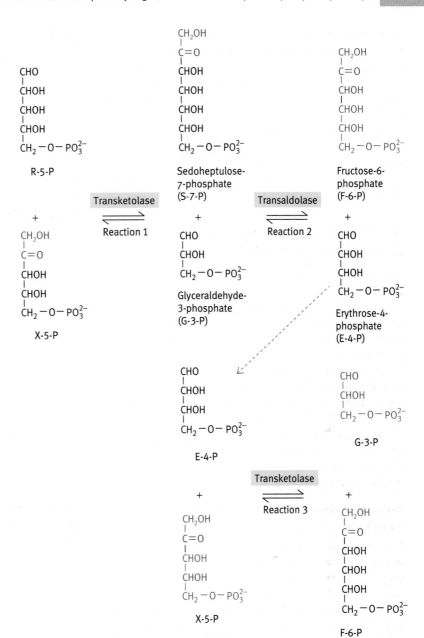

Fig. 17.4 Conversion of ribose-5-phosphate to glyceraldehyde-3-phosphate. Glyceraldehyde-3-phosphate is converted to glucose-6-phosphate by gluconeogenesis reactions. Together, these are the reactions by which six C_5 sugars (three ribose-5-phosphates and three xylulose-5-phosphates) are converted to five glucose-6-phosphates.

The following transformations then occur, reaction 1 being between xylulose-5-phosphate and the remaining ribose-5-phosphate. (Fig. 17.4 gives these reactions in full, so that you can understand what is happening).

(1)　　$2\ C_5 \rightleftharpoons C_3 + C_7$　Transketolase
　　　Reaction 1, Fig.17.4

(2)　$C_7 + C_3 \rightleftharpoons C_4 + C_6$　Transaldolase
　　　Reaction 2, Fig.17.4

(3)　$C_5 + C_4 \rightleftharpoons C_3 + C_6$　Transketolase
　　　Reaction 3, Fig.17.4

(4)　$2\ C_3 \rightarrow 1\ C_6$

The net effect of the first three reactions is that three molecules of C_5 (indicated in red) are converted into 2.5 molecules of C_6 (blue). The final C_3 compound is glyceraldehyde-3-phosphate, two molecules of which are converted to glucose-6-phosphate by the gluconeogenesis pathway. Note that this set of reactions can also convert dietary ribose to glucose-6-phosphate following its conversion to ribose-5-phosphate by an ATP-requiring kinase.

Two rounds of reactions 1–3 will produce two molecules of glyceraldehyde-3-phosphate, which are converted to fructose-6-phosphate by the gluconeogenesis pathway.

$$2 \begin{array}{l} CHO \\ | \\ CHOH \\ | \\ CH_2-O-PO_3^{2-} \end{array} \xrightarrow[\text{(Gluconeogenesis reactions)}]{\longrightarrow\longrightarrow\longrightarrow\longrightarrow} 1\ \text{F-6-P} + P_i.$$

G-3-P

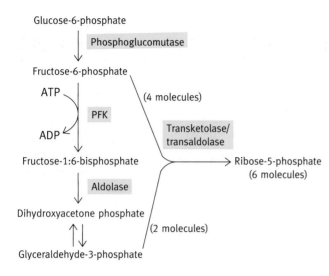

Fig. 17.5 Scheme by which glucose-6-phosphate is converted to ribose-5-phosphate without the production of NADPH. The oxidative part of the pentose phosphate pathway is not involved. PFK, phosphofructokinase.

Fructose-6-phosphate is converted to glucose-6-phosphate by phosphohexose isomerase.

The net effect of these reactions is:

$$6 \text{ Ribose-5-phosphate} \rightarrow 5 \text{ glucose-6-phosphate} + P_i$$

Conversion of glucose-6-phosphate to ribose-5-phosphate without NADPH generation

The pentose phosphate pathway is extremely flexible. Consider another situation where a cell needs to make ribose-5-phosphate for nucleotide synthesis, but has little demand for NADPH. The following overall process caters for this:

$$5 \text{ Glucose-6-phosphate} + ATP \rightarrow 6 \text{ ribose-5-phosphate} + ADP + H^+$$

The mechanism is summarized in Fig. 17.5. In this, glucose-6-phosphate is converted by the glycolytic pathway partly to fructose-6-phosphate and partly to glyceraldehyde-3-phosphate. These are the C_6 and C_3 products of reaction 3. Reversal of the three steps produces xylulose-5-phosphate, which is isomerized into ribose-5-phosphate. This scheme does not involve the oxidative part of the pentose phosphate pathway at all.

Generation of NADPH without net production of ribose-5-phosphate

The oxidative part of the pentose phosphate pathway (see Fig. 17.1) is sometimes referred to as being capable of the direct oxidation of glucose to CO_2, and $NADPH + H^+$. The stoichiometry of the overall process is:

$$6 \text{ glucose-6-phosphate} + 12 \text{ NADP}^+ + 7 \text{ H}_2O \rightarrow$$
$$5 \text{ glucose-6-phosphate} + 6 \text{ CO}_2 + 12 \text{ NADPH} + 12 \text{ H}^+ + P_i$$

This set of events generates NADPH without net production of ribose-5-phosphate – a situation required in a cell with rapid fat synthesis but no cell division. In fact, a *single* molecule of glucose-6-phosphate is *not* converted to six CO_2 molecules by the pathway. What happens is that six molecules of glucose-6-phosphate can *each* give rise to one molecule of CO_2 and one molecule of ribose-5-phosphate plus one molecule of NADPH by the reactions given above. Now, if nonoxidative reactions convert the six ribose-5-phosphate molecules back to five glucose-6-phosphate molecules, as described, then on a balance sheet it looks as if six glucose-6-phosphate molecules have been converted to five molecules of glucose-6-phosphate plus six molecules of CO_2 and six molecules of NADPH. However, as stated, it has not really oxidized one molecule of glucose completely.

BOX 17.1 Why do red blood cells have the pentose phosphate pathway?

Mature erythrocytes do not divide so they have no need for ribose-5-phosphate for nucleic acid synthesis, nor do they synthesize fat. Their energy is derived from anaerobic glycolysis – anaerobic, for they have no mitochondria.

Nonetheless, in patients whose red blood cells lack the first enzyme in the pentose phosphate pathway (glucose-6-phosphate dehydrogenase), a massive haemolytic anaemia may be induced by the antimalarial drug, pamaquine. The reason for this relates to the fact that NADPH, generated by the pentose phosphate pathway, is needed for a protective mechanism described on page 490, namely the reduction of glutathione, a thiol compound, which maintains haemoglobin in its reduced (ferrous) state, and protects against reactive oxygen species. These are increased by pamaquine oxidation.

An interesting sidelight on glucose-6-phosphate dehydrogenase deficiency is that mutations leading to a defective enzyme confer a selective advantage in areas where a lethal type of malaria is endemic. Possible explanations for this are that the parasite has a requirement for the products of the pentose phosphate pathway and/or that the extra stress caused by the parasite causes the deficient red blood cell host to lyse before the parasite completes its development. It is interesting to compare the selective advantage conferred in this case with that conferred by the sickle cell trait (page 272), where a potentially lethal disease can provide a survival advantage because it gives protection against a more lethal disease, malaria.

Another medical condition related to inherited deficiency in glucose-6-phosphate dehydrogenase in red blood cells is favism, which results in haemolysis of red blood cells in certain individuals, and is caused by the toxin in faba beans (*Vicia faba*). See Further reading reference.

Summary

The pentose phosphate pathway should not be viewed as a pathway to oxidize glucose; it does not generate ATP nor does it take a molecule of glucose and oxidize it completely. It is a versatile pathway, which supplies three main needs. It produces riibose-5-phosphate for nucleotide synthesis; it supplies NADPH for fat synthesis and other reductive systems; and it provides a route for the metabolism of excess pentose sugars from the diet.

The pathway has an oxidative section converting glucose-6-phosphate to ribose-5-phosphate and produces NADPH. The nonoxidative section manipulates ribose-5-phosphate according to the needs of the cell. If a cell requires equal amounts of ribose-5-phosphate and NADPH, only the oxidative section is required. If it is synthesizing fat and requires a lot of NADPH but little ribose-5-phosphate the nonoxidative section takes care of excess of the latter by converting it back to glucose-6-phosphate. The pathway is needed in red blood cells to generate NADPH needed to keep the protective molecule glutathione in a reduced state. Without this, certain drugs produce anaemias.

The reaction sequences involved can be complex, involving the interconversion of several sugars. The key reactions in these manipulations are catalysed by transaldolase and transketolase, which between them can effect a range of sugar interconversions.

Further reading

McMillan, D. C., Bolchoz, L. J. C., and Jollow, D. J. (2001). Favism: effect of divicine on rat erythrocyte sulfhydryl status, hexose monophosphate shunt activity, morphology, and membrane skeletal proteins. *Toxocol. Sci.*, **62**, 353–9.

Horecker, B. L. (2002). The pentose phosphate pathway. *J. Biol. Chem.*, **277**, 47965–71.
Historical perspective in Reflections series to celebrate the centenary of the Journal of Biological Chemistry in 2005.

Veech, R.L. (2003). A humble hexose monophosphate pathway metabolite regulates short- and long-term control of lipogenesis. *Proc. Natl. Acad. Sci. U.S.A.*, **100**, 5578–80.
Emphasizes the relationship with fatty acid synthesis.

Problems

1 What are the functions of the pentose phosphate pathway?

2 What is the oxidative part of the pentose phosphate pathway?

3 What enzymes are involved in the nonoxidative part?

4 A nondividing fat cell requires large amounts of NADPH for fat synthesis but very little ribose-5-phosphate. However, the oxidative section produces equal amounts of the two products. Explain how the nonoxidative reactions cope with this problem. (The answer need not give actual reactions.)

5 The pentose phosphate pathway was sometimes referred to as the direct oxidation pathway for glucose. Why was this and why was it a somewhat misleading term?

6 Mature red blood cells have no need for nucleotide or fat synthesis. Why then do they have glucose-6-phosphate dehydrogenase?

Chapter 18

Raising electrons of water back up the energy scale – photosynthesis

Chapter 12 describes how ATP generation in aerobic cells depends on transporting electrons of **high energy potential**, present in food, down the energy scale to end up as the electrons present in the hydrogen atoms of water.

Since the amount of food on Earth is limited, if life in general is to continue indefinitely, a way must exist to kick those electrons back up the energy scale. A minor qualification to this statement is that life forms have been discovered in deep oceans around cracks in the Earth's crust from which compounds such as H_2S emerge. H_2S is a strong reducing agent (that is, of low redox potential) and its electrons could be transported down the energy gradient releasing energy, provided appropriate biochemical systems are there. Such life could presumably exist as long as H_2S and other such agents are generated in the Earth's crust but, for continuation of the vast majority of living organisms, electron recycling is necessary.

Overview

Photosynthesis is the biological process that recycles electrons from water, to produce oxygen and carbohydrate. It has several crucial advantages – water is an inexhaustible source of electrons, sunlight an inexhaustible source of energy, and, in releasing oxygen, an inexhaustible supply of electron sinks (acceptors) is provided to allow the energy to be extracted back from the high-energy electrons of food by life forms in general. The onset of photosynthesis was arguably the most important biological event following the establishment of life. The global energy cycle is shown in Fig. 18.1.

You are probably familiar with the concept of photosynthesis producing carbohydrate (usually starch or sugar) from CO_2 and H_2O; the overall equation (written for glucose production) is

$$6CO_2 + 6H_2O \rightarrow C_6H_{12}O_6 + 6O_2 \; \Delta G^{0'} = 2820 \text{ kJ mol}^{-1}$$

To synthesize glucose from CO_2 and H_2O there are two basic essentials, looked at from the point of view of energy. First, there

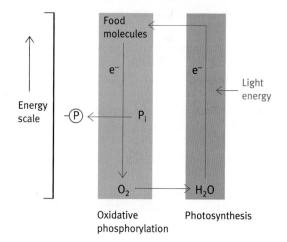

Fig. 18.1 Global 'electron cycling' by oxidative phosphorylation and photosynthesis. Note that the fixation of CO_2 is a process secondary to the raising of electrons from water to a higher energy potential.

must be a reducing agent of sufficiently low **redox potential** (**high energy**). If you need to refresh your memory on redox potentials, turn to page 118. In photosynthesis, the reducing agent is NADPH. (Gluconeogenesis in animals, as described in Chapter 15, uses NADH as the reductant, but, in photosynthesis, note that it is NADPH.) Secondly, there must be ATP to drive the synthesis of carbohydrates.

Light energy is directly involved *only* in transferring electrons from water to $NADP^+$ and in the generation of the proton gradient that drives ATP production.

Site of photosynthesis – the chloroplast

Photosynthesis occurs in the **chloroplasts** of the cells of green plants. They are reminiscent of mitochondria in being membrane-bounded organelles in the cytoplasm with an outer

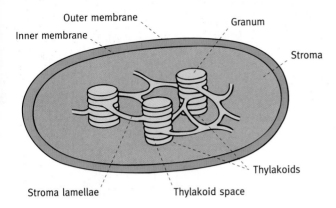

Fig. 18.2 A chloroplast. Grana are stacks of thylakoids.

permeable membrane and an inner one impermeable to protons. Like mitochondria they have their own DNA coding for part of their proteins. Their protein-synthesizing machinery is prokaryotic in type and it is believed that they arose by a symbiotic colonization of eukaryotic cells by primitive prokaryotic photosynthetic unicellular organisms.

Unlike mitochondria, however, chloroplasts contain yet another type of membrane-bounded structure – the thylakoids – membrane sacs within the chloroplast. The **thylakoids** are flattened sacs piled up like stacks of coins into grana, these being connected at intervals by single-layer extensions. Inside the thylakoids is the thylakoid lumen; outside is the chloroplast stroma (Fig. 18.2).

All of the light-harvesting chlorophyll and the electron transport pathways are in the thylakoid membranes. The conversion of CO_2 and H_2O into carbohydrate molecules is not itself light dependent and occurs in the chloroplast stroma. The latter processes are referred to as 'dark reactions', not to imply that they only occur in the dark, but rather that light is not involved in them. In fact, the dark reactions occur mainly in the light, when NADPH and ATP generation is occurring in the thylakoid membranes. This is summarized in Fig. 18.3.

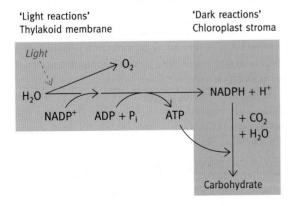

Fig. 18.3 The processes in photosynthesis.

The light-dependent reactions of photosynthesis

The photosynthetic apparatus and its organization in the thylakoid membrane

It would be useful for you to refresh your memory of the electron transport chain in mitochondria for there are considerable similarities between this and the photosynthetic machinery. In the inner mitochondrial membrane there are four complexes (see Fig. 12.17) that transport electrons. Connecting these complexes by ferrying electrons between them are ubiquinone, a small lipid-soluble molecule, and cytochrome c, the small water-soluble mobile protein.

In the thylakoid membrane there are three complexes (Fig. 18.4); connecting the first two is the electron carrier plastoquinone the structure of which is very similar to that of ubiquinone (page 201). Connecting the second two complexes is plastocyanin, a small water-soluble protein that has a bound copper ion as its electron-accepting moiety; this oscillates between the Cu^+ and Cu^{2+} states as it accepts and donates electrons.

The three complexes are named photosystem II (PSII), the cytochrome bf complex, and photosystem I (PSI). PSII comes before PSI in the scheme of things; the numbers refer to the order in which they were discovered rather than to their place in the scheme. The function of the whole array shown in Fig. 18.4 is to carry out the overall reaction:

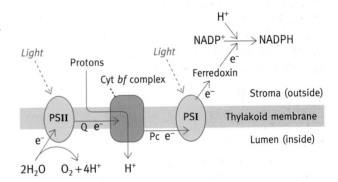

Fig. 18.4 The diagram gives an overall view of the light-dependent part of the photosynthetic apparatus, without reaction details. The feature to note is that the purpose is to take electrons from water and transfer them to NADP$^+$ and in the process to create a proton gradient across the thylakoid membrane that can drive ATP synthesis by the chemiosmotic mechanism. The gradient is created from two sources – the splitting of water (photosystem II, PSII) and the proton pumping of the plastoquinone (Q) – cytochrome bf complex. Electrons are transported from PSII to the cytochrome bf complex by plastoquinone (Q). The cycle is analogous to that in mitochondria (where 'Q' is used for ubiquinone) that results in four protons translocated per plastoquinol (QH$_2$) molecule oxidized. Note also that plastocyanin (Pc) is a mobile carrier analogous in this respect to cytochrome c in mitochondria and that ferredoxin is located on the opposite side of the membrane.

$$2H_2O + 2NADP^+ \rightarrow O_2 + 2NADPH + 2H^+$$

This involves a very large increase in free energy. What is unique in photosynthesis, so far as biochemical reactions are concerned, is that this energy is supplied by light. For each molecule of NADPH produced, four photons are absorbed. During the reduction process a proton gradient is established that is used to generate ATP. The arrangement is rather beautiful for it supplies the two requirements for carbohydrate synthesis from CO_2 and water – a reducing agent and ATP. Added to this, as mentioned, it generates an electron sink – oxygen – which makes it possible for living organisms to recover the energy entrapped in the carbohydrate produced. The logic of it all is magnificent.

We now turn to the light-harvesting machinery present in PSII and PSI.

How is light energy captured?

In green plants, the light receptor is chlorophyll. Other receptor pigments exist in bacteria and algae.

Chlorophyll is a tetrapyrrole similar to haem except that it has a magnesium atom at its centre instead of iron and the substituent side groups are different. One of these side groups is a very long hydrophobic group that anchors it into the lipid layer. As in haem, there is a conjugated double-bond system (alternate double and single bonds right round the molecule) resulting in strong absorption of certain wavelengths of light and thus a strong green colour from wave lengths not absorbed.

In green plants there are two chlorophylls (*a* and *b*) differing in one of the side groups. They both absorb light in the red and blue ranges, leaving the intermediate green light to be reflected. The two chlorophylls have slightly different absorption maxima, which complement each other so that in the red and blue ranges between them they absorb a higher proportion of the incident light. Chlorophyll *a* is shown below. (Chlorophyll *b* is the same except for the $-CH_3$ side chain (in red), which is $-CHO$ in chlorophyll *b*.)

Structure of chlorophyll *a*

When a chlorophyll molecule absorbs light, it is excited so that one of its electrons is raised to a higher energy state; it

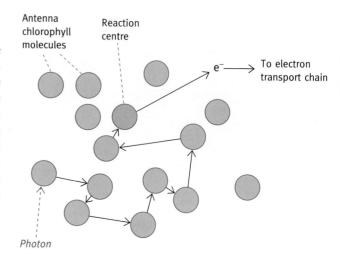

Fig. 18.5 Activation of reaction-centre chlorophyll molecule (red circle) by resonance transfer of energy from activated antenna chlorophyll molecules (green circles). The reaction centre of PSII is called P680 and of PSI, P700. Close packing of the chlorophyll molecules is needed for efficient resonance energy transfer.

jumps into a new atomic orbit. In an isolated chlorophyll molecule, after such excitation, the electron drops back to its ground state liberating energy as heat or fluorescence in doing so (and nothing is thereby achieved). But, when chlorophyll molecules are closely arranged together, a process known as **resonance energy transfer** transfers that energy from one molecule to another. In green plants, chlorophyll molecules are packed in functional units called **photosystems**, such that this resonance transfer occurs readily. Thus when a chlorophyll molecule is excited by absorption of a photon, its energy is transferred to another molecule and it drops back to the ground state itself. The excitation wanders at random from one chlorophyll molecule to another (Fig. 18.5).

This has a function, for among large numbers of 'ordinary' chlorophyll molecules there is a special **reaction centre** (an arrangement of a pair of chlorophyll molecules in association with proteins). The properties of this reaction centre are such that the excitation of a constituent chlorophyll molecule by resonance transfer results in the excited electron being at a somewhat lower energy level, as compared with that of other excited chlorophyll molecules, so that resonance energy transfer *from* this molecule to other chlorophyll molecules does not occur. The excitation energy is, in this sense, trapped in an energetic hole – a shallow hole because the 'trapped' electron is still at a higher energy level than an unexcited electron, sufficient for the excited reaction centre to hand on an electron to an electron acceptor of appropriate redox potential. The latter is the first carrier of an electron transport chain in photosynthesis (to be described shortly). The chlorophyll molecules feeding excitation energy to the centres are called **antenna chlorophylls** (Fig. 18.5).

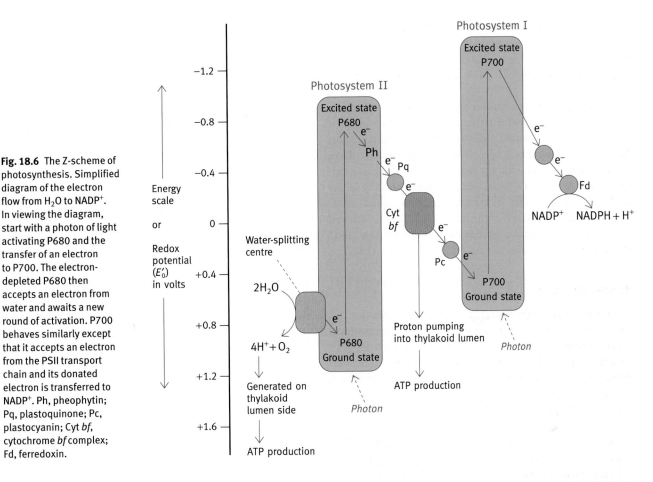

Fig. 18.6 The Z-scheme of photosynthesis. Simplified diagram of the electron flow from H_2O to $NADP^+$. In viewing the diagram, start with a photon of light activating P680 and the transfer of an electron to P700. The electron-depleted P680 then accepts an electron from water and awaits a new round of activation. P700 behaves similarly except that it accepts an electron from the PSII transport chain and its donated electron is transferred to $NADP^+$. Ph, pheophytin; Pq, plastoquinone; Pc, plastocyanin; Cyt *bf*, cytochrome *bf* complex; Fd, ferredoxin.

A photosystem is therefore a complex of light-absorbing chlorophylls, reaction-centre chlorophyll, and an electron transport chain. In the case of PSII (the first system) the reaction-centre chlorophyll is called P680 because it absorbs light up to that wavelength (in nanometres) and in PSI is called P700 for an equivalent reason.

Mechanism of light-dependent reduction of $NADP^+$

Figure 18.6 shows the 'Z' arrangement of the two photosystems II and I, with the redox potentials of the components indicated. Why two photosystems? A familiar analogy might help at the outset. An electric torch with a bulb (globe) requiring 3 V to light it up often uses two 1.5 V batteries in series. In photosynthesis, lighting of the bulb is represented by $NADP^+$ reduction, and the batteries by the two photosystems operating in series. In fact, as stated, the latter supply more energy than is needed and some of it is sidetracked into ATP generation (see below).

Photosystem II

Let us start with a P680 chlorophyll of a reaction centre of PSII. In the dark, it is in its ground, unexcited state in which it has no tendency to hand on an electron. When energy of a photon

reaches it *via* antenna chlorophylls, it is excited such that it has a strong tendency to hand on its excited electron. It is, in fact, a reducing agent and it reduces the first component (a chlorophyll-like pigment lacking the Mg^{2+} atom, called **pheophytin**) of the PSII electron transport chain.

Two molecules of reduced pheophytin then hand on an electron each (one at a time) to reduce **plastoquinone**, the lipid-soluble electron carrier between PSII and the **cytochrome** *bf* **complex**:

$$\text{Plastoquinone (Q)} \xrightarrow[2H^+]{2e^-} \text{Plastoquinol (QH}_2)$$

R = a long hydrophobic group

The latter complex contains two cytochromes and an iron-sulphur centre (see page 201 if you need to be reminded what this is); the complex transports electrons from **plastoquinol** (QH_2 – reduced plastoquinone) to **plastocyanin** to give the reduced form of the latter (see Fig. 18.4). Plastocyanin is a copper-protein complex in which the copper ion alternates between the Cu^{2+} (oxidized) and Cu^+ (reduced) forms. At this point we will leave PSII and move on to PSI but we will return to the reduced plastocyanin very soon.

Photosystem I

The chlorophyll at the reaction centre of PSI is P700; when activated by a photon arriving from the antenna chlorophylls it becomes a reducing agent. It passes on its electron to a short chain of electron carriers (details of this not given), which reduces **ferredoxin**, a protein with an iron-cluster electron acceptor. Ferredoxin is a water-soluble, mobile protein residing in the chloroplast stroma (that is, outside of the thylakoid membrane). It reduces $NADP^+$ by the following reaction, catalysed by the FAD enzyme called **ferredoxin-NADP reductase** (FAD or flavin adenine dinucleotide was discussed on page 179):

$$2 \text{ Ferredoxin}_{RED} + NADP^+ + 2H^+ \rightarrow 2 \text{ ferredoxin}_{OX} + NADPH + H^+$$

If we summarize what has taken place in PSI, an electron has been excited out of P700 and transported to ferredoxin which, in turn, has reduced $NADP^+$. However, this has left P700 an electron short; it is now $P700^+$, an oxidizing agent. It accepts an electron from plastocyanin (Pc), which, you remember, we left after it had been reduced by PSII. The reaction is:

$$P700^+ + Pc_{Cu^+} \rightarrow P700 + Pc_{Cu^{2+}}.$$

To go back further, remember that we started with light exciting an electron out of P680 (Fig. 18.6), the reaction-centre pigment of PSII; this leaves $P680^+$, which must have its electron restored so that it can revert to the ground state, ready for another photon to start a new round of reactions. The electron comes from water.

The water-splitting centre of PSII

$P680^+$ is a very strong oxidizing agent – it has a very strong affinity for an electron (greater than that of oxygen) so that it can even extract electrons from water. Four electrons are extracted from two molecules of H_2O with the release of O_2 and four H^+ into the thylakoid lumen. It is necessary to extract all four electrons so as not to release any intermediate oxygen free radicals, which are dangerous to biological systems, just as, in mitochondria, the addition of electrons to oxygen to form H_2O must be complete. In PSII there is a complex of proteins with Mn^{2+}, known as the **water-splitting centre**, which extracts the electrons from water, with the release of oxygen and protons, and passes them on to $P680^+$ molecules, thus restoring them to the ground state (Fig. 18.8(a)). The P680 is now ready to be activated by another photon.

How is ATP generated?

The **cytochrome *bf* complex** of PSII, which uses plastoquinol (QH_2) to reduce plastocyanin, resembles complex III of

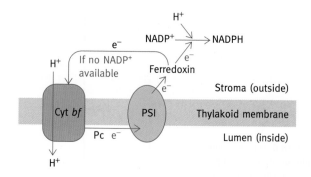

Fig. 18.7 Cyclic electron flow. When all of the $NADP^+$ is reduced, ferredoxin transfers electrons back to the cytochrome *bf* complex. Flow of electrons through this leads to increased proton pumping and hence increased ATP synthesis. Pc, plastocyanin.

mitochondria (see Fig. 12.17) in that electron transport through the complex causes translocation of protons from the outside of the thylakoid membrane to the inside. In addition, the water-splitting centre generates protons inside the thylakoid lumen, the two effects producing a proton gradient that reduces the pH of the thylakoid lumen to about 4.5. The uptake of a proton in the reduction of $NADP^+$ by ferredoxin in the stroma further contributes to the proton gradient across the membrane.

There is a device that under certain circumstances leads to an increased proton translocation and therefore a greater potential for ATP synthesis. When virtually all of the $NADP^+$ has been reduced by ferredoxin, ferredoxin donates electrons to the cytochrome *bf* complex (Fig. 18.7) instead. The passage of these through the complex to plastocyanin leads to increased proton pumping by that complex. The extra ATP production is referred to as **cyclic photophosphorylation** driven by **cyclic electron flow**.

The proton gradient is used to generate ATP from ADP and P_i by the chemiosmotic mechanism described for mitochondria. The whole process of the events described so far is summarized in Fig. 18.8(b).

In mitochondria, the proton gradient is from the outside (high) to the inside (low). The thylakoid membrane is formed from invaginations of the inner chloroplast membrane (cf. the inner mitochondrial membrane), which explains why the proton gradients and the ATP synthases of mitochondria and thylakoid discs look as if they are in opposite orientation.

The 'dark reactions' of photosynthesis – the Calvin cycle

How is CO_2 converted to carbohydrate?

As already emphasized, the aspect of photosynthesis that is fundamentally different from other biochemical processes is the harnessing of light energy to split water and reduce $NADP^+$.

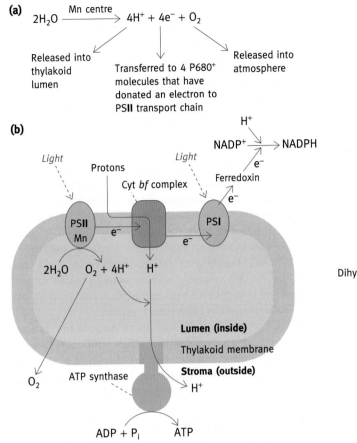

Fig. 18.8 Processes in thylakoid sacs. **(a)** Sum of reactions at the Mn water-splitting centre. **(b)** Routes followed by protons and electrons in thylakoids.

From there, although the actual metabolic pathways by which glucose and its derivatives are synthesized are unique to plants, it is nonetheless 'ordinary' enzyme biochemistry and quite secondary to the light-dependent process.

Getting from 3-phosphoglycerate to glucose

Glucose is formed from the metabolite 3-phosphoglycerate by a series of steps that are the same as the process of gluconeogenesis in the liver (Fig. 15.2) except that NADPH is used instead of NADH as the reductant (Fig. 18.9). This leaves the question of how 3-phosphoglycerate is produced in photosynthesis.

3-Phosphoglycerate is formed from ribulose-1:5-bisphosphate

The most plentiful single protein on Earth is the enzyme called 'Rubisco' for short, which stands for **ribulose-1:5-bisphosphate carboxylase**. It is the enzyme that utilizes CO_2 to produce 3-phosphoglycerate.

To remind you of terminology, ribose is an aldose sugar and ribulose is its ketose isomer. Ribulose-1:5-bisphosphate is cleaved by Rubisco into two molecules of 3-phosphoglycerate; this fixes one molecule of CO_2.

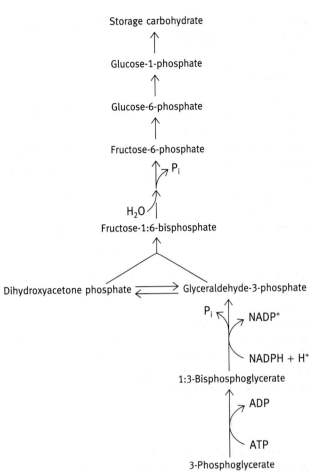

Fig. 18.9 Pathway of starch synthesis in photosynthesis, starting with 3-phosphoglycerate. The process is the same as in gluconeogenesis in liver except that NADPH is the reductant rather than NADH. In starch synthesis, the activated glucose is ADP-glucose rather than the UDP-glucose involved in glycogen synthesis. The special question in photosynthesis is the mechanism by which 3-phosphoglycerate is produced (see text for this).

The 3-phosphoglycerate is converted to carbohydrate as already described. This leads to the next question.

Where does the ribulose-1:5-bisphosphate come from?

The answer to this is very simple, in principle. From six molecules of ribulose bisphosphate (30 carbon atoms in total) and six molecules of CO_2 (six carbon atoms) we get 12 molecules of 3-phosphoglycerate (36 carbon atoms) produced. Two of these latter molecules ($2C_3$) ('the profit') are used to make storage carbohydrate (C_6) by the pathway already outlined in Fig. 18.9.

The remaining 10 phosphoglycerate molecules (30 carbon atoms in total) are manipulated to produce six molecules of ribulose bisphosphate (30 carbon atoms in total). The manipulations involve C_3, C_4, C_5, C_6, and C_7 sugars, and aldolase and transketolase reactions (page 268) reminiscent of the pentose phosphate pathway. The reactions are rather involved and are not presented here; instead they are summarized below.

$$C_3 + C_3 \rightarrow C_6 \quad \boxed{\text{Aldolase}}$$
$$C_6 + C_3 \rightarrow C_4 + C_5 \quad \boxed{\text{Transketolase}}$$
$$C_4 + C_3 \rightarrow C_7 \quad \boxed{\text{Aldolase}}$$
$$C_7 + C_3 \rightarrow C_5 + C_5 \quad \boxed{\text{Transketolase}}$$

In summary,

$$5\,C_3 \rightarrow 3\,C_5.$$

The outcome of it all is that six molecules of ribulose bisphosphate, plus six molecules of CO_2 and six of H_2O, are converted to 12 molecules of 3-phosphoglycerate. From there, the six molecules of ribulose bisphosphate are regenerated, plus the dividend of one molecule of fructose-6-phosphate, which is converted to the storage carbohydrate. This whole process, known as the Calvin cycle after its discoverer, is shown in Fig. 18.10. The stoichiometry of the whole business is given (for reference purposes only) in the following rather daunting equation:

$$6\,CO_2 + 18\,ATP + 12\,NADPH + 12H^+ + 12\,H_2O \rightarrow$$
$$C_6H_{12}O_6 + 18\,ADP + 18\,P_i + 12\,NADP^+.$$

Rubisco has an apparent efficiency problem

At the dawn of photosynthesis there was no oxygen and, it is thought, much higher CO_2 levels than now, but, as photosynthesis occurred, oxygen accumulated. It so happens that Rubisco, in addition to using CO_2, can also react with oxygen, the two competing with one another, so that strictly the enzyme should be called **ribulose-bisphosphate carboxylase/oxygenase**. The oxygenation reaction, so far as is known at present, serves no useful purpose and is apparently wasteful in the sense that it destroys ribulose-1:5-bisphosphate and wastes ATP in a reaction pathway known as **photorespiration**.

A molecule of ribulose bisphosphate is broken down to one molecule of 3-phosphoglycerate and one molecule of glycolate + CO_2 + P_i. This release of CO_2 is wasteful to the fixation process and it sacrifices a high-energy phosphoryl group. The glycolate is salvaged by conversion to glycine.

In high temperatures the wasteful oxygen reaction is maximized to the detriment of photosynthetic efficiency. This can reduce CO_2 assimilation by about 30%. Presumably, in earlier evolutionary times, when there was little oxygen and higher CO_2 levels, this would not have been significant, but today in the presence of high oxygen levels the situation is different. It might have been expected that a new Rubisco that excluded the oxygen use reaction would have evolved but this has not occurred. There may be good reasons for the apparent inefficiency that are not yet appreciated. However, in some plants, a biochemical device has evolved to raise the CO_2 level in cells where Rubisco operates, and thus minimize the oxygenase reaction. This occurs in species of plants that live in high-light and

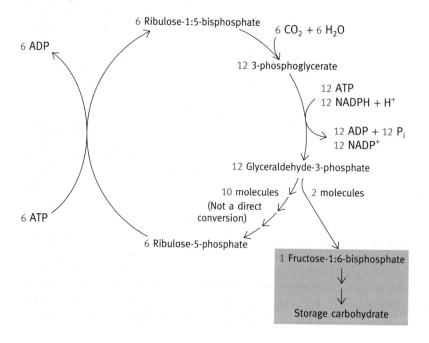

Fig. 18.10 Net effect of the reactions in the Calvin cycle. Note: the diagram is simplified in that the conversion of 3-phosphoglyceraldehyde to ribulose-5-phosphate involves part of it being converted to fructose bisphosphate as an intermediate. The presentation is intended to show the net effect of all the reactions involved. Dihydroxyacetone phosphate, which is in equilibrium with glyceraldehyde-3-phosphate, is also omitted for simplicity.

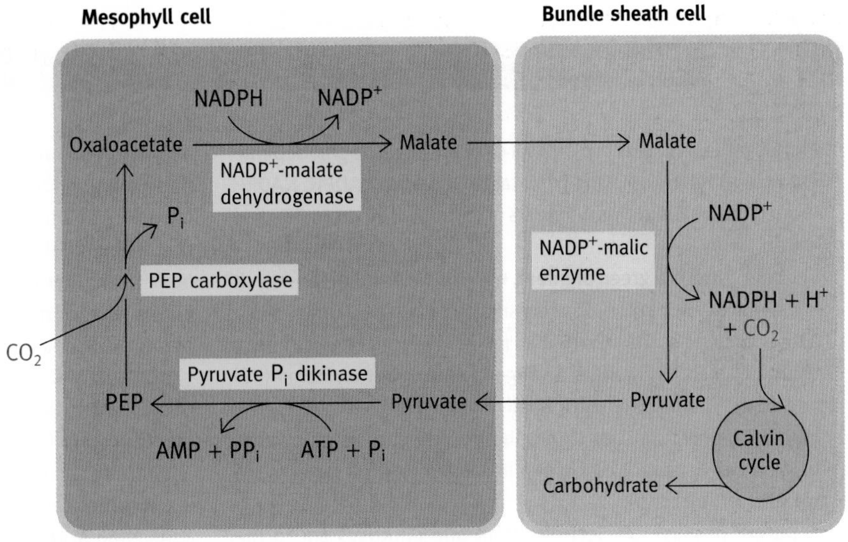

Fig. 18.11 The C_4 pathway for raising the CO_2 concentration for photosynthesis in bundle sheath cells. Important note: the pathway of oxaloacetate formation from pyruvate is quite different from that in animals. Direct conversion of pyruvate to phosphoenolpyruvate (PEP) does not occur in animals and nor does the carboxylation of PEP. It is important to be clear on this for it has a major effect on metabolic regulation in animals. Note also that the malate dehydrogenase that reduces oxaloacetate uses NADPH, unlike that in the citric acid cycle which is NAD^+-specific.

high-temperature environments where the problem of photorespiration would be maximized – in plants such as maize (corn) and sugar cane.

The C_4 pathway

C_3 plants are so called because the first stable labelled product that is experimentally detectable, if they are allowed to photosynthesize in the presence of $^{14}CO_2$, is the C_3 compound, 3-phosphoglycerate, produced by the Rubisco reaction. Some plants, however, *initially* fix CO_2 from the atmosphere into oxaloacetate (Fig. 18.11). The latter is a C_4 compound, the process is referred to as C_4 photosynthesis, and the plants as C_4 **plants**.

The anatomical or cellular structure of C_4 plant leaves differs from that of C_3 leaves. In the former, the mesophyll cells just below the epidermis cells at the surface, which are exposed to the atmospheric CO_2, do not contain the Rubisco enzyme but do fix CO_2 very efficiently into oxaloacetate by the carboxylation of phosphoenolpyruvate (PEP) by the enzyme **PEP carboxylase**. (*Note that this is not found in animals.*)

$$CO_2 + H_2O + PEP + NADP^+ \rightarrow oxaloacetate + CO_2 + NADPH + H^+$$

PEP carboxylase has a high affinity for CO_2, and there is no competition from oxygen. The oxaloacetate is reduced to malate, which is transported into neighbouring bundle sheath cells where the Calvin cycle occurs. Here the CO_2 is released from malate by the 'malic enzyme' (which you have met before in fatty acid synthesis).

$$Malate + NADP^+ \rightarrow pyruvate + CO_2 + NADPH + H^+$$

This raises the concentration of CO_2 in the bundle sheath cells 10–60-fold, resulting in a more efficient operation of the Rubisco reaction. The pyruvate returns to the mesophyll cell where it is reconverted to PEP by **pyruvate phosphate dikinase**.

This is an unusual reaction (*also absent from animals*) in which two —Ⓟ groups of ATP are liberated:

$$CH_3{-}CO{-}COO^- \quad + \ ATP \ + \ P_i$$

$$\downarrow$$

$$CH_2{=}C{-}COO^- \quad + \ AMP \ + \ PP_i + H^+$$
$$| \atop OPO_3^{2-}$$

In animals, PEP can be made from pyruvate only *via* oxaloacetate by a quite different route (see page 237). Once phosphoglycerate is made in the bundle sheath cells, the Calvin cycle operates exactly as in C_3 plants.

The C_4 route incurs an energy price for raising the CO_2 concentration in bundle sheath cells, since ATP is consumed in making PEP and transporting acids. However, at higher temperature and light levels, the C_4 route becomes a considerable advantage. C_4 plants such as corn or sugar cane are prolific producers of carbohydrates.

There is great diversity in the biochemical strategies used by different C_4 plants. For example, the C_4 acid labelled in the presence of $^{14}CO_2$ may be aspartate in some species, and not malate. These plants have high levels of aspartate aminotransferase (Chapter 19) instead of $NADP^+$-malate dehydrogenase. C_4 plants have also evolved three distinct options for decarboxylating C_4 acids in bundle sheath cells, two being located in the mitochondria unlike the $NADP^+$-malic enzyme shown in Fig. 18.11, which is located in the cytoplasm. They are all mechanisms to achieve the same end – increase of CO_2 levels where Rubisco operates.

Succulent plants living in arid areas use the C_4 pathway to save water during the day by closing stomata on leaves, which means that CO_2 cannot be taken in except at night when they fix the CO_2 as malate, which is stored until the sun shines. In light the CO_2 is then released by decarboxylation for fixation by the Calvin cycle.

Summary

Photosynthesis occurs in plant cell chloroplasts. The part dependent on light is the splitting of water to generate NADPH. NADPH is used for the reductive synthesis of carbohydrate from CO_2 and water.

Chlorophyll is a green pigment, which receives light energy. It is present in the membrane of organelles called thylakoids. When activated by photons, chlorophyll molecules donate electrons to chains of electron carriers arranged in two photosystems (PSI and PSII). The electrons are finally used to reduce $NADP^+$. The loss of the electrons by chlorophyll makes it a very powerful oxidizing agent capable of accepting electrons from water in the water-splitting centre.

During passage of electrons from one photosystem to the other, ATP is generated by the chemiosmotic mechanism. Thus both NADPH and ATP are produced. Carbohydrate is synthesized using these by the Calvin cycle. The key reaction in this is catalysed by ribulose-1:5-bisphosphate carboxylase/oxygenase (Rubisco) which generates 3-phosphoglycerate from which carbohydrate synthesis proceeds by reversal of glycolytic reactions but using NADPH as reductant.

Rubisco works less efficiently at low CO_2 levels because it can also react apparently wastefully with oxygen in a process known as photorespiration. Oxygen and CO_2 compete for the Rubisco and at higher temperatures the photorespiration reaction is maximized. At the dawn of photosynthesis it may be speculated that this did not matter because the ratio of CO_2/oxygen would have been very much higher. But today, especially in tropical plants such as maize (corn) and sugar cane grown in high temperatures, it becomes a very significant factor.

Such plants, known as C_4 plants, have developed a means of combating this in the form of a device that greatly elevates the concentration of CO_2 available for photosynthesis, thus minimizing the oxygenase reaction of Rubisco. The CO_2 is first incorporated into oxaloacetate (a C_4 acid) in mesophyll cells by a reaction with phosphoenolpyruvate (a reaction that does not occur in animals or *Escherichia coli*). The oxaloacetate is reduced to malate, which migrates into the bundle sheath cells containing the Calvin cycle. There the malate is decarboxylated to pyruvate and CO_2 by the malic enzyme thus greatly raising the CO_2 concentration, which competes more effectively with oxygen in the Rubisco reaction and minimizes photorespiration. Analogous mechanisms using different acids operate in other C_4 plants. Temperate plants that fix CO_2 initially into 3-phosphoglycerate are known as C_3 plants.

Further reading

The light-harvesting pathway

Slater, E. C. (1983). The Q cycle, an ubiquitous mechanism of electron transport. *Trends Biochem. Sci.*, **8**, 239–42.
Describes the proton-pumping system present in both mitochondria and chloroplasts.

The C_4 pathway

Rawsthorne, S. (1992). Towards an understanding of C_3—C_4 photosynthesis. *Essays Biochem.*, **27**, 135–46.
Discusses the distribution of the biochemical systems among plant species as well as the pathways themselves.

Problems

1 Explain, in general terms, what is meant by the terms 'light' and 'dark' reactions in photosynthesis.

2 What is meant by the term 'antenna chlorophyll'?

3 What is meant by:
 (a) photophosphorylation?
 (b) cyclic photophosphorylation?

4 The oxidation of water requires a very powerful oxidizing agent. In photosynthesis what is this agent?

5 Proton pumping due to electron transport in photosystem II (PSII) causes movement of protons from the outside of thylakoids to the inside. In mitochondria protons are pumped from the inside to the outside. Comment on this.

6 If a photosynthesizing system is exposed for a very brief period to radioactive CO_2, in C_3 plants the first compound to be labelled is 3-phosphoglycerate. Explain why this is so.

7   Explain the Calvin cycle in simplified terms.

8   The enzyme Rubisco can react with oxygen as well as CO_2, the oxygen reaction being, so far as we know, an entirely wasteful one. At low CO_2 concentrations such as can occur particularly in intense sunlight, the wasteful oxygenation reaction is maximized.

What mechanisms have evolved to ameliorate this problem?

9   In C_4 plants, pyruvate is converted to phosphoenolpyruvate by an ATP-requiring reaction. Have you any comments on this, bearing in mind the corresponding process in animals?

Chapter 19

Amino acid metabolism

Digestion of proteins in a normal diet results in relatively large amounts of the 20 different amino acids being absorbed from the intestine into the portal blood, which goes directly to the liver. All cells, except non-nucleated ones such as mature erythrocytes, use amino acids for protein synthesis and for the synthesis of a variety of essential molecules such as membrane components, neurotransmitters, haem, and the like. All cells take up amino acids by selective transport mechanisms since, in their free, ionized form, they do not readily penetrate the membrane lipid bilayer.

In the body there is no dedicated storage of amino acids in the sense that there is no polymeric form of amino acids whose function is simply to be a reserve of these compounds to be called upon when needed. The only 'reserves' are in the form of functional proteins, and since the biggest mass is in muscle proteins, these are the main reserves. But these muscle proteins are part of the contractile machinery and if broken down, for instance to provide amino acids for gluconeogenesis in the liver, muscle wasting ensues.

This biochemical situation where dietary inadequacies exist can be an unfortunate one because, during evolution, humans have lost the ability to synthesize 10 of the amino acids. Interestingly enough, the ones that we cannot synthesize are those with many steps in their synthesis and therefore require many enzymes and many genes to code for them – they are 'expensive' to manufacture.

Why this evolutionary loss occurred is an interesting question that can be answered only in speculative terms. One possibility is that it was simply a measure of economy – that it was more advantageous to obtain 10 of the more complicated amino acids ready-made in the diet and to make only the simpler ones. When the human diet was from the 'wild', if sufficient food could be obtained to sustain life, energy-wise, it would probably contain all the necessary amino acids. In this situation, loss of the ability to synthesize certain amino acids would be pure gain, since you would probably have been more likely to die from insufficient energy sources than from lack of essential amino acids in the food that was available.

With the development of agriculture, however, vast production of chemical energy in the form of the carbohydrate of cereals permitted large populations to survive but did not necessarily supply adequate amounts of the essential amino acids, for some plant proteins, such as those in corn (maize), have low levels of lysine and tryptophan. What exacerbates the situation is that when humans have a rich source of protein, containing enough essential amino acids to provide for an extended period, there is no method for storing them. After immediate needs are satisfied, the surplus amino acids are simply oxidized or converted to glycogen or fat and the nitrogen excreted as urea. In some circumstances, it is somewhat like burning diamonds to keep warm.

The result of a diet with adequate energy but inadequate protein in even a single essential amino acid is the condition known as kwashiorkor. This leads to susceptibility to infections, wasting, apathy, inadequate growth, deleterious brain effects, and lowered levels of serum proteins which, by reducing the osmotic pressure of the blood, is a cause of oedema of the tissues, giving the child a plump appearance. The condition is a vicious circle since the cells lining the intestine are constantly renewed and, in kwashiorkor, this may be inadequate, as might also apply to the production of digestive enzymes. The result is that what food is available may be inadequately digested and absorbed. The disease affects developing children more than adults because of their greater demand for protein to support growth. This evolutionary loss of the ability to synthesize certain amino acids and the lack of provision for their storage have turned out to be, apparently, considerable human disadvantages. The conclusion is speculative, for it may be that unknown compelling reasons dictated the evolutionary loss of synthetic capability. One speculation has been made that the intermediates in the synthesis of essential amino acids may have been toxic to higher organisms (brain development or impairment of its function has been suggested), but there is little evidence on the matter.

Nitrogen balance of the body

The nutritive aspects of amino acids can be treated on a generalized level through the concept of **nitrogen balance**. (This term refers to organic nitrogen; it has nothing to do with N_2.) If the total intake of nitrogen (mainly as amino acids) equals the total excretion, the individual is in a state of nitrogen balance. During growth or repair, more is taken in than is excreted – this is positive nitrogen balance. Negative nitrogen balance occurs in wasting, where excretion exceeds intake, when, for example, amino acids of muscle proteins are converted to glucose and the nitrogen excreted. The proteins of animals are continually 'turning over'; they are continuously broken down and resynthesized. Although many of the derived amino acids are recycled back into proteins, something of the order of about 0.3% of total body protein nitrogen per day is converted to urea and excreted. To a much lesser extent, in mammals nitrogen excretion also occurs as ammonia, creatinine, and uric acid.

An essential amino acid is one that, when omitted from an otherwise complete diet, results in negative nitrogen balance or fails to support the growth of experimental animals. Some amino acids are essential without qualification, such as lysine, phenylalanine, and tryptophan (see Table 19.1 for a list of essential and nonessential amino acids). The essential amino acids include a subclass known as conditionally essential. For example, tyrosine is not essential provided sufficient phenylalanine is available, since phenylalanine is convertible to tyrosine; similarly, cysteine synthesis in mammals requires the availability of methionine, another essential amino acid. The nutritive picture, in this respect, is not completely neat and tidy. 'first-class' proteins are rich in all of the essential amino acids. Plant proteins may be poor in one or two essential amino acids. Since the amino acid compositions of proteins vary, a mixture of plant proteins is needed to ensure adequate amino acid nutrition in vegetarian diets.

General metabolism of amino acids

The general situation of amino acid metabolism is shown in Fig. 19.1 essentially linking the broad aspects of catabolism in which the integration of amino acid, carbohydrate, and fat metabolism is emphasized. A physiologically vital aspect is that in starvation survival depends on the release of amino acids by muscle wasting. These travel to the liver to be used for glucose synthesis (page 237).

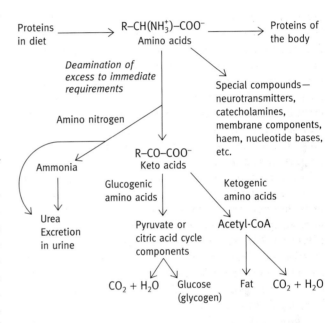

Fig. 19.1 The overall catabolism of amino acids. Note that some amino acids are partly ketogenic and partly glucogenic (see page 285 for explanation of these terms).

Table 19.1 Classification of dietary amino acids in humans.

Nonessential	Essential
Alanine	Arginine[‡]
Asparagine	Histidine
Aspartic acid	Isoleucine
Cysteine*	Leucine
Glutamic acid	Lysine
Glutamine	Methionine
Glycine	Phenylalanine
Proline	Threonine
Serine	Tryptophan
Tyrosine[†]	Valine

* Cysteine is produced only from the essential amino acid methionine.
[†] Tyrosine is produced only from the essential amino acid phenylalanine.
[‡] Arginine is required only in the growing stages.

Aspects of amino acid metabolism

These include the following questions.

- How are the amino groups of amino acids removed? In other words, how does deamination occur? Although there are 20 different amino acids most of them are deaminated by a common mechanism and this is dealt with.

- What happens to the keto acids – the carbon-hydrogen 'skeletons' – of the amino acids after deamination? In this case, each amino acid has its own special metabolic route, and we will give only a limited treatment of this aspect, dealing with features of special biochemical or medical relevance.

- How is the amino-group nitrogen, which is removed from amino acids, converted to urea? This is an area of central importance and will be dealt with.

- How are amino acids synthesized? In the animal body this concerns only the nonessential amino acids. If the carbon skeleton (the keto acid) is available, synthesis is often the reverse of deamination. We will deal with only a few examples of these. In bacteria and plants, all amino acids are synthesized by individual pathways using general metabolic intermediates. Animals depend on this synthetic activity to supply their essential amino acids. The pathways for the synthesis (and breakdown) of all amino acids have been elucidated, and constitute a formidable amount of detailed information. We will deal only with particular cases of amino acid biosynthesis of special biochemical or medical interest.

There are aspects of amino acid metabolism involving methyl group transfer, haem synthesis and ammonia transport which are also covered.

There are also major topics of biochemistry, such as protein synthesis, and nucleotide synthesis, that have amino acids as their reactants. These are dealt with in the relevant later chapters.

Glutamate dehydrogenase has a central role in the deamination of amino acids

What has dehydrogenation to do with deamination? To explain this we will briefly digress to a piece of simple chemistry. It concerns **Schiff bases**. A Schiff base results from a spontaneous (noncatalysed) equilibrium between a molecule with a carbonyl group (aldehyde or ketone) and one with a free amino group.

$$\overset{|}{\underset{|}{C}}=O \;+\; H_2N-R \;\rightleftharpoons\; \overset{|}{\underset{|}{C}}=N-R \;+\; H_2O$$

The reaction is freely reversible so a Schiff base can readily hydrolyse. If R = H in the above equation, the Schiff base will hydrolyse to give NH_3. It shows how the removal of a pair of hydrogen atoms from an amino acid can result in **deamination**, and the supply of a pair of hydrogen atoms can result in the synthesis of an amino acid from a keto acid and ammonia by the reverse reaction (nonionized structures are used below for clarity).

$$\overset{|}{\underset{|}{CHNH_2}} \xrightarrow{-2H} \overset{|}{\underset{|}{C}}=N-H \xrightarrow{H_2O} \overset{|}{\underset{|}{C}}=O \;+\; NH_3$$

Now back to biological deamination. Glutamic acid is an amino acid of central importance in metabolism. It is deaminated by **glutamate dehydrogenase**, unusually working with either NAD^+ or $NADP^+$. (The NH_3 is protonated to NH_4^+ at physiological pH.)

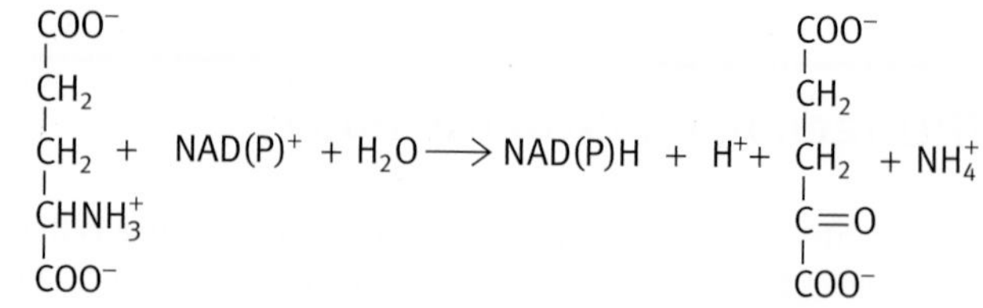

$$\begin{array}{l} COO^- \\ | \\ CH_2 \\ | \\ CH_2 \\ | \\ CHNH_3^+ \\ | \\ COO^- \end{array} + NAD(P)^+ + H_2O \longrightarrow NAD(P)H + H^+ + \begin{array}{l} COO^- \\ | \\ CH_2 \\ | \\ CH_2 \\ | \\ C=O \\ | \\ COO^- \end{array} + NH_4^+$$

Glutamate .... α-Ketoglutarate

As is described below, glutamate dehydrogenase plays a major role in the deamination, and therefore in the oxidation, of many amino acids. In keeping with this, the enzyme is allosterically inhibited by ATP and GTP (indicators of a high energy charge) and activated by ADP and GDP which signal that an increased rate of oxidative phosphorylation is needed.

The α-ketoglutarate produced can feed into the citric acid cycle. Since α-ketoglutarate is converted to oxaloacetate in the cycle (Fig. 12.10), glutamic acid can give rise to glucose synthesis in appropriate physiological situations. It is a glucogenic amino acid. However, there are no corresponding dehydrogenases for the other amino acids. How are they deaminated? A few have special individual mechanisms (see below) but many of them have their amino groups transferred enzymically to α-ketoglutarate, forming glutamate and the corresponding keto acid. The glutamate is then deaminated by the glutamate dehydrogenase reaction given above. The amino acids are thus deaminated by a two-step process.

The first reaction is called **transamination:**

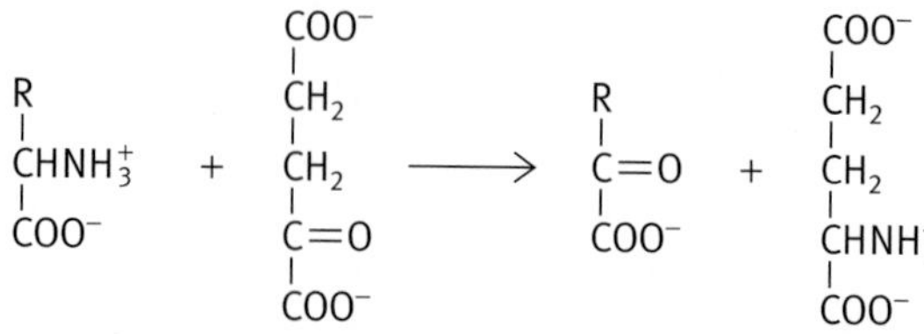

$$\begin{array}{l} R \\ | \\ CHNH_3^+ \\ | \\ COO^- \end{array} + \begin{array}{l} COO^- \\ | \\ CH_2 \\ | \\ CH_2 \\ | \\ C=O \\ | \\ COO^- \end{array} \longrightarrow \begin{array}{l} R \\ | \\ C=O \\ | \\ COO^- \end{array} + \begin{array}{l} COO^- \\ | \\ CH_2 \\ | \\ CH_2 \\ | \\ CHNH_3^+ \\ | \\ COO^- \end{array}$$

If, in the above reaction, R = CH_3, the amino acid is alanine. Deamination of alanine proceeds as follows:

1. Alanine + α-ketoglutarate → pyruvate + glutamate
2. Glutamate + NAD^+ + H_2O → α-ketoglutarate + NADH + $NH4_4^+$

Net reaction: Alanine + NAD^+ + H_2O → pyruvate + NADH + NH_4^+

The two-step process involving transamination and then deamination of glutamate is called **transdeamination** for obvious reasons. Enzymes of the type involved in reaction 1 above are called **transaminases** or **aminotransferases**. A number of these exist with specific substrate specificities and most amino acids can be deaminated by this route. The specific one shown is called **alanine transaminase**. Liver damage can cause its level in the blood to rise and this may be used as a clinical diagnostic tool.

The reversibility of transamination means that, provided a keto acid is available, the corresponding amino acid can be synthesized. (The keto acids for essential amino acids are not, however, synthesized in the body.) An example of the importance of this is in the malate – aspartate shuttle (page 212), in which the following reversible reaction occurs.

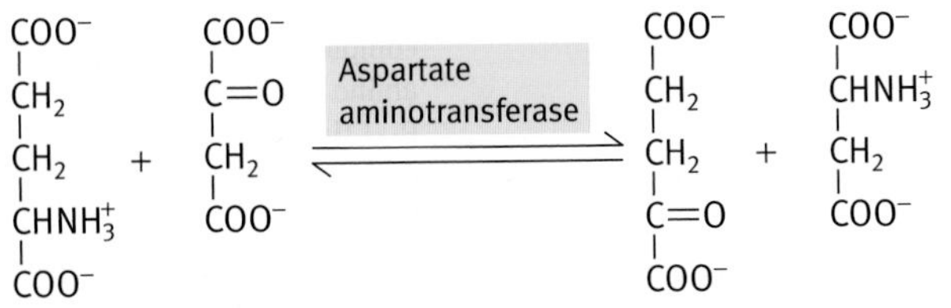

Glutamate   Oxaloacetate   α-Ketoglutarate Aspartate

Mechanism of transamination reactions

All transaminases have, tightly bound to the active centre of the enzyme, a cofactor, pyridoxal-5′-phosphate (**PLP**), that participates in the transaminase reaction.

PLP is a remarkably versatile cofactor; it participates as an electrophilic agent in a wide variety of reactions involving amino acids. In simple terms, PLP acts as an intermediary, accepting the amino group from the donor amino acid and then handing it on to the keto acid acceptor, both phases occurring on the same enzyme. As so often is the case, the cofactor is a B vitamin derivative. Vitamin B_6 in the diet consists of three interrelated compounds, pyridoxin, pyridoxal, and pyridoxamine (Fig. 19.2). They can all be converted to PLP (Fig. 19.2) in the cell. The business end of the molecule is the —CHO group.

The general reaction catalysed in transamination occurs in two steps:

$$ENZ—PLP + amino\ acid\ 1 \rightleftharpoons ENZ—PLP—NH_2 + keto\ acid\ 1$$

$$ENZ—PLP—NH_2 + keto\ acid\ 2 \rightleftharpoons ENZ—PLP + amino\ acid\ 2$$

where PLP represents pyridoxal phosphate and PLP—NH$_2$, pyridoxamine phosphate. The mechanism of the reactions is as shown in Fig. 19.3. Both parts of the reaction occur at the active site of the transaminase, the pyridoxamine phosphate remaining attached to the enzyme.

Special deamination mechanisms for serine and cysteine

In addition to the general transdeamination reactions, certain amino acids have their own particular way of losing their amino groups.

Serine is a hydroxy amino acid; cysteine is the corresponding thiol amino acid. Serine can be deaminated by a PLP-requiring dehydratase enzyme (Fig. 19.4). Cysteine can be deaminated by a somewhat analogous reaction in which H$_2$S is removed instead of H$_2$O, but this occurs only in bacteria. In animals, two more complex routes are used involving direct oxidation of the sulphur in one, or transamination and then desulphuration in the other (not shown). The product in both cases is pyruvate.

Fate of the keto acid or carbon skeletons of deaminated amino acids

Amino acid metabolism links up to the major metabolic pathways of carbohydrate and fat metabolism as already outlined in Fig. 11.10.

As far as general metabolism is concerned, some amino acids are **glucogenic** and some are **ketogenic**. The term 'ketogenic' might seem to imply that an amino acid giving rise to acetyl-CoA results in formation of ketone bodies in the blood, which would conflict with the earlier explanation that ketone bodies arise only in conditions of excess fat metabolism. Acetyl-CoA in normal metabolic situations does *not* give rise to ketone bodies.

The universally used term ketogenic for certain amino acids is an old one, resulting from the use of fasting animals as a test system for the metabolic fate of amino acids since, in

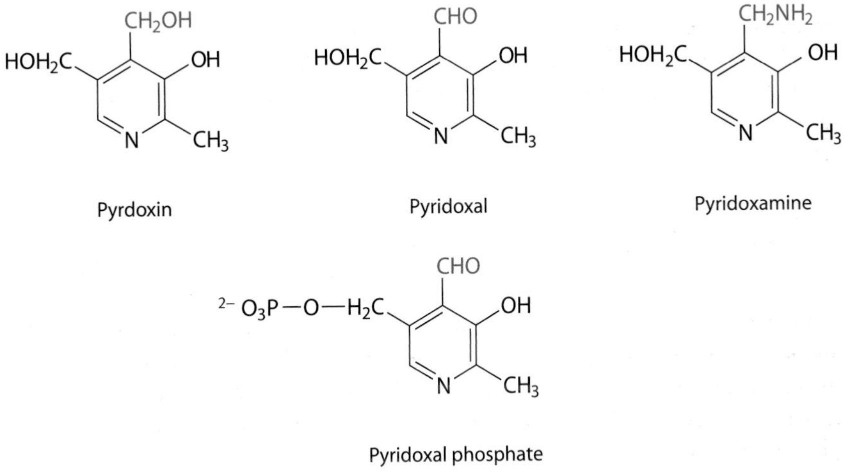

Pyrdoxin   Pyridoxal   Pyridoxamine

Pyridoxal phosphate

Fig. 19.2 Structures of vitamin B_6 components and of the transaminase cofactor, pyridoxal phosphate.

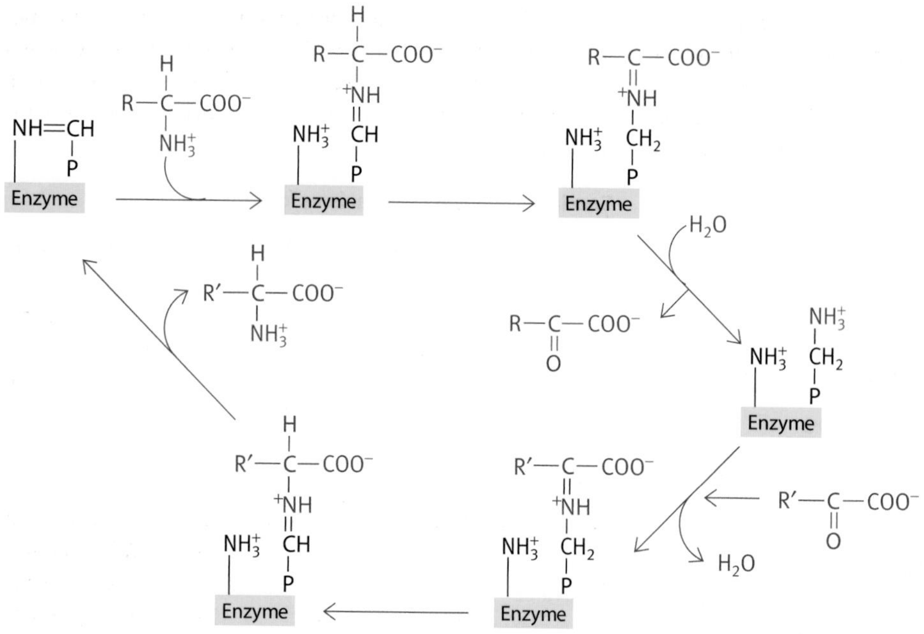

Fig. 19.3 Simplified diagram of the mechanism of transamination. P—CH=NH— in the first structure represents pyridoxal phosphate complexed with a lysine-amino group of the protein. P—CH$_2$—NH$_3^+$ represents pyridoxamine phosphate. The forward reaction (red arrows) results in the conversion of an amino acid to a keto acid and of pyridoxal phosphate to pyridoxamine phosphate. The reverse reaction (blue arrows) reacting with a different keto acid results in transamination between the (red) amino acid and the (blue) keto acid.

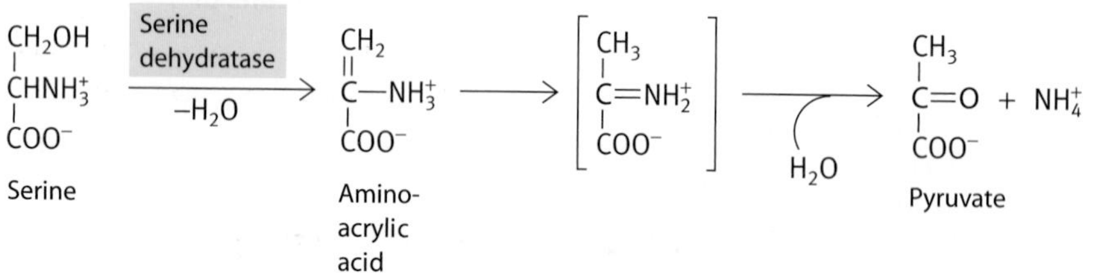

Fig. 19.4 Conversion of serine to pyruvate.

such animals, any increase in acetyl-CoA production causes increases in blood ketone bodies that are easily measured. It does not mean that ketogenic amino acids produce these exclusively in normal animals, but simply that they *can* give rise to acetyl-CoA but not to pyruvate. Glucogenic amino acids were detected by increased levels of blood glucose or glucose excretion when administered to diabetic animals, and similar qualifications apply to this term. Whether, in normal animals, the pyruvate goes to glucose formation or is oxidized depends on metabolic controls.

Aspartate, like glutamate, is converted to a metabolite of the citric acid cycle (oxaloacetate) by the transamination reaction already described and is therefore glucogenic. Alanine and glutamate, producing pyruvate and α-ketoglutarate respectively on deamination, are also glucogenic, as are serine, cysteine, and others.

Some amino acids are both ketogenic and glucogenic – for example, phenylalanine produces fumarate (an intermediate in the citric acid cycle) and acetyl-CoA. Of the 20 amino acids only two (leucine and lysine) are solely ketogenic. The degradation of four amino acids (isoleucine, methionine, threonine, and valine) is referred to on page 287 because of its special interest there. The product is propionate, which is metabolized by the route for oxidation of odd-numbered fatty acids, a process involving vitamin B$_{12}$ (page 219).

Genetic errors in amino acid metabolism cause diseases

Phenylketonuria

Phenylalanine is an aromatic amino acid, an excess of which is normally converted to tyrosine by an enzyme, phenylalanine hydroxylase (Fig. 19.5). This enzyme is interesting in that a pair of hydrogen atoms are supplied by an electron donor – a coenzyme molecule called **tetrahydrobiopterin** (RH$_4$ in Fig. 19.5; structures of RH$_4$ and dihydrobiopterin (RH$_2$) are given below for reference purposes). Oxygen is also needed.

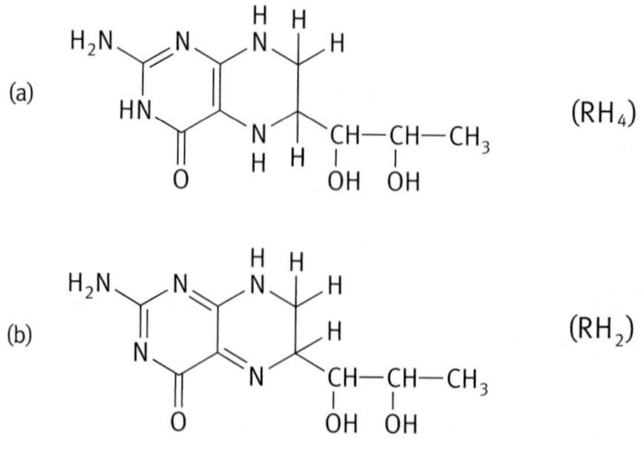

It may seem odd that a reaction requires *both* oxygen and a reducing agent but it is a mechanism used in other reactions also, as you will see later. The trick is that one *atom* of an oxygen molecule is used to form an —OH group on the aromatic ring, but this leaves the other oxygen atom to be taken care of.

It is reduced to H_2O by the two hydrogen atoms donated by the tetrahydrobiopterin. A separate enzyme system reduces the **dihydrobiopterin** formed back to the tetrahydro form, using NADPH as the reductant, and so the cofactor acts catalytically.

The phenylalanine hydroxylase belongs to a class of enzymes known as **monooxygenases** (because one atom of O appears in the product) or, alternatively, **mixed-function oxygenases** because two things are oxygenated – the amino acid and a pair of hydrogen atoms.

Phenylalanine as such is not normally deaminated, being converted to tyrosine, and only then does deamination occur (Fig. 19.5). However, there is a genetic abnormality, occurring in around 1 in 10,000, in which the phenylalanine conversion to tyrosine is impaired or blocked because of enzyme deficiency or, rarely, to lack of tetrahydrobiopterin. This causes excess phenylalanine to accumulate and, in this situation, it abnormally participates in transamination producing phenylpyruvate (an abnormal metabolite), which spills out into the urine (Fig. 19.5). The disease is called **phenylketonuria** or **PKU**. The consequences of phenylpyruvate in babies are irreparable mental impairment. If diagnosed at birth (by blood analysis, for phenylalanine level) a child with the disease can be given a diet limited in phenylalanine (but adequate in tyrosine) and development is normal. The urine of patients has a mousy colour

due to the phenylacetate formed from the phenylpyruvate. Mass screening programmes of neonatals, giving early detection, avoids the tragic consequences of the abnormality. It is not known how phenylpyruvate causes such deleterious brain damage.

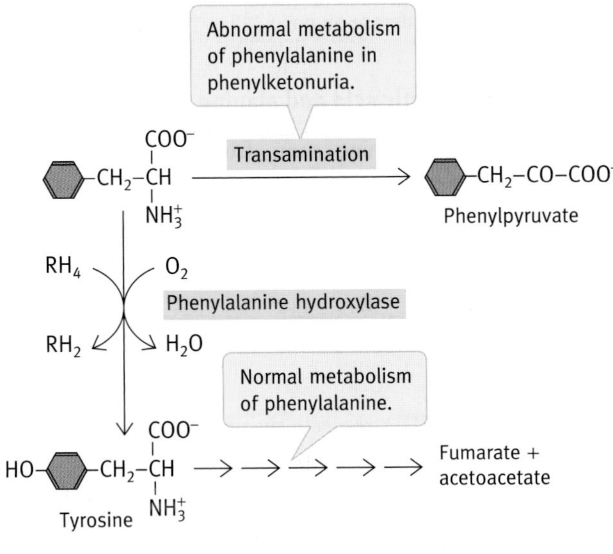

Maple syrup disease

Curiously enough, another genetic condition, called **maple syrup disease**, involves accumulation of the keto acids of three aliphatic amino acids, valine, isoleucine, and leucine, and this also involves brain impairment. (The name of the disease comes from the keto acids in the urine having a characteristic smell.) This disease is much rarer than PKU.

Alcaptonuria

Another much-quoted genetic condition is **alcaptonuria** in which the urine turns black on exposure to air. This is a relatively benign condition but in later years may cause problems in connective tissue. It is due to a block in the tyrosine-degradation pathway in which a diphenol intermediate metabolite, homogentisate, is excreted. The diphenol in air oxidizes to form a dark pigment. It was the first metabolic disease for which the genetic pattern of inheritance was worked out.

Methionine and transfer of methyl groups

Methionine is one of the essential amino acids. It has the structure:

$$CH_3—S—CH_2—CH_2—\underset{\underset{NH_3^+}{|}}{CH}—COO^-.$$

The interesting part is the **methyl group**. Methyl groups are very important in the cell – a variety of compounds are methylated, and methionine is the source of methyl groups that are transferred to other compounds. Methionine is a stable molecule – the methyl group has no tendency to leave; however, if the molecule is converted into **S-adenosylmethionine (SAM)**, a sulphonium ion is created and the methyl group is 'activated' – it has a strong group transfer potential making it thermodynamically favourable for it to be transferred (by transmethylase enzymes) to O and N atoms of other compounds (it cannot form C—C bonds). The reason for the high energy is that when the methyl group is transferred, the sulphonium ion reverts to an uncharged thioether. ATP supplies the energy for SAM synthesis – in this case three —P groups are converted to $PP_i + P_i$ and then the PP_i is cleaved to two P_i (Fig. 19.6).

Transfer of the methyl group of SAM to other compounds generates **S-adenosylhomocysteine**. The latter is hydrolysed

Fig. 19.5 Normal and abnormal metabolism of phenylalanine.

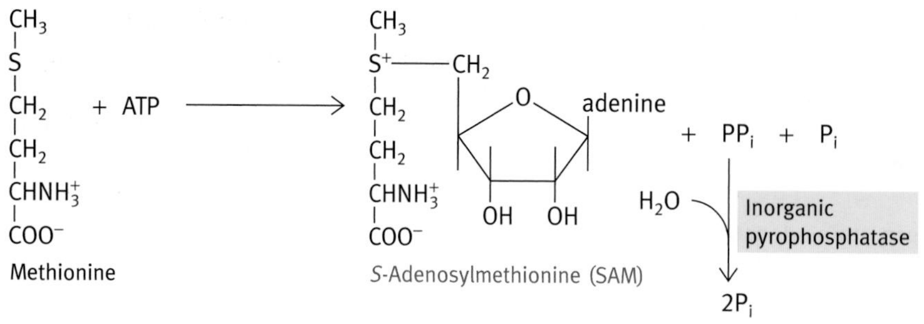

Fig. 19.6 The synthesis of S-adenosylmethionine (SAM) from methionine.

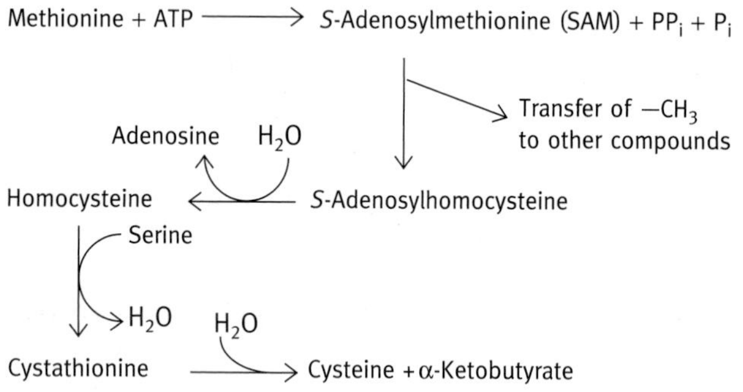

$$\text{Methionine} + \text{ATP} \longrightarrow S\text{-Adenosylmethionine (SAM)} + PP_i + P_i$$

Fig. 19.7 Breakdown of s-adenosylhomocysteine to form cysteine and α-ketobutyrate. Note that the recycling of homocysteine to methionine is dealt with on page 308.

to produce **homocysteine** – this is methionine with —SH instead of —S—CH_3 (it is not one of the 'magic' 20 used in protein synthesis). The homocysteine is complexed with serine to form cystathionine, which is hydrolysed to cysteine and α-ketobutyrate (Fig. 19.7). Deficiency in cystathionine formation leads to accumulation of homocysteine which, for unaccounted reasons, is associated with a variety of childhood pathologies including mental retardation.

Recycling of homocysteine back to methionine requires a folate derivative (page 308). Hence folate deficiency can cause a build-up of homocysteine, and folate therapy is being tried for patients with raised homocysteine levels. Elevated homocysteine levels are also believed to be somehow associated with development of blood vessel blockages and coronary heart disease, and is currently of considerable medical interest.

What are the methyl groups transferred to?

In the body the methyl groups of creatine (page 125), phosphatidylcholine (page 103), and epinephrine (adrenaline; page 292) come from S-adenosylmethionine, and so do the methyl groups attached to the bases of nucleic acids. Note, however, that the latter do not include that of thymine, which is separately synthesized and is an important topic to be dealt with later (page 306), as also is the subject of nucleic acid base methylation.

Synthesis of amino acids

In the body, as explained, only the nonessential amino acids can be synthesized, but all are made in plants and bacteria. The pathways of synthesis of all have been long established. Our aim here is to deal only with aspects of special interest and to illustrate how some amino acids are synthesized from glycolytic and citric acid cycle intermediates. In fact, five of these intermediates (3-phosphoglycerate, phosphoenolpyruvate, pyruvate, oxaloacetate, and α-ketoglutarate) together with two sugars of the pentose phosphate pathway are the precursors of all 20 amino acids (in plants and bacteria).

Synthesis of glutamic acid

As explained earlier, deamination of glutamic acid occurs *via* glutamate dehydrogenase, an $NADP^+$ enzyme of central importance. This is reversible, and is the route by which glutamate can be formed from α-ketoglutarate and ammonia. Glutamate formation also occurs in animals *via* transamination of α-ketoglutarate using other amino acids such as alanine or aspartate as the donor of the amino group. See the reaction for aspartate aminotransferase on page 285.

Synthesis of aspartic acid and alanine

Aspartic acid and alanine come from transamination of oxaloacetate

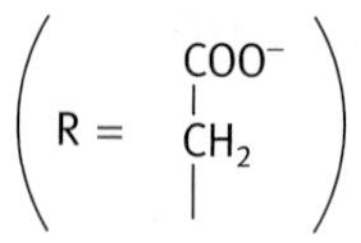

$$\left(R = \begin{array}{c} COO^- \\ | \\ CH_2 \\ | \end{array} \right)$$

and pyruvate ($R = CH_3$), respectively, with glutamate.

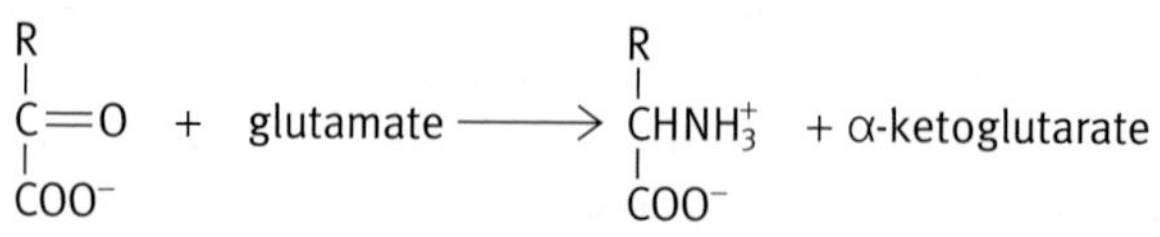

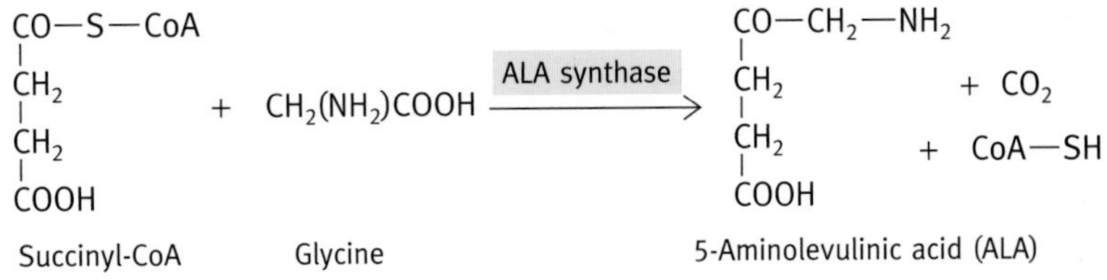

Fig. 19.9 The first step in haem synthesis, catalysed by aminolevulinate (ALA) synthase. For clarity of illustration, nonionized structures are given.

Synthesis of serine

Serine is formed from the glycolytic intermediate, 3-phosphoglycerate, which is first converted to a keto acid, 3-phosphohydroxypyruvate:

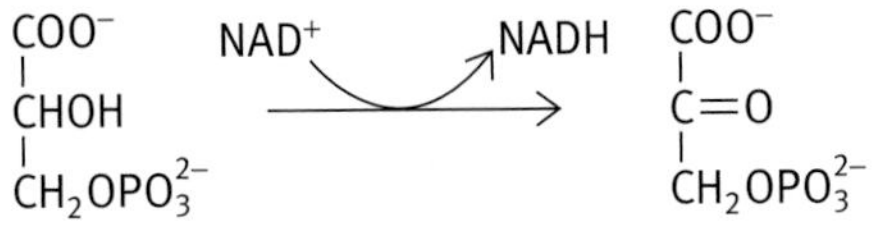

This keto acid is transaminated by glutamic acid to give 3-phosphoserine, which is hydrolysed to serine and P_i.

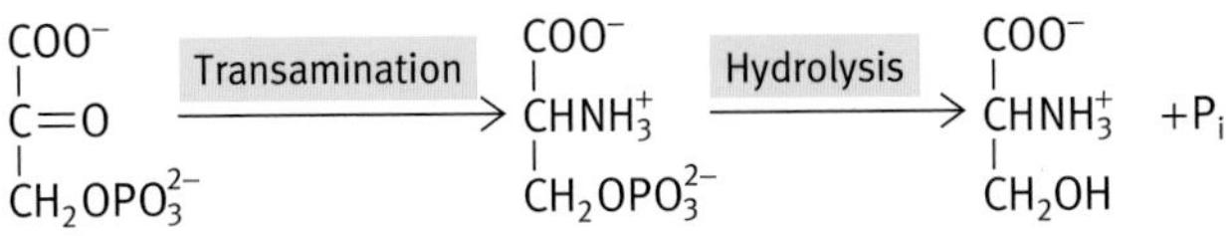

Synthesis of glycine

Glycine is the simplest amino acid of all ($CH_2NH_3^+ COO^-$). It is formed by a reaction that is completely new, so far as this book is concerned, involving withdrawal of a hydroxymethyl group ($-CH_2OH$) from serine and adding it to a coenzyme, tetrahydrofolate (again, not yet mentioned in this book), the function of which is to act as a one-carbon-unit carrier. The one-carbon-unit transfer area is of importance in nucleotide synthesis. It will be more appropriate to go into this thoroughly later (page 301). (Other sources of glycine exist.)

Haem and its synthesis from glycine

The full structure of haem is given for reference purposes in Fig. 4.19, where it is described in relation to its role in haemoglobin, but Fig. 19.8 gives an outline structure. It is a ferrous iron complex with protoporphyrin. Protoporphyrin is a tetrapyrrole, the four substituted pyrroles being linked by methene ($=CH-$) bridges such that a **conjugated double-bond system** exists (that is, you can go right round the molecule *via* alternating single and double bonds). This gives protoporphyrin and haem their deep red colour.

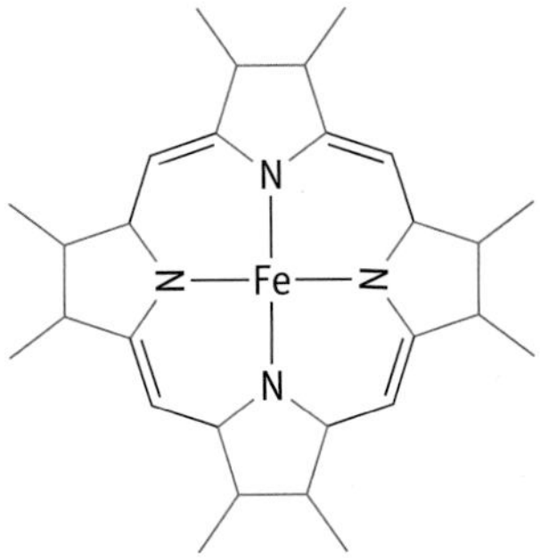

Fig. 19.8 Outline of the haem molecule. (The full structure is given in Fig. 4.19.)

In haem, the four pyrrole N atoms are bound to Fe^{2+} as shown in Fig. 4.20, leaving two more of the six ligand positions of the Fe^{2+} available for other purposes.

Red blood cells have the vast majority of the body's haem content, though haem is found in all aerobic cells as the prosthetic group for cytochromes and other proteins.

The synthesis of haem appears to be a formidable task but the essentials are surprisingly simple, requiring in animals only two starting reactants, glycine and succinyl-CoA. You have met succinyl-CoA in the citric acid cycle. An enzyme, **aminolevulinate synthase**, or **ALA synthase** (ALA-S), carries out the reaction shown in Fig. 19.9. 5-Aminolevulinic acid (ALA) is the precursor solely committed to porphyrin synthesis. Two molecules of ALA are used to form the pyrrole, **porphobilinogen** (**PBG**), the two molecules being dehydrated by ALA dehydratase (Fig. 19.10). The remainder of the haem biosynthetic pathway consists of linking four PBG molecules together,

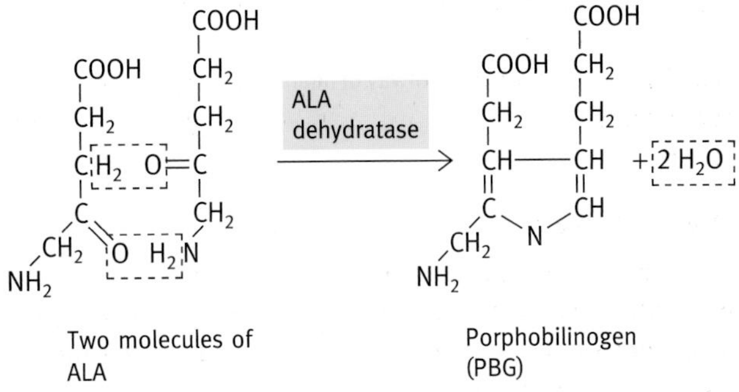

Fig. 19.10 Synthesis of porphobilinogen (PBG) – the aminolevulinate (ALA) dehydratase step. Nonionized structures are given for simplicity. PBG is a monopyrrole; haem is a tetrapyrrole. The conversion of PBG to haem is shown in Fig. 19.11.

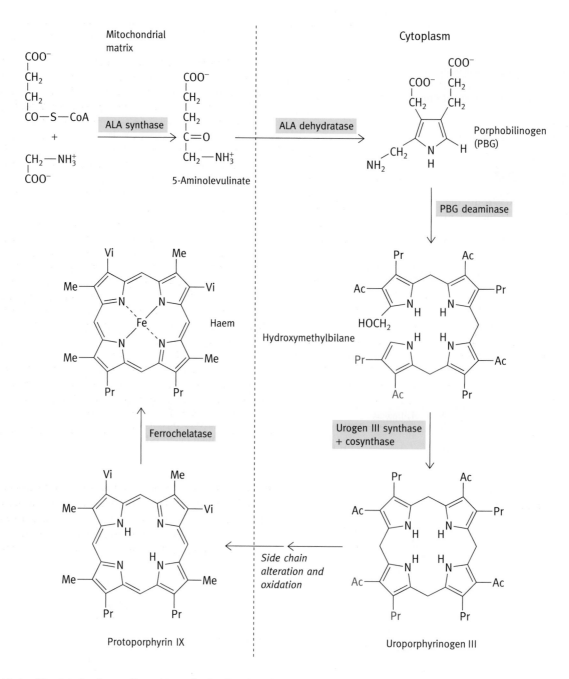

Fig. 19.11 An abbreviated pathway of haem biosynthesis, given for reference purposes. Me, Pr, Ac, and Vi indicate methyl, propyl, acetyl, and vinyl groups, respectively.

modifying the side groups, and chelating an atom of ferrous iron to form haem. The intermediate tetrapyrroles between PBG and haem are the colourless **uro-** and **copropor-phyrinogens** (in which PBG units are linked by methylene bridges) and the red **protoporphyrin** (in which they are linked by methene bridges). The pathway is given in Fig. 19.11; have a quick look at this.

The haem synthesis pathway has a curious feature. The first reaction, synthesis of ALA, occurs inside mitochondria, after

which the ALA moves out to the cytoplasm, but the final three steps also occur in the mitochondria. Why this should be so is not clear.

A group of porphyria diseases is known, each of which is associated with a deficiency of one of the enzymes of the haem biosynthetic pathway (Box 19.1). Each deficiency can cause the accumulation of the metabolite(s) preceding the deficiency and these can have deleterious effects.

Acute intermittent porphyria (AIP), the most common type of porphyria disease, though not encountered frequently, is clinically important because it can be life threatening. It is due to a deficiency (about 50%) of the enzyme porphobilinogen deaminase. Most of the time the patient is normal but acute attacks can be precipitated by a variety of drugs such as barbiturates. The triggering agents appear to have in common the ability to induce the synthesis of hepatic cytochrome P450. This is a haem protein that is massively induced by some drugs, barbiturates being the classic ones in this respect. This causes a demand for increased haem synthesis, and in response to this the level of ALA synthase is increased to meet the demand. In acute intermittent porphyrics, the haem biosynthetic pathway cannot handle the increased supply of ALA, resulting in its accumulation and that of PBG, the next metabolite, which spill over into the urine. Such accumulation is associated with the onset of the symptoms of the disease. These are neurological in nature and result in severe abdominal pain and psychiatric abnormalities. It is not known how these effects are caused.

The control mechanisms that underlie the disease are now understood. When cytochrome P450 is induced it causes a reduced haem level. It has been shown that haem controls ALA synthase in three different ways. First it represses transcription of the mRNA for the enzyme in the nucleus; secondly it destabilizes the mRNA resulting in a shorter half-life and reduced synthesis of the enzyme; thirdly, the enzyme is synthesized in the cytoplasm as a precursor protein, which is transported into the mitochondria where it functions. Haem inhibits this transport. Thus in a normal person, haem biosynthesis maintains a haem level that balances haem production with demand. In a patient with AIP, the impaired haem biosynthesis pathway is inadequate to cope with the extra demand so the ALA synthase is excessively induced, and the ALA and PBG accumulate due to the block.

AIP is known as a hepatic porphyria because its effects originate in the liver. In red blood cells the ALA synthase is coded for by a different gene and haem-synthesis control is different in mechanism. The enzyme blockages in erythropoietic porphyrias, as the associated diseases are called, result in accumulations of porphyrins in the skin, giving rise to distressing photosensitivity damage.

For the past few decades the literature has implied that King George III (the 'mad king') had acute intermittent porphyria (and speculated that the attacks may have had some relevance to the American War of Independence), but more recently, opinion has favoured the view that he had variegate porphyria, also of hepatic origin. It has also been implied that Vincent Van Gogh possibly had acute intermittent porphyria (see the **May** *et al.* review at the end of this chapter for a fuller account).

In general, metabolic diseases are inherited recessively, because most metabolic pathways can operate on the 50% enzyme level in heterozygotes with one gene deficient. However the porphyria diseases involve rate-limiting enzymes, and a 59% deficiency is sufficient in some circumstances to cause the disease.

Destruction of haem

Red blood cells are destroyed mainly by **reticuloendothelial cells** of spleen, lymph nodes, bone marrow, and liver. Removal of sialic acid groups from the red-cell-membrane glycoproteins is a signal that the cell is aged and ready for destruction. The degraded carbohydrate attaches to cell receptors, which leads to endocytosis of the erythrocyte. The enzyme, haem oxygenase opens up the tetrapyrrole ring, releasing the iron for reuse and forming **biliverdin**, a linear tetrapyrrole (Fig. 19.12). Biliverdin

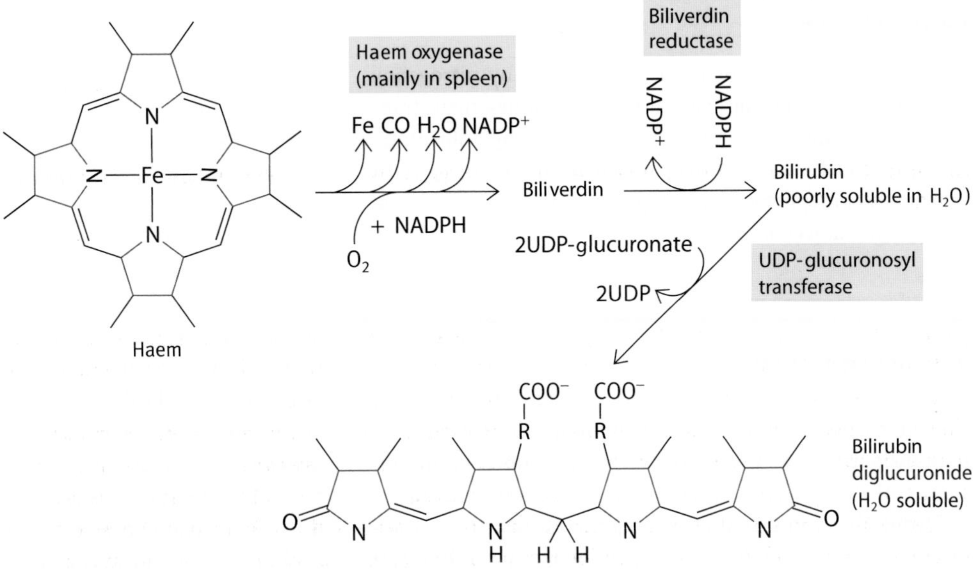

Fig. 19.12 Simplified representation of haem breakdown, the oxygenase reaction, and the reduction of biliverdin (see text). See Fig. 29.5 for UDP-glucuronate formation.

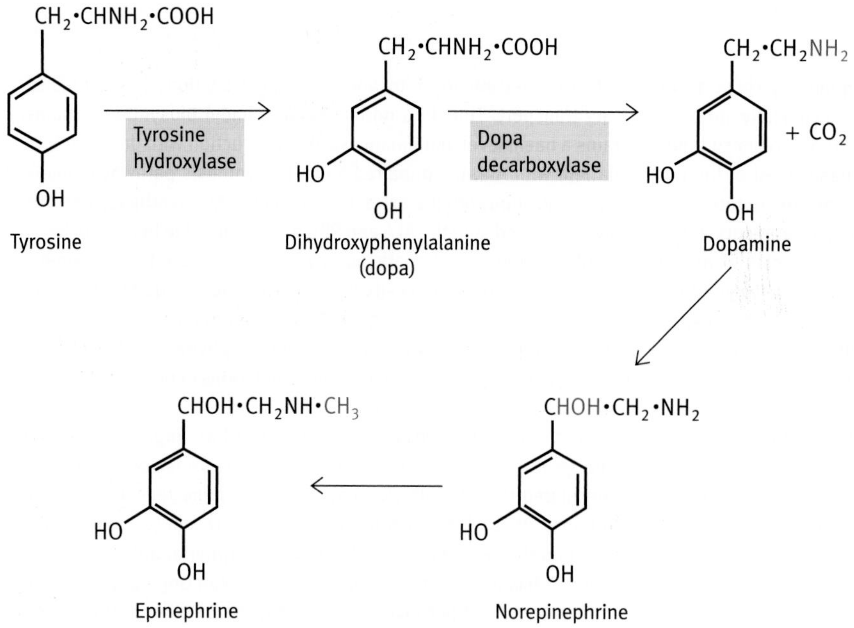

Fig. 19.13 Intermediates in the pathway for the synthesis of the catecholamines.

is reduced to **bilirubin**. This is water insoluble but is transported in the blood, attached to serum albumin, to the liver, where it is rendered much more polar by the addition of two glucuronate groups (Fig. 19.12) and then excreted, in the bile, into the gut where bacteria convert it to **stercobilin**, giving the brown colour to faeces; modification and partial reabsorption of some bile compounds leads to the yellow colour of urine. **Jaundice** may result from excessive red blood cell breakdown, lack of the glucuronidation enzyme, or blockage of the bile duct.

Synthesis of epinephrine and norepinephrine

Amino acids are used for the synthesis of many nitrogenous molecules including long chain amines, hormones, neurotransmitters, creatine, nucleotides and others. A single illustrative example (Fig. 19.13) gives the pathway for the synthesis of the hormones **epinephrine** and **norepinephrine**. (Nucleotide synthesis is the subject of Chapter 20.)

The urea cycle

The amino groups of catabolized amino acids are excreted in mammals mainly as urea, a highly water-soluble, inert, non-toxic molecule. Its essential role is to prevent accumulation of the toxic NH_4^+ ion produced from amino acids. Urea is produced only in the liver from the guanidino group of arginine by the hydrolytic enzyme **arginase**. The other product is ornithine, an amino acid not found in proteins.

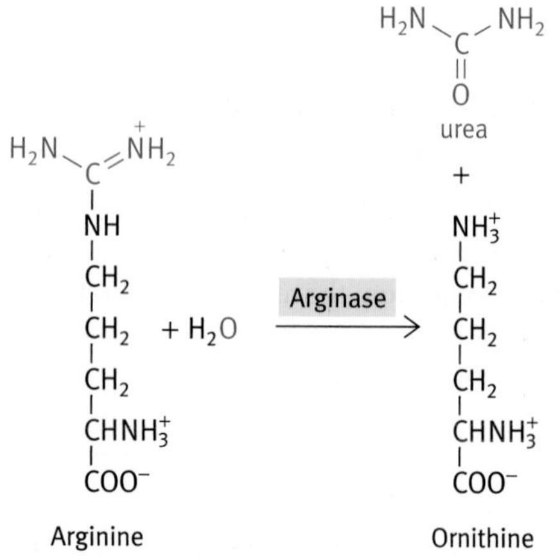

The amino nitrogen of the catabolized 20 amino acids is used to convert ornithine back to arginine. The extra carbon comes from CO_2 (as HCO_3^-). Krebs (who also discovered the citric acid cycle) observed together with Henseleit that, when arginine was added to liver cells, the increased urea formation caused by this addition far exceeded in amount that of the arginine added – that is, it was acting catalytically and ornithine did the same, suggesting a cyclical process. It was discovered that citrulline, an amino acid intermediate between ornithine and arginine, is involved, since it also acts catalytically on urea synthesis when added to liver cells. This discovery led to the famous **urea cycle**, the first biochemical cycle. Citrulline is not one of the 20 amino acids used for synthesizing proteins.

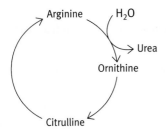

Fig. 19.14 Outline of the arginine–urea cycle. The input of CO_2 and nitrogen into the cycle is dealt with in the section on mechanism of arginine synthesis.

H₂N group structure labeled:

$$H_2N \!-\! \underset{NH}{\overset{O}{\underset{|}{C}}} \!-\! ...$$

NH
|
CH₂
|
CH₂
|
CH₂
|
CHNH₃⁺
|
COO⁻

Citrulline

An outline of the cycle is given in Fig. 19.14 . The whole cycle is given in Fig. 19.15.

Mechanism of arginine synthesis

We need first to see how ornithine is converted to arginine. Ammonia, CO_2, and ornithine are the reactants for the first step, that of citrulline synthesis which occurs in the mitochondrial matrix. Energy is needed and ATP supplies it. Firstly, ammonia and CO_2 are converted to a reactive intermediate, **carbamoyl phosphate**, which then combines with ornithine to give citrulline. Carbamic acid has the structure NH_2COOH and carbamoyl phosphate is therefore $NH_2\!-\!\overset{O}{\overset{\|}{C}}\!-\!PO_3^{2-}$. It is a high-energy phosphoryl compound, being an acid anhydride. It is synthesized by an enzyme **carbamoyl phosphate synthetase**, catalysing the following reaction:

$$NH_4^+ \;+\; HCO_3^- \;+\; 2ATP$$

$$\downarrow$$

$$NH_2\!-\!\overset{O}{\overset{\|}{C}}\!-\!O\!-\!\overset{O}{\underset{\underset{O^-}{|}}{\overset{\|}{P}}}\!-\!O^- \;+\; 2ADP \;+\; P_i \;+\; 2H^+$$

Two ATP molecules are used, the first ATP being broken down to ADP and P_i to drive the production of carbamate

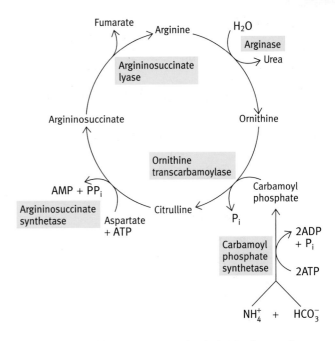

Fig. 19.15 The enzymes of the urea cycle. The levels of urea cycle enzymes are coordinated with the dietary intake of protein. The cycle is allosterically controlled at the carbamoyl phosphate synthetase step. The positive allosteric effector of the enzyme is N-acetyl glutamate. The conversion of ornithine to citrulline takes place inside mitochondria while the rest of the cycle occurs in the cytoplasm.

from ammonia and CO_2 and the second used to phosphorylate the carbamate. It all happens on the surface of the one enzyme. The carbamoyl group of carbamoyl phosphate is now transferred to ornithine by **ornithine transcarbamoylase** giving citrulline. If we represent ornithine as $R\!-\!NH_3^+$ the reaction is:

$$R\!-\!NH_3^+ \;+\; NH_2\!-\!\overset{O}{\overset{\|}{C}}\!-\!O\!-\!\overset{O}{\underset{\underset{O^-}{|}}{\overset{\|}{P}}}\!-\!O^- \longrightarrow HN\!-\!\overset{O}{\overset{\|}{C}}\!-\!NH_2 \;+\; P_i$$

Ornithine   Carbamoyl phosphate   Citrulline

Conversion of citrulline to arginine

The final step in arginine synthesis is to convert the C=O group of citrulline to the C=NH of arginine. This occurs in the cytosol. Ammonia is *not* used here, but instead the amino group of aspartate is added directly. Ammonia can be converted to glutamate and this can generate aspartate by transamination with oxaloacetate. In this way, ammonia and also the amino groups of most amino acids can be converted to urea *via* this second stage of the cycle.

First, the enzyme **argininosuccinate synthetase** condenses aspartate with citrulline as follows:

The molecule formed is **argininosuccinate**. The name derives from the fact that the molecule structurally is like an arginine derivative of succinate. The molecule is now 'pulled apart' by **argininosuccinate lyase** to yield arginine and fumarate (not arginine and succinate).

Note that the urea cycle is linked to the citric acid cycle because fumarate is converted to oxaloacetate. Transamination of this forms aspartate which is available for arginosuccinate synthesis.

Control of the urea cycle

The major control of the urea cycle is by variation in the level of enzymes catalyzing each step. When there is a high level of free amino acids to be metabolized, enzymes are synthesized, and the reverse applies at low levels. High amino acids levels occur after a rich intake and in starvation when muscle proteins are breaking down to supply the gluconeogenic pathway in liver. In addition, carbamoyl phosphate synthetase is inactive without its allosteric activator, N-acetyl glutamate, whose cellular level reflects that of amino acids.

Transport of the amino nitrogen from extrahepatic tissues to the liver

Transport of ammonia in the blood as glutamine

There are two main mechanisms. Ammonia produced from amino acids is toxic, and blood ammonia levels are kept very low since abnormally high levels can impair brain function and cause coma. Free ammonia is not therefore transported, as such, from peripheral tissues to the liver, but is first converted to the nontoxic amide, glutamine. This is synthesized from glutamate by the enzyme, **glutamine synthetase**.

The reaction involves the intermediate formation of an enzyme-bound g-glutamyl phosphate. This is a high-energy phosphoryl anhydride compound with sufficient energy to react with ammonia.

The glutamate for this synthesis can be formed from α-ketoglutarate generated in the citric acid cycle followed by transamination with other amino acids. The glutamine is carried in the blood to the liver where it is hydrolysed to release ammonia which is used for urea synthesis.

Glutaminase in the kidney also liberates ammonia to be excreted along with excess acids from the blood. Glutamine is one of the 20 amino acids found in proteins and is involved in the synthesis of several other metabolites.

Transport of amino nitrogen in the blood as alanine

From muscle, as much as 30% of the amino nitrogen produced by protein breakdown is sent to the liver as alanine (and, to a lesser extent, as glutamine). The released amino acids from protein breakdown transaminate with pyruvate to yield alanine, which is released into the blood. This is taken up by the liver; the amino group is used to form urea (*via* ammonia and/or aspartic acid). The released pyruvate is converted to blood glucose which can go back to the muscle. The sequence of events is referred to as the glucose-alanine cycle (Fig. 19.16).

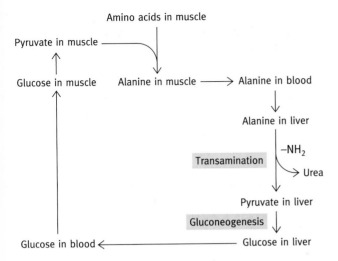

Fig. 19.16 The glucose–alanine cycle for transporting nitrogen to the liver as alanine, and glucose back to the muscles.

This alanine transport from muscle to liver has another important physiological role in starvation. As already explained, after glycogen reserves are exhausted the liver *must* make glucose to supply the brain and other cells with an obligatory requirement for the sugar. The main sources of metabolites for this hepatic gluconeogenesis are the amino acids derived from muscle protein breakdown. Many of the amino acids can give rise to pyruvate in the muscle which is sent to the liver as alanine. However, the glucose-alanine cycle itself gives no net increase in glucose and in starvation the amino acids liberated from muscle protein breakdown must be converted to alanine without utilizing glucose as a source of pyruvate. In that situation, the pyruvate comes from breakdown of glucogenic amino acids.

Diseases due to urea cycle deficiencies

Since the role of the urea cycle is to prevent accumulation of toxic levels of ammonium ions, deficiencies in pathway enzymes cause diseases. They are fortunately rare, since the diseases are very serious.

A variety of such diseases have been identified because of the deficiencies in the synthesis of the activator of carbamoyl phosphate synthetase, N-acetyl glutamate, and to deficiencies of the enzyme. Other conditions are associated with separate deficiencies in argininosuccinate synthetase and lyase, and rarely with arginase deficiency.

The diseases vary in severity according to the site of the blockage in the urea synthesis pathway, and according to the fractional loss of the enzyme in question. Many are only partially lost. Severe disease typically is associated with mental retardation and early death. The reason for the toxic effect of ammonia on the brain is not known.

Various treatments have been devised in different forms of disease to cause alternative routes of nitrogen excretion. Feeding of benzoate and phenylacetate results in excretion of these acids coupled to glycine and glutamine respectively. A more complex metabolic strategy in the case of arginosuccinate lyase deficiency is to provide excess arginine (and a low-protein diet). Arginine is converted to urea and ornithine. Ornithine is converted to citrulline (using carbamoyl phosphate), which, with aspartate, forms arginosuccinate. This accumulates and is excreted, carrying with it two nitrogen atoms from carbamoyl phosphate and aspartate.

Alternatives to urea formation exist in different animals

In humans, some nitrogen is excreted as uric acid, ammonium ions, and creatinine. However birds excrete it as a white paste of solid uric acid rather than urea; fish and other animals living in water excrete it as ammonia. There is an evolutionary logic here. In an aqueous environment, unlimited water means that ammonia can be constantly excreted and dispersed; mammals are intermediate in this respect, and urea, being nontoxic and highly soluble, is the route of choice. Birds have the problem that chicks develop in a closed egg, and accumulation of any soluble form of excretory product could be deleterious. They therefore excrete nitrogen as uric acid which, being almost insoluble, causes no problems as it accumulates.

Summary

Amino acids are supplied in the diet from protein hydrolysis in the gut. Proteins in the body are also constantly degraded and resynthesised. The body can synthesize about 10 of the amino acids but the rest must be obtained from the diet. All 20 are needed for protein synthesis. Amino acids are also used to synthesize a wide variety of other molecules.

Amino acids in excess of immediate requirements are deaminated; the amino nitrogen is mainly converted to urea (in mammals) and excreted. The carbon-hydrogen skeletons are oxidized to release energy or converted to fat or glycogen according to the metabolic controls operating at the time and the particular amino acid. Most amino acids are deaminated via transamination with α-ketoglutarate. The glutamate so formed is deaminated by glutamate dehydrogenase, releasing ammonia. Transaminases are pyridoxal phosphate-containing proteins, the cofactor being essential in the transamination reaction.

A few amino acids such as serine, cysteine, and glycine have different metabolic pathways of their own. Phenylalanine is converted to tyrosine before deamination. If the hydroxylating enzyme is missing, phenylalanine is deaminated to produce phenylpyruvate, which causes mental impairment in children with the disease phenylketonuria. If the disease is detected early, dietary strategies to restrict phenylalanine intake results in normal development. The keto acids of branched-chain amino acids also accumulate in a rare genetic disease (maple syrup disease), which also results in mental impairment. Methionine has the special role of providing methyl groups. The latter must be first activated by the formation of S-adenosylmethionine.

Haem is synthesized from glycine, the first step being the formation of 5-aminolevulinate by condensation with succinyl-CoA. Haem is destroyed by haem oxygenase, mainly in the spleen, resulting in bilirubin formation, which is excreted as the diglucuronide into the bile. Blockage of the bile duct leads to jaundice.

The ammonia formed by deamination of amino acids is excreted mainly as urea. This is formed in the liver by the Krebs urea cycle. In this, arginine is converted to urea + ornithine by arginase. Arginine is resynthesized from ornithine, using HCO_3^-, ammonium ions, and the amino group of aspartate. Citrulline is an intermediate in the process. The urea cycle is metabolically linked to the citric acid cycle.

Urea synthesis is controlled in two ways. The urea cycle is allosterically activated at the carbamoyl synthetase step by N-acetyl glutamate whose level reflects that of the amino acids available. In addition, high levels of amino acids cause an increase in the enzymes of the cycle.

Free ammonia from deamination in the tissues is potentially toxic and is transported to the liver as the amide group of glutamine. There the glutamine is hydrolysed by glutaminase to release ammonium ion to be used in urea synthesis.

An alanine cycle is responsible for transporting amino nitrogen from muscles resulting from muscle protein breakdown to the liver.

Several diseases exist in which urea cycle enzymes are deficient in amount. In some cases these can be treated by dietary strategies.

Alternatives to urea formation exist in different animals. In humans, some nitrogen is excreted as uric acid, ammonium ions, and creatinine. However birds excrete it as a white paste of solid uric acid rather than urea; fish and other animals living in water excrete it as ammonia.

Further reading

Felig, P. (1975). Amino acid metabolism in man. *Annu. Rev. Biochem.*, **44**, 933–55.
A useful review, which discusses amino acid metabolism at the organ level.

Snell, K. (1979). Alanine as a gluconeogenic carrier. *Trends Biochem. Sci.*, **4**, 124–8.
Reviews the role of muscle in supplying the liver with metabolites for glucose synthesis.

Walford, M. (1991). The urea cycle: a two compartment system. *Essays Biochem.*, **26**, 49–58.

Straka, J. G., Rank, J. M., and Bloomer, R. (1990). Porphyria and porphyrin metabolism. *Annu. Rev. Med.*, **41**, 457–69.
This review deals with the defects in the enzymes of haem biosynthesis which lead to porphyrias, and the therapeutic uses of porphyrins.

May, B. K., et al. (1995). Molecular regulation of heme biosynthesis in higher vertebrates. *Prog. Nucleic Acid Res.*, **51**, 1–47.
A comprehensive review of all aspects including the hereditary diseases.

Problems

1   Explain how an oxidation can result in the deamination of an amino acid.

2   Which amino acid is deaminated by the mechanism referred to in question 1?

3   How are several of the amino acids deaminated, where the reaction in question 2 is involved? Use alanine as an example.

4   What is the cofactor involved in transamination? Give its structure and explain how transamination occurs.

5   Explain how serine is deaminated.

6   What is meant by the terms glucogenic and ketogenic amino acids? Which amino acids are purely ketogenic?

7   Explain the genetic disease phenylketonuria.

8   What is the role of tetrahydrobiopterin in phenylalanine hydroxylation?

9   Methionine is the source of methyl groups in several biochemical processes. Explain how methionine is activated to donate such groups.

10 Describe the first two steps in haem biosynthesis in animals.

11 Outline the reactions of the urea cycle.

12 Why should the level of urea cycle enzymes be increased both in the situation of a high intake of amino acids and in starvation?

13 How are (a) ammonia and (b) amino nitrogen in peripheral tissues transported to the liver for conversion to urea?

14 Several genetic diseases involving the urea cycle are known. What are these and what strategies have been used to ameliorate them?

15 What is the medical relevance of ALA synthase control in liver?

Chapter 20

Nucleotide synthesis and metabolism

Several of the preceding chapters have been mainly concerned with energy release from food. In this, ATP occupies the central position, but GTP, CTP, and UTP are also involved in aspects of food components metabolism. However, the involvement of these nucleotides described so far has all been concerned with the phosphoryl groups of the molecules. The nature of the bases, whether A, G, C, or U, has been important only for recognition by the appropriate enzymes but otherwise has not been directly relevant to the metabolic processes. For this reason we have not previously given information about the bases themselves.

We are now about to start, in the next chapter, on a new main area in which **information transfer** is a primary purpose – we refer to nucleic acids and protein synthesis and for this the structures, synthesis, and metabolism of nucleotides are important and a prerequisite for the subsequent chapters.

The metabolic routes in nucleotide synthesis and conversions are also important in the treatment of several diseases, notably cancer, by drugs.

Structure and nomenclature of nucleotides

The term 'nucleotide' originates from the name of nucleic acids, originally found in nuclei; you are reminded that a **nucleotide** has the general structure:

phosphate—pentose sugar—base.

A **nucleoside** has the structure:

pentose sugar—base.

Thus, AMP and the corresponding nucleoside, adenosine, have the structures shown:

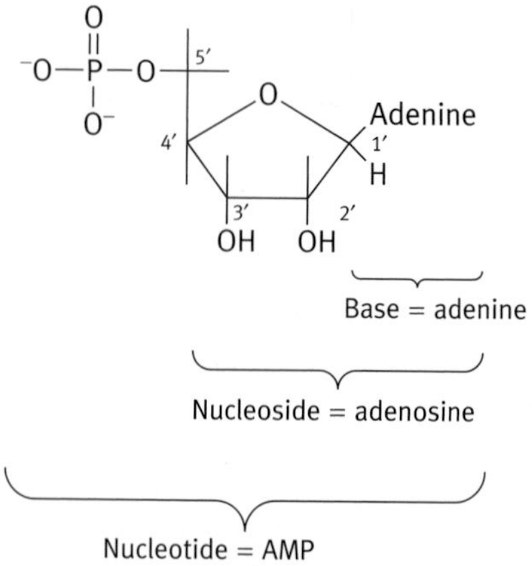

Base = adenine

Nucleoside = adenosine

Nucleotide = AMP

Strictly speaking, the AMP shown here should be written as 5'AMP. The prime (') indicates that the number refers to the position on the ribose sugar ring, to which the phosphate is attached, rather than to the numbering of atoms in the adenine ring. It is a common practice to assume that the phosphate is 5' unless specified, since this is the most usual position. Thus 5'AMP is often called AMP, whereas if the phosphate is on the carbon atom 3 of the ribose, this is always specified as 3' AMP. Both are different from cyclic AMP (cAMP, page 249).

The sugar component of nucleotides

The sugar component of a nucleotide is always a pentose, ribose, or 2'-deoxyribose, which are always in the D-configuration, never the L-form.

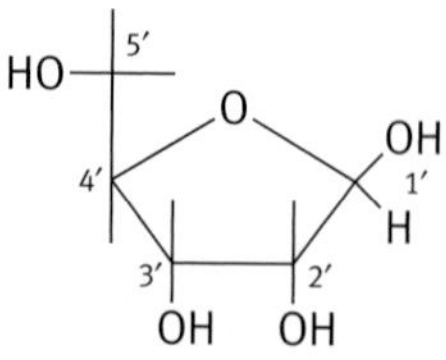

D-Ribose

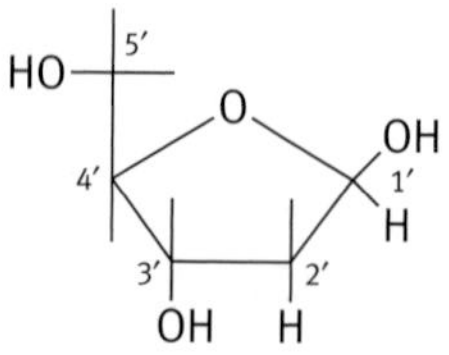

D-2'-Deoxyribose

In **RNA** the sugar is always ribose (hence the name, **ribonucleic acid**) and in **DNA**, deoxyribose (hence, **deoxyribonucleic acid**). A nucleotide containing ribose is a ribonucleotide but this is not usually specified; unless otherwise stated, a named nucleotide such as AMP is taken to be a ribonucleotide. A deoxyribonucleotide *is always* specified; for example, deoxyadenosine monophosphate or dAMP, etc. (with the one occasional exception mentioned below).

The base component of nucleotides

Nomenclature

We are primarily concerned with five different bases – adenine, guanine, cytosine, uracil, and thymine, all often abbreviated to their initial letter.

A, G, C, and U are found in RNA;

A, G, C, and T are found in DNA.

The ribonucleotides are AMP, GMP, CMP, and UMP, but older and still used terms are adenylic, guanylic, cytidylic, and uridylic acids, respectively (or adenylate, guanylate, cytidylate, and uridylate for the ionized forms at physiological pH). The deoxyribonucleotides are dAMP, dGMP, dCMP, and dTMP. The latter often is called TMP, or thymidylate, without the d-prefix because T is found only in deoxynucleotides.

When the intention is to indicate a nucleotide without specifying the base, the abbreviations NMP, and 5′ NMP, or dNMP, and 5′ dNMP for deoxynucleotides are used.

Deoxy UMP exists only as an intermediate in the formation of dTMP; it does not occur in DNA (except as a result of chemical damage to the DNA, when it is promptly removed – see page 345).

Other so-called 'minor' bases exist and are found in transfer RNA (page 383).

Structure of the bases

The first point is that:

A and G are purines

C, U, and T are pyrimidines

These names originate from their being formally related to purine and pyrimidine, respectively (neither of which occur in nature).

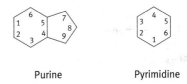

Purine   Pyrimidine

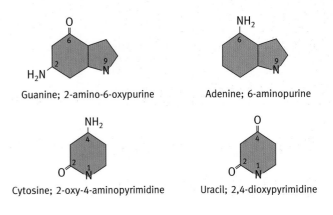

Guanine; 2-amino-6-oxypurine   Adenine; 6-aminopurine

Cytosine; 2-oxy-4-aminopyrimidine   Uracil; 2,4-dioxypyrimidine

Thymine; 2,4-dioxy-5-methylpyrimidine

Fig. 20.1 Diagrammatic representation of structures of purine and pyrimidine bases found in nucleic acids (full structures in Chapter 22). The oxy and amino groups exist largely, or entirely, in the form shown, rather than the —OH and=NH tautomeric forms. Other minor bases are found in transfer RNA, described in Chapter 24.

These rings in future structures will be represented by the simplified forms below.

Purine   Pyrimidine

The structures of the nucleotide bases are represented in Fig. 20.1.

Of special importance, note that *T is simply a methylated U*. It will be useful to fix in your mind that T is essentially the same as U except that it is 'tagged' by a methyl group. T is found only in DNA; U only in RNA. The significance of this will be apparent later (page 345).

Attachment of the bases in nucleotides

The bases are attached to the pentose sugar moieties of nucleotides at the N-9 position of purines and the N-1 position of pyrimidines. The glycosidic bond is in the β-configuration, that is, it is above the plane of the pentose ring. The structures of AMP and CMP are given in Fig. 20.2.

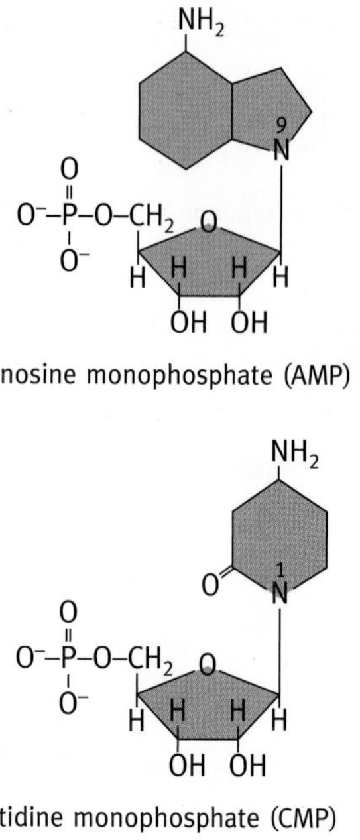

Adenosine monophosphate (AMP)

Cytidine monophosphate (CMP)

Fig. 20.2 Structures of a purine and a pyrimidine nucleotide.

Ribose-5-phosphate

PRPP

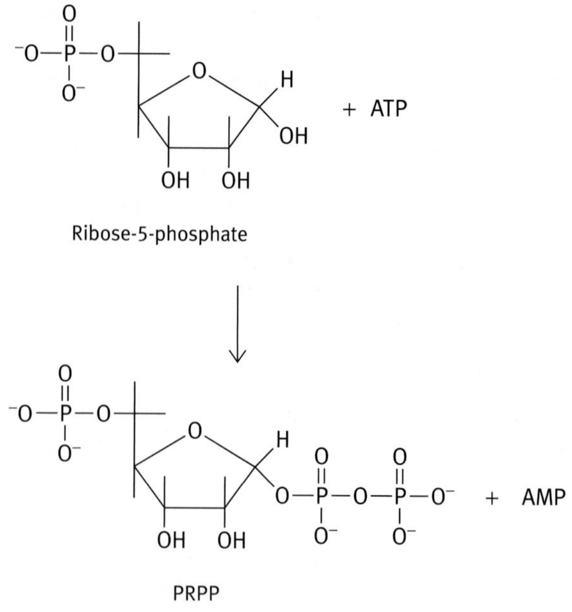

The PRPP is an 'activated' form of ribose-5-phosphate; appropriate enzymes can donate the latter to an amino group or a whole base forming a nucleotide, and splitting out —P—P. Hydrolysis of the latter to $2P_i$ drives the reaction thermodymically.

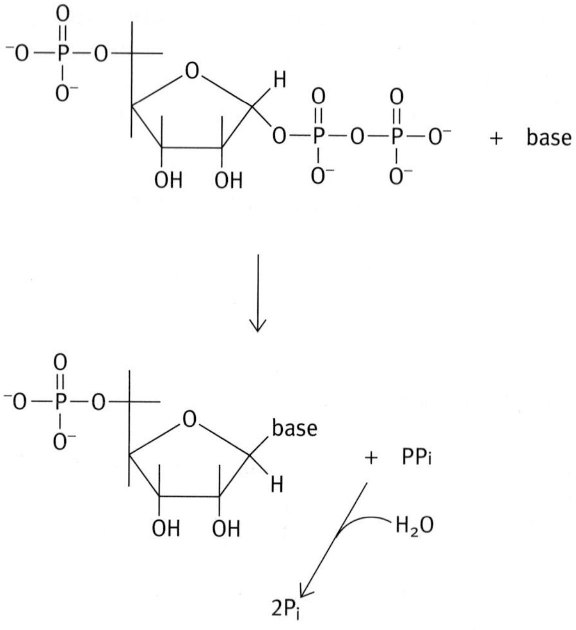

Synthesis of purine and pyrimidine nucleotides

Purine nucleotides

Most cells can synthesize purine bases *de novo* from smaller precursor molecules. In the *de novo* synthesis of purine nucleotides, bases are not synthesized in the free form but rather the purine ring is assembled piece by piece as a nucleotide starting with an amino group on ribose-5-phosphate. (This refers only to the *de novo* synthesis of purines because free purine bases released by degradation of nucleotides *are* utilized for nucleotide synthesis by the separate **salvage pathway** to be described later.) The mechanism of **ribotidation** (the addition of ribose-5-phosphate to an amino group or to a whole base) is the same in both pathways, as well as in pyrimidine nucleotide synthesis. This brings us to PRPP, the metabolite that is involved in all ribotidation.

PRPP – the ribotidation agent

PRPP is **5-phosphoribosyl-1-pyrophosphate**. It is formed from ribose-5-phosphate (produced by the pentose phosphate pathway; page 267) by the transfer of a pyrophosphate group from ATP by the enzyme **PRPP synthetase**.

In this reaction the configuration at carbon atom 1 is inverted so that the base is in the required β position. Note that in the *de novo* pathway (see reaction 1 of this pathway in Fig. 20.4) the 'base' of the starting nucleotide is simply —NH_2, derived from glutamine and PRPP giving 5-phosphoribosylamine; in the purine *salvage* pathway it is a purine. We usually associate the term nucleotide with a purine or pyrimidine base but it can be applied to any base attached in the appropriate manner to the sugar phosphate.

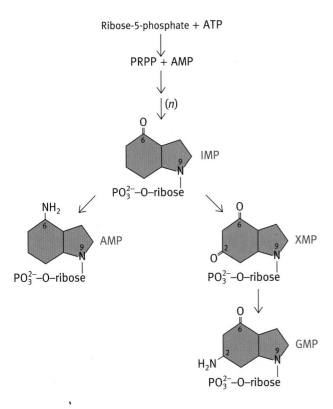

Fig. 20.3 Diagram of the purine *de novo* pathway of GMP, AMP, and XMP synthesis. The base in IMP (inosine monophosphate) is hypoxanthine. The base in XMP (xanthosine monophosphate) is xanthine. PRPP, 5-phosphoribosyl-1-pyrophosphate. The complete pathway of AMP and GMP synthesis can be seen in Figs 20.4 and 20.5.

The *de novo* purine nucleotide synthesis pathway

To return from this general point to the purine *de novo* pathway in particular, after the formation of 5-phosphoribosylamine, there follows a series of nine reactions resulting in the assembly of the first purine nucleotide in which hypoxanthine is the base (see Fig. 20.3). For historical reasons this nucleotide is called IMP or inosinic acid.

IMP is a branch-point since its hypoxanthine base may be converted either to adenine or guanine yielding AMP and GMP, respectively. The overall pathway is summarized in Fig. 20.3, but all of the reactions of the pathway are set out in Figs 20.4 and 20.5.

The daunting *de novo* pathway in Fig. 20.4 is given so that we can refer to some reactions of specific interest. Six molecules of ATP are consumed in the synthesis of one purine nucleotide molecule. *The ATP utilization refers only to —℗ groups*; there is no loss of the adenine nucleotide of ATP so that the pathway results in a net synthesis of AMP.

Reactions 3 and 9 of this pathway are an important class of reaction and have a general interest and we need to divert to deal with these in some detail. This concerns a type of reaction not dealt with in this book before, and one that is medically important – **one-carbon transfer**.

The one-carbon transfer reaction in purine nucleotide synthesis

Reactions 3 and 9 of the pathway involve the addition of a one-carbon formyl (HCO—) group to intermediates in the pathway (Fig. 20.4). The donor molecule in both cases is N^{10}-**formyltetrahydrofolate**, a molecule not mentioned before in this book. **Tetrahydrofolate** (FH_4, or sometimes **THF**) is the carrier in the cell of formyl groups. It is a coenzyme derived from the vitamin **folic acid** (F) or **pteroylglutamic acid**. We suggest that you need to learn only the relevant part of this and related structures below, given in blue (not the whole molecules).

Folic acid (pteroylglutamic acid or F)

The vitamin (F) is reduced to FH_4 by NADPH in two stages.

Dihydrofolate reductase

The abbreviated structures of **dihydrofolate** (FH_2, or sometimes **DHF**) and FH_4 are given below:

Dihydrofolate (FH_2)

Tetrahydrofolate (FH_4)

Have a look at the structure of FH_4 and notice that the N-5 and N-10 atoms are placed such that a single carbon atom can neatly bridge the gap between them. For our present purposes we can therefore represent FH_4 as:

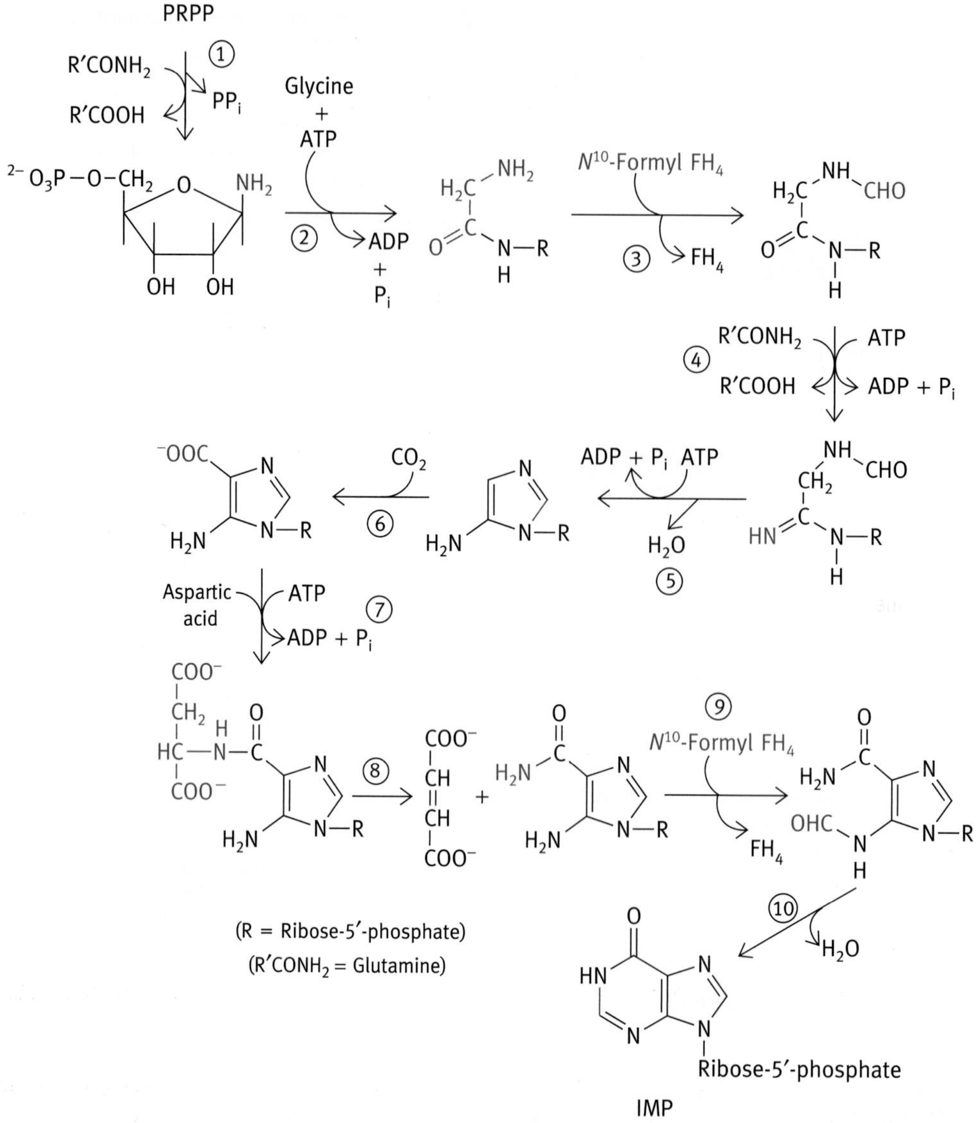

Fig. 20.4 Details of the pathway for the *de novo* synthesis of the purine ring from 5-phosphoribosyl-1-pyrophosphate (PRPP) to inosinic acid (IMP), given for reference purposes. (The circled reaction numbers are referred to in the text.) Blue indicates the structural change resulting from the latest reaction. N^{10}-formyl FH$_4$, N^{10}-formyltetrahydrofolate.

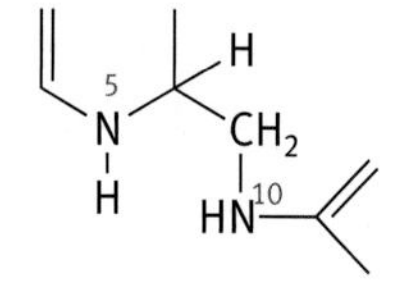

and N^{10}-formyl-FH$_4$ as

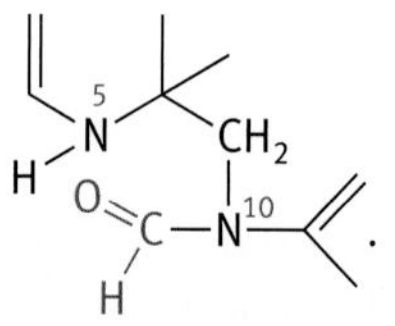

N^{10}-formyltetrahydrofolate
(N^{10}-formyl FH$_4$)

This is the donor of the formyl group in reactions 3 and 9 of the purine biosynthesis pathway; specific formyl transferase enzymes catalyse the reactions.

Where does the formyl group in N^{10}-formyl FH$_4$ come from?

The answer is the amino acid serine (which is readily synthesized from the glycolytic intermediate 3-phosphoglycerate). An enzyme, **serine hydroxymethylase**, transfers the hydroxymethyl group (—CH$_2$OH) to FH$_4$ leaving glycine and forming N^5, N^{10}-methylene FH$_4$.

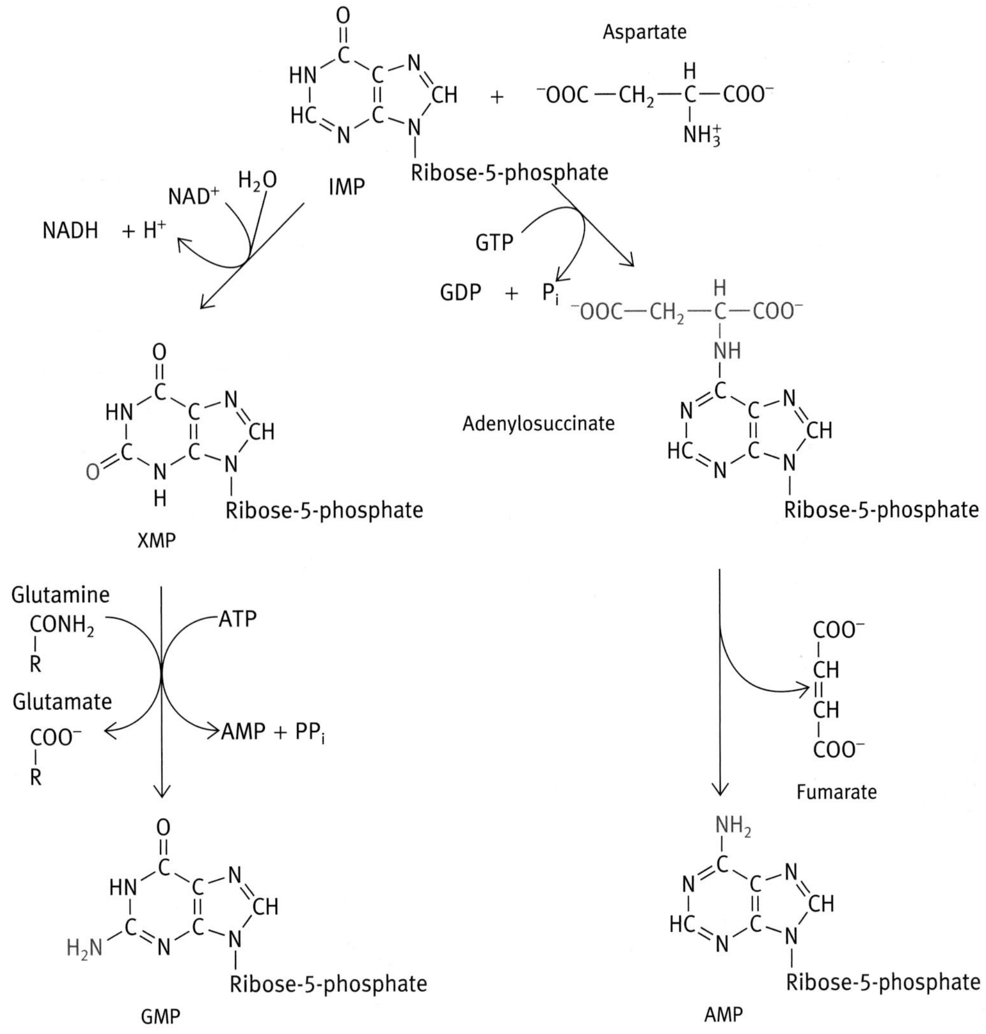

Fig. 20.5 Details of the pathways for the synthesis of GMP and AMP from inosinic acid (IMP), given for reference purposes. The colour shows the change resulting from each reaction. The pathways are summarized in Fig. 20.3. XMP, xanthosine monophosphate.

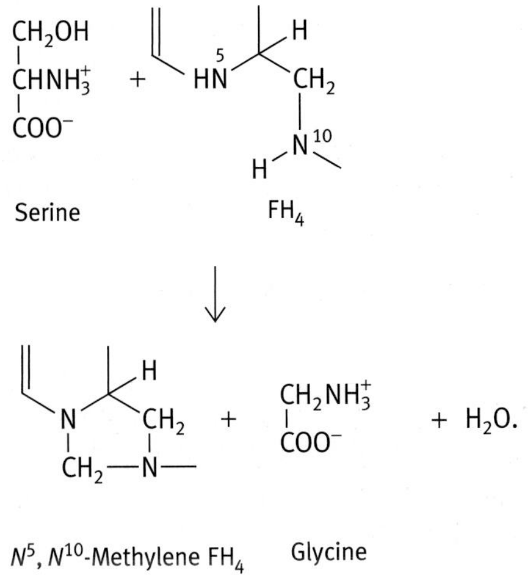

than a formyl group. It is therefore oxidized by an $NADP^+$-requiring enzyme, forming the methenyl derivative, which is hydrolysed to N^{10}-formyl-FH_4, the formyl group donor.

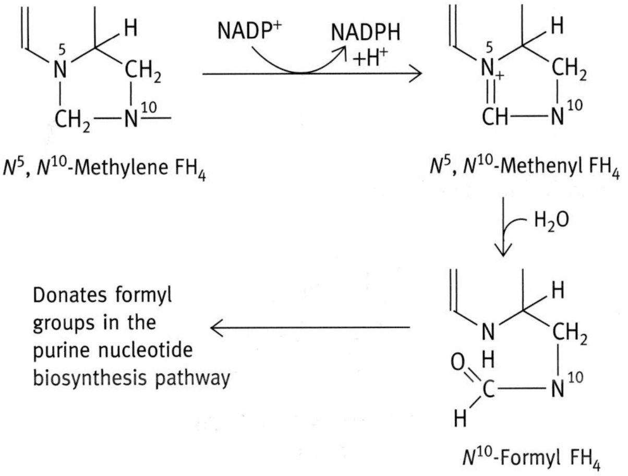

The product, N^5, N^{10}-methylene FH_4 is not quite what we want for formylation because the $-CH_2-$ group is more reduced

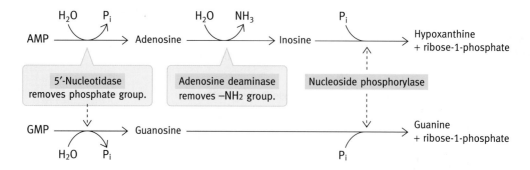

Fig. 20.6 Production of free purine bases hypoxanthine and guanine by nucleotide breakdown. Patients lacking adenosine deaminase in lymphocytes have an immune deficiency that formerly could be treated only by keeping the affected child in a sterile plastic bubble. The disease was the first to be successfully treated by gene therapy in which the normal gene for adenosine deaminase was inserted *in vitro* into bone marrow stem cells and returned to the patient. (See page 476.)

How are ATP and GTP produced from AMP and GMP?

Most of the synthetic reactions of the cell involve nucleoside triphosphates. As you will see later, these are needed for nucleic acid synthesis. It is a simple but especially important concept that enzymes (kinases) exist in the cell to transfer —P groups between nucleotides *at the high-energy level*. There is little free-energy change involved. The main source of —P is, of course, ATP, for remember that the energy-generating metabolism constantly regenerates ATP from ADP and P_i. Newly formed AMP and GMP are phosphorylated by kinase enzymes as shown:

$AMP + ATP \rightleftharpoons 2\,ADP$	Adenylate kinase
$GMP + ATP \rightleftharpoons GDP + ADP$	Guanylate kinase
$GDP + ATP \rightleftharpoons GTP + ADP$	Nucleoside diphosphate kinase

The nucleoside diphosphate kinase has a wide specificity and can use any pair of nucleoside di- and triphosphates.

The purine salvage pathway

We have emphasized that the *de novo* synthesis of purines does not involve free purine bases – purine nucleotides are produced. However, as already indicated, there is a separate route of purine nucleotide synthesis in which **free bases** are converted to nucleotides by reaction with PRPP. The free bases originate from degradation of nucleotides mainly in the liver and supplied to other tissues in the blood – they are salvaged (recycled) and hence the name of the pathway. Two enzymes are involved – these are phosphoribosyltransferases, one of which forms nucleotides from adenine and the other from hypoxanthine or guanine. The latter enzyme, known as **HGPRT** (for **hypoxanthine-guanine phosphoribosyltransferase**), catalyses the reaction:

$$\text{Guanine or hypoxanthine} + \text{PRPP} \rightarrow \begin{array}{c}\text{GMP}\\ \text{or}\\ \text{IMP}\end{array} + \text{PP}_i.$$

The enzyme salvaging adenine may be of lesser importance than that dealing with guanine and hypoxanthine in humans, for the main routes of nucleotide breakdown produce the free bases hypoxanthine (from AMP) and guanine (from GMP) as shown in Fig. 20.6.

What is the physiological role of the purine salvage pathway?

Since purines are energetically 'expensive' to make, a mechanism for re-utilizing free purine bases is economical since it can reduce the amount of *de novo* synthesis a cell has to carry out. Moreover, certain cells such as erythrocytes have no *de novo* purine synthesis pathway and must rely on the salvage pathway.

The physiological importance of purine salvage is underlined by the rare genetic disease of infants called the **Lesch-Nyhan syndrome**, in which the enzyme HGPRT is missing. This results in neurological problems including mental retardation and self-mutilation. Brain possesses the *de novo* pathway only at low levels, so purine nucleotide synthesis is very sensitive to the salvage defect. Lack of the salvage reaction leads to a hepatic *overproduction* of purine nucleotides by the *de novo* pathway in these patients because the level of PRPP rises (due to lack of utilization by the salvage reaction) and stimulates the *de novo* pathway. This explains why excessive uric acid production occurs as in gout (see below), which may result in kidney failure caused by the urate crystals. The connection between the biochemical defect and the neurological symptoms is not clear in the Lesch-Nyhan patients. While uric acid overproduction is treatable with allopurinol (see below), this does not relieve the neurological problems. Nor do patients with gout develop the neurological symptoms.

The recycling of preformed purine bases has the obvious advantage of energy-saving provided, of course, that the *de novo* pathway synthesis is correspondingly reduced. This is achieved in two ways:

- salvage reduces the level of PRPP and hence of the *de novo* pathway
- the AMP and GMP produced by salvage exert feedback inhibition on the pathway (see below).

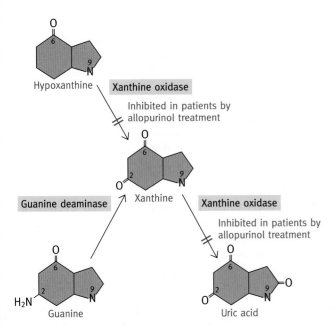

Fig. 20.7 Conversion of hypoxanthine and guanine to uric acid. The drug allopurinol is closely related in structure to hypoxanthine.

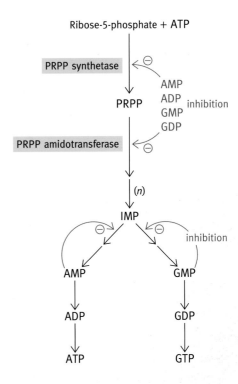

Fig. 20.8 Simplified scheme of the control of the purine nucleotide biosynthesis pathway. PRPP, 5-phosphoribosyl-1-pyrophosphate.

Formation of uric acid from purines

Nucleotide degradation leads to the production of free hypoxanthine and guanine. Part of this is salvaged back to nucleotides but part is oxidized to produce **uric acid** (Fig. 20.7). The enzyme, xanthine oxidase, that produces uric acid is present mainly in the liver and intestinal mucosa. **Gout** is due to a raised level of urate in the blood, leading to the deposition of crystals in tissues. Although gout is traditionally associated with rich living, the main source of uric acid is excess *de novo* production of purine nucleotides due, in some patients, to a high level of PRPP synthetase activity. Also, as described above, deficiency of the HGPRT enzyme leads to the overproduction of purine nucleotides. The drug **allopurinol**, used in the treatment of gout, mimics the structure of hypoxanthine. It is a potent xanthine oxidase inhibitor. This inhibition results in xanthine and hypoxanthine formation rather than that of uric acid (Fig. 20.7). These products are more water-soluble than uric acid and more readily excreted, thus preventing the deposition of insoluble uric acid crystals in tissues that results in the clinical symptoms of **gout**.

Control of purine nucleotide synthesis

As with all metabolic pathways, there must be regulation or chemical anarchy would prevail. The *de novo* pathway is a classical example of **allosteric feedback control** (see page 245). The first step of a pathway is a logical place for control. In the *de novo* pathway this is the PRPP synthetase. This enzyme is negatively controlled by AMP, ADP, GMP, and GDP. The next enzyme, which catalyses the first *committed* step to synthesis of purine nucleotides (reaction 2, Fig. 20.4), is inhibited also, as shown in Fig. 20.8. However, this is not quite the end of the story, because the *de novo* pathway produces IMP, and then the IMP goes in two directions – to AMP and GMP. These latter feedback-control their own production. The regulatory loops serve to ensure a balanced production of ATP and GTP since both are required for nucleic acid synthesis (Chapter 22).

Synthesis of pyrimidine nucleotides

Most cells of the body synthesize pyrimidine nucleotides *de novo* but, unlike bacteria, mammals do not appear to have significant pyrimidine salvage pathways for free bases, analogous to those for purines. The nucleoside thymidine, however, is readily phosphorylated to TMP by **thymidine kinase** and, in that sense, salvage of this nucleoside does occur.

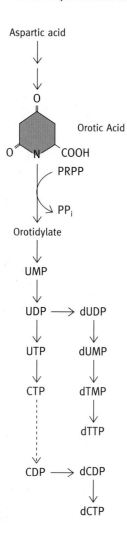

Fig. 20.9 Summary of pyrimidine nucleotide synthesis. (It is assumed here that CTP is converted to CDP.) The formation of deoxynucleotides is described later in this chapter. The complete pathway is given in Fig. 20.10.

The **pyrimidine pathway** is summarized in Fig. 20.9 and given in full in Fig. 20.10. It starts with aspartic acid and produces a ring structure compound, orotic acid. Orotic acid is converted to the corresponding nucleotide by the PRPP reaction and this is converted to UMP. UTP is produced by kinase enzymes much as in the purine pathway. CTP is produced by amination of UTP.

In *Escherichia coli*, control of pyrimidine nucleotide synthesis is mainly at the aspartate transcarbamoylase step. In mammals the pathway is controlled at the carbamoyl phosphate synthase step; it is inhibited by pyrimidine nucleotides and activated by purine nucleotides. The latter control serves to keep the supply of all the nucleotides required for nucleic acid synthesis in balance.

How are deoxyribonucleotides formed?

For DNA synthesis dATP, dGTP, dCTP, and dTTP are required (Chapter 23). The reduction of ribonucleotides to deoxy

compounds occurs at the diphosphate level with NADPH as the reductant. The ribonucleotide reductase has a complex mechanism involving formation of a stable radical. There are different, but related, reductases in different organisms.

ADP		dADP	
GDP	Ribonucleotide reductase	dGDP	
CDP	$\xrightarrow{\hspace{2cm}}$	dCDP	$+ H_2O$
UDP	NADPH NADP$^+$	dUDP	
	+ H$^+$		

The resultant dADP, dGDP, dCDP, and dUDP are converted to the triphosphates by phosphoryl transfer from ATP.

However, dUTP is not used for DNA synthesis since DNA, you will recall, has thymine (the methylated uracil) as one of its four bases, but never uracil. The dUTP is converted to dTTP. This is done in three steps: first, dUTP is hydrolysed to dUMP:

$$dUTP + H_2O \rightarrow dUMP + PP_i$$

The dUMP is converted to dTMP (see below) and then to dTTP by phosphoryl transfer from ATP. An appropriate system of allosteric feedback controls exists to keep the production of the four deoxynucleotide triphosphates in balance.

Thymidylate synthesis – conversion of dUMP to dTMP

The methylation of dUMP is carried out by the enzyme **thymidylate synthase**; it utilizes N^5, N^{10}-methylene FH$_4$. In purine synthesis, the methylene group of the latter is oxidized to produce a formyl group. In thymidylate synthesis, the methylene group is transferred and, at the same time, *reduced* to the methyl group of thymine. The reducing equivalents for the reduction come from FH$_4$ itself, leaving it as FH$_2$ (note how versatile this coenzyme is). In the scheme below, only the relevant part of N^5, N^{10}-methylene FH$_4$ is shown.

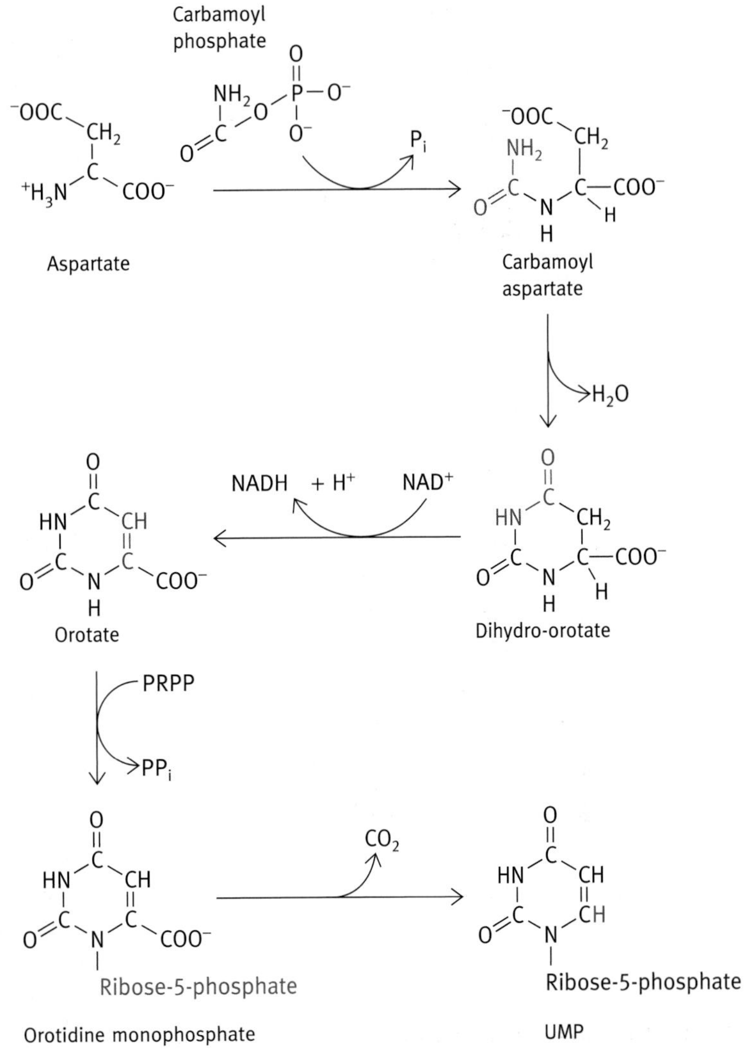

Fig. 20.10 Details of the pathway for the *de novo* synthesis of pyrimidine nucleotides, given for reference. Blue indicates the structural change resulting from the latest reaction.

The FH_2 produced in this reaction is reconverted to FH_4 by **dihydrofolate reductase**. The FH_4 is reconverted to methylene FH_4 by reaction with serine (page 303).

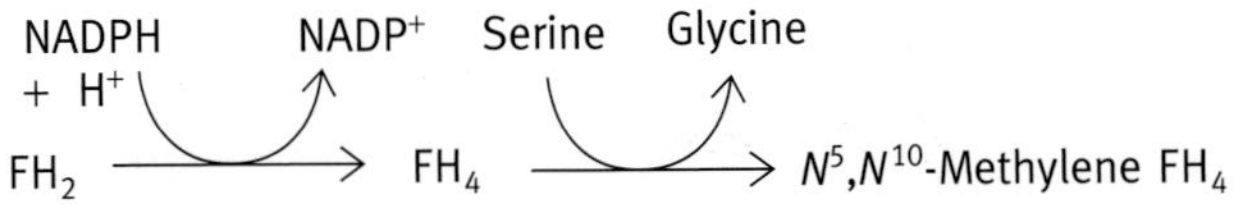

Medical effects of folate deficiencies

Cells must synthesize DNA if they are to divide and obviously for this they need supplies of all four deoxynucleotides; failure to produce any of them in adequate quantities will impair cell division. Some, such as red blood precursor cells and cancer cells, which divide rapidly, are particularly sensitive to anything that restricts nucleotide availability.

FH_4 is involved in several metabolic syntheses such as glycine, serine, and methionine formation but these components are usually available from the diet. Nucleotide synthesis, however, is sensitive to folate deficiency in the diet and it leads to **anaemias** typified by large fragile red blood cells. Folate deficiency during pregnancy has also been associated with birth of babies with **spina bifida**.

Thymidylate synthesis is targeted by anticancer agents such as the antifolate, methotrexate

In the synthesis of thymidylate from dUMP and N^5, N^{10}-methylene FH_4, described above, the products are dTMP and FH_2 since the FH_4 supplies the two H atoms needed for the reduction of the methylene group to a methyl group. The FH_2 must be reduced to FH_4 by dihydrofolate reductase for the molecule to be used catalytically. If this reduction is inhibited, then the FH_4 becomes trapped as unusable FH_2 resulting in an effective folate deficiency. The antileukaemic drugs **methotrexate** (**amethopterin**) and the related **aminopterin** (both known as antifolates) competitively inhibit the dihydrofolate reductase, by mimicking the folate structure.

The structure of methotrexate (amethopterin) is shown here for interest; aminopterin is similar but lacks the N^{10}-methyl group.

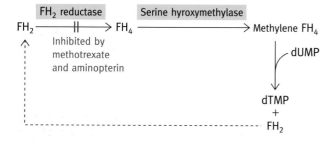

Methotrexate (amethopterin)

Cancer cells, like those of leukaemia, require rapid dTMP production to synthesize DNA and are thus selectively inhibited. The scheme is outlined in Fig. 20.11.

Fluorouracil is another agent attacking folate reduction. It is converted in cells to the corresponding deoxynucleotide, a potent FH_2 inhibitor used in some cancer therapies.

Fig. 20.11 Site of action of the anticancer agent, methotrexate. The relationship of the methotrexate structure to folic acid can be seen by comparing the structure of methotrexate with that of folic acid on page 301.

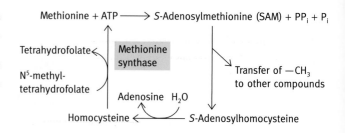

Fig. 20.12 Recycling methionine back from homocysteine. The methyl group of methionine is activated to form S-adenosylmethionine (SAM), a major donor of methyl groups. Homocysteine can be recycled to methionine by the transfer of a methyl group from N^5-methyltetrahydrofolate, a reaction that needs a coenzyme derived from vitamin B_{12}.

Vitamin B_{12} deficiency in cells and the folate methyl trap

N^5, N^{10}-Methylene FH_4 may be further reduced to N^5-methyl FH_4, which supplies the methyl group to convert homocysteine to methionine. A coenzyme derived from vitamin B_{12} is needed in the transfer of the methyl group of N^5-methyl FH_4 to homocysteine. This reaction is catalysed by **methionine synthase** (Fig. 20.12). In the disease **pernicious anaemia**, a gastric glycoprotein called the extrinsic factor, needed for absorption of the vitamin, is missing so that cells of the body lack the coenzyme irrespective of the adequacy of the diet. Vitamin B_{12} occurs in most diets but is not present in plants so that vegan diets are often deficient.

The vitamin has no role in nucleotide synthesis but if it is deficient in cells, FH_4 becomes trapped in its methylated form; the methyl group cannot be transferred to homocysteine which is the only metabolic route for releasing the FH_4 for reuse in other reactions in nucleotide synthesis. This **methyl trap hypothesis** explains why the red blood cell abnormalities in megaloblastic anaemia resemble those found in folate deficiency. Vitamin B_{12} is involved in other reactions (see page 219) which possibly account for the neurological abnormalities found in pernicious anaemia. The latter are not alleviated by folate administration.

Summary

Nucleotides are synthesized *de novo*. Purine nucleotides are also synthesized by a salvage pathway, by ribotidation of free bases released by the degradation of nucleotides. In mammals the pyrimidine salvage pathway appears to be unimportant.

In the *de novo* synthesis of purine nucleotides, the purine ring is assembled by a series of steps in which the intermediates are already joined to ribose-5-phosphate. 5-Phosphoribosyl-1-pyrophosphate (PRPP) is the universal ribotidation agent. The

product of the synthesis pathway is inosine phosphate (IMP), which is converted to form both GMP and AMP.

Synthesis of purine nucleotides is allosterically controlled.

The pathway involves two additions of formyl groups, reactions depending on the coenzyme tetrahydrofolate (FH_4). This molecule is formylated using serine as the formyl donor forming glycine and formyltetrahydrofolate.

Dividing cells require deoxynucleotides and so are sensitive to deficiencies in the synthesis pathway. Folate deficiency in the diet can produce anaemias. Deficiency during pregnancy is associated with birth of babies with spina bifida.

The salvage pathway involves two enzymes, one of which catalyses the ribotidation of free adenine, the other of hypoxanthine or guanine, the latter transferase (HGPRT or hypoxanthine-guanine phosphoribosyl transferase) being the more important in humans.

In the Lesch-Nyham syndrome, infants lack the HGPRT and this leads to mental retardation and self-mutilation; *de novo* synthesis is low in amount in brain, which therefore is sensitive to deficiency in the salvage pathway. In liver, HGPRT deficiency causes PRPP accumulation possibly because it is not used as much as is normal. This stimulates *de novo* synthesis, leading to overproduction of purines. The excess is converted to uric acid. The uric acid problem does not appear to be related to the neurological symptoms of the Lesch-Nyham syndrome.

Excessive synthesis of PRPP is a factor causing gout in other patients. Allopurinol is used clinically to inhibit uric acid formation. Pyrimidine nucleotides are formed *de novo* via a different synthesis pathway resulting in UMP synthesis.

The formation of deoxynucleotides needed for DNA synthesis occurs at the diphosphate level, catalysed by a reductase using NADPH. The dNDPs are converted to the triphosphates by phosphoryl transfer from ATP to provide the substrates for DNA synthesis. dUTP is not used in DNA synthesis. dTTP is used. The methylation to form thymidylate occurs at the dUMP level catalysed by thymidylate synthase. The conversion requires FH_4, which, during the formation of the thymine methyl group, is oxidized to dihydrofolate (FH_2). The antileukaemic drugs, methotrexate and aminopterin, inhibit the recycling of the FH_2 back to FH_4 and inhibit cell multiplication.

Vitamin B_{12} is required for the formation of a cofactor involved in the conversion of homocysteine to methionine. If this vitamin is absent, such as in pernicious anaemia, FH_4 accumulates in the methylated form. This creates a shortage of FH_4 for nucleic acid synthesis. This **methyl trap hypothesis** explains why the red blood cell abnormalities in megaloblastic anaemia resemble those found in folate deficiency.

Further reading

Antifolate drugs

Huennekens, F. M. (1994). The methotrexate story: a paradigm for development of cancer therapeutic agents. Adv. *Enzyme Regul.*, **34**, 392–419.

Gives history of the antifolate drug, the structure of dihydrofolate reductase, and development of methotrexate resistance in cells.

Problems

1 Describe the method of ribotidation.

2 What is the cofactor involved in the two formylation reactions involved in purine nucleotide synthesis? Give the essential structure of its formyl derivative.

3 What is the origin of the formyl group? Explain how this is generated.

4 What is the function of hypoxanthine-guanine phosphoribosyltransferase (HGPRT)?

5 The Lesch-Nyhan syndrome is a severe genetic disease in children. Discuss its biochemistry.

6 How does the drug allopurinol reduce uric acid formation?

7 Draw a diagram illustrating the main allosteric controls on purine nucleotide synthesis.

8 Several compounds in the body are methylated using methionine as a methyl group source. Is this true of thymidine monophosphate synthesis? Explain your answer.

9 How does the antileukaemic drug methotrexate inhibit cancer cell reproduction?

10 The symptoms of pernicious anaemia resemble, in some respects, those of folate deficiency. What is a possible reason for this?

Part 4

Information storage and utilization

Chapter 21

The genome

A brief overview

In this chapter we will deal with what genomes are and their typical structure with the human genome central to our discussion. This includes the chemistry of the DNA and its associated molecules, its physical state in the cell, how this varies during a life cycle, the size of the genomes of different organisms and its relationship to the complexity of the particular organism it codes for. Structures of genes are covered but aspects of the genome – how genes work and are controlled – come in the subsequent Chapters 22–24 on DNA synthesis, gene transcription and protein synthesis respectively.

In the past decade, a great advance in our knowledge of the genome has been the complete determination of the base sequence of the human genome. Following this, the genomes of several other species have been sequenced. Recently the discovery of large numbers of hitherto unrecognized nonprotein-coding genes has revolutionized our concept of what was formerly referred to as 'junk DNA'.

Knowledge of the components of the genome has been crucial in elucidating how genes work and in the development of technologies. Examples of the latter are methods for the location and isolation of disease-producing genes and DNA fingerprinting in forensic science. However, we will leave a description of these and other technologies to Chapter 28, because it is a big topic in its own right. The information in this chapter will provide a foundation for understanding the technologies.

The term genome usually refers to all the information coded by the DNA of the cell but there are related usages of the term. In many viruses such as influenza and HIV, RNA is the genetic material, not DNA. In eukaryotes we distinguish between the nuclear genome and those in mitochondria and chloroplasts.

The prokaryotic genome

A typical prokaryote such as *Escherichia coli* is **haploid**, which means it has only a single set of genes. The single chromosome consists of a closed circle of double stranded DNA There is no nuclear membrane surrounding it so that the DNA is in direct contact with the cytoplasm, being visible in the electron microscope as an ill-defined tangled fine thread called a nucleoid (page 17). The cell grows continuously in suitable nutritive conditions because the chromosome remains in an active state throughout the life cycle of the cell. After the DNA has been duplicated and a critical size reached, cell division occurs with each daughter cell receiving one chromosome. There are no separate phases in the life cycle – cell growth and replication of DNA, and transcription go on continuously.

The eukaryotic genome

Eukaryotic genomes are quite different. The cells are usually **diploid** (sperm and eggs excepted) which means that they have two sets of chromosomes, one derived from each parent. *The chromosomes are linear, not circular.* This has relevance to cancer and ageing (page 347). To take the human as an example, there are 46 chromosomes consisting of 24 **homologous pairs** known as **autosomes** and 2 **sex chromosomes** X–Y in the male and X–X in the female. Homologous chromosomes have the same genes in the sense that each gene of a pair control the same phenotypic characteristic but the two genes at a given locus (location in the chromosome), referred to as **alleles**, are often slightly different in their base sequence. For example a gene for human haemoglobin codes for the normal protein but in some humans its allele codes for an abnormal one, resulting in the disease sickle cell anaemia (page 68). At meiosis, homologous chromosomes pair up and crossovers occur, in which genes are exchanged before the two separate (see below, page 325).

Unlike prokaryotes, eukaryotes contain their chromosomes within a nuclear membrane. When a gene is active, a messenger RNA (mRNA) (see page 11 for a simple description or page 353 for a fuller one) is transcribed (copied) from the DNA, processed, transported through the nuclear membrane and is then translated by ribosomes into proteins. This separation of gene transcription from mRNA translation in time and space has profound repercussions. By contrast, in *E. coli* the mRNA is translated immediately, beginning even before it has been completely synthesized and released from the DNA.

The human cell nucleus contains about 2 metres of DNA in its chromosomes, and this is packed into a sphere about 10 mm (10^{-5} m) in diameter, so that an immense degree of packing is necessary. (Due to the vast number of cells, usually estimated at around 10^{13} (10 trillion), the total length of DNA in a human, put end to end would span the earthly orbit in the solar system several times). The compact X-shaped chromosomes of dividing cells, known as **metaphase chromosomes** (page 323), are actually newly formed pairs of chromosomes, each one known as a chromatid. They are very condensed, which facilitates movement of a set of chromosomes into each daughter cell.

The structures of DNA and RNA

A basic essential for the existence of life is that organisms must carry genetic information that is replicated and given to their offspring so that they are copies of themselves. At the origin of life this was almost certainly in the form of RNA (see below), but in all cellular life it is DNA. DNA and RNA have basically the same structure. Some viruses today have RNA as their genetic material but viruses are not cells (page 23). Also, RNA has been retained in cells as the intermediary between genes and the ribosomes in the form of messenger RNA. Other important roles of RNA will be described in Chapters 23 and 25.

DNA is chemically a very simple molecule

This may seem a surprising statement considering the immense size of DNA molecules and their central role in determining the characteristics of life forms. DNA has only four different types of 'units', known as nucleotides (see below). Great numbers of these are linked together to form immensely long thin threads, so that there is a monotonous succession of nucleotide after nucleotide. So far as defining amino acid sequences, the succession of the four bases is a form of code based on triplets of bases called **codons** that correspond to individual amino acids in polypeptide chains. The information that ultimately specifies the characteristics of complex organisms like humans is carried in this way.

DNA (like RNA) has the essential characteristic of being able to direct its own replication. Since at the origin of life there could not have been any of the elaborate machinery that cells use today to replicate nucleic acids, life had to start with a basically simple molecule and a basically simple replicative mechanism that would operate before the development of cells. However, although much of the detail has been refined by evolution to produce the present day highly controlled replication machinery, the basic principle remains. Life was locked into it at the beginning.

DNA and RNA are both nucleic acids

DNA was first isolated from cell nuclei. It is an acid because of its phosphate groups, which at physiological pH are dissociated to liberate hydrogen ions – hence the term **nucleic acid**. Its nucleotide subunits contain a pentose sugar, **2-deoxy-D-ribose** and therefore is called **deoxyribonucleic acid**, or DNA for short. As already stated, RNA has a similar structure, and it is useful to mention briefly here how it differs from that of DNA. The pentose sugar in the nucleotide subunits of RNA is **D-ribose, not deoxyribose**. It is therefore called **ribonucleic acid** or RNA for short. The two sugars are shown below. 2-Deoxy-D-ribose lacks the oxygen on the carbon 2 position; it is usually simply referred to as **deoxyribose**.

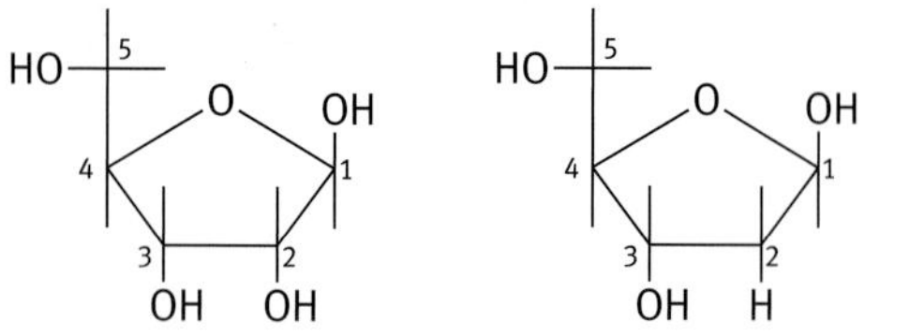

D-Ribose   2-Deoxy-D-ribose

In eukaryotic cells, the bulk of the DNA is confined inside the nuclear membrane, the remainder being in the mitochondria and chloroplasts. Most of the RNA is found in the cytoplasm although, with the exception of the small amounts made in mitochondria and chloroplasts, RNA originates in the nucleus. Although we have already discussed nucleotides in Chapter 19, which dealt with their synthesis, we will repeat some of the material here both for convenience and because of its importance.

The primary structure of DNA

DNA is a **polynucleotide**. A single nucleotide has the structure:

$$phosphate-sugar-base$$

The structure of a deoxyribonucleotide is shown in the nonionized form for simplicity:

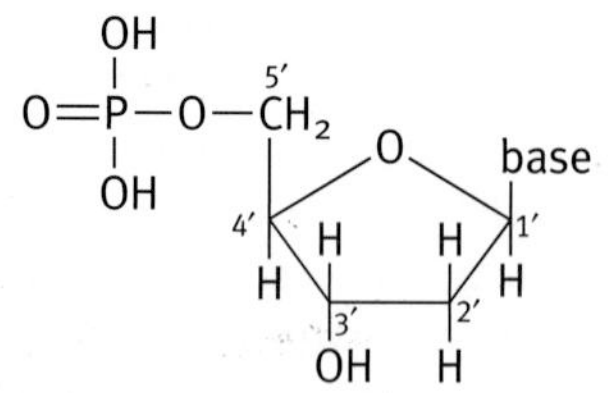

To specify a position in the deoxyribose moiety, a prime (′) is added to distinguish it from the numbering of the base ring atoms. Thus the sugar carbon atoms are 1′, 2′, 3′, 4′, and 5′ (pronounced 'five prime', etc.) and indicated outside the ring. The sugar is in the furanose, five-membered ring form. The nomenclature of nucleotides is described on page 298.

There are four different nucleotide bases in DNA

The bases are **adenine, guanine, cytosine, and thymine** – abbreviated to **A, G, C,** and **T. A** and **G** are **purines,** C and T, **pyrimidines.** The numbering of atoms in the bases is given inside the ring structures.

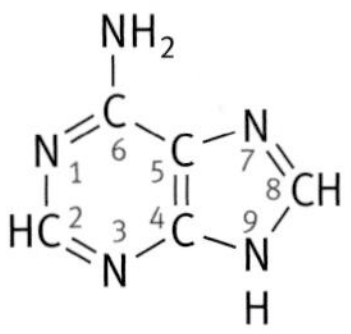

Adenine (A) 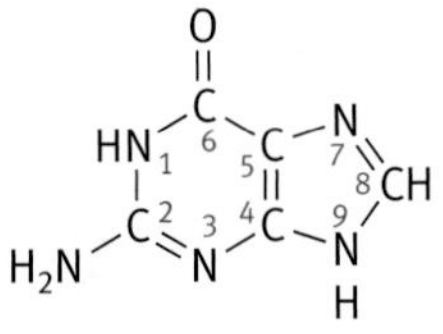   Guanine (G)

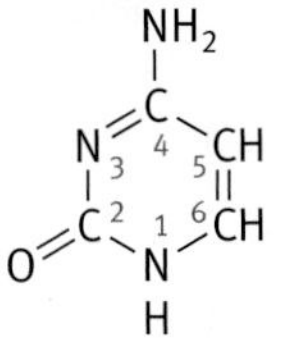

Cytosine (C) 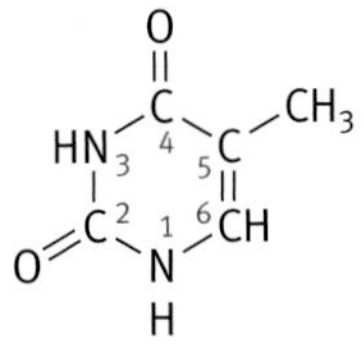   Thymine (T)

Attachment of the bases to deoxyribose

The bases are attached to deoxyribose *via* a glycosidic link between carbon atom 1 of the deoxyribose and nitrogen atoms at positions 9 and 1, respectively, of the purine and pyrimidine rings. The linkage is β (i.e. above the plane of the sugar ring).

The structure, base—sugar, is called a **nucleoside**; if the sugar is deoxyribose it is a **deoxyribonucleoside**. All the deoxyribonucleosides have specific names. Where the base is adenine it is deoxyadenosine; the guanine derivative is deoxyguanosine. Deoxycytidine and deoxythymidine are the deoxynucleosides of cytosine and thymine respectively. The structures are shown below in diagrammatic form in which the heterocyclic ring structures are not shown in detail but the characteristic side groups are given.

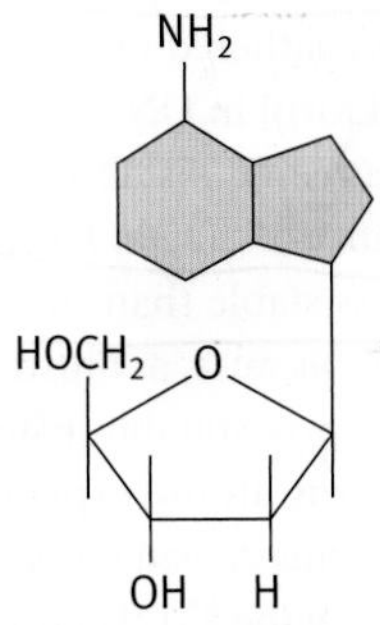

Deoxyadenosine 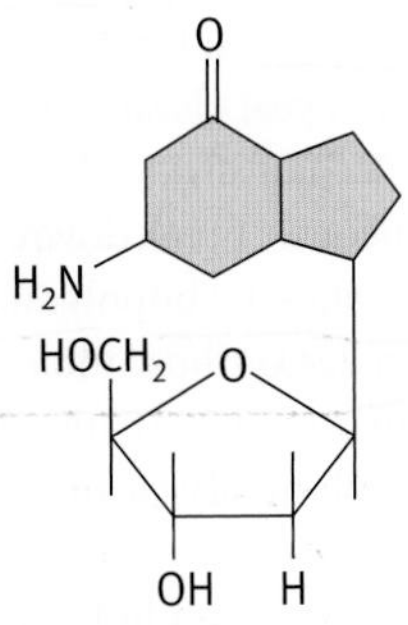   Deoxyguanosine

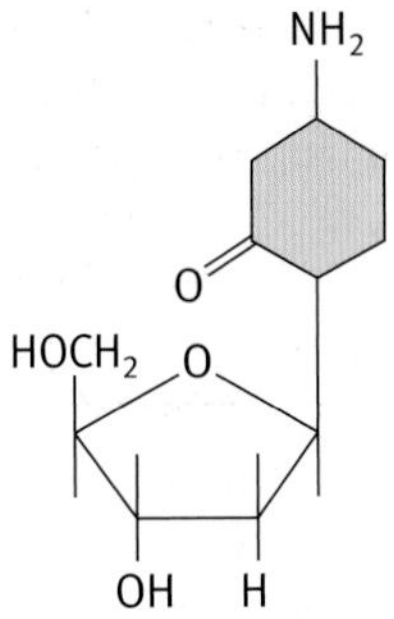

Deoxycytidine   Deoxythmidine
(or, simply, thymidine)

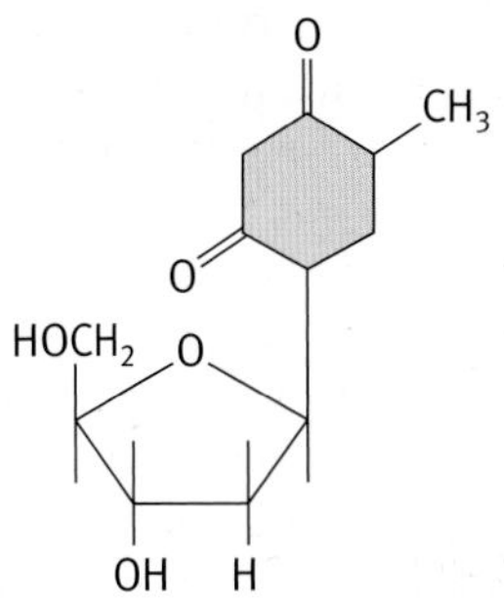

The physical properties of the polynucleotide components

The nucleotides in DNA are the 5′ phosphate compounds, dAMP, dGMP, dCMP, and dTMP (or TMP, since the ribose analogue does not occur, also known as thymidylate).

As noted, phosphoric acid —OH group of nucleotide components in DNA is ionized at physiological pH and thus has a negative charge. This, together with the hydroxyl groups of the deoxyribose, makes the exterior of the DNA double helix strongly hydrophilic. The bases are different – and this is important – in that they are relatively water-insoluble, guanine almost completely so. Their flat faces are essentially hydrophobic so they have a tendency to bind face-to-face because of hydrophobic forces. At the edge of each, however, there are polar groups with hydrogen-bonding potentiality so they can bind edge-to-edge by hydrogen bonds in the core of the double helix to form base pairs.

Structure of the polynucleotide of DNA

A **dinucleotide** consists of two nucleotides linked together by a phosphate group between the 3′-OH of one and the 5′-OH of the second (the nonionized forms are shown here for clarity; see page 337 and Fig. 22.9 for a description of the reaction).

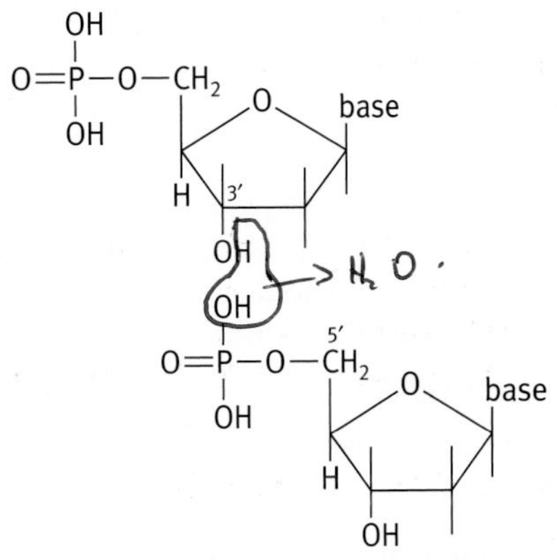

Two nucleotides

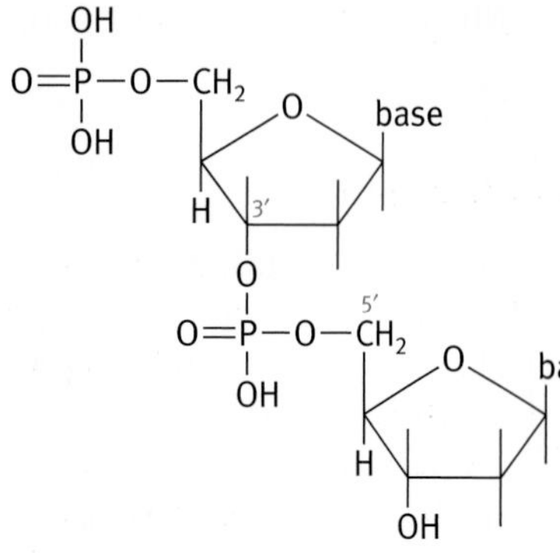

A dinucleotide

or, in structural terms,

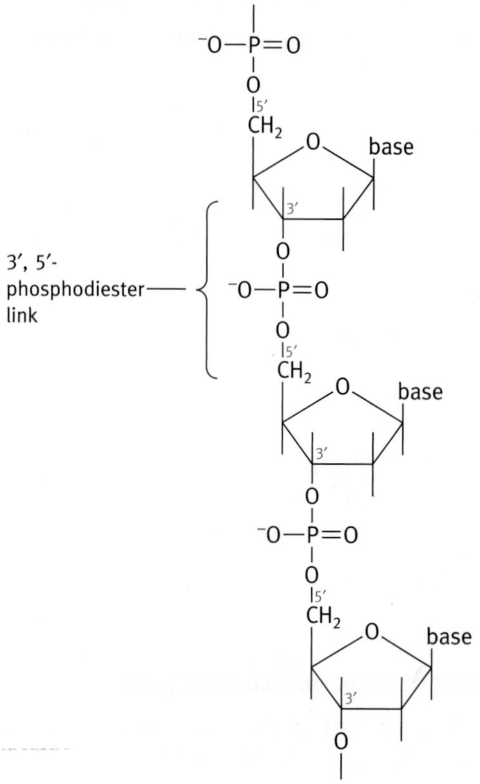

3′, 5′-
phosphodiester-
link

In a mononucleotide, the phosphate group is a **primary phosphate ester**, which means that there is only a single ester bond. In the polynucleotide structure, **phosphodiester links** are formed – the phosphate being linked to two deoxyribose moieties by two ester bonds.

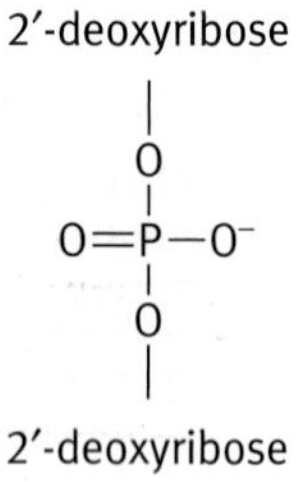

In the dinucleotide shown above it is a 3′, 5′-phosphodiester. Nucleotides can be added (by energy-requiring reactions) in the same way indefinitely, giving a polynucleotide. DNA is in its primary structure a polynucleotide of immense length. In dealing with proteins you may recall that there is a **polypeptide** backbone with amino acid side-chains attached. By analogy a polynucleotide has a backbone of alternating sugar—phosphate—sugar groups with a base attached to each sugar residue on the 1′ position; coded information is carried in the sequence of the bases. DNA therefore has the primary structure.

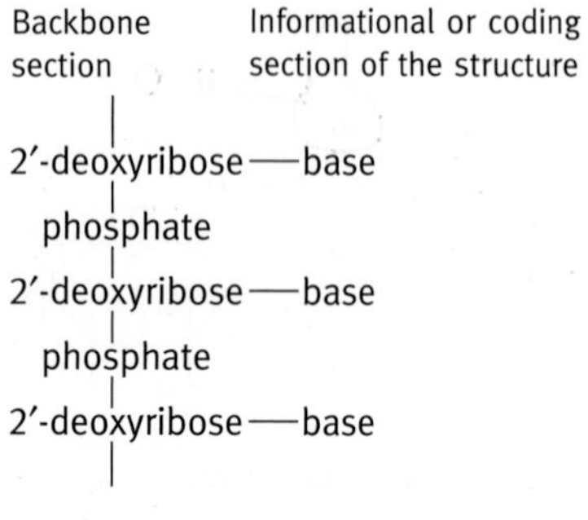

Why is deoxyribose used in DNA rather than ribose?

The cell has several nucleotides (for example, AMP) with ribose as the sugar component, and the nucleic acid, RNA (described in Chapter 23), has ribose as its sugar. One might then ask, why then did we end up with deoxyribose in DNA?

In evolution it is highly probable that ribonucleotides predated deoxyribonucleotides, and RNA predated DNA. The cell nevertheless goes to considerable energetic expense to convert ribonucleotides to the deoxyribonucleotides required for DNA synthesis. The reason for this is that DNA is chemically more stable than RNA. Genetic information gathered over countless millions of years is stored in chemical form in DNA molecules, but molecules always have some degree of instability – they spontaneously break down. The presence of the 2′-OH group of ribose makes a ribopolynucleotide less stable than the corresponding deoxyribose molecule. This is because, as illustrated in the structures below, the 2′-OH group is suitably placed for a nucleophilic attack on the phosphorus atom (explained on page 95) in the presence of OH⁻ ions, thus causing breakage of the phosphodiester link. In DNA, lacking the 2′-OH group, this does not happen.

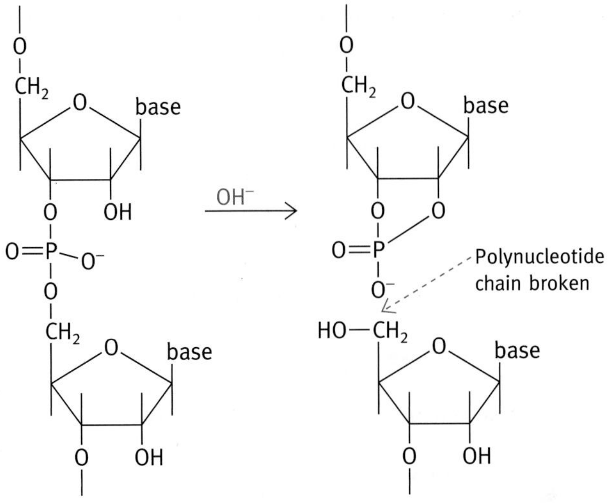

The 2'-OH of ribose facilitates the reaction because it can generate a 2'-O⁻, which attacks the phosphorus atom and converts the phosphodiester group into a 2', 3'-cyclic nucleotide, thus breaking the polynucleotide chain. Hydrolysis of the cyclic nucleotide produces a mixture of 2' and 3' nucleotides at the breakpoint.

The difference in stability is illustrated by the fact that dilute NaOH will completely destroy RNA at room temperature while DNA is unaffected. DNA is therefore a more stable repository of genetic information than is RNA. (The fact that some viruses can get away with RNA for this role does not contradict this concept, as is explained later.) Nevertheless chemical damage continually occurs in DNA. Repairing of the damage is the essential role of DNA repair mechanisms.

Why does RNA have uracil and DNA thymine?

The base pairing properties of uracil and thymine are the same so there has to be a reason for the two nucleic acids to be different in this respect. It is that DNA has to be constantly repaired. A commonly occurring spontaneous damage is the hydrolysis of cytosine bases to uracil. A DNA repair enzyme recognizes uracil as abnormal and replaces it with cytosine. If uracil occurred normally it couldn't make the distinction between uracil groups that should be there and those generated from cytosine. Use of thymine disposes of this problem. RNA is not repaired, probably because viral genomes are so small and suffer less damage per genome, and also because rapid mutation can be an advantage to viruses.

The DNA double helix

There will be few readers who have not heard of the double helix. DNA almost always exists as a double strand – only in a few viruses is it not double-stranded (see below). In other

words, you have two polynucleotide molecules paired together. What holds them together? The answer is **complementary base-pairing**.

Complementary base-pairing

Complementarity refers to the bases. A / T, and G / C in DNA chains are complementary in structure so that, when they are opposite one another in the two chains, hydrogen bonds form between them, two between A and T and three between G and C, attaching the double helices together. Only A—T and G—C pairing takes place in DNA, this being known as **Watson–Crick base-pairing**. This is a spontaneous process between closely positioned atoms, requiring no catalysis. Because hydrogen bonds are weak, they are easily broken, for example, by heat, which causes unpairing; cool the solution and re-association occurs.

The geometry of base-pairing is shown in Fig. 21.1. The base pairs always include one purine (larger molecule) and one pyrimidine (smaller) so that the pairs are essentially the same size. The reality of base-pairing as a spontaneous process is shown by the phenomenon of **hybridization**. If a molecule of DNA in solution is cut up into double-stranded pieces, each say, about 20 to any number of nucleotides long and then the mixture is heated to an appropriate temperature (about 95°C), the two strands of each piece of DNA will separate – referred to as **DNA melting**, due to heat disruption of hydrogen bonds and associated weak forces. This results in many pieces of single-stranded DNA of different base sequence. However, if the solution is cooled, the pieces will find their original partners and reassociate (Fig. 21.2). This hybridization technique, sometimes called **annealing**, lies at the heart of gene molecular biology (Chapter 28). At low temperatures there is a thermodynamic driving force for hybridization of the pieces, since the formation of hydrogen bonds releases energy. The most stable state in which free energy is minimized is that in which the bases

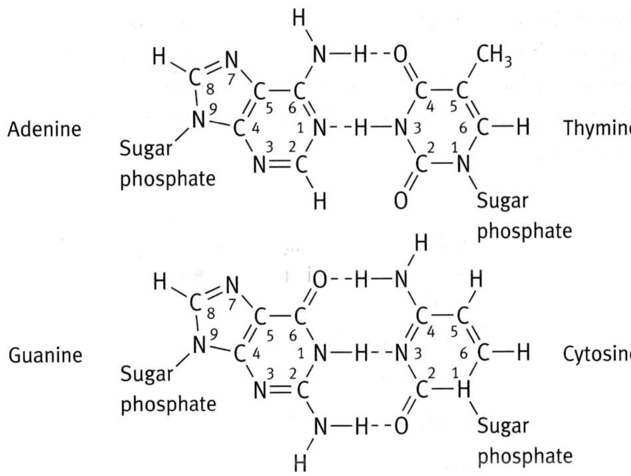

Fig. 21.1 Hydrogen bonding in the Watson–Crick base pairs.

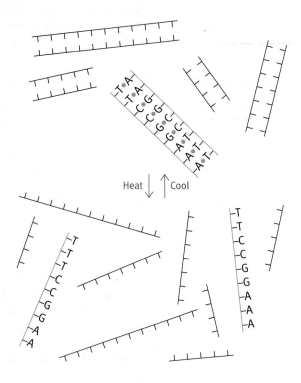

Fig. 21.2 Spontaneous hybridization of pieces of complementary DNA. Base sequences are shown on a single piece of DNA to illustrate the fact that hybridization depends on them.

are paired, since this gives the maximum number of hydrogen bonds.

Because of base complementarity, if you analyse different double-stranded DNAs, the amount of G equals that of C, and that of A equals T. Discovery of this by Erwin Chargaff in New York was a vital clue in the elucidation of DNA structure. However, DNAs from different organisms vary in their percentages of [A + T], and of [G + C], reflecting their different genetic information. For the first approximation then, a stretch of DNA is commonly represented as shown below, the long solid line representing the sugar—phosphate backbones and the attached bases interacting by hydrogen bonding.

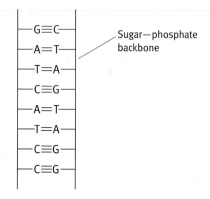

In a piece of DNA rich in G + C, the two strands will be more strongly held together than in a piece rich in A + T.

However, this straight ladder structure does not occur under normal solution conditions, for it violates a structural requirement as shown in the diagram below.

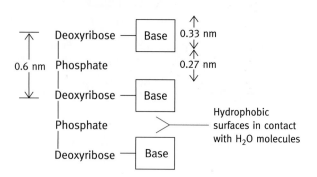

The length of the phosphodiester link is 0.6 nm (1 nanometer (nm) = 10^{-9} m) while the bases are about 0.33 nm thick, so that in the straight ladder-like structure there would be a gap between them. The faces of bases are, as mentioned, hydrophobic, and in the 'straight' structure above, they would be exposed to H_2O molecules, an unstable situation. Instead, each of the two DNA chains forms a helix, with the bases inside and the hydrophilic sugar and phosphate groups outside (Fig. 21.3(a), (b)). In the double helix the sloping of the chains as they follow a double helical pathway collapses the bases together and minimizes the exposure of the hydrophobic faces to water, as illustrated in Fig. 21.3(c).

The base pairs still lie almost flat, stacked on top of one another – a phenomenon known as **base stacking**. The hydrogen-bonding face at the edge is still free so that it can bond to its partner strand. In the DNA double helix, the stacking of bases is not exactly vertical as shown in Fig. 21.3(c). Successive base pairs rotate slightly relative to one another, as illustrated in Fig. 21.3(d), such that approximately 10 base pairs are required to rotate through one complete turn, and there is a slight cross-helical slope called 'tilt'.

The helices are right-handed – as you move along a strand or a groove you continually turn clockwise; alternatively, imagine a right-handed person driving in a screw. The turning motion gives the direction of twist. The structure of the double helix is such that there are major and minor grooves (see Fig. 21.3(a)). Any base pair can be viewed from both the major and the minor grooves but only their edges are visible. The major grooves have more atoms accessible for bond formation and hence provide easier access for proteins to 'recognize' (by which we mean attach to) the base pair edges. A given base pair 'looks' quite different when viewed from the two grooves (Fig. 21.4), the significance of which will become apparent in Chapter 23, where gene regulation is discussed.

The DNA conformation described above is known as the **B form** (Fig. 21.5) and is the normal form that exists in cells. However, DNA *can* adopt different configurations in special

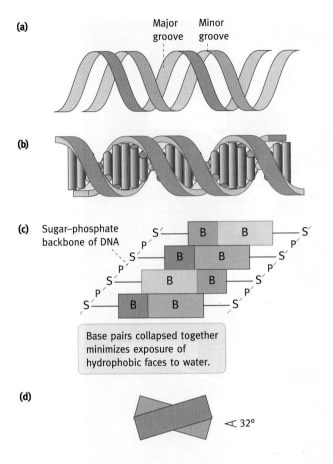

(a)

Major groove   Minor groove

(b)

(c) Sugar–phosphate backbone of DNA

Base pairs collapsed together minimizes exposure of hydrophobic faces to water.

(d)

◁ 32°

circumstances. When dehydrated, the double helix is more squat and the bases are more tilted; this is known as the **A form**. Another form is known as Z (because the polynucleotide backbone zigzags); in this, the double helix is left-handed (*cf.* right-handed in B-DNA and A-DNA). It has been observed to occur in short synthetic DNA molecules with alternating purine and pyrimidine bases provided the solution is of high ionic strength. Whether either of the A or Z forms has biological significance is not known, but the possibility of DNA adopting modified configurations in localized sections of the chromosome is not excluded.

An important property of the double helix structure is that it can bend. A molecule of DNA is vastly longer than the widest dimension of a nucleus; to pack it in, flexibility is necessary to allow for all the coiling and folding required.

As mentioned, while DNA is almost always in the double helix form, single-stranded DNA does occur, for example, in certain bacterial viruses. In such situations the molecule takes up very complex internally folded structures to satisfy thermodynamic considerations. However, even in such cases the life cycle involves a double helix form of DNA so that the basic principles of genetic information with complementary

Fig. 21.3 (*opposite*) **(a)** Outline of the backbone arrangements in the DNA double helix. **(b)** As (a), but showing the base pairs in the centre of the helix. (Note that each coloured band is a base *pair.*) **(c)** How a skewed arrangement of the ladder collapses the bases together. **(d)** Diagram of two successive base pairs in a double helix showing the twist imposed by the double helix. A more realistic model corresponding to (b) is shown in Fig. 21.5.

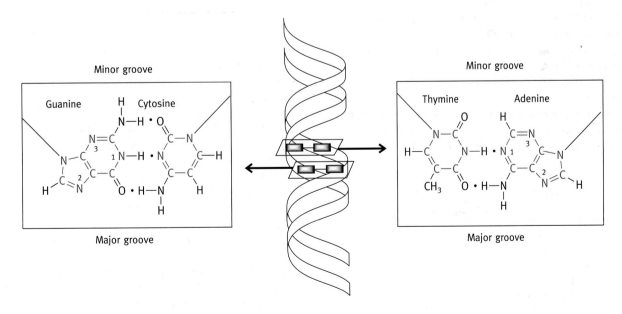

Minor groove

Guanine   H   Cytosine

Major groove

Minor groove

Thymine   Adenine

Major groove

Fig. 21.4 The edges of a given base pair in DNA look different when viewed from the major and minor grooves. DNA-binding proteins designed to recognize specific sequences of base pairs in DNA can identify (bind to) the characteristic chemical groupings of different base pairs without unwinding the DNA. The importance of this is discussed in Chapter 23, which deals with gene control.

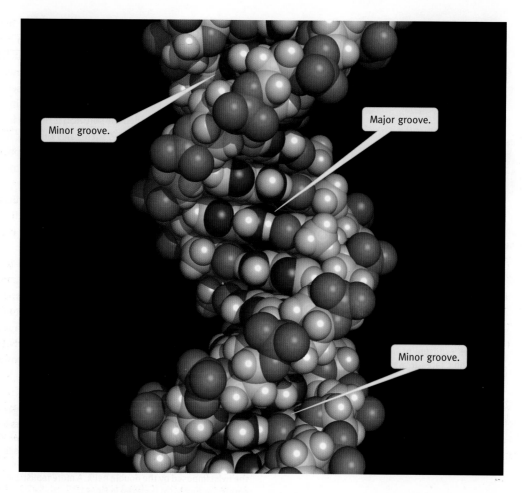

Fig. 21.5 A model of B DNA, Protein Data Bank code **1BNA**. Space-filling atomic model of a DNA segment with one major groove and part of two minor grooves.

base-pairing are the same in all cases. The existence of single-stranded forms is in this sense a specialized idiosyncrasy – not a fundamental difference.

DNA chains are antiparallel; what does this mean?

By antiparallel we mean that the two chains of a double helix have opposite polarity – they run in opposite directions. It may not be immediately clear what is meant by the polarity or direction of a DNA strand. In DNA we are talking about the direction in which a sequence of nucleotides is read, and not polarity in the sense of a bond between oppositely charged ions. It is worth spending a little time on this so that you are comfortable with the concept, because a lot of biochemistry requires an understanding of it. Two antiparallel strands of DNA are illustrated in Fig. 21.6.

Any single linear strand of DNA has (obviously) two ends. One has a 5′-OH group on the sugar nucleotide that is *not* connected to another nucleotide. (It may have a phosphate on it.) This is the 5′ end. The other end has a 3′-OH group that is *not* connected to another nucleotide (though it also may have a phosphate group on it). This is the 3′ end. At one end of a linear piece of DNA double helix there is always one 5′ end and one 3′ end. If a piece of DNA is circular, as in bacteria, there are no free ends but there is still inherent polarity in the individual strands, as you can see from the deoxyribose moieties.

Thus the structure on the left runs 5′ → 3′ down the page and that on the right 5′ → 3′ up the page.

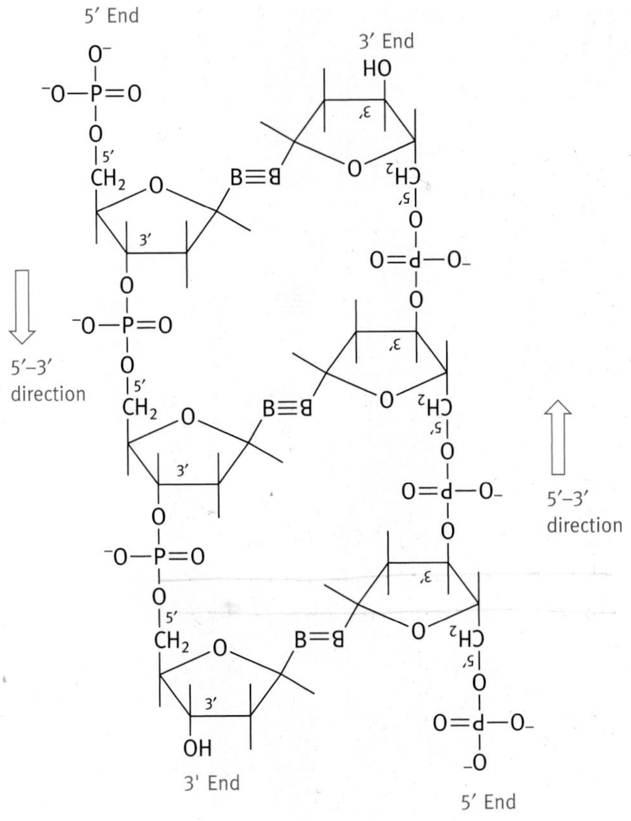

Fig. 21.6 Two antiparallel strands of DNA. B, base.

Incidentally, and confusingly, as you move down a DNA strand in the 5′ → 3′ direction the phosphodiester links that you traverse are 3′ → 5′ (see above diagram). *Remember that a 5′ → 3′ direction in a DNA strand means that you are proceeding (in a linear molecule) from a terminal 5′-OH group towards a terminal 3′-OH group* or, if you prefer it put it another way, you are travelling in the direction from the 5′ carbon of the deoxyribose towards the 3′ one of the same sugar, as illustrated in the diagram above.

There are conventions in writing down the base sequence of DNA. It is usual to represent a polynucleotide structure simply by a string of letters representing the bases of component nucleotides, but sometimes the phosphodiester link is indicated by the letter p inserted between the bases (for example, CpApTpGp, etc.). Suppose we have a piece of double-stranded DNA whose base sequence is:

5′ CATGTA 3′

3′ GTACAT 5′.

Sometimes it is useful to write both strand sequences, but usually it is not necessary to write both sequences since, given one, the complementary sequence is automatically specified. So you will find that the structure of a gene is often given as a single

base sequence, despite there being two strands. There is a convention that a single base sequence is written with the 5′ end to the left and there is no need therefore to specify the 5′ and 3′ ends of a sequence. Thus, if the structure illustrated above is part of a gene, it would be written as CATGTA.

How is the DNA packed into a nucleus?

There are many cases in biochemistry where a problem facing life is so staggering that the solution boggles the imagination to a point where it would be reasonable to wonder if it could possibly work – except for the fact that it obviously does. The packing of DNA is one of these.

In a human cell, there are 46 chromosomes giving a total length of DNA per cell of about 2 metres, packed into a nucleus about 10 μm in diameter. A very elaborate packing procedure is needed. DNA in eukaryotic cells exists as **chromatin** – a DNA–protein complex. The main proteins are **histones** in somatic cells and **protamine** in sperm cells; these are small basic proteins rich in arginine and/or lysine, giving them positive charges that form ionic bonds with the negative charges on the phosphate groups on the outside of the double helix of DNA. The amino acid sequences of eukaryotic histones are highly conserved throughout evolution, although the protamines are not. One of the histones differs only in two amino acids in the entire molecule between peas and cows and the changes are conservative (valine for isoleucine, lysine for arginine). The extreme conservation means that changes in their structure would be lethal or sufficiently deleterious for natural selection to eliminate.

We will start with the four histones called H2A, H2B, H3, and H4. These form an octamer protein complex called a **nucleosome core** around each of which two turns of DNA are wrapped (160 base pairs) with intervening stretches linking successive nucleosomes, arranged like beads zig-zagging on a string (Fig. 21.7(a), (b)). Histone H1 is different from the other four. It is larger and less well conserved. It plays a role in condensing the nucleosomes together (Fig 21.7(a)). The arrangement condenses (packages) the DNA somewhat, but nowhere near enough. The nucleosomes are roughly discs about 10 nm in diameter and 5 nm thick. They associate together to form the 10 nm fibre shown in Fig. 21.7(b). The histone H1 protein is bound to the DNA between the nucleosomes and is not evolutionarily conserved. This 10 nm fibre is condensed further to form a fibre, known as the 30 nm fibre, illustrated in Fig. 21.7(c). The nucleosomes in this are now arranged in a zigzag manner. An electron micrograph of such a fibre is shown in Fig. 21.8. The condensation of the DNA achieved so far is about 100-fold. The 30 nm fibres now form long loops that are attached to a central chromosomal protein scaffolding (Fig. 21.7(d)). This looped structure forms yet more densely packed structures, not fully understood, involving folding and/or coiling and achieving a 10,000-fold packing of the original DNA.

(a)

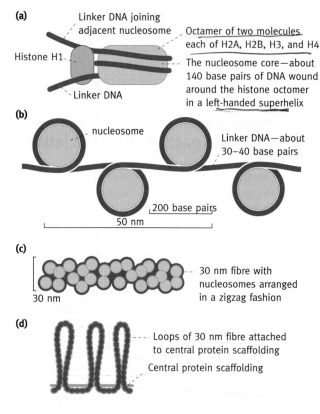

Linker DNA joining adjacent nucleosome

Histone H1

Linker DNA

Octamer of two molecules each of H2A, H2B, H3, and H4

The nucleosome core—about 140 base pairs of DNA wound around the histone octomer in a left-handed superhelix

(b)

nucleosome

Linker DNA—about 30–40 base pairs

200 base pairs

50 nm

(c)

30 nm

30 nm fibre with nucleosomes arranged in a zigzag fashion

(d)

Loops of 30 nm fibre attached to central protein scaffolding

Central protein scaffolding

Fig. 21.7 Order of chromatin packing in eukaryotes. **(a)** Diagram of a nucleosome. **(b)** Beads on a string form, the 10 nm fibre. **(c)** A 30 nm fibre of chromatin (see Fig. 21.8 for an electron micrograph of a 30 nm fibre). **(d)** Loops of the 30 nm fibre are attached to a central protein scaffold in a 360° array. It is believed that these loops are yet further condensed, perhaps by supercoiling and ultimately into the extremely compact metaphase chromosome. The latter condensation stage is not illustrated.

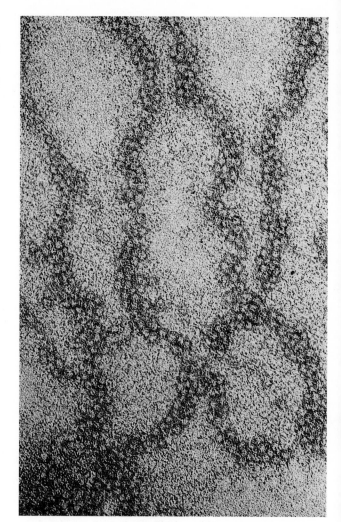

Fig. 21.8 Electron micrograph of a 30 nm fibre of chromatin.

We thus have several levels of organization:

1 Winding around nucleosomes.

2 Nucleosomes packed into a 30 nm fibre about 3–5 nucleosomes thick.

3 The fibres form loops, thousands of nucleosomes long, attached to a central scaffolding.

4 The loops form yet other coils and/or folds (see Fig. 21.7(a)–(d)).

Stages 1, 2 and 3 are probably found in most chromatin but stage 4 is probably restricted to the highly condensed metaphase chromosomes and areas known as heterochromatin.

As will be described in later chapters, to be functionally active, all this packed DNA has to be accessible to enzymes that read its informational content and to enzymes that replicate the entire DNA, so that there are changes to the degree of packing during the cell cycle of DNA.

The packing of the prokaryotic genome is different from that in eukaryotes

As already mentioned, prokaryotes have no nuclear membrane and there is no well-established structure corresponding to the eukaryotic nucleosome organization. Proteins reminiscent of histones have been postulated to play a somewhat analogous role but less is known of the physical structure of prokaryotic chromosomes in the cell.

The bacterial DNA is visible in electronmicrograph cross-sections as a 'nucleoid' mass (see Fig. 2.2). At cell division, as described, the duplicated circular chromosomes become separated into daughter cells without the condensation that occurs at the metaphase stage of mitosis in eukaryotes, so that the compact X-shaped pair of chromatids are not seen in bacteria. This difference is related to the very much smaller size of the prokaryote genome.

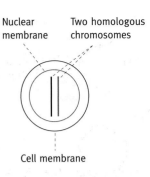

Nuclear membrane   Two homologous chromosomes

1. **Interphase.** Chromosomes are in the form of unravelled fine threads. To keep the diagram simple only a single chromosome is depicted for the illustration. Note that the red and blue lines each represent **double-stranded** DNA of homologous chromosomes. Genes are expressed and cell size increases.

Cell membrane

Duplicated homologous chromosomes

2. DNA is replicated ready to enter the mitotic phase but the duplicated chromosomes remain connected at the centromeres.

3. Metaphase. The nuclear membrane has disappeared and the chromosomes have condensed into compact X shapes and the genes shut down. The two unseparated halves of the X shape are complete chromosomes but at this stage are known as **chromatids**. Each chromatid becomes attached at the mid line of the cell to a kinetochore fibre.

4. The chromatids are pulled apart as chromosomes by kinetochore fibre shortening.

5. The chromosomes move apart. See Fig. 8.19 for the mechanism by which this happens.

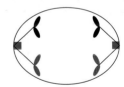

6. Cell division occurs, the nuclear membranes form, the DNA unravels giving two copies of the original cell in interphase.

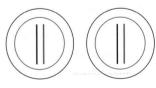

Fig. 21.9 A simplified diagram of mitosis.

The packing of DNA in the eukaryotic nucleus changes during the cell life cycle

Chromatin is unpacked whenever it is read during RNA synthesis and DNA synthesis to allow access to enzymes and regulatory molecules. It is most tightly packed when chromosomes are segregating, to aid separation and avoid tangling and is not genetically active during this period.

Eukaryotic cells have a strictly controlled cell cycle (Chapter 31) which typically occupies about 24 hours (see Fig. 22.4). During this time the cell doubles in size and its DNA is replicated. Cell division, in all but germ line cells which give rise to sperm and eggs, involves **mitosis** (Fig. 21.9). In this, the replicated chromosomes are equally partitioned into the two daughter cells. The period between one mitosis and the next is known as **interphase**, when the DNA of the chromosomes exists as

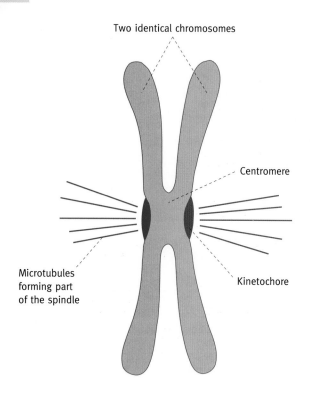

Two identical chromosomes

Centromere

Microtubules forming part of the spindle

Kinetochore

Fig. 21.10 Chromosome at metaphase consisting of two daughter chromatids.

diffuse long threads though not completely unraveled. In this state the genes are active.

As the cycle enters mitosis (prophase), duplicated chromosomes become condensed into compact X shaped **metaphase** chromosomes. At this stage the chromosomes in the pairs are referred to as **chromatids** (Fig. 21.10). The chromatids are made of a pair of identical, daughter chromosomes (chromatids) which have not separated completely. (Each chromosome is double-stranded DNA, one of the strands of each being newly synthesized, and the other being the parental strand.) They are held together at their **centromeres**, by **cohesin** proteins. In this state the genes are in a shut down condition. The cell now undergoes a major structural change; the nuclear membrane disintegrates, allowing the **mitotic spindle** to invade the area. The spindle is a group of microtubules (page 136) originating in the **centrosomes**, a complex of proteins at each pole of the nucleus. At metaphase the double chromosomes are arranged at the spindle equator with each chromatid attached to spindle **kinetochore fibres** at the **kinetochore**, a protein complex assembled at the chromatid centromere. In **anaphase** the chromatids separate due to the activation of an enzyme (a protease) that destroys the cohesin proteins holding the chromatids together. The separated chromatids, now called chromosomes, move towards the spindle poles and the latter separate further apart. This process is completed at **telophase** when the chromosomes reach the spindle poles. The mitotic apparatus is dis-

assembled and nuclear membranes appear. Cell division occurs with each daughter cell receiving one set of chromosomes. The chromosomes decondense into the active interphase state. The physical movements are caused by molecular motors that attach to spindle fibres (see Fig. 8.19).

Meiosis

Germ line cells divide to produce gametes (eggs and sperm in mammals) which are haploid – they have only a single set of chromosomes – while somatic cells are diploid. *Mitosis produces diploid cells; meiosis produces haploid cells.* Mitosis and meiosis have much in common so far as mechanism goes with spindle fibres and centrosomes. The difference is that, in mitosis, DNA is replicated once and a **single** cell division occurs. In meiosis the DNA is also replicated once but **two** cell divisions occur. Exactly as in mitosis, in meiosis DNA replication and chromosome condensation produce two pairs of chromatids (Fig. 21.11(a)); at this stage, as in mitosis, the nuclear membrane disappears. Up till now the process has been the same as in mitosis but now meiosis is different in that the two homologous pairs of chromatids align themselves together in what is called **synapsis** and they swap small sections of DNA by **crossing-over** (Fig. 21.11(b)).

Now cell division occurs but of a type that does not occur in mitosis. Instead of the chromatids separating, the **two pairs** of homologous chromatids (the X-shaped structures) are separated in the first meiotic cell division (Fig. 21.11(c)), one pair going into each daughter cell (Fig. 21.11(d)). These immediately undergo a second cell division (telophase I) (*without prior DNA replication*) with the chromatids separating (Fig. 21.11(e)), the nuclear membranes form as in mitosis giving four haploid cells then called the **gametes** (Fig. 21.11(f)).

On fertilization, combination of sperm and egg produces a diploid cell which by mitotic divisions produces the embryo.

The tightness of DNA packing is the initial control on gene activity

As briefly alluded to before, the degree of 'unpacking' of DNA in a eukaryotic cell is vitally important. To use information encoded in the base sequence of DNA, the molecule must be accessible to enzymes and other proteins and not concealed in tightly condensed structures. Certain parts of eukaryotic chromosomes contain DNA believed not to be involved in gene structure – that is, DNA that does not carry coded information for the synthesis of proteins. The DNA in these regions even in interphase chromosomes remains highly condensed in regions known as **heterochromatin**. Heterochromatin is particularly found in the chromosomal centromeres and telomeres. Centromeres are the sites of attachment of the filaments that are involved in chromosome separation in cell division, while telomeres protect the ends of the linear chromosomes. This may seem like a mundane role but telomeres are now known to be of fundamental importance and of relevance to ageing and

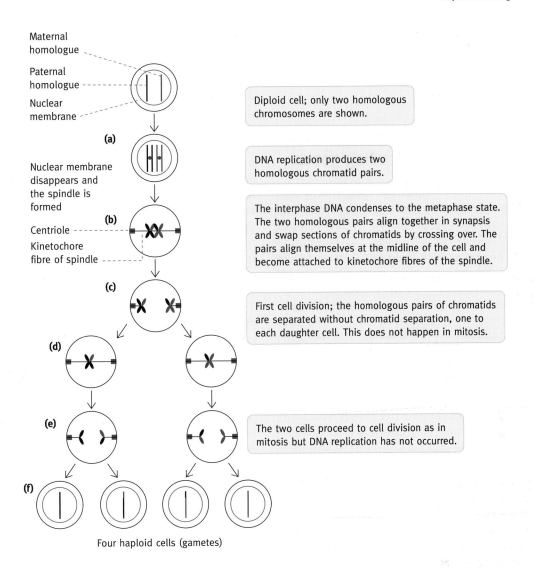

Maternal homologue

Paternal homologue

Nuclear membrane

(a)

Nuclear membrane disappears and the spindle is formed

(b)

Centriole

Kinetochore fibre of spindle

(c)

(d)

(e)

(f)

Four haploid cells (gametes)

Diploid cell; only two homologous chromosomes are shown.

DNA replication produces two homologous chromatid pairs.

The interphase DNA condenses to the metaphase state. The two homologous pairs align together in synapsis and swap sections of chromatids by crossing over. The pairs align themselves at the midline of the cell and become attached to kinetochore fibres of the spindle.

First cell division; the homologous pairs of chromatids are separated without chromatid separation, one to each daughter cell. This does not happen in mitosis.

The two cells proceed to cell division as in mitosis but DNA replication has not occurred.

Fig. 21.11 A simplified diagram of meiosis. The process of meiosis to generate haploid germ cells occurs mechanistically by stages similar to those of mitosis. The diagram therefore just shows the changes that occur to the chromosomes ignoring spindle formation and the chromatid separation process. (See Fig. 8.19 for this mechanism.) The main difference between mitosis and meiosis is that, in meiosis, the chromosomes are replicated once but two cell divisions occur. Thus the daughter cells (the gametes) receive only one homologous chromosome. In the figure a single pair of homologous chromosomes are used for illustrative purposes.

cancer (discussed on page 347). The active regions are less condensed and are known as **euchromatin** (the prefix 'eu' means 'true'). Within this, eukaryotic genes in the default state (where nothing is done to activate them) are in the 'shut down' state. When sections of chromosomes, such as individual genes, become active, the chromatin becomes less tightly packed. This is the initial event in gene 'switch on' (described in Chapter 23).

The mitochondrial genome

It has been explained (page 20) that mitochondria originated in the evolutionary sense from the engulfment of a prokaryote cell by the precursor of modern eukaryotic cells. Mitochondria reproduce by division.

Most mitochondrial proteins today are coded for by genes in the nucleus, synthesised in the cytoplasm and transported into the organelle but a small proportion are still made in the organelle. The mitochondrial DNA is circular with, in humans, 16,569 base pairs. The genome reflects its prokaryote origin with protein-coding sequences tightly packed along the length of the DNA (page 326). Gene transcription and mRNA translation are also prokaryote like. There are other differences from nuclear DNA referred to in Chapters 23 and 25, where transcription of eukaryotic genes and protein synthesis are discussed.

Mitochondrial inheritance is different to that of nuclear genes. It simply follows the cytoplasm through cell division, and does not get transmitted by sperm, which has little cytoplasm. Thus the mitochondrial genome's inheritance is maternal (down the 'female line').

The structure of protein-coding genes

A **gene** is the unit of heredity. In molecular terms it has no independent existence – it is simply a stretch of DNA, part of a huge molecule, and carries coded information for the sequence of amino acids of (usually) one polypeptide chain, though some genes produce noncoding RNA molecules. A more detailed description of protein-coding genes is given on page 356 and of noncoding genes on page 374. For a protein with a single polypeptide chain, a single gene therefore codes for that protein. Where there are more polypeptide chains, more genes are usually needed (one exception is insulin, with two chains derived from a single initial one (page 54). The information resides in the base sequence of the DNA – there is no physical or molecular discontinuity between one gene and the next. For a more familiar analogy, a chromosome can be compared with an entire magnetic tape, a gene with a piece of music recorded on it, and the base sequence with the magnetic signals. Associated with the coding region are gene promoters that regulate them, and untranslated regions at each end.

Protein-coding regions of genes in eukaryotes are split up into different sections

In prokaryotes the genes are arranged close together on the DNA with little 'spacer segments' between them. The coding region that specifies the amino acid sequence of a protein is continuous. In eukaryotes this is not the case. The coding region is interrupted by segments of DNA that do not code for amino acid sequences. The interrupting sequences are called **introns** while the coding sections are called **exons** (see Fig. 23.12). There can be 1–500 introns in a gene, which can each vary from 50 to 20,000 base pairs in length. Exons are smaller, usually around 150 base pairs. Thus, typically, the sum of the coding regions of a eukaryotic split gene is very much smaller than the total of its introns. In the human genome the coding regions total only about 1.6% of the DNA, while the total including introns is around 25%. When split genes are transcribed into RNA, the sections of the latter corresponding to the introns are removed and the coding region corresponding to the exons joined together to produce a continuous messenger. The process is called 'splicing'. It is complex and best left to Chapter 23.

What is the origin of split genes?

There are two views on this question. One is that the prokaryotic type, non-interrupted genes, are primitive and that introns were inserted into eukaryotic genes later in evolution ('introns late' model).

The alternative view is the exon theory of genes, or 'introns early' model. This postulates that primitive genes were small minigenes with untranscribed flanking regions alongside their coding section. Modern genes are postulated to have formed by the fusion of several of these **minigenes**. Introns would then represent the fused untranscribed flanking regions of adjacent minigenes. On this hypothesis, prokaryotic uninterrupted genes are regarded as a later development in which introns were thrown overboard to achieve rapid replication. An argument against the introns early model would seem to be that to synthesize proteins, the primary RNA gene transcripts have to be spliced and this requires a large multiprotein complex (page 361), which would not have been available in primitive times. The biochemical world was shocked when it was found that the RNA transcripts from certain genes of the organism *Tetrahymena*, remarkably can self-splice (page 361) without extra help so the objection does not necessarily apply. The jury is still out on the issue of split gene origin.

What explains the evolutionary survival of introns?

Since prokaryotes manage perfectly well without introns it might seem surprising that they have survived so abundantly in eukaryotic protein-coding genes. They must have been some survival advantage, or natural selection would presumably have eliminated them.

An important biological aspect of the current existence of introns is that they may have facilitated evolution of proteins (though not necessarily the reason why introns occurred in the first place). At this point you might re-read page 55 on protein domains. As research on protein structure identified domains, and gene research identified exons, it was found that the exons often code for discrete protein domains. New proteins are believed to have evolved particularly rapidly by '**domain-shuffling**' – the concept is that the same domain is used repeatedly, combined with various other domains, so that new families of proteins are assembled from pre-existing domains. This would be similar to the principle of assembling a variety of electronic machines using different combinations of plug in boards. The separation of the parts of a gene into exons coding for discrete protein domains could facilitate domain-shuffling during evolution.

Domain shuffling is dependent on the assembly of new genes by **exon-shuffling** involving genetic recombination between genes. Introns could help in the process because they are zones where breaks and joins could take place safely without disrupting exons and hence without destroying protein domains. The process would provide a more rapid means of producing novel proteins by recombination events than would point mutations in DNA leading to single amino acid changes. Note, however, that it is an assumption that introns have no function, a view now being challenged (see Chapter 25).

BOX 21.1 Size of genomes related to complexity of organisms

When we come to discuss quantitative aspects of the genome, it is customary always to refer only to the haploid genome (a single copy of each chromosome) whether the organism in question is haploid, diploid or polyploid. This makes comparisons possible. In prokaryotes and viruses the amount of DNA roughly correlates with the complexity of the organism. However in eukaryotes this is not the case. A frog has almost as much DNA as a human; a newt or a broad bean has more. A much-cited example is the amoeba Dubia, which has 200 times more base pairs than humans. Given that such a small proportion of the DNA is devoted to protein-coding genes it is not surprising perhaps that base pair numbers can show little correlation with coding gene numbers. Nor does the eukaryotic gene number obviously correlate with complexity very well.

The smallest known cellular genome is that of *Mycoplasma genitalium* with 580,000 base pairs (per haploid genome) and 468 protein-coding genes. *Escherichia coli* has about 4.5 million base pairs and just over 4000 coding genes. Yeast, a single cell eukaryote, has 12 million base pairs and about 6000 genes. The fruit fly, *Drosophila*, has 137 million base pairs and 14,000 genes, but the much simpler nematode roundworm *C. elegans* with only about a thousand cells has 18,000 genes. Humans, with about 10 trillion cells, have over 3 billion base pairs and an estimated 25,000 protein-coding genes, only about double the number of protein-coding genes that the roundworm has. *Arabidopsis*, a small cress plant, has about 142 million base pairs and 26,000 genes, only about four thousand fewer than humans, or, possibly, about the same number. It seems clear that there is no close correlation between the number of protein-coding genes and the complexity of the organism. Current thinking is that the more complex organisms make more efficient use of genes to generate additional complexity to which the discovery of the large number of microRNA genes, which exert control over conventional genes has lent weight.

It should be noted that the cited numbers of protein-coding genes in eukaryotes are estimates, with a considerable degree of uncertainty despite the elucidation of the sequence of whole genomes. This is because of the split structure of eukaryotic genes; it is not necessarily a simple matter to deduce gene numbers from the genome sequence. Bacterial genes are easier to identify and count from the sequenced genomes because all that need be looked for is a stretch of DNA long enough to code for a protein before running into a stop codon (page 381).

Multiple gene copies facilitate evolution of new genes

Sometimes there are multiple copies of the same gene. This can arise from gene duplication by accident during evolution. (Gene duplication is also observed medically in human cells, particularly during cancer treatment by antifolates – see page 307). Multiple copies of genes provide for more rapid production of the cognate proteins; there are multiple copies of histone genes that occur in clusters in some organisms – hundreds of repeating copies in the case of the sea urchin. This also gives the evolutionary potential for modification of copies to develop new genes coding for new proteins. During evolution it often has happened that new genes are created by modification of pre-existing genes. Duplicate genes allow one copy to be modified while the other retains the essential function of the original gene.

Views on so-called junk DNA have changed dramatically

One might have expected that to code for an organism such as a human, the genome would have an organized look about it. However, when the human genome project was completed and the base sequence of the DNA available, the reverse was found. It looks more like a disorganized mess thrown together haphazardly. As one distinguished worker in the field put it: 'The general arrangement of the genome provides another startling jolt. In some ways it may resemble your garage/bedroom/refrigerator/life – highly individualistic and unkempt, with little evidence of organization and much accumulated clutter. Virtually nothing is ever discarded and valuable items are scattered indiscriminately, apparently carelessly'.

The valuable items referred to in this quote are the protein-coding genes and the clutter is DNA that until recently was presumed to have no function and was called 'junk' DNA. It was presumed to be useless DNA that could not be discarded. It constitutes over half of the total in humans; bacteria, whose genomes look far better organized, have only 1–2%. Genes occupy only a fraction of the human genomes, and the actual protein-coding regions constitute only 1.6% of the total DNA excluding introns.

The newly discovered microRNA genes of junk DNA are revolutionizing important concepts of gene control

It is now believed that much or most of the junk DNA contains large numbers of hitherto unknown **microRNA genes**. These are transcribed into small RNA molecules that do not code for the synthesis of proteins.

One important known aspect of the microRNAs is that they are capable of causing the specific destruction of individual selected messenger RNA molecules and thus control gene expression. The phenomenon is known as **RNA interference (RNAi)**. Apart from the biological effects of RNAi it also promises to be of immense medical importance, because it provides a relatively simple way to selectively inhibit the expression of virtually any known gene. It is discussed in Chapter 25.

The microRNAs are believed to cause *heritable changes* in the phenotype of cells but *without any changes in DNA sequence*. This is known as **epigenetic control**. In simplistic terms, it is a type of genetic control superimposed on the protein-coding gene-based one. Although understanding of the mechanisms are in their infancy the microRNAs are postulated to be involved in a wide range of regulatory controls, including development and differentiation. RNAi also protects against RNA virus infection. It is reasonable to speculate that this development is likely to have repercussions on our concepts on the genetic inheritance and the manner in which complex eukaryotic genomes function (Chapter 25).

Transposons

DNA sequences are not necessarily fixed in position; chromosomes have sections that are mobile – they move from one place in the DNA to another. They are called **transposons** or jumping DNA sequences. They have no known direct functions, but, like gene duplication, may produce new sequences for evolution to operate on.

Different classes of transposons exist. **DNA-only transposons** are those in which the transposon sequence is cut out of the chromosome and then becomes inserted in some other place in the chromosome. The transposon codes for a transposase enzyme, which facilitates the process. They don't transfer frequently and it is not known what causes them to jump.

The more common class in the human genome is called **retrotransposons**. These are also sections of DNA inserted into the chromosome but they never jump out of it. They replicate by their DNA being transcribed into RNA copies by an enzyme called **reverse transcriptase** (see Fig. 22.23) coded for by the transposon and then converted to double-stranded DNA versions. The new DNA transposons then insert themselves into the genome elsewhere. They are called retrotransposons because retroviruses such as HIV also insert into host chromosomes and replicate in the same way, also using reverse transcriptase (page 347). (A retrovirus is very like a retrotransposon, with a protective coat, so that it can survive outside an organism). Transposon replication in this manner has, in evolutionary terms, produced a large part of the repetitive DNA sequences mentioned below. In humans and other hominids almost all of the transposons are inactive fossilized versions of previously active mobile elements though some are still active in other species such as the mouse. A retrotransposon may be thousands of bases in length.

When a transposon inserts at random into the genome there is the danger that it might disrupt an essential gene or cause a gene to be dangerously overactive. RNA interference (Chapter 25) has been postulated to destroy RNA intermediates in transposon replication and guard against this danger.

Chromosomes can also acquire new resident genes from retroviruses that infect the cell.

Repetitive DNA sequences

About half the DNA of the genome is made up of repetitive sequences of different types. We will mention three here – others are given in Box 28.1 on page 471.

- **LINES (Long Interspersed Repeated Sequences)** are a few thousands of base pairs long; there are a few thousand of these scattered throughout the genome, and probably are mostly from transposons in the evolutionary sense.

- **SINES (short interspersed repeated sequences)** are much the same but only 100–500 base pairs long with typically several thousands scattered throughout a chromosome. There are different families of them. The best known in humans is the *Alu* one repeated almost a million times.

- **Tandemly repeated sequences**: the other main category of repetitive DNAs consists of sequences of short nucleotide units arranged one after the other. An example is five bases in tandem, TTCCA/TTCCA/TTCCA (with their complementary strands) repeated dozens or thousands of times. Repeats of this style are found particularly in the tightly packed **heterochromatin** of centromeres, in patches near the centromeres and near the ends of the chromosomes. They constitute perhaps 10% of the total DNA. Nothing is known of their function, if any, though it has been suggested that they may originate from transposons. Most tandem repeated sequences are highly polymorphic. This gives them some important practical applications in human genetics. The term '**polymorphism**' refers to the situation in which, within a population, there are several or many variants, of a sequence at a given locus. Tandem repeats are important in forensic science and as genetic markers. We will deal with this aspect in Chapter 28, page 471.

Pseudogenes

Pseudogenes are another major component of junk DNA. They are 'fossilized' copies of genes that are no longer active or produce nonfunctional proteins or fragments of proteins.

Summary

DNA is a polynucleotide consisting of strands of deoxynucleotides linked by $5' \rightarrow 3'$ phosphodiester links between the sugar residues. Although RNA probably preceded DNA in early life as the genetic material, DNA is chemically more stable because of the lack of the oxygen atom on the $2'$ position of the deoxyribose and this is probably the reason for its present dominance in life, though RNA genomes exist in viruses.

DNA is a double helix of two antiparallel strands (they run in opposite directions) held together by complementary base pairing. It contains four bases, A, T, G, and C, but no U. T is the same as U except that it has a methyl group. Complementary base pairing by hydrogen bonding occurs between A and T, and G and C; this holds the two strands together. High temperatures cause strand separation due to hydrogen bond breakage. Reversal occurs on cooling, the phenomenon being known as strand hybridization or annealing. This is highly specific for pieces longer than about 15–20 nucleotides and is at the centre of DNA technologies (Chapter 28).

The bases in a double helix point to the inside of the molecule and the phosphate – sugar backbone to the outside with the edges of the bases visible in the two grooves of the double helix. This is known as the B form. The bases themselves have flat hydrophobic faces and are stacked on one another. Most DNA is in the B form but A and Z forms are possible in certain circumstances, their biological significance, if any, being unknown.

Each eukaryotic cell has about 2 m of DNA packed into the microscopic nucleus. There it is in the form of chromatin in which the double helix is wrapped around nucleosomes; these are octets of histone proteins. Between nucleosomes and linking them together are stretches of DNA 30–40 base pairs in length, forming a 'beads on a string' structure. This nucleosome – DNA fibre is further packed in complex looping arrangements. The maximum condensation occurs in mitotic chromosomes during cell division in which form the DNA is genetically inert. In *Escherichia coli* the chromosome is circular and the packing is not so highly structured as in chromatin and lacks nucleosomes.

In eukaryotes cell division occurs by mitosis (except for germ line cells). The process involves prophase, metaphase, anaphase and telophase with the resulting cells segregating by cytokinesis.

Germ line cells divide so that the resulting cells are haploid and after fertilisation a diploid cell is reformed. In meiosis there are two cell divisions, but only one replication (Fig. 21.11) and exchange of DNA between chromosomes occurs by crossover.

The role of DNA is to store information on the amino acid sequences of proteins, in the form of base sequences of genes. A gene is a small section of DNA, which is part of the large molecule of DNA constituting a chromosome and has no physical independence. The base sequences distinguish one gene from another. In addition to protein-coding genes, there are mobile genes in chromosomes as well as genes coding for ribosomal and transfer RNAs.

The mitochondrial genome is small and circular. It codes for only a small number of mitochondrial proteins, the rest being coded for by nuclear genes.

The sizes and organization of genomes have produced surprises. The amount of DNA is not proportional to the complexity of an organism. Amphibia have more DNA than humans per cell. Humans have about 30,000 genes, (or, by more recent estimates, 25,000) not the 100,000 previously estimated. *E. coli* has 4000, while the cress plant has 26,000 (close to the human figure).

The human genome has been sequenced. Only about 1.5% of the DNA actually codes for protein sequences and genes in total occupy only 20%, most of this being due to noncoding introns.

Half the DNA is repetitive and was regarded as 'junk' DNA of no known function. A change has occurred in thinking about junk DNA for large numbers of microRNA-coding genes have been discovered in it. There is evidence that the small RNA transcripts play essential parts in the determination of the phenotypes of complex organisms. Thus instead of protein-coding genes being the be all and end all in this determination, it seems that the small RNAs constitute an extra layer of epigenetic control. The view has been expressed by some that it may be the start of a revolution in molecular genetics.

Further reading

Earnshaw, W. C. and Pluta, A. F. (1994). Mitosis. *BioEssays*, **16**, 639–3.

Desai, A. (2000). Kinetocores. *Curr. Biol.*, **10**, R508. *A one-page guide.*

Sharp, D. J., Rogers, G. C., and Scholey, J. M. (2000). Microtubule motors in mitosis. *Nature*, **407**, 41–7.

Wittman, T., Hyman, A., and Desai, A. (2001). The spindle: a dynamic assembly of microtubules and motors. *Nat. Cell Biol.*, **3**, E28–E34.

Moens, P. B. (1994). Molecular perspectives and chromosome pairing at meiosis. *BioEssays*, **16**, 101–6.

Cooper, G. M. (1997). *The Cell: a Molecular Approach.* ASM Press.

Callandine, C. R. and Drew, H. R. (1992). *Understanding DNA: the molecule and how it works.* Academic Press.
A discussion of DNA structure, supercoiling, and DNA organization in the cell.

Rennie, J. (1993). DNAs new twists. *Sci. Am.,* **266**(3), 88–96.
An account of the 'unorthodox' genes – jumping genes, fragile chromosomes, expanded genes, edited mRNA, and nonstandard genetic code in mitochondria and chloroplasts.

Chambers, D. A., Reid, K. B. M., and Cohen, R. L. (1994). DNA: the double helix and the biomedical revolution at 40 years. *FASEB J.,* **8**, 1219–26.
Reviews a meeting to mark the 40th anniversary of the double helix. Biochemical nostalgia, but which summarizes the landmarks in the area and looks to the future.

Pennisi, E. (2000). And the gene number is . . . ? *Science,* **288**, 1146–7.
Betting on human gene numbers in 2000. Amusing, but gives perspectives on DNA research.

Dennis, C. and Gallagher, R. (Eds.) (2001). *The Human Genome.* Nature Publishing Group.

Levine, M. and Tjlan, R. (2003). Transcription regulation and animal diversity. *Nature,* **424**, 147–51.
Transcriptional complexity may be the answer to why humans have relatively few genes.

Klose, R. J. and Bird, A. P. (2006). Genomic DNA methylation: the mark and its mediators. *Trends in Biochem. Sci.,* **31**, 89–97.

Nucleosomes

Hewish, D. R. and Burgoyne, L. A. (1973). Chromatin substructure. The digestion of chromatin DNA at regularly spaced sites by an nuclear deoxyribonuclease. *Biochem. Biophys. Res. Commun.,* **52**, 504–10.
Chromatin was found to be digested in a nonrandom manner.

Kornberg, R. D. and Lorsch, Y. (1999). Twenty-five years of the nucleosome, fundamental particle of the eukaryote chromosome. *Cell,* **98**, 185–94.
Reviews the nucleosome story.

Junk DNA

Riddihough, G. (2002). Three 'Viewpoint' articles, and a review on The other RNA world. *Science,* **296**, 1259–1273.
Papers cover noncoding RNAs and RNA silencing.

Gibbs, W. W. (2003). Hidden genes. *Sci. Am.,* **289**, 28–33.
Genes that work through RNA rather than protein.

Mattick, J. S. (2003). Challenging the dogma: the hidden layer of nonprotein-coding RNAs in complex organisms. *BioEssays,* **25**, 930–9.
An early prediction of the importance of RNA in gene control.

Nudler, E. and Mironov, A. S. (2004). The riboswitch control of bacterial metabolism. *Trends Biochem. Sci.,* **29**, 11–17.
Summary of the recent progress identifying small RNA aptamers, called 'riboswitches'. Found in the leader sequences of many metabolic genes they can repress or activate their cognate genes at both transcriptional and translational levels.

Gusella, J. F. and MacDonald, M. E. (2006). Huntington's disease: seeing the pathogenic process through a genetic lens. *Trends Biochem. Sci.,* **31**, 533–40.

Greally, J. M. (2007). Genomics: encyclopaedia of humble DNA. *Nature,* **447**, 782–3.
Nature News and Views article.

No authors quoted. (2007). Identification and analysis of functional elements in 1% of the human genome by the ENCODE pilot project. *Nature,* **447**, 799–816.

Pennisi, E. (2007). DNA study forces rethink of what it means to be a gene. *Science,* **316**, 1556–7.
A commentary on the ENCODE project results which shows that genes may be far from compact assemblies but have regions scattered around the genome. Highly readable.

Problems

1 Write down the structure of a dinucleotide.

2 Ribonucleic acid (RNA) almost certainly evolved before deoxyribonucleic acid. Why do you think DNA evolved?

3 The flat faces of the bases of DNA are hydrophobic. Explain the structural repercussions of this fact on the structure of double-stranded DNA.

4 What is the main form of double-stranded helical DNA called? Is it a right- or left-handed helix? Approximately how many base pairs are there in a stretch of DNA that completes one rotation of the helix?

5 Explain what is meant by DNA chains in a double helix being antiparallel.

6 Explain in everyday language what is meant by a $5' \rightarrow 3'$ direction in a linear DNA molecule.

7 If you see a DNA structure simply written as CATAGCCG, what exactly does this means in terms of a double-stranded structure and the polarity of the two chains? Explain your answer.

8 (a) What is a nucleosome?
   (b) Describe an experiment that was an important clue to their existence.

9   What are Alu sequences?

10 All genes are sections of chromosomes that code for the amino acid sequences of proteins. Comment briefly on how this statement has to be qualified, especially in the light of recent discoveries.

11 Summarize the differences between mitosis and meiosis.

12 Which of the following is out of place – adenine, guanine, thymine, cytosine, uracil?

13 Describe the broad differences between the typical prokaryotic and eukaryotic genomes.

Chapter 22

DNA synthesis, repair, and recombination

DNA synthesis is simple in concept, but complex in practice. The mechanism was initially studied mainly in *Escherichia coli* but sufficient is now known of synthesis in eukaryotic cells to be sure that the processes are basically the same in both, if not in absolute detail.

Each time a cell divides, its DNA must be duplicated or, as is more usually stated, the chromosome(s) must be replicated, so that a complete complement of DNA can be given to each daughter cell. A human cell has about 6 billion base pairs in its total DNA (3.2 billion per haploid genome). The magnitude of the task of faithfully replicating these needs no emphasis. Even a single incorrect base in a gene may cause a protein with impaired function to be produced. The unavoidable minute proportion of errors that are not repaired are the feedstock of evolution, and, unfortunately, of genetic diseases.

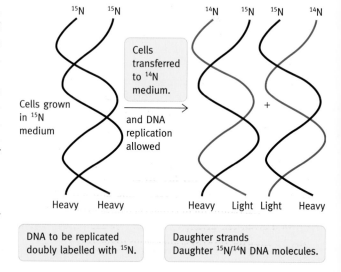

DNA to be replicated doubly labelled with ^{15}N.

Daughter strands
Daughter ^{15}N/^{14}N DNA molecules.

Fig. 22.1 Demonstration of semiconservative DNA replication by Meselson and Stahl. The DNA of cells was labelled by growing them in a medium in which the nitrogen source was ^{15}N, so that both strands of DNA were 'heavy'. They were then transferred to ^{14}N medium so that all subsequent DNA chains synthesized would be 'light'. The density gradient analysis indicated that, one generation after the transfer, each DNA molecule contained one 'heavy' and one 'light' strand. This is known as semiconservative replication. Continuation of the experiment for further generations confirmed the result. The red strands are newly synthesized.

Overall principle of DNA replication

We will go into the question of *how* DNA is synthesized in due course, but for the moment let us look at it at a general level.

A chromosome is double-stranded DNA. Its replication is described as **semiconservative** in that the two original strands, called parental strands, are separated and each acts as a template to direct the synthesis of a new complementary strand; each new double helix has one old and one new strand. This was established in the classic experiment shown in Fig. 22.1.

The basis of the replication is that of complementarity in that a G will base pair with C, and A with T, so that a base on the parental strand specifies which base is to be incorporated into the new strand as its partner. Since this copying process depends on Watson–Crick hydrogen bonding of base pairs, it follows that strand separation is essential to unpair the bases in double-stranded DNA, and make them available for base-pairing with incoming nucleotides.

The E. coli chromosome is circular. It contains about 4.6 million base pairs. The strands are initially separated at one particular point called the **origin of replication** and *two* **replication forks**, moving in opposite directions, synthesize DNA at the rate of up to a maximum of about 1000 base pair copies per second (the rate will vary with conditions in the growth medium, and the growth rate), with separation of parental DNA and synthesis of

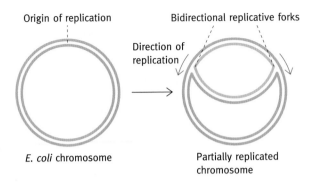

Fig. 22.2 Bidirectional replication of the *Escherichia coli* chromosome. Parental strands are blue; newly synthesized strands are red.

new DNA occurring at the same time (Fig. 22.2). The two forks meet at the opposite side of the circle.

Eukaryotic chromosomes are linear. In eukaryotes, a replication fork synthesizes DNA at about an average 50 base pairs copied per second – too slow by a huge margin for a single replicon (explained below) to synthesize the vast lengths of DNA in a chromosome in the time allowed for it in cell division. To cope with this, there are hundreds of origins of replication along the chromosome from which replicative forks work in both directions (Fig. 22.3). A **replicon** is a section of DNA where replication is initiated by a centre of origin. The multiple replicons of a chromosome fire off at different times. Within one cell cycle there are early-replicating and late-replicating sections. A vital requirement is that each does so once, and once only, in a given cell division cycle. Except for the number of initiation sites, the process is analogous to the *E. coli* situation.

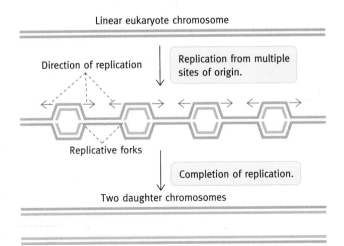

Fig. 22.3 Diagram of multiple bidirectional replicative forks in a eukaryotic chromosome.

Control of initiation of DNA replication in *E. coli*

Before cell division occurs, there must be a complete duplication of the chromosome. In *E. coli* cell division under optimum conditions follows about 20 minutes later. Exactly how cell division and DNA replication are coordinated is not understood. Protein synthesis and a critical enlargement of the cell are required. As already seen (Fig. 22.2), in *E. coli* there is a single point of origin of DNA synthesis called *oriC* at which replication commences bidirectionally.

The origin of replication has a specific base sequence, very rich in A—T pairs, presumably to facilitate strand separation. (We remind you that A—T pairs have two hydrogen bonds and G—C pairs three and, therefore, the former are less tightly bound together.) At the time of initiation, a protein referred to as DnaA binds in multiple copies to this region and causes strand separation. This permits the main unwinding enzyme (helicase), which works at each replicative fork, to attach and begin progressive unwinding of the strands in both directions. The **helicase** (or DnaB, the protein coded for by the gene *dnaB*) is referred to later when we describe the mechanism of DNA synthesis. The two strands of DNA are held together by numbers of hydrogen bonds, and the ATP-driven helicase enzyme is necessary to separate them for replication to proceed. The helicase is believed to move along one strand using ATP hydrolysis as energy source to displace the other strand and unwind the DNA. There is a mechanism (see below) to stop them coming together again prematurely.

Initiation and regulation of DNA replication in eukaryotes

As outlined in Chapter 2, in eukaryotic cells there is a strictly controlled cell cycle. DNA synthesis is confined to a period of time in the cell cycle called the S (for synthesis) phase. Different cells vary a great deal but, in cultured animal cells, the **S phase** typically takes about 8 hours out of a total cycle of 24 hours. Before the S phase is the G_1 phase, (G for gap in DNA synthesis), and afterwards the G_2 phase (Fig. 22.4). To proceed to cell division, a mammalian cell requires a mitogenic (mitosis- or cell division-producing) signal from other cells. This takes the form of protein signalling molecules that attach to surface receptors and transmit the signal to the interior of the cell. The latter process (that of cell signalling) is the subject of Chapter 27, and eukaryotic cell-cycle control is dealt with more fully in Chapter 31.

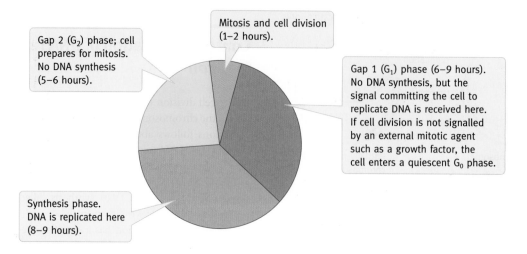

Fig. 22.4 The eukaryotic cell cycle. The duration of the cell cycle varies greatly between different cell types. The times given here are for a rapidly dividing mammalian cell in culture (24 hours to complete the cycle). See Chapter 31 for a more detailed account of the cell cycle.

Unwinding the DNA double helix and supercoiling

DNA strand separation by helicase presents topological problems. (Topology refers to arrangement in space of components relative to each other.) To explain these we must deal with the subject of DNA supercoiling.

Since duplex DNA has two strands in the form of right-handed helices, it has an inherent degree of twist, there being one turn of the helix per approximately 10 base pairs. A short piece of linear DNA that is free to rotate on its own long axis adopts this strain-free configuration, known as the **relaxed state**. Suppose instead that you clamped one end of the duplex so that it was not free to rotate and you gave an extra twist to the other end, so that the coil of the double helix is tightened – the number of turns of a given length of DNA is increased; that is, the number of base pairs per turn is decreased. It is now **positively supercoiled** or **overwound**. If you twisted in the opposite direction, the coil would be opened up – the number of turns per unit length would be reduced, or the number of bases per turn increased. The DNA would be **negatively supercoiled** or **underwound**. Both the underwound and overwound states are under tension and one way of accommodating the strain is for the DNA double helix to coil upon itself forming a **coiled coil** or **supercoil** (Fig. 22.5). You can illustrate supercoiling with a piece of double-stranded rope. Have someone hold one end or clamp it somehow, so that the rope cannot rotate freely, and twist the rope on its axis. Coils will form to take up the twisting strain. (Alternatively, look at a telephone cord which has been inadvertently twisted every time the phone has been picked up – it is a mass of supercoils.) If you release the end of the supercoiled rope it will spin back to the relaxed state. To determine

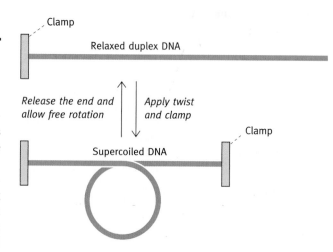

Fig. 22.5 The twisting of a piece of DNA that is not free to rotate induces supercoiling to accommodate the twisting strain. Cellular DNA is effectively clamped and is not free to rotate. If, somehow, free rotation is allowed, the supercoil will relax. If the applied twist is in the direction of unwinding, the supercoil will be negative; it will be positive if the applied twist is in the opposite direction.

whether a coil is positive or negative, look along the coil from either end. If the uppermost strand is turning to the left it is positive, if to the right it is negative (see Fig. 22.7).

What has this to do with DNA replication? DNA in the cell is not free to rotate on its own long axis; in *E. coli* the closed-circle chromosome effectively 'clamps' the DNA. In eukaryotes, the DNA is of such vast length, arranged in fixed loops (see Fig. 22.7) and attached to protein structures, that once again free rotation is impossible. But, separation of DNA strands demands that the duplex rotates. This causes overwinding – it generates positive supercoils ahead of the replicative fork and, as the helix tightens, further strand separation is resisted. If unrelieved, the tension would bring strand separation and DNA replication to a halt.

A simple experiment will convince you of this. If you take a short piece of double-stranded rope or string and pull the ends apart, the rope will spin, thus preventing the accumulation of positive supercoils, and the rope strands will separate completely. Now take a long piece of the same rope coiled on the floor, or have someone hold one end of a reasonably long piece, so that it cannot freely rotate, and try to pull the strands apart. Positive supercoils will snarl up and oppose the separating process and prevent any further separation. This would be the situation in DNA replication in the cell if something were not done about it.

It follows that, for DNA synthesis to proceed, the positive supercoils ahead of the replicative fork must be relieved and this, of necessity, involves the transient breakage of the polynucleotide chain.

How are positive supercoils removed ahead of the replicative fork?

A group of enzymes, known as **topoisomerases**, catalyse the process. They act on the DNA and isomerize or change its topology. There are two classes of topoisomerase, types I and II. We will deal with the principles of their mechanisms first, and then explain their roles in DNA replication.

In type I, the enzyme breaks *one* strand of a supercoiled double helix, which permits the duplex to rotate on the single phosphodiester bond of the partner strand, effectively introducing a swivel, or point of free rotation, into the DNA. After rotation has occurred, the enzyme reseals the duplex (see Fig. 22.6). The enzyme *does not hydrolyse the phosphodiester bond it attacks* – it transfers the bond from the deoxyribose-3′-OH to the —OH of one of its own tyrosine side chains. Since little energy change is involved, the process is freely reversible. The enzyme does not use ATP. A **topoisomerase type I** can relax only supercoiled DNA.

A **type II topoisomerase** breaks *two* strands of the DNA double helix transferring the bonds to itself, making the breakage of the polynucleotide chains freely reversible. The enzyme physically transfers the DNA duplex of a supercoil through the gap. (You can imagine that untangling string or a fishing line would be helped if you could cut a loop, transfer a strand through the cut and magically rejoin the ends without a knot.) Conformational change in the protein is involved, generated by ATP hydrolysis. In *E. coli*, the topoisomerase II is called gyrase; it introduces negative supercoils into the DNA. The mechanism of this is illustrated in Fig. 22.7. In this figure we use the removal of a single positive supercoil from a circular DNA molecule for purposes of illustration. In Fig. 22.7(b) the supercoil is positive. The gyrase cuts both strands of the lower duplex to form a gap (Fig. 22.7(c)) through which the front strand is transferred. It then reseals the cut, now at the front (Fig. 22.7(d)), creating a negative supercoil. ATP hydrolysis supplies the energy required in the physical transfer process. Gyrase actively inserts negative

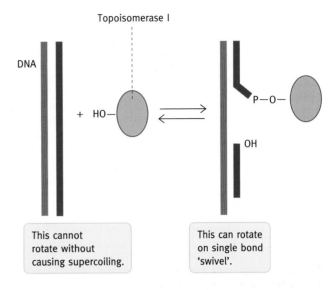

Fig. 22.6 Mechanism of topoisomerase I action. The enzyme breaks a phosphodiester link in the backbone of one strand of DNA by transferring the bond to a tyrosine —OH group in its own protein. The DNA can then be allowed to rotate on the single-bond 'swivel' in the partner strand. After rotation, the enzyme restores the original bond. Note that the broken bond is transferred to the enzyme; it is not hydrolysed so the process is freely reversible.

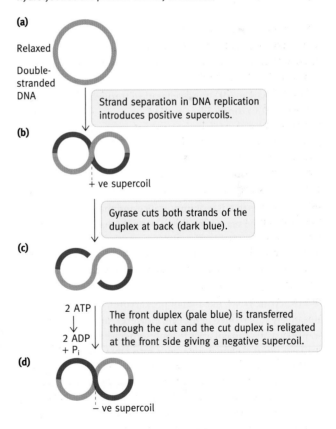

Fig. 22.7 Simplified mechanism by which gyrase in *Escherichia coli* neutralizes positive supercoiling by insertion of negative supercoils. Unlike topoisomerase I, ATP is required to provide energy for transfer of the duplex through the cut.

(a) DNA of chromosome (zero supercoil);
note that this strand is not free to rotate.

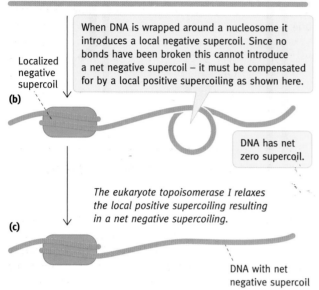

When DNA is wrapped around a nucleosome it introduces a local negative supercoil. Since no bonds have been broken this cannot introduce a net negative supercoil – it must be compensated for by a local positive supercoiling as shown here.

Localized negative supercoil

(b)

DNA has net zero supercoil.

The eukaryote topoisomerase I relaxes the local positive supercoiling resulting in a net negative supercoiling.

(c)

DNA with net negative supercoil

Fig. 22.8 A mechanism by which eukaryotic DNA becomes negatively supercoiled despite the absence of any enzyme capable of actively inserting negative supercoils such as the prokaryotic gyrase. Steps (a)–(c) are referred to in the text.

supercoils and, since this will relax positive supercoils, permits DNA synthesis to proceed.

Thus, in prokaryotic and eukaryotic DNA replication, the potential snarl-up of strand separation through the accumulation of positive supercoils is averted.

When DNA is carefully isolated from cells it is found to be negatively supercoiled. In relaxed DNA, the double helix has one turn per 10.5 base pairs; in cellular DNA it has about one turn per 12 base pairs – it is underwound. The degree of supercoiling is roughly comparable in the DNA of different cells, which suggests that it is of importance and that its generation is controlled. The reason for this is possibly that, in such a state, DNA strand separation occurs more readily than in the relaxed or positively supercoiled state. In *E. coli* the degree of underwinding will be a balance between topoisomerase I relaxing negative supercoils, and gyrase inserting them.

Eukaryotic DNA, like that of prokaryotes, is underwound or negatively supercoiled in the cell. However, unlike the situation in prokaryotes, no eukaryotic topoisomerase is known that can actively insert negative supercoils into DNA. How then is the underwinding achieved?

When chromatin is assembled, it is believed that the DNA winds around nucleosomes in such a manner that, in the local region in contact with the protein, it is in an underwound state. This is achieved by a left-handed coil. Since this nucleosome winding does not involve any bond breakage and since the chromosomal DNA cannot freely rotate, it follows that there cannot have been any *net* change in the supercoiling of the

DNA. Therefore, the local negative supercoiling at the nucleosome must be compensated for by positive supercoiling elsewhere so that the net change in the structure is zero (Fig. 22.8(b)). The eukaryotic topoisomerases I or II now relax the positively supercoiled section, thus achieving the insertion of a negative supercoil (Fig. 22.8(c)). A prokaryotic type of gyrase is thus not needed; the fact that prokaryotes do not have the nucleosome structures correlates with the need for their own type of gyrase. The antibiotic, **nalidixic acid**, inhibits bacterial gyrase, and is used to treat certain infections resistant to other antibiotics. Since humans do not have gyrase, nalidixic acid can be used to treat patients.

So far we have dealt with the broad aspects of DNA replication – its semiconservative nature based on Watson–Crick base pairing, with the cell cycle, with the initiation of replication, and with the mechanism of unwinding. We now turn to the mechanism of DNA synthesis. The enzyme(s) that catalyse this are called **DNA polymerases**; they polymerize nucleotides into DNA, using deoxynucleotide triphosphates as substrates.

The basic enzymic reaction catalysed by DNA polymerases

A series of facts first:

- There are three DNA polymerases in *E. coli*, called Pol I, II, and III – named in order of their discovery.

- The DNA synthesis occurring in the replicative fork is catalysed by Pol III or its eukaryotic equivalents, but Pol I also plays an essential role in DNA replication as well as in repair. Less is known of Pol II, but it is believed to be associated with certain types of DNA repair.

- The substrates for DNA polymerases are the four deoxyribonucleoside triphosphates dATP, dCTP, dGTP, and dTTP. These are synthesized in the cell as described in Chapter 20. The regulatory mechanisms in their synthetic pathways ensure that they are produced in adequate and coordinated amounts.

- The polymerase must have a DNA template strand to copy. 'Copy' is used in the complementary sense – a G on the template strand is 'copied' into a C in the new strand, and likewise A into T, C into G, T into A.

- A most important fact to fix in your mind: *a DNA polymerase can only elongate (add to) a pre-existing strand called a primer*. This **primer** may only be 20 nucleotides long but without it nothing happens. **DNA polymerases cannot start a chain – they cannot join together two free nucleotides**. The priming mechanism is described below.

- As illustrated in Fig. 22.9, the polymerase attaches a nucleotide to the 3′ free OH group of the end of the primer (or to the last nucleotide added), liberating inorganic

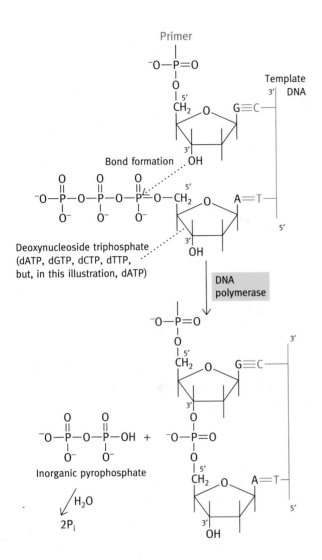

Fig. 22.9 The elongation reaction catalysed by DNA polymerase. The diagram shows the addition of an adenine deoxynucleotide from dATP to the 3′ end of the primer DNA strand, the base selected for addition being determined by the base on the template strand. Note that the synthesis is in the 5′→ 3′ direction; the chain is being lengthened in the 5′→ 3′ direction. The dotted line with arrow shows the attack of the 3′-OH on the α-phosphate.

pyrophosphate (PP$_i$). Hydrolysis of PP$_i$ increases the negative $\Delta G^{0'}$ value for the synthesis thus helping to drive the reaction. Incorporation of a nucleotide into the new strand of DNA involves the formation of hydrogen bonds with its template partner, with the liberation of energy, thus adding to the thermodynamic drive of the process.

- Which of the four deoxyribonucleoside triphosphates is accepted by the DNA polymerase is determined by the base on the parental strand being copied.

DNA synthesis always proceeds in the 5′ → 3′ direction with respect to the growing strand. Be sure that you know what this

means – that the growing DNA chain is being elongated in the 5′ → 3′ direction – a nucleotide is added to the free 3′-OH of the preceding terminal nucleotide. At the risk of overemphasizing the point, for it is important, when we talk of synthesis being in the 5′ → 3′ direction we always refer to the direction of elongation – the polarity of the *new* strand. We are *not* referring to the template strand, which has the opposite polarity. The polarity of DNA strands has been explained in the previous chapter (page 320).

How does a new strand get started?

DNA polymerases cannot initiate new chains and yet, at each origin of replication, new chains must be initiated. The solution to the question in the above heading is rather surprising in that DNA chains are initiated (primed) by RNA, whose structure and synthesis are described in the next chapter. However, for the moment, RNA has the same structure as single-stranded DNA except that the sugar is ribose and the base thymine (T) is replaced by uracil (U). RNA is synthesized by RNA polymerases by essentially the same basic chemical mechanism as outlined above for DNA except that ATP, CTP, GTP, and UTP, are used. But, in the present context, the vital difference is that RNA polymerases *can initiate new chains*. They can take two nucleotides and link them, a template being required here also; DNA polymerase cannot do this.

When a small piece of RNA primer (perhaps 10–20 nucleotides) has been synthesized by a special RNA polymerase called **primase**, DNA polymerase takes over and extends the chain. Primers are later removed.

We now come to yet another problem due to the antiparallel nature of DNA.

The polarity problem in DNA replication

It might be useful to turn back again to page 355 to make sure you understand DNA polarity. DNA is synthesized at each replicative fork which steadily progresses along the chromosome. In *E. coli*, there are two DNA polymerase III molecules involved in each fork, one for each strand, the two enzymes being linked together into a single asymmetric holoenzyme dimer. As shown in Fig. 22.10, the polymerase dimer molecules that are replicating the two strands must physically move in the same direction (that is, up the page as it were). However, one parent strand runs 5′ → 3′ in the reverse direction of fork movement and the other runs 5′ → 3′, the opposite way, in the direction of fork movement. Since DNA polymerase can synthesize only in the 5′ → 3′ direction, the template strand must run 5′ → 3′ in the *opposite* direction to that of synthesis. This is fine for the synthesis of one new strand (the left-hand one in Fig. 22.10),

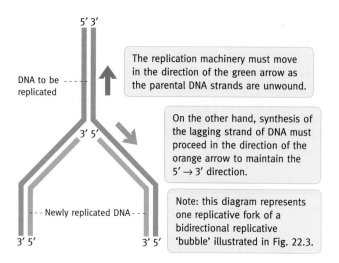

Fig. 22.10 The polarity problem in DNA replication.

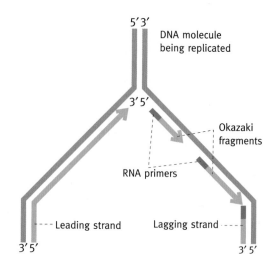

Fig. 22.11 Diagram of a replicative fork. The leading strand is synthesized continuously, while the lagging strand is synthesized as a series of short (Okazaki) fragments.

but what about the other? The other template strand runs $5' \rightarrow 3'$ in the direction of fork movement and hence DNA synthesis must run the opposite way. A seemingly impossible problem is illustrated in Fig. 22.10 in which the polymerase on the right-hand strand *must* physically move up the page, as it were, but synthesize DNA in the direction of down the page. The left-hand strand, with no problems, is called the **leading strand** and the other the **lagging strand**. (As a brief aside, relevant to Fig. 22.10, be careful not to fall into the trap of imagining that a linear piece of DNA is replicated from the end. It does not happen in that way but starts from the centre of a replicative 'bubble' in both directions as shown in Fig. 22.3. This misconception may easily arise because, to avoid having to show both replicative forks of a 'bubble', diagrams of the replicative fork usually show an inverted Y-shaped structure.)

How can the lagging strand initiate? In the case of the leading strand, primase lays down a single primer at the origin of initiation and the DNA polymerase proceeds from there until replication is complete, but the same will not suffice for the lagging strand. The solution is that, as the DNA unwinds, there is repeated initiation by primase, each primer being extended by the polymerase into a short stretch of DNA synthesis 1000–2000 bases long in *E. coli* and 100–200 in eukaryotes. The net result is illustrated in Fig. 22.11 but do not be worried about how this is achieved, for it still looks to be an impossible situation!

The short stretches of DNA attached to RNA primers on the lagging stand are called **Okazaki fragments** after their discoverer (Reiji Okazaki). This still leaves the original problem of how a DNA polymerase can synthesize DNA backwards while moving forwards – a physical or topological problem. It also leaves the lagging strand not as a single piece of DNA but a series of disconnected short pieces attached to RNA primers that must be made into uninterrupted DNA. Let us deal with the physical problem first.

Mechanism of Okazaki fragment synthesis

The basic principle in solving the physical or topological problem of how the lagging strand is synthesized in the $5' \rightarrow 3'$ direction, without the replicative machinery moving backwards, is simple. The lagging template strand is looped, so that for a short distance it is oriented with the same polarity as the leading strand template. The replicative machinery can therefore proceed in the direction of the fork and synthesize both new strands.

Although the principle is simple, the mechanical problem of how the loop system can move along the entire length of the parental strand and permit the synthesis of Okazaki fragments is not simple because it requires that the loop is reformed and enlarged at regular intervals, and a new RNA primer laid down at each stage.

This model, put forward by Arthur Kornberg, is shown in Fig. 22.12. As the loop enlarges and the replicative machinery moves forward, the polymerase will meet the 5'-RNA end of the previous Okazaki fragment. At this point the polymerase detaches and the loop falls away and a new one started. To understand this model we must first discuss the replicative machinery at the replication fork.

Enzyme complex at the replicative fork in *E. coli*

The functional complex of proteins and protein subunits at the replication fork is illustrated in Fig. 22.13. The key enzymes are the helicase to unwind the double helix, attached to which is the primase, which synthesizes RNA primers at intervals on the

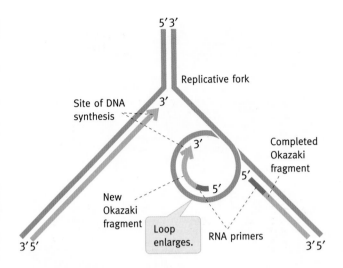

Fig. 22.12 Principle of Kornberg's 'loop' model for Okazaki fragment synthesis. (Pink arrowheads indicate DNA synthesis.) This model requires the loop to fall away when the new Okazaki fragment meets the old one. A new loop has then to be made. Both strands can be synthesized in this model in the required $5' \rightarrow 3'$ direction as the replication machinery moves in the direction of the fork. It is not possible to specify the precise mechanical details of the looping mechanism. A more detailed model is shown in Fig. 22.13.

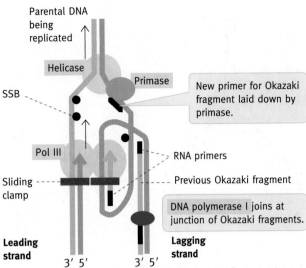

Fig. 22.13 A more detailed loop model. This is to solve the problem of how the polymerase can move forward (upwards on the page) but still synthesize DNA in the required $5' \rightarrow 3'$ direction when the template strand polarity demands that synthesis is in the opposite direction (downwards on the page). The action of DNA polymerase I is described in Fig. 22.16 and in the text below. The sliding clamps are described in the text and shown in Fig. 22.14. SSB, single strand binding protein.

lagging strand as it moves along, and the extremely complex Pol III. The helicase unwinding activity is ATP driven and moves along a DNA strand and, in doing so, separates the two strands of the double helix. The primase and helicase (*E. coli*) form a complex in the replicative fork known as a **primasome**. Finally, a single-strand binding protein (**SSB**), which has a high affinity for single-stranded DNA but with no base sequence specificity, binds to the separated DNA strands and stabilizes the single strands. As indicated, in *E. coli* there are two connected molecules of Pol III in the replicative fork, one synthesizing the leading strand and the other the lagging strand. They have the same core enzyme, but the holo-enzyme dimer is asymmetric with extra subunits present on the lagging strand side. Pol III has high processivity – once it locks on to a DNA strand it does not fall off but can go on replicating its template strand indefinitely without the risk of dissociating. A special mechanism has been devised to guard against premature dissociation. We will now describe this.

The DNA sliding clamp and the clamp loading mechanism

The sliding clamp is a ring-shaped protein structure surrounding the DNA. The annulus has a hole big enough for double-stranded DNA to slide through it easily but it cannot fall off the DNA (Fig. 22.14). This clamp structure is found both in *E. coli* where it is known as the β **protein** and in eukaryotes, known as **proliferating cell nuclear antigen** (**PCNA**), or in other words a protein in the nucleus necessary for proliferation. (The name derives from the manner of its discovery, antigen being an

immunological term, in the present context simply meaning a protein.) Although the structures of the two clamps look at first sight to be the same (Fig. 22.14), the *E. coli* one is a dimer whereas the PCNA has a three-subunit structure and the two have little sequence homology. The convergent evolution of different structures for the same function emphasizes its vital role. An additional protein complex is needed to load the clamps onto the DNA. These have been found in *E. coli* and in eukaryotes. In *E. coli* it is known as the γ **complex**. As explained, at the initiation of replication, a special RNA polymerase, the primase, lays down a short stretch of RNA against the template DNA strand. The γ complex, with a bound molecule of ATP, recognizes the short stretch of RNA primer/DNA hybrid and attaches a sliding clamp around it. It does this by seizing a circular clamp from solution which it opens and places around the RNA primer/DNA hybrid. The ring then snaps shut, this step being associated with ATP hydrolysis and release of the loading protein. The face of the clamp has a site for binding the Pol III DNA polymerase which is recruited from solution. The Pol III is now firmly attached to the DNA by the clamp so that it cannot fall off but is free to move along the DNA and replicate the template strand (Fig. 22.15). The clamp-loading complex places a clamp wherever there is a primer RNA laid down.

When the Pol III on the lagging strand meets the primer for the next Okazaki fragment, it must detach and re-initiate at the next primer RNA laid down by the primase. We turn now to the question of how the Okazaki fragments are tidied up into continuous DNA.

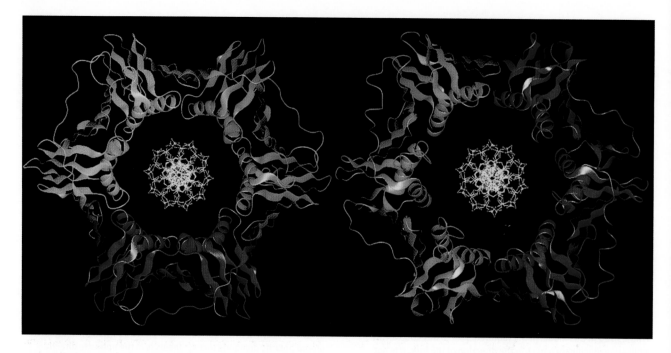

Fig. 22.14 Ribbon representations of the yeast and *Escherichia coli* sliding 'clamps'. **(a)** The yeast clamp (PCNA) that confers processivity on DNA polymerase δ is a trimer. **(b)** The *E. coli* clamp (β protein) that attaches to DNAPol III is a dimer. The individual subunits within each ring are distinguished by different colours. Strands of β sheet are shown as flat ribbons and α helices as spirals. A model of B DNA is placed in the centre of each structure to show that the rings can encircle duplex DNA.

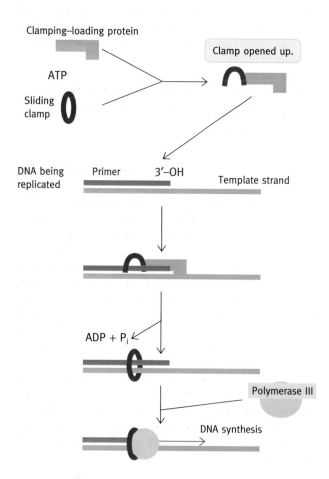

Processing the Okazaki fragments

In *E. coli*, when the Pol III reaches the RNA primer of the preceding Okazaki fragment, it disengages from the DNA, leaving a nick at the DNA/RNA junction. This is where DNA polymerase I (Pol I) comes in. Pol I is an astonishing enzyme with three separate catalytic activities on the same molecule.

If we look at the problem, as illustrated in Fig. 22.13, the separate pieces of DNA, the Okazaki fragments, must be converted into a continuous DNA molecule. Each piece starts with RNA which has to be removed, replaced with DNA, and the separate DNA pieces joined up. The Pol I attaches to the nicks, or breaks, between successive Okazaki fragments, and adds nucleotides to the 3'-OH of the preceding fragment, moving in the 5' → 3' direction; as with Pol III (or any DNA synthesis), nucleotide additions are always in the 5' → 3' direction. Since, as Pol I moves, it encounters the RNA of the next Okazaki fragment, the nucleotides of this are hydrolysed off. Thus, as it were, the front end of Pol I removes RNA nucleotides, and a site further back adds DNA nucleotides to fill the gap with DNA. The 'front' activity is a 5' → 3' **exonuclease** activity – 'exo' because it works on the

Fig. 22.15 (*opposite*) The steps in loading a sliding clamp for DNA synthesis in *Escherichia coli*. The clamps exist in solution as complete rings. The clamp-loading protein, in the presence of ATP opens up the ring, binds to the DNA wherever a primer laid down by primase awaits elongation, and snaps the clamp around the DNA to which a Pol III molecule attaches.

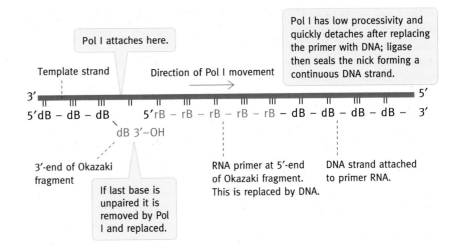

Fig. 22.16 Pol I actions in processing Okazaki fragments. dB, deoxynucleotide; rB, ribonucleotide. Removal of the last base, if unpaired, is described on page 342.

end of the molecule, 'nuclease' because it hydrolyses nucleic acids, and 5′ → 3′ because it nibbles away at the 5′ end of the RNA and moves in the direction of the 3′ end of the molecule. Note that Pol III does not have a 5′ → 3′ exonuclease activity like Pol I. Thus, Pol III cannot chop out the RNA primer when it meets the preceding Okazaki fragment. As stated, it disengages from the DNA at this point and hands over the job to Pol I.

The DNA Pol I has (unlike Pol III) *low processivity* – it does not hold on to the DNA template strand firmly and detaches relatively soon after the RNA has been replaced. It does not have the annular clamp to hold it on to the DNA. This is essential for otherwise it would go on replacing long stretches of the newly synthesized Okazaki fragments. When it detaches, a nick is left in the chain, which is healed by a separate enzyme, called **DNA ligase**.

This enzyme synthesizes a phosphodiester bond between the 3′-OH of one DNA fragment and the 5′-phosphate of the next, a process requiring energy. In some prokaryotes and all eukaryotes, ATP supplies this. The mechanism is that the enzyme (E) accepts the AMP group of ATP, liberating pyrophosphate, and transfers AMP to the 5′-phosphate of the DNA. Finally the DNA-AMP reacts with the DNA-3′-OH, releasing AMP and sealing the break.

$$E + ATP \rightarrow E—AMP + PP_i$$
$$E—AMP + \text{(P)}—5′ \, DNA \rightarrow E + AMP—\text{(P)}—5′ \, DNA$$
$$DNA\text{-}3′\text{-}OH + AMP—\text{(P)}—5′ \, DNA \rightarrow$$
$$DNA\text{-}3′\text{-}O—\text{(P)}—5′ \, DNA + AMP.$$

The AMP is linked to the enzyme and to the DNA *via* its 5′-phosphate group during the process.

In *E. coli*, instead of ATP, NAD⁺ is used. This is most unusual role for NAD⁺, which you have met only as an electron carrier. However, NAD⁺ can donate an AMP group just like ATP and, for some reason, *E. coli* uses this route.

What happens then, in summary, is the following. The DNA Pol I binds at the attachment site shown in Fig. 22.16 – at the nick. The polymerase adds DNA nucleotides to the 3′ end of the fragment on the left hand of the diagram, and moves in the 5′ → 3′ direction. The RNA, and some DNA, is nibbled away and the gap replaced by DNA. The Pol I detaches and a ligase joins the two fragments of DNA together. Thus, a series of Okazaki fragments becomes a continuous new DNA strand. Pol I also proofreads the nucleotide additions (see page 342).

The machinery in the eukaryotic replicative fork

The general principles of DNA replication in *E. coli* apply to eukaryotes, but there are differences in detail. Nine eukaryotic DNA polymerases are known, identified by Greek letters. The one corresponding to *E. coli*'s Pol III, responsible for synthesizing both strands of DNA, is polymerase δ. This has a 3′ → 5′ exonuclease activity for proofreading and, as with Pol III, it is held on to the template strand by an annular clamp. It therefore has high processivity. However, the initiation of strand replication is different in eukaryotes. The primase activity for both strands is contained in the DNA polymerase α subunit which lays down the RNA primer for the leading-strand synthesis and for Okazaki-fragment synthesis on the lagging strand. The same polymerase adds a short stretch of DNA, about 30 nucleotides long, to the primer and then hands over to polymerase δ which carries on the synthesis of both strands. Polymerase γ is responsible for mitochondrial DNA replication. The other polymerases are variously employed in Okazaki fragment processing and repair though their precise roles are not all fully elucidated.

The nucleosomes (page 321) must be displaced or otherwise coped with at the replicative fork as the polymerase moves along. Immediately behind the fork the nucleosomes reform so that the replicated DNA is immediately reassembled into the normal chromatin structure.

How is fidelity achieved in DNA replication?

An *E. coli* cell has to copy about 4.6 million base pairs each time it divides and this, at maximum growth rates, can occur every 20 minutes. A human diploid cell has to replicate 6.4 billion base pairs for each cell division. Because of the large numbers of centres of origin and the longer time of a cell cycle the polymerases can work at a slower rate than is the case for *E. coli* but the size of the problem is obvious. Even single mistakes, in which an incorrect base is incorporated, if at a critical point in a critical gene, can cause a genetic disease. The cells achieve an error rate of < 1 in a billion. How is this achieved?

When a deoxynucleotide triphosphate (dNTP) enters the active site of a DNA polymerase, it has to pair with the template nucleotide in a Watson–Crick fashion. If it does not it is incorrect. The free-energy difference between the pairing of a correct base and an incorrect one is not large enough to give sufficient discrimination. There has to be something else selecting the correct dNTP. The structure of the polymerase protein plays a large part, as in enzyme-substrate binding in general. The two main factors which operate in the polymerases are **geometric selection** and **conformational changes** in the polymerases.

Geometric selection. The geometry of nucleotide Watson–Crick base pairs (A/T and G/C) are almost identical in their shape, the distance apart, and angle of their glycosidic links. Given the emphasis that is placed on the specificity of base pairing, it may be somewhat of a surprise to learn now that other hydrogen bonding between non-Watson–Crick pairs can, in principle, be formed. They are not known to occur in DNA. The 'illegitimate' base pairs have a different geometry from Watson–Crick pairs. This aspect is further discussed on page 384 where it becomes relevant to protein synthesis. An incorrect dNTP pairing with the base in the template strand will therefore not have the correct geometric shape to fit the polymerase active site.

Conformational changes. When a correct dNTP pairs with the template nucleotide a large conformational change occurs in the polymerase active site, as happens with many enzymes. The DNA polymerase has an open structure in the absence of substrate. The entry of a correct dNTP causes the enzyme to close it up around the base pair. The conformational changes are 10,000 times slower with the entry of an incorrect dNTP. The selectivity achieved by this is high, with an error rate of one in a million, but this would still give an unacceptable mutation rate. Further improvement in selectivity is needed.

The next stage is for the polymerase itself to check the correctness of each addition.

Exonucleolytic proofreading

Many DNA polymerases including *E. coli* polymerases I and III and their eukaryotic counterparts have a catalytic activity that we have not mentioned so far. They have $3' \rightarrow 5'$ exonuclease (backward) activity which can chop off the last added nucleotide from the growing DNA chain. Note that this is quite different from the forward-acting $5' \rightarrow 3'$ exonuclease by which Pol I removes the RNA portions of Okazaki fragments. However, this backward chopping activity only occurs if the last added nucleotide was incorrect. Each time the enzyme adds a nucleotide it checks if it was correct; if not, it is removed and it has another try to replace it with a correct one. It is a proofreading mechanism that many DNA polymerases have.

The mechanism of this has been examined in DNA Pol I. The synthesis site and exonuclease sites are sufficiently close on the enzyme surface for the newly formed end of the DNA chain to slide from one site to another (Fig. 22.17). If this swings the last

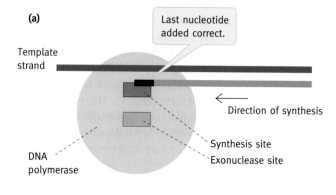

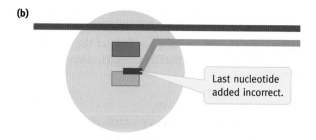

Fig. 22.17 Simplified diagram of the mechanism of exonucleolytic proofreading by DNA polymerase I. **(a)** Situation if last addition was correct. **(b)** Situation if last addition was not correct. An incorrect base on the growing DNA chain (mauve) in the lower diagram increases the chance of it detaching from the template base and swinging to the exonuclease site so that the error is removed. The polymerase can then replace it with the correct nucleotide when the chain swings back. If the last added base is correct, it is less likely to detach from the synthesis site.

added nucleotide from the synthesis site into the exonuclease site it is removed. An incorrect base is more likely to become detached from its template partner and slide from the synthesis into the hydrolysis site than is an end with the correct base. This **proofreading** mechanism occurs in the synthesis of the leading and lagging strands and in the processing of the Okazaki fragments. There is yet another fidelity check.

Methyl-directed mismatch repair

The number of mismatches that slip through into the newly synthesized DNA would still give an unacceptable rate of mutation. The cell therefore has a backstop mechanism to replace faulty bases even after the newly synthesized DNA has been released from the polymerase. The final fidelity mechanism in *E. coli* is called **methyl-directed mismatch** repair. It increases fidelity. If a mismatch has escaped the polymerase proofreading correction, the error will cause a distortion in the duplex chain illustrated diagrammatically below.

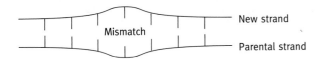

The error will be in the new strand and the repair system has to recognize which strand this is. The important point is that, in such a mismatch, the base on the parental (template strand), by definition, is correct and it is the complementary base on the new strand that is incorrect. The repair system must discriminate between the two strands, for, if it replaced, the template strand base it would confirm the mutation. It must remove the base from the new strand and correct it. How is this strand discrimination made? In *E. coli*, wherever there is a GATC sequence in the DNA, the adenine of this sequence is **methylated** by an enzyme in the cytoplasm. This does not affect base pairing or DNA structure. It takes some time after its synthesis for the new strand to be methylated and so, for this brief period, it is unmethylated. Thus the parental strand is methylated but the just-synthesized new strand is not. The repair system is little less complex than that for DNA synthesis. First the protein **Mut S** recognizes the mismatch distortion in the double helix; **Mut H**, which is an *endo*nuclease, is linked to Mut S by **Mut L**; then Mut H finds an *unmethylated* GATC on one of the duplex strands that identifies it as being newly synthesized and nicks it at this site (Fig. 22.18). This may be thousands of bases from the error. Then helicase, SSB, and an exonuclease cooperate to remove the entire section from the nick to beyond the error and pol III replaces it with new DNA and, in doing so, corrects the mismatch. DNA ligase completes the repair by sealing the nick. To correct one base, thousands of nucleotides may be replaced (Fig. 22.18). This error-correction system increases the fidelity of replication so that there is a final error rate of less than or equal to 10^{-10}. There is evidence that a mismatch repair also

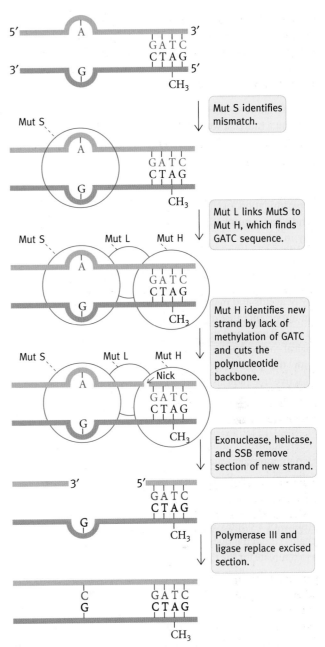

Fig. 22.18 Methyl-directed pathway for mismatch repair. If the GATC sequence is distant from the error, bending of the DNA could bring the two into proximity. Mut proteins are coded for by mutator genes whose inactivation increases DNA synthesis error rates.

occurs in eukaryotes. In humans, proteins corresponding to Mut S and Mut L proteins of *E. coli* are known, but nothing corresponding to Mut H has been found, suggesting that something other than methylation is used for strand identification. As might be expected, mutations in the genes for these in humans are associated with increased risk of colon cancer.

Repair of DNA damage in *E. coli*

The mechanisms described above ensure that DNA is replicated with the degree of accuracy needed to ensure continuity of cellular life. However, there is still a major problem. Chemical changes occur at a rate that would result in large numbers of mutations per day, in each cell, if there were not constant repair. Some of the damage change is spontaneous. The glycosidic link that binds the purine and, to a lesser extent, pyrimidine bases to the deoxyribose moieties is somewhat unstable so that depurination and depyrimidation occur spontaneously – numbers of purines (and a lesser number of pyrimidines) break off the DNA every day in a human cell. In addition, cytosine and adenine occasionally chemically deaminate to become uracil and hypoxanthine, respectively (see page 345 and Fig. 20.7 for structures).

The DNA of all cells is also subject to 'insults' by a wide variety of agents and many of these cause carcinogenic mutations. An important cause of damage is oxygen free radicals generated in cells. These, and the mechanism by which they damage biological molecules, have been described on page 488, together with the protective mechanisms developed against them. Reactive free radicals are also generated by ionizing radiation. UV light is well known to cause cancers by cross-linking adjacent pyrimidine bases. The best known are thymine dimers, but all four types of pyrimidine dimer can be formed. A variety of other abnormal molecules can be produced by UV light. Certain molecules such as aflatoxins and other carcinogens are activated by the P450 system to form reactive molecules that attack and modify DNA.

It will be relatively infrequent for damage to happen to *both* strands of a duplex DNA molecule at the same place in both chains (though it does happen), so that the other strand can act as a template for the repair of the damaged part. A variety of systems exist in cells to repair DNA damage for, of all things, this is the area where maintaining the integrity of a molecule is of paramount importance. When other molecules such as proteins are damaged they are simply destroyed, but DNA must be repaired at all costs (or, if not, in a complex animal the whole cell must be destroyed by apoptosis before it develops into a cancer; see Chapters 32, 33).

- **Direct repair** Exposure of DNA to UV light can result in the covalent linking of two adjacent thymine bases (on the same strand), forming a **T dimer** and other pyrimidine dimmers. (The structure is shown in two dimensions, side by side for clarity, rather than one on top of the other.)

T dimer

In *E. coli*, the abnormal bonds are cleaved by a light-activated mechanism that restores the two thymine moieties to their original form. A photo-activated enzyme called photolyase is involved. Another direct repair system removes alkyl groups on bases (which are formed by some mutagenic agents). The 'suicide enzyme' that accepts the alkyl group in doing so destroys its own action. It is more of a specific protein reagent than an enzyme since it is changed in the process, unlike a true catalyst.

- **Nucleotide excision repair** Lesions that distort the double helix such as a T dimer are also repaired by the excision of a short stretch of nucleotides, including the lesion, followed by its correct replacement, the opposite strand serving as the template for this. In *E. coli*, an unusual endonuclease called **exinuclease**, (or the uvrABC complex (unrelated to ABC transporters) after the three genes coding for the enzyme), cuts the DNA on both sides of the lesion. It then removes a single-stranded section of 12–13 nucleotides (Fig. 22.19). DNA polymerase I adds nucleotides to the 3' end of the nicked chain and ligase heals the nick. The system depends on it being possible to recognize which strand of the DNA is faulty.

- **Base excision repair and AP site repair** Deamination converts cytosine to uracil and adenine to hypoxanthine. DNA glycosylases recognize the abnormal bases and hydrolyse them off, leaving AP (apurine or apyrimidine) sites in which the deoxyribose moiety has no base attached to it (Fig. 22.20). AP sites can also be formed spontaneously, since the purine-deoxyribose link especially is somewhat unstable. Repair of AP sites involves nicking of the polynucleotide chain adjacent to the lesion followed by replacement of the section containing the latter by DNA polymerase I and sealing by ligase.

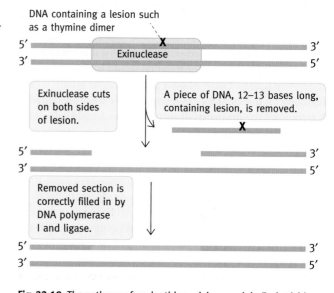

Fig. 22.19 The pathway of nucleotide excision repair in *Escherichia coli*.

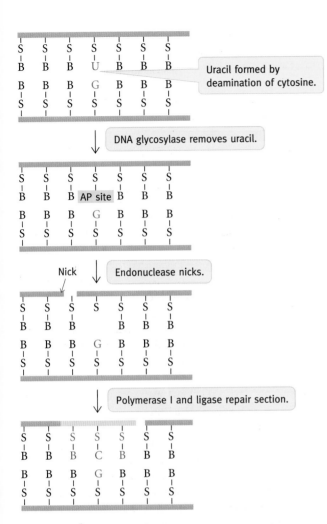

Fig. 22.20 AP site formation and repair. In the example given, the site is created by removal of a uracil by a glycosylase, but sites are also formed by spontaneous hydrolysis of purine bases (and, to a lesser extent, of pyrimidine bases) from the nucleotide. S, sugar; B, base.

The need to remove uracil formed from cytosine explains why DNA has T, instead of U. Remember that T is, in essence, a U that is tagged in DNA for identification purposes with a methyl group. If DNA *normally* contained U, it would be impossible to distinguish between a U that should be there and an 'improper' U, formed by deamination of C.

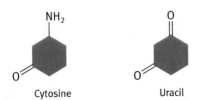

Cytosine   Uracil

Using T in DNA, instead of U, solves the problem. (As described in the next chapter, U can be used in RNA, because RNA has a relatively short lifetime, is much smaller, and errors

do not have the same long-term consequences. Hence RNA is not repaired.)

Repair of double-strand breaks

Potentially disastrous forms of DNA damage are double-strand breaks, which can be due to ionizing radiation and other agents. This type poses an extra repair problem because there is no partner strand to act as a template for the repair as occurs in the systems so far described. Two methods have evolved. In the end-joining mechanism the cut ends are simply ligated together. This may change the sequence slightly due to a nucleotide or so being trimmed off at the cut ends, but unless it occurs at a critical site this will not matter (only 1.5% of human DNA actually codes for protein).

The other method is more accurate. It uses homologous recombination with the other homologous undamaged chromosome of a diploid organism to direct the repair. This involves a complex procedure (not described here) in which resynthesis of excised sections occurs.

Telomeres solve the problem of replicating the ends of eukaryotic chromosomes

Eukaryotic chromosomes are linear and this poses a problem not encountered in the replication of circular chromosomes of *E. coli.*

Consider the replication of the chromosome represented in Fig. 22.21 as a very short one for diagrammatic convenience. It is shown as being replicated (in a bidirectional manner) from a single initiation site in its centre (but remember that a real eukaryotic chromosome has many such sites). The 5′ ends of each strand are fully replicated by leading-strand synthesis. This is not true of the lagging 5′ ends, because the synthesis of the end Okazaki fragments requires RNA primers to be laid down, as shown, on the 3′ ends of the template strands. When the primers are removed (as believed to occur), it leaves these ends unreplicated and no mechanism exists by which it could be replicated by the DNA synthesizing machinery that we have described so far, since all DNA synthesis requires a starting primer. To fill in the missing parts would require RNA primers but there are no templates against which they could be laid down. This means that, at each cell division, chromosomes would become progressively shorter, on average in a vertebrate, by about 100 nucleotides per cell division. A more potentially disastrous situation could hardly be imagined.

The mechanism of DNA synthesis means that incomplete replication of linear double-stranded DNA *cannot* be avoided. The solution adopted is that, at the ends of eukaryotic chromosomes, stretches of special DNA are attached which have no informational role, called **telomeric DNA**; the ends of the

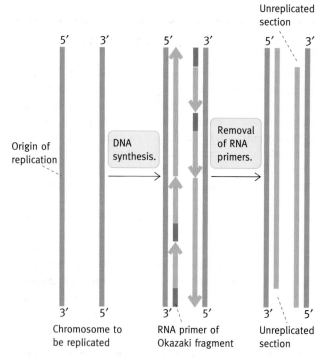

Fig. 22.21 The shortening of linear chromosomes by replication. For diagrammatic convenience the bidirectional replication of a very short piece of DNA is represented. It should be noted that primer removal from Okazaki fragments is a continuous process – it is represented here, for clarity, as occurring as a separate event. A typical chromosome will have multiple origins of replication. The pink lines represent new DNA synthesis; the green lines the RNA primers of Okazaki fragments.

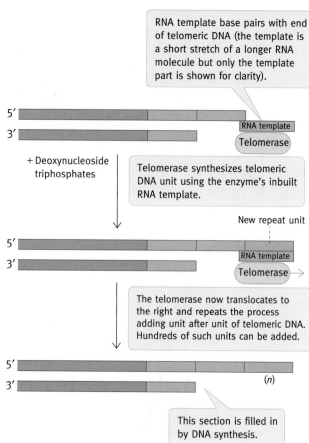

Fig. 22.22 Mechanism by which telomerase synthesizes telomeric DNA. The blue lines indicate the chromosomal or informational DNA (the 'real' chromosome) and the red lines the pre-existing telomeric DNA at one end of a chromosome. The telomerase has an inbuilt short RNA molecule that contains the sequence complementary to the repeating unit characteristic of the species. The enzyme becomes positioned with the RNA pairing with the terminal bases of the pre-existing telomere and adds one repeating unit of TTAGGG (in the case of humans) one base at a time to the G-rich strand. Synthesis is, as always, in the $5' \rightarrow 3'$ direction. The enzyme moves so that the RNA template is now paired with the end bases of the new repeating unit and a further unit is added, and so on. The newly synthesized telomeric DNA acts as the template for filling in the opposite strand, using conventional RNA priming, so that the telomere is double-stranded.

chromosomes containing it are called **telomeres** (Fig. 22.22). The lagging strand ends will still not be replicated by the DNA synthesis machinery so far described, but it no longer matters for it is only 'sacrificial' DNA that is lost. The 'real' chromosome is fully replicated with the primer at the 3' end of the template strand being laid down against telomeric DNA. In many rapidly dividing cells, the telomere is added to so that the chromosome is never at risk. This, however, does not occur in human somatic cells (see below).

How is telomeric DNA synthesized?

A telomere consists of repeating short stretches of bases – the repeated sequences vary between species. In humans there are hundreds of repeats of the TTAGGG sequence. The enzyme **telomerase** adds these sequences one after the other to the 3' end of pre-existing telomeric DNA and so extends the latter (all chromosomes have telomeres to start with). Telomerase has two remarkable features:

- it uses RNA as the template for DNA synthesis – it is a **reverse transcriptase** (page 347)
- it carries its own RNA template in its structure.

This RNA carries the sequence complementary to several repeats of the TTAGGG sequence (in humans). It hybridizes to the end of the overhang (Fig. 22.22), which positions the template bases for the repeating unit correctly, and a DNA unit is then added, one base at a time. When this is done the enzyme moves along and hybridizes to the end of the new repeating unit and thus the telomere is constructed in a discontinuous manner. The enzyme extends only one strand of the DNA. This is made double stranded by conventional lagging-strand DNA synthesis; in this an RNA primer is laid down and the DNA filled in as an Okazaki fragment, which is processed. Removal of the RNA primer results in the strand being shorter than its partner and explains the overhang referred to earlier.

The necessity for telomeres has been demonstrated by the use of **yeast artificial chromosomes (YACs)**. These contain the three types of DNA essential for chromosome replication – centromeres, sites of origin, and telomeres. It was shown that YACs are correctly maintained for generations when inserted into yeast cells but that, when they lack telomeric ends, they disappear in time from the cells. YACs are used as expression vehicles to produce wanted proteins (page 467).

Telomeres are not present in prokaryotes, for in a circular chromosome there is always a DNA template available for priming so chromosome shortening does not occur.

Telomere shortening correlates with ageing

The existence of telomeres in eukaryotes has profound significance. Telomerase occurs in germ cells, which are continuously dividing so that lengthening of the telomeres is necessary to compensate for the incomplete replication. However, in somatic cells where cell division occurs only to replace dead cells or wound healing, additions to the telomeres does not occur and, indeed, telomerase is believed to be absent in such cells. An implication is that somatic cells receive their initial 'ration' of telomeric DNA to suffice for the lifetime of that cell and its progeny, and, during life, the telomeres of most cells become shorter with age. In specific cases, this has been observed. It may be a factor in determining the lifespan of species.

Somatic cells can be transformed to the rapidly dividing state if they become cancerous; such cells in effect become immortal – there is no limit to their ability to divide in culture. It is significant that telomerase is found in most cancer cells. However, some immortal cancer cells do not have telomerase; it turns out that these have an alternative means of lengthening chromosomes, a point of importance in considering telomerase as a likely target for cancer therapy.

Telomeres stabilize the ends of linear chromosomes

Quite apart from the problem of replicative shortening of chromosomes, linear chromosomes face another problem. At the telomeric ends of eukaryotic chromosomes the 3′ end of one strand overhangs the 5′ end of the complementary strand. (See above for an explanation of this.) Free DNA ends are likely to be mistaken by the cell for damaged DNA and maybe attacked by repair systems or nucleases. The ends have to be protected. It is now believed that a long stretch of the double-stranded telomere loops back on itself and the single-stranded overhang is tucked into the double-stranded DNA by single-strand invasion of the telomeric duplex forming a short triple-stranded structure and a D loop (see Fig. 22.25 for an explanation of these terms). While this neatly explains how the telomeric ends are securely tied up into a protective loop it creates the (unanswered) problem of exactly how telomerase is able to access the free end to elongate it.

DNA damage repair in eukaryotes

We have described above the errors that can arise in DNA after the synthesis is completed resulting from the inherent chemical instabilities of the molecule, and damage from radiations and chemical attacks. A human cell would acquire unacceptably high rates of mutations were it not for the DNA repair mechanisms. Analogous mechanisms to those in *E. coli* are present in eukaryotic cells. The dealkylating 'suicide enzymes' also occur in cells of higher organisms. In humans with the genetic disease **xeroderma pigmentosum**, normal excision repair of UV-generated pyrimidine dimers is faulty and exposure to sunlight causes cancerous skin lesions.

Genes for proteins corresponding to the mismatch-repair proteins, Mut S and Mut L of *E. coli* (page 343), have been found, and their mutations are associated with one of the most common cancers called **hereditary nonpolyposis colorectal cancer (HNPCC)**, also known as **Lynch syndrome**.

Replication of mitochondrial DNA

In humans each mitochondrial genome has 16,600 base pairs that code for 24 RNAs and 13 proteins. (Most mitochrondrial proteins are imported into the organelle.) Mitochondria have multiple copies of double-stranded circular DNA. Replication of individual genomes within an organelle seems to occur at random, and is not synchronized with the cell cycle.

There are few or no DNA-repair systems in mitochondria, presumably reflecting the fact that the chromosome is very much smaller with fewer chances for errors. Also, since there are large numbers of mitochondria per cell, a small proportion of faulty ones has less significance. The apparent absence of repair, however, means that mutations in mitochondrial DNA may accumulate with time and it has been suggested that this is another component of the process of ageing.

Inheritance of mitochondrial genes is non-Mendelian and maternal, since mitochondria are donated *via* the ovum only. A large number of genetic diseases are now known to be caused by mitochondrial mutations.

DNA synthesis by reverse transcription in retroviruses

Some viruses use a protein instead of RNA for DNA chain-synthesis priming, and, as already indicated, retroviruses have a fundamentally different mechanism in that the RNA genome is copied into DNA, which we will now discuss.

RNA **retroviruses** are of much current interest; that causing AIDS is one example. They have a single-strand RNA genome.

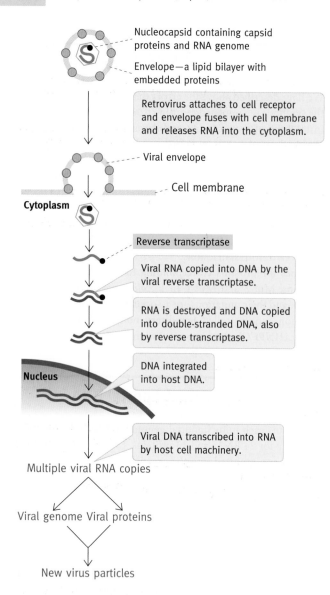

Fig. 22.23 Replication of a hypothetical retrovirus.

Typically the retrovirus particle or virion carries within itself a few molecules of an enzyme whose discovery caused initial disbelief, to be followed by the award of a Nobel Prize in 1975 to its discoverers, David Baltimore, Renato Dulbecco, and Howard Temin; it is called **reverse transcriptase**. Before its discovery it was, of course, known that DNA directs RNA synthesis, but the accepted dogma was that the reverse never happened. Viral reverse transcriptase is an amazingly versatile enzyme; when a retrovirus infects a cell, the viral reverse transcriptase copies the RNA strand into DNA. As with all DNA synthesis, a primer is necessary. Retroviruses use a host cell transfer RNA (tRNA) molecule (page 383), which hybridizes to the start of the template strand, and it carries the appropriate tRNA in the virion particle. The RNA/DNA hybrid formed by reverse transcription is converted to single-strand DNA by RNA hydrolysis,

an enzyme activity also present in the reverse transcriptase molecule. The single-stranded DNA is then copied by the same enzyme to form double-stranded DNA (called **proviral DNA**) (Fig. 22.23). It is incorporated into host DNA by an integrase, also carried in the virus. Once in, it is replicated along with the host DNA chromosome. For the production of new retrovirus particles, the proviral genes (that is, viral genes in the host chromosomes) are transcribed into RNA transcripts that direct synthesis of the proteins needed for new virus particle assembly.

Homologous recombination

Genetic recombination involves the *in vivo* rearrangement of the DNA of chromosomes within cells. This is different from recombinant DNA technology described in Chapter 28, in which the recombinant molecules are created artificially in the test tube. Genetic recombination occurs naturally in living cells. The main type is **general or homologous recombination** between separate chromosomes (DNA duplexes) at a point where there are homologous sections of DNA – where sections of the base sequences of the two are largely the same. There is another quite different type of recombination known as **site-specific recombination** involved in chromosomal rearrangements involved in antibody production (see page 496).

The continual reassortment of genes by homologous recombination (coupled with selection) means that new combinations of genes may be assembled to be tested by evolution. A most important homologous recombination to produce genetic diversity occurs in bacteria during conjugation and in eukaryotic meiosis.

The result of homologous recombination is illustrated in Fig. 22.24; in this, you see two chromosomes (both are DNA duplexes) with a stretch of DNA homologous in base sequence in the two. They come together at this point and exchange sections of duplex DNA as shown in the diagram. The recombinant process produces a very small patch of heteroduplex DNA if the two homologous pieces are slightly different at the crossover point. This hybrid DNA patch could indeed cause recombination within this area, but the main event is that the two arms on either side of the cross-over point are exchanged, producing extensive swapping of genes between the two parent chromosomes.

Mechanism of homologous recombination in *E. coli*

We can conveniently divide the process of homologous recombination into two parts:

- the linking together of two chromosomes by DNA strand exchange
- the separation of the pair in such a way as to produce genetic recombinants.

Fig. 22.24 Homologous recombination which occurs *via* cross-over junctions as described in the text. Genetic recombination is the resultant exchange of chromosome arms or sections. This is caused by the formation of hybrid DNA sections at the site of cross-over junctions (described later). It is important not to confuse the exchange of DNA strands at the cross-over junction with the exchange of chromosome arms. The possibility of this confusion is heightened because, for space reasons, subsequent diagrams are confined to the cross-over mechanism with the long arms not shown. A, a, B, and b represent allelic genes on the two chromosomes. (Gene alleles represent the same gene but differ in base sequence such that the proteins expressed are slightly different, or are expressed differently.)

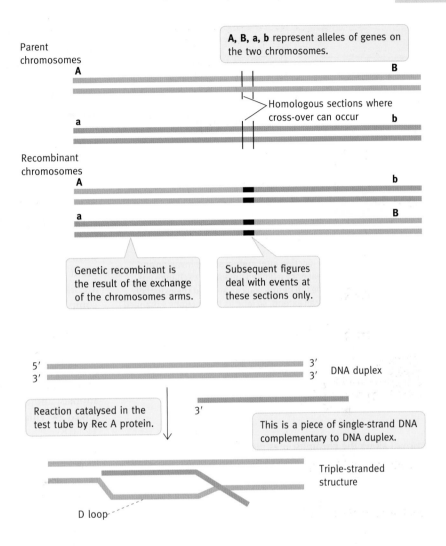

Parent chromosomes

A, B, a, b represent alleles of genes on the two chromosomes.

A   B

Homologous sections where cross-over can occur

a   b

Recombinant chromosomes

A   b

a   B

Genetic recombinant is the result of the exchange of the chromosomes arms.

Subsequent figures deal with events at these sections only.

Fig. 22.25 Single-strand invasion. This is to illustrate a somewhat remarkable reaction that can happen *in vitro* between a free piece of single-stranded DNA and a piece of duplex DNA, one strand of which is homologous with the single strand. The single-stranded DNA base pairs with one strand of the duplex, displacing it, and forming the D-shaped loop. The process requires that the single-stranded invading molecule is homologous to the duplex DNA, otherwise base pairing could not occur. This reaction requires the presence of *Escherichia coli* Rec A protein. ATP is hydrolysed in the process of invasion.

5′   3′   DNA duplex
3′   3′

Reaction catalysed in the test tube by Rec A protein.

3′

This is a piece of single-strand DNA complementary to DNA duplex.

Triple-stranded structure

D loop

The molecular mechanism of homologous recombination is more fully established in *E. coli* than in eukaryotic cells, but sufficient is known to make it likely that the two classes of cells have much in common.

We will first describe a basic reaction in all homologous recombination – **single-strand invasion**. This has been established in experiments *in vitro* using proteins isolated from *E. coli* and pieces of DNA. It involves the invasion of a DNA duplex molecule by a single strand of DNA complementary in base sequence to one of the duplex strands. The 3′ end of the single strand inserts itself into the duplex and, by hybridizing to the complementary strand, displaces the original duplex strand forming a triple-stranded D loop structure, so called because of its resemblance to the shape of the letter (Fig. 22.25).

Single-strand invasion requires energy and is thus catalysed in *E. coli* by an ATP-using enzyme, RecA, though other proteins are involved. RecA-mutants, lacking the gene for this protein, are deficient in recombination. Multiple molecules of RecA, each with an attached ATP molecule, bind to the single invading DNA strand. The RecA with bound ATP has a high affinity

for DNA. The single strand now invades the duplex and searches for a complementary sequence with which it base pairs, forming the D loop structure. ATP hydrolysis by the RecA causes loss of affinity of the enzyme for duplex DNA from which it dissociates. We now have to see how this single-strand invasion reaction participates in recombination within cells.

Formation of cross-over junctions by single-strand invasion

A model to account for homologous recombination was put forward by Robin Holliday. The first stage is the pairing of homologous sections of DNA alongside each other. Since single-strand invasion requires a strand of DNA with a free 3′ end, one strand of each DNA duplex is nicked by an endonuclease and the nicked strands mutually invade the homologous duplexes leading with their 3′ ends. This results in the formation of limited stretches of hybrid DNA where the invading strands have displaced the original partners. The nicks are sealed resulting in the formation of the cross-over junction shown in Fig. 22.26, known as a **Holliday junction**. As explained in the figure, the cross-over junction randomly

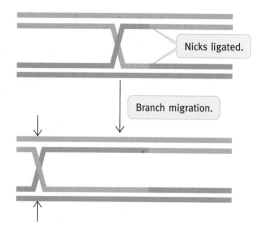

Fig. 22.26 A cross-over junction after mutual strand invasion and ligation of nicks. Note that once a limited amount of exchange has occurred, as shown in the top figure, you can see that the extent of the cross-hybridization can extend, so long as the sections exhanged are homologous so that hybridization can occur. In doing so, the crossover junctions are moved; this is known as branch migration. The migration can occur to the limits of the homologous sections. It should be emphasized that the diagram represents only the small area which will form the hybrid DNA section shown in Fig. 22.24, not the long arms of the chromosome.

moves along the chromosomes just as long as there is homology. This increases the length of the hybridized sections of the invading strands (see lower part of Fig. 22.26 and the legend) and has the result of moving the cross-overs along (known as **branch migration**).

Separation of the duplexes

The DNA duplexes bound together by the Holliday junction must now be separated. If you examine Fig. 22.26 it may seem that the obvious thing to do is to cut the crossed-over strands and ligate the breaks; this will separate the duplexes and produce a heteroduplex patch at the junction site, but the genes on the arms will be exactly as before on the two chromosomes. Genetic recombination is not achieved. (If the base sequences in the homologous patches were not identical then there will be some gene conversion or alteration but this is confined only to the local heteroduplex area.) To obtain the wanted genetic recombination, the duplexes are separated by cutting the *noninvading* strands, which has the result of swapping the chromosome arms (see Fig. 22.24). To achieve this the cell rearranges the Holliday junction in three dimensions, which involves turning over half of the molecule to uncross the strands, a process known as **isomerization**. The result is a structure in the form of a cross. The noninvading strands are cut at the arrowed point (Fig. 22.26) and then the cut ends on the upper strand are ligated to their opposite ones on the lower strand. (The complex manipulations involved are not described here.) This separates the duplexes and effects exchange of the arms.

Recombination in eukaryotes

In **meiosis**, in which haploid sperm and eggs are produced in animals, and corresponding gametes in plants, chromosomes linked by **chiasmata** (strand cross-overs) are seen (see Fig. 21.11(b)). The mechanism has not been resolved fully. However, a eukaryotic homologue of *E. coli* RecA, known as RAD51 protein, has been identified in humans. It is similar to RecA structurally and in its biochemical properties. It is believed to be involved both in homologous recombination and in double-strand break repair. From this, and the finding of RecA homologues in yeast, it would seem that much of the mechanism of homologous recombination and chromosome break repair have been strongly conserved throughout evolution, which underlines their importance.

Summary

DNA synthesis is catalysed by DNA polymerases that require four deoxynucleoside triphosphates, a template or parental strand to copy, and a primer because they cannot initiate new chains. The primer is a short RNA molecule to which deoxynucleotides are added. Synthesis starts at a site of origin on the chromosome where strand separation occurs.

E. coli has a single site of origin and eukaryotic chromosomes have hundreds. As the polymerase proceeds, a helicase separates parental strands, which produces supercoiling ahead of it. This is removed by topoisomerases. The problem of replicating the antiparallel strands by the asymmetric polymerase dimer while maintaining the $5' \rightarrow 3'$ direction of synthesis is achieved by a looping mechanism and discontinuous synthesis of one strand followed by processing of the separate Okazaki fragments into a single chain.

Fidelity of replication is achieved by several means. The polymerase selectively accepts triphosphates that form Watson–Crick base pairs with the template nucleotides; the free-energy difference between the formation of a correct and an incorrect pair is not sufficient to give sufficient discrimination. Correct base pairs have a shape, which differs from incorrect ones, and this plays an important part in the selection process. The DNA polymerase responds to correct pairs by an allosteric change. They also proofread by removing and replacing the last added nucleotide if it is incorrect.

DNA of cells is subject to continual insults from radiation and chemical instability, for which there is a variety of repair mechanisms.

The mechanism of synthesis of DNA inevitably means that linear eukaryotic chromosomes are shortened on each replicative round. To prevent genetic damage and to avoid them being mistaken for double-stranded breaks and 'repaired', the ends are protected by telomeres, consisting of repetitive DNA added by the enzyme telomerase. This enzyme is not present in somatic cells so their telomeres shorten with age and this limits the number of cell divisions possible. Stem cells and cancer cells that replicate indefinitely have telomerase to maintain telomere length or some cancer cells may develop an alternative telomere-lengthening mechanism.

Retroviruses such as HIV have an RNA genome but on infection replicate their RNA as DNA by a reverse transcriptase.

Genetic recombination involves the *in vivo* rearrangement of the DNA of chromosomes within cells. The main type is **general or homologous recombination** between separate chromosomes (DNA duplexes) at a point where there are homologous sections of DNA (Fig. 22.24). The continual re-assortment of genes by homologous recombination (coupled with selection) means that new combinations of genes may be assembled to be tested by evolution.

Further reading

Replication

Li, J. J. (1995). Once and once only. *Curr. Biol.*, 5, 472–5.
Discusses the essential link between the eukaryotic cell cycle and initiation of sites of origin of DNA replication.

Wyman, C. and Botcham, M. (1995). A familiar ring to DNA polymerase processivity. *Curr. Biol.*, 5, 334–7.
Reviews the sliding clamps in DNA replication.

Baker, T. A. and Bell, S. P. (1998). Polymerases and the replisome: machines within machines. *Cell*, 92, 295–305.

O'Donnell, M. (1999). Processivity factors. *Curr. Biol.*, 9, R545.
A one-page quick guide to essential facts on the circular clamps that ensure that DNA polymerases can progress for long distances.

Hubscher, U., Nasheuer, H.-P., and Sybaoja, J. E. (2000). Eukaryotic DNA polymerases, a growing family. *Trends Biochem. Sci.*, 25, 143–7.
Describes the various enzymes and their roles.

Kunkel, T. A. and Bebenek, K. (2000). DNA replication fidelity. *Annu. Rev. Biochem.*, 69, 497–529.
An advanced comprehensive review.

Mendez, J. and Stillman, B. (2003). Perpetuating the double helix: molecular machines at eukaryotic DNA replication origins. *BioEssays*, 25, 1158–67.
Molecules and processes involved in initiation of DNA replication.

Von Hippel, P. H. (2003). Macromolecular complexes that unwind nucleic acids. *BioEssays*, 25, 1168–77.
Review on how helicases are 'coupled' to the macromolecular machines of gene expression.

Repair

Demple, B. and Karran, P. (1983). Death of an enzyme: suicide repair of DNA. *Trends Biochem. Sci.*, 8, 137–9.
Reviews de-alkylation DNA repair enzymes.

Critchlow, S. E. and Jackson, S. P. (1998). DNA end-joining: from yeast to man. *Trends Biochem. Sci.*, 23, 394–8.
Reviews double-strand break repair.

Prolla, T. A. (1998). DNA mismatch repair and cancer. *Curr. Opin. Cell Biol.*, 10, 311.

Featherstone, C. and Jackson, S. P. (1999). Double strand break repair. *Curr. Biol.*, 9, R759–61.
A 'primer' concise summary.

Jiricny, J. (2002). An APE that proofreads. *Nature*, 415, 593–4.
Short summary on correction of base excision-repair mistakes.

Sancar, A., *et al.* (2004). Molecular mechanisms of mammalian DNA repair and the DNA damage checkpoints. *Annu. Rev. Biochem.*, 73, 39–85.
A review of DNA damage response reactions.

David, S. S. (2005). DNA search and rescue. *Nature*, 434, 569–70.
A News and Views article on the mechanism by which DNA repair enzymes locate a fault in the DNA among all the correct ones.

Telomeres

Greider, C. W. and Blackburn, E. H. (1996). Telomeres, telomerase and cancer. *Sci. Am.*, 274(2), 80–5.
Discusses the problem of DNA end replication, DNA shortening, and how telomerase protects chromosomal end segments. Possible relevance to ageing and cancer discussed.

Johnson, F. B., Marcniak, R. A., and Guarente, L. (1998). Telomeres, the nucleolus and ageing. *Curr. Opin. Cell Biol.*, 10, 332–8.
Discusses the relationship between telomere lengthening and replicative senescence. Includes discussion of the relationship to human ageing.

Nagley, P. and Wei, Y.-H. (1998). Ageing and mammalian mitochondrial genetics. *Trends Genet.*, **14**, 513–17.
An account of mitochondrial mutations and their possible relationship to ageing.

Greider, C. W. (1999). Telomeres do D-loop-T-loop. *Cell*, **97**, 419–422.
Minireview, explains how telomere ends form terminal loops.

Dunham, M. A., et al. (2000). Telomere maintenance by recombination in human cells. *Nat. Genet.*, **26**, 447–50.
Describes the ALT mechanism of telomere synthesis which is an alternative to telomerase in cancer cells.

Battacharya, M. K. and Lustig, A. J. (2006). Telomere dynamics in genome stability. *Trends Biochem. Sci.*, **31**, 114–22.
An advanced review dealing with the consequences of telomerase deficiency.

Mitochondrial DNA abnormalities

Hammans, S. R. (1994). Mitochondrial DNA and disease. *Essays Biochem.*, **28**, 99–112.
A discussion of diseases arising from abnormalities in mitochondrial DNA.

Problems

1 What is a replicon?

2 In separating the strands of parental DNA during replication, what topological problem occurs?

3 How is the problem referred to in the preceding question solved, both in *E. coli* and eukaryotes?

4 By means of diagrams, explain the actions of topoisomerases I and II.

5 Eukaryotes have no topoisomerase capable of inserting negative supercoiling into DNA and yet eukaryotic DNA is negatively supercoiled. Explain how this is brought about.

6 What are the substrates for DNA synthesis?

7 Why is TTP used in DNA synthesis – why not UTP as in RNA?

8 (a) Can a DNA chain be synthesized entirely from the four deoxytriphosphate substrates? Explain your answer.
   (b) In which direction does DNA synthesis proceed? Explain your answer so as to be totally unambiguous.

9 What are the thermodynamic forces driving DNA synthesis?

10 *E. coli* polymerase I is a complex enzyme. Describe its different activities and explain their roles in DNA synthesis.

11 Discuss the mechanism by which DNA polymerase III of *E. coli* achieves a high standard of fidelity in DNA synthesis.

12 The proofreading activity of *E. coli* polymerase III is important, but insufficient to give a sufficiently high fidelity rate. If an improperly paired nucleotide is incorporated giving a mismatch, it has to be replaced. This demands that the repair system recognizes which of the two bases in the mismatch is wrong. How is this done and how is the problem fixed? Does this mechanism exist in humans?

13 Explain what a thymine dimer is, how it is formed, and how it is repaired.

14 Explain how eukaryotic chromosomes become shortened at each round of replication.

15 Explain how the DNA-shortening problem in replication is coped with.

16 What is meant by processivity? Which DNA polymerase does not have it and why?

17 Which of the following components does not belong to the series? CTP, UTP, DNA, ATP, GTP, RNA.

Chapter 23

Gene transcription and control

Information in DNA, encoded in the sequence of the four bases, is used to direct the assemblage of the 20 standard amino acids in the correct sequence to produce the protein for which a given gene is responsible, or, in the case of noncoding genes, to produce the correct RNA sequence. A gene does not participate directly in protein synthesis; in eukaryotes the DNA is enclosed inside the nuclear membrane while the protein-synthesizing machinery is outside in the cytoplasm, so the two never meet. How then does a gene direct protein synthesis? It does so by sending out RNA copies of its coded information to the cytoplasm, although in *Escherichia coli* the copies are immediately in contact with the cytoplasm because there is no nuclear membrane. There are several types of RNA and this one is called **messenger RNA (mRNA)**. It carries the message of the gene to the ribosome.

Messenger RNA

The structure of RNA

RNA stands for ribonucleic acid. It is a polynucleotide essentially the same as DNA but with these differences:

- The sugar is D-ribose, not the deoxyribose of DNA. Ribose has an —OH in the 2′ position.

D-Ribose   2′-Deoxy-D-ribose

- mRNA is single stranded, not a duplex of two molecules as is DNA. mRNA is a copy of only one of the two strands of the DNA of a gene.

- Its four bases are A, C, G, and U. There is no T. U and T have identical base-pairing properties and thus they both pair with A.

Were it not for these differences the structure of a single strand of DNA, shown on page 316, could be that of single-stranded RNA. There are the same 3′ → 5′ phosphodiester bonds between successive nucleotides.

How is mRNA synthesized?

The building-block reactants for RNA synthesis are ATP, CTP, GTP, and UTP, which are produced in all cells. In *E. coli*, all RNA is synthesized from these by a single enzyme called DNA-dependent RNA polymerase, or **RNA polymerase**. In eukaryotes three RNA polymerases exist, known as polymerases I, II, and III, or, often, as Pol I, II, or III. mRNA is synthesized by polymerase II, which transcribes all protein-coding genes, the other two being for making ribosomal RNA (rRNA) and transfer RNA (tRNA; page 383).

The synthesis first requires that the duplex DNA strands are separated to provide a single-stranded template for directing the sequence of nucleotides to be assembled into mRNA. The two strands are transitorily separated over a short sequence at the site of mRNA synthesis, and then come together again after the polymerase has passed. In effect, a separation 'bubble' moves along the DNA. The basic process of synthesis, called **gene transcription**, is much the same as in DNA synthesis in that the base of the incoming ribonucleotide is complementary to the base on the DNA template (Fig. 23.1) but, unlike DNA synthesis, only one strand is formed.

The RNA polymerase works its way along the template, joining together the nucleotides in the correct order as determined by the DNA template. *RNA synthesis is always in the 5′ → 3′ direction.* That is, new nucleotides are added to the 3′-OH and so the chain elongates in the 5′ → 3′ direction. The template is antiparallel, running in the opposite (3′ → 5′) direction. The chemical reaction catalysed by the polymerase is very much like that of DNA synthesis in that it involves the transfer of the

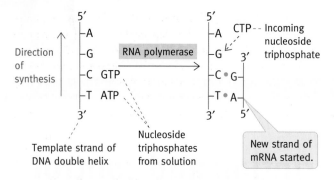

Fig. 23.1 Copying mRNA from a DNA template strand. The nontemplate strand is not shown. Note that the separation of the two strands is transitory. A bubble of DNA strand separation moves along the DNA as the polymerase progresses along it.

why nucleotide triphosphate .

α-phosphoryl group (the first one attached to the ribose) of the nucleotide triphosphates to the 3′-OH of the preceding nucleotide, splitting off inorganic pyrophosphate (PP$_i$). PP$_i$ is hydrolysed to two P$_i$ molecules, making the reaction, shown in Fig. 23.2, strongly exergonic, again like DNA synthesis.

An important point of difference, however, is that RNA polymerase *can initiate new chains* – it does not need a primer; it can synthesize the entire mRNA molecule from the four nucleoside triphosphates, provided a DNA template is there. This is quite different from the situation in DNA synthesis where primer is always required for DNA polymerase activity, which can only elongate existing chains.

Some general properties of mRNA

In a typical chromosome there are thousands of different genes (see Chapter 21, page 356, for gene definition). An mRNA molecule is coded for by a single gene (or, in prokaryotes, often by a small group of genes) and, therefore, large numbers of different mRNA molecules are formed in the cell. While the DNA molecule is vast in length, an mRNA molecule is minute in comparison. In the cytoplasm the mRNAs, as stated, direct the synthesis of proteins for which their respective genes code.

DNA is immortal in cellular terms, but mRNA is ephemeral, with a half-life of perhaps 20 minutes to several hours in mammals and about 2 minutes in bacteria. Thus, for expression of a gene (the term 'expression' means that the protein coded for is actually being synthesized), a continuous stream of mRNA molecules must be produced from that gene. The gene, as it were, 'stamps out' copy after copy, RNA polymerase being the stamping machinery. This might seem wasteful but it gives the important benefit of permitting positive control of the expression of individual genes. Destruction of mRNA is the main 'off switch' in protein synthesis, once a gene ceases to produce mRNA.

In eukaryotic cells, mRNA almost always corresponds to single genes but in prokaryotes it may carry the coded instructions for the synthesis of several proteins, all joined together in a single RNA molecule. This is called **polycistronic mRNA** after the genetic term **cistron**, which is more or less synonymous with the term gene. Such clustered genes giving rise to a single mRNA results in the coordinated expression of several genes, the proteins of which function as a unit, such as the enzymes of a metabolic pathway.

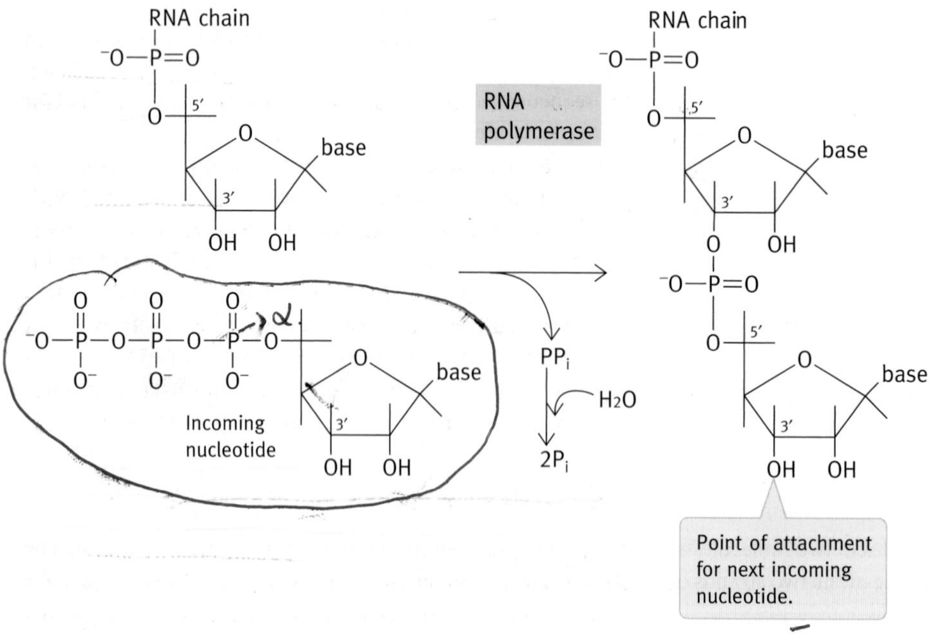

Fig. 23.2 The reaction catalysed by RNA polymerase.

Some essential terminology

The flow of **information** in gene expression is:

(transcription)   (translation)
DNA — — — → mRNA — — — → protein.

(The broken arrows represent *information* flow, not chemical conversions. DNA cannot be converted into RNA nor RNA into protein.)

The 'language' in DNA and RNA is the same – it consists of the base sequences. In copying DNA into RNA there is transcription of the information. Hence mRNA production is called **transcription**, and the DNA is said to be transcribed. The RNA molecules produced are called transcripts, and, in the case of those which are yet to be modified, primary transcripts. The 'language' of the protein is different – it consists of the amino acid sequence the structures and chemistry of which are quite different from those of nucleic acid. The synthesis of protein, directed by mRNA, is therefore called **translation**. If you copy this page in English you are transcribing it. If you copy it into Mandarin characters, you are translating it.

We have so far talked of mRNA synthesis as 'copying' the DNA. The template DNA strand is 'copied' only in the complementary sense as you have seen from its method of synthesis, which is dependent on Watson–Crick base pairing of incoming ribonucleotides to the template bases. A in the template becomes U in the copy and so on. The DNA strand being copied into mRNA is called the **template strand**, the other one the **nontemplate strand**. mRNA has the information for the sequence of amino acids in a protein; it therefore carries the sense or message of the gene to the translational machinery. The base sequence of the mRNA is the same (apart from the T → U switch) as that of the nontemplate DNA strand, which therefore is often called the **coding or sense strand**. The template strand therefore is the noncoding strand (also called the nonsense strand in an older terminology; Fig. 23.3).

So far we have dealt with gene expression only in general terms in that we have referred to the gene only as supplying template DNA. However, there are many important questions such as how the genes that are to be expressed are selected out of the huge collection available in a cell. In transcribing a gene,

which is part of a huge chromosome, how does the RNA polymerase 'know' where to start copying the DNA and where to stop? How is the rate of gene expression controlled? A protein coded for by one gene may be produced in large amounts and another in tiny amounts or not at all; and some genes may be expressed at one time and not at other times. How are these situations achieved? To answer these questions we now must look at the structures of genes.

Apart from the basic chemistry already described, gene transcription and control in prokaryotes and eukaryotes are very different. We will deal with the process first in *E. coli* and then in eukaryotic cells.

Gene transcription in *E. coli*

A gene functions only by acting as a template for the transcription of RNA molecules the base sequence of which corresponds to the nontemplate DNA strand. A gene is a specific section of DNA that is transcribed into RNA. There are some noncoding genes the function of which is to code for RNA molecules that are not messengers – these are the ribosomal and transfer RNAs whose role in protein synthesis is described in Chapter 24. Also recent discoveries have revealed large numbers of genes which produce microRNA transcripts believed to be involved in control (Chapter 25). The rest of the vast array of genes produce mRNA molecules in which the base sequences code for the amino acid sequences of specific proteins. An mRNA molecule has sections at each end that are not translated into protein. The 5′ untranslated regions (UTRs) contain encoded signals necessary for initiation of translation and the 3′ UTR signals for its termination (Chapter 25), so that a gene includes sections of DNA that are transcribed into these regions as well as that for protein coding.

This does not complete the list of DNA regions associated with the gene, for, in addition, there is a region of DNA adjacent to the 5′ end of the gene, called the **promoter** that is essential for transcription of the gene, but is not itself transcribed into RNA. At the opposite (3′) end is a **terminator region** necessary for termination of transcription (which is not transcribed into RNA). A typical gene is illustrated in Fig. 23.4. The first template nucleotide is given the number +1 and the nucleotide 5′ to this −1 and so on. The start site is illustrated by an arrow (→) that indicates the direction of transcription. Nucleotides 5′ to this are referred to as 'upstream' and 3′ to this as 'downstream'. You will notice that Fig. 23.4 includes one end of the gene being labelled 5′. This brings us to the next question.

What do we mean by the 5′ end of a gene?

A gene has two strands of DNA of opposite polarity, and **duplex DNA** has therefore no intrinsic polarity. The 5′ end of a gene is the end containing the promoter. Thus the conventional polarity

DNA ⟨

5′ C G A T G C A T 3′   Nontemplate strand
(coding or sense)

3′ G C T A C G T A 5′   Template strand
(noncoding or nonsense)

5′ C G A U G C A U 3′   mRNA strand

Fig. 23.3 Relationship of transcribed mRNA to template and nontemplate strands. The mRNA is the sense of the information (see text). In viruses a frequently used terminology is: template, minus (–) strand; nontemplate, plus (+) strand.

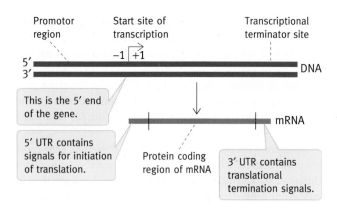

Fig. 23.4 Geography of a prokaryotic gene and its mRNA. Note that the 5′ end of a gene is referring to the nontemplate strand or sense strand.

of a gene is the 5′ → 3′ direction in which the RNA is transcribed from the promoter (but note that there are no physical ends to the gene – the DNA strand continues to the next gene).

Phases of gene transcription

There are three phases – **initiation**, **elongation**, and **termination**.

Initiation of transcription in *E. coli*

The gene initiates at a 'promoter' sequence and this is the sequence the RNA polymerase locks on to. In the promoters there are short stretches of bases that are called 'boxes' or 'elements'. These are accepted terms but the stretches are not boxes nor do they have anything to do with atomic elements. Probably the term 'box' derives from the practice of drawing a rectangle around a sequence of bases to specify it. In a typical *E. coli* promoter there are two boxes – the **Pribnow box** (named after its discoverer, David Pribnow), centred at nucleotide −10, and the other centred at −35. The consensus sequences of the boxes are shown in Fig. 23.5. A **consensus sequence** is obtained by determining the sequence of, for example, the Pribnow box in a number of different genes, for they are often not exactly the same. You then look at the first nucleotide position in the box and count up which base is most often used by the different genes and so on for all the other positions. In fact, the consensus sequence itself might never occur in any gene but the variation from it will be small.

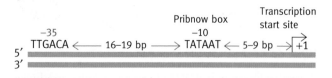

Fig. 23.5 Consensus sequences of *Escherichia coli* promoter elements.

Single sequences are always given in the 5′ → 3′ direction but, of course, in a gene there is the second DNA strand. Thus, although we say the Pribnow box has the sequence TATAAT, it really is:

$$5'\text{——}TATAAT\text{——}3'$$
$$3'\text{——}ATATTA\text{——}5'.$$

It is the *double strand* that is recognized by the proteins that control transcription (see below).

Correct initiation of transcription is obviously important. Synthesis of an mRNA needs to commence at exactly the correct nucleotide on the template and on the correct strand, though in a few cases 'sloppy' start sites with close multiple sites occur. The question is how the RNA polymerase is positioned in the correct place to start transcribing a gene. The −35 and Pribnow boxes are the signals for this. RNA polymerase of *E. coli* is a large protein complex. The 'core' enzyme has an affinity for DNA, but it cannot recognize the correct initiation site until it is joined by another protein from the cytoplasm, the sigma protein (σ) or **sigma factor**. With this attached, the polymerase binds to the −35 and Pribnow boxes and initiation of transcription can start. (We remind you that, although the DNA bases are Watson–Crick paired in the centre of the duplex and largely concealed, their edges are 'visible' in the DNA grooves and can be recognized, that is to say, contacted, by proteins designed to do so; see page 318 and Fig. 21.4.) This aligns the enzyme in the correct starting position, and the correct orientation.

Separating the DNA strands

The polymerase can now separate the DNA strands as it moves, thus making the template strand bases available for pairing with incoming bases of NTPs. The enzyme synthesizes the first few phosphodiester bonds from nucleoside triphosphates and initiation is thus achieved. At this point sigma factor protein flies off (to be used again) and the polymerase, now released, moves down the gene synthesizing mRNA. It is not known what causes the sigma factor protein to detach. The polymerase moves at the rate of about 40 nucleotides per second and unwinds the DNA ahead. The DNA rewinds behind it forming a temporary unwound 'bubble' (about one and a half turns of the helix in length), which passes along the gene with the polymerase (Fig. 23.6).

Termination of transcription

At the end of many transcribed prokaryote genes are sequences that result in the transcribed RNA having a **stem loop structure**. This needs to be explained. Although mRNA is single-stranded, it maximizes base-pairing within itself to achieve the lowest free energy. The newly synthesized mRNA is attached to the DNA template strand by base-pairing but only for about 12 nucleotides because the polymerase deflects the mRNA from the template. Near the end of the gene, the sequence of bases in the mRNA produced is such that the stem loop structure shown

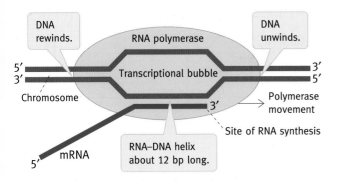

Fig. 23.6 DNA transcription by *Escherichia coli* RNA polymerase. The polymerase unwinds a stretch of DNA about 17 base pairs in length forming a transcriptional bubble that progresses along the DNA. The DNA has to unwind ahead of the polymerase and rewind behind it. The newly formed RNA forms an RNA–DNA double helix about 12 base pairs long.

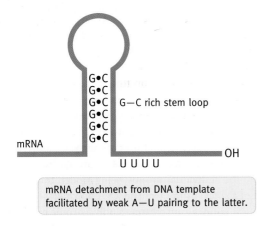

Fig. 23.7 The stem loop structure of an RNA transcript involved in the Rho-independent termination of gene transcription (see text).

in Fig. 23.7 forms because the base sequence here permits formation of G—C pairs, a stable structure due to triple hydrogen bonding between G and C. The stem loop structure prevents binding of the mRNA to the template at this point, since, if its bases are preferentially internally paired, they cannot pair with the template DNA. Immediately following the stem loop structure in the mRNA transcript is a string of U residues, giving weak bonding of the RNA to DNA because of weak A—U hydrogen bonding. This facilitates detachment of the mRNA and terminates transcription.

There is a second method of termination of transcription in many prokaryote genes. This requires a protein called the **Rho** factor which attaches to the newly transcribed mRNA and moves along it behind the RNA polymerase. At the termination site, the polymerase pauses, possibly because of a difficult to separate G—C-rich sections of the DNA, and allows the Rho factor to catch up with the polymerase. The Rho factor has an unwinding (helicase) activity for unwinding the RNA–DNA duplex formed by transcription. ATP breakdown is involved.

This releases mRNA and terminates transcription. In *E. coli* the mRNA starts to direct protein synthesis immediately before the full mRNA molecule is completed because the transcription occurs in contact with the cytoplasm. In fact, the ribosomes follow closely behind the polymerase.

However, we are not finished with *E. coli* gene transcription for there is still the question of gene control.

The rate of gene transcription initiation in prokaryotes

Genes that are **constitutively expressed** are those that are 'switched on' all the time and whose rates of expression are not selectively modulated. Such genes code for enzymes and other proteins that are needed at all times and in amounts that do not vary from time to time. However, among these constitutive proteins some will be required in larger amounts than others. The major control on the rate of gene expression in bacteria is the rate of mRNA production and this is largely determined by the frequency of initiation of transcription of a given gene. This varies because genes have promoters of different 'strengths'. A 'strong' promoter will initiate transcription frequently and cause many mRNA transcripts of the gene to be made and hence a lot of the specific protein. A 'weak' promoter has the reverse effect. The strength of a promoter is a function of the precise base sequence of the Pribnow and −35 boxes, the distance between them, and the nature of the bases in the −1 to −10 region. The greater the affinity of these regions for the polymerase, the stronger the promotion, though this may not be the sole determinant.

Control of transcription by different sigma factors

A particularly neat method of controlling blocks of genes in prokaryotes is by using different sigma factors. Under certain conditions, the usual sigma protein is replaced by one that causes the RNA polymerase to initiate at a different set of genes that have different promoter sequences.

- In sporulating bacilli a new sigma factor is produced in response to adverse conditions in the environment – this causes expression of a set of genes leading to sporulation.

- In nitrogen starvation a special sigma factor is produced.

- After a heat shock (a sudden rise in temperature), *E. coli* transitorily increases the synthesis, stability, and activity of a different protein (σ^{32}) that normally is present at a nonfunctional level. This factor directs the transcription of genes for a set of 'heat-shock proteins' that protect the cell against the consequences of the heat shock (page 395).

Gene control in *E. coli*: the *lac* operon

Bacteria live in continually changing environments to which they must adapt for survival. A typical *E. coli* cell under optimal conditions will divide in about 20 minutes. If a strain lops 1

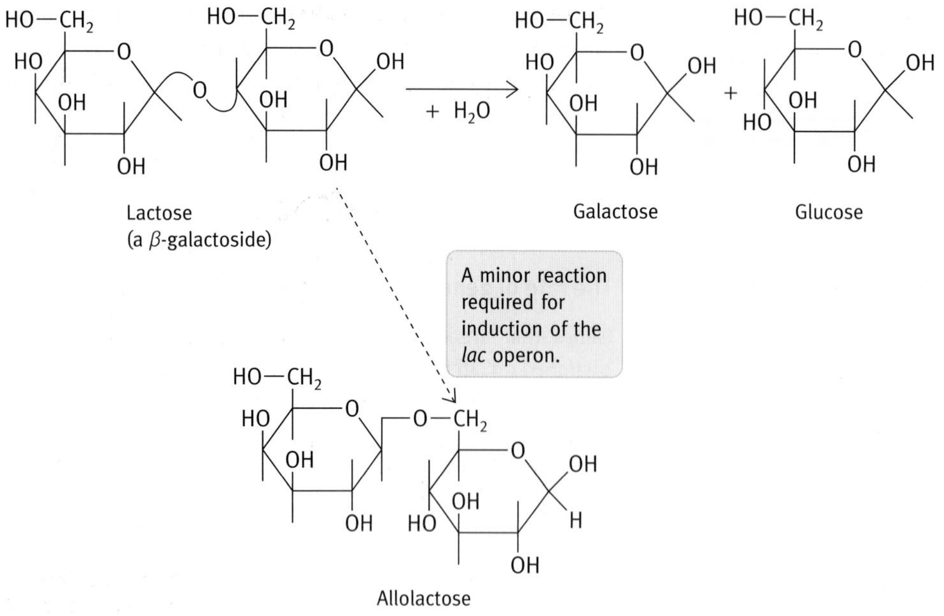

Lactose
(a β-galactoside)

Galactose

Glucose

A minor reaction required for induction of the *lac* operon.

Allolactose

Fig. 23.8 The reactions catalysed by β-galactosidase. The term lactase is often applied to the enzyme catalyzing this reaction in digestion. See page 359 for explanation of allolactose formation.

minute off that time, it will outgrow other strains very quickly. Wastage and inefficiency in biochemical processes is not tolerated in the face of savage natural selection.

Production of enzymes consumes resources and energy, and it would not do if the *E. coli* cell produced enzymes that were unnecessary at the time. Glucose-metabolizing enzymes are **constitutive**, for glucose is the most common sugar, and other sugars are shunted on to the glucose pathways. These enzymes are always needed so the promoters of the genes coding for them have no 'on' and 'off' switches – they are always 'on'.

However, the *E. coli* cell may encounter other sugars sporadically – the disaccharide lactose present in milk is an example. If it is the sole source of carbon available, ability to utilize it would be necessary for survival. The enzyme needed to utilize this sugar is β-galactosidase, so called because lactose is a β-galactoside, and it must be hydrolysed to free galactose and glucose before it can be metabolized (Fig. 23.8). An additional transport protein, **β-galactoside permease**, is needed to transport the lactose into the cell. A third protein, **galactoside transacetylase**, is believed to be involved in protection of the cell against nonmetabolizable, potentially toxic β-galactosides that may be imported, though less is known of this. The three proteins are normally made in minute amounts (basal levels) because they are not required unless lactose is encountered. When lactose is encountered as the sole energy source, there is an almost instant burst of synthesis of the three proteins. The cell can then use the lactose as a carbon and energy source. However, if, in addition to lactose being present, there is also glucose, then production of the three enzymes would be wasteful since this merely leads to production of more glucose inside the cell when there is plenty of it available anyway. The cell therefore 'ignores' the lactose signal and does not produce the enzymes. The regulation that produces this situation is at the level of gene transcription initiation.

Structure of the *E. coli lac* operon

First a few terms: production of a protein in response to a chemical signal is called **induction** of that protein, and the responsible chemical, the **inducer**. Prevention of the production of a protein is called **repression**. Many prokaryotic genes are grouped together, the individual groups being under transcriptional control of a single promoter. The RNA polymerase transcribes through the entire group creating a **polycistronic mRNA** molecule with the coding instructions for several proteins.

Such a group of genes, with its single promoter control, is called an **operon** (the promoter being part of the operon). β-Galactosidase, lactose permease, and transacetylase genes belong to such an operon. The three genes are often referred to as *z*, *y*, and *a*, respectively (capital letters may now be used). There is also a separate *i* gene (*i* for inducibility) that codes for a protein called the *lac* repressor and there is a stretch of DNA called the **operator region** to which the *lac* repressor protein can bind. Finally, there is a stretch of DNA to which a cyclic AMP (cAMP) receptor protein (**catabolite gene-activator protein – CAP**) can bind. It is given this general name because it is involved in the induction of other enzymes involved in catabolism of substrates. The approximate arrangement of these elements in the operon is shown Fig. 23.9. We can now systematically go through the control mechanism noting the following points. The *lac* promoter *on its own* is a weak one so that the RNA polymerase does not readily bind to it and initiate transcription. Without extra help the *lac* operon is not transcribed except at a low basal level. The extra help in polymerase binding is given by the attachment of the protein, CAP, to the adjacent site. When CAP is attached to the DNA, the promoter is a strong one.

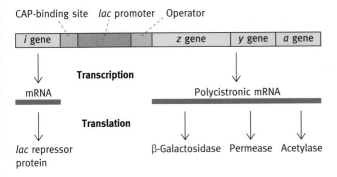

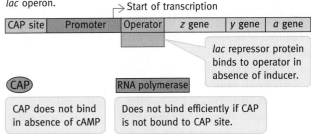

Situation (a). High glucose; no cAMP; no lactose; no transcription of *lac* operon.

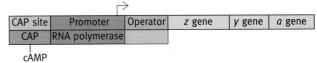

Situation (b). Low glucose; high cAMP; CAP–cAMP complex binds CAP site; RNA polymerase can now bind to promoter; no lactose; repressor protein blocks operator; no transcription.

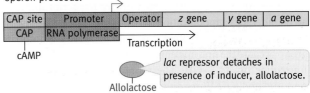

Fig. 23.9 Diagram of the *lac* operon. Note that the *i* gene is an independent gene that codes for the *lac* repressor protein. Similarly, there is a completely independent gene producing the catabolite gene-activator protein (CAP) to which cyclic AMP can bind.

Situation (c). Low glucose; high cAMP; lactose present; repressor protein–allolactose complex detaches from operator; transcription of operon proceeds.

Fig. 23.10 Expression of the *lac* operon. **(a)** In the presence of high glucose there is no cyclic AMP (cAMP) to cause catabolite gene-activator protein (CAP) to bind, and this binding is necessary for the attachment of RNA polymerase to the promoter. **(b)** With low glucose but no lactose, although CAP binds and assists the RNA polymerase to bind, transcription still does not occur because the *lac* repressor is bound to the operator, blocking polymerase movement. In **(c)** the inducing allolactose binds to the repressor, causing its release from the operator, and transcription can proceed. (In the presence of lactose, a small amount of a lactose isomer, allolactose, is produced, which is the actual inducer – see text.)

However, CAP does not attach unless cAMP is bound to it – it is an allosteric protein and *cAMP is present in the cell only when glucose levels are low*. The binding of cAMP to CAP is freely reversible. When glucose is at a high level, cAMP is scarce, CAP does not bind, the RNA polymerase therefore does not bind effectively, and *lac* operon transcription is minimal and so is the production of β-galactosidase. This situation is illustrated in Fig. 23.10(a).

Does this mean that the *lac* operon is always transcribed when glucose is scarce? The answer is no. As explained, there is no point in doing so unless lactose is present. In the absence of lactose the *lac* repressor protein is attached to the operator and blocks the RNA polymerase from transcribing the genes (Fig. 23.10(b)). The *lac* repressor is also an allosteric protein. In the absence of lactose in the environment, it has strong affinity for the operator stretch of DNA, but, if lactose is present, a small amount is able to enter the cell via the basal level of permease. It is hydrolysed; during the course of this a small amount is converted to the lactose isomer, **allolactose**, which is the true inducer of the *lac* operon (see Fig. 23.8). This binds reversibly to the repressor protein, causing an allosteric change which causes the repressor to dissociate from the operator and unblocks the operon.

It is not known why allolactose is the inducer rather than lactose. It is formed by β-galactosidase catalysing a small amount of transglycosylation in which a galactosyl residue is reversibly transferred to the 6-OH of glucose instead of to water (Fig. 23.8). Conceivably it is a mechanism for preventing the firing off of wasteful induction when only uneconomic amounts of lactose are present. With the unblocking of the operator, the RNA polymerase is now free to move down the operon, producing the polycistronic mRNA (Fig. 23.10(c)); this is translated to produce the three enzymes.

The *lac* operon was the first understood example of pro-karyote operon control, but its mechanism applies to many pathways. In *E. coli* the **tryptophan operon** (*trp* **operon**), which contains the five structural genes needed to produce enzymes involved in tryptophan synthesis, is similarly controlled by a *trp* **repressor protein**. In this case, tryptophan bound to the latter causes it to block the operator site. *Operons do not occur in eukaryotes.* In addition, the *trp* operon is regulated by a process known as **attenuation**. We describe this on page 393.

Gene transcription in eukaryotic cells

The basic enzymic reaction by which RNA is synthesized in eukaryotes is the same as in prokaryotes (Fig. 23.1). The DNA-dependent RNA polymerase which transcribes genes coding for proteins in eukaryotes is **RNA polymerase II**, a large multi-subunit enzyme. This is highly sensitive to the deadly toxin

α-amanitin, a complex octapeptide produced by the death cap mushroom, *Amanita phalloides*.

In prokaryotes, genes are transcribed and the RNA molecules produced in this way (the primary transcripts) are fully functional mRNAs which are immediately translated. The situation in eukaryotic cells is very different, because *the primary transcript is greatly modified before it is a functional mRNA*. The processes of initiation of the transcription of a gene and its control are inseparable from one another in eukaryotes and very different from those in prokaryotes. We think it will be more easily understood if we first deal with the production of RNA transcripts and their processing to become mRNA molecules in eukaryotes and then return to what is the really major topic, initiation and control of transcription. So, for the moment just assume that transcription of RNA from the DNA template has started and we will first describe what happens to these transcripts.

Capping the RNA transcribed by RNA polymerase II

The RNA of the eukaryotic gene primary transcript immediately undergoes a modification at its 5′ end, called **capping**. At the 5′ end of the primary RNA transcript there is a triphosphate group because the first nucleotide triphosphate simply accepts a nucleotide on its 3′-OH. The terminal phosphate of this is removed and a GMP residue is added from GTP (Fig. 23.11). The 5′—5′ triphosphate linkage is unusual. The G is then methylated in the N-7 position as also is the 2′-OH of the

second nucleotide. The cap protects the end of the mRNA from exonuclease attack and it is involved in initiation of translation as described in the next chapter.

These are not the only differences between mRNA production in prokaryotes and eukaryotes, because, with a few exceptions, eukaryotic genes are split genes.

Split genes

The DNA coding for a given protein is split up into several parts, linked together by intervening stretches that do not code for amino acid sequences. The latter sections with no protein-coding content are called **introns**, and the coding stretches, **exons**. There can be 1–500 or so introns in a gene (Fig. 23.12(a)), which can typically vary in length from about 50 to 20,000 base pairs (or sometimes longer). Exons vary in size but usually are smaller than introns, around 150 base pairs in length on average. The primary transcript is processed to eliminate the introns and link together the exons into one mRNA molecule. This is known as **RNA splicing** (Fig. 23.12(b)).

Mechanism of splicing

Removing the unwanted RNA introns of a primary transcript and joining up the exons into mRNA looks a formidable task but the key to it is the **trans-esterification** reaction. In this, a phosphodiester bond is transferred to a different —OH group. There is no hydrolysis and no significant energy change during the bond rearrangements.

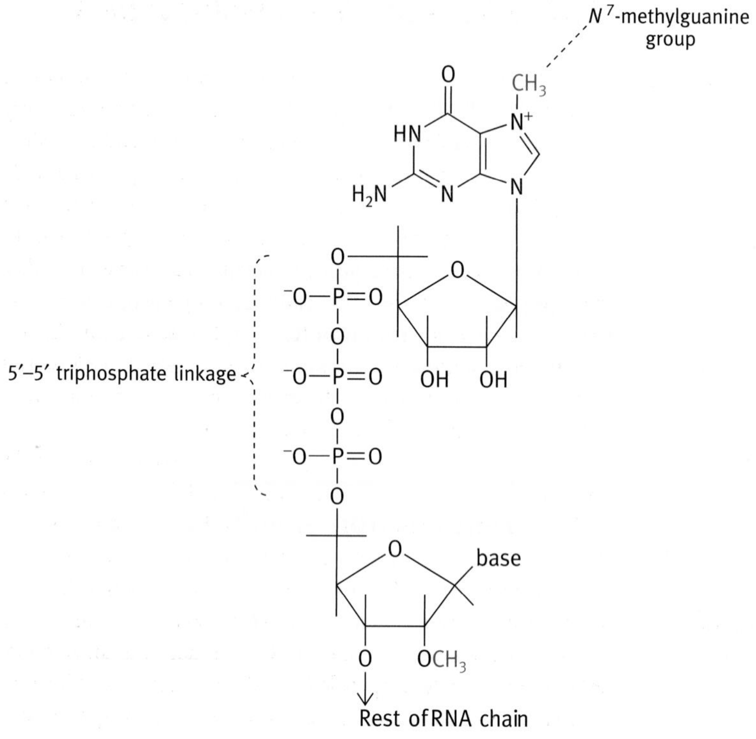

Fig. 23.11 Structure of the 5′ cap in eukaryotic mRNA. The terminal nucleoside triphosphate of the primary RNA transcript is converted to a diphosphate followed by a reaction with GTP in which pyrophosphate is eliminated. This is followed by methylation reactions. (A third methyl group may be added to the 2′-OH of the next nucleotide of the primary transcript.) The capped primary transcript is then processed to mRNA – see text for details.

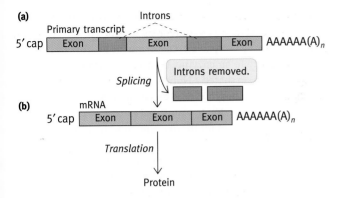

Fig. 23.12 Primary polymerase II transcript of a eukaryotic gene: **(a)** introns after capping and addition of polyA tail; **(b)** excision of introns to form the mature mRNA is called splicing.

If X—Y is an RNA chain, it would be broken.

In RNA splicing, the exon–intron junctions are 'labelled' by consensus sequences; all introns begin with GU and end with AG (though the full consensus sequences are longer than this). The ROH of the diagram in Fig 23.13 is the 2′-OH of an adenine nucleotide in a short sequence (seven bases long in yeast) of the intron chain, known as the **branch site**. The reason for this name will be seen from the structure in Fig. 23.13. The 2′-OH group attacks the 5′ phosphate of the G nucleotide at the splice site, forming a lariat structure. This breaks the chain

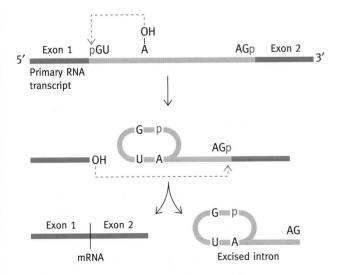

Fig. 23.13 Mechanism of RNA splicing.

at the 3′ end of exon 1, thus producing a free 3′-OH and the 3′-OH attacks the 5′ end of exon 2, joining the two exons.

In most cases, the splicing reaction in the nucleus is catalysed by very complex protein-RNA structures called **spliceosomes**. They are complexes of about 300 different proteins and also five RNA molecules, 100–300 bases long in higher eukaryotes, called **small nuclear RNAs (snRNAs)**. These are associated with proteins in structures known as **small ribonucleoprotein particles (snRNPs)**, each containing multiple protein subunits. There are five snRNPs in a spliceosome, known as **U1, U2, U4, U5**, and **U6**, comprised of about 40 proteins (see **Nilsen** review in the Further reading). Additional proteins known as **splicing factors** are also needed. The U1 and U2 snRNPs bind by base pairing to the 5′ splice site and the branch site respectively and then associate with each other. A trimer of U4, U5, and U6 then associates with the complex to form the spliceosome. U6 catalyses the transphospho-esterification reactions but this has to be triggered by the release of U4, which masks the catalytic centre. Release of U4 requires ATP hydrolysis. The various interactions between the snRNPs themselves and with the RNA to be spliced are believed to cause the reaction sites to be appropriately aligned and also to create the catalytic site, which brings about the reactions. The catalytic site is probably an RNA structure.

Faulty splicing can lead to genetic diseases. In β-thalassaemia (page 68), the β subunit of haemoglobin is not produced in normal amounts, because the G at the 5′ splicing sequence of an intron is mutated to an A, and primary transcripts are therefore not properly processed to mRNA.

Ribozymes and self-splicing of RNA

The spliceosome is, as described, a complex with scores of protein components and several RNA components and a very elaborate mechanism. The biochemical world was shocked when it was discovered that a few RNA gene transcripts accurately self-spliced without any help from proteins. It was the first case known of a specific biochemical reaction occurring as the result of catalytic activity brought about by a macromolecule other than a protein. In the protozoan *Tetrahymena*, one of the rRNAs is made as a precursor transcript containing an intron which has to be spliced out to produce the mature rRNA. The two exons on either side of the intron become joined together by the splicing to form the mature molecule.

During a study of the splicing of the isolated precursor RNA in the laboratory of Tom Cech, it was found that the intron was spliced out with the two exons properly joined without any protein being needed. A divalent metal ion (Mg^{2+} is used *in vitro*) and guanosine (or a 5′ guanine nucleotide) are essential. The mechanism of the self-splicing in *Tetrahymena* is shown in Fig. 23.14. The guanosine 3′-OH (G—OH) attacks the phosphodiester bond thus releasing the left-hand exon with a 3′-OH group. The latter now makes an attack on the second phosphodiester

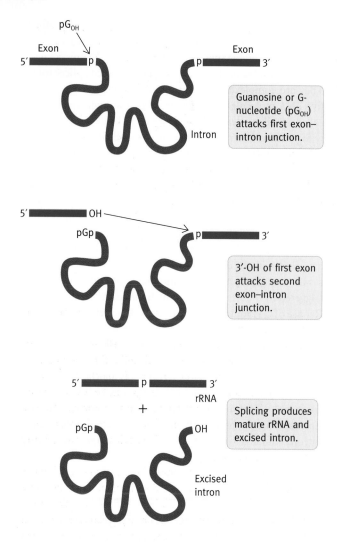

Fig. 23.14 Mechanism of the self-splicing reaction of the *Tetrahymena* ribosomal RNA (rRNA) precursor. G_{OH} represents guanosine, GMP, GDP, or GTP. A metal ion is needed; Mg^{2+} is usually used for *in vitro* experiments.

bond releasing the intron and splicing the two exons. The chemistry of the two reactions involved is as illustrated in the diagram of trans-esterification. The intron is fairly large (414 nucleotides) and has a complex three-dimensional structure, the integrity of which is essential for the self-cleaving. It is believed that the structure provides a binding site for the guanosine (G—OH). This type of catalytic activity was not anticipated to occur in RNA since it does not have the range of dissociable groups of proteins which are involved in enzyme catalysis nor was it expected to be able to form binding sites. Both of these expectations have proved to be wrong.

This self-splicing is not a true catalytic reaction because the molecule itself is changed. However, the laboratory of Sydney Altman showed that RNA can also act as a true catalyst. An enzyme called **ribonuclease P**, found in *E. coli*, and elsewhere

processes tRNA precursors by a specific hydrolytic reaction. This enzyme has an RNA component attached to a protein, but the RNA by itself is capable of catalysing the hydrolysis. Because of its similarity to an enzyme, the term **ribozyme** was coined. Cech and Altman shared a Nobel Prize for these discoveries.

A number of ribozymes are known but they are not common and seem to occur for the most part unpredictably. Enzyme reactions still account for the overwhelming majority of all biochemical reactions. RNA transcripts in the newt and certain mitochondria have ribozyme capability. Most spectacularly, a reaction of central importance in protein synthesis is catalysed by an RNA component of the ribosomes (see page 389) of all species so far as is known.

Why should most splicing occur by the elaborate spliceosome apparatus and others self-process? In the case of the ribosomal ribozyme a most likely answer is that the reaction was needed early in the 'RNA' world before proteins existed; it is a relic that has been retained in a few cases.

Alternative splicing or two (or more) proteins for the price of one gene

There is one well-established advantage that split genes confer – alternative splicing. In typical splicing, all the exons of the primary RNA transcript are linked together to form the mature mRNA leading to the formation of a single protein. However, there are many known cases where the splicing can occur in different patterns so that a particular group of exons form one mRNA, whereas a different group from the same gene transcript forms another leading to different proteins. The mechanism is often employed to produce variant forms (isoforms) of a protein required in different tissues or at different times such as occurs in antibody production (page 498). Nineteen different human kinesins formed in this way have been identified. Alternative splicing may partly explain why the human genome has a surprisingly small number of protein-coding genes.

We will now return to initiation of transcription in eukaryotes.

Mechanism of initiation of eukaryotic gene transcription and its control

Unpacking of the DNA for transcription

We have already described the packing of eukaryotic DNA necessary to fit its huge length into the nucleus (page 321). Before (and/or during) transcription, this packing structure is first loosened up specifically at those parts of the DNA that are to be transcribed. We know that selective loosening up occurs from two lines of evidence.

Salivary gland cells of the larval form of the fruit fly, *Drosophila*, are very large. Their chromosomes (known as polytene chromosomes) are replicated about 1000 times without cell division occurring and the elongated chromosomes, still

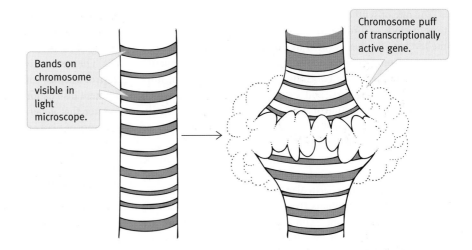

Bands on chromosome visible in light microscope.

Chromosome puff of transcriptionally active gene.

Fig. 23.15 Chromosome puffs at transcriptionally active chromatin on insect salivary gland polytene chromosomes.

with much condensation packing, lie side by side precisely aligned lengthways. This allows bands to be visible in the light microscope, probably representing different zones of packing. During larval development successive banks of genes are expressed and it is seen that during transcription the DNA of specific bands becomes loosened and expands, forming what are called **chromosome puffs** (Fig. 23.15). These are the sites of mRNA formation.

Transcriptionally active regions of chromatin are more susceptible to attack by added DNase – an enzyme that hydrolyses DNA. Globin genes, for example, become hypersensitive to DNase only in chromatin from those cells that synthesize globin and only at the developmental time at which the genes are actively transcribed.

A general overview of the differences in the initiation and control of gene transcription in prokaryotes and eukaryotes

In prokaryotes, as described, the RNA polymerase, attached to a sigma factor, binds to the DNA at a gene promoter (page 356). The elements of the different prokaryotic gene promoters are essentially the same, differences in their affinity for the polymerase being due to subtle variations in the base sequence of elements and their spacings. In the case of the *lac* operon described earlier, the rate of gene transcription is a function of the affinity of the polymerase for the promoters.

Eukaryotic gene control is also largely effected at the level of transcriptional initiation. The process looks less orderly, indeed at first sight almost chaotic. The gene promoter has control elements (boxes) but gene control regions vary greatly in the elements they contain. *The polymerase does not bind by recognizing sites on the DNA, as happens in prokaryotes.* Instead, a large number of other proteins assemble on the promoter and the polymerase is recruited by them from the surrounding medium. As a result it becomes correctly positioned on the DNA

but, again it is emphasized, it does not itself initially join to the DNA as in prokaryotes.

The control requirements in eukaryotic cells are vastly different from those of prokaryotes. An *E. coli* cell has about 4000 genes. During the lifetime of a single cell it may be necessary to transcribe all of those genes. For the most part what is required is an 'on-off' switch for controlled genes or, for constitutive genes, a permanent 'on' condition with the actual rates of initiation of the latter being determined in both cases by the affinity of the polymerase for the promoter.

In a eukaryote such as a mammal, the situation is much more complex; a human cell has about 25,000 protein-coding genes. There are many different types of cell – liver, muscle, brain, epithelial, blood, bone, and so on. Many proteins are common to all – those for glycolysis (Chapter 12) are an example, coded for by what are known as **housekeeping genes** – but each type of cell also has its own cohort of proteins needed for specific cell functions. Liver cells have liver-specific proteins not present in brain and muscle cells, and *vice versa*. However, the DNA of all cells in the animal is the same (ignoring gametes and special cases such as the gene rearrangements in B and T cells of the immune system described on page 496). In a liver cell, genes for liver-specific proteins must be activated while those coding for brain-specific proteins must be ignored by transcription apparatus in that cell. All cells of the body arise from a single fertilized egg cell and differentiation into specific cell types involves gene control.

Even in mature cells when the differentiation into cell types has been achieved, a different set of control problems exist in mammals, which have no counterpart in prokaryotic cells. An animal cell cannot make all decisions itself. Its activities must be such that they correspond to the needs of the animal as a whole. An obvious instance of this is that cell division should not proceed independently (as happens in cancer) – the cell must receive one or more signals from other cells before it proceeds to division.

The rate of synthesis of individual proteins varies over short time periods according to needs. For example, after feeding, the level of enzymes devoted to storage of foodstuffs increases resulting from hormonal activation of specific genes. The activities of many cells are controlled by a whole battery of hormones, growth factors, and cytokines (Chapter 27), which bombard them and control appropriate genes. A given gene in a cell may be simultaneously instructed by a multiplicity of signals from hormones or other factors.

The key to eukaryotic gene control is that there can be any number of short sequences called 'control elements' on the DNA associated with a particular gene. Some control elements, with their cognate proteins (transcription factors, see below) determine tissue-specific expression – why certain genes are activated only in the appropriate tissues; others respond to the different signals arriving at the cell from other cells of the body both positive and negative. Many different control elements exist around the genes in a typical cell and a given single gene maybe given many and even contradictory signals by them; it all adds up to a balanced control. It is a mind-boggling problem when you think about it for, to put it in a more general way, the cell has to make sense out of multiple signals as to what the rate of transcription of a given gene should be. There is nothing remotely as complex as this in prokaryotes.

To anticipate what we shall shortly be describing more fully, the important concept you need to become familiar with is that of **transcription factors** (**TFs**), the proteins that specifically bind to control elements, sometimes known as **gene activators** (they usually, but not in all cases, activate). A plethora of different TFs in eukaryotic cells control a multitude of different genes. It has been estimated that more than 5% of human genes code for TFs. TFs are proteins that attach to control elements of eukaryotic genes and are responsible for controlling transcription at the level of initiation. Tissue-specific expression of genes depends on the presence of the cognate TFs in the cells. (The term 'cognate' implies a relationship – a cognate TF is one related to, or having affinity with, the element in question – the particular TF designed to be capable of binding to that control element sequence.) Some TFs are activated only when the cell receives the appropriate signal such as from a hormone; a wide variety of external signals work in this way, by activating specific TFs. The system gives some of the very great flexibility to gene control in eukaryotes which is needed to respond to the many different instructions that may be showered on individual genes.

After that general preview, the first thing to discuss is the anatomy of a eukaryotic gene promoter.

Types of eukaryotic genes and their controlling regions

In eukaryotes there are three different RNA polymerases, designated I, II, and III (nothing to do with order of importance). They are specific for transcribing types of genes also known as types I, II, and III. Type I and type III genes code for rRNA and tRNA molecules respectively which are not translated into proteins but are involved in protein synthesis. We will refer to transcription of these latter genes later in this chapter. The important ones in our present context are the type II genes which include all nuclear genes coding for proteins and are therefore numerically dominant. (Nuclear genes are those residing in the cell nucleus rather than in mitochondria or chloroplasts.) The rest of this chapter is devoted largely to these.

Type II eukaryotic gene promoters

Figure 23.16 shows the DNA components of a type II promoter. There are two sections.

The term **basal elements** refer to the **initiator** (**Inr**), a short sequence at the start site, and the **TATA box**, centred at −25 base pairs from the start site, with a consensus sequence TATAAAA reminiscent of the Pribnow box of prokaryotes. It is the only element with a *fixed* position relative to the start site. Many, but not all, type II genes have the TATA box. Where there is no TATA box, the Inr and a short section called the **downstream promoter element** (**DPE**) position the polymerase. The DPE is about +25 base pairs downstream from the start site.

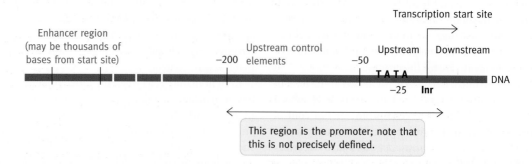

Fig. 23.16 Eukaryotic type II gene control elements. Examples of upstream common control elements are the CAAT box and the GC box. Different promoters can have any mixture of the general transcription elements including multiple copies of individual ones and no one element is present in every gene (see Fig. 23.17 for an illustration of this). Inr, initiator (a pyrimidine-rich stretch on one of the strands).

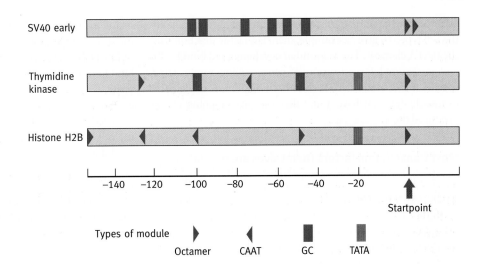

Fig. 23.17 The elements of three eukaryotic gene promoters. Promoters contain different combinations of TATA boxes, CAAT boxes, GC boxes, and other general control elements.

The **upstream control elements** are found at variable positions within the range of about −50 to −200 or so base pairs. The most commonly found ones are the CAAT (pronounced CAT) box with a consensus sequence GGCCAATCT, and the GC box (GGGCGG). It seems that eukaryotic genes usually have at least one of such common control elements but different gene promoters are quite different in which are present. It seems almost undisciplined that a given promoter may have any one or a mixture of them and there may be several copies of one or more of them; there is no obvious standard 'recipe'. Figure 23.17 shows examples of the variety of control elements found in promoters. In addition to these, there may be any number (as appropriate to the particular gene) of control elements concerned with tissue-specific expression, hormonal control, and control by many other factors.

Enhancers

The eukaryotic gene promoter is usually taken to be roughly in the region of 200 base pairs upstream (5′) of the start site which includes the upstream control elements. However, many genes have another section of DNA called the **enhancer**, another cluster of elements, which can greatly increase the expression of the gene, sometimes by as much as 200 times. The remarkable thing is that a given enhancer may be thousands of base pairs distant from the gene that is affected and may be upstream or downstream of the gene. It doesn't matter which orientation the enhancer section is in. The enhancer does not have any promoter of its own and its operation is often tissue specific. It exerts its effects only in a particular tissue.

The obvious problem is how the enhancer can act at such a distance from the gene it affects. This is achieved by looping of the DNA between the gene and its enhancer so that the two can be brought into proximity with one another. The enhancer is a cluster of control elements, many of which bind the same activators (TFs) as on the gene promoter. A complex

of proteins called the **enhanceosome** assembles on the enhancer in a stepwise manner. Some of the proteins that bind to the DNA cause bending of the DNA and thus bring the cluster of protein factors on the enhancer to a position of interacting with the transcriptional complex on the gene promoter, probably via the mediator (see below for an explanation of this term).

Insulators

Another curious feature of enhancers is that they can be quite promiscuous in their effects. They can influence the expression of any gene that comes within their range. To guard against this the genome is apparently divided into sections or loci which confine their action to where their activity is appropriate. This is achieved by having sections of DNA called **insulators** to which proteins attach and limit the range of genes which enhancers can affect.

Transcription factors

We now turn to the role of the specific DNA control elements within the enhancer and promoter. In all cases they are sites of attachment for specific proteins, the transcription factors. There are two broad categories of these.

- The **general TFs** are present in all cells and are components of the **basal transcription machinery**, which form a large complex attached to the basal elements (sometimes called the pre-initiation complex). These are essential for all eukaryotic gene transcription.

- The **sequence-specific TFs (activators)** bind to their cognate upstream control and enhancer elements. These are at the heart of eukaryotic gene control.

There are common elements already mentioned such as the CAAT box and the GC box. In addition there are TFs for a wide variety of control elements found as appropriate only in specific cells to bring about patterns of gene activation appropriate to the cell type. To give an example, there is a factor that binds to the GATA element. This is found in developing red blood cells and is required for activation of genes coding for proteins specific to red blood cells, such as haemoglobin. A tissue such as muscle does not have it and therefore haemoglobin is not produced there.

Most transcription factors themselves are regulated

The TFs that bind to common elements such as the CAAT and GC boxes are present in cells in an active form. Those that exist in an inactive form in the cell cannot stimulate transcription until they are activated often by phosphorylation (or dephosphorylation) or other change causing a conformational change

in the protein which then can bind to the DNA sequence in question. Often the activation is associated with the movement of the TF from the cytoplasm to the nucleus where it can then bind to its cognate DNA elements. The activation is usually the result of signals arriving at the cell from other cells. Figure 23.18 gives a few examples, in outline, of the activation mechanisms involved: in Fig. 23.18(a) a steroid hormone is shown to enter the cell directly (due to its lipid solubility) and on binding to a soluble receptor protein causes a conformational change in the latter so that it is now an active TF for cognate genes; Fig. 23.18(b) shows that cAMP, which is elevated as a result of the action of certain hormones, activates protein kinase A, which phosphorylates an otherwise inactive transcription factor – the latter activates genes appropriate to the hormone signal; Fig. 23.18(c) shows the general concept of the way in which

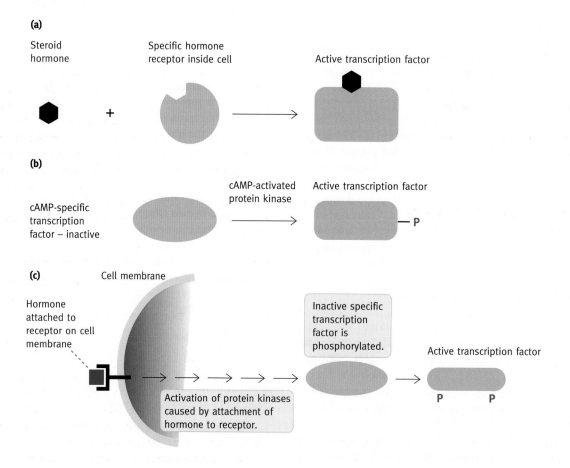

Fig. 23.18 Examples of transcription factor activation. **(a)** A1 steroid hormone enters the cell; it attaches to a receptor specific for that hormone and causes a conformational change in the receptor protein which is now an active transcription factor. This activates the gene(s) which are controlled by the particular hormone. A whole family of steroid hormone receptors exists. **(b)** Cyclic AMP (cAMP) is produced as a result of epinephrine binding to cell receptors. The cAMP activates a protein kinase, which phosphorylates the inactive transcription factor, which is activated. **(c)** Protein hormones such as insulin do not enter the cell but bind to receptors on the cell surface. This results in a sequence of events which ends in the phosphorylation of the appropriate inactive transcription factors and activates them. Note that there are many different hormones which bind to different specific receptors and activate different transcription factors. In many cases the activation involves transport from the cytoplasm into the nucleus. The transcription factors bind to specific response elements of different genes. Thus each hormone can exert control over appropriate genes (activation of transcription factors inside cells by steroid binding and receptor-mediated signal transduction is dealt with in Chapter 27 on cell signalling).

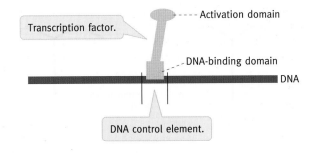

Fig. 23.19 Transcription factor domains.

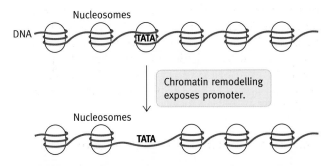

Fig. 23.20 Chromatin remodelling. The principle is that the promoter in chromatin is blocked by nucleosomes. Gene activation requires exposure of the promoter; this may require the physical removal of one or more nucleosomes, or it could be some change in the relationship of the nucleosome(s) to the DNA which effectively gives accessibility to the promoter. Use of the term 'chromatin remodelling' reflects the current uncertainty about exactly what happens at the molecular level.

many hormones bind to membrane receptors and induce a signal cascade inside the cells. This results in activation of specific TFs often by their phosphorylation. The regulation is effected largely by signals from other cells in the form of hormones, cytokines, and growth factors. This type of cell signaling control lies at the heart of cell regulation and is much more fully dealt with in Chapter 27. Inappropriate activation of TFs is important in the generation of cancer because genes are activated when they should not be. The same is true of overproduction of certain TFs, which leads to improper stimulation of cell growth as covered in Chapter 31.

An active TF has two domains (Fig. 23.19) – the DNA-binding domain and the activation domain, which is a binding site for other initiation proteins.

How do transcription factors promote transcriptional initiation?

Before we can answer that we must first deal with another important aspect.

The role of chromatin in eukaryotic gene control

Eukaryotic genes *in vivo* are in the form of the protein–DNA complex known as **chromatin**, not as naked DNA, as discussed earlier. In this, two turns of DNA are wrapped around nucleosomes made of octamers of histone proteins. Individual nucleosomes are separated by linker DNA, which varies somewhat in length in different species but averages about 50 base pairs, so the whole length of DNA per nucleosome is about 200 base pairs (see Fig. 21.7).

Chromatin used to be regarded as an inert structure with the sole function of condensing the DNA to fit into the nucleus. However, as mentioned, when a gene is activated, the chromatin opens up, making the DNA more accessible. *The 'default' state of chromatin (the state in the absence of any action to counteract it) is a 'shut-down' condition – the genes are inactive.* The reason is that gene promoters are blocked by nucleosomes that prevent assembly of the basal initiation machinery on the promoters.

Gene control in eukaryotes involves 'opening up' or unblocking of the promoters. It requires modification of the chromatin structure known as **chromatin remodelling** (Fig. 23.20). This term means the *effective* removal of nucleosomes from the promoter site of the gene to be activated. It is not known whether a nucleosome physically leaves the DNA or just changes its attachment so as to permit the transcription complex to assemble on the promoter. The term 'remodelling' avoids implication of what exactly is happening in molecular terms. It is not known how many nucleosomes need to be remodelled – a nucleosome and its linker is about 200 bases in length, which is about the size of a promoter, so one might suffice.

How do transcription factors open up gene promoters?

First, one (or more?) transcription factor(s) attach(es) to cognate elements on the promoter and/or the enhancer (in Fig. 23.21(a) only a single one is illustrated). A TF has two domains, one to attach to the DNA element and the other, the activation domain, being free to bind to other proteins. The role of TF binding is to cause the opening up of the promoter by the mechanism described below so that all of the factors required for initiation can bind. There is obviously a chicken-and-egg dilemma here – a TF has to bind to open up the promoter so that factors can bind. The probable answer is that some TFs are able to bind to their DNA elements even before chromatin remodelling has occurred. This leads to remodelling of the promoter, which allows all the other factors to access and assemble on the promoter.

The coactivator

One of the latter factors so far not mentioned now enters the picture – the **coactivator**. This is a protein complex that binds to the activation domain of the TF. The coactivator is not a TF and does not itself bind to DNA but is essential for gene transcription. Its attachment to the DNA-bound TF positions

(a)

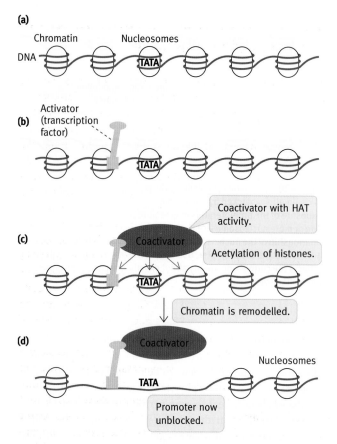

Fig. 23.21 Remodelling of chromatin – the first step in gene transcription. **(a)** Chromatin with a promoter blocked by nucleosomes. **(b)** An activator (a transcription factor) attaches to its site. **(c)** A coactivator with histone acetyltransferase (HAT) activity attaches to the bound activator and acetylates the histones of the blocking nucleosome(s). **(d)** The blocking nucleosomes are removed or remodelled. Several different activators are involved in control of a typical eukaryotic gene (see Fig. 23.16 for illustration of this) and several or all may be involved in recruiting the coactivator. In case of the activation of the thyroid hormone gene, three or four nucleosomes are removed from the DNA.

it close to the promoter and its blocking nucleosomes (Fig. 23.21(b)). A family of coactivators have been identified, two well-known ones being CBP, and PCAF. (To explain the naming of only one of these, **CBP** stands for CREB-binding protein and **CREB** in turn stands for **cAMP-response-element-binding protein**. CREB is a transcription factor activated by a cAMP-induced phosphorylation (Fig. 23.18(b)). It has been shown that when CREB is activated it binds to the coactivator CBP.

An exciting discovery was that several coactivators have an enzymic activity known as **histone acetyltransferase** (HAT; Fig. 23.22). It catalyses the transfer of the acetyl group of acetyl-CoA to the ε-NH$_2$ group of lysine residues in the N-terminal domains of the histone octamer subunits which form the nucleosomes (Fig. 23.21(c)). These domains are exposed on

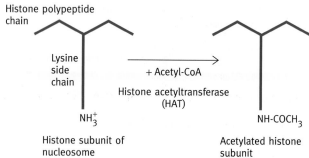

Fig. 23.22 Acetylation of the lysine residues of the N-terminal tails of subunits of the histone octamers of nucleosomes. The acetylation reduces the positive charge on the protein and is believed to result in lessening the attachment of DNA to the nucleosome leading to the chromatin remodelling described in the text.

the surface of the nucleosomes like short tails so that they are accessible to HAT activity. Acetylation eliminates the positive charge on the amino groups and is believed to loosen the attachment of the negatively charged DNA to the nucleosome and nucleosome-nucleosome interactions. It appears to be a factor in chromatin remodelling (Fig. 23.21(d)) and therefore for opening up the gene prior to initiation.

It is not the only mechanism known for remodelling chromatin; separate ATP-dependent remodelling machines have been discovered first in yeast and later in humans.

How is transcription initiated on the opened promoter?

The next step is to assemble the **basal initiation machinery** on the promoter. Let us look at the components of this first. In all cells there are **general transcription factors** required for initiation of all genes. One of these is a large complex called **TFIID** (transcription factor for type II genes, the D indicating which of several it is; about 10 different proteins are involved in the complex). The heart of this complex is the **TATA-box-binding protein (TBP)**, which attaches the TFIID to the TATA box. The other components are known as **TAFs** (TBP-Associated Factors; Fig. 23.23). RNA polymerase II is a large multisubunit enzyme that exists in the nucleoplasm (essentially the nuclear equivalent of cytoplasm). It joins up to the TFIID complex on the TATA box; several other general initiation factors bind such as TFIIB which plays a role in linking up the polymerase to the TFIID. Since the TATA box is in a fixed position relative to the start site the position of TFIID is fixed and therefore the polymerase in joining up to it is automatically correctly positioned at the start site and pointing in the right direction. As already mentioned, rather disconcertingly some genes do not have a TATA box; in these cases it is the **Inr** and the **DPE** (page 364) that positions the basal transcription machinery.

The opening up of the promoter now permits the collection of sequence-specific TFs to attach to their appropriate boxes.

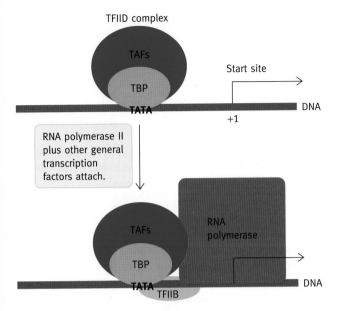

Fig. 23.23 Diagrammatic representation of the components of the basal initiation complex. TFIID is a complex of the TATA-box-binding protein (TBP) and a number of TAFs (TBP-associated factors). RNA polymerase attaches to the preinitiation complex and forms the basal transcription complex. This figure is to illustrate the components of the basal complex but this does not exist in isolation; it is part of a very large assembly and interacts with transcription and other factors (see Fig. 23.24).

Somehow the gabble is integrated to give the intensity of transcription that should occur on the gene.

Transcription repressors

We have so far described transcriptional factors in terms of them being activators of transcription. As emphasized in the introduction to eukaryotic gene control, transcription regulation is achieved by a variable number of signals to a given promoter which may be both negative and positive. This gives stable and finely balanced control. The negative control is achieved by transcription factors that are repressors. The repression may be due to simple competition with activators for their binding sites on the DNA. Or they may bind near the activator and block the activation site of the latter. Another possibility is that they may cause gene switch off by preventing the opening up of the chromatin structure around the gene (such as by recruiting histone deacylase; this will be explained on page 371).

Finally it may be that repressors work in a manner equivalent to that of activators – they combine with regulatory sites on the promoter and contact the transcription complex assembled on the gene. This is particularly relevant to a model of repression that suggests that repressors prevent the mediator forming a connecting link between transcriptional factors and the polymerase (page 370).

Discovery of the mediator

The mediator was discovered by Roger Kornberg. He was awarded the Nobel Prize in 2006. He studied yeast transcription because it is somewhat simpler than in higher eukaryotes. On the basis of what was known of the mechanism of transcription of genes by polymerase II, it was possible to assemble *in vitro* the known essential components for transcription in a partially purified form. He found that the assembled components transcribed genes in what is known as basal transcription but did not respond to addition of activators (TFs). Something was missing. Addition of crude (unpurified) yeast extracts raised the basal level of transcription and the system now responded to activators, suggesting that there was a factor that permitted TFs to control Pol II. This factor was given the name of **mediator** because it apparently mediated the transmission of regulatory information between TFs and the polymerase within the initiation complex (Fig. 23.24). A more mechanistic model of the mediator-pol II complex is given in Fig. 23.25.

The discovery was viewed at first with some scepticism because when isolated, the yeast mediator fraction turned out to be an immense complex of many protein subunits. Since the initial discovery, many studies have decisively confirmed the reality of the mediator. It is present in all eukaryotes, for the transcription of Pol II specific genes. The protein components of the mediator have been identified and many of their

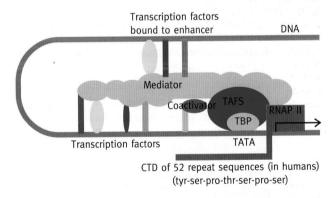

Fig. 23.24 Cartoon of final initiation complex. The sizes, positions and interactions of the various components are conjectural. They represent the binding of the general transcription factors (TBP and associated TAFS of the TFIID complex) to the TATA box which positions the RNA polymerase II (RNAPII) at the correct site. The upstream sequence-specific transcription factors of the promoter interact with the mediator as do those of the enhancer. The mediator is a complex of 30 proteins in humans; its representation is based on the fact that it does not bind to DNA but forms a physical connection between the polymerase and other components of the initiation complex. By an undefined mechanism it is believed to convey instructions from the various regulatory elements to the enzyme. The CTD (C-terminal domain) of the polymerase becomes phosphorylated on the serine residues before elongation (see text).

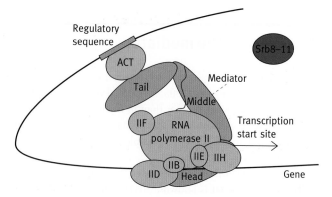

Fig. 23.25 A model of the believed association of activators with the mediator on the one hand and the Pol II transcriptional complex. The mediator (blue) functions as a bridge between gene-specific activators (ACT, red) and the general Pol II transcription machinery (purple) at the promoter. Activator interactions mainly take place within the tail region of the mediator, whereas contacts with Pol II are localized to the head and middle region. A subgroup of mediator components forms a module (Srb8–11, pink) that is involved in negative regulation of transcription. Only mediator lacking the Srb8–11 module can associate with Pol II. A model for transcriptional repressor action is that the Srb8–11 module associates with the mediator and causes the active mediator-pol II complex connection to dissociate.

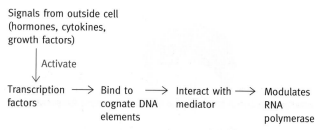

Fig. 23.26 Sequence of events believed to control RNA transcription in eukaryotic cells following opening up of the promoter and attachment of the general transcription complex.

Fig. 23.26 summarizes what currently appear to be the association between mediator components and Pol II in the initiation complex assembly leading to the control of RNA polymerase. A major question is the nature of the information transmitted by the mediator to the polymerase.

The RNA polymerase II of eukaryotic cells

The polymerase for mRNA production is a complex of sub units one of which has a long **C-terminal domain** (CTD) made of, in mammals, 52 repeating peptide heptads with the sequence Tyr-Ser-Pro-Thr-Ser-Pro-Ser (Fig. 23.24). For synthesis of the RNA transcript (elongation) the polymerase must detach from the initiation complex. This is associated with phosphorylation of the serine residues of the CTD – two serines on every repeat. During transcription, the template strand of the DNA is positioned in a groove in the polymerase with the non-template strand outside. When transcription begins, the first eight nucleotides of the RNA transcript forms a duplex with the template strand after which a lobe of the enzyme diverts it (the RNA) along a groove so that it exits in the direction of the CTD. The template DNA strand exits behind and then reassociates with the nontemplate strand (Fig. 23.27).

The groove of the enzyme in which the template strand resides is a claw-like structure which closes and firmly attaches the DNA to the enzyme. This is essential because the gene transcribed may be enormous in length and if it prematurely detached there is no mechanism for it to reattach. After 20–30 nucleotides have been added, there is a pause during which the RNA is capped (page 360). The function of the phosphorylated CTD may be to position the appropriate capping enzymes and possibly other factors so that they can act on the emerging RNA transcript. The polymerase then proceeds along the template strand, melting the DNA as it moves forward. The mRNA resulting from the spliced transcripts is finally 'packaged' for transport out of the nucleus by attachment of proteins (for nuclear-cytoplasmic transport, see page 421).

interactions mapped. The mediator isolated in pure form from mammalian cells has 30 protein subunits, 22 of which are homologous with yeast mediator subunits.

The mediator is required both for activator and repressor control of transcription. This was proved in a compelling way. Previous screening by other workers had identified many yeast mutants that were deficient in transcriptional control. In 13 of these, the deficient genes were found to code for mediator protein subunits. The mutants included ones deficient in activation and others in repression control confirming that the mediator is involved in both negative and positive control.

Homologues of most of the yeast mediator subunits are found in humans. The yeast and human mediators are similar in their arrangement in the complexes and have similar shapes.

The mediator interacts with Pol II and with regulatory element binding proteins on the DNA

In three cases there is direct evidence that the mediator binds to TFs including those specific for the thyroid and sterol regulatory elements. The mediator connects TFs and Pol II in the transcriptional complex. Structural studies have shown that it forms a crescent shape at one end that envelops the polymerase. A model of the believed association is shown in Fig. 23.25. The compact mediator structure unfolds into the open crescent shaped structure in the presence of activator. (See **Bjorkland and Gustafsson** in Further reading.) A model for the mechanism of repression has also been proposed in which a transcription repressor prevents association between the mediator and the polymerase complex.

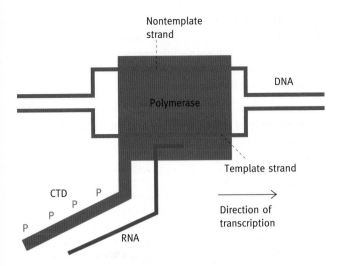

Fig. 23.27 Diagrammatic representation of RNA polymerase II transcribing a gene. It is designed to represent the enzyme melting the DNA as it progresses; the template strand is in a groove of the enzyme containing the active centre while the other strand takes a separate path. The two re-associate behind the enzyme. The RNA forms a hybrid with the template strand for eight or nine bases but is then diverted to a separate exit from the enzyme in the direction of the phosphorylated C-terminal domain (CTD). It has been postulated that the latter has enzymes attached for capping and splicing the RNA transcript as it is synthesized, as well as packing the mRNA for transport into cytoplasm.

As the polymerase moves along reading the template strand it faces a problem not found in prokaryotes, namely how to negotiate the nucleosomes, but this is not fully understood yet.

Termination of transcription in eukaryotic cells

Termination of transcription in eukaryotic cells is less understood than the prokaryotic mechanism (page 357). At the end of the RNA there is a polyadenylation signal (AAUAAA). The polymerase transcribes a short distance beyond this and then, somehow, terminates. The RNA is cleaved near the polyadenylation signal and another enzyme, not dependent on a template, adds as many as 200 adenine nucleotides (using ATP as their source) to form a polyA tail. (For the role of the polyA tail, see below.) Histone mRNA does not have this tail.

After transcription is completed, the CTD is dephosphorylated and the polymerase returns to the promoter for another round of transcription.

Switching off the gene

Gene control requires switching off genes when their transcription should cease. This is dependent on the initial activating signal from hormones and cytokines no longer arriving. The TFs which have been activated are deactivated, often by dephosphorylation. The full breaking up of the initiation complex may be

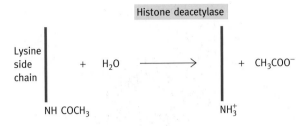

Fig. 23.28 Reaction catalysed by histone deacetylase.

a function of histone deacetylases (Fig. 23.28), which reverse the initial activation. The nucleosomes can then re-establish themselves on the promoter, which is thereby closed down.

DNA methylation affects gene transcription

Methylation of bases of DNA occurs in bacteria and in eukaryotes. In bacteria it is concerned with preventing restriction enzymes from hydrolysing the cell's own DNA; the restriction enzymes are a protection against bacteriophage infection as discussed on page 459.

In mammals, which this section is concerned with, it has a different role, not well understood. A few general remarks: It is essential in vertebrates; mice lacking the ability to methylate DNA are nonviable. DNA methylation is *associated* with inhibition of gene expression but the relationship between the two is not a simple one. It cannot be said that methylation necessarily itself results in gene silencing but rather that it appears to reinforce or confirm gene silencing by other means. Heterochromatic areas of the genome whose genes are inactive are methylated while the areas with active genes tend not to be. Thirdly, its effect is believed to be associated with chromatin remodeling, but there may be other mechanisms too. We will now discuss methylation in more detail.

In higher vertebrates only cytosine is methylated; it occurs at the N-5 position as shown in the diagram.

It occurs only where there is a CG sequence. There are so called 'C—G islands' near gene promoters which remain unmethylated. There are two classes of methyltransferases both using S-adenosylmethionine as methyl donor (page 288). There are the *de novo* **methyltransferases**, which methylate previously unmethylated DNA, and there are **maintenance methyltransferases**. After fertilization of a vertebrate oöcyte, the DNA is largely demethylated and during development the de novo enzymes establish a new labelling pattern which can change during development. It is not certain how the new methylation pattern is determined.

What is the function of the maintenance methyltransferases? It is to copy the methylation pattern of double-stranded DNA when cell replication occurs. To explain this more fully, when the two strands of chromosomal DNA replicates, the daughter double strands each have a parent strand which is methylated and a new strand which is not methylated. The maintenance methyltransferases copy the parental strand's methylation onto the daughter strand's complementary CG pair, so that in cell multiplication the methylation pattern is inherited, but the pattern is not usually inherited by the next generation due to demethylation after fertilisation. However in a few instances the methylation pattern is locked in and is somewhat unstably inherited by offspring – known as **genetic imprinting**. It is a case of epigenetic control (heritable characteristics not due to changes in the nucleotide sequence of the DNA).

Possible roles of DNA methylation in gene silencing

In one case it is clear that DNA methylation is involved in preventing the transcription of ribosomal RNA genes. A single GC dinucleotide present in the promoter is methylated by de novo transmethylase. Other possibilities are that methylation of gene promoters may block TFs from binding or may recruit chromatin remodelling complexes to 'close down' the chromatin.

mRNA stability and the control of gene expression

Although regulation of gene transcription is of overriding importance in the control of gene expression, the stability of individual mRNAs is also of significance. In most situations, the rate of synthesis of a protein is a reflection of the mRNA level for that protein except where translation of a messenger is controlled. The level of an mRNA in a cell is a function of its synthesis and breakdown rates. At a given rate of mRNA production, a messenger with a longer half-life will be present at a higher steady-state level within the cell than a less stable one, resulting in a higher rate of synthesis of the cognate protein. Mechanisms that determine the half-life of an mRNA can thus provide a way of regulating gene expression.

Prokaryotic mRNAs in general have an ephemeral existence, with half-lives of 2–3 minutes. In mammals, the half-life of individual mRNAs ranges from about 10 minutes to 2 days. The mRNA for globin, a stable one, has a half-life of about 10 hours. Regulatory proteins tend to be coded for by short-lived messengers so that changes in the rate of transcription of their genes

have rapid effects on their synthesis; the relevant messengers usually have half-lives of less than 30 minutes. A given cell thus contains mRNAs with widely different rates of degradation, and this may change in individual mRNAs according to the prevailing conditions in the cell.

Determinants of mRNA stability and their role in gene expression control

Role of the polyA tail

Almost all eukaryotic mRNAs have a polyA tail added to their 3′ ends before they emerge from the nucleus (Fig. 23.29(a)), histone mRNAs being exceptions. There are still some uncertainties about the polyA tail but it seems that it protects the messengers from rapid destruction. A polyA-binding protein binds to the tail and protects the mRNA from destruction by a 3′ → 5′ exonuclease. It also may protect the other end of the mRNA; presumably it loops around and brings the polyA-binding protein into contact with the 5′-methylguanosine cap and shields the latter from attack by a decapping enzyme. Removal of this cap exposes the 5′ end of the mRNA to attack by a 5′ → 3′ exonuclease. mRNA destruction is preceded by deadenylation. When the tail is reduced to less than 15 adenylate residues the polyA-binding protein cannot attach and the cap site is removed by the decapping enzyme, thus triggering destruction from both ends of the mRNA molecule.

Structural stability determinants of mRNAs

Histone mRNAs lack a polyA tail and their stability is determined by a stem loop at the extreme 3′ end of the molecules

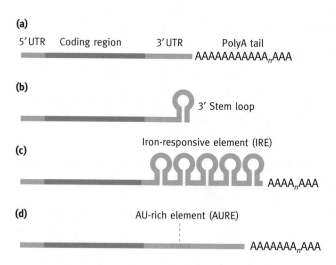

Fig. 23.29 Some structures present in the 3′ untranslated regions (UTRs) of mammalian mRNAs that influence the half-lives of the molecules in the cell: **(a)** the polyA tail found in most eukaryotic mRNAs; **(b)** the G—C rich 3′ stem loop found in histone mRNAs; **(c)** the iron-responsive element (IRE) of transferrin mRNA; **(d)** the A—U-rich element (AURE) found in a large number of unstable eukaryotic mRNAs.

(Fig. 23.29(b)). Synthesis of histones is required only in the S phase of the eukaryotic cell cycle (page 507) when DNA synthesis and nucleosome assembly are occurring. Histone genes are transcribed during S phase but this ceases in the G2 phase when the level of histone mRNAs rapidly falls. The latter is partly due to cessation of transcription but, in addition, the half-life of the mRNAs decreases from 40 to 10 minutes, a change dependent on the presence of the stem loop referred to above. The experimental transfer of this feature to globin mRNA destabilized it at the end of the S phase as if it were a histone mRNA. The destabilization of the histone messengers requires the presence of free histone (protein) monomers, which accumulate as soon as DNA synthesis (and therefore nucleosome assembly) ceases at the end of the S phase. Prompt switch-off of histone synthesis is necessary because free histone is toxic to cells.

The synthesis of the **transferrin receptor protein** is another case where mRNA stability regulates synthesis of the protein. The receptor is responsible for the transport of iron into cells (page 398). In the 3′ untranslated region of the mRNA is a group of five stem loops called an **iron-responsive element** (**IRE**; Fig. 23.29(c)). In the absence of iron, an IRE-binding protein attaches to the IRE and stabilizes the mRNA, thus increasing receptor synthesis with a consequent increase in the import of iron into the cell. In iron abundance, the metal complexes with the protein, which then no longer binds to the IRE. It is believed that this exposes the region to attack by a ribonuclease, resulting in reduction of the import of iron into the cell.

Some short-lived mRNAs contain 'instability' sequences in their nucleotide sequence that mark the molecules for rapid destruction. A structural feature of many unstable mRNAs is the **adenine/uracil-rich element** (**AURE**) found in the 3′ untranslated regions of the molecules (Fig. 23.28(d)). These elements vary from one messenger to another but all contain AU sequences at least nine bases in length. They cause messenger destabilization, possibly by forming recognition (attachment) sites for endonucleases which hydrolyse the messenger.

It is also known that several hormones modulate the stability of specific messengers in target cells. It seems that many different controls determine different mRNA lifetimes.

Gene transcription in mitochondria

Mitochondria are replicating organelles of eukaryotic cells, with their own DNA and protein-synthesizing machinery (the same is true of plant chloroplasts). They divide to maintain the number appropriate to the cell – about 1000 in rat liver and 10^7 in a frog's oöcyte. The DNA of mitochondria is a circular duplex – less than 20,000 nucleotides in mammals, but there can be 5 or 10 copies in each mitochondrion.

In mitochondria *both* strands of the DNA are completely transcribed into single long transcripts (the polymerase molecules moving in opposite directions). The single primary transcripts are then processed into mRNAs, tRNAs, and rRNAs.

The mitochondrial transcriptional system has prokaryotic and eukaryotic features. In mammalian mitochondria, the mRNAs are polyadenylated (eukaryotic) but not capped (prokaryotic), and there are no introns in the genes (prokaryotic).

Editing of mRNAs

In trypanosome mitochondria, mRNAs are produced which then have to be 'edited'. RNA transcripts have extra U residues inserted at specific places to produce the correct coding sequence for proteins. These Us are *not* coded for in the DNA. Editing occurs even in nuclear transcripts. Thus a gene for mammalian apolipoprotein B, coded for by a nuclear gene, has two mRNA transcripts. One of these has to have a DNA-coded C converted to a U to form a translational stop codon (see next chapter).

Transcription of noncoding genes

Noncoding genes are those which do not code for proteins. The RNA molecules they give rise to are stable end products. Most of the RNA is ribosomal RNA (rRNA). This forms the core of ribosomes.

Most eukaryote cells have several hundred such genes in tandem clusters. The oöcyte of the South African toad (*Xenopus*) is very large and has about 2 million rRNA genes. This enables large numbers of ribosomes to be made.

In animals, the gene clusters are located near the tips of five chromosomes and these are associated with a region of the nucleus called the **nucleolus**. There are four different rRNA components in eukaryotic ribosomes (three in *E. coli* ones). These are described in the next chapter (page 386). Three of the four rRNAs are transcribed from a single gene giving a primary transcript. This is broken down into the mature rRNA components and also is chemically modified. **Small nucleolar RNAs** (**snoRNAs**) are involved in the precise processing of the primary transcripts, possibly acting as **guide RNA molecules**, which hybridise to the transcript by specific base pairing and guide the processing enzymes to the correct sites. The fourth rRNA molecule is derived from the processing of a different cluster of rRNA genes.

There is yet another type of RNA molecule involved in protein synthesis – a smaller type called **transfer RNA** (**tRNA**) (described in the next chapter, page 383). These are processed from various primary transcripts of rRNA genes.

In eukaryotes these noncoding RNAs are produced by polymerases I and III, different from the polymerase II which produces messenger RNAs. Pol I and III do not have the C-terminal extension which is believed to be the site where mRNA capping and **polyadenylation** occurs. None of the noncoding RNAs are capped or polyadenylated.

DNA-binding proteins

The term DNA-binding proteins is reserved for the control proteins that bind to specific sequences. Histones, positively charged proteins, bind DNA tightly too but we don't call them DNA-binding proteins because they are not sequence-specific. From what has been said in this chapter, it is clear that gene-regulatory proteins that bind to specific sites on DNA play a central role in control. There are numerous repressors and transcriptional factors involved in differentiation, embryonic development, and gene control in general and all depend on their ability to recognize and bind to sequence elements like those in promoters. The gene-controlling proteins bind to double-stranded DNA in which the bases are already Watson–Crick hydrogen bonded to each other so that the binding proteins have to recognize the exposed edges of the bases visible in the grooves of the double helix (see Fig. 21.4). Most proteins bind in the major groove.

Sequence-specific binding to a double helix involves non-covalent bond formation between the amino acid side chains of the protein and the DNA bases, though additional stabilizing bonding to the sugar-phosphate-sugar backbone may occur. In many cases, the contact is made by a 'recognition α helix' (see page 50 for α helix) that fits into the major groove of the DNA. The edge view of attachment groups is more variable here than in the minor groove and thus has more potential recognition specificity.

There are large numbers of different TFs but they can be grouped into a small number of families on the basis of characteristic structures of the protein 'motifs' involved in the recognition. The motifs are small parts of the whole recognition proteins; the latter usually attach as dimers with the attachment site of each monomer located adjacently in the major groove. The sites that the DNA binding proteins recognise are usually sections of about 20 base pairs in length. We will now describe four different classes of DNA binding proteins on the basis of their attachment motifs.

Helix–turn–helix proteins

This was the first type of DNA-binding protein to be identified and the most studied; it occurs commonly in prokaryotes and eukaryotes. We will use as an example the bacteriophage **CRO** whose binding mechanism was first elucidated. It is involved in controlling the lysogenic/lysis switch in the phage life cycle. The lambda repressor protein and the cAMP catabolite gene-activator protein of *E. coli* are essentially similar.

The helix–turn–helix (HTH) motif is a small section of the protein that makes the binding contact to the operator site on the DNA. It is not a protein domain for it could not exist as such in isolation. The motif has two α helices linked by a β turn; one of the two (the recognition helix) sits in the major groove of DNA (Fig. 23.30(a)).

The CRO protein is a dimer, the recognition helices of the two HTH motifs fitting into adjacent binding sites (Fig. 23.30(b)).

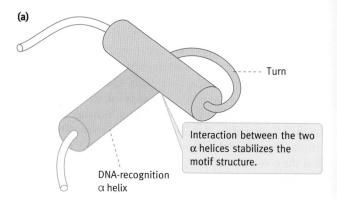

(a)

Turn

Interaction between the two α helices stabilizes the motif structure.

DNA-recognition α helix

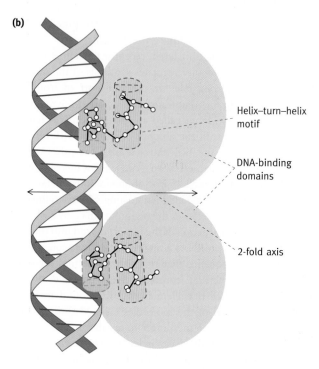

(b)

Helix–turn–helix motif

DNA-binding domains

2-fold axis

Fig. 23.30 (a) The helix–turn–helix motif of a DNA-binding protein monomer. **(b)** Dimer helix–turn–helix protein binding to DNA in major grooves.

The use of a dimer rather than a monomer presumably gives tighter binding, for the free-energy fall is greater than the sum of the two separately. In eukaryotes the **homeodomain proteins** are HTH proteins controlling **homeotic genes** that are of great importance in development. (The homeodomain proteins themselves are encoded by homeotic genes.)

Leucine zipper proteins

These are found in many eukaryotic transcription factors, and are really protein–protein recognition motifs. Note that the name does *not*, in this case, refer to the DNA-recognition motif as in HTH proteins, but rather to the characteristic structure found in these proteins that causes dimerization of two subunits. The whole domain is an α helix but, in the 'leucine zipper' motif, every seventh amino acid is leucine. Since there are 3.6 residues per turn, the leucines all appear on the same side of

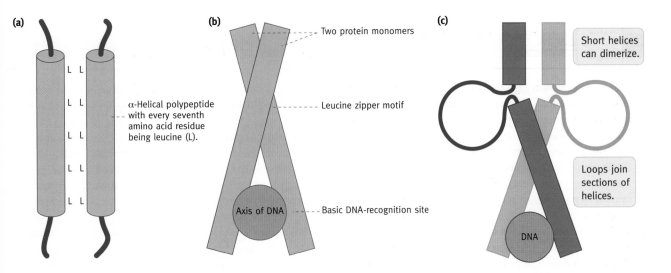

Fig. 23.31 **(a)** The leucine zipper motif. The hydrophobic leucine residues are opposed, not interdigitated as in a zipper. **(b)** The leucine zipper protein attached to DNA, looking down the axis of DNA. The two arms of the protein lie in adjacent sections of the wide groove. The attachment sites are rich in basic amino acids. **(c)** The helix–loop–helix is a different type of dimerization in that the flexibility of the loop allows the short helical sections to bind to each other.

the α helix forming a row of hydrophobic faces. Two such subunits attach by hydrophobic forces between the leucine side chains (Fig. 23.31(a)). The term 'zipper' is a misnomer resulting from the initial belief that the leucine residues interdigitated. The DNA-binding region of each monomer is a region rich in positively charged arginine and lysine residues. The dimer attaches to the DNA in a 'scissor' grip, the two arms being in adjacent major groove sections of the duplex (Fig. 23.31(b)). The dimers may be of the same or different monomers; heterodimers recognize different adjacent sites thus giving flexibility of control.

Helix–loop–helix proteins

A type of dimerization occurs in the **helix–loop–helix** (HLH) that is different from the HTH. In this the α helix of each monomer is interrupted by a polypeptide loop that gives flexibility for the short helices of the dimer to associate (Fig. 23.31(c)).

Zinc finger proteins

The zinc finger motif is found very commonly in eukaryotic DNA-binding proteins; over 200 are known and about 2% of human genes code for them. The name derives from a finger-like projection of the polypeptide chain which binds inside the wide groove of DNA. The first discovered, 'classic' type, is a finger structure of about 30 amino acid residues stabilized by a zinc atom bound to four residues two cysteines and two histidines, but in some, the zinc-binding residues are four cysteines. Several of these fingers often occur as a group so that successive fingers can bind into the major grooves of the DNA giving firm attachment of the protein. The classic type is illustrated in Fig. 23.32. In all zinc fingers one side of the structure is an α helix that recognizes the binding site on the DNA.

An important class of regulatory protein that contains zinc fingers with four cysteines are the steroid receptor family (dealt with in Chapter 27; page 433). These proteins dimerize or heterodimerize. They then bind to their gene elements by zinc fingers. The family includes, as well as steroid binders, those for thyroid hormone, retinoic acid, and vitamin D. The transcription factor TFIIIA which regulates the 5 S RNA gene has nine zinc fingers.

In addition to these zinc finger-binding proteins there are other classes based on the presence of one (or sometimes two) zinc atoms. Although it was thought that zinc fingers were only for binding of proteins to DNA, it has been shown that protein-protein interactions can be involved (see **Gamsjaeger** in Further reading).

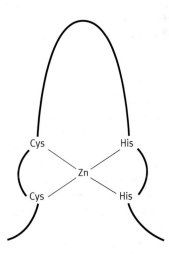

Fig. 23.32 Diagram of one type of zinc finger structure. This motif inserts into the major groove of DNA and binds to five base pairs by its recognition α helix. Different proteins have variable numbers of zinc fingers arranged sequentially which bind into successive regions of the major groove in the DNA giving firm attachments.

Summary

Genes are sections of huge DNA molecules (viral genes may be RNA) which are the chromosomes. To be expressed (to cause production of a protein) one strand of the DNA of a gene, the template strand, is copied (transcribed) into single-stranded RNA. The coding region of a prokaryotic gene is a continuous stretch of DNA so that the primary transcript is a messenger RNA (mRNA).

Each gene has adjacent to its 5′ end a promoter region, which controls its transcription at the level of initiation. The control requirements are different in prokaryotes and eukaryotes. While the basic mechanism of RNA synthesis is the same in both, the mechanism of initiation and its control in the two are very different.

In prokaryotes, RNA is transcribed by a single RNA polymerase using ATP, CTP, GTP, and UTP as building blocks. RNA polymerases do not require a primer. The promoter has a DNA sequence known as a TATA box and a Pribnow box further upstream. The RNA polymerase directly attaches at these sites, which position the enzyme correctly. The rate of mRNA production is determined by the affinity of the polymerase for the promoter. As the polymerase moves along the DNA it separates the two DNA strands forming a transcription bubble and copies the template strand into RNA. Termination of transcription at the end of the coding region of the gene can occur in either of two ways. A stem loop in the mRNA formed by G—C base pairing followed by a string of U residues causes detachment from the DNA template. Alternatively a Rho factor, which is a helicase, unwinds the mRNA from the DNA, causing detachment.

In *E. coli*, groups of genes called operons often occur and are transcribed together forming polycistronic messengers. In the *lac* operon, comprising three genes, a repressor protein effects control by blocking an operator region at the initiation site of transcription. The repressor is an allosteric protein and in the presence of lactose it detaches and allows transcription of genes required to form enzymes needed to utilize the sugar. The actual inducer is allolactose formed from lactose. This type of operon control is prevalent in other metabolic systems.

Eukaryotic gene control is also at the level of transcription initiation but the process is more complex. Eukaryotic cells of an animal have to respond to batteries of signals from hormones, growth factors, and cytokines, so that a given gene is likely to be instructed by a multiplicity of signals which the initiation process interprets.

mRNA production in eukaryotes is complicated because most genes are split; the protein-coding region of each gene is broken up into segments known as exons separated by noncoding introns. The transcript is 'capped' by the addition of a GMP moiety to the end of the RNA, followed by addition of methyl groups. The transcript is spliced by spliceosomes to remove exons and link the introns into a continuous mRNA molecule. In rare cases, transcribed RNAs splice themselves without the aid of proteins, the first case of catalytic RNAs. The eukaryotic mRNAs are transported into the cytoplasm *via* pores in the nuclear membrane.

Eukaryotic genes are in a closed-down default state – they are inactive until something is done to change this. The gene promoters are blocked by nucleosomes (Chapter 21). The key to eukaryotic gene control are transcription factors. These are proteins that bind to gene regulatory sequences upstream of the start site; they open up specific promoters and recruit coactivator proteins that acetylate nucleosome histones and unblock the promoter (known as DNA remodelling since it is not clear exactly what happens). Some genes have enhancers, large distances away from the gene, to which transcription factors also attach and enhance gene activity. Sections of DNA, called insulators, confine enhancer effects to their intended target gene(s).

Eukaryotic transcription is carried out by three RNA polymerases; polymerase II is the one dealing with protein-coding enzymes. Transcription by it depends on a host of transcription factors. They attach to the DNA at upstream DNA sites (boxes). A protein complex known as the mediator completes the initiation complex. The initiation complex recruits the polymerase from the nucleoplasm. We therefore have a large complex of transcription factors, coactivators, mediator, and the polymerase all interacting in not yet fully defined ways and somehow this gives control over the rate at which new messengers are initiated. The gene has a termination signal that causes release of the transcript, which is usually polyadenylated by a separate enzyme.

Controls have to be reversible. Transcription factors are inactivated. Before they function many transcription factors require activation by phosphorylation, so dephosphorylation inactivates them. Deacetylases restore nucleosomes to their original state to give the shut-down condition.

Other genes code for ribosomal and transfer RNAs essential for protein synthesis (Chapter 24). Further gene expression control may be exerted by mRNAs having different stabilities.

Transcription of the small circular genome in mitochondria has basically the same synthesis mechanism but the transcription is different in detail.

The recognition of specific DNA sequences by the various transcription factors is the crucial aspect of transcription. There are a number of specific patterns of DNA-binding proteins. These include helix–turn–helix proteins, leucine zipper proteins, helix–loop–helix proteins, and zinc finger proteins. They usually are dimers that recognize two DNA sequences visible in the broad grooves of the double helix.

Further reading

mRNA editing

Hodges, R. and Scott, J. (1992). Apolipoprotein B mRNA editing; a new tier for the control of gene expression. *Trends Biochem. Sci.*, **17**, 77–81.
How mRNA editing leads to the production of two forms of apolipoprotein B from one mRNA transcript.

Proudfoot, N. J., Furger, A. and Dye, M. J. (2002). Integrating mRNA processing with transcription. *Cell*, **108**, 501–12.
Reviews eukaryotic transcription and subsequent modification of transcripts.

Introns and exons

Gilbert, W. (1987). The exon theory of genes. *ColdSpring Harbor Symp. Quant. Biol.*, **LII**, 901–5.
Describes the 'introns early' theory of assemblage of genes from mini-genes.

Go, M. and Nosaka, M. (1987). Protein architecture and the origin of introns. *Cold Spring Harbor Symp. Quant. Biol.*, **LII**, 915–24.
Considers the two theories of the origins of introns.

Mattick, J. S. (2003). Challenging the dogma: the hidden layer of nonprotein-coding RNAs in complex organisms. *BioEssays*, **25**, 930–9.

Protein modules or domains

Doolittle, R. F. (1995). The multiplicity of domains in proteins. *Annu. Rev. Biochem.*, **64**, 287–314.
A fascinating discussion of domains, domain shuffling, exons, and introns.

Splicing

Breitbart, R. E., Andreadis, A., and Nadal-Ginard, B. (1987). Alternative splicing: a ubiquitous mechanism for the generation of multiple protein isoforms from single genes. *Annu. Rev. Biochem.*, **56**, 467–95.
Fairly detailed but gives an overview of the biological role.

Orgel, L. E. (1994). The origin of life on earth. *Sci. Am.*, **271**(4), 52–61.
Growing evidence supports the idea that the emergence of catalytic RNA was a crucial early step.

Nilsen, T. W. (2003). The spliceosome: the most complex macromolecuar machine in the cell? *BioEssays*, **25**, 1147–9.
A short informative review on how introns are excised rom transcripts.

RNA self-cleavage

Fedor, M. J. (1998). Ribozymes. *Curr. Biol.*, **8**, R441–3.
Concise summary.

Chromatin remodelling

Pollard, K. J. and Peterson C. L. (1998). Chromatin remodelling: a marriage between two families. *BioEssays*, **20**, 771–80.
Extensive review of two families of remodelling enzymes.

Gregory, P. D. and Horz, W. (1998). Life with nucleosomes: chromatin remodelling in gene regulation. *Curr. Opin. Cell Biol.*, **10**, 339–42.

Kornberg, R. D. (1999). Eukaryote transcriptional control. *Trends Cell Biol., Trends Biochem. Sci.* and *Trends Genet.* (joint issue), **24**, M46–8.
Millenium review including chromatin remodelling. A short overview of the subject including an account of what still needs to be done.

Sudarsanam, P. and Winston, F. (2000). The Swi/Snf family: nucleosome-remodeling complexes and transcriptional control. *Trends Genet.*, **16**, 345–50.

Lusser, A. and Kadonaga, J. T. (2003) Chromatin remodeling by ATP-dependent molecular machines. *BioEssays*, **25**, 1192–200.
How nucleosome structure may be re-organized during chromatin remodelling.

Svejstrup, J. Q. (2003). Histones face the FACT. *Science*, **301**, 1053–4.
This Perspectives article summarizes new evidence on how histones cope with transcription.

Control of eukaryotic transcription

Kiermaier, A. and Eilers, M. (1997). Transcriptional control: calling in histone deacetylase. *Curr. Biol.*, **7**, R505–7.

Siegfried, Z. and Cedar, H. (1997). DNA methylation: a molecular lock. *Curr. Biol.*, **7**, R305–7.
Modulation of gene expression by DNA methylation.

Kuo, M.-H. and Allis, C. D. (1998). Roles of histone acetyltransferases and deacetylases in gene regulation. *BioEssays*, **20**, 615–26.

Ogbourne, S. and Antalis, T. M. (1998). Transcriptional control and the role of silencers in transcriptional regulation in eukaryotes. *Biochem. J.*, **331**, 1–14.
A review article.

Struhl, K. (1999). Fundamentally different logic of gene regulation in eukaryotes and prokaryotes. *Cell*, **98**, 1–4.
Minireview giving a good overview.

Fiering, S., Whitelaw, E., and Martin, D. I. K. (2000). To be or not to be active: the stochastic nature of enhancer action. *BioEssays*, **22**, 381–7.

Ahringer, J. (2000). NuRD and SIN3 histone deacetylase complexes in development. *Trends Genet.*, **16**, 351–6.

Brown, C. E., Lechner, T., Howe, L., and Workman, J. L. (2000) The many HATs of transcription coactivators. *Trends Biochem. Sci.*, **25**, 15–18.
A discussion of the multiple histone acetyltransferases.

Malik, S. and Roeder, R. G. (2000). Transcriptional regulation through mediator-like coactivators in yeast and metazoan cells. *Trends Biochem. Sci.*, **25**, 277–83.

Myers, L. C. and Kornberg, R. D. (2000). Mediator of transcriptional regulation. *Annu. Rev. Biochem.*, **69**, 729–49. *More suitable for instructors.*

Orphanides, G. and Reinberg, D. (2002). A unified theory of gene expression. *Cell*, **108**, 439–51. *Wide-ranging review including concept that DNA transcription in eukaryotes is one continuous smooth process up to transport of mRNA from the nucleus.*

Freiman, R. N. and Tjian, R. (2003). Regulating the regulators: lysine modifications make their mark. *Cell*, **112**, 11–17. *Reviews the concept that covalent modifications fine-tune transcription in different organisms. Also discusses the relevance to lack of the correlation between complexity and gene numbers.*

Levine, M. and Tjian, R. (2003). Transcription regulation and animal diversity. *Nature*, **424**, 147–51. *Transcriptional complexity may be the answer to why humans have relatively few genes.*

Bjorklund, S. and Gustafsson, C. M. (2005). The yeast mediator complex and its regulation. *Trends Biochem. Sci.*, **30**, 240–4.

Kornberg, R. (2005). Mediator and the mechanism of transcriptional activation. *Trends Biochem. Sci.*, **30**, 235. *Introduces a special edition on mediators.*

Berger, S. L. (2007). The complex language of chromatin regulation during transcription. *Nature*, **447**, 407–11. *An Insight review.*

Ferguson-Smith, A. C. and Greally, J. M. (2007). Perceptive enzymes. *Nature*, **449**, 148–9. *Adding methyl groups to DNA is a way of regulating some genes and genomic sequences – a window on epigenetic printing. A News and Views article.*

Paik, W. K., Paik, D. C., and Kim, S. (2007). Historical review: the field of protein methylation. *Trends Biochem. Sci.*, **32**, 101–52.

Reik, W. (2007). Stability and flexibility of epigenetic gene regulation in mammalian development. *Nature*, **447**, 425–32, *An Insight review. An advanced review comparing the concepts of gene regulation by transcription factors and by methylation. Probably more suitable for instructors.*

Merger, S. L. (2007). The complex language of chromatin regulation during transcription. *Nature*, **447**, 407–12. *An advanced Insight review dealing with broad concepts. Probably more suitable for instructors.*

Termination of transcription

Reeder, R. H. and Lang, W. H. (1997). Terminating transcription in eukaryotes: lessons learned from RNA polymerase I. *Trends Biochem. Sci.*, **22**, 473–7.

Richardson, J. P. (2003). Loading Rho to terminate transcription. *Cell*, **114**, 157–9. *Minireview on the Rho mechanism.*

RNA polymerase machinery

Landick, R. (2001). RNA polymerase clamps down. *Cell*, **105**, 567–70. *Discusses mechanism of RNA polymerase holding on to DNA being transcribed.*

Woychik, N. A. and Hampsey, M. (2002). The RNA polymerase II machinery: structure illuminates function. *Cell*, **108**, 453–63. *Comprehensive review of eukaryotic transcriptional machinery.*

Transcription in mitochondria

Clayton, D. A. (1984). Transcription of the mammalian mitochondrial genome. *Annu. Rev. Biochem.*, **53**, 573–94. *If differs from nuclear transcription. General review of the topic.*

Asin-Cayuela, J. and Gustafsson, C. M. (2007). Mitochondrial transcription and its regulation in mammalian cells. *Trends Biochem. Sci.*, **32**, 111–7.

DNA-binding proteins

Brennan, R. G. and Matthews, B. W. (1989). Structural basis of DNA-protein recognition. *Trends Biochem. Sci.*, **14**, 287–90. *Reviews the structures of proteins involved in the control of gene transcription and the way they interact with DNA.*

DNA-binding proteins – zinc fingers

Klevit, R. E. (1991). Recognition of DNA by Cys_2, His_2 zinc fingers. *Science*, **253**, 1367, 1393. *Concise summary.*

Rhodes, D. and Klug, A. (1993). Zinc fingers. *Sci. Am.*, **263**(2), 56–65. *Discusses structures, functions, and distribution of these transcription factors.*

Klug, A. and Schwabe, J. W. R. (1995). Zinc fingers. *FASEB J.*, **9**, 597–604. *A complete account of the structure of this protein motif and its role in DNA binding.*

DNA-binding proteins – leucine zippers

McKnight, S. L. (1991). Molecular zippers. *Sci. Am.*, **264**(4), 32–9.

Ellenberger, T. E., Brandl, C. J., Struhl, K., and Harrison, S. C. (1992). The GCN4 basic region of leucine zipper binds DNA as a dimer of uninterrupted α helices: crystal structure of the protein-DNA complex. *Cell*, **71**, 1223–37. *A research paper, with molecular illustrations.*

Gamsjaeger, R. *et al.* (2007). Sticky fingers as protein-recognition motifs. *Trends Biochem. Sci.*, **32**, 63–70.

Problems

1 In what ways does RNA synthesis differ from DNA synthesis?

2 By means of notes and diagrams, describe the components of an *E. coli* single gene and its associated flanking regions.

3 Describe the process of initiation of transcription of a gene in *E. coli*.

4 Describe two methods by which, in *E. coli*, gene transcription is terminated.

5 What factors determine, in a constitutive gene of *E. coli*, the strength of a promoter?

6 Describe how the *lac* operon is controlled.

7 Describe, in broad terms, the main ways in which the formation of mRNA in eukaryotes differs from that in prokaryotes.

8 By means of a diagram, explain the mechanism of splicing. What possible biological significance does the existence of introns have?

9 In what ways does eukaryotic initiation of transcription differ from that in prokaryotes?

10 What part does acetyl-CoA play in the initiation of eukaryotic gene transcription?

11 What is the experimental evidence that chromatin loosening is involved in activation of eukaryotic genes?

12 What is one probable way by which an activated eukaryotic gene is inactivated?

13 The control of gene initiation by the acetylase and deacetylase implies that these enzymes are somehow targeted to specific gene promoters. How is this done?

14 Discuss the two viewpoints on the nature of introns.

15 In terms of structural motifs, there are several families of transcription factors. What are these?

16 What are proteomes and genomes? Are they fixed in size?

17 There are estimated to be about 25,000 genes in humans. Does this mean that the maximum number of human proteins is also about 25,000?

Chapter 24

Protein synthesis and controlled protein breakdown

In Chapter 23 we dealt with production of messenger RNA (mRNA). In this chapter we deal with the way in which mRNA directs the assembly of amino acids into the finished protein coded for by the gene from which the mRNA was transcribed.

Protein synthesis occurs on **ribosomes**. These are present in cells in large numbers in the cytoplasm, mitochondria, and chloroplasts; an *Escherichia coli* cell has about 20,000 and they constitute about 25% of the dry weight. Prokaryotic ribosomes and those in mitochondria and chloroplasts are smaller than eukaryotic ones, with 55 proteins and >80 proteins respectively. However, their structures, more details of which will be given in due course, are basically similar, each having a small and a large subunit, each made of many molecules.

Ribosomes are the molecular machines that move along a tape of instructions from genes (mRNA), and assemble the amino acids into the sequences that constitute proteins. Ribosomes work quickly; an *E. coli* ribosome adds about 20 amino acid residues per second at 37°C. They do it with a high degree of accuracy, for an incorrect amino acid residue in a protein could mean a non-functional molecule. The accuracy is about one mistake in the addition of 10^4 amino acids. An error does not have the long-term consequences as in DNA synthesis because the faulty protein is simply destroyed. A few faulty molecules are of no consequence. Evolution seems to have selected an error rate that allows most proteins of a few hundred amino acids to be mainly error-free, and allows an acceptable error-rate in the largest proteins. A greater accuracy might have made protein synthesis too slow.

We have earlier referred to the 'RNA world', now believed to have predated the DNA world. In a sense reminiscent of the way in which astronomers view the earlier state of the universe through long-distance telescopes, in viewing the ribosome we are looking at earlier stages of life. Ribosomes were probably originally only RNA, and proteins were a refining addition later in evolution. As pointed out earlier, one of the essential properties of nucleic acid molecules is that they can, due to specific base-pairing properties, direct their own replication. A 'useful' RNA molecule could have been directly replicated in primitive circumstances but nothing analogous to this is known for proteins. While protein-containing ribosomes today synthesize proteins, if this were true at the origin of life a logical impasse arises – how can a structure dependent on proteins be required for the synthesis of proteins? An ancient RNA-only ribosome explains this.

Protein synthesis is basically the same in prokaryotes and eukaryotes but there are sufficient differences to require some separate treatment. We will first outline the concepts that apply to both, then deal with protein synthesis in *E. coli*, followed by a discussion of differences in eukaryotic systems.

At the end of the chapter we will deal with the targeted breakdown of proteins. It may seem that the breakdown of proteins is a mundane subject analogous to the simple digestion of proteins. Nothing could be further from the truth, for targeted breakdown inside cells is fundamental to many aspects of cellular mechanisms, including the control of cell division in eukaryotes. The mechanisms involved are elegant and have been conserved throughout much of evolution. The key player in this endgame is a cellular structure known as the **proteasome**, yet another astonishing molecular machine.

Essential basis of the process of protein synthesis

mRNA is a long molecule with four different species of bases in its component nucleotides; A, U, G, and C. Proteins are synthesized from 20 different species of amino acids ignoring a relatively rare additional one – selenocysteine (see page 393). They are listed in Table 4.1. The sequence of bases in the messenger specifies the sequence of amino acids in the protein. A one-base code (in which a single base represents an amino acid) could code for only four amino acids; a two-base code could code for 16; still not enough. A three-base code will do it with some excess capacity left over, and this is the situation. A triplet of bases called a **codon** represents each amino acid on the RNA. With four possible bases at each site, 64 different triplets or codons ($4 \times 4 \times 4$) are possible.

The genetic code

The identification of which codon triplets correspond to which amino acids constitutes the **genetic code**. This is almost universal, but not quite. In mitochondria, derived from symbiotic early prokaryotes, there are one or two minor variations with corresponding variations in their translation apparatus; the same is true of some protozoans, but apart from these the code is universal. Although there may originally have been a simpler code, or many codes, one is now fixed, because a single alteration in the code would mean that many, or most, proteins of a modern cell could not be made. The genetic code has been described as 'a frozen accident' but the final 'frozen' genetic code that we have today is not entirely a collection of randomly acquired 'accidental' triplets, but has apparently evolved with a certain logic to it as described below.

With 64 different triplet codons available for the 20 amino acids, you might think that evolution would have picked out 20 codons to be used and ignored the rest. However, this would mean that chance mutations of codons would frequently result in unusable triplets corresponding to no amino acid, which would thus render genes inoperative since the synthesis of the protein from its messenger would halt when an unusable triplet in the mRNA was encountered.

An alternative policy has been adopted. Three codons have been reserved as 'stop' signals, which indicate to the protein-synthesizing machinery that the protein is complete. These three codons (UAA, for example, is one) have no amino acids assigned to them in the genetic code. If, by chance, a mutation produces a stop signal in an mRNA coding region (known as a **nonsense mutation**), the completed protein will not be produced from that gene. However, with only three stop triplets, the chance of this is much less than if there were 44 of them. Of the remaining 61 codons, all code for amino acids, which means that an amino acid is likely to have several different codons, giving what is known as a **degenerate code**. The assignment of codons to amino acids, the genetic code, is shown in Table 24.1.

Only two amino acids, methionine and tryptophan, have single codons (AUG for methionine). The rest have more than one – leucine, arginine and serine have six. Codon assignment is not random. Where several codons exist for one amino acid, they tend to be closely related and vary mainly in the third base. For example, those for isoleucine are AUU, AUC, and AUA. Not only that, but codons for similar amino acids tend to be similar. For example isoleucine and leucine are very similar aliphatic hydrophobic amino acids; their codons include CUU (leucine)

Table 24.1 The genetic code

| 5′ base | Middle base | | | | 3′ base |
	U	C	A	G	
U	UUU Phe	UCU Ser	UAU Tyr	UGU Cys	U
	UUC Phe	UCC Ser	UAC Tyr	UGC Cys	C
	UUA Leu	UCA Ser	UAA Stop*	UGA Stop*	A
	UUG Leu	UCG Ser	UAG Stop*	UGG Trp	G
C	CUU Leu	CCU Pro	CAU His	CGU Arg	U
	CUC Leu	CCC Pro	CAC His	CGC Arg	C
	CUA Leu	CCA Pro	CAA Gln	CGA Arg	A
	CUG Leu	CCG Pro	CAG Gln	CGG Arg	G
A	AUU Ile	ACU Thr	AAU Asn	AGU Ser	U
	AUC Ile	ACC Thr	AAC Asn	AGC Ser	C
	AUA Ile	ACA Thr	AAA Lys	AGA Arg	A
	AUG Met†	ACG Thr	AAG Lys	AGG Arg	G
G	GUU Val	GCU Ala	GAU Asp	GGU Gly	U
	GUC Val	GCC Ala	GAC Asp	GGC Gly	C
	GUA Val	GCA Ala	GAA Glu	GGA Gly	A
	GUG Val	GCG Ala	GAG Glu	GGG Gly	G

* Stop codons have no amino acids assigned to them.
† The AUG codon is the initiation codon as well as that for other methionine residues.

and AUU (isoleucine). This has important genetic consequences, since it means that many mutations involving single-base-changes have relatively little effect on the protein synthesized (changing AUU to AUC still represents isoleucine) or else substitute a very similar amino acid (changing CUU to AUU substitutes isoleucine for leucine). Isoleucine and leucine are so similar in size and hydrophobic properties that the substitution may not impair the function of the protein. Thus the arrangement of the genetic code provides a 'genetic buffering' action whereby the effects of many single-base-change mutations on the proteins synthesized are minimized.

If you know the base sequence of an mRNA you can work out the amino acid sequence of the protein it codes for. The reverse is not completely true because we cannot be sure in the case where an amino acid has several codons which of these was in the mRNA.

The process by which protein is synthesized is called *translation* because the language of nucleic acid bases is translated into the different language of protein amino acids.

A preliminary simplified look at the chemistry of peptide synthesis

Some (or many) students find protein synthesis a difficult subject to learn. It might help if we give a very simplistic description of the essence of the chemistry and energy needs of the synthesis of a polypeptide devoid of other complicating aspects. If you are completely familiar with this central chemistry (which is slightly counter-intuitive), it will be easier to understand all the other activities going on in the ribosome. The latter are to organize the process on the ribosome and ensure that the genetic code of the messenger RNA being translated is accurately followed; that is where the real complexity lies.

To form a peptide bond two amino acids must become joined together by a CO—NH link. This requires energy so the first step is that each amino acid that will participate in protein synthesis is activated. An activated amino acid has sufficient energy to form a peptide bond. The activation takes place in the cytoplasm at the expense of ATP hydrolysis. There is a family of enzymes called **aminoacyl synthetases**, each of which activates a specific amino acid. Each also recognizes specific transfer RNA molecules (tRNA). As will be described later, these are adaptor molecules that ensure that the amino acid added to the growing polypeptide corresponds with the codon on the messenger the tRNA is attached to. The activation process consists of attaching an amino acid to a cognate (correct) tRNA molecule by an ester bond between a hydroxyl group on the ribose molecule at the 3′ terminus of the tRNA and the —COOH of the amino acid. Each amino acid has its own specific tRNA which its synthetase recognizes. The overall activation reaction is shown below using nonionized structures for simplicity (tRNA—OH signifies the 2′– or 3′–OH of the 3′ terminal ribose of a tRNA molecule).

$$tRNA—OH + HOOC—\underset{\underset{R}{|}}{CH}—NH_2 + ATP \longrightarrow$$

Amino acid

$$tRNA—O—OC—\underset{\underset{R}{|}}{CH}—NH_2 + AMP + PP_i.$$

Aminoacyl—tRNA

The inorganic pyrophosphate (PP_i) is hydrolysed to two inorganic phosphate (P_i) molecules so that the process has a large negative ΔG value.

Protein synthesis always involves these activated amino acids (called **aminoacyl-tRNAs**), never free amino acids. For simplicity we will in the scheme below represent them as tRNA-AA. *Note that the amino end of the activated amino acid is free and the carboxyl group is bound to the tRNA. The growing polypeptide chain also has a free —NH$_2$ group. Protein synthesis consists of extending the —COOH end of the peptide not the amino terminal end.* One might reasonably assume that additional activated amino acid would be added to the free —NH$_2$ groups of the peptide attached to the tRNA, but it does not happen like that. To illustrate what happens let us assume that we have a partially formed growing polypeptide on the ribosome attached to tRNA and we are going to add the amino acid specified by the next codon on the messenger. Fig. 24.1 shows that the *peptide (in red) is transferred from its tRNA to the —NH$_2$ of the incoming*

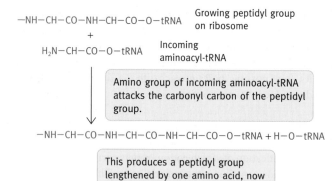

Amino group of incoming aminoacyl-tRNA attacks the carbonyl carbon of the peptidyl group.

This produces a peptidyl group lengthened by one amino acid, now attached to the incoming tRNA.

Fig. 24.1 Illustration of the reaction by which the growing peptide chain on the ribosome is elongated by one amino acid unit. Note that the peptidyl group is transferred from its transfer RNA (tRNA) attachment to the newly arrived aminoacyl-tRNA; the amino group remains free This extends the carboxyl terminal end of the peptide chain. The same process is repeated over and over again until the protein chain is completed. Note particularly that after each round the peptide is attached to the most recent incoming aminoacyl-tRNA.

aminoacyl—tRNA (in blue). This gives us a peptide, one amino acid longer, attached to the last arrived tRNA.

$$tRNA—peptide + tRNA—AA \rightarrow tRNA + tRNA—AA—peptide$$

The transfer is effected by the —NH$_2$ of the incoming aminoacyl —tRNA attacking the carbonyl carbon of the peptide—ester linkage to its tRNA; the *peptide* is transferred to the —NH$_2$ of the incoming activated amino acid.

This mechanism means that in the synthesis of a protein 200 amino acids in length, the last peptide bond synthesis involves a polypeptide of 199 amino acids being transferred to the amino end of the 200th aminoacyl—tRNA. One might intuitively expect that each incoming aminoacyl group would simply be added to the peptide amino end but, as stated, it's not like that. To use a fanciful image, to complete the dog, you don't add the tail to the dog, you add the dog to the tail, or, if you prefer it, in completing a wall you don't add the last brick to the wall, you add the wall to the last brick. The amino terminal end of the protein emerges from the ribosome first.

The above is the essence of the *chemistry* of polypeptide synthesis. The actual process on the ribosome is more complicated but, as already implied, the complications are to ensure that the process is organized to ensure that translation of messenger starts at the right place, that the codons on the mRNA are faithfully translated, that the ribosome moves along the mRNA one triplet at a time for each amino acid inserted, and that the finished polypeptide is successfully released.

One other general concept that may help concerns energy. We have explained that ATP is used to activate each amino acid, and no further energy input is needed for the reaction forming each peptide bond. You are familiar with this type of energy utilization for bond formation. However on the ribosome, each time a peptide bond is made, two molecules of GTP are hydrolysed to GDP and P$_i$, but no bonds are made as a result. It looks like a waste of energy, but what happens is that the hydrolysis causes a conformational change in the protein that the GTP is attached to. In each case the protein attaches to the ribosome in the GTP form, and can't be released except in the GDP form. Hydrolysis occurs only after a slight delay which allows an essential step to happen. This GTP/GDP switch mechanism is a crucial part of the organization of events on the ribosome, as will become evident. It is a vital mechanism in much of life.

We will now turn back to dealing with the complexities of protein synthesis but from this preliminary account you should understand what it is all aimed at – simply to synthesize one peptide bond after another.

How are the codons translated?

It is important at this stage to remember that protein synthesis occurs only on **ribosomes** – we will see shortly what these are and how they function. But for now, we can concentrate on the concept of how codons on mRNA are translated into the amino acid residues of proteins.

There is no physical or chemical resemblance or relationship between an amino acid and its codon that could lead to their direct association. It was therefore predicted that there must be **adaptor molecules** to associate amino acids with particular codons and, since the hydrogen-bonding potential of codons was most likely to be of importance, the adaptors were postulated to be small RNA molecules. Almost at the same time these were discovered as the **transfer RNA (tRNA)** molecules.

Transfer RNA

These are small RNAs, less than 100 nucleotides in length. Diagrammatically they have a cloverleaf structure (Fig. 24.2(a)). Internal base pairing forms the stem loops. The important parts (from our present viewpoint) are the three unpaired bases, which form the **anticodon** (explained below), and the 3'-CCA terminal trinucleotide flexible arm to which an amino acid can be attached. Two of these tRNA molecules at a time, with attached amino acids, have to be positioned side by side on the ribosome with their anticodons hydrogen-bonded to adjacent codons on the mRNA. In real life the tRNA molecules are folded up into quite a narrow shape. Figure 24.2 (b) shows the backbone structure of a tRNA.

The anticodon is a triplet of bases complementary to a codon (Fig. 24.3). It is located at a hairpin bend of the tRNA so that the three bases are unpaired and available for hydrogen bonding. Thus if a codon on the mRNA is 5'UUC3' (coding for phenylalanine), the anticodon corresponding to this on a tRNA molecule will be 5'GAA3'. This particular tRNA molecule must

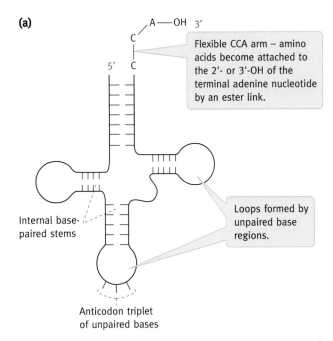

(a)

Flexible CCA arm – amino acids become attached to the 2'- or 3'-OH of the terminal adenine nucleotide by an ester link.

Loops formed by unpaired base regions.

Internal base-paired stems

Anticodon triplet of unpaired bases

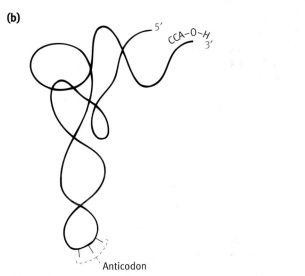

(b)

Anticodon

Fig. 24.2 (a) The cloverleaf structure of transfer RNA. **(b)** The folded structure of tRNA molecules. Minor addition to labeling of (a) *viz.*

have put on to it *only* phenylalanine and not any other amino acid; we will come to how this is done shortly. Since there are 61 codons each representing an amino acid (the other three are the stop codons), to translate these, it might be expected that there are 61 different tRNA molecules each with its own anticodon complementary to one codon and each accepting the one amino acid represented by its codon. In fact, there are fewer than 61 tRNA species – obviously there must be at least one for each of the 20 amino acids, but some tRNA molecules can recognize several codons. In these cases each of the codons recognized by a single tRNA must, of course, represent the same amino acid. The arrangement means that the cell needs to make

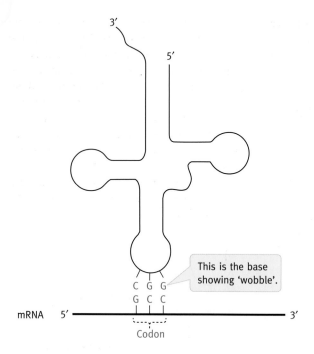

Fig. 24.3 Base pairing of an anticodon of a transfer RNA (tRNA) molecule to a messenger RNA (mRNA) codon. To achieve antiparallel pairing the tRNA molecule is flipped over. This is why a tRNA structure is presented in one way in Fig. 24.2 (with 5′ by convention to the left) but in this paired form in the reverse way; the mRNA is written with the 5′ end to the left. (In the example shown, the tRNA can base pair with codons GCC and GCU, both of which code for alanine.)

fewer tRNA molecules. How is this achieved? The answer is **wobble pairing**.

The wobble mechanism

In view of the importance of complementarity in DNA replication and transcription where Watson–Crick base pairing is a vital essential, it may be disconcerting that, in codon-anticodon base pairing, the rules are bent a little. This applies *only* to the first base of the anticodon; that is to say, the *third* base of the codon. It is known as wobble pairing; a U in this position will pair with A or G on the codon, and a G with C or U. The other two bases have the dominant role in codon-anti-codon binding.

The term 'first base of the anticodon' needs explanation. It is a convention that nucleic acid sequences are always written in the 5′ to 3′ direction but nucleic acids always base-pair in an antiparallel manner (with the two strands having opposite directionality). This is why, in diagrams, a tRNA molecule *on its own* as in Fig. 24.2 is shown with the 5′ end to the left, but when it is base-paired on a codon, it is flipped over as in Fig. 24.3. Thus, the anticodon CGG will base pair in the normal Watson–Crick fashion as shown, the first (5′) base of the latter being printed in red:

3′ CG**G** 5′ anticodon

5′ GC**C** 3′ codon

The wobble mechanism permits the same anticodon to pair 'improperly', as follows:

3′ CG**G** 5′ anticodon

5′ GC**U** 3′ codon

Since GCC and GCU both code for the amino acid alanine, wobble pairing does not alter the amino acid sequence of the protein synthesized, but it enables a single tRNA to translate both. In short, as stated, it permits the cell to manage with fewer species of tRNA molecules.

Evolutionary tinkering has produced other ways of enabling more flexible codon-anticodon interactions without jeopardizing translational accuracy. One is by using the nonstandard base hypoxanthine (Fig. 20.7) as the 5′ base of the anticodon. This will pair with C, U, or A in codons.

How are amino acids attached to tRNA molecules?

The system depends on a tRNA molecule having attached to it the particular amino acid specified by the codon which is complementary to the anticodon on that tRNA molecule. Thus a tRNA molecule with the anticodon GAA must be 'charged' only with phenylalanine, since UUC and UUU are the codons for that amino acid. If any other amino acid were to be attached to that tRNA, a mistake would be made in the synthesis of protein molecules since phenylalanine would be replaced by the other amino acid. This has been proved by experiments which showed that a noncognate ('incorrect') amino acid experimentally attached to a tRNA was inserted into a protein as if it was the correct one specified by that tRNA. Everything depends on the synthetase recognizing its own (cognate) tRNA. If it gets this wrong, then an incorrect amino acid could be inserted (if the proofreading failed). The latter is described below.

tRNA specific for phenylalanine is depicted as tRNA^{Phe}, and so on for each of the 20 amino acids; most amino acids use the first three letters of their name as an abbreviation (see Table 4.1). (Note that tRNA^{Phe} specifies only the tRNA; it does not mean that it has Phe attached. For the latter situation the term Phe-tRNA^{Phe} is used.) Enzymes that attach amino acids to tRNAs are called **aminoacyl-tRNA synthetases** or, sometimes, **aminoacyl-tRNA ligases**. Each cell must have at least 20 different species of these enzymes, which each can attach a specific amino acid to a cognate tRNA. This means that each enzyme recognizes both a specific tRNA and its cognate amino acid, and joins them together. There is no general pattern for the way in which a synthetase recognizes its tRNA – the anticodon is recognized in some cases, but in others a specific base or several bases elsewhere in the molecule are recognized.

The overall reaction is repeated here for convenience:

$$\text{tRNA—OH} + \text{HOOC—CH—NH}_2 + \text{ATP} \longrightarrow$$
$$\underset{\text{R}}{|}$$
$$\text{Amino acid}$$

$$\text{tRNA—O—OC—CH—NH}_2 + \text{AMP} + \text{PP}_i.$$
$$\underset{\text{R}}{|}$$
$$\text{Aminoacyl—tRNA}$$

$$\text{Amino acid} + \text{tRNA} + \text{ATP} \rightarrow \text{aminoacyl-tRNA} + \text{PP}_i + \text{AMP}.$$

The process is also referred to as **amino acid activation.**

The PP_i is hydrolysed to 2P_i and this drives the reaction to the right.

Proofreading by aminoacyl-tRNA synthetases

However, there is more to the above reaction. The synthetase, as well as selecting a correct tRNA, must also select the correct amino acid. As stated, the accuracy of the translation of mRNA into protein depends on the accurate selection by these enzymes of the correct amino acid. After an amino acid is loaded on to a tRNA, the aminoacyl-tRNA enters into the protein-synthesizing machinery but the latter has no known means of checking that a particular tRNA is carrying the correct amino acid. It 'assumes', as it were, that it is correct, so it is very important that the enzyme attaching the amino acid does not have a high error rate. It is relatively easy for an enzyme to distinguish between amino acids with markedly different structures, but much harder to do this between amino acids such as valine and isoleucine that are very similar.

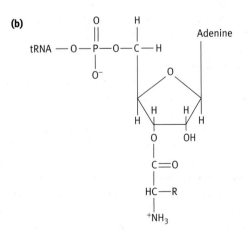

Valine   Isoleucine

The difference in binding energies due to a single —CH_2 group is not enough to give a sufficiently high degree of selectivity between isoleucine and valine. The aminoacyl-tRNA synthetase specific for isoleucine would attach valine to tRNA^{Ile} at a rate that would result in an unacceptable rate of errors in mRNA translation unless there was a corrective mechanism. A 'proofreading' mechanism has been evolved. The overall reaction above occurs in two stages as follows:

$$\text{Amino acid} + \text{ATP} \rightarrow \text{aminoacyl-AMP} + \text{PP}_i;$$

$$\text{Aminoacyl-AMP} + \text{tRNA} \rightarrow \text{aminoacyl-tRNA} + \text{AMP}$$

The aminoacyl-AMP does not leave the enzyme. The initial selection is on the active site of the synthetase selecting its correct amino acid. Active sites are usually highly selective, but as stated, and where amino acids are very similar, unacceptable rates of error would occur. A proofreading mechanism is present on most aminoacyl-tRNA synthetases. The enzymes have an extra editing site that hydrolyse incorrect aminoacyl-tRNAs. Thus the isoleucyl-tRNA synthetase hydrolyses valyl-tRNA. Another example is that threonyl-tRNA synthetase hydrolyses seryl-tRNA. The tyrosine specific enzyme has no such mechanism because tyrosine is sufficiently different from all other, and the initial selection is adequate.

The tRNA molecule has a terminal 3′ trinucleotide sequence of CCA, the terminal adenosine having free 2′ and 3′-OH groups on the ribose moiety (Fig. 24.4 (a)). It is to one of these

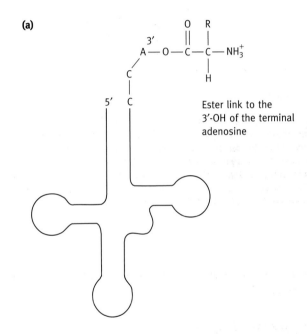

Fig. 24.4 **(a)** Diagram of transfer RNA molecule showing the CCA base sequence at the 3′ end, where the amino acid is attached by ester link, **(b)** Structure of the terminal nucleotide with the attached amino acid. In some cases the ester link is on the 2′OH of the ribose.

that the amino acid is attached by an ester bond (Fig. 24.4 (b)). (There are two classes of synthetases, each class having its own group of tRNAs. One class attaches its amino acid to the 3′ of the ribose, and the other class to the 2′.) This CCA trinucleotide forms a flexible arm, which can position the aminoacyl group on the appropriate reactive site on the ribosome (see below). The ester bond formed by the aminoacyl group is at a slightly higher energy level than that of a peptide bond so there is no thermodynamic problem in transferring the aminoacyl-ester group to the —NH_2 of another aminoacyl group to form a peptide bond. In other words, the energy required for the formation of a peptide bond is inserted into the process by the aminoacyl-tRNA synthetase using ATP as the energy source.

Ribosomes

Ribosomes derive their name from the content of RNA or ribonucleic acid, which accounts for about 60% of the dry weight. A ribosome consists of two subunits – in *E. coli*, a large one containing two molecules of RNA (23 S and 5 S; see below for an explanation) and a small subunit with one RNA molecule (16 S). Eukaryotic ribosomes, which are somewhat larger, are described later. You have already met mRNA and tRNA; **ribosomal RNAs (rRNAs)** are different again. Because of internal base pairing, they assume highly folded compact structures. Fig. 24.5 illustrates the complex secondary structure of an RNA molecule of an *E. coli* ribosome. The rRNAs are associated with many proteins, forming solid particles – 34 proteins in the case of the large subunit and 21 for the small.

With very large structures such as a ribosome, their sizes are measured in terms of the rate at which they sediment in an ultracentrifuge – expressed as Svedberg units or S values. (An ultracentrifuge spins so fast and therefore generates such a high G force that large molecules in solution move towards the bottom of the tube.) An *E. coli* ribosome is 70 S, the subunits being 50 S and 30 S (the S values are not simply additive, since S depends both on size and shape).

The overall *principle* of protein synthesis can be given very simply. The ribosome becomes attached to the mRNA near the 5′ end and then moves down the mRNA towards the 3′ end, assembling the aminoacyl groups of charged tRNA molecules into a polypeptide chain according to the sequence of codons in the mRNA which is read in the 5′ → 3′ direction. The N-terminal amino acid of the polypeptide is the first to emerge from the ribosome's exit tunnel through the large subunit and the C-terminal one is the last. At the end of the mRNA, the ribosome meets a stop codon, the protein is released, the ribosome detaches from the mRNA, and dissociates into its two subunits. Note that there are no 'special' ribosomes. In a given cell, any ribosome can use any mRNA just as a tape player can play any recorded tape; a slight qualification is that mitochondrial and

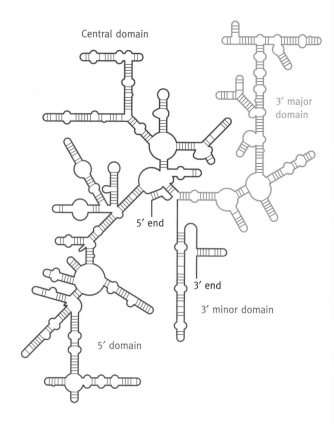

Fig. 24.5 16 S RNA

chloroplast ribosomes are different from those in the cytosol of eukaryotic cells, the former being prokaryotic in type.

That is the principle – to understand the process we need to give more detail. Synthesis of a protein molecule can be divided into three phases – **initiation**, **elongation**, and **termination**.

Initiation of translation

It is important that a ribosome begins its translation of an mRNA at exactly the correct point – in other words, that it initiates translation correctly. mRNAs have 5′ and 3′ untranslated regions and the coding region lies in between. The ribosome therefore must recognize the *first* codon of the coding sequence, which is not at the beginning of the mRNA, and start translating at that site. Absolutely precise initiation is essential, for this puts the ribosome in the correct **reading frame**.

This may need explanation. Suppose the *coding region* of an mRNA starts with the sequence:

5′ AUGUUUAAACCCCUG— — — — —3′.

The first five amino acids are specified by the codons AUG, UUU, AAA, etc. There is nothing to indicate what constitutes a codon, other than that the first three bases of the coding part of the mRNA encountered constitute codon 1, the next three are codon 2, and so on. There are no commas or full stops between them. The message therefore depends on starting to read *exactly* at AUG. Suppose an error of one base is made and a U is deleted from the codon UUU at the second base. The codons then read would be:

$$5'\ \text{AUG,UUA,AAC,CCC,UG}- - - - - -3'$$

The codons translated would be totally different and the amino acids inserted would correspondingly be totally different. This error is known as a **reading frameshift**; mutations deleting or adding one or two bases from a mRNA also cause reading frameshifts and result in the amino acid sequence of the polypeptide synthesized after the frameshift being nonsense, the process being halted when a stop codon arising randomly as a result of the frameshift is reached. The resultant polypeptide is useless garbage instead of the one specified by the gene. This contrasts with the loss of a complete codon in which case one amino acid would be missing from an otherwise accurate sequence.

Initiation of translation in *E. coli*

First the ribosome must be positioned at the correct starting place on the message. As shown in Fig. 24.6, there is a puirine-rich sequence of 3–8 nucleotides long 5′ to the translational start site on the mRNA known as the **Shine-Dalgarno sequence**, which is complementary to a section of the 16 S rRNA. Binding of the two by base pairing correctly positions the mRNA on the small ribosomal subunit. A ribosome has three sites on it, each of which can accommodate a tRNA molecule. These are the A, P, and E sites (for **acceptor**, **peptidyl**, and **exit** sites) the roles of which will become apparent shortly. The sites extend into both ribosome subunits. The mRNA is positioned such that the first and second codons are aligned with the P and A sites respectively (Fig. 24.6). Many bacterial mRNA molecules are polycistronic; the *lac* mRNA is one such example, with three regions in the one mRNA molecule coding for three different proteins. Each cistron has a Shine-Dalgarno sequence adjacent to it so that each can be initiated independently (Fig. 24.7).

Since initiation of translation is *not* at the 5′ end of the mRNA but at a distance along the molecule after the Shine-Dalgarno sequence, the first codon must be identified in some way. The start site is the codon **AUG**, but there is a problem. An AUG triplet can occur anywhere in the mRNA, since it is needed to code for each internal methionine. So how is the absolutely crucial step of identifying the correct initiating AUG carried out? The answer is that the Shine Dalgarno sequence centred approximately 10 nucleotides upstream of the AUG base-paired with the 3′ end of the 16S RNA of the small ribosomal subunit (30S ribosome), so positioning the AUD in the P

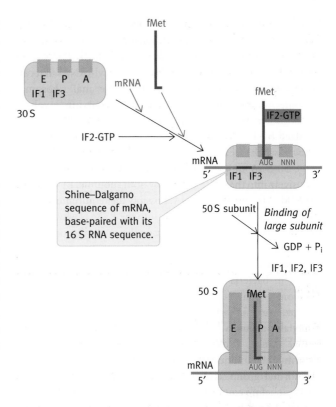

Fig. 24.6 Initiation of translation in *Escherichia coli*. The initiating transfer RNA, tRNA$_f$, is represented by the blue line, the anticodon being the horizontal short line. The binding of fMet-tRNA$_f$ to the 30 S subunit requires IF2. NNN represents any codon (N for any nucleotide).

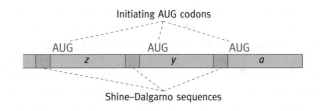

Fig. 24.7 The structure of a polycistronic prokaryote messenger RNA. *z*, *y*, and *a* refer to coding regions of the *lac* mRNA (see Fig. 23.10).

site that is present in the 30 S ribosome. Thus the recognition signal is effectively longer than just the AUG.

Now that the AUG is in the correct position, the initiating tRNA can be placed so that its anticodon 3′ UAC 5′ can base pair with the AUG initiation codon on the mRNA. There are two different types of tRNAs both specific for methionine and both with the same anticodon, but one is used exclusively for initiation and the other exclusively for inserting methionine internally into the growing polypeptide during the elongation process. The two differ in that the initiating methionyl-tRNA has structural features that are recognized by an **initiation protein** factor (**IF2**). Binding of this to the methionyl-tRNA makes it the only aminoacyl-tRNA of all the aminoacyl-tRNAs that can directly enter the P site that is present on the small ribosomal subunit. All the other aminoacyl-tRNAs that are involved in *elongation* of the polypeptide following initiation are recognized by a different cytoplasmic factor (EF-Tu; described below), which delivers them to the complete 70 S ribosome (large 50 S subunit joined with the small 30 S subunit) but exclusively into the A site formed by both subunits. EF-Tu does not bind to the initiating tRNA. Possibly the requirement to get the first aminoacyl-tRNA directly into the P site is why the first step of initiation always takes place on the small ribosomal subunit alone.

There is another difference between the two methionyl-tRNAs in bacteria that does not occur in eukaryotes – the methionine that becomes attached to the initiating tRNA in prokaryotes is formylated on its —NH_2 group by a transformylase using N^{10}-formyltetrahydrofolate (page 303) as formyl donor. Prokaryotic proteins are synthesized with N-formylmethionine (fMet) as the first amino acid residue. The formyl group, and frequently the methionine also, are removed before completion of the synthesis.

That then is how the initiating methionyl-tRNA is distinguished from the *N*-formylmethionyl-tRNA used in elongation, the most important factor being that they are recognized by different protein factors, IF2 and EF-Tu respectively. The initiating tRNA is called tRNA$_f$ (f for formyl), and the charged version fMet-tRNA$_f$, while the tRNA for methionine involved in elongation is called tRNA$_m$; the nomenclature practice in this varies somewhat.

To summarize the initiation procedure, in the cytoplasm there is a pool of 30 S and 50 S ribosomal subunits. There are three cytosolic initiating factors that are involved in initiation – they bind, participate in the process, and then are released for further use. Initiating factors IF1 and IF3 in the cytoplasm bind to the 30 S subunit. IF3 prevents premature reassociation with the 50 S subunit at this stage. One function of IF1 is apparently to prevent an aminoacyl-tRNA entering the A site before initiation is complete. **IF2**, as explained above, binds to fMet-tRNA$_f$.

In the presence of mRNA, fMet-tRNA$_f$, and GTP a complex of these with a 30 S subunit (with its attached IFs 1 and 3) is formed. The fMet-tRNA$_f$ is delivered to the P (for peptidyl) site on the subunit with its anticodon paired with the AUG codon, by IF2 (Fig. 24.6). This complex now associates with a 50 S subunit. The event is accompanied by the hydrolysis of GTP and the release of GDP, P_i, IF1, IF2, and IF3 (Fig. 24.6).

We now have a complete 70 S ribosome positioned on the mRNA with the fMet-tRNA$_f$ in the P site with its anticodon base paired with the initiating AUG codon. The A site is vacant, awaiting delivery of the second amino acid on its tRNA. Initiation is complete.

Once initiation is achieved, elongation is the next step

Cytoplasmic elongation factors in *E. coli*

Two soluble protein elongation factors in the cytoplasm are involved in this. They bind alternately, perform their task for each round of peptide synthesis and detach. They cannot both be attached at once – they alternate in this. In both cases they can attach only when bound to a molecule of GTP. Both are latent GTPases, active only when ribosome-bound. Hydrolysis of GTP to GDP causes them to detach. In the cytoplasm their GDP is exchanged for GTP so that the factors are then ready to participate in a new round of elongation.

The two factors are **EF-Tu** (elongation factor, temperature unstable) and **EF-G**, also known as **ribosomal translocase**. EF-Tu has the task of delivering the incoming aminoacyl-tRNA to the ribosome, EF-G of moving the ribosome along the mRNA in the $5' \rightarrow 3'$ direction to the next codon, once an aminoacyl group has been added to the growing peptide (see below).

It is useful to keep in mind this central concept of a pair of factors *alternately* hopping on to the ribosome in their GTP form, performing their tasks, and detaching in their GDP form to be recycled for subsequent rounds of elongation.

Mechanism of elongation in *E. coli*

We have already given the chemistry of elongation in principle (page 382). We suggest that you follow the steps in Fig. 24.8 as you read the next part. Starting with the initiation complex (state a), we have an fMet-tRNA$_f$ in the P site and the A site is vacant. It might be worth re-emphasising that *only* in the initiation process does the P site accept a tRNA charged with an amino acid – in this case N-formylmethionine; *all* subsequent aminoacyl-tRNAs enter the A site.

Aminoacyl-tRNAs (other than the initiating species) are complexed in the cytosol with the elongation factor EF-Tu, carrying a molecule of GTP bound to it. This factor attaches to the ribosome only if both GTP and an aminoacyl-tRNA molecule are bound to it. The EF-Tu-GTP-aminoacyl-tRNA complex binds to the ribosome such that the aminoacyl-tRNA occupies the A site with its anticodon positioned at the mRNA codon (Fig. 24.8, state b). EF-Tu exists in high concentration in the cytosol, sufficient to bind all of the aminoacyl-tRNA. Hydrolysis of its bound GTP molecule by EF-Tu releases the latter in its GDP form due to a conformational change, and frees the ribosome to proceed with the next step in elongation (Fig. 24.8, state c).

The aminoacyl groups on the two tRNA molecules on the P and A sites are in the vicinity of the catalytic site of **peptidyl transferase**, which transfers the fMet group from the tRNA in the P site to the free amino group of the incoming aminoacyl-tRNA

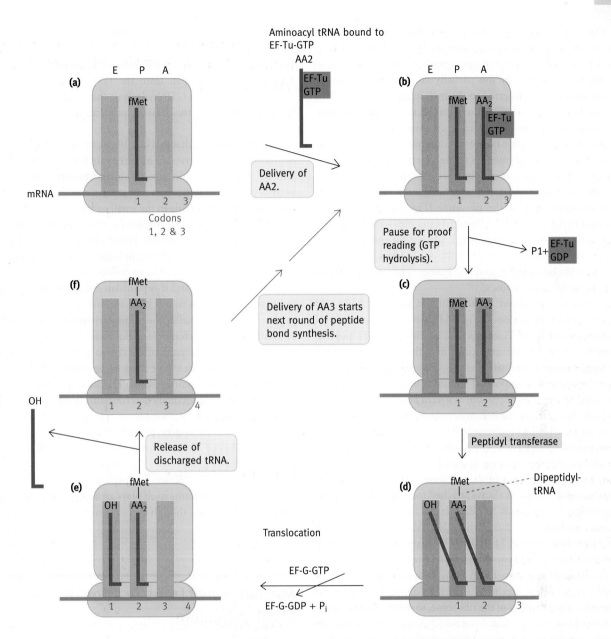

Fig. 24.8 The elongation process in protein synthesis in *Escherichia coli* following translational initiation. Transfer RNAs are shown as blue lines, the anticodon being represented by the short horizontal section; AA2 and AA3 represent amino acids and fMet represents formylmethionine. The positioning of EF-Tu-GTP on the tRNA is purely diagramatic but see Fig. 24.9(c). The reason for the naming of the enzyme peptidyl transferase is not evident from the diagram, but if you do the next round of synthesis you will see that in all subsequent rounds of synthesis it is a peptide that is transferred to the incoming aminoacyl-tRNA as is also explained in the text. E, exit site; P, peptidyl site; A, acceptor site. Note movement of mRNA, like a tape through a reader.

in the A site, producing a dipeptide attached to the tRNA (state d). Despite its name, *'peptidyl transferase' is an RNA molecule with catalytic properties*, part of the 23S RNA of the large ribosomal subunit, not a protein; it is a ribozyme. The peptidyl transferase reaction is shown, using the first peptide bond synthesis as an example:

$$tRNA—O—CO—fMet + NH_2—CH—CO—O—tRNA \longrightarrow$$
$$\text{(P site)} \qquad \qquad | \qquad \text{(A site)}$$
$$R'$$

$$tRNA—OH + fMet—CO—NH—CH—CO—O—tRNA.$$
$$\text{(P site)} \qquad \qquad | \qquad \text{(A site)}$$
$$R'$$

The aminoacyl group on the tRNA in the P site is transferred to the free amino group of the aminoacyl-tRNA in the A site. As the synthesis of the polypeptide proceeds, the aminoacyl group in the P site is actually the partially completed polypeptide chain and this is transferred to the incoming amino acid. This is why the process is called the **peptidyl transferase reaction**. Adjacent to the peptidyl transferase site there is the opening to a tunnel through the large ribosomal subunit. As the polypeptide is synthesized it is fed into this tunnel to emerge from the sub-unit, the amino terminal end first (see Fig. 24.11).

During the peptidyl transfer process, each tRNA becomes bound to two sites on the ribosome (Fig. 24.8(d)). The dis-charged tRNA straddles the P and E (exit) sites – the anticodon end of the molecule is still in the P site but the other end is in the E site. Similarly, the tRNA in the A site (now carrying the peptide) straddles the A and the P sites. This was shown by chemical protection studies.

The model proposed to account for this attachment of tRNAs to two sites at once (Fig. 24.8) envisages that one end of the aminoacyl-tRNA swings over as shown (c→d), peptide transfer occurs at the same time, and the discharged tRNA swings one end to the E site.

In this model an important feature is that, during the pro-cess, the *nascent peptide remains in a fixed position relative to the large subunit*, as shown; it is always in the P site. This would eliminate the problem of how a tRNA physically moves with a polypeptide attached to it. The model also removes the problem of how the tRNAs can move on the ribosome during transloca-tion without the danger of them diffusing away, since at least one end of the tRNAs is always attached to a site. Translocation (see below) now causes the discharged tRNA to move with respect to the large subunit, the tRNA exits, and we have the situation shown in Fig. 24.8 (state f). This moves the ribosome to the next codon and completes the first round.

An interesting comment on elongation comes from one of the leading laboratories in this field, to quote 'Perhaps most impressive is the ability of the ribosome, together with the elon-gation factor, EF-G, to translocate mRNA and tRNAs over very large molecular distances with high speed and accuracy, while maintaining the correct reading frame' (see **Korostelev and Noller** reference in Further reading). They further add that our understanding of translocation is only rudimentary despite decades of experimentation.

How is accuracy of translation achieved?

The basis of this is due to codon-anticodon interaction select-ing the correct aminoacyl-tRNA, but exactly how this selection is effected is a problem. The binding-energy difference between a correct and an incorrect codon-anticodon base-pairing alone is insufficient to account for the translational fidelity achieved. The EF-Tu does not 'know' which aminoacyl-tRNA has to react next, and it must deliver them to the A site randomly. It is believed that an incorrect aminoacyl-tRNA diffuses away

before it reacts since it will not be held in the A site as strongly by hydrogen bonding to the codon as a correct one would be. Peptide synthesis cannot occur until EF-Tu-GDP is released and this cannot happen until the GTP is hydrolysed (Fig. 24.8(c)). The delay in GTP hydrolysis that occurs is postulated to provide time for this kinetic 'proofreading' process to be effective.

It seems, however, that this is unlikely, by itself, to achieve the degree of fidelity observed of an error rate between 1 in 10^3 and 1 in 10^4. It is believed that the ribosome may play a part in the accuracy of codon-anticodon recognition and this is a function of the 16 S RNA molecule in the small subunit of the ribosome. There are three bases, two adenines and one guanine, which, if altered, render *E. coli* nonviable. When a cognate tRNA binds to the anticodon, it is believed that the bases flip out of their orien-tation in the rRNA helix. When a codon binds to an anticodon it forms a mini-helix and it is postulated that the three flipped bases act like fingers which bind to base pairs in the minor groove of this mini-helix. (See **Ramakrishnan**, and also **Ogle and Ramakrishnan** references in Further reading). They check that the first two base pairs of the codon-anticodon action are *bona fide* Watson–Crick base pairs rather than aberrant pairs. The third (wobble) pair is not so checked; it is the first two which are the main specificity determinants and this agrees with the wobble situation.

If the codon-anticodon pairing is correct it triggers the hydrolysis of GTP attached to the EF-Tu which delivered the aminoacyl-tRNA to the A site. This causes the release of EF-Tu-GDP and in turn this allows transpeptidation to occur. It thus appears that the ribosome itself plays an active role in achieving fidelity of translation, though there is not a full understanding of the mechanism.

Mechanism of translocation on the *E. coli* ribosome

Translocation of peptidyl-tRNA from the A site to the P site is catalysed in *E. coli* by **EF-G** (**ribosomal translocase**). It attaches to the ribosome after each peptide synthesis is completed and catalyses the movement of the tRNA straddling the A/P sites (now carrying the peptide) completely into the P site, leaving the A site vacant. At the same time the discharged tRNA strad-dling the P/E sites moves completely to the exit site and leaves. During translocation mRNA moves with the newly formed peptidyl-tRNA, bringing the just translated codon into the Psite and the next untranslated codon into the A site.

EF-G has attached to it a molecule of GTP and the trans-location is associated with the hydrolysis of this and release of the EF-G-GDP complex. The mechanism by which EF-G acts involves **protein mimicry**. It mimics the shape of tRNA. Figure 24.9 shows space-filling models of (a) tRNA^Phe, (b) the protein

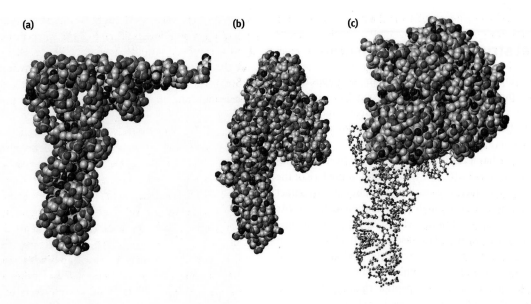

Fig. 24.9 Space-filling models of **(a)** yeast transfer RNA^{phe} (Protein Data Bank code 4TNA), **(b)** EF-G in complex with GDP (1DAR), and **(c)** the ternary complex of Phe-tRNA^{phe}, EF-Tu, and a GTP analogue (1TTT), showing the protein tRNA mimicry described in the text.

EF-G in complex with GDP, and (c) the ternary complex of Phe-tRNA^{Phe}, EF-Tu, and a GTP analogue. You can see that a domain of EF-G resembles in shape that of the tRNA stem helix

carrying the anticodon, and that the shape of the whole EF-G molecule is a close match to the complex between EF-Tu and tRNA. Possibly the EF-G-GTP inserts itself into the A site and physically pushes the peptidyl-tRNA into the P site, and the discharged tRNA from the P site into the exit site. During translocation GTP is hydrolysed into GDP, and, in this form, the EF-G-GDP is able to detach. It is not fully understood, however, how translocation is achieved.

Termination of protein synthesis in *E. coli*

At the end of each mRNA coding section there is a stop codon (Table 24.1). In *E. coli* there is a menu of three of these and two release factors, **RF1** and **RF2**, respectively, both of which recognize UAA. There is another case of protein mimicry here for the primary amino acid sequence of RF1 and RF2 suggested that these proteins also mimic the shape of the tRNA as occurs in EF-G. The release factors carry a molecule of water to the ribosome that hydrolyses the ester bond between the now compleatred polypeptide chain and the final tRNA. This releases the polypeptide from the ribosome.

After this the remaining ribosome complex is disassembled into the two subunits and the bound tRNA and mRNA released. The dissociation procedure requires a **ribosomal recycling factor** (RRF) and IF3, and is associated with GTP hydrolysis. The ribosome recycling factor is a protein which also accurately mimics the shape of tRNA. IF3 attaches to the small subunit, and prevents reassociation with the large one.

Physical structure of the ribosome

In the preceding diagrams we have used simple shapes for the ribosome for the purpose of clarity, but a good deal is known of the structure, visible in the electron microscope and recently elucidated through the crystal structure of the prokaryotic 50 S and 30 S subunits and the intact 70 S ribosome.

The ribosome has a distinctive shape that reflects its function, as shown in Fig. 24.10. The tRNAs bind to the face of the large subunit, which faces the small subunit, but at the contact faces there is a hollow big enough to accommodate the tRNAs and form a channel. The aminoacyl-tRNAs enter the channel on the right of the diagram and the discharged tRNAs emerge on the left. The peptidyl transferase catalytic region is located on the face of the large sub unit within the cavity, and a tunnel for the exit of the synthesized polypeptide chain opens next to this. The chain is fed into it and emerges from the large subunit, —NH_2 end first. The mRNA binds to the face of the small subunit facing the cavity. Fig. 24.11 shows a cross-section of the large ribosomal subunit with the nascent peptide in the exit tunnel.

What is a polysome?

It takes about 20 seconds for a ribosome to synthesize an average protein in *E. coli*, which adds about 20 amino acid residues per second. If only a single ribosome at a time moved along the mRNA molecule, the latter could direct the synthesis of the pro-

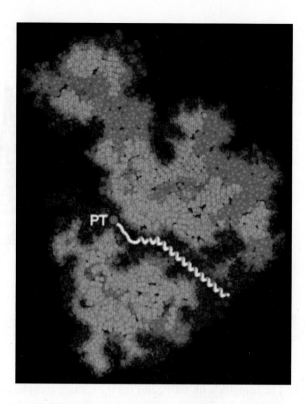

Fig. 24.11 The modeled nascent polypeptide shown schematically in the exit tunnel of the large ribosomal subunit. PT, peptidyl transferase centre. Buried RNA atoms are shown in grey, and protein atoms in green.

tein at the rate of 1 molecule per 20 seconds. However, as soon as an initiated ribosome has got under way and has moved along about 30 codons, another initiation can occur. One ribosome after another hops on to the mRNA. They follow one another down the mRNA, each independently synthesizing a protein molecule – a typical case would be about five ribosomes per mRNA molecule, but this varies with the length of the mRNA. This greatly increases the rate of protein synthesis. The term **polysome** is thus a shortened version of **polyribosome**.

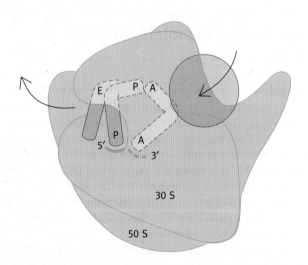

Fig. 24.10 Diagram of a ribosome with plausible locations for ribosome-bound transfer RNAs in the A/A, P/P, and E states (see Fig. 24.8 for the terminology). The shaded area at the right shows the approximate site of interaction of EF-Tu. The polarity of a fragment of messenger RNA containing the A- and P-site codons is shown. The arrows indicate the likely path of a tRNA as it transits the ribosome.

Riboswitches

Riboswitches are the most recently discovered example of the ability of RNA molecules to regulate aspects of metabolism without the assistance of proteins. They can influence gene expression both by effects on translation and transcription. It is speculated that they may represent an ancient regulatory mechanism in bacteria that modulates the synthesis of certain metabolites according to its prevailing level in the cell. The principle is that mRNAs coding for the synthesis of enzymes involved in the production of the metabolite contain a sequence in the untranslated leader sequence known as an **aptamer** (from the

Latin *aptus* to fit), which specifically binds its cognate metabolite. The aptamer is acting as a metabolite sensor. When the metabolite binds to its aptamer the RNA is believed to undergo a conformational change, which inhibits the expression of the gene(s) coding for enzyme(s) involved in the synthesis of the metabolite. The control may be achieved in one of two ways.

- The leader sequence is synthesized first and the conformation of the aptamer can influence the synthesis of the rest of the messenger by forming a terminator helix which aborts the transcription of the messenger.
- Alternatively it may cause the masking of the Shine–Dalgarno sequence on the mRNA and so prevent initiation of translation.

Riboswitches have been found to include ones specific for flavin mononucleotide, thiamine pyrophosphate, certain amino acids, purines and S-adenosylmethionine.

This is not the only case of the conformation of the leader sequence RNA controlling the gene it is attached to. It has long been known in *E. coli* that expression of operons responsible for synthesis of certain amino acids is controlled by the abundance of the amino acid in question. In this case, however, the control by the metabolite is not direct attachment to the RNA as in riboswitches but an indirect one involving the translation of the leader sequence that affects the conformation of the RNA. This in turn results in transcription of the mRNA being completed if the specific amino acid level is low or if high, the transcription is aborted by the formation of a terminator structure in the mRNA. The mechanism is known as **attenuation**. The **Wakeman** *et al.* reference in Further reading gives further information on this.

Protein synthesis in eukaryotes

The major difference in eukaryotes from that in prokaryotes is in initiation of translation. It is more complicated but it is schematically similar in that it requires the initiating methionyl-tRNA to bind into the P site of a small ribosomal subunit that positions itself correctly so that the initiating AUG is in that site. Following the formation of this complex the large subunit joins it and elongation proceeds as in bacteria. Initiation in eukaryotes involves over 12 eiFs (**eukaryotic initiation factors**) and one of these (**eiF 3**) the largest and earliest discovered has 13 subunits.

The process starts with the assembly of a 43 S **ternary complex** involving free 40 S ribosomal subunits arising in the cytoplasm from the separation of the large and small subunits of ribosomes that have just completed the synthesis of a protein. First, **eiF2** (corresponding to prokaryote EF2) combines with GTP and Met-tRNA$_i$ forming a ternary complex. This, aided by numerous other initiation factors, combines with the small 40 S ribosomal subunit near the cap site at the 5′ end of the mRNA

(Fig. 24.12). Yeast studies suggest that the poly A tail of the mRNA at the 3′ end is also involved. The polyA binding protein attached to this is needed for binding of the 40 S subunit into the **preinitiation complex** (PIC). The next stage is to position the PIC at the AUG start codon of the mRNA with the Met-tRNA$_i$ (i for initiation) anticodon base-paired with it (Note, not formylated fMet as in bacteria). There is no equivalent of the Shine–Dalgarno sequence which positions the 30 S subunit correctly at the start site in prokaryotes. Instead, the PIC scans the message by moving along it until it finds the correct AUG. This is usually the first one encountered but there is an adjacent base sequence recognized by the PIC. At this point the large 60 S ribosomal subunit joins to the small subunit forming the 80 S complex, GTP on the eiF2 is hydrolysed, and the eiF-GDP and all the other initiation factors are released and initiation is complete. The eiF-GDP is recycled back to the GTPstate in the cytoplasm ready for further use.

The next stage is **elongation** – the synthesis of the complete polypeptide chain of the protein. This proceeds as in prokaryotes involving two cytosolic **elongation factors**. The counterpart of EF-Tu in eukaryotes is called **EF1a**, and that of EF-G (translocase) is **EF2**. Termination is similar to that in *E. coli* except that a single release factor, eRF1, is used.

There can, with the scanning mechanism for selecting the AUG initiation codon, be only one initiating site per mRNA molecule so that eukaryotic mRNAs must be monocistronic, coding for a single polypeptide only, unlike polycistronic prokaryotic mRNA.

Termination is similar to that in prokaryotes (page 391).

Incorporation of selenocysteine into proteins

Francis Crick invented the phrase, the 'magic 20', to indicate the amino acids for which there are codons in the mRNAs and hence are incorporated into proteins. However there is a 21st amino acid called **selenocysteine** that is incorporated by a 'freak' mechanism into a few specific proteins (specified by specific mRNAs) during their synthesis on the ribosome.

$$H—Se—CH_2—CH(NH_2)—COOH$$

There is no codon for this amino acid but a specific UGA *stop codon* is used as a substitute. A messenger RNA for a selenocysteine protein has a long stem loop section near to a UGA codon, which identifies it as the one to be used. The stem loop is called the **selenocysteine insertion sequence** (SECIS). A special tRNA with an anticodon complementary to the UGA codon is used in the incorporation. The aminoacyl synthetase that recognizes this particular tRNA attaches *serine* to it. The seryl tRNA is then converted by a reaction with **selenophosphate** (produced enzymically from the metal and ATP) to **selenocysteinyl-tRNA**. This is delivered to the ribosome by a protein factor which recognises only this aminoacyl-tRNA. The incorporation involves a SECIS binding protein and an associated SECIS specific

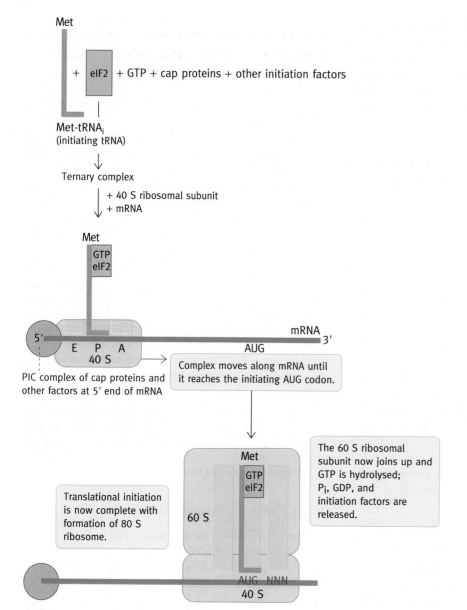

Fig. 24.12 Simplified diagram of intiation in eukaryotes. Note that several eukaryotic initiation factors besides eIF2 are involved. eIF2 is the eukaryotic initiation factor corresponding to IF2 in prokaryotes. It selects the initiating met-tRNA$_i$ and delivers it to the P site. PIC, preinitiation complex. NNN represents the second codon, N representing any nucleotide.

elongation factor. While the mechanisms of selenocysteine incorporation in eukaryotes and in *E. coli* are similar, the details of the process are less well elucidated in eukaryotes.

Selenium is toxic in large amounts, but is still an essential trace metal. There are not many selenocysteine proteins – the family of glutathione peroxidase enzymes (page 490) is an example. They occur in both prokaryotes and eukaryotes; about 25 are known in humans. The naturally occurring selenocysteine proteins should not be confused with selenomethionine proteins produced by incorporation of artificial selenomethionine in place of methionine during protein synthesis in *E. coli* cDNA expression systems. Selenomethionine-labelled proteins are now often used in determining the three-dimensional struc-

tures of the proteins using synchrotron radiation. The selenium replaces the heavy metal isomorphous labelling necessary in traditional X-ray diffraction (page 79).

Protein synthesis in mitochondria

Mitochondria contain DNA of their own and have their own protein-synthesizing machinery. The ribosomes of mitochondria are prokaryote-like and use fMet-tRNA$_f$ for initiation. They have other features of interest, such as a slightly different genetic code, and codon-anticodon interactions are simplified so that mammalian mitochondria can manage with only 22 tRNA species. The possibility of such simplification is related to the fact that mitochondria synthesize only a handful of different proteins, most being synthesized in the cytoplasm and then transported in. The same is true of chloroplasts which are likewise believed to have originated from incorporated (in this case, photosynthetic) prokaryotes.

Folding up of the polypeptide chain

The newly synthesized polypeptide chain emerges from the channel in the ribosome in a denatured (unfolded) state. Proteins are not active in this state; they must acquire the three-dimensional structure specified by the amino acid sequence. It is not fully understood how proteins fold up so rapidly. It takes the cell anything from one second to two minutes. To achieve the correct conformation by trying all variations until the lowest free energy is found would take millions of years. It is now believed that certain sections, called 'molten globules' very rapidly fold to give the main features of a secondary structure, which is followed by side chain adjustments to give the final tertiary structure. The folding is probably a stepwise procedure in which correct foldings of sections are preserved, and nonprofitable ones avoided or allowed to correct themselves with the help of chaperones.

Chaperones (heat shock proteins)

As a newly synthesized polypeptide emerges from the ribosome, hydrophobic groups, which will ultimately be buried in the interior of the folded native protein, are exposed to the aqueous medium of the cytosol. Unless something is done to prevent it, these groups will form nonspecific hydrophobic associations to whatever other hydrophobic groups are available, either on the same or adjacent polypeptides. Such random, 'improper' associations could prevent the polypeptide from folding (unless corrected).

A family of proteins collectively referred to as **chaperones** guard against this. They are highly conserved in evolution and are present normally in all cells. They were discovered when it was observed that in cells subjected to temperatures higher than normal, certain proteins increased in amounts and were called **heat shock proteins (Hsps)**. Chaperones are Hsps. What does heat shock have to do with protein synthesis? Heat denatures (unfolds) native proteins; newly synthesized polypeptides also are unfolded as they emerge from the ribosome. The heat-denatured proteins have improperly exposed hydrophobic groups, and if they are to be salvaged must be refolded. So the problem is similar to that of newly synthesized proteins.

Mechanism of action of molecular chaperones

The Hsps are classified into three groups. The two best known are the **Hsp 70** and the **Hsp 60** groups.

Hsp 70 attaches to the hydrophobic groups of nascent polypeptides (Fig. 24.13). It has a molecule of attached ATP, in which form it has a high affinity for the unfolded chain. Hsp 70 is a slow ATPase. After the chaperone attaches to a polypeptide, the ATP is hydrolysed after a short period; in the ADP form it has a low affinity for the polypeptide and so releases it, conformational change of the protein being responsible for this. The ATPase is thus a timing mechanism to determine how long the chaperone remains attached to the unfolded polypeptide. By attaching, the Hsp prevents incorrect hydrophobic associations occurring during the time when polypeptide synthesis is likely to be completed. When it detaches, the correct folding is given the opportunity to occur. Probably the Hsp 70 type of chaperone assistance is all that is required for many proteins; even large ones probably have several essentially structurally independent domains which fold sequentially as the polypeptide emerges. (Page 55 has a description of what a domain is.) A number of proteins use a different form of assistance for completion by a molecular chaperone of the Hsp 60 class (sometimes known as **chaperonins**). The best known of these

Nascent polypeptide with exposed
hydrophobic groups

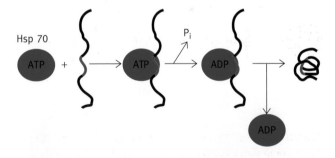

Fig. 24.13 Simplified diagram of *Escherichia coli* chaperone Hsp 70 action in assisting folding of a polypeptide. The participation of other cochaperone proteins in the process has been omitted for simplicity. The essence of the process is that the chaperone is a slow ATPase and alternates between an ATP-bound form and an ADP-bound form, the former having a high affinity for the polypeptide and the latter a low affinity. The red line represents a hydrophobic patch on the polypeptide. Hsp 70 can attach to nascent proteins emerging from the ribosome.

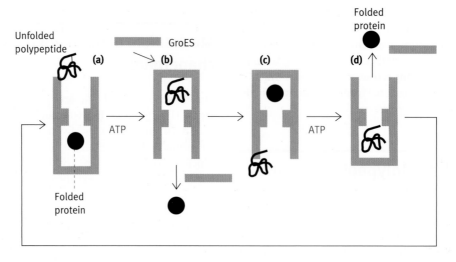

Fig. 24.14 Simplified diagram to illustrate the principle of GroEL action in *E. coli*. The 'lid' structure is known as GroES. The chaperonin has two folding chambers, which in this model are postulated to work alternately. In **(a)** an unfolded polypeptide has attached *via* its hydrophobic groups to GroEL. The lower cavity has a folded polypeptide represented as a solid circle waiting for release (this corresponds to the situation in the upper chamber in (c)). In **(b)** the unfolded polypeptide has entered the hydrophilic cavity and a 'lid' seals it in. Meanwhile the lower cavity has released its folded protein. In **(c)** the situation is just the same as (a) but upside down. Note that the diagram does not show the ring structure of the chaperonin nor the conformational changes that occur. These involve the changing of the lining of the cavity accepting the protein from hydrophobic to hydrophilic. The mechanism involves the hydrolysis of seven molecules of ATP at the steps indicated. The next figure shows a more realistic structural representation. The mammalian HSP counterpart may function as a single ring.

is a protein complex in *E. coli* known as **GroEL**, together with a 'lid' structure known as **GroES**. GroEL is a multi-subunit structure with two back-to-back rings of seven subunits. In Fig. 24.14 all such details of subunit structure are omitted for simplicity. A space-filling model of the chaperonin, together with a vertical cross-section, is given in Fig. 24.15.

We mentioned earlier that in Anfinsen's experiment, in which ribonuclease was refolded *in vitro*, the denatured protein was at low concentration which favoured the refolding because

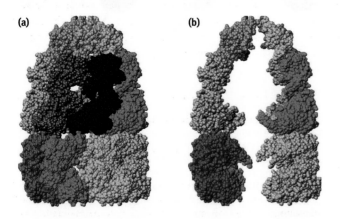

Fig. 24.15 (a) Space-filling model of GroEL with a GroES cap (Protein Data Bank code 1AON). This state corresponds with state (b) in the previous figure. **(b)** A cross-section of the same.

there is less chance of aggregation with other molecules. GroEL provides what is sometimes known as an 'Anfinsen cage' because it literally encloses a polypeptide so that it can fold in a hydrophilic box secluded from all the other proteins in the cell. It gives the protein a private folding chamber and then ejects it. If it has failed to fold it can try again because its hydrophobic groups are still exposed and will reattach to the GroEL. Hsp 60 is needed typically for proteins imported, in an extended form, into the mitochondrial matrix where the concentration of proteins is extremely high. As illustrated in Fig. 24.14 the unfolded, or partially unfolded, protein attaches to the Hsp 60 entry point by its hydrophobic groups. At this stage the chamber has a lining of exposed hydrophobic groups to which the unfolded protein can attach. ATP attaches to the central domain of the barrel and this results in a dramatic allosteric conformational change in the lining of the cavity containing the unfolded protein. This causes hydrophilic groups to be exposed instead of the hydrophobic ones, the cavity changes its shape and enlarges and the GroES cap attaches and seals in the protein. The hydrophilic box is an ideal environment for refolding to occur. Hydrophobic groups become hidden and its hydrophilic ones exposed, as is required in a folded protein. After a short interval the ATP is hydrolysed, the cap detaches, and the protein is ejected and the conformational changes reverse. If the folding was not successful hydrophobic groups will still be exposed and it can have another try. Note that the GroEL provides no folding guidance to the protein but just optimal conditions for it to fold itself. It is the amino acid sequence which determines the folded configuration. A main driving force for the folding is the hiding of hydrophobic groups from a hydrophilic environment. GroEL has two cavities and it has been suggested that they act alternatively (see **Rye** *et al.* in Further reading). The yeast HSP chaperone is believed to also act using two rings alternately, but the mammalian one is thought to function as a single ring.

Enzymes involved in protein folding

The chaperone assistance in folding is not the end of the folding story; there are two additional problems this time requiring enzymic intervention.

The first is by **protein disulphide isomerase (PDI)**. This enzyme 'shuffles' S—S bonds in polypeptide chains (page 54); it rearranges them. If an incorrect S—S bond were to be formed, being covalent it would not break spontaneously and would lock the polypeptide in an incorrect configuration. PDI, by breaking and reforming S—S bonds between different cysteine residues, permits the folding to correct itself. It transfers an S— bond from one disulphide bridge to another, using itself as the intermediary attachment site. There is little or no free energy change in the process and so it is freely reversible. High concentrations of PDI are found inside the endoplasmic reticulum involved in folding proteins destined for secretion, many of which have disulphide bridges.

Another enzyme is **peptidyl proline isomerase (PPI)**. Wherever a proline group occurs in peptide linkage two conformations are possible – *cis* and *trans*. The PPI plays the role of 'shuffling' proline residues between the conformations in order to permit the whole protein to fold correctly.

Protein folding and prion diseases

A group of fatal neurological degenerative diseases exist that affect humans and animals, known as **prion** diseases. These include **Creutzfeldt–Jakob disease** and **kuru** in humans. Kuru, known as the **laughing disease** because of the facial grimaces it causes, used to be transmitted in certain New Guinea tribes by cannibalism. In sheep, a related disease is known as **scrapie** because the animals scrape off their wool by rubbing against fence posts; in cattle there is **bovine spongiform encephalopathy (BSE)**, commonly known as **mad cow disease**. The diseases can be transmitted by consumption of infected tissue or, rarely, can be an inherited trait. The only known case of a prion disease being transmitted from animals to humans is the mad cow disease. The diseases were first believed to be caused by 'slow viruses' because they are infectious and can take years to develop. It is now known that the diseases are associated with an abnormal form of a normal protein of unknown function found in brain. The *disease-producing unit* is called a **prion**

(for **proteinaceous infectious particle**) and the protein itself as PrP^{Sc} (for prion protein, scrapie; it applies to human prion protein not just sheep) while its *normal* counterpart is called PrP^c (for the normal constitutive prion protein). The human proteins (PrP^C and PrP^{Sc}) have identical polypeptide amino acid sequences *and are coded for by the same gene*, but their folded conformations are different. PrP^c has four α helices and no β sheet content, is soluble and protease sensitive. In PrP^{sc} two of the four α helices are folded instead into four β strands that make up a single β sheet, which make it resistant to proteases.

The question arises as to how an improperly folded protein can be infectious and reproduce itself. No mechanism is known by which a protein molecule can direct its own replication and considering their complex folded structures such a possibility is improbable. However, PrP^{Sc} somehow causes PrP^c (the normal protein) to convert to the abnormal form. This has been demonstrated *in vitro* by incubating the two together. It is believed that it occurs by a 'seeding' mechanism, which is not yet understood. It is generally accepted that the induced conversion is the explanation for the infectivity of a prion protein (though final proof has remained elusive in that it has not been shown that PrP^{Sc} produced from PrP^c in vitro is infectious). The conversion *in vivo* of PrP^c to PrP^{Sc} is a rare event in the absence of infection by PrP^{Sc}, so spontaneous occurrence of the disease is rare. It is believed that mutations in the gene for the normal PrP^c may increase the probability of this, which might explain the hereditary origin of some cases of the disease. Once some PrP^{Sc} is formed, it would then trigger the autocatalytic formation of more.

It is not known how prions cause disease (see **Taubes** reference in Further reading) though it is associated with the accumulation of amyloids in the brain. These are insoluble accumulations of PrP^{Sc} in the form of elongated structures composed of the molecules stacked up on top of one another known as cross β-filaments. It is not known what effects these have on the brain.

It is now known that other proteins are capable of forming other deposits also, often, called amyloid and some are known to be associated with mental diseases. However in the case of Huntingdon disease (page 471) the abnormal protein aggregates are associated with an abnormal gene inherited genetically.

Translational control mechanisms

Regulation of globin synthesis

The two components of haemoglobin, the protein **globin** and the porphyrin haem, need to be produced in the correct relative amounts if excess of one or the other is to be avoided. A coordinating regulatory link exists. In reticulocyte lysates, in the absence of haem, a protein kinase phosphorylates one of the factors (eIF2) involved in initiation of protein synthesis.

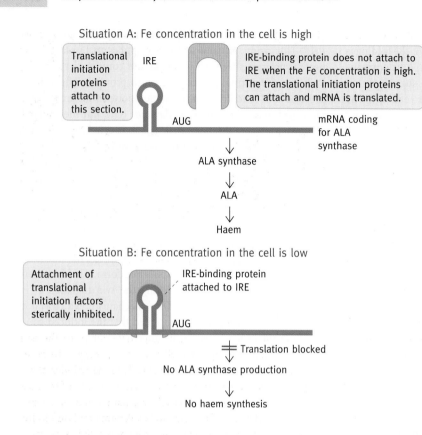

Situation A: Fe concentration in the cell is high

Translational initiation proteins attach to this section.

IRE

IRE-binding protein does not attach to IRE when the Fe concentration is high. The translational initiation proteins can attach and mRNA is translated.

AUG

mRNA coding for ALA synthase

↓
ALA synthase
↓
ALA
↓
Haem

Situation B: Fe concentration in the cell is low

Attachment of translational initiation factors sterically inhibited.

IRE-binding protein attached to IRE

AUG

Translation blocked
↓
No ALA synthase production
↓
No haem synthesis

Fig. 24.16 Haem synthesis control by iron (Fe) levels in the red blood cell. The regulatory mechanism presumably depends on the ALA synthase having a short half-life in the cell so that, once synthesis of the enzyme stops, haem synthesis stops shortly afterwards. The enzyme is known to be unstable in reticulocyte lysates *in vitro* and presumably is so in the reticulocytes. The question of how Fe levels in the cell are controlled is dealt with separately. IRE, iron-responsive element, a stem loop structure in the mRNA. When the IRE-binding protein attaches to the IRE, it probably sterically inhibits the attachment of the translational initiation proteins. Binding of Fe to the IRE-binding protein abolishes its affinity for the IRE.

This halts the initiation of translation in the red blood cell and prevents globin synthesis. In the presence of haem, the kinase is inactivated and a phosphatase restores the normal activity of the initiation factor. Thus globin synthesis can proceed only when haem is present. The mechanism by which haem synthesis is regulated by translational control of the enzymes involved will now be described in more detail.

Translational control of proteins involved in haem synthesis and iron metabolism

The rate-limiting enzyme in haem production is **aminolevulinate synthase** (**ALA synthase**) (Fig. 19.9). The control of haem biosynthesis is of unusual interest and is intimately tied up with regulation of iron uptake and storage in the cell.

The situation is as follows.

- ALA synthase activity is the rate-limiting step in haem synthesis.
- ALA synthase *synthesis* is controlled by iron levels in the cell.
- Iron levels are controlled by transferrin receptors.
- Iron levels feedback-control the synthesis of transferrin receptor protein at the translational level.

In erythrocytes the mRNA for the enzyme ALA synthase has, at its 5′ untranslated end (that is, the translational initiation

end), a hairpin loop called an **iron-responsive element** or **IRE**. At low iron levels, an **IRE-binding protein** attaches to the IRE and prevents initiation and translation by steric hindrance (Fig. 24.16). At higher cellular levels of iron, the IRE-binding protein is prevented from attaching. Translation can now occur, thus coordinating the rate of haem synthesis with the iron supply.

The question now follows as to what controls the cellular iron level. Iron is transported in the blood plasma as a complex with the protein, **transferrin**, produced in the liver. This is taken up by receptor-mediated endocytosis (page 170). Uptake is regulated by translational control of the synthesis of the receptor protein by cellular iron. The mRNA for the receptor protein has iron-responsive elements at the 3′ end of the molecule. At low levels of iron, the IRE-binding protein binds to the mRNA, which protects the latter from degradation. In conditions of high iron levels, the binding protein detaches and removes the protection and the mRNA is destroyed. Thus, iron causes downregulation (reduction in the number) of transferrin receptors which, in turn, reduces iron intake.

Iron is stored in liver cells as **ferritin** – a complex of the protein, apoferritin, and inorganic iron. Apoferritin synthesis is regulated by a mechanism essentially the same as that for ALA synthase.

The area of haem biosynthesis is dealt with in Chapter 19. It has medical importance, since intermediates of haem biosynthesis, if allowed to accumulate in excess of needs, cause diseases, the best known of which is **acute intermittent porphyria** (see Box 19.1). The porphyria diseases are summarized, and a summary of the control area given, in the **May *et al.*** reference in the Further reading list at the end of this chapter.

Programmed destruction of protein by proteasomes

Introduction

There are three main ways in which proteins are broken down in the human body. The most obvious is in digestion, which is not an intracellular process. The second is intracellular, in the lysosome system (page 416) in which material to be destroyed is enclosed in a vesicle to which are delivered vesicles of destructive enzymes for all classes of materials. The third is by the **ubiquitin–proteasome system**, which is our present topic. This destroys individual selected protein molecules in cells. It is a tightly controlled in an elaborate and sophisticated way.

Controlled protein breakdown in eukaryotic cells is of almost astonishing importance. For example, the cell cycle control (page 508) depends on the specific breakdown of cyclin proteins at precisely the correct time to allow the cycle to progress from one phase to the next. Failure to destroy them as appropriate would cause an improperly controlled cell cycle with disastrous consequences. Enzymes and other proteins are destroyed in a selective manner; some proteins have, in humans, a half-life of hours and others of days. Despite chaperones and the other methods of guarding against faulty folding of polypeptides errors still occur and unless these faulty proteins were removed they would accumulate and cells would be loaded down with them. As will be described in the next chapter many proteins are transported into the endoplasmic reticulum (ER) as a linear polypeptide where they fold up into the mature form. Mistakes happen and unfolded proteins are sensed and transported back out of the ER lumen into the cytoplasm where they are degraded in proteasomes. As a final example of its importance, as will be explained later (page 500), a major part of the immune protection against viruses is dependent on the proteasomal destruction of proteins.

The structure of proteasomes

Proteasomes are organelles which provide a cavity in which proteins destined for destruction are segregated from the rest of the cell in an unfolded form and degraded to small peptides. These organelles are large protein structures about 2 million daltons in size. They are present in all eukaryotic cells in both the cytoplasmic and nuclear compartments in large numbers and are visible in the electron microscope. A model of the structure is shown in Fig. 24.17. There is a central **20 S core** (see page 386 for S values) consisting of a barrel-shaped cylin-

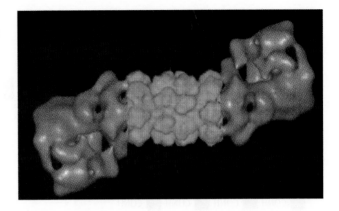

Fig. 24.17 Model of the proteasome. The yellow represents the structure of the 20 S proteasome core and the blue, the 19 S caps.

der made of four annular rings of protein subunits; the end ones, known as α rings, sandwich the two β rings. The latter contain proteolytic enzymes on the inside of the cylinder; the α rings have no known enzyme activity. At both ends of the 20 S core are **19 S caps**, also known as the regulatory units since they control the selection of proteins to be admitted and unfold the proteins so that they can enter the cavity of the 20 S core where the actual hydrolysis takes place. The unfolding by the caps is ATP-driven. The dimensions of the proteasome cavity are such that extended polypeptide chains can be accommodated but not folded proteins. Here the polypeptide is cut into small peptides which emerge into the cytoplasm. It is more or less like a tree trunk being sliced up into short logs. This may seem a trivial detail, but the correctly 'sized' peptides are vital in the immune system.

Proteasomes occur in a limited number of modern bacteria (eubacteria) but are found in the archaebacteria, which are organisms living in hostile environments such as sulphur hot springs at 80°C and pH 2. These proteasomes are different in that they lack the end caps, and have only the 20 S core, virtually identical in appearance to the core of those in yeast, but with fewer types of subunits in the rings. It is remarkable that the basic 20 S core structure has been conserved from archaebacteria through yeast down to humans, spanning billions of years of evolution.

Proteins destined for destruction in proteasomes are marked by ubiquitination

Ubiquitin is a small protein of 76 amino acids found universally in eukaryotes but not in prokaryotes. It has been highly conserved throughout evolution. Its amino acid sequence is identical in all animal species and differs from that in yeast and plants by only three, very conservative, amino acid changes.

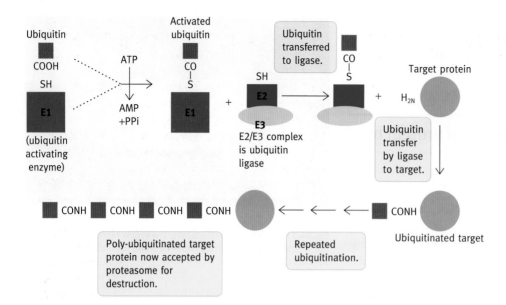

Fig. 24.18 Sequence of events in ubiquitinating target proteins destined for proteasomal destruction. Ubiquitin is first activated by attachment to E1 by a thioester link at the expense of ATP breakdown. It is then transferred to ubiquitin ligase which is a complex of E2 and E3. The ligase transfers it to a lysine amino group on the target protein. This is repeated to give a poly-ubiquitinated target which is then accepted by a proteasome for destruction.

Mechanism of ubiquitination of proteins

Three enzymes are involved (Fig. 24.18).

- Step 1: the ubiquitin is activated, its terminal —COOH becomes attached by a thioester link to the —SH group of enzyme 1 (**E1**) the **ubiquitin activating enzyme**. This requires ATP breakdown to AMP and PP$_i$ because the thioester is a high-energy bond.

- Step 2: E1 transfers its attached ubiquitin to an —SH group on **E2**. (The latter exists in the cell complexed with **E3**.)

- Step 3: E3 is the final enzyme. It exists as a complex with E2. The E2/E3 complex is the enzyme **ubiquitin ligase** (the active form of E3). It transfers the ubiquitin from E2 to an ε-amino group of a lysine residue on the target protein.

- Step 4: The ubiquitin attachment is the 'death ticket' for a protein but rather oddly the attached ubiquitin itself becomes ubiquitinated. About four molecules are optimal for proteasomal destruction. The multi-ubiquitinated protein molecule binds to the proteasome caps, is unfolded, and fed into the proteasome core containing the proteases. Prior to this, the attached ubiquitin molecules are released by enzymes and recycled.

Selection of target proteins for ubiquitination

This is a critical question for unless there are rigorous selection criteria there would be mayhem from random destruction. The selection system is not fully understood but a good deal is known. One criterion is the N-terminal amino acid of a protein. Some proteins with very short half-lives in yeast (less than 1 hour) are characterized by arginine, lysine and the aromatic ones. However, there are other criteria. Some destabilizing signals are masked and revealed by conformational changes induced by a ligand. Phosphorylation by a protein kinase is also believed in some cases to trigger recognition of a target protein by the E3 ligase. Denaturation may reveal destabilizing signals that are normally hidden. With all the different signals to recognize there needs to be a variety of different ligases and there are. Several hundred different E3 enzymes and a variety of E2s exist so that there are a great many different ubiquitin ligases. There must also be a special mechanism whereby for example cyclins in the cell cycle are triggered to be degraded at the end of each cell cycle phase (page 508). The complexity of the organization that must go into the safe running of this system is somewhat mind-boggling.

The role of proteasomes in the immune system

Somatic cells are at risk of virus infection. The body defends itself by destroying infected cells thus aborting virus multiplication, but it must detect those cells carrying foreign proteins and ignore normal ones. Their destruction is brought about by cytotoxic killer T cells (page 499), which recognize infected cells by a remarkable mechanism in which proteasomes play an essential part. All somatic cells continually hydrolyse 'samples' of cytosolic proteins into short peptides and display these on their surface for inspection by the killer cells, which ignore displayed peptides originating from normal (self) proteins but recognize those from a foreign, usually virus, protein synthesized in the cytosol of an infected cell. The cell is then attacked and destroyed. A more mechanistic account of this system can be found in Chapter 30 (page 500).

Summary

Protein synthesis is the joining together of amino acids in the correct sequence to form a polypeptide chain. It involves three phases, initiation, chain elongation, and termination.

Messenger RNA (mRNA) contains information of the amino acid sequence of the protein in the form of triplets of bases known as codons. Each codon corresponds to a specific amino acid, the correlation between them being the genetic code.

The mRNA codons are translated by cytoplasmic ribosomes into amino acid residues of polypeptide chains. Ribosomes have a large and a small subunit each containing RNA and many proteins.

As well as ribosomes, many cytoplasmic proteins are needed, and also a group of transfer RNAs (tRNAs). The tRNAs act as adaptors between the amino acids and the codons on the messenger being translated by the ribosome. Each tRNA has an unpaired triplet of bases known as the anticodon. There are tRNAs specific for each amino acid; protein synthesis depends on amino acids being placed on their own cognate tRNA – the one whose anticodon is complementary to the codon specifying it on the messenger. The amino acids are attached to the tRNAs by ligases that recognize their appropriate tRNA and the cognate amino acid, and join them together by ester linkage. ATP is broken down to supply the energy to link the amino acid to the tRNA. This supplies the energy for the subsequent formation of a peptide bond.

The initiation codon is AUG, which specifies methionine but it is also used for other methionine residues in the protein. Initiation has to distinguish between them. There is a special tRNA specific for methionine that is recognized by a protein that delivers methionyl-tRNA only to the initiation codon. (In *E. coli* the methionine is formylated.) The essence of the process is that the next amino acid on its tRNA is delivered to the second codon on the ribosome and the formylmethionine (in the case of *E. coli*) is transferred to the incoming amino acyl group to form a dipeptide.

The ribosome has three tRNA sites, A, P, and E (acceptor, peptide, and exit). Initiation in *E. coli* involves formation of a preinitiation complex (PIC) of formyl-methionyl-tRNA, mRNA and initiation factors on the small subunit. The formyl-methionyl-tRNA is delivered to the P site by a cytoplasmic protein, which recognizes only the initiation tRNA. Other methionine-specific tRNAs involved only in chain elongation are different and require a different delivery protein. Initiation is completed by joining of the large ribosomal subunbit to the small one.

Chain elongation involves delivery of aminoacyl-tRNAs to the A site where they bind by codon-anticodon hydrogen bonding. The P site is already occupied by the peptidyl-tRNA. The peptidyl group is transferred to the amino group of the latter thus elongating the nascent peptidyl chain in the amino to carboxyl direction.

Translocation transfers the elongated peptidyl-tRNA to the P site while the discharged tRNA moves to the exit site. The vacant A site accepts the next incoming aminoacyl-tRNA and a new round of elongation ensues. Translocation requires the cytoplasmic translocase (EF-G in *E. coli*) to attach to the ribosome and GTP hydrolysis is involved. The mechanism is uncertain and it seems that the ribosome itself may play an active part in the process. The synthesis of new peptide bonds is effected by peptidyl transferase which is a ribozyme.

At the end of the mRNA is the termination site comprised of stop codons where a protein factor releases the polypeptide. GTP hydrolysis plays a role in inducing conformational changes in proteins attached to ribosomes during the various phases.

To achieve fidelity of translation, the ligase attaching the amino acid to tRNAs in some cases hydrolyses incorrect amino acid loadings. Additionally the small subunit ribosomal RNA (rRNA) monitors that the codon – anticodon interaction is *via* Watson–Crick base pairing before the peptidyl transfer reaction occurs. A pause mechanism allows incorrect aminoacyl-tRNAs to leave the ribosome.

It now appears that the rRNA is involved in protein synthesis rather than being an inert scaffold for proteins, as was believed for so long. Probably the primitive ribosome was entirely RNA and proteins a refining addition. This avoids the origin of life chicken-and-egg problem of how proteins could be essential for the synthesis of proteins.

Initiation of eukaryotic ribosomes differs in that in initiation, protein factors and the small subunit assemble at the mRNA cap site. The preinitiation complex moves down the mRNA until it encounters the AUG initiation codon; initiation is then completed by attachment of the large subunit. The initiating methionyl-tRNA is not formylated.

The final stage is the folding of polypeptide chains into their native form. This is assisted by chaperones, proteins that provide maximum opportunities for the polypeptide to fold correctly either by preventing unprofitable hydrophobic associations or providing an optimum environment for correct folding. Deficiencies of folding cause prion diseases such as mad-cow disease (bovine spongiform encephalopathy) and probably others.

Selenocysteine is the 21st amino acid. It is incorporated into a few proteins during polypeptide synthesis by a special mechanism using a stop codon instead of a more normal one as used by the other amino acids.

Targeted protein degradation is of central importance in many aspects of the life of the cell. Prominent among these is the degradation of cyclins (Chapter 31), which control the eukaryotic cell cycle.

The selection of proteins for destruction is effected by attachment of ubiquitin proteins, which direct the protein to proteolytic destruction chambers known as proteasomes. The signals for ubiquitination are not fully elucidated but involve specific amino acid sequences. The ubiquitination process is complex.

Further reading

Transfer RNA

Schimmel, P. and de Pouplana, L. R. (1995). Transfer RNA: from minihelix to genetic code. *Cell*, **81**, 983–6.
This review addresses the question of why particular nucleotide triplets correspond to specific amino acids. tRNA can be thought of as comprising of two informational domains: the operational RNA code for amino acids, and the anticodon-containing domain with the trinucleotides of the genetic code.

Trifonov, E. N. (2000). Consensus temporal order of amino acids and evolution of the triplet code. *Gene*, **261**, 139–51.

Ogle, J. M., Carter, A. P., and Ramakrishnan, V. (2003) Insights into the decoding mechanism from recent ribosome structures. *Trends Biochem. Sci.*, **28**, 259–66.

Initiation

Preiss, T. and Hentze, M. W. (2003). Starting the protein synthesis machine. *BioEssays*, **25**, 1201–11.
The ribosome as the ultimate protein-synthesis machine in initiation.

Hinnebusch, A. G. (2006). eIF3: a versatile scaffold for translation initiation complexes. *Trends Biochem. Sci.*, **31**, 553–62.

The ribosome

Powers, T. and Noller, H. F. (1994). The 530 loop of 16S rRNA – a signal to EF–Tu? *Trends Genet.*, **10**, 27–31.
A penetrating discussion of how fidelity in protein synthesis is achieved.

Stadtman, T. C. (1996). Selenocysteine. *Annu. Rev. Biochem.*, **65**, 83.

Bock, A. (2000). Biosynthesis of selenoproteins – an overview. *Biofactors*, **11**, 77.

Wilson, K. S. and Noller, H. F. (1998). Molecular movement inside the translational engine. *Cell*, **92**, 337–49.
An authoritative review of how the ribosome works. It gives more information than most students would require but is listed for anyone taking a special interest in the area.

Rodnina, M. V. and Wintermeyer, W. (2001). Fidelity of aminoacyl-tRNA selection on the ribosome: kinetic and structural mechanisms. *Annu. Rev. Biochem.*, **70**, 415–35.
Advanced research-level review.

Dever, T. E. (2002). Gene-specific regulation by general translation factors. *Cell*, **108**, 545–56.
Reviews initiation of translation in eukaryotes.

Doudna, J. A. and Rath, V. L. (2002). Structure and function of the eukaryotic ribosome: the next frontier. *Cell*, **109**, 153–6.
A mini-review on the unique properties of the ribosome.

Ramakrishnan, V. (2002). Ribosome structure and the mechanism of translation. *Cell*, **108**, 557–72.
Readable review of all aspects of protein synthesis and ribosome formation.

Joseph, S. (2003). After the ribosome structure: How does translocation work? *RNA*, **9**, 160–4.

Maden, B. E. H. (2003). Historical review: Peptidyl transfer, the Monro era. *Trends Biochem. Sci.*, **28**, 619–24.
Review of developments that have led to our current knowledge of the peptide-bond-forming reaction.

Moore, P. B. and Steitz, T. A. (2003). The structural basis of large ribosomal subunit function. *Annu. Rev. Biochem.*, **72**, 813–50.
Crystal structures of the ribosome discussed in relation to its peptide-bond-forming activity, the way antibiotics inhibit large subunit function, and the ribosome as an RNA enzyme.

Ogle, J. M. and Ramakrishnan, V. (2005). Structural insights into translational fidelity. *Annu. Rev. Biochem.*, **74**, 129.

Mankin, A. S. (2006). Nascent peptide in the 'birth canal' of the ribosome. *Trends Biochem. Sci.*, **31**, 11–3.

Rodnina, M. V., Beringa, M., and Wintermyer, W. (2007). How ribosomes make peptide bonds. *Trends Biochem. Sci.*, **32**, 20–6.

Caban, K. and Copeland, P. R. (2006). Size matters: a view of selenocysteine incorporation from the ribosome. *Cell Mol. Life Sci.*, **63**, 73–81.

Korostelev, A. and Noller, H. F. (2007). The ribosome in focus: new structures bring new insights. *Trends Bochem. Sci.*, **32**, 434.

Riboswitches

Wakeman, C. A., Winkler, W. C., and Dann, C. E. (2007). Structural features of metabolite-sensing riboswitches. *Trends Biochem. Sci.*, **32**, 415–24.

Molecular chaperones and protein folding

Hortl, F.-U., Hlodon, R., and Langer, T. (1994). Molecular chaperones in protein folding: the art of avoiding sticky situations. *Trends Biochem. Sci.*, **19**, 20–5.
A general review of protein folding.

Horwich, A. L. (1995). Resurrection or destruction? *Curr. Biol.*, **5**, 455–8.
Discusses how chaperones are involved both in rescuing proteins and directing them towards proteolytic destruction.

Netzer, W. J. and Hartl, F. U. (1998). Protein folding in the cytosol: chaperonin-dependent and independent mechanisms. *Trends Biochem. Sci.*, **23**, 68–73.
A comprehensive account of protein folding and the mechanism of chaperone action.

Bukau, B. and Horwich, A. L. (1998). The Hsp 70 and Hsp 60 chaperone machines. *Cell*, **92**, 351–66.
An authoritative clear review, with structural models.

Pfanner, N. (1999). Protein folding: who chaperones nascent chains in bacteria? *Curr. Biol.*, **9**, R722.
Summarizes the actions of Hsp 60 and Hsp 70.

Rye, H. S. *et al.* (1999). GroEL-GroES cycling ATP and non-native polypeptide direct alternation of folding-active rings. *Cell*, **97**, 325–38.

Gottesman, M. E. and Hendrickson, W. A. (2000). Protein folding and unfolding by *E. coli* chaperones and chaperonins. *Curr. Opin. Microbiol.*, **3**, 197–202.
Summarizes the structures and functions of these molecular machines.

Prion diseases

Weissmann, C. (1995). Yielding under the strain. *Nature*, **375**, 628–9.
A concise summary of the molecular basis of prion diseases.

Prusiner, S. B. (1995). The prion diseases. *Sci. Am.*, **272**(1), 30–7.
General review of the field.

Thomas, P. J., Qu, B.-H., and Pedersen, P. L. (1995). Defective protein folding as a basis of human disease. *Trends Biochem. Sci.*, **20**, 456–9.
Discusses the possibility that a large number of diseases, in addition to prion diseases, may be due to protein-folding abnormalities.

Taubes, G. (1996). Misfolding the way to disease. *Science*, **272**, 1493–5.
A research news item presents the general hypothesis that protein misfolding may cause amyloid diseases such as Alzheimer's disease as well as the prion diseases. Introduces the concept that aggregation of proteins into insoluble complexes may be more prevalent and more important than hitherto suspected.

Dobson, C. M. (1999). Protein misfolding, evolution and disease. *Trends Biochem. Sci.*, **24**, 329–32.
Reviews diseases in which protein misfolding leads to amyloid or fibrillar aggregates.

Hunter, N. (1999). Prion diseases and the central dogma of molecular biology. *Trends Microbiol.*, **7**, 265–6.
A 'Comment' article summarizing the nature of prions.

Manson, J. C. (1999). Understanding transmission of the prion diseases. *Trends Microbiol.*, **7**, 465–7.
A 'Comment' article summarizing this topic.

Hope, J. (1999). Prions. *Curr. Biol.*, **9**, R763–4.
A quick guide to the essential facts.

Butler, D. (2001). Unfolding issues. *Nature*, **414**, 577.
Briefly discusses prions and gives three-dimensional structures.

Dobson, C. M. (2002). Getting out of shape. *Nature*, **418**, 729–30.
Short article on protein-misfolding diseases.

Buxbaum, J. N. (2003). Diseases of protein conformation: what do *in vitro* experiments tell us about *in vivo* diseases? *Trends Biochem. Sci.*, **28**, 585–92.
Review on human diseases associated with misfolded proteins with decreased solubility.

Soto, C., Estrada, L., and Castilla, J. (2006). Amyloids, prions and the inherent infectious nature of misfolded proteins. *Trends Biochem. Sci.*, **31**, 150–5.

Arolas, J. L., and Aviles, F. X., Chang, J., and Ventura, S. (2006). Folding of small disulphide rich proteins: clarifying the puzzle. *Trends Biochem. Sci.*, **31**, 292–301.

Chiti, F. and Dobson, C. M. (2006). Protein misfolding, functional amyloid and human disease. *Annu. Rev. Biochem.*, **75**, 333–66.

Translational control

Russell, J. E., Morales, J., and Liebhaber, S. A. (1997). The role of mRNA stability in the control of globin gene expression. *Prog. Nucleic Acid Res. Mol. Biol.*, **57**, 249–87.

Gebauer, F. and Hentze, M. W. (2004). Molecular mechanisms of translational control. *Nature Reviews*, **5**, 827–35.
A comprehensive detailed account.

Proud, C. G. (2006). Regulation of protein synthesis by insulin. *Biochem. Soc. Trans.*, **34**, 213–6.

Translational control of proteins involved in haem synthesis and iron metabolism

Chen, J.-J. and London, I. M. (1995). Regulation of protein synthesis by heme-regulated eIF-2α kinase. *Trends Biochem. Sci.*, **20**, 105–8.
A comprehensive review of the control of haemoglobin synthesis by this mechanism.

May, B. K., *et al.* (1995). Molecular regulation of heme biosynthesis in higher vertebrates. *Prog. Nucleic Acid Res.*, **51**, 1–47.
A comprehensive review of all aspects including the hereditary diseases.

Proteasomes

Coux, O., Tanaka, K., and Goldberg, A. L. (1996). Structure and function of the 20S and 26S proteasomes. *Ann. Rev. Biochem.*, **65**, 801–44.
A comprehensive review of the subject suitable for advanced students.

Stuart, D. I. and Jones, E. Y. (1997). Cutting complexity down to size. *Nature*, **386**, 437–8.
A News and Views summary of proteasomes and their structure.

Ciechanover, A. (1998). The ubiquitin-proteasome pathway: on protein death and cell life. *EMBO J.*, **17**, 7151–60.

De Mot, R., Nagy, I., Walz, J., and Baumeister, W. (1999). Proteasomes and other self-compartmentalizing proteases in prokaryotes. *Trends Microbiol.*, **7**, 88–92.

Kirschner, M. (1999). Intracellular proteolysis. *Trends Cell Biol., Trends Biochem. Sci.* and *Trends Genet.*, Joint issue. **24**, M42–5.
Part of a special millennium issue of the journals, this article reviews the development of an area of biochemistry from what was a relatively dull subject to one of the exciting ones today.

Scheffner, M. and Whitaker, N. J. (2001). Proteolytic relay comes to end. *Nature*, **410**, 882–3.

Hooper, N. M. Ed. (2002). Proteases in biology and medicine. *Essays Biochem.*, **36**, 1–167.
A complete issue with 12 reviews on many aspects of proteases, including caspases, cancer, proteasomes, and blood clotting.

Gille C., *et al.* (2003). A comprehensive view on proteasomal sequences: implications for the evolution of the proteasome. *J. Mol. Biol.*, **326**, 1437–48.

Problems

1   There are 64 codons available for 20 amino acids. Why do you think 61 codons are actually used to specify the 20 amino acids?

2   Despite the facts stated in question 1, there are fewer than 61 tRNA molecules. Explain why this is so.

3   In diagrams, when a tRNA molecule is shown base paired to a codon, the molecule is shown flipped over as compared with the same tRNA shown on its own. Why is this so?

4   At which points in protein synthesis do fidelity mechanisms operate?

5   Describe the participation and, where known, the role of GTP in protein synthesis.

6   Protection studies have indicated that, in *E. coli*, tRNA molecules (with their aminoacyl or peptidyl attachments) straddle A, P, and E sites on the ribosome. Explain why this occurs.

7   The mechanism of initiation of translation in eukaryotes is not compatible with polycistronic mRNA. Explain why.

8   Explain the role of chaperones in protein synthesis.

9   What diseases are associated with improper protein folding?

10   (a)   If you are given the base sequence of the coding region of a mRNA can you deduce the amino acid sequence of the protein it codes for?

   (b)   If you are given the amino acid sequence of a protein can you deduce the sequence of the coding region of the mRNA that directed its synthesis? Explain your answers.

11   Consider an mRNA that codes for a protein 200 amino acid residues in length. What would the resultant polypeptide be from translation of this messenger if codon number 100 was mutated so that its first base was deleted or if the first and second bases were deleted? What if all three bases were deleted? Explain your conclusions.

12   In a ribosome, is the RNA there simply as an inert scaffold on which to hang proteins? Discuss two pieces of evidence that bear on this problem.

13   Explain the difference between transcription and translation.

14   Explain how the red blood cell ALA synthase level is coordinated with the availability of iron.

15   Describe a proteasome. What are its roles in cells? Give some examples of the latter. What is the evidence for their great importance in cells?

16   How are proteins targeted to the proteasomes?

17   Codons and anticodons specifically base pair on the ribosome. Is the formation of hydrogen bonds itself sufficient to account for the observed fidelity of protein synthesis?

18   Explain briefly how an Hsp 70 chaperone works. Why should chaperones be given the prefix Hsp in their names?

Chapter 25

The RNA world – RNA microgenes and RNA interference

There are few developments that have had the impact of RNA interference (RNAi) on biology; as one reviewer put it, 'it has taken the world by storm'. It is a natural process that has probably always been essential for eukaryotes to survive but which also provides a method of silencing (i.e. inhibiting) gene expression of unprecedented simplicity and width of potential application in medicine and research.

However, this is not the only aspect of the importance of the new RNA world. As will be described more fully below, it has been discovered that almost all of the genome is transcribed. It is believed that large numbers of RNAs that do not code for proteins (noncoding RNAs) are transcribed from previously unrecognized microRNA genes existing in the areas once believed to be useless 'junk' DNA. This is causing a widespread rethink of many of our ideas on how the genomes of complex life are organized and controlled. It is to be noted that these remarks apply primarily to eukaryotes; it seems that it is in the generation of complexity of eukaryotic organisms that the 'new RNA world' concepts mainly apply, so that its relevance to the most complex organism, the human being, is especially great.

A general overview

We have long known of the existence in biochemical processes of noncoding species of RNAs other than messenger RNAs (mRNAs), such as ribosomal, transfer and small nucleolar RNAs (snoRNAs) (see page 374). That life could not function without them is evident, but their known roles were always subsidiary – they were all involved in other major processes, more or less as infrastructure needed to allow the protein-coding genes to be expressed. The protein-coding genes were the all important centre of attention for the very good reason that they apparently determined most things about the heritable characteristics of living organisms. The broad concept, which still is not questioned, is that if a fertilized egg produces the correct proteins, at the correct time and in the correct amount then the correct organism will self-assemble from them. It was confidently believed that all the necessary information to achieve this is present in nucleotide sequences of the formal control regions of genes to which a multiplicity of control proteins can bind and determine gene transcription. In the informational sense, protein-coding genes determined life. The **noncoding genes**, genes which are transcribed into RNAs as the *terminal* product, such as ribosomal RNAs, it was believed, determined nothing in the sense of phenotypic characteristics and development of the organism. That, the dogma said, was the job of protein-coding genes, their adjunct control regions and associated protein factors.

In the past 10 years there has been a dramatic change in our concepts of how the complex genomes of eukaryotes may function.

The human genome project astonished everybody when it revealed that the human genome has a disorganized look, with genes scattered around like jewels in a cluttered 'desert' of seemingly useless DNA debris, called 'junk' DNA (see page 327 for a fuller description).

Now the situation has changed. Large numbers of hitherto unrecognized noncoding RNA genes exist widely spread in the eukaryotic genome, including the 'junk' DNA areas devoid of protein-coding genes, and even in the introns that divide up eukaryotic protein-coding genes hitherto presumed by most to have no function. They produce small RNA molecules (**microRNAs**) that are believed to be regulatory RNAs and quite different from the hitherto known noncoding species of RNAs referred to earlier.

MicroRNAs and the functioning of the human genome

Before the discovery of microRNA genes, most thought that the basic rules of genome function had been established despite gaps in our understanding. We still accept that protein-coding genes determine more or less all of the heritable characteristics, since the proteins that are produced in turn determine the chemistry and assembly of organisms. There was abundant evidence for this since phenotypic characteristics were, for example, altered by mutations in protein-coding genes, but the situation has proved to be more complex than that. There are apparent inconsistencies, such as the fact that protein-coding genes occupy so little of the DNA (about 1.5%), and the number of such genes do not correlate with complexity (see page 327 and **Mattick** reference). To re-emphasize with two examples, the small roundworm, *Caenorhabditis elegans* (a nematode), with only 1000 cells has 18,000 genes, and the human, estimated to have between 10 and 100 trillion cells, has an estimated 25,000 genes; the rice plant has more protein-coding genes than a human (although, because of alternative splicing, it doesn't necessarily mean that rice has more proteins).

Research on genes hitherto has unavoidably tended to be on individual genes, their components, and how they work. It has been very successful and has given necessary insight into the fundamentals of gene molecular biology. It seems that what is now needed is a genome-wide insight. This task is comparable in scale to the sequencing of human DNA. To this end an international collaborative research effort has been set up in which 35 different research teams aim to identify the functional elements in the human genome, and to identify their transcriptional activities. It is called ENCODE (Encyclopaedia of DNA Elements).

In its initial phase, the consortium set out to examine a total of 1% (about 30,000 kilobases) of the human genome in great detail. To make up the total it sampled 44 different sections of the genome taken from different representative types of areas, such as ones rich in protein-coding genes and areas hitherto believed to be transcriptionally inert (junk DNA areas). It was found (see Nature and Science references under ENCODE in Further reading), remarkably, that of the bases in the genome about 80% are found in RNA primary transcripts. It could have been that the RNA transcription was background 'transcriptional noise' that went on without meaning but it was found that a number of the noncoding transcripts were conserved across mice and humans, showing that they almost certainly have an important function.

There is an increasing body of evidence pointing to microRNAs as regulators of a wide range of the most fundamental processes of life, including development, cell growth and apoptosis. The view has been expressed that the RNAs may represent a layer of epigenetic control that regulates and coordinates gene expression. (**Epigenetic control** means alteration of hereditary or phenotypic characteristics without altering the nucleotide sequence of the DNA.)

RNA interference (RNAi) is a method of gene silencing triggered by double-stranded RNA

RNA interference (RNAi) is a natural process mediated by microRNAs that has probably always existed in eukaryotic organisms. The process is called **posttranscriptional gene silencing (PTGS)** in plants, **quelling** in fungi and **RNAi** in animals. We will use the latter form in the rest of this chapter. *The double-stranded RNA that triggers it arises from several sources. It can come from the outside in the form of double-stranded RNA virus infections or chemically synthesized ds RNA; it can be generated endogenously by transcription of **microgenes** producing the small RNA molecules known as **microRNAs (miRNAs)**.*

Gene silencing triggered by RNA was first described in plants when it was observed that RNA virus infection of plants led to them developing an immunity to the virus. After this, it was discovered in the small roundworm *C. elegans* that a gene *essential for its development* (called *lin-4*) did not code for a protein but for a small noncoding RNA that silenced the expression of the gene *lin-14* that produces a protein essential for the development of the worm. Thus, the cellular concentration of *lin-14* protein, which regulates the timing of certain developmental events, is itself regulated by the *lin-4* noncoding RNA. Evolutionary conservation occurs only when a gene is important so that natural selective pressure prevents it being 'mutated away'. It is now known that a number of human genes producing miRNAs are conserved in the mouse.

A paper by **Fire *et al.*** in 1998 (see Further reading) showed that silencing of genes in the roundworm required **double-stranded RNA**. Andrew Fire and Craig Mello were awarded the Nobel Prize for this work in 2006. They were initially studying the possibility of silencing a protein-coding gene using **antisense RNA** chemically synthesized in the laboratory. The concept here is that an RNA molecule *complementary* to a messenger RNA (an antisense RNA), present in excess would bind to the mRNA by Watson–Crick pairing and block the translation of the latter, and thus inhibit expression of the gene. The studies used worms containing a green fluorescent reporter protein (GFP) gene. An antisense preparation when injected into the worm was effective, but paradoxically so was a sense strand preparation which, with the same base sequence as the messenger, could not have hybridized with the latter. It was obvious that a different explanation for the silencing must be involved. The explanation was that both preparations were contaminated with *double-stranded RNA (dsRNA)* which was what did the silencing. The roundworms were genetically engineered, to express in their tissues GFP which is easily detected by its fluorescence. It was expected that when dsRNA homologous with GFP mRNA was injected into the body of the worm only the injected area would be silenced but in fact it happened to the whole worm showing that it can be transmitted from cell to cell. It was also transmitted for several generations *via* the germ line cells, suggesting that the effect was amplified in the organism (see **Mello and Conte** in Further reading).

MicroRNAs are double-stranded hairpin molecules

When a gene is transcribed by RNA polymerase, a single strand transcript is produced. The RNA transcripts are typically about 75 nucleotides in total length and have base complementary sections on either side of the centre so that they automatically fold back on themselves by base pairing to produce double-stranded hairpins that can trigger silencing. The same strategy is used in chemical synthesis of miRNAs. (Peter Waterman, who invented the chemical method, commented 'We thought we'd come up with something unique and man-made, but it has now become clear that plants and animals have been producing their own hairpin dsRNAs for at least 400 million years.)

Protection against retroviruses (page 348) may have been the evolutionary origin of the RNAi mechanism. There is another important and related reason. We mentioned in an earlier chapter (page 328) that much of the 'junk' DNA consisted of **retrotransposons** (also called **transposons**). These are essentially the same as retroviruses in their replication. They transpose, and multiply promiscuously, inserting copies of themselves elsewhere in the genome. This potentially could cause mayhem in the genome because insertion of 'foreign' DNA sections into genes could inactivate them or if adjacent to their promoters, cause their normal control to become aberrant and possibly produce tumours. Retrotransposon transposition proceeds *via* an RNA intermediate, and RNAi is postulated to prevent excessive retrotransposon multiplication by destroying this intermediate. For this reason the RNAi mechanism has been called the guardian of the genome.

Molecular mechanism of gene silencing by RNAi

The various double strand RNAs which trigger silencing have to be processed into the actual effector molecules. As stated already, injection of double-stranded RNAs into roundworms, and a wide variety of invertebrates, silenced genes to which the injected RNAs were homologous. Irrespective of the source of the triggering double-stranded RNA (natural or experimental), they all enter the same processing machinery and are converted to short double strand RNAs 22 nucleotides long, *which are the effectors of gene silencing* and called **short interfering RNAs** (**siRNAs**). These are found in all eukaryotic species. However, when dsRNAs were first injected into *vertebrate* cells in culture they did not elicit the expected specific gene silencing. This was because the tested RNAs were too long. If they are longer than 30 base pairs they induce interferon protection. This is an antiviral response to double-stranded RNA that silences a broad range of genes and obscures RNAi responses. However,

double-stranded RNAs shorter than 30 base pairs only induce the expected silencing of expressed complementary genes in vertebrates as for other organisms.

The processing of microRNAs and viral RNAs into molecules 22 base pairs in length occurs in several steps (Fig. 25.1). The hairpin double-stranded miRNAs emerge from the nucleus into the cytoplasm (in some species after trimming of the initial transcript in the nucleus by an enzyme called **Drosha**) where it meets another enzyme called **Dicer**, which processes them into siRNA 22 base pairs in length. The strands of the siRNAs are separated by a **helicase** and one, the **guide strand**, becomes associated with a complex of proteins called **RISC** (**RNA-induced silencing complex**). The other, known as the **passenger strand** is discarded and degraded. The guide strand has to be complementary in base sequence to the target messenger RNA of the target gene. It is not understood how the strand selection is made. The guide strand associated with the RISC guides the complex to the target messenger RNA so that the RISC complex in this way identifies its target messenger RNA. (It is possible that the helicase is a component of RISC.) RISC is an astonishing molecular machine.

What happens then appears to be determined by the degree of complementarity between the guide strand and the mRNA. If it is perfect or nearly so it base pairs with the translatable section of the mRNA. This determines where an enzyme in the RISC complex called **argonaute** starts to cut up the mRNA, which is thereby inactivated. In other species the cutting enzyme is called **slicer** – it makes a single cut. (We give the quaint names of these enzymes because it is helpful to have met them if you read a review on RNAi in the Further reading.) In other cases where the complementarity match between the guide strand and the mRNA is only partial, the RISCs attach to the 3′ untranslated section (UTR) of the mRNA and prevents its translation by ribosomes. The attachment site is downstream of the stop signal where the ribosomes detach and translation ceases and before the polyadenylation site so it is not clear how translation is blocked in this way. Either way the gene is silenced, completely or partially; partial silencing is called **gene knockdown**. The silencing by blocking of the 3′UTR requires only about a 7 base pair complementarity, which might imply that genes other than the targeted one presumably could be silenced by an miRNA. There is much yet to be understood about the precise mechanism of the process of gene silencing *in vivo*.

The potential medical and practical importance of RNAi

With the discovery that easily prepared siRNAs were effective in silencing specific genes in vertebrates we now have a peerless method, in principle, of silencing any known gene in the human body. The siRNAs are synthesized in the laboratory in

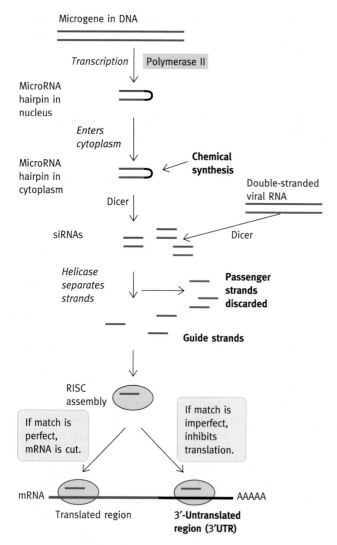

Fig. 25.1 Simplified mechanism of RNA interference. The diagram shows that short interfering RNAs (siRNAs) can be produced in the cell either from microgene transcription or from processing of exogenous double-stranded RNAs, such as viral RNAs or chemically synthesized double-stranded RNAs. The latter used for experimental purposes are conveniently synthesized as short hairpins as described in the text. The hairpin RNAs are processed by Dicer into short double-stranded interfering RNAs. The strands of these are separated by helicase action and one of the strands, the passenger strand, is discarded. The mechanism of strand selection is not understood. The remaining guide strand associates with the RISC complex of proteins and targets the RISC onto a specific mRNA. If the guide strand sequence matches that of the target messenger perfectly, or nearly so, the messenger is cut in its coding region and inactivated. If the match is imperfect, it may attach to the 3′UTR of the messenger downstream from the stop signal and this in some unknown way blocks messenger translation.

automated machines as hairpins as explained earlier. The synthetic ones are effective in tissue cultures. Since the human gene project gives the sequence of human DNA this implies that, in principle, any identified gene could be silenced by making an siRNA complementary to the mRNA of the target gene. For example, it has been established in cultured cells that oncogenes can be successfully silenced by 80%. Oncogenes are cancer-inducing genes (see page 520). However, as stated, the field is in its infancy and there are wide-ranging ideas as to how and where RNAi might be used. Even viral infections of eukaryotic cells are potential targets, since these depend on mRNA production from their genes, 50 siRNAs can be designed to silence them. Another important use of gene silencing could be the production of mouse strains as models of human diseases caused by specific gene deficiency.

MicroRNAs may orchestrate expression of protein-coding genes in eukaryotes

Perhaps the most profound outcome of this new field of research is the possibility referred to earlier that the miRNAs appear to constitute another layer of control that may orchestrate the function of the genomes of humans and complex organisms generally. It may be that evolution has increased complexity by having large numbers of regulatory RNAs that somehow network a limited number of protein-coding genes into producing greater phenotypic complexity than could be otherwise achieved, though it is not understood how this might be organized.

An exciting recent discovery has shown that metastasizing breast cancer cells have a higher level of a specific miRNA. Experimental overexpression of this miRNA causes non-metastasizing cancer cells to metastasize when implanted into mice (see **Ma et al.** in Further reading). Control experiments without the overexpression did not do so (see page 518 for more detail). (Also see **Zhongxing** in Further reading for blockage of breast cancer cell metastasis by an miRNA.)

As a research tool the potentialities of RNAi seem unlimited and almost unrivalled. Therapeutic applications face the problem of delivery to the tissues and cells, and novel ways are being developed such as by inserting DNA, which is transcribed into appropriate siRNAs.

Bacteria don't need the complexity required in eukaryotes; their survival mechanism apparently centres on rapid reproduction for which simplicity is a presumed advantage, so that the new RNA-world concepts and RNAi are not applicable there so far as is believed at present (though some participation of RNA in control mechanisms does occur in a few cases, for example in riboswitches, page 393). This correlates well with the bacterial genome being an orderly array of tightly-packed protein-coding genes largely controlled by regulatory proteins. There is little or no 'junk' DNA.

Many people have been referring to what has happened in the past decade as the RNA revolution. Mello and Conte (see

Further reading) commented that 'it was more apt to call it an RNA revelation – that RNA has been in control all along. We just didn't realise it until now'. It is believed that RNA was supreme in the origin and earliest stage of the evolution of life, before DNA or proteins existed. The dogma of DNA's absolute supremacy in control of eukaryotic life is now being challenged by RNA.

Summary

RNA interference (RNAi) is an ancient natural mechanism of silencing protein-coding genes by destroying or preventing the translation of specific mRNAs. It is widely found in eukaryotes, but not in prokaryotes. Its function was probably initially to protect against invasion by single-stranded RNA viruses, most likely against retroviruses. In mammals it has the believed role of preventing excessive proliferation of retrotransposons and, in this role, is known as the guardian of the genome.

RNAis start with microRNAs (miRNAs) transcribed from microRNA genes, which are found almost everywhere in the genome, including so-called 'junk DNA', and even in introns.

The miRNAs are hairpin transcripts, about 72 bases long. They are processed into short interfering double-stranded RNAs (siRNAs) 22 bases long. Chemically synthesized siRNAs are also effective. The strands are separated, and one, the guide strand, attaches to the complex, RISC, which guides the latter to target and destroy the complimentary mRNA, if the match is perfect or blocks translation if less perfect.

The microRNAs are believed to orchestrate the expression of protein-coding genes. They are believed to constitute an epigenetic control. One microRNA has been shown to promote the ability of breast tumour cells to metastasize. It seems that microRNAs enable increasing complexity of life forms without a corresponding increase in protein-coding genes, though much remains to be learnt about how this works.

RNAi has the potential to become one of the great advances in medicine and research.

Further reading

ENCODE

Greally, J. M. (2007). Genomics: encyclopaedia of humble DNA. *Nature*, **447**, 782–3.
Nature News and Views article.

(No authors quoted). (2007). Identification and analysis of functional elements in 1% of the human genome by the ENCODE pilot project. *Nature*, **447**, 799–816.

Pennisi, E. (2007). DNA study forces rethink of what it means to be a gene. *Science*, **316**, 1556–7.
A commentary on the ENCODE project results which shows that genes may be far from compact assemblies but have regions scattered around the genome. Highly readable.

MicroRNA

Fire, A. *et al.* (1998). Potent and specific genetic interference by double-stranded RNA in *C elegans*. *Nature*, **391**, 806–11.

Nishikura, K. (2001). A short primer on RNAi: RNA-directed RNA polymerase acts as a key catalyst. *Cell*, **107**, 415–18.
Minireview.

Zamore, P. D. (2001). RNA interference: listening to the sound of silence. *Nat. Struct. Biol.*, **8**, 746–50.
Review on gene silencing.

Martinez, J. (2002). Single-stranded antisense siRNAs guide target RNA cleavage in RNAi. *Cell*, **110**, 563–74.

Shi, Y. (2003). Mammalian RNAi for the masses. *Trends Genet.*, **19**, 9–12.
The siRNAs (small interfering RNAs) are easily synthesized chemically. Their sequence specificity means that, in principle, any mRNA can be targeted to silence (or knockdown) expression of specific genes. Deals with potential medical applications.

Lavorgna, G. *et al.* (2004). In search of antisense. *Trends Biochem. Sci.*, **29**, 80–7.
Natural antisense transcripts have been implicated in many aspects of eukaryotic gene expression.

Hannon, G. J. (2002). RNA interference. *Nature*, **418**, 244–51.
An Insight review article.

Nelson, P., Kiriakodou, M., Sharma, A., Maniataki, E., and Mourelatos, Z. (2003). The microRNA world: small is mighty. *Trends Biochem. Sci.*, **28**, 534–40.

Ambros, V. (2004). The functions of animal microRNAs. *Nature*, **431**, 350–5.
Useful introduction to the subject of microRNAs.

Bartel, D. P. (2004). MicroRNAs. Genomics, biogenesis mechanism and function. *Cell*, **116**, 281–97.
Very comprehensive, more suitable for instructors.

Meister, G. and Tuschl, T. (2004). Mechanisms of gene silencing by double-stranded RNA. *Nature*, **431**, 343–9.
A general review.

Mello, C. C. and Conte, D. (2004). Revealing the world of RNA interference. *Nature*, **431**, 338–42.

Eckstein, F. (2005). Small non-coding RNAs as magic bullets. *Trends Biochem. Sci.*, **30**, 445–52.
Very good summary of all aspects.

Lu, J. *et al.* (2005). MicroRNA expression profiles classify human cancers. *Nature*, **435**, 834–8.

Storz, G., Alluvia, S., and Wassarman, K. M. (2005). An abundance of RNA regulators. *Annu. Rev. Biochem.*, **74**, 199–217.

Mattick, J. S. and Makunin, I. V. (2006). Noncoding RNA. *Human Molecular Genetics*, R17–R29.

Kloosterman, W. P. and Plasterk, R. H. A. (2006). The diverse functions of microRNAs in animal development and disease. *Devel. Cell*, **11**, 441–50.

Parker, S. and Barford, D. (2006). Argonaute: a scaffold for the function of short regulatory RNAs. *Trends Biochem. Sci.*, **31**, 622–30.

Plasterk, R. H. A. (2006). MicroRNAs in animal development. *Cell*, **124**, 877–81.

Hernando, E. (2007). MicroRNAs and cancer: role in tumorigenesis, patient classification and therapy. *Clinical and Translational Oncology*, **9**, 155–60.
This review examines the role of miRNAs in the pathogenesis of cancer as well as miRNA-profiling studies performed in human malignancies.

Ma, L., Teruya-Feldstein, J., and Weinberg, R. A. (2007). Tumour invasion and metastasis initiated by microRNA-10b in breast cancer. *Nature*, **449**, 682–8.

Mattick, J. (2007). A new paradigm for developmental biology. *J. Exptl. Biol.*, **210**, 1526–47.
Proposes a model for RNA regulatory networks.

Zhao, Y. and Srivastava, D. (2007). *Trends Biochem. Sci.*, **32**.

Zhongxing. (2007). Blockade of invasion and metastasis of breast cancer cells via targeting CXCR4 with an artificial microRNA. *Biochem. Biophys. Res. Commun.*, **363**, 542–6.
CXCR4 is a chemokine receptor.

Amaral, P. P. *et al.* (2008). The eukaryotic genome as an RNA machine. *Science*, **319**, 1787–9.
A Science perspective article.

Nowacki, M., Vijayan, V., Zhou, Y., Schotanus. K., Doak, T. G., and Landweber, L. F. (2008). RNA-mediated epigenetic programming of a genome-rearrangement pathway. *Nature*, **451**, 153–60.

 ## Problems

1   What is a noncoding RNA? Give examples.

2   Where are microgenes located in the genome?

3   What is the evidence that miRNAs represent meaningful transcription and not just background transcriptional 'noise'?

4   What is believed to be the significance of miRNAs?

5   How do miRNAs exert epigenetic control over protein coding genes?

6   Describe briefly the mechanism of RNA interference (RNAi).

7   What is one medical potential of RNAi?

Chapter 26

Protein sorting and delivery

As explained in Chapter 24 there are two types of ribosome in the eukaryotic cell, those in the cytoplasm and the prokaryotic-like ones of the mitochondria and chloroplasts. Those in the mitochondria and chloroplasts produce only a handful of proteins; all the rest are synthesized by cytoplasmic ribosomes. Which protein a given ribosome synthesizes at any one time is solely a function of the mRNA that it happens to be translating, but, once synthesized, proteins have a number of different destinations.

So far as delivery goes, cytoplasmic proteins present no problems – they are synthesized in the cytoplasm, released from the ribosome, and stay there. However, proteins destined for other compartments present intriguing problems. How do the integral proteins of the plasma membrane and other membranes get to be there? How are blood serum proteins selectively released by liver cells to their exterior? The same applies to release of any of the many extracellular proteins such as the digestive enzymes, insulin from the pancreas, and the connective tissue proteins from fibroblasts to name only a few. Most mitochondrial proteins are synthesized in the cytoplasm coded for by nuclear genes. How are these selectively transported into the mitochondria? Lysosomes and peroxisomes are membrane-bounded vesicles full of enzymes specifically found in those vesicles, but they cannot synthesize proteins. Again, how are the different enzymes transported into the correct compartment? The nucleus has its own cohort of proteins such as the enzymes responsible for synthesizing and transcribing the DNA, but these are synthesized in the cytoplasm. There is also traffic of proteins (and RNA) out of the nucleus. How is the two-way traffic across the nuclear membrane orrganised? As you will learn from Chapter 27 on cell signalling, many gene-control proteins exist in the cytosol but, on receipt of extracellular signals, enter the nucleus to regulate transcription of genes. It is not just a question of how proteins are able to cross membranes but also how specific proteins are selected from the whole mixture of proteins in the cell to be delivered to, and transported across, or into the correct membrane. Much of the mechanisms of protein targeting have been substantially elucidate, though much detail of the processes remains to be understood.

A preliminary overview of the field

An overview summary without any details may be useful.

- **Cytoplasmic proteins** are released from the ribosome on completion of their synthesis and they stay there (Fig. 26.1).
- Proteins destined to go into **mitochondria, peroxisomes, or the nucleus** are released from the free cytoplasmic ribosomes that have synthesized them and are then

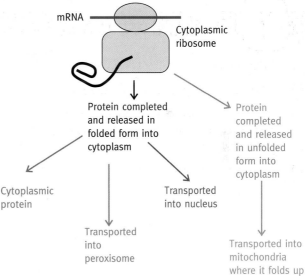

(The method of transport into the three organelles is different in each case.)

Fig. 26.1 Preliminary overview summary of events in posttranslational targeting of proteins to the cytoplasm, peroxisomes, nucleus, and mitochondria. The essence of the process is that free ribosomes synthesize complete polypeptide chains, release them in the cytoplasm, and then these proteins are transported into their targeted destinations but by different mechanisms in each case, as will be described. This contrasts with cotranslational transport depicted in Fig. 26.6 in which proteins are transported across the endoplasmic reticulum lipid bilayer as they are synthesized.

transported into the appropriate organelle, but by a different mechanism in each case; this is known as **post-translational transport** (Fig. 26.1).

- The synthesis of **extracellular (secreted) proteins, lysosomal proteins, proteins inside the lumen of the endoplasmic reticulum (ER)**, and all **integral membrane proteins** *commences* on free ribosomes, but these become transitorily attached to the ER membrane and the proteins are transported into the lumen (or into the ER membrane in the case of membrane proteins) *as they are synthesized.* This is known as **cotranslational transport.**

Once inside the ER lumen, proteins move along to the smooth ER and then are transported to the Golgi apparatus. In the ER and Golgi, proteins have carbohydrates added. The Golgi sort out and package proteins into transport vesicles, which deliver their cargo to appropriate target membranes or compartments. Secretory vesicles eject their contents from the cell by exocytosis (page 107), and transport vesicles for lysosomal enzymes deliver their contents to endosomes (see below) to form lysosomes. Vesicles carrying integral proteins in their membrane fuse with their target membrane to produce new membrane complete with proteins (Fig. 26.2).

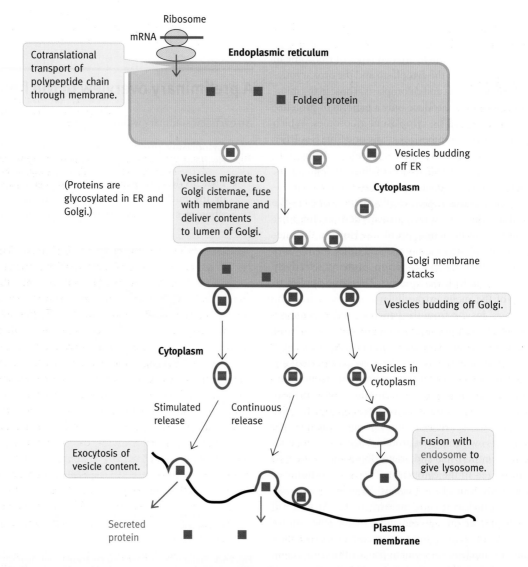

Fig. 26.2 Preliminary overview of how proteins are secreted from cells and how enzymes are delivered to lysosomes. The essence of the process is that the initial transport of proteins through the lipid bilayer into the lumen of the endoplasmic reticulum (ER) is cotranslational – the polypeptide traverses the membrane as it is synthesized. This is quite different from the transport into peroxisomes, mitochondria, and nuclei which occurs posttranslationally (see Fig. 26.1). The targeting of new membrane proteins to their appropriate sites is basically by a similar mechanism; the proteins are inserted into the ER membrane as they are synthesized and then sections of membrane, complete with the new proteins, are packaged as vesicles. These migrate to and fuse with the target membrane, thus delivering both new lipid bilayer and membrane proteins. Note that, as described later, proteins are glycosylated as they pass through the ER and Golgi. The many questions as to the mechanism of all this are dealt with below.

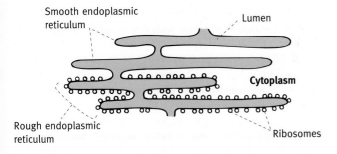

Fig. 26.3 Diagrammatic representation of the endoplasmic reticulum.

Some information on the ER and the Golgi apparatus will be useful since these remarkable organelles play such a central role.

Structure and function of the ER and Golgi apparatus

The ER is a membranous structure that pervades eukaryotic cells to varying degrees (erythrocytes excepted). It is a complex of linked sacs so that its lumen is one continuous cavity separated from the cytoplasm by the single ER membrane. Its size varies enormously in different cells, depending on the metabolic functions and state of the cell. Part of the ER when seen in the electron microscope is studded with attached ribosomes and is the **rough ER**; the rest, **smooth ER**, has no attached ribosomes, but the two types merge into each other on the same membrane (Fig. 26.3). They are not physically discrete membrane structures but areas that have different functions. The ER lumen is continuous with the space enclosed by the double membrane surrounding the nucleus.

The ribosomes are not a permanent fixture on the rough ER but are those which just happen, at the time, to be translating an mRNA coding for a protein destined for secretion, inclusion in a lysosome, or incorporation as an integral membrane protein. When the ribosome on the ER has completed the synthesis of a protein molecule it detaches and its subunits re-enter the general cytosolic pool to be replaced on the ER membrane by other ribosomes. There are no 'special' ribosomes for this role. The polypeptides are transported into the ER lumen as they are synthesized and are folded in the lumen. The proteins progress through the smooth ER, during which they often have carbohydrates added. To leave the ER smooth section, proteins are enclosed in small transport vesicles which bud off the ER and transport them to the **Golgi apparatus** (Fig. 26.4).

The ER and the Golgi are completely closed structures, there being no physically evident entry or exit sites. The cytoplasmic face of the smooth ER is the major site for synthesis of new lipid bilayer, a function which correlates with the formation of transport vesicles.

The Golgi apparatus consist of 4–6 (more in plant cells) membranous flattened structures enclosing spaces known as **cisternae**. They resemble a stack of large plate-like vesicles placed near the nucleus. The side facing the ER is the transport

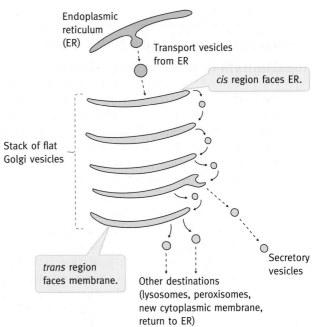

Fig. 26.4 The central role of the Golgi apparatus in posttranslational sorting and targeting of proteins. In addition, newly synthesized membrane lipids are transported to the appropriate destination by transport vesicles.

vesicle reception area known as the *cis* cisternae (cis = near to; cisternae = chambers). Transport vesicles carrying newly synthesized proteins are budded off from the smooth ER, move to fuse with the *cis* membranes, and deliver their contents to the Golgi cysternae. Proteins move through the Golgi stacks in the same way via transport vesicles; there are no direct connections between the plates. Proteins are modified in successive cisternae, progressing towards the final one on the other or *trans* (trans = distant) side where they are sorted out and packaged into vesicles and despatched to their destinations. The Golgi thus takes newly synthesized proteins arriving from the ER, modifies them, packages them into membrane vesicles addressed to their proper destinations, and finally despatches them. It is what a mail-sorting office is to posted mail. There is even a 'return-to-sender' service; as proteins move through the ER, proteins that are needed in the ER such as chaperones and enzymes involved in polypeptide folding are swept along with them. These need to be sent back in appropriately addressed transport vesicles.

That is the end of the general overview, and we can now turn to the molecular processes. We start with the ER-mediated processes. This will be followed by the transport of proteins into mitochondria and peroxisomes (in which the ER and Golgi are not involved), and finally with transport of proteins across the nuclear membrane, which is a quite different story from all the rest. However, there is a concept in some of the molecular processes that is worth spending a little time on first.

The importance of the GTP/GDP switch mechanism in protein targeting

As we go through the protein-targeting mechanisms, GTP hydrolysis to GDP + P$_i$ will be encountered several times. You are used to ATP breaking down to perform chemical or other work, but GTP is broken down on single **GTPase** subunit proteins, apparently without any useful work being performed. This is not a waste of energy. The hydrolysis of the GTP to GDP causes an allosteric conformational change in the protein. This acts as a switch to allow the next step in a process to occur, and, since the hydrolysis is irreversible, it confers a unidirectionality on the process. Note that the GTP hydrolysis by these proteins is believed to occur only after a slight delay, which gives time for a stage in a process to happen, and then it moves on to the next stage. After the GTP hydrolysis, special mechanisms exchange the GDP on the protein for GTP, restoring the original form. This type of switching is important in many processes. It occurs in protein synthesis (page 383), cell signalling (page 438), nuclear–cytoplasmic transport (page 424) and vesicle transport (page 418).

Translocation of proteins through the ER membrane

For his work on this, Gunther Blobel at the Rockefeller Institute, New York, was awarded a Nobel Prize in 1999. When a protein is to be secreted or has to end up in the external plasma membrane, inside lysosomes, or in the lumen of the ER itself, as it is synthesized on the ribosome it has on its N-terminal end a **signal peptide sequence** of about 29 amino acids. If the coding information for such a signal sequence is artificially added to the mRNA for a protein that normally stays in the cytoplasm, the protein will be transported into the ER as it is synthesized. The signal peptide amino acid sequences of different proteins have a pattern, rather than a fixed sequence, as shown diagrammatically in Fig. 26.5. There is a short, positively charged N-terminal section and a central hydrophobic region 10–15 amino acid residues in length. Substitution of a single charged residue into the hydrophobic central region is enough to inactivate the signal sequence. When the polypeptide traverses the ER membrane the signal sequence is cleaved off by a **signal peptidase** on the inner face of the membrane, so the mature protein does not have the sequence.

Mechanism of cotranslational transport through the ER membrane

It will help if you follow each numbered step in Fig. 26.6 as you read the following section. A free ribosome in the cytoplasm

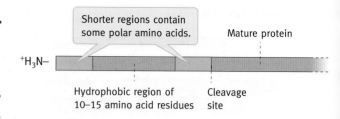

Fig.26.5 A typical signal sequence attached to the N-terminal end of a protein destined to be transported through the endoplasmic reticulum membrane. Such signal peptide sequences show the same general pattern of amino acids but no specific amino acid sequence. Acidic amino acids are not present.

translating an mRNA for an ER-targeted protein synthesizes the signal sequence first (since it is at the N-terminal end of the polypeptide). This is recognized by a **signal-recognition particle** (**SRP**), an RNA-protein complex in the cytoplasm that binds to the nascent signal peptide when it emerges from the ribosomal tunnel and arrests further elongation of the polypeptide chain (step 1 in Fig. 26.6).

The loaded ribosome migrates to the ER membrane (step 2 in Fig. 26.6) which has on it **SRP receptors** or **SRP docking proteins** to *which GDP is bound*. The receptors are found only on the ER membrane. The SRP of the cytosolic ribosome-SRP complex attaches (step 3 of Fig. 26.6), and positions the ribosome on the membrane. In the ER membrane are protein assemblies known as **translocons**. These are channels formed by several subunits of a protein complex which span the membrane, but in the absence of a ribosome these are in an effectively closed condition. Possibly it is a gated channel (page 113) which opens on the contact with the signal peptide and this is the possibility we have used for illustrative purposes in the diagrams. The essential point is that a ribosome attaching to the SRP docking protein on the membrane becomes associated with a translocon channel. This attachment triggers step 4, the exchange of GDP on the SRP receptor for GTP catalysed by an exchange enzyme present in the cytoplasm. *In this form, the receptor (the docking protein) is postulated to bind more tightly to the SRP and allow the SRP to detach from the signal peptide*, all believed to be the result of the conformational change brought about by the binding of GTP in place of GDP.

The signal peptide is now free to insert into the translocon (step 5 in Fig. 26.6), and SRP is released into the cytosol for further use (step 6).

The ribosome recommences synthesis of the polypeptide which traverses the membrane via the translocon channel as it is synthesized. The signal peptide is postulated , in this model, to remain in the channel but the cleavage site is exposed at the internal face of the membrane as illustrated in step 6. (Other models show the signal peptide going straight through.) The signal peptidase, which is responsible for the cleavage (step 7 in Fig. 26.6), has a hydrophobic patch which attaches it on the

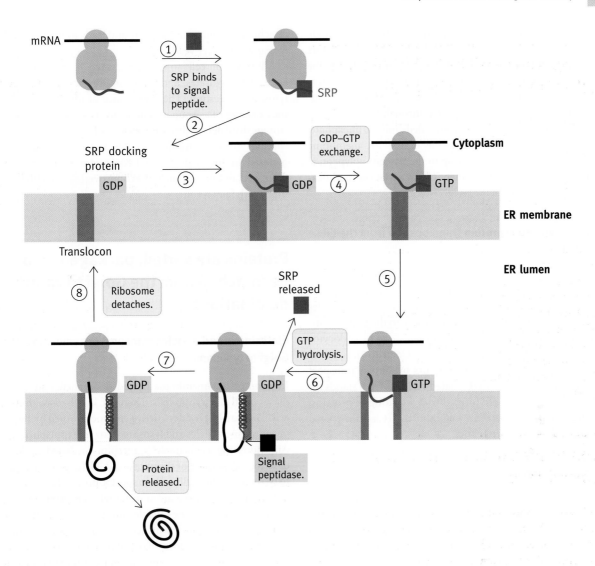

mRNA

① SRP binds to signal peptide.

SRP

② SRP docking protein

③

GDP

④ GDP–GTP exchange.

GDP   GTP

Cytoplasm

ER membrane

Translocon

ER lumen

⑧ Ribosome detaches.

SRP released

⑤

GTP

⑦

GDP

GTP hydrolysis.

GDP ⑥ GTP

Protein released.

Signal peptidase.

Fig. 26.6 Sequence of events by which proteins are cotranslationally transported into the lumen of the endoplasmic reticulum (ER). The numbered steps are referred to in the text. The signal peptide is shown in red and the polypeptide that constitutes the mature protein is shown in black. SRP, signal-recognition particle.

membrane so that as the signal peptide cleavage site emerges from the membrane it encounters the peptidase.

On completion of the polypeptide synthesis, the polypeptide is released into the lumen of the ER (step 8 in Fig. 26.6). The ribosome dissociates into its subunits to rejoin the cytoplasmic pool for further use. The signal peptide is presumed to be destroyed and the translocon, it is postulated, becomes closed.

Folding of the polypeptides inside the ER

The lumen of the ER contains **chaperones** (page 395) that attach to unfolded and partially folded polypeptides. Their function is to hold the chain in a conformation that prevents the polypeptides from going down an unproductive folding route leading to aggregation. A key component in this system is the ER isoform of the chaperone **Hsp 70**, which interacts with the polypeptide chain as it emerges from the translocon channel so that the incoming polypeptide folds correctly. To assist with this process the lumen contains **protein disulphide isomerase** and **peptidylproline *cis-trans*-isomerase** whose role in assisting correct folding has already been dealt with. Proteins which misfold, and are not corrected, are not allowed to accumulate. They are transported back into the cytoplasm, and degraded in proteasomes.

Glycosylation of proteins in the ER lumen and Golgi apparatus

In Chapter 4 (page 56) we described proteins, particularly membrane and secreted proteins, which have complex oligosaccharides added to them. The attachment points are either the amide —NH$_2$ of asparagine side groups (*N*-glycosylation) or the —OH of serine and threonine residues (*O*-glycosylation). Inside the ER, *N*-glycosylation is carried out. The first step of this is interesting in that a 'core' oligosaccharide of 14 sugar units is assembled in the cytoplasm and transported through the membrane attached to a long hydrophobic chain called **dolichol phosphate**. A transferase enzyme on the inside of the ER membrane transfers the oligosaccharide group to the nascent polypeptide chains as they enter the ER lumen. *O*-glycosylation modification of proteins occurs in the Golgi cisternae.

Vesicles involved in protein translocation from the ER and Golgi

As already stated, there is no physical connection between the ER and Golgi or between the Golgi stacks. Movement of proteins between them occurs via membrane transport vesicles budded off from the organelles. These move to the target membranes to which they fuse and deliver their cargo by endocytosis. The transport of proteins from the Golgi to the plasma membrane for secretion likewise occurs by similar transport vesicles. These, known as **COP-coated vesicles** (COP stands for **c**oat **p**rotein complex), are formed with a coating of proteins; their formation will be dealt with shortly. (Clathrin-coated vesicles are different – see below).

Proteins for lysosomes

In the case of enzymes destined for inclusion in lysosomes, the signal is on the carbohydrate part of the glycoprotein. In the Golgi the carbohydrate attachment is enzymically modified so that it terminates in a mannose-6-phosphate residue which attaches to membrane receptors and leads to their inclusion in lysosomal delivery vesicles.

All types of cell components, such as proteins, nucleic acids, carbohydrates, and lipid components, are destroyed by lysosomal digestion. The importance of this is underlined by the existence of a group of genetic diseases known as **lysosomal storage disorders** which arise because of the lack of one or more specific lysosomal enzymes (see Box 26.1). Their cognate target material in the lysosomes scheduled for destruction is not destroyed and the organelles become overloaded with it, sometimes with fatal results.

Proteins are sorted, packaged, and despatched from the Golgi to various destinations

Clathrin-coated vesicles transport enzymes from the Golgi to lysosomes

Lysosomes are membrane-bounded organelles in the cytoplasm of all eukaryotic cells. There are hundreds within a liver cell. They are bags of destructive enzymes formed by the fusion of **endosome vesicles** and **lysosomal enzyme transport vesicles**. An endosome is formed by receptor-mediated endocytosis described in outline for the uptake of LDL (page 170). A particle to be delivered to a cell binds to specific protein receptors on the cell membrane. These are transmembrane proteins with the receptor exposed to the outside and their cytoplasmic domain exposed on the inside. A protein known as **adaptin** combines with the cytoplasmic domains of a number of the receptors and clusters them together forming a depression of the membrane. A basketwork-like coating made up of the protein **clathrin** attaches to the adaptin molecules the depressions being known as **clathrin-coated pits** (Fig. 26.7). The pits invaginate and another protein called **dynamin** attaches to the neck of the invagination causing a vesicle to be nipped off as a **clathrin-coated vesicle**. Inside the liver cell, the coated vesicle is uncoated and the coat molecules recycled back to the membrane. The interiors of the vesicles (now called **endosomes**) are acidified by proton pumps in the membrane, forming late endosomes. Vesicles budded off from the Golgi membranes by the same clathrin-coated mechanism fuse with the endosomes and deliver the hydrolytic enzymes. The result is a **lysosome** in which digestion of the material initially endocytosed is hydrolysed into its component parts.

The cell is protected from destruction by the lysosomal enzymes because they are segregated by the lysosomal membrane, and also because the enzymes require a pH between 4.5 and 5.0 for activity maintained inside the vesicle by the ATP-dependent proton pumps in the membrane. If a lysosome were to rupture, the buffering of the cytoplasm would

---

> **BOX 26.1 Lysosomal storage disorders**
>
> In one family of these genetic diseases, called sphingolipidoses, specific lysosomal enzymes are missing, which impair degradation of gangliosides (page 105) of the cell membranes. A classic example is Tay–Sachs disease.
>
> In I-cell disease all of the lysosomal enzymes are missing so that all manner of molecules accumulate within the vesicles. It is caused by a deficiency in the tagging of lysosomal enzymes with mannose-6-phosphate so they are not packaged into transport vesicles, and are secreted from cells into the plasma, rather than ending up in lysosomes. The disease is often fatal before 10 years of age.
>
> In Pompe's disease, one of a series of glycogen storage diseases, the deficiency is of a lysosomal enzyme which hydrolyses the $(1 \rightarrow 4)$-α links of glycogen. In the absence of the enzyme there is a massive accumulation of glycogen in the cells usually causing death in infancy.

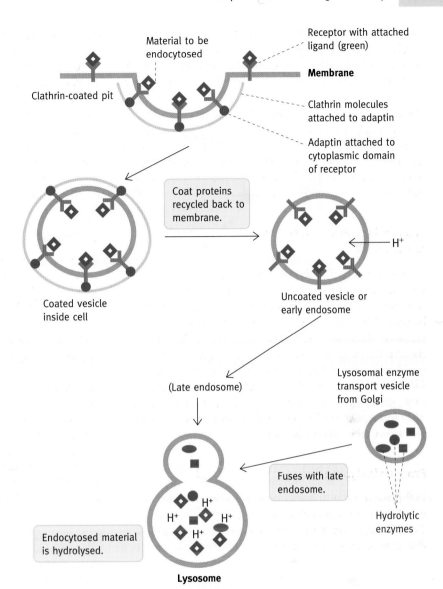

Fig. 26.7 Formation of a lysosome by receptor-mediated endocytosis involves clathrin-coated vesicles, not COP vesicles. Ligand molecules (such as low-density lipoprotein) bind to membrane receptors. Adaptin molecules attach to receptor domains inside the cell and cluster them into a coated pit. Clathrin molecules bind to adaptin and the pit invaginates into a coated vesicle in the cytoplasm. The latter is uncoated and the coat molecules are recycled. The uncoated vesicle, now an endosome, is acidified. Acidification releases the receptors (which in many cases are recycled back to the cell membrane) producing a late endosome. A Golgi transport vesicle containing hydrolytic enzymes fuses with the latter, forming a lysosome.

maintain the pH at 7.3 or so, at which lysosomal enzymes are inactive. Intracellular components such as aged mitochondria are also destroyed by enveloping them in endosomal vesicles known as **autophagosomes** followed by lysosomal enzyme delivery.

Proteins to be returned to the ER

In most cases the sorting mechanism must include membrane receptors that recognize certain structural features or specific sequences of the proteins to be sorted. These are not fully eluci-dated but in those proteins (such as protein disulphide iso-merase) that have to be returned to the ER, the 'return address label' is a peptide sequence of four amino acids in the protein, Lys–Asp–Glu–Leu (KDEL in the single-letter abbreviation sys-tem for amino acids; see Table 4.1). (The vesicles involved are COP-coated – see below).

Proteins to be secreted from the cell

The Golgi packages proteins for secretion into COP-coated (see below) vesicles which migrate towards the plasma membrane (Fig. 26.4). There are two types of secretion. Some proteins are released continuously as produced, while others are released periodically as required. Serum proteins for example are released continuously from the liver without any signal being required. The transport vesicles fuse with the cell plasma mem-brane as they arrive there and release their contents by exocyto-sis. In the case of digestive enzymes, however, these are released from the pancreas only when food enters the gut. In this case the vesicles from the Golgi containing these enzymes are larger secretory vesicles also known as **secretory granules**. These store the enzymes until wanted, at which point a neuronal, or hor-monal, stimulus causes a release by exocytosis. There are many other examples – release of insulin is a familiar one.

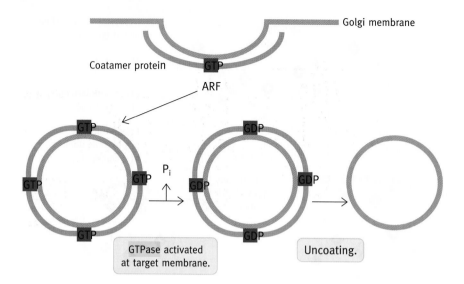

Fig. 26.8 Formation of transport vesicles from the Golgi membranes. The membrane at the site of vesicle formation becomes coated with subunits of a coatamer protein and of an ARF–GTP protein. When the vesicle arrives at its target membrane the ARF–GTPase is activated to form ARF–GDP. This triggers the uncoating of the vesicle followed by fusion with its target membrane. In the case of a secretory vesicle the contents are ejected by exocytosis.

Mechanism of COP-coated vesicle formation

Most transport vesicles are of this type, the exception being the clathrin-coated type already described for receptor-mediated endocytosis. Two varieties of COP-coated vesicle exist, COP I and II, used in forming transport vesicles from the Golgi and ER respectively. In COP I a protein known as **ARF** with attached GTP binds to the membrane and recruits **coatamer proteins** to cover the site of vesicle budding (Fig. 26.8). The name ARF comes from another role of the protein, not relevant in the present context. COP II contains different proteins.

How does a vesicle find its target membrane?

The vesicle has a molecule called a v-SNARE (v for vesicle) on it that binds to a complementary t-SNARE (t for target) on the target membrane (Fig. 26.9). These SNAREs are long helical proteins, which can associate and bring the two membranes together, ready for fusion.

When a coated vesicle reaches its target membrane the vesicle is uncoated due to the activation of the GTPase of the ARF protein which hydrolyses GTP to GDP. This results in the coatamer and ARF proteins dissociating from the vesicle. Following uncoating, other proteins then join and bring about the fusion of the vesicle and target membrane. The mechanism of this is complicated, involving several proteins; only an outline of the process is given here.

If the target membrane is the plasma membrane then vesicle contents are secreted by exocytosis. This method of targeting vesicles to its destination is a general one, not just confined to protein secretion. In nerve-impulse transmission, when a signal arrives at a synapse, vesicles containing neurotransmitter

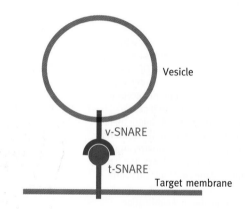

Fig. 26.9 Binding of v-SNAREs and t-SNAREs leads to uncoating of the vesicles followed by fusion of the two membranes. This is a general mechanism for vesicle targeting.

substances fuse with the presynaptic membrane and eject their contents into the synapse. This occurs also by courtesy of v-SNAREs attaching to membrane t-SNAREs. The **tetanus** and **botulinus** neurotoxins, two of the most deadly substances known, have protease activity which snip off these SNARES and interfere with nerve-impulse propagation, since neurotransmitters are not then released.

Synthesis of integral membrane proteins and their transport

Integral membrane proteins are synthesized on the rough ER, integrated into the membrane and transported *in situ* by vesicles, as new membrane, to the plasma or other target membranes. Why are the proteins fixed in the ER membrane instead of going right through it? One model is that a **stop transfer**

Fig. 26.10 Insertion of integral membrane proteins into the membrane of the endoplasmic reticulum (ER). The initial steps are as depicted up to step 7 in Fig. 26.6 but the ribosome synthesizes another sequence, shown in blue, which is a stop transfer or anchor sequence. This arrests the further movement of the polypeptide so that when the protein is fully synthesized and exits the channel, the protein is left as an integral membrane protein. As described in the text, models have been proposed for a mechanism by which the integral protein may be oriented across the membrane in the opposite orientation, with the N-terminus in the cytoplasm. There must presumably be a mechanism for releasing the protein laterally from the translocon.

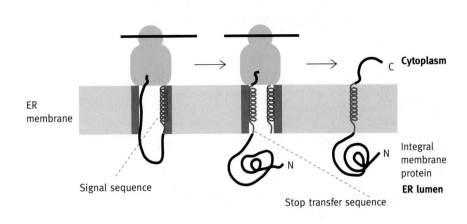

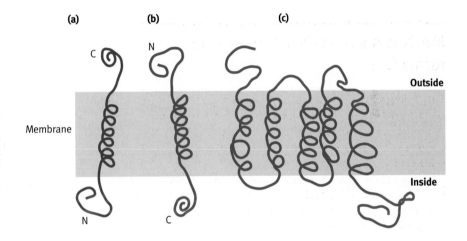

Fig. 26.11 Different orientations of integral membrane proteins. See text.

sequence or **anchor sequence** is translated which anchors the polypeptide into the membrane (Fig. 26.10). After completion of the protein it is presumed to exit laterally from the translocon into the lipid bilayer.

This sequence of events produces a transmembrane protein with its N-terminus inside the ER (Fig. 26.11(a)). Other orientations occur (Fig. 26.11(b),(c)).

A possible mechanism for achieving the reverse orientation (b) is shown in Fig. 26.12. In this, there is a **noncleavable signal peptide** positioned within the polypeptide that also acts as a stop transfer signal and is therefore called a **signal anchor sequence**. It is envisaged that the signal sequence inserts into the channel in a hairpin-looped fashion, resulting in the situation shown in Fig. 26.12. Release of the protein from the channel would produce an integral protein with the C-terminal end inside the ER and the N-terminal end in the cytosol. The synthesis of serpentine proteins, which criss-cross the membrane several times, is believed to involve a succession of stop transfer sequences.

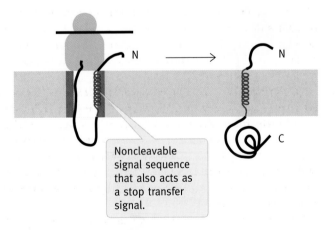

Noncleavable signal sequence that also acts as a stop transfer signal.

Fig. 26.12 How a transmembrane protein may be synthesized with the orientation shown. See text for explanation. In this the signal sequence is also a stop transfer signal. It inserts into the channel in a looped fashion in this model.

Posttranslational transport of proteins into organelles

To remind you, all of the membrane transport so far described has been cotranslational. However, transport into mitochondria, (and chloroplasts in plants), peroxisomes and the nucleus is posttranslational. The polypeptides are completely synthesized, released by cytoplasmic ribosomes and then transported. Different mechanisms are used for transport into the different organelles.

Transport of proteins into mitochondria

A mitochondrion contains hundreds of proteins, only a 13 of which (in humans) are synthesized within the mitochondrion, coded for by mitochondrial genes. All of these are subunits of larger complexes of the oxidative phosphorylation components of the inner mitochondrial membrane, such as cytochrome oxidase and ATP synthase. The rest are coded for by genes in the nucleus, the mRNAs for which are translated on free ribosomes and the polypeptides released into the cytoplasm. There they have to be selected from other cytoplasmic proteins, delivered to receptors on the mitochondrial membrane and transported to their destinations.

The transport of proteins into mitochondria from the outside is quite complex because of the several compartments to which the proteins have to be targeted. In some cases alternative routes to the one compartment have been evolved. The nascent peptide of a mitochondrial protein as it emerges from the ribosome becomes attached to chaperones of the Hsp70 type and delivered in this form to the receptors. Proteins entering the matrix have to cross both membranes; this happens at points where the inner and outer membranes become close together.

Mitochondrial matrix proteins are synthesized as preproteins

Preproteins have targeting sequences, one of the best characterized of which is an N-terminal sequence of 15–35 amino acids, which is not present in the mature mitochondrial proteins. These sequences are variable but usually consist of patterns of hydrophobic, hydroxylated, and basic amino acids, which form an α helix with hydrophobic side chains on one side and basic, positively charged groups on the other. They are thus amphipathic. However other proteins have internal recognition signals instead. As the polypeptide is synthesized on the ribosome, it is held in the unfolded form by chaperones attached to the extended chain (Fig. 26.13). It is not known if there are special chaperone attachment signals in

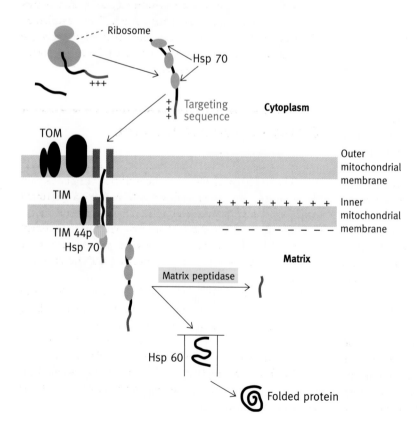

Fig. 26.13 Targeting of proteins to the mitochondrial matrix. Proteins are synthesized as preproteins with an amphipathic targeting sequence at the N-terminal end (red). They are released into the cytoplasm but held in the unfolded state by chaperones (yellow). The preprotein attaches to the TOM complex receptors (the chaperone is involved in this also). It is transported through to the TIM complex and into the matrix where a peptidase cleaves off the targeting sequence and the protein folds up assisted by the chaperone Hsp 70 and the chaperonin Hsp 60. The Hsp 60 provides an isolated cage amongst the densely packed matrix proteins inside which the transported polypeptide can fold up (see page 395 for the mechanism of action of Hsp 60 and Hsp 70). The TOM and TIM complexes contain a variety of protein subunits. Import of the polypeptide into the matrix is assisted by TIM 44p, Hsp 70, and other subunits that form the translocation motor. The inner membrane must have a charge potential for protein transport through it.

this context. The polypeptide–chaperone complex docks with a mitochondrial receptor, which is part of the *translocase* of the *outer mitochondrial membrane* (TOM). This is a multi-protein complex forming a channel through which the pre-protein traverses the outer membrane. *The TOM complex is involved in the transport of virtually all proteins that enter the mitochondrion.*

In the case of proteins destined for the mitochondrial matrix, the preprotein is transported in an extended form (but without the chaperones) through the outer membrane, and then across the **intermembrane space**. It is delivered to the *translocase* complex of the *inner mitochondrial* membrane (TIM). For transport by TIM the inner membrane must have a charge potential (negative inside) generated by electron transport. As the preprotein enters the matrix it meets an ATP-driven translocation motor, composed of the mitochondrial isoform of Hsp 70, plus other TIM subunits which drives the import process to completion. A **matrix peptidase** hydrolyses off the targeting peptide sequence and the protein becomes fully folded, a process involving the Hsp 70 chaperone and/or the Hsp 60 chaperonin. The latter forms a secluded chamber amongst the extremely densely packed proteins of the matrix inside which the protein can fold properly. (A description of chaperones, chaperonins and their mechanisms can be found on page 395.)

Delivery of proteins to mitochondrial membranes and intermembrane space

In the case of integral proteins of the inner membrane there are three different routes but all involve the TOM complex. In the first, the proteins are delivered as matrix proteins but the removal of the preprotein signal sequence exposes another leader sequence which targets the protein to the inner membrane via one of two other TIM complexes. (One of these complexes does the same for the inner membrane proteins synthesized *in* the mitochondrion). Two other routes involve the delivery of proteins by the TOM complex to TIM complexes which release the proteins laterally into the inner membrane. Proteins for the outer membrane have internal signals which the TOM complex recognizes. Different classes of outer membrane proteins exist and appear to be inserted by different mechanisms by the TOM complex.

Proteins are delivered to the intermembrane space by a variety of routes. One of these is that the protein is first inserted into the inner membrane and then snipped off to release the external part into the space.

Plant **chloroplasts** also import about 90% of their total proteins from the cytoplasm. The mechanism is similar to that for mitochondria with the targeted proteins synthesized on cytoplasmic ribosomes as preproteins but the targeting peptide, known as the **transit peptide**, is 30–100 amino acids long and is not positively charged as is the mitochondrial targeting sequence. Possibly correlating with this there is no requirement

for a negative membrane potential on the inner membrane as there is for mitochondria.

Targeting peroxisomal proteins

Peroxisomes and their metabolic functions have been described earlier (page 21). They are small organelles with a single membrane and have no DNA or protein synthesising machinery. Their content of about 50 different enzymes are synthesized on free cytoplasmic ribosomes and transported into the organelle. The proteins become fully folded up in the cytoplasm and, unusually, they are transported into the peroxisome in this form. This is in marked contrast to mitochondrial import, described above, where the polypeptides are kept in an extended form.

There are two known **peroxisome-targeting signals** (**PTS1 and PTS2**) on proteins to be transported. PTS1, the most common, is a C-terminal tripeptide Ser–Lys–Leu (SKL in the single-letter abbreviations) which is not removed. PTS2 is nonapeptide located near the N-terminus but sometimes internally. A variety of cytoplasmic proteins called **peroxins** are needed for transport of the targeted proteins. Since folded proteins are transported into the peroxisome, a very large membrane pore is implied, but how the transport occurs is not known nor whether there is a single mechanism only. However, it is known that soluble peroxins recognize candidate proteins and dock them to receptors on the peroxisome.

The interest in the field is heightened by the existence of genetic diseases that are due to defects in peroxisome biogenesis. In one of these, the **Zellweger syndrome**, death often occurs before 6 months of age. Yeast genetic studies have revealed 23 PEX genes involved in peroxisome biogenesis, 13 of which are conserved in humans. Of these, 10 have been shown to be associated with peroxisomal disorders.

Nuclear–cytoplasmic traffic

The eukaryotic nucleus is surrounded by inner and outer membranes, the spaces between them being continuous with the ER lumen. The existence of a separate nuclear compartment means that there is intensive traffic across the membrane between the nuclear and cytoplasmic compartments different from anything in the sections already dealt with. Moreover, we are dealing with two-way traffic of proteins and protein-RNA complexes.

The nuclear membrane is studded with pores of huge size and elaborate construction. Proteins, such as those synthesizing and transcribing DNA, function in the nucleus and have to be delivered to it. During DNA replication the traffic is two-way and intense; when a cell is synthesizing DNA, histones are supplied from the cytoplasm to allow nucleosome assembly. It has been calculated that each pore must transport 100 histone molecules per minute.

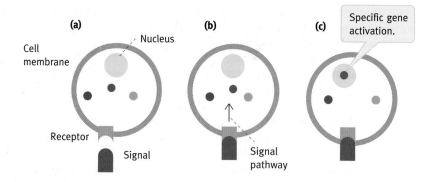

Fig. 26.14 Outline of the role of nuclear import in the control of genes by cell signalling. **(a)** A hormone or other chemical signal arrives at a cell surface and combines with its specific receptor. **(b)** The receptor triggers a series of events which selects a specific cytoplasmic protein (blue) to migrate into the nucleus (this is a major subject dealt with in Chapter 27). **(c)** The protein inside the nucleus causes specific gene activation (also a major subject dealt with in Chapter 23). It should be noted that this is an exceedingly simplified scheme to give the bare outlines of gene control by hormones etc. as related to nuclear transport. Although it applies to most situations, lipid-soluble hormones such as steroids, thyroxine, and vitamin D diffuse through the membrane and bind to cytoplasmic receptors or to receptors already in the nucleus. Nonetheless most signalling agents do not enter the cell. (All of these matters are dealt with in Chapter 27.)

The nucleus manufactures all the RNA of the cell (apart from that in mitochondria and chloroplasts) most of which is required in the cytoplasm so that it has to be transported out. mRNA in particular is 'packaged' for transport out into the cytoplasm. rRNA is not transported as such; insteadthe ribosomal proteins are transported in, the ribosomal subunits assembled and transported out. By contrast, **snRNP**, a ribonucleoprotein, is a nuclear component involved in mRNA splicing (page 360). It has small nuclear RNAs (snRNAs) bound to protein, but the snRNA component is transported out to be equipped with its protein in the cytoplasm and then the complete snRNP is transported back into the nucleus, the reverse of what happens with ribosomal subunit assembly.

Why is there a nuclear membrane?

This is a slight, but relevant, diversion from the mechanism of nuclear-cytoplasmic traffic.

An interesting question is why, in eukaryotes, evolution has chosen to go to the trouble of having its DNA inside a nuclear membrane. An *E. coli* cell manages perfectly well without one and has none of the associated transport problems. The eukaryotic nucleus separates the act of transcription of DNA in both time and space from translation of the mRNA in the cytoplasm. This provides a time-gap for the mRNA to be modified before translation; the most important of these may have been the splicing which, in allowing the existence of split genes, facilitated exon shuffling (page 390) a factor likely to have been important in fostering evolution. In addition, differential splicing allows the production of different proteins from a single gene (page 362). In *E. coli*, ribosomes begin translating mRNA even before the synthesis of the latter has been completed and detached from the gene. It is not easy to see how splicing could occur in this circumstance.

Much of eukaryotic gene control is the result of hormones and cytokines arriving at the cell. This almost always involves specific proteins being transported from the cytoplasm into the nucleus on arrival of a signal (Fig. 26.14). Transcriptional control and cell signalling are major topics dealt with in Chapters 24 and 27 respectively.

After that diversion we will turn to the mechanism of nuclear–cytoplasmic traffic.

The nuclear pore complex

Each nucleus has several thousand pores forming aqueous channels between the cytoplasm and nucleoplasm (Fig. 26.15). Nuclear pores are huge structures with a total size of 10^8 daltons and built up of over 30 species of protein subunits in multiple copies. The proteins are called **nucleoporins**. Pores have been isolated free of the membrane by detergent treatment and their structure studied. They consist of an annular ring of protein subunits at each end connected by 8 spokes with gaps between them. The outer 'knobs' anchor the pore into the nuclear membrane, the inner and outer nuclear membranes being continuous at the site of the pore (Fig. 26.16). In addition to the main pore structure spanning the membrane, there is a basket-like projection into the nucleoplasm and fibrils extend into the cytoplasm, giving it a rather science-fiction appearance. The fibrils are the site of receptors to which bind the complexes due for transport.

Molecules up to about 40,000 daltons can enter the nucleus simply by diffusion, though proteins in the upper ranges of this do so more slowly than smaller ones. Anything greater than this size must be specifically accepted by the transport machinery of the pore.

Fig. 26.15 Spread *Xenopus* oöcyte nuclear envelopes (NEs) prepared for transmission electron microscopy (TEM). **(a)** Electron micrograph of chemically unfixed and unstained *Xenopus* oöcyte nuclear pore complexes (NPCs) embedded in a thick (i.e. ~250 nm) amorphous ice layer. **(b)** Nuclear face of quick-frozen, frozen-hydrated, and metal-shadowed NPCs revealing well preserved nuclear baskets (see arrows). Scale bars, 200 nm. Photographs courtesy Professor Ueli Aebi, University of Basel, Switzerland.

(a)

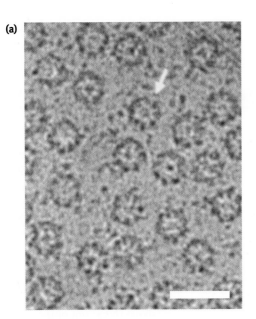

(b)

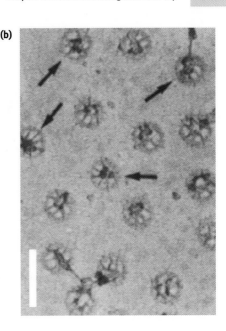

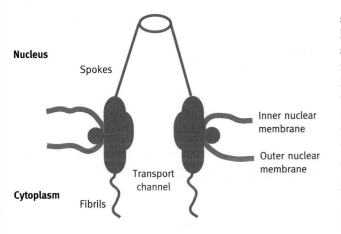

Fig. 26.16 Simplified diagram of the main body of a nuclear pore. See text for the mechanism of transfer through the pore.

Nuclear localization signals

Large (> 40,000 daltons), nuclear-localized proteins need a **nuclear localization signal (NLS)**. This is a sequence of amino acids that may be located anywhere in the polypeptide chain. A number of these have been identified ranging from 6 to 18 amino acid residues in length, rich in arginine and lysine, sometimes with a central spacer of noncritical residues. The 'prototype' of a class of NLS, known as **conventional** because it signals only the inwards transport into the nucleus, is that found in the T antigen of the SV40 virus (a protein first identified immunologically); it is transported into the nucleus as part of the infective process. The NLS is PKKKRKV (using

single-letter abbreviations of amino acids). Another class is known as **bipartite**; an example is found on nucleoplasmin, a chromatin assembly protein; the sequence of this is KR, followed by a 10 amino acid (unspecified) spacer, then by PAAIKKAGQAKKKK (given only to illustrate the type). Mutation of an NLS results in a normally nuclear-located protein remaining in the cytoplasm, whereas addition of such a signal to a cytoplasmic protein results in it being transported into the nucleus. Most transcription factors (page 364) are examples of proteins with NLS sequences. As well as conventional (import) NLS signals, **nuclear export signals (NESs)** have been identified (see below). However, variations occur. Some proteins shuttle in and out of the nucleus, perhaps the best known being the protein which binds mRNA in the nucleus and transports it *via* the nuclear pore into the cytoplasm as a **hetero-ribonucleoprotein complex (hnRNP)**. mRNAs are relatively short-lived in the cytoplasm (20 minutes to a few days in mammals) so the mRNA binding protein shuttles back into the nucleus to pick up more. It has a signal of 38 amino acid residues, which can act as both an import signal and an export signal.

Importins combine with nuclear localization signals on proteins to be transported into the nucleus

Importins are also known as **nuclear import receptors** or **karyopherins** (*we will use 'importin' in the account below*) are soluble cytoplasmic proteins which recognize the NLS on 'cargo' molecules (step 1 in Fig. 26.17). There is a family of different ones. Each has two binding sites; one binds to the NLS on the 'cargo' proteins to be transported and the other to receptor nucleoporins on the fibres of the pore complex. They are adaptors attaching the cargo to sites on nucleoporins (step 2). Many of the nucleoporins of the pore (about 30% of the total) contain

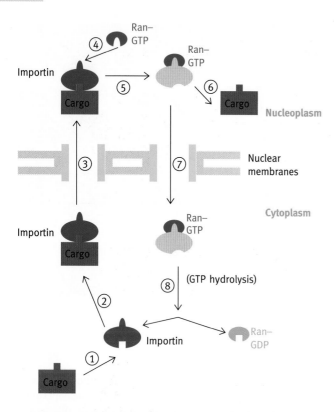

Fig. 26.17 Mechanism of protein import into the nucleus. See text for explanation. The deep red colour represents importin in its form capable of attaching to a protein to be transported; the pale red form cannot do so. Note that Ran–GDP is transported back into the nucleus (not shown) to complete the cycle.

short stretches of clustered hydrophobic amino acids known as FG-repeats (F and G are one letter abbreviations for phenylalanine and glycine respectively) interspersed by hydrophilic regions. The proteins have an extended conformation. *The FG-repeats are binding sites for the importins (carrying the cargo).* It is believed that movement through the pore (step 3) involves progressive binding and detachment of the importin/cargo complex from the FG-repeats. It moves from one to the next, but the precise mechanism is the subject of investigation still.

Where does the energy for nuclear transport come from?

The nuclear pore has no ATPases or other known direct energy input mechanisms. The driving force for the transport comes indirectly from a gradient created by GTP hydrolysis as will now be described.

The importin, which is a heterodimer of α/β subunits, undergoes conformational change when a small protein, **Ran**, a **guanine nucleotide binding protein** combines with it. **Ran** *exists in the nucleus combined with GTP*; when this attaches to the arriving importin–cargo complex (step 4) it causes a conformational change in the importin which releases its cargo

(steps 5 and 6) The **Ran–GTP**, coupled to the importin, recycles back to the cytoplasm *via* the nuclear pore (step 7). Ran has GTPase *activity but this requires activation by a* **GTPase activating protein (GAP)**, *which occurs only in the cytoplasm, not in the nucleus.* Thus, *in the cytoplasm Ran–GTP is hydrolysed to Ran–GDP* (step 8), resulting in release of the importin in a conformation able to pick up a new cargo molecule to transport into the nucleus. The Ran–GDP is recycled back into the nucleus by a nuclear transport factor (NTF2) where the GDP–GTP exchange enzyme converts it to Ran–GTP. *The exchange enzyme does not occur in the cytoplasm. The system depends on the Ran protein being in the GTP form in the nucleus and in the GDP form in the cytoplasm.* To reemphasize the crucial point, this is achieved because of the asymmetric compartmentalization of the exchange factor and the GTPase activating protein.

Exportins transfer out of the nucleus

The reverse transport of proteins carrying a **nuclear export signal (NES)** out of the nucleus into the cytoplasm occurs by a cycle (Fig. 26.18) that is the mirror image of the import cycle. In the nucleus there is a group of **exportin** proteins (also known as **nuclear export receptors** or **karyopherins**), each specific for proteins carrying the appropriate NES. Attachment of the exportin to the NES of its target cargo (step 1) occurs only when Ran–GTP is attached to the exportin (steps 2 and 3). Note that this is the reverse of the situation with importin, which releases

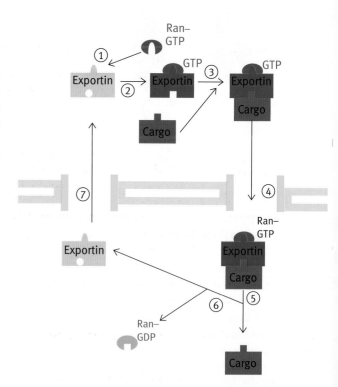

Fig. 26.18 The mechanism of export of proteins from the nucleus. See text for explanation. The deep red colour represents exportin in a form capable of carrying a cargo; the pale red form cannot do so.

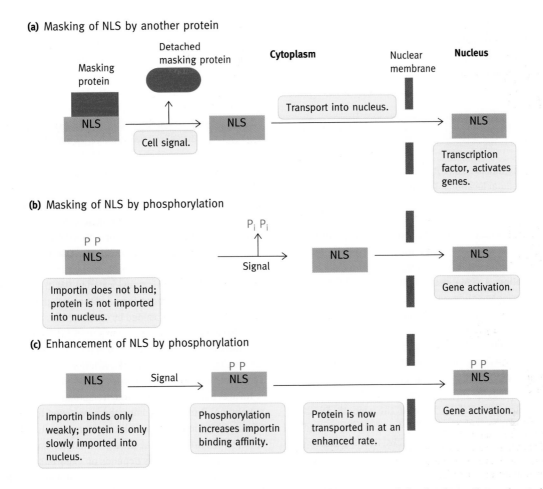

(a) Masking of NLS by another protein

(b) Masking of NLS by phosphorylation

(c) Enhancement of NLS by phosphorylation

Fig. 26.19 Mechanisms of signal-mediated regulation of import of proteins into the nucleus. Extracellular signals to cells are of central importance in gene control, and nuclear import regulation plays a vital role in this. The essence of it is that proteins required for specific gene control reside in the cytoplasm until a signal causes them to be imported into the nucleus. This requires that the proteins in question are not transported into the nucleus (or at a slow rate) until a signal causes them to be transported. There are several mechanisms for this, three of which are illustrated. NLS, nuclear localization signal.

its cargo when Ran–GTP attaches. The Ran–GTP–exportin–cargo is transported through the pore (step 4) into the cytoplasm where, as before, the GAP activates the Ran–GTPase (step 5) causing hydrolysis of the GTP and dissociation of the complex into Ran–GDP and exportin. This releases the cargo (steps 5 and 6). The Ran–GDP is recycled back to the nucleus (step 7). The exportin is also recycled back to the nucleus.

The establishment of the gradients of RAN–GTP and RAN–GDP across the nuclear membrane energetically drives nuclear transport, and the contrasting properties of importin and exportin give the opposing directionality.

Regulation of nuclear transport by cell signals and its role in gene control

As already explained, in much of eukaryotic gene control, proteins are transported into the nucleus on receipt by the cell of signals (Chapter 27). The proteins must each have an NLS so the question arises of why they are not carried into the nucleus

in the absence of any signal. Such proteins have their NLS rendered ineffective until an extracellular signal arriving at the cell causes changes in the protein which now make it available for transport.

There are several strategies employed to bring this about. One is to mask the NLS on a given protein by the binding of another protein molecule (Fig. 26.19(a)). An example of this is to be found in the chapter on cell signalling (page 433), describing how certain steroid hormones combine with their cognate protein to be transported, causing a conformational change that results in detachment of the masking protein. The NLS is thereby made available for combination with the importin, resulting in import of the protein into the nucleus where it is a transcription factor for those genes that are activated by the steroid hormone.

An alternative strategy is to mask the NLS by phosphorylation (Fig. 26.19(b)). Some NLSs have a low binding affinity for importins, so that the proteins carrying them are transported at a slow rate; phosphorylation of the protein may increase the binding affinity of the two proteins and increase the rate of transport into the nucleus (Fig. 26.19(c)). Dephosphorylation by protein phosphatases provides a way of reversing the process when the cell signal is terminated.

The realization that the nuclear-cytoplasmic transport traffic plays a major role in cell signalling and gene control enormously extends its importance and interest.

Summary

Proteins are synthesized on cytoplasmic ribosomes but function in different cellular locations. They are targeted to their destinations by a number of mechanisms.

Some proteins traverse the membrane of the rough endoplasmic reticulum (ER) and enter the lumen as they are synthesized. Free cytoplasmic ribosomes begin translating mRNAs for these proteins. The initial sequence of polypeptide produced is called a signal peptide. A signal-recognition particle (SRP) binds to this, and halts polypeptide synthesis.

The SRP docks the ribosome to a translocon site in the ER membrane. The signal peptide is postulated to open the translocon channel and leads the nascent polypeptide through it as it is synthesized. On reaching the ER lumen the signal peptide is cleaved off, the synthesis of the protein is completed and released into the lumen where it folds up. This is known as cotranslational transport.

The proteins move along through the smooth ER, being processed en route by attachment of carbohydrates and are transported in vesicles to the Golgi stacks. Here the proteins are sorted, addressed, and transported by vesicles to lysosomes and the plasma membrane or return to the rough ER lumen. The synthesis of integral membrane proteins also occurs in the ER. In this case, in addition to the signal peptide, the polypeptide chain contains a stop transfer or anchor signal that fixes the protein in the membrane. Vesicles transport new membrane sections complete with proteins to specific existing membranes.

Membrane transport vesicles are of two types; clathrin-coated and COP-coated depending on their site of origin and destination. Vesicles are targeted to their destination by v-SNARES, protein complexes that bind to complementary t-SNARES on their target membrane.

GTP hydrolysis plays an important part in these processes, the GTP-binding proteins acting as molecular switches. The proteins undergo allosteric changes when GTP replaces GDP or bound GTP is hydrolysed.

Transport of proteins into mitochondria is posttranslational. On completion of synthesis, the polypeptide after release is held in an extended form by a chaperone. This delivers it to the receptor of the transport complex, known as TOM (translocase, outer membrane) receptor on the outer membrane of the mitochondrion. The polypeptide is translocated to TIM (translocase, inner membrane) in an extended form and then into the mitochondrial matrix where it folds up aided by chaperones.

Peroxisomal proteins are transported in a fully folded form into the organelle by mechanism(s) not understood. Two peroxisomal targeting signals are known in the form of amino acid sequences in the proteins to be transported.

Nuclear transport occurs through elaborate pores in the nuclear membrane. Proteins are imported complexed to importin proteins, which bind to nuclear localization signals (NLSs) in proteins, which are essential to target them for import. Export of proteins from the nucleus depends on exportins that recognize nuclear export signals (NESs) on target proteins and transport them through the pores.

Nuclear traffic depends on the hydrolysis of GTP attached to an allosteric protein, Ran, which has latent GTPase activity. Compartmentation of enzymes results in Ran being in the GTP form inside the nucleus and in the GDP form in the cytoplasm. Ran–GTP causes cargo release from importin in the nucleus while hydrolysis of GTP causes cargo release from exportin in the cytoplasm. Exchange of GDP for GTP occurs only in the nucleus and activation of GTP hydrolysis occurs only in the cytoplasm. The Ran–GTP/Ran–GDP gradient so established across the nuclear membrane gives directionality to nuclear pore transport and is the energy source for the process.

Further reading

von Heijne, G. (1998). Life and death of a signal peptide. *Nature*, **396**, 111–12.
A News and Views article on signal peptide cleavage as the transported polypeptide emerges into the lumen of the ER.

Matlack, K. E. S., Mothes, W., and Rapoport, T. A. (1998). Protein translocation: tunnel vision. *Cell*, **92**, 381–90.
Refers to protein transport through the ER membrane.

Bernstein, P. (2000). Molecular trafficking. *Essays Biochem.*, **36**, 1–129.
A complete issue devoted to 10 reviews on protein targeting.

Eichler, J. and Irihimovitch, V. (2003). Move it on over: getting proteins across biological membranes. *BioEssays*, **25**, 1154–7.
Two distinct modes mediated by 'translocons' – the macro molecular complex that translocates proteins across membranes.

Golgi

Short, B. and Barr, F. A. (2000). The Golgi apparatus. *Curr. Biol.*, **10**, R583–5.
Short review.

Lysosomes

Neufeld, E. F. (1991). Lysosomal storage diseases. *Annu. Rev. Biochem.*, **60**, 257–80.
Reviews general principles and specific diseases.

Bonifacino, J. S. and Traub, L. M. (2003). Signals for sorting of transmembrane proteins to endosomes and lysosomes. *Annu. Rev. Biochem.*, **72**, 395–447.
A research-level review.

COP-coated vesicle formation and targeting

Edwardson, J. M. (1998). Membrane fusion: all done with SNARE pins. *Curr. Biol.*, **8**, R390–3.
All about SNARE proteins and transport vesicle targeting.

Lowe, M. (2000). Membrane transport: Tethers and TRAPPs. *Curr. Biol.*, **10**, R407–9.
Brief summary of v-SNARES and t-SNARES.

Mallabiabarrena, A. and Malhotra, V. (1995). Vesicle biogenesis: the coat connection. *Cell*, **83**, 667–9.
A minireview that discusses why different coat proteins are used for the transport of vesicles.

Schmid, S. L. and Damke, H. (1995). Coated vesicles: a diversity of form and function. *FASEB J.*, **9**, 1445–53.
Reviews the different ways in which coated vesicles are budded off from the Golgi and other membranes.

Schekman, R. and Orci, L. (1996). Coat proteins and vesicle budding. *Science*, **271**, 1526–33.
A review of vesicle transport and their role in sorting of proteins.

Wieland, F. and Harte, C. (1999). Mechanisms of vesicle formation: insights from the COP system. *Curr. Opin. Cell Biol.*, **11**, 440–6.
Discussing formation of transport vesicles.

Clathrin

Marsh, M. and McMahon, H. T. (1999). The structural era of endocytosis. *Science*, **285**, 215–19.
A concise review of clathrin coat assembly.

Peroxisomes

Shimozawa, N., et al. (1992). A human gene responsible for Zellweger syndrome that affects peroxisome assembly. *Science*, **255**, 1132–4.
Shows that the Zellweger syndrome primary cause is a defect in the synthesis of a peroxisome assembly protein.

McNew, J. A. and Goodman, J. M. (1995). The targeting and assembly of peroxisomal proteins: some old rules do not apply. *Trends Biochem. Sci.*, **21**, 54–8.
Reviews evidence that folded proteins are imported into peroxisomes as such.

Subramani, S., Koller, A., and Snyder, W. B. (2000). Import of peroxisomal matrix and membrane proteins. *Annu. Rev. Biochem.*, **69**, 399–418.
Comprehensive review of peroxisomal biogenesis.

Mitochondria

Pfanner, N. and Meijer, M. (1997). Mitochondrial biogenesis: the TOM and TIM machine. *Curr. Biol.*, **7**, R100–3.
Reviews the field.

Koehler, C. M., Merchant, S., and Schatz, G. (1998). How membrane proteins travel across the mitochondrial intermembrane space. *Trends Biochem. Sci.*, **24**, 428–32.
Deals with the special proteins needed to conduct hydrophobic proteins across the aqueous intermembrane space. Deficiencies in these lead to blindness and deafness.

Paschen, S. A. and Neupert, W. (2001). Protein import into mitochondria. *IUBMB Life*, **52**, 101–12.

Neupert, W. and Herrmann, J. M. (2007). Translocation of proteins into mitochondria. *Annu. Rev. Biochem.*, **76**, 723–49.

Schatz, G. (2007). The Magic Garden. *Annu.* Rev. Biochem., **76**, 673–678.
A highly readable personal review of mitochondrial assembly and how the field, in which the author has played a major part, has developed .

Nucleus

Mattaj, I. W. and Englmeir, L. (1998). Nucleocytoplasmic transport: the soluble phase. *Annu. Rev. Biochem*, **67**, 265–306.
A comprehensive review.

Mattaj, I. W. and Conti, E. (1999). Snail mail to the nucleus. *Nature*, **399**, 208–10.
Concise News and Views article describing the mechanism of nuclear-cytoplasmic transport.

Blobel, G. and Wozniak, R. W. (2000). Proteomics for the pore. *Nature*, **403**, 835–6.
News and Views article describing work by Rout and colleagues, which gives a complete structure of the nuclear pore.

Weis, K. (2003). Regulating access to the genome: nucleocytoplasmic transport throughout the cell cycle. *Cell*, **112**, 441–51.

Burke, B. (2006). Nuclear pore complex models gel. *Science*, **314**, 766–7.
A postulated mechanism for nuclear pore transport.

Beck, M., Lucic, V., Forster, F., Baumeister, W., and Medalia, O. (2007). Snapshots of nuclear pore complexes in action captured by cryoelectron tomography. *Nature*, **449**, 611–5.

Protein glycosylation

Abeijon, C. and Hirschberg, C. B. (1992). Topography of glycosylation reactions in the endoplasmic reticulum. *Trends Biochem. Sci.*, **17**, 32–6.
Reviews the different protein glycosylation reactions occurring in the endoplasmic reticulum.

Problems

1   What is meant by cotranslational and posttranslational transport of a protein across a membrane? Give examples.

2   In protein targeting, GTP hydrolysis is often involved although this is not associated with any chemical synthesis or performance of other obvious work. What is the function of this?

3   What are the purposes of lysosomes?

4   What are lysosomal transport vehicles and where are they produced?

5   What evidence is there to indicate that lysosomal digestion is an essential process?

6   Why is Pompe's disease somewhat of a biochemical puzzle?

7   The transport of proteins through nuclear pores is critically dependent on the asymmetric distribution of two specific protein-catalysed activities between the nucleoplasm and cytoplasm. Explain this.

8   What are FG repeats of the nuclear pore?

9   Briefly summarize the basic differences between the targeting of proteins for secretion, for import into mitochondria, nucleus, and peroxisomes respectively.

Chapter 27

Cell signalling

Cell signalling brings together several major biochemical concepts that have been dealt with separately elsewhere in the book. Important in this category are conformational changes in proteins (page 248), protein kinases (page 436), second messengers (page 249), gene control in eukaryotes (page 364), transcription factors (page 364), and the cell cycle (page 507). In this chapter there are brief recapitulations of essentials where these seem useful and there are page references to fuller treatments of individual topics.

Overview

There is hardly a topic in the molecular aspects of life more important than cell signalling in an organism as complex as a mammal. A human being is a society of about 10^{13} individual cells, which have to be coordinated so that their activities correspond to those which are in the interests of the body as a whole. This is achieved by chemical signalling between cells. There is no room for independent cellular enterprise. Cancer is the end result of a cell replicating, irrespective of whether or not it should do so. It is a disease involving aberrant signalling systems (Chapter 33). The number of cells in a human body far exceeds the total human population so the organizational task is large and in recent years it has emerged that the controls operating on mammalian cells are far more complex than was ever imagined. There is no parallel to this in prokaryotes.

The most fundamental controls are those on cell replication, differentiation, and programmed cell death (apoptosis; page 231); apoptosis is a normal part of life. Metabolic events must also be regulated to serve the interests of the body as a whole (Chapter 16). Cells in the body are bombarded with instructions on all manner of things by cell signalling. As well as coordination of cell growth, replication, differentiation, and death, there are ever-changing physiological requirements such as responding to the state of nutrition, inflammatory signals, and adapting to a changing environment. Cell signalling by

hormones and other factors is the mechanism by which cells communicate with each other about these metabolic needs.

There are, in the broadest sense, three avenues of signal delivery.

Neurotransmitter substances are released by nerve endings to signal the next neuron, or to activate muscle contraction. They usually are ligands controlling gated ion channels present on all cells but of special importance in nerve-impulse conduction. We have described the mechanism of these in Chapters 7 and 8.

The second delivery system is *via* **hormones** (Chapter 16). These are produced and secreted into the bloodstream by cells specialized for their production and aggregated into endocrine glands. The insulin-secreting cells of the pancreas are a typical example. Hormones flood the body to reach distant **target cells** – those designed to receive the signal.

Neurological control by transmitters and hormonal control issued by 'central authorities' (the brain and the endocrine glands) are easy concepts to grasp. The control is hierarchical – how human societies are organized.

The third broad class of signalling control involves a more elusive concept of mutual control by cells. The signalling molecules are known as **cytokines** or **growth factors**; there is no absolute distinction between the two terms and what they are called often depends on how they were discovered. Factors involved with blood cells are usually called cytokines and growth factors were often detected by their effects on cell growth and replication but they may have other effects on different cells or the same cells at different times. Here are a few examples of what they do.

- Normal (noncancerous) cells will not grow on nutritive plates unless supplied with growth factors (like those in foetal calf plasma).
- When stem cells differentiate into specific somatic cells, cytokines/growth factors direct the process, and production of various classes of blood cells from adult stem cells is a typical example of a battery of different cytokines controlling the process.

- When a wound heals, growth factors direct the cells to replicate.
- In the immune system, cytokine signals are needed to trigger B cells to differentiate into antibody-producing cells.
- Interferons produced by virus-infected cells signal neighbouring cells to take protective actions.

From these few examples you can see that they are of fundamental importance and are intensely researched, especially as they are of major medical interest.

Identification of the many growth factors and cytokines and elucidation of their signalling pathways has revealed much about the nature of cellular controls and of diseases such as cancer. It is not understood, however, how local collections of cells can between themselves sort out the problems of coordination of the activities of all the cells in, for example, a tissue. Tissues and organs, such as the liver, essentially remain the same size in adults because cell division and cell death balance each other. Cell signalling somehow achieves this. For example, if as much as two-thirds of the liver of a rat is removed, cell division is increased until – in about 2 weeks – normal liver size is again achieved. It then reverts to the normal slow rate needed to compensate for cell death. How this is achieved is not known, nor is it understood how each organ reaches its appropriate size, and then remains constant.

All eukaryotic signalling molecules combine with protein receptors. Most of these are on the outside of cells, though a handful of lipid-soluble signals enter the cell before they combine with **intracellular** receptors (Fig. 27.1(a)). A cell without a receptor for a given signal cannot respond to it. The chemistry of the signalling molecules has no known intrinsic significance because they do not enter into any reactions. All they do is to accurately combine with their cognate receptors. It is the relationship between the receptor and the signal, which is important. Nothing happens to the signalling molecule apart from its ultimate destruction. When a signal combines with a **transmembrane** receptor, conformational changes occur in the receptor that results in its cytoplasmic domain changing its shape. The receptor proteins have evolved to respond to their cognate signals. The signal is thus conveyed (transduced) across the membrane by the transmembrane receptor protein.

In the case of gene-activating signals, the change in the receptor domain results in a chain of events which triggers the relay of the message to the nucleus. In the case of hormones affecting metabolism, the action may be primarily in the cytoplasm (Fig. 27.1(b)). In both cases, the whole chain of events from the membrane inwards are called **signal transduction pathways**. There are various types of these. Some, like the Ras pathway (see below), involve a chain of proteins; others produce a **second messenger**, which passes on the message to other proteins. The second-messenger term (page 249) is used for a transduction pathway component, which is a small molecule rather than a peptide, but it is still its shape and binding

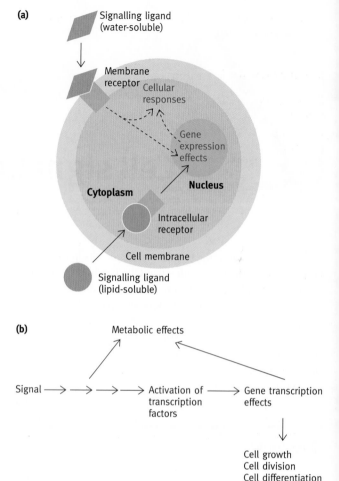

Fig. 27.1 (a) Outline of receptor-mediated signalling. Water-soluble signalling molecules cannot pass through the lipid bilayer. They bind to external domains of receptors. The lipid-soluble signalling molecules such as steroids and thyroxine enter the cell directly and bind to intracellular receptors. Note that some intracellular receptors reside in the nucleus as described later; cytoplasmic steroid receptors move into the nucleus upon ligand binding. **(b)** A more detailed account of cellular responses to signals.

properties that matters. It fits to its target component and relays the signal.

With so many signals there have to be different receptors and signal transduction pathways, just as a landline telephone system depends on separate lines to take signals to the correct house. Variation in the structures of the protein components involved gives a multiplicity of receptors and pathways.

It adds up to a mind-boggling complexity but fortunately for our understanding there is a relatively small number of reasonably direct pathways used for individual receptors and transduction pathways. For example, the different Ras pathways have variations in individual components of the pathway can route signals from different receptors to their correct molecular destinations.

Organization of this chapter

In the sections that follow, the patterns of receptor design are first presented. First there is the class of **intracellular** receptors, the prototype of which is the glucocorticoid-specific one. Then, of the **extracellular** receptors, most fall into two main categories – these are **tyrosine kinase receptors** and **G-protein receptors**. A number of examples of the rest of the transduction pathways are then given.

The prototype of the tyrosine kinase pathways is the Ras pathway, an ancient highly conserved signalling pathway present in all animals from fruit flies to humans. Its relative, the JAK/STAT pathway, that follows has a different type of tyrosine kinase receptor and pathway, used by many cytokines. How insulin works (involving another tyrosine kinase receptor) is also a topic of obvious medical interest.

Turning to the second major class, the G-protein-associated signal transduction pathway, of which epinephrine (adrenaline) signalling is the prototype. This illustrates the different strategy of pathway activation which does not involve tyrosine phosphorylation of receptors. This is a very large group (about half of all pharmaceutical drugs target these) but all have a common basic mechanism. They are known as G-protein receptors because in all cases the signal is transduced by a GTP-binding protein. The phosphatidylinositide cascade pathway is a more complicated example of a G-protein pathway of importance.

Other examples given are to illustrate various signalling strategies; for example, how light signals are handled in the visual process and how the simplest signal of all, nitric oxide (NO), works, a process associated with Viagra® action.

What are the signalling molecules?

As stated in the overview, from the viewpoint of cell signalling mechanisms, the signalling molecules do not enter into chemical reactions – they are simply molecules of the right shape and properties for binding with great specificity by noncovalent bonds to their receptors. Briefly, to illustrate their variety, in chemical terms they include **proteins, large peptides, small peptides, steroids, eicosanoids** (page 231), **thyroid hormone, epinephrines, nitric oxide**, and **derivatives of vitamins A and D**. Examples of the first five groups respectively are **insulin, glucagon, vasopressin, sex hormones**, and **prostaglandin**.

A biological classification is as follows (nitric oxide being in a class of its own):

- neurotransmitters
- hormones

- growth factors and cytokines
- vitamin A and D derivatives.

This classification is important in terms of nomenclature and physiology, but they are all signalling molecules that bind to cellular receptors and elicit responses; so that they have basic roles in common and conceptually are all the same. With that qualification we will now describe the basis of that classification.

Neurotransmitters

A variety of neurotransmitters are involved in nerve function but we will mention only a few relevant to our immediate topic. The sympathetic (involuntary) nervous system, which innervates fat cell depots for example, secretes epinephrine (adrenaline) and norepinephrine (noradrenaline) on receipt of a nerve impulse. Thus, a fat cell may be regulated by epinephrine from the adrenal glands *via* the blood (Chapter 16) or from nerve endings. The difference is that the latter delivery route is faster and the signal is released precisely at the target cell site. The motor neurons innervating voluntary striated muscle trigger contraction by the release of acetylcholine from the nerve endings. All of these work by binding to external cell receptors which usually control the opening of ion channels.

The transmitters are rapidly destroyed to prepare the neurons for the next impulse.

Hormones

These are the 'classic' signalling molecules, most of which have been known for a long time. They are important both in metabolic control and in control of the expression of specific genes. Hormones are produced by endocrine glands which secrete the hormones into the circulation. They reach their target cells by flooding the whole body and so reach cells distant from the secreting gland. The system works because only the **target cells** have **receptors** capable of picking up the signal. A large number of hormones are known and their biological effects well documented. Table 27.1 lists some of the principal hormones, for reference purposes.

Much of the endocrine system is under a hierarchical control system with the **hypothalamus** being at the top of the chain of command. The hypothalmus is a small part of the brain that produces hormones that stimulate release of anterior pituitary hormones, known as tropic hormones because they cause target endocrine glands (thyroid, adrenal cortex, and the gonads) to release *their* hormones, the 'final' ones in the chain, which elicit responses from cells in general. Feedback loops in which end products (the final cell-targeted hormones) inhibit the first step (release of hypothalamic hormones) maintain appropriate levels of circulating hormones. There are exceptions to this control system, a notable one being release of the pancreatic hormones, insulin and glucagon, which is more directly controlled by the level of blood glucose.

Table 27.1 The principal hormones

Secreting organ	Hormone	Target tissue	Function
Hypothalamus	Hormone-releasing factors	Anterior pituitary	Stimulation of circulating hormone secretion
	Somatostatin (also from pancreas)	Anterior pituitary	Inhibits release of somatotrophin
Anterior pituitary	Thyroid-stimulating hormone (TSH)	Thyroid	Stimulates T_4 and T_3 release
	Adrenocorticotrophic hormone (ACTH)	Adrenal cortex	Stimulates release of adrenocorticosteroids
	Gonadotrophins (luteinizing hormone [LH]) and follicle-stimulating hormone (FSH)	Testis and ovary	Stimulates release of sex hormones and cell development
	Somatotrophin (growth hormone)	Liver	Stimulates synthesis of insulin-like growth factors, IGFI and IGFII
	Prolactin	Mammary gland	Required for lactation
Posterior pituitary	Antidiuretic hormone (ADH) or vasopressin	Kidney tubule	Promotes water resorption
	Oxytocin	Smooth muscle	Stimulates uterine contractions
Thyroid	Thyroxine (T_4), triiodothyronine (T_3)	Liver, muscles	Metabolic stimulation
Parathyroid	Parathyroid hormone	Bone, kidney, intestine	Maintains blood Ca^{2+} level, stimulates bone resorption and dietary Ca^{2+} uptake
	Calcitonin (also from thyroid)	Bone, kidney	Inhibits resorption of Ca^{2+} from bone
Adrenal cortex	Glucocorticosteroids(cortisol)	Many tissues	Promotes gluconeogenesis
	Mineralocorticosteroids(aldosterone)	Kidney, blood	Maintains salt and water balance
Adrenal medulla	Catecholamines (epinephrine, norepinephrine)	Liver, muscles, heart	Mobilize fatty acids and glucose into bloodstream
Gonads	Sex hormones (testosterone from testes, oestradiol, progesterone from ovaries)	Reproductive organs, secondary sex organs	Promote maturation and function in sex organs
Liver	Somatomedins (insulin-like growth factors: IGFI, IGFII)	Liver, bone	Stimulate growth
Pancreas	Insulin	Liver, muscles	Stimulates gluconeogenesis, lipogenesis, protein synthesis
	Glucagon		Stimulates glycogen breakdown, lipolysis

Cytokines and growth factors

It has been suggested that cytokines and growth factors should be regarded as developmental regulatory factors. **Growth factors** for example are more understandable if regarded as signals that, depending on the cell type and the circumstances, may induce cell division or inhibit it. They may control differentiation or instruct the cell to undergo programmed cell death (**apoptosis**). The cytokines and growth factors control such fundamental processes because they control specific gene transcription. Many such factors exist; this is an intensively researched area and of great medical interest. There are no 'authorized' definitions that distinguish between cytokines and growth factors, and the terms are sometimes used interchangeably. The many factors controlling blood cell development, including those involved in immune responses (page 500) together with the interferons (page 443), are referred to as **cytokines**. Cytokines and growth factors are regulatory proteins or peptides secreted by many cell types that, unlike the cells producing hormones, are not specialized for producing signals, but are typical of whichever tissue they belong to, such as hepatocytes and lymphocytes. Most cytokines/growth factors are **paracrine** in their action – they diffuse short distances to act only on local cells and are rapidly destroyed – while some are **autocrine** in action (Fig. 27.2). These, such as **interleukin 2**, which stimulates T cell proliferation (page 500), act on the same type of cells that secrete them. The localization of their action to closely neighbouring cells may be a biologically important function of their action. The converse can also be the case; the cytokine **erythropoietin** produced by the kidney medulla controls the proliferation of erythrocytes. It is released when the oxygen tension in the blood is low and is different in that it is released into the general circulation of the blood.

The names given to cytokines and growth factors usually depend, as stated, on the way they were discovered. One of the earliest known growth factors was **platelet-derived growth factor (PDGF)**. Blood platelets lyse at the sites of damage in blood vessels to initiate clotting and the released PDGF stimulates cell division and repair. However, other cells also produce the factor so its name does not fully define its role. **Epidermal growth factor (EGF)** stimulates the growth of skin cells. **Interleukins**

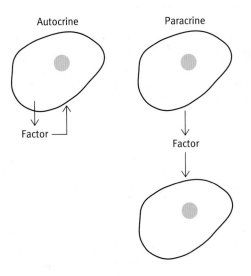

Fig. 27.2 Autocrine signals affect the cell producing them; paracrine signals diffuse only a short distance to affect nearby cells.

are produced by leucocytes (white blood cells) to affect other leucocytes. **Colony-stimulating factors (CSFs)** are so named because they were discovered in experiments in which they stimulated the growth of colonies of white cells on culture plates. Some are used clinically to control white cell production. For example, people with leukaemia given bone marrow transplants first have their bone marrow cells ablated (number reduced) by radiation and chemotherapy. After the transplant, there is a period during which there are insufficient neutrophils, with the risk of infections, since these white cells combat bacteria. Treatment with **granulocyte-colony-stimulating factor (G-CSF)** stimulates neutrophil (granulocyte) production, reducing the risk period.

Growth factors/cytokines and the cell cycle

The eukaryotic cell cycle is described in Chapter 31 but, briefly, for present purposes, eukaryotic cell division involves a progression of phases from one division to the next. DNA synthesis is confined to a definite period of time called the S (for synthesis) phase (see Fig. 22.4) typically lasting about 7 hours out of the total cycle of 24 hours. Before and after the S phase are the G_1 and G_2 phases respectively (G denoting gap in DNA synthesis). The crucial checkpoint known as the **restriction point** (in mammals) controls the transition from the G_1 to the S phase. Once past this checkpoint the cell is committed to duplicate its chromosome and proceed to mitosis. If a mitogenic (replication) signal is not received in the G_1 phase, the cell is shunted into a quiescent (G_0) phase until a signal arrives. This is the state of most cells in a mature tissue. The mitogenic signal is delivered by a growth factor or cytokine which thus assumes critical importance in cell division control.

Vitamin D and retinoic acid

We usually think of vitamins as being enzyme cofactors, or components of these, but vitamins A and D are different. Retinoic acid, derived from vitamin A, is important as a signalling molecule in embryonic development and normal cell growth, and in vision. Calcitriol (1,25-dihydroxyvitamin D, the active form of Vitamin D) has an important role in control of genes involved in calcium absorption from the intestine.

Having considered many of the 'signals' and their target processes, we will now move on to the question of how cells detect signals and transduce signals into the cell.

Responses mediated by intracellular receptors

As already mentioned, a handful of hormones are lipid-soluble and easily traverse the cell membrane. These are steroid hormones, thyroid hormone, Vitamin D, and retinoic acid. They meet their receptors *inside the cell* in contrast to most hormones and other signals, which are hydrophilic and do not enter the cell but meet their receptors exposed on the outside surface. The lipid-soluble ones regulate expression of specific genes in target cells at the level of initiation of gene transcription. Examples of the steroids include **glucocorticoids**, **oestrogen**, and **progesterone** (see list of hormones on page 432.) A number of the lipid-soluble signalling molecules are shown in Fig. 27.3 for reference. A superfamily of related steroid/thyroxine receptors exists, suggesting an ancient, common ancestor protein.

We will use the glucocorticoid receptor as an example. Glucocorticoid hormones have diverse effects on metabolism (see Box 27.1), including increasing gluconeogenesis (page 237). The receptor exists in the cytoplasm, attached to a complex of heat shock proteins (Hsps, described on page 395) which mask its nuclear localization signal (NLS), a peptide sequence on the protein (page 423). When a glucocorticoid hormone binds to a specific site on the receptor, the receptor undergoes a conformational change and the Hsps dissociate from it, revealing the NLS so that the receptor is transported into the nucleus. There, it combines as a dimer at the specific glucocorticoid response element (Fig. 27.4) on the DNA causing activation of appropriate genes.

In the case of the steroid family of receptors, which reside in the nucleus (those for thyroid hormones, and retinoic acid), the nuclear localization signal is not obscured by the Hsp complex so that the receptor–Hsp complex is transported as a unit into the nucleus soon after it is synthesized. However, until the Hsp is released by hormone attachment, it does not become an active transcription factor and does not modulate genes. The principle of the control is the same as with glucocorticoids despite these differences.

BOX 27.1   The glucocorticoid receptor and anti-inflammatory drugs

Several synthetic glucocorticoids are among the most effective suppressors of inflammation, their activity being mediated by the glucocorticoid receptor described below. The inflammatory response provides immediate protection against infections and assists in tissue repair after injury, but inappropriate or unregulated inflammation is a key element in the development of **rheumatoid arthritis** and other autoimmune diseases such as **inflammatory bowel disease**, the skin disorder **psoriasis**, and **multiple sclerosis**.

One important element in inflammation is the release by phagocytic white blood cells (neutrophils and macrophages) of protein mediators called chemokines (they attract white cells) and cytokines (page 433), of which tumour necrosis factor-α (TNF-α) and interleukin-1 (IL-1) are two important examples. These attach to cell-surface receptors and stimulate the inflammatory response. They do so by initiating a cascade of intracellular events, leading ultimately to the activation of the transcription factor family **NF-κB**. This is present in the cytoplasm of all cells but in an inactive form due to being complexed with **IκB** inhibitory proteins. The TNF-α signal causes the activation of kinases that phosphorylate the inhibitor protein marking it for destruction by proteasomes. This exposes the nuclear localization signal (NLS) on the NF-κB, which is then transported into the nucleus where it activates many genes, leading to the inflammatory response.

The anti-inflammatory glucocorticoid drugs attach to the glucocorticoid receptor in the cytoplasm as shown in Fig. 27.4 and the receptor also moves into the nucleus. There it attaches to and inhibits the activated NF-kB protein, thus helping to inhibit the inflammatory response.

Glucocorticoids remain important drugs in the treatment of inappropriate inflammation. Recently, however, new classes of anti-inflammatory drugs have been developed. These include inhibitors of the inflammation-associated enzyme cyclooxygenase 2 (**COX-2 inhibitors**), which acts on membrane-derived lipids to release inflammatory prostaglandins (page 274), and the use of monoclonal antibodies or soluble cytokine receptors, which mop up cytokines such as TNF before they have a chance to bind their cell-surface receptors (TNF inhibitors). Finally, small-molecule drug candidates targeting components of the intracellular signalling cascades are also under development.

This is a large and complex system, not fully elucidated, and has been given here only in brief outline. It is an area of intense medical interest. See Further reading list at the end of this chapter. See Box 14.2 on page 232 for more on COX-2 inhibitors, which are now of restricted use because of deleterious side effects.

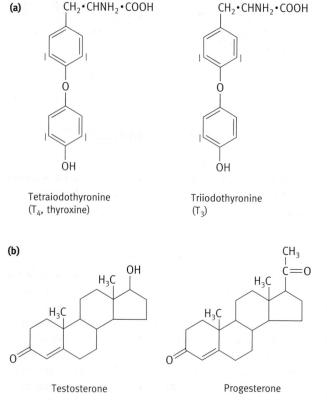

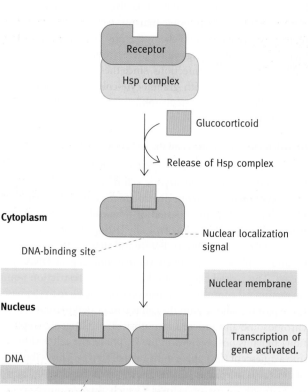

Fig. 27.3 (a) The structures of the thyroid hormones; **(b)** the structures of two steroid hormones.

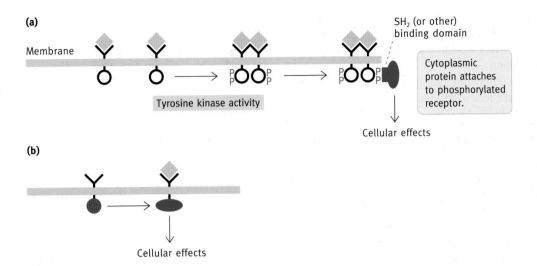

Fig. 27.5 **(a)** Diagram illustrating the tyrosine kinase type of receptors; the phosphorylation of the receptors causes attachment of the appropriate cytoplasmic protein by its SH2 domain; the attached protein then activates a signalling pathway. (SH2 domains are common, but other binding domains exist – see below.) **(b)** In this type of signalling, the receptor is not phosphorylated but attachment of the signal molecule causes a conformational change in the cytoplasmic domain of the receptor. This leads to the activation of one or more signalling pathways. The G-protein-coupled receptors are of this type.

Another signalling molecule, nitric oxide, is also lipid-soluble with an intracellular receptor but is in quite a different category and for convenience will be dealt with later (page 451).

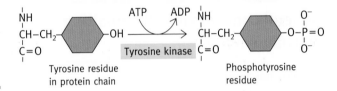

Fig. 27.6 Tyrosine phosphorylation by tyrosine kinase.

Responses mediated by receptors in the cell membrane

For the water-soluble hormones, most of the signalling in the body occurs *via* membrane receptors. There are large numbers of them, but all have an external domain to which the signal binds, a transmembrane domain, and a cytoplasmic domain. When its cognate signal molecule (cognate means 'having affinity to') is bound to a receptor, the cytoplasmic domain undergoes a conformational change that activates a signalling pathway. This in turn activates specific genes and/or modulates metabolic systems.

Fig. 27.4 (*opposite*) Gene activation by steroid hormones. The example shown is the glucocorticoid receptor, one of a superfamily of steroid/thyroxine receptors, all of which involve complexes with heat-shock proteins (Hsps) and binding to DNA sites by zinc fingers (page 376). In the case of those members of the superfamily that reside in the nucleus, the Hsp complex does not obscure the nuclear localization signal. The locations of the latter and the various binding sites are drawn arbitrarily in the diagram.

There are two types of membrane-bound receptors

The biggest class of receptors are called **G-protein-coupled receptors** because they are associated with GTP-binding proteins. The other main group is the **tyrosine kinase-coupled receptors**.

The tyrosine kinase-coupled receptors

When cognate ligands bind to these, they dimerize by lateral movement in the lipid bilayer (Fig. 27.5(a)). This brings the cytoplasmic domains of a pair together. These domains are themselves tyrosine protein kinases and they phosphorylate each other on tyrosine groups by the reaction shown in Fig. 27.6.

In the cytoplasm are different proteins involved in control processes, which bind to phosphorylated receptors and act as adaptor molecules that link the phosphorylated receptors and the signalling pathways that convey the signals, usually to the nucleus. Around the phosphorylation sites on a receptor are amino acid sequences that are recognized by these proteins, so that the correct adaptor proteins bind to the correct receptor and thus the correct signalling pathway(s) are activated (see section on binding domains below). A given receptor may be connected to a number of signalling pathways and thus have multiple control effects in the cell.

The G-protein-coupled receptors

We have earlier described the control system of the GTP/GDP switch (page 414), which is important in protein synthesis and protein delivery mechanisms and elsewhere. The principle, to recapitulate very briefly is that the G-control proteins have GTP bound to them. They are monomeric **GTPases** and hydrolyse the attached GTP to GDP and phosphate causing a conformational change in the protein. In the case of G-coupled receptors this occurs as a result of the binding of the cognate hormone or other signal to the receptor which causes an allosteric change in the cytoplasmic domain. In turn this leads to a response to the signal (Fig 27.5b). A very wide range of hormones have receptors of this type. We will describe later in detail how the 'prototype' of G-protein receptors, the epinephrine-specific one operates.

General concepts in cell signalling mechanisms

Protein phosphorylation

The interconversion of phosphorylated and dephosphorylated forms of proteins plays a crucial role both in metabolic control (Chapter 16) and in signalling and there are large numbers of different protein kinases in eukaryotes for phosphorylation of proteins. The principle is simple. Addition of a highly charged phosphate group to a protein can have a major effect on the conformation of a protein, which result in the protein participating in the relevant signalling pathways. Reversal of this by protein phosphatases terminates the action. Fig. 27.7 illustrates how phosphorylation is used in signalling both for metabolic and gene control. The figure shows that the protein kinase and the phosphatase are themselves controllable by some sort of signal so that highly integrated control is possible. In this type of control, phosphorylation is usually on serine and/or threonine —OH groups. (This is different from the way tyrosine kinase phosphorylation of receptors works for there it acts as an attachment point for other proteins.)

Binding domains of signal transduction proteins

Cell signalling in mammals is complex and includes proteins recognizing other proteins or small molecules and joining together to form chains or clusters which relay the signals, one

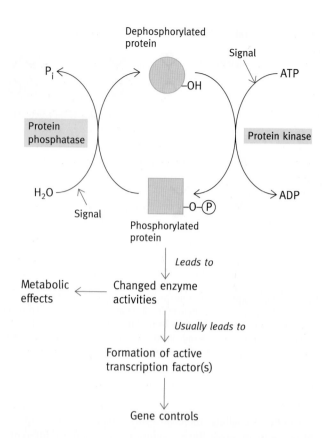

Fig. 27.7 Central principle of control by many extracellular signals. The diagram illustrates protein kinase and protein phosphatase action. Phosphorylation is on serine or threonine residues of proteins. The phosphorylation brings about a conformational change in the protein that changes its activity in some way, resulting in a cellular response. The process is reversed by removal of the phosphoryl group by a protein phosphatase action. In the case of some membrane receptors, described later in the text, phosphorylation of their cytoplasmic domains on tyrosine —OH groups of protein side chains is involved in their activation. Note that an arrow may represent several steps.

to another, to their destinations. The same recognition domains have been used in different proteins to build up signalling pathways reminiscent of the sections in a child's building blocks that click together. On page 55 we described how domains of proteins are found over and over in different proteins, an evolutionary concept known as domain shuffling.

Such a domain, called SH2, is found in a wide variety of regulatory proteins. The name SH2 means Src (pronounced sarc) homology domain, region 2. This refers to a kinase found in the Rous sarcoma virus which causes tumours. One of the Src kinase domains recognizes a small amino acid sequence. SH2 proteins typically bind to phosphorylated tyrosines on the cytoplasmic face of membrane receptors. As mentioned, the phosphorylated tyrosines have adjacent to them different amino acid sequences in different receptors, and different SH2 domains of signalling proteins recognize these, as well as the tyrosine phosphate group. In the human genome many hundreds of genes coding for hundreds of proteins with variant SH2 domains have been identified.

In addition to SH2, the prototype Src kinase has an **SH3 domain** which binds to proline-rich sequences around phosphorylated tyrosines and to proline-rich regions of other proteins. These are found in many signalling proteins often in addition to the SH2. Thus an adaptor protein may bind to its phosphorylated receptor by the SH2 domain and also connect to other proteins in signalling pathways *via* its SH3 domain.

The existence of variants of these domains provides for the specific recognition of large numbers of different activated receptors by adaptor proteins.

Another class of proteins involved in signalling pathways binds to membrane areas enriched in inositol-containing second messengers (page 249) by the **pleckstrin homology** or **PH domain** first found in a blood platelet protein (pleckstrin). It has since been found in over 60 signalling proteins downstream from the receptor in signalling pathways.

In summary, the use of a few main types of recognition domains with large numbers of variants giving many signalling proteins allows evolution of great flexibility in receptor recognition and the assembly of transduction pathways in a modular fashion.

Terminating signals

Regulatory systems in the body must be reversible, otherwise a signal once given could not be switched off. Cancers are often the result of runaway signalling pathways (Chapter 33). Much signalling involves phosphorylations and families of protein phosphatases exist to reverse these. It is estimated that 2–3% of human genes code for these. Some are specific for serine/ threonine phosphates and others for tyrosine phosphates and may themselves be regulated by phosphorylation. In addition, some signalling pathways have individual restraining mechanisms which will be dealt with when we describe examples.

The rest of this chapter illustrates examples of the main classes of signalling pathways. To help you keep track of the text as we deal with them, they are listed below as a reference guide – it does not matter if, at this stage, some terms do not mean much to you.

Examples of signal transduction pathways

- Pathways from tyrosine kinase-associated receptors:
  - the Ras pathway
  - the phosphatidylinositide 3-kinase (PI 3-kinase) pathway (used by insulin)
  - JAK/STAT pathways.

- G-protein-associated pathways:
  - cyclic AMP (cAMP) pathway
  - phosphatidylinositol cascade pathway
  - vision – the light-transduction pathway.
- Signalling pathways mediated by cyclic GMP as second messengers:
  - membrane receptor-mediated pathways
  - nitric oxide signaling.

Signal transduction pathways from tyrosine kinase receptors

The Ras pathway

This widespread signalling pathway from membrane receptor to genes in eukaryotes is used by a wide variety of growth factors. **Ras**, the protein that gave its name to this pathway, is found in all eukaryotic cells. It is part of a major signalling pathway from growth factor-specific tyrosine kinase receptors, to modulation of gene transcription factors. *There are no low-molecular-weight second messengers in this pathway – all of the components are proteins.*

Ras, and other components of the pathway, were known to be important in cell regulation, because mutations resulting in abnormal forms of some of the proteins are **oncogenic** – they are associated with cancer. Ras was discovered as the oncogenic protein coded for by the **rat sarcoma virus** that causes muscle tumours (sarcomas) in rats. Its normal counterpart is found in all eukaryotic cells and a mutated form of Ras (not of viral origin) occurs in many human cancers.

The signalling pathway involves the receptor stimulating a cascade of three cytoplasmic protein kinases, the final one of which causes gene activation. Fig. 27.8 gives a simplified outline of the pathway to provide a preliminary orientation.

Mechanism of the Ras signalling pathway

We will use the receptor for the **epidermal growth factor (EGF)** to illustrate this (Fig. 27.9). The receptor for EGF is a monomer with external binding, single transmembrane and cytoplasmic domains. The latter is a tyrosine kinase. When EGF binds, the receptors dimerize in the membrane, bringing their kinase domains together and they phosphorylate each other as shown in Fig. 27.5(a).

In the cytoplasm is a **growth receptor-binding protein (GRB)**, an adaptor protein. It binds to the phosphorylated receptor *via* its SH2 domain, but not to the nonphosphorylated form. The next component in the pathway is **SOS**, initially found in a *Drosophila* (fruit fly) genetic mutant where it was named 'son of sevenless'. SOS is activated by the receptor-bound GRB (but not by free GRB). The activated GRB-SOS complex in turn activates Ras, the mechanism of which we will now describe.

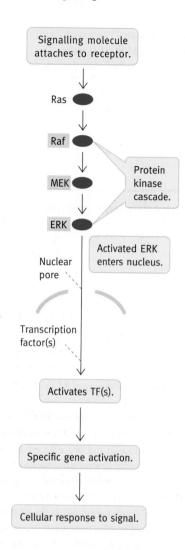

Fig. 27.8 Simplified overview of the Ras signal tranduction pathway. Raf, MEK, and ERK are protein kinases, each of which phosphorylates the next component of the pathway. The nomenclature and further details of the pathway are given in the text and subsequent figures. TF, transcription factor. Arrows represent activation.

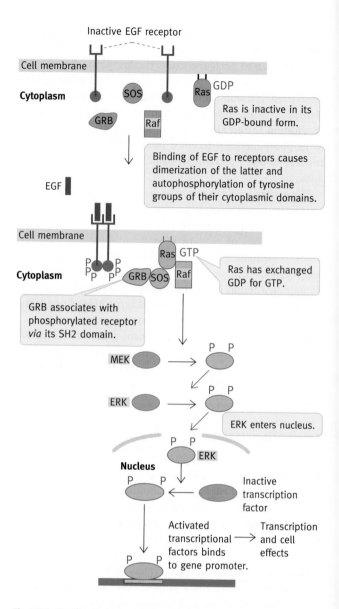

Fig. 27.9 The Ras pathway of signal transduction. See text for explanation. In this diagram signalling by epidermal growth factor (EGF) is used as an example. Activation of the various proteins involves changes in conformation but these are not indicated to keep the presentation reasonably simple. The nomenclature is explained in the text. GRB, growth-receptor binding protein.

Concept of the GTP/GDP switch mechanism, as used by the Ras pathway

Ras has a switch mechanism widely used in control pathway proteins; it is a GTPase that belongs to a family known as **small monomeric GTPase proteins**. (The term **G-protein**, which is reserved for the trimeric GTPase components of G-protein receptors, is described later (see Fig. 27.19).) They have a molecule of GDP *or* GTP bound to them. *Ras is active in the GTP form and inactive in the GDP form* (Fig. 27.10), the

nucleotides causing conformational changes in the protein. To activate Ras (for Ras to activate the next component) the GDP must be exchanged for GTP, but this exchange occurs only when Ras is in contact with GRB-SOS bound to the receptor (which occurs only when a signal is attached to the receptor).

When the EGF is no longer present (due to its destruction) the signal has to be terminated. The Ras protein is a slow GTPase that switches itself off by hydrolysing its attached

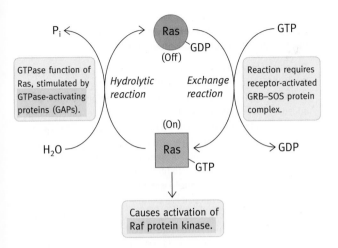

Fig. 27.10 The control of the Ras protein.

activating GTP molecule to GDP + P_i. If the receptor is still activated, Ras will be reactivated by a repeat of the GDP/GTP exchange process; if not, Ras will be inactive until more EGF arrives. One gets a glimpse here of the complexity of signalling controls, because other proteins called **GTPase-activating proteins (GAPs)** stimulate the rate of GTP hydrolysis by the Ras-type proteins and speed up the switch-off. Why have this complication? GAPs provide a modulation of the control. This accounts for the fact that GAP mutations can be oncogenic. If the GTPase activity of Ras is not activated by a functional GAP, Ras does not switch itself off.

The MAP kinase cascade in the Ras pathway

Ras in its GTP form activates a cascade of three protein kinases, collectively called **mitogen-activated protein kinases (MAP kinases)**. It activates **Raf** by combining with it and inducing a conformational change. Raf is the first protein kinase in the Ras pathway downstream from Ras. The name Raf derives from a viral oncogene; after activation by Ras-GTP it phosphorylates **MEK**, the second protein kinase, which in turn phosphorylates **ERK**, the last of the three protein kinases (nomenclature explained below). The cascade is as shown in Fig. 27.9, the first kinase (Raf) activating the second by phosphorylation and the second phosphorylating the third which migrates into the nucleus. Why so many components? This strategy amplifies the signal. The use of amplifying cascades is discussed in glycogen breakdown (page 253) and blood clotting (page 483).

When the phosphorylated ERK is transported into the nucleus it phosphorylates target transcription factors, which results in the transcription of specific genes, the synthesis of their cognate proteins, and the desired cellular response to the EGF signal such as cell proliferation. Raf and ERK are serine/threonine-type-specific kinases. MEK, uniquely, has dual specificity for these and tyrosine.

To summarize, the activation sequence, also illustrated in Fig. 27.9, is as follows:

- EGF attaches to its receptor; the receptor dimerizes and is autophosphorylated on its tyrosine —OH groups
- GRB-SOS attaches to the phosphorylated receptor by the SH2 domain of GRB
- receptor-bound GRB-SOS activates Ras; Ras activates Raf, a protein kinase
- Raf phosphorylates and thus activates MEK, also a protein kinase
- MEK phosphorylates ERK, the final protein kinase
- ERK migrates into the nucleus and phosphorylates target transcription factor(s); the latter activates gene transcription; specific proteins are synthesized that promote cell proliferation.

Nomenclature of the protein kinases of the Ras pathway

Raf, MEK, and ERK are collectively known as **mitogen-activated protein kinases (MAP kinases)**, a mitogen being something that stimulates cell division. Since there are several parallel pathways involving MAP kinase cascades (see below), *individual* MAP kinases in a cascade are given identifying names. In the case of the Ras pathway these are Raf, MEK, and ERK. MEK stands for MAP kinase/ERK; that is a MAP kinase whose substrate is ERK. **ERK** stands for **extracellular signal-regulated protein kinase**. Since related, but different, MAP kinase variant cascades have kinases analogous to Raf, MEK, and ERK but with different specificities, these also have individual names.

It is useful to be able to collectively refer to all kinases of the Raf type, and similarly to all protein kinases of the MEK type, and also to all of the ERK type found in the different signalling pathways. The terminology seems daunting at first sight, but is really quite simple. In the literature the terms **MAPKKK**, **MAPKK**, and **MAPK** are used; to understand them it is best to start with the last and work backwards. Thus ERK is a MAP kinase or MAPK. MEK is a kinase that phosphorylates ERK (a MAP kinase), and therefore is a MAP kinase-kinase or MAPKK. Raf is a kinase that phosphorylates MEK (a MAP kinase-kinase), and therefore is a MAP kinase-kinase-kinase or MAP-KKK. (See terms in brackets in Fig. 27.12; these are the generic names for MAP kinases in Ras-type parallel pathways.)

Inactivation of the Ras pathway

The removal of EGF or other factors from receptors terminates the activation of the receptor but the whole pathway needs to be switched off also. Ras is inactivated by its intrinsic GTPase activity. How are the activated MEK and ERK kinases inactivated? A family of **protein phosphatases** is known which inactivate MAP kinases and other signalling proteins by dephosphorylating them. Somewhat surprisingly, the protein kinases and their cognate phosphatases are physically bound together

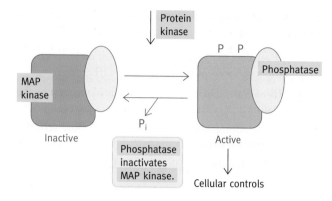

Fig. 27.11 The association of a MAP kinase with a phosphatase to provide a rapid molecular switch mechanism. The MAP kinase is activated by phosphorylation and the associated phosphatase reverses this by removing the phosphoryl groups. The implication is that MAP kinase pathway signalling involves the rapid oscillation of the kinase components between the active and inactive states. This may be to give stability of control.

(Fig. 27.11), implying that no sooner has a phosphorylation taken place than it is reversed. A competing push-pull situation is less likely to get out of control – perhaps analogous to continually touching the brakes of a car going down a steep hill, and, as stated, a signalling pathway out of control is characteristic of cancer cells.

Some of the protein phosphatases are induced by the same signal that activates the pathway, so that negative feedback loops are established. The importance of protein phosphatases is demonstrated by the toxins that increase or decrease protein dephosphorylation (Box 27.2).

Scaffold proteins prevent cross-talk between pathways

There are multiple Ras type pathways operating in parallel using protein kinases corresponding to Raf, MEK, and ERK in other pathways and targeted at different transcription factors (Fig. 27.12) or cytoplasmic effectors.

These multiple pathways convey signals from different receptors to the nucleus so that activation of specific genes appropriate to the particular signal occurs. It is necessary that the different pathways do not 'cross-talk' *inappropriately* just as you don't want telephone lines to different houses to allow cross-talk when they should be conveying separate messages to their different destinations.

However, different MAP kinase cascades may contain a component common to two or more pathways. This would seem to allow a mix up of signals from different receptors and negate the whole point of having specific receptors transmitting their signal to specific endpoints. A solution to the problem is found in scaffold proteins, which bind together in a series, the components of specific cascades. This is illustrated in Fig. 27.13. It is thus possible to use the same MAP kinase in two pathways since the scaffold protein does not allow cross-talk between pathways.

BOX 27.2 **Some deadly toxins work by increasing or inhibiting dephosphorylation of proteins**

The crucial importance of protein phosphatases is emphasized by the fact that some biological toxins target them. One of current environmental interest is the phosphatase-inhibitory toxin produced by blue-green algae. This is a cyclic peptide, called microcystin, containing seven residues. It is a powerful liver toxin and a dangerous cancer-producing agent. Inhibition of dephosphorylation would prevent switch-off of signalling pathways. As the nitrification of rivers occurs, due to fertilizer run off from agricultural areas, coupled with the reduced water flows caused by irrigation, toxic algal blooms develop. It is a worldwide problem.

One of the poisons produced by shellfish, called okadaic acid, is an inhibitor of serine/threonine-specific phosphatases of the type found in the Ras pathway. This toxin is of medical concern, as well as a problem for the shellfish industry.

The other side of the coin is a protein that has killed vast numbers of people in earlier centuries. It is a protein of the bubonic plague *Bacillus*, which is a tyrosine-specific protein phosphatase. This causes signalling mayhem since much of receptor-mediated signalling involves phosphorylation of tyrosines of receptor proteins (see the **Cohen** reference at the end of this chapter for further literature).

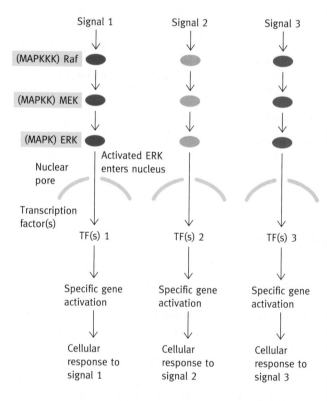

Fig. 27.12 Multiple signal pathways of the Ras type. Raf, MEK, and ERK are shown as green shapes. The orange and blue shapes represent protein kinases corresponding in function to Raf, MEK, and ERK but are protein kinases of other signalling pathways. The terms in brackets are the generic names for MAP kinases in the Ras-type parallel pathways. TF, transcription factor.

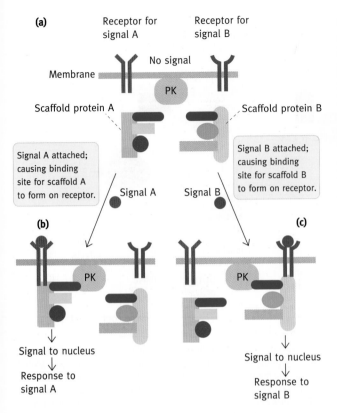

(a)
Receptor for signal A   Receptor for signal B

No signal

Membrane

PK

Scaffold protein A   Scaffold protein B

Signal A attached; causing binding site for scaffold A to form on receptor.

Signal A   Signal B

Signal B attached; causing binding site for scaffold B to form on receptor.

(b)   (c)

PK   PK

Signal to nucleus   Signal to nucleus
↓   ↓
Response to signal A   Response to signal B

Fig. 27.13 Simplified diagram to give the overall concept of how two receptors can route signals to two signalling cascades *via* a common protein kinase (PK). The mechanism has been established for two yeast receptors but is presented here without yeast specific details and in a generalized way. (See text for explanation and the **Park** *et al.*, and **Ptashne and Gann** references for the yeast situation.) In some cases a domain of the scaffold protein itself constitutes the middle MAP kinase. The figure also illustrates how scaffold proteins allow independent pathways to have components in common (here seen in dark green) without cross-talk occurring.

Signal sorting

A more difficult problem exists in that it is known that two independent signals may activate the same component of different signalling pathways and yet the signals have to remain separate and reach different specific final targets. For example it is known that Ras can be activated both by growth promoting mitogens and by hormones, and yet their effects on the cell are quite different. As an example, EGF and insulin can both activate Ras, but the responses to the two signals are quite different. Somehow the cell sorts out signals from receptors to the correct end points.

A likely solution to such a problem was found in a simple signalling system in yeast. Two receptors were studied, one activated by a mating pheromone and the other by high salt concentrations. The responses to the two signals are totally

different. They both activate different signal cascades, each attached to a scaffold protein, but are activated by single protein kinase, which interacts with a MAP kinase common to both pathways. The single activating kinase is attached to the cell membrane. In the absence of a signal to either receptor, the MAP kinase cascade on their scaffold proteins are detached from the membrane protein kinase so that neither signal pathway is activated. It is postulated that when a cognate signal binds to a receptor there is conformational change in its cytoplasmic domain which creates a binding site for its cognate scaffold protein. The latter binding brings the first MAP kinase on the scaffold into contact with the kinase which activates it thus firing off the signal pathway. When the signal leaves the receptor the conformational change reverses and the scaffold protein detaches and the signal is terminated. Thus signals from the different receptors in this way are thus routed down separate pathways to the correct endpoints. The mechanism is presented in Fig. 27.13 in a generalized way to give the overall concept without details specific to the yeast situation.

The problem cited above, of different signals activating Ras but the correct messages getting to the intended destinations seem much more difficult in mammalian systems and the mechanism by which the cell sorts the signals so efficiently is not understood. However, there are aspects to Ras that may ultimately be found to be relevant. In mammals, four largely homologous Ras isoforms are known, which activate different signalling pathways within the cell, and there is considerable speculation on their compartmentation.

It is speculated that **lipid rafts** may play a role in this area. These are small islands of membrane lipids rich in sphingolipids and cholesterol, which do not mix with the rest of the lipid bilayer and remain discrete as slightly thickened areas due to the length of the hydrocarbon tails of the sphingolipids. It is thought that possibly these lipid rafts are sites for the specific attachment of regulatory proteins of signalling pathways, though here again no definite examples of how this may function in signal sorting are known.

The phosphatidylinositide 3-kinase (PI 3-kinase) pathway and insulin signalling

This is another example of the class of signalling pathways involving tyrosine kinase-associated receptors. The PI 3-kinase pathway has an almost bewildering range of control roles involved in cell proliferation, differentiation, inhibition of apoptosis, and other cellular activities including metabolic control; it is activated by many different receptors responding to a range of hormones, growth factors, and neurotransmitters. We will use **insulin** to illustrate the pathway.

Insulin's effects have been discussed earlier but briefly it controls glucose uptake into muscle and adipose cells, increases protein synthesis, and controls the synthesis of specific enzymes. Of special metabolic interest, it controls **glycogen synthase** (page 255).

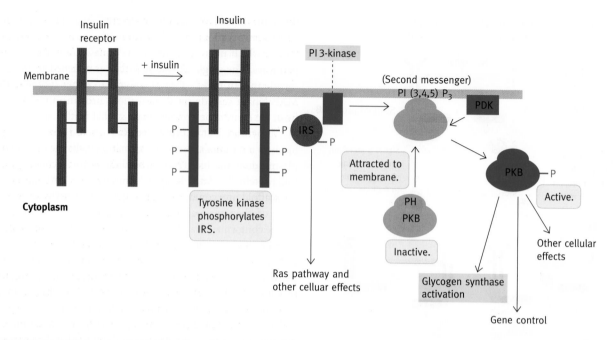

Fig. 27.14 Simplified insulin signalling pathway. The attachment of insulin to its receptor activates tyrosine kinase of the receptor's cytoplasmic domain which phosphorylates itself. Insulin receptor substrate (IRS) attaches to the phosphorylated receptor and itself becomes phosphorylated. PI 3-kinase binds to the phosphorylated form of IRS, thus coming into contact with the membrane where it converts phosphatidylinositol to a second messenger $PI(3,4,5)P_3$ by the reactions shown in Fig. 27.15. This attracts inactive protein kinase B (PKB) to attach to the membrane by its pleckstrin homology (PH) domain which brings it into contact with the kinase PDK. PDK activates PKB by phosphorylation. PKB has many target proteins; it leads to the activation of glycogen synthase and to other cellular and gene control effects. PH, pleckstrin homology domain.

The **insulin receptor** is unusual in structure in that it resembles two receptors of the EGF type, covalently dimerized (Fig. 27.14). On binding of insulin, the cytoplasmic domain of the receptor, *which is a tyrosine kinase*, becomes autophosphorylated on tyrosine —OH groups. The specific adaptor protein that binds to this phosphorylated receptor by its SH2 domain is called the **insulin receptor substrate** (IRS) and this is also phosphorylated by the receptor tyrosine kinase. An enzyme, **PI 3-kinase**, binds to the phosphorylated IRS and is thereby brought in an activated form close to the plasma membrane. The substrate of PI 3-kinase is a membrane component, **phosphatidylinositol-4,5-bisphosphate** (**PI(4,5)P$_2$ or PIP$_2$**). As a result of the PI 3-kinase action, a **second messenger**, **PI(3,4,5)P$_3$ (or PIP$_3$)**, is formed (Fig. 27.15). The second messenger causes activation of **protein kinase B (PKB)**, also known as **Akt**. (PKB is the mammalian homologue of the kinase of retrovirus AKT 8.) The membrane section enriched with PI(3,4,5)P$_3$ attracts *inactive* cytoplasmic PKB to attach to the membrane because it has the pleckstrin homology domain, which binds to PI(3,4,5)P$_3$. There it encounters a membrane-located kinase called **PDK1**, which phosphorylates the PKB, thus activating it. PKB has many protein targets and is involved in gene control, as well as inactivation of glycogen synthase kinase 3 (page 255).

A summary may be useful here.

- Insulin attaches to its receptor.
- The receptor is autophosphorylated on its tyrosine —OH groups.
- The IRS attaches by its SH2 domain to the phosphorylated receptor and is itself phosphorylated.
- PI 3-kinase binds to the phosphorylated IRS and is activated. Its action on PI(4,5)P$_2$, a membrane phospholipid, leads to the production of a second messenger (PI(3,4,5)P$_3$).
- Cytoplasmic inactive PKB is attracted to bind to the membrane enriched with the second messenger.
- The PKB is phosphorylated by the kinase PDK1 located in the membrane and is thus activated.
- PKB phosphorylates other proteins and has widespread cellular effects; specifically it leads to the activation of glycogen synthase by inactivating GSK3 (described on page 255).

Insulin has many effects in addition to those on glycogen metabolism control; IRS in the presence of insulin can activate Ras and thus affect gene control by the pathway described in the previous section. The IRS thus acts as an adaptor to connect the receptor to different signalling pathways.

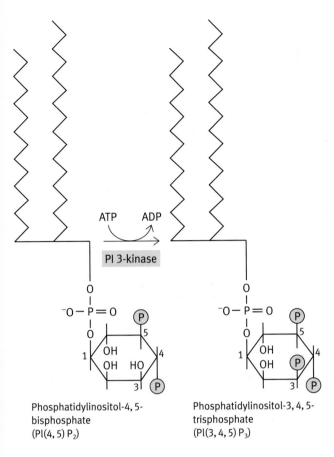

Fig. 27.15 Production of the second messenger PI(3,4,5)P$_3$ from the membrane component PI(4,5)P$_2$ by PI 3-kinase, the latter being insulin-activated.

The strategy of creating a membrane site which entices other signal transducers to bind to particular locations where they can be phosphorylated by kinases lying in wait is a very interesting one which may have wide application. It introduces the possibility of cellular controls involving micro-localization of components. A large variety of proteins with PH domains are known to bind to membranes in response to signals leading to activation of PI 3-kinase. Possibly mechanisms, not yet elucidated, operate to route the signals to the pathway appropriate to the particular signal activated. Mutations of PI 3-kinase are known to be oncogenic and indeed a large proportion of advanced human cancers have been reported to be associated with these.

The JAK/STAT pathways: another type of tyrosine kinase-associated signalling system

A wide variety of genes are controlled by JAK/STAT receptors; they occur in *Drosophila* and humans but not in yeast. The first

discovered was involved in the control of genes by the antiviral agent **interferon** (produced in response to an infection) but they are now known to be used by a wide variety of cytokines and growth factors. The remarkable feature of JAK/STAT signalling pathways is that instead of the multiplicity of intermediates in the pathway the signal is carried from receptor to nucleus by a single protein dimer (Fig. 27.16).

The receptors in the cell membrane exist as monomers. The cytoplasmic domains have no tyrosine kinase activity themselves but associated with each is a JAK, or Janus kinase. Janus was the two-faced Roman god; JAKs have two kinase sites. In the absence of an activating signal the kinases are inactive. On binding of a cytokine, the receptors in the membrane dimerize and their associated kinases are activated. They phosphorylate tyrosine groups on the receptor. In the cytoplasm are **signal transducer and activator of transcription (STAT) proteins**. These have SH2 domains that bind to the phosphorylated tyrosine receptor sites and are themselves phosphorylated by the JAKs on tyrosine sites. The phosphorylated STATs leave the receptor and dimerize, each dimer being an active transcription factor, which is transported into the nucleus. There it binds to its target gene-control elements and switches on genes that, in the case of interferon activation, protect the cell against viruses.

The pathways have great versatility. There are three classes of receptors and four different kinases (JAKs 1, 2, and 3 and TYK2) that bind to different phosphorylated receptors with high specificity. There are in mammals seven different STAT proteins and these can form homodimers or heterodimers. Many different signalling pathways can thus be assembled.

Negative control of JAK/STAT pathways

As emphasized previously, signalling pathways have to have a mechanism for switch-off as well as a switch-on for otherwise a signal would last forever. In keeping with this, the JAK/STAT pathways have a multiplicity of checks. There is a phosphatase that reverses the activation by removing phosphate groups from proteins. Cytokines attached to receptors may be internalized and destroyed. There are also **suppressors of cytokine signalling (SOCS)** proteins, which bind to and inhibit the JAKs attached to the receptors. The SOCS proteins inhibit the protein kinase domain by blocking its active site.

The STAT dimers activate the genes needed to respond to the cytokine signal but also activate SOCS genes giving a feedback-control system. The cytokine thus switches on the positive and negative responses at the same time. Gene-knockout mice (page 472) lacking SOCS, when given interferon, show adverse physiological effects due to excessive signal. Additionally there are protein inhibitors called **protein inhibitors of activated STATs (PIAS)**, which bind to activated STAT proteins and prevent them from switching genes on.

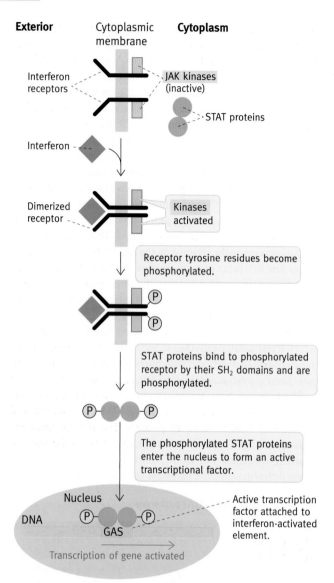

Fig. 27.16 The signalling pathway by which γ-interferon activates specific gene transcription. STAT, signal transducer and activator of transcription; GAS, γ-interferon-activated response element. Note that, unlike the epidermal growth factor (EGF) receptor (Fig. 27.9), which is a tyrosine kinase, in this pathway the dimerized receptor activates attached tyrosine kinases; these are known as Janus kinases (JAKs) after the Roman god with two faces. (JAKs have two kinase sites.) γ-Interferon and several cytokines have analogous signalling pathways but may differ somewhat in detail.

G-protein-coupled receptors and associated signal transduction pathways

Overview

We now turn to the second major class of signal transduction pathway – which does not involve tyrosine kinases. The latter was the first major class. This is now the second major class. The G-protein-coupled receptors constitute a very large super-family of receptors; examples are found in organisms from yeast and insects to humans. The signals include hormones, growth factors, odours, and light.

The general principle can be stated simply. Each receptor has associated with it on the cytoplasmic side a **heterotrimeric G-protein**, made up of three different subunits, α, β, and γ. The G-proteins all have a common pattern. The α subunit of the trimeric *unactivated* G-protein has bound to it a molecule of GDP (hence the name, G-protein). On binding of the signal molecule to the receptor a conformational change in the associated G-proteins occurs and GDP is replaced by GTP. This in turn causes activation of a signal pathway involving the formation of second messengers which bring about the cellular responses as outlined earlier (see Fig. 27.7(b)).

A note on terminology: Ras also is a GTP/GDP-binding protein and has a switch mechanism the same as that to be described for G-proteins; as stated, Ras belongs to the class known as **small monomeric GTPase proteins** rather than G-proteins, *a term used only for the heterotrimeric class*. After that overview, we will now give more details of this class of signalling system.

Structure of G-protein-coupled receptors

All are proteins whose polypeptide chain crosses the membrane seven times (Fig. 27.17); this type of protein is called serpentine or polytopic. In the case of epinephrine, the hormone binds to a cleft formed by the transmembrane helices while the hetero-trimeric G-proteins associate with a polypeptide loop on the cytosolic side of the membrane. In real life, the transmembrane domains are clustered together, not extended as in the figure.

cAMP as second messenger: epinephrine signalling – a G-protein pathway

Earlier (page 249) we dealt with control of metabolism by cAMP activation of protein kinases, but not with how the cAMP level is modulated; in this section we deal only with this and how, importantly, cAMP regulates gene expression. A wide array of hormones use cAMP as *second messenger*, including adrenocorticotrophic hormone (ACTH; or corticotrophin), antidiuretic hormone (ADH; or vasopressin), gonadotroph-ins, thyroid-stimulating hormone (TSH), parathyroid hormone,

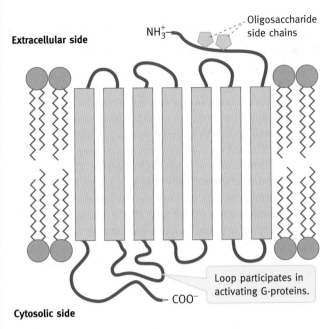

Extracellular side

NH$_3^+$ — Oligosaccharide side chains

Loop participates in activating G-proteins.

COO$^-$

Cytosolic side

Fig. 27.17 The structure of the β_2 adrenergic receptor.

glucagon, the catecholamines epinephrine and norepinephrine, and somatostatin (the latter negatively controls cAMP levels, see below). This list, which is not exhaustive, illustrates that cAMP has different effects in different cells. The system works because the response to cAMP in a given cell is appropriate to the signal that that cell recognizes *via* its receptors. Thus, in cell A, signal X elevates cAMP levels. This produces cellular responses appropriate to signal X. Cell B does not have receptors for X, but does for signal Y, which also elevates cAMP. In cell B this causes responses appropriate to signal Y. For example, cAMP mobilizes glycogen breakdown in liver cells and triacylglyceride breakdown in fat cells.

Control of cAMP levels in cells

cAMP is produced from ATP by adenylate cyclase (Fig. 27.18(a)), an enzyme which is an integral cell membrane protein (Fig. 27.19(a)). A typical example of a pathway that activates adenylate cyclase is that using the β_2-**adrenergic receptor**. Associated with the cytoplasmic face of the receptor is a G-protein. The α subunit has a site on it that can have bound to it *either* GTP or GDP. When the site on the α subunit is occupied by GDP, nothing happens. This is the situation in the absence of hormone, illustrated in Fig. 27.19(a).

When a molecule of epinephrine binds to the receptor, the receptor undergoes a conformational change, causing conformational change in the α subunit of the G-protein bound to it, which exchanges its GDP for GTP. *The G-protein cannot make this exchange unless it is attached to a receptor to which the hormone is bound.* The α-GTP complex detaches, migrates to, and activates by binding to an adenylate cyclase enzyme molecule which now produces cAMP (Fig. 27.19(b),(c)) by the reaction shown in Fig. 27.18(a).

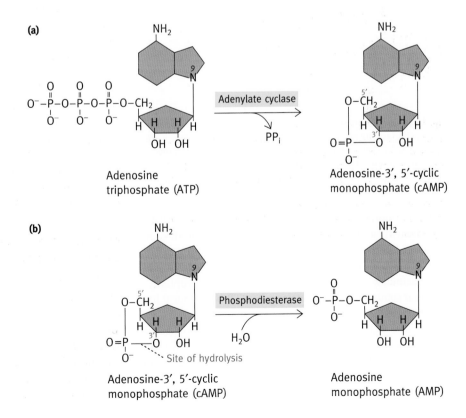

(a)

Adenosine triphosphate (ATP) → Adenylate cyclase → Adenosine-3′, 5′-cyclic monophosphate (cAMP) + PP$_i$

(b)

Adenosine-3′, 5′-cyclic monophosphate (cAMP) → Phosphodiesterase, H$_2$O → Adenosine monophosphate (AMP)

Site of hydrolysis

Fig. 27.18 (a) Formation of cyclic AMP (cAMP) by adenylate cyclase; **(b)** hydrolysis of cAMP by phosphodiesterase.

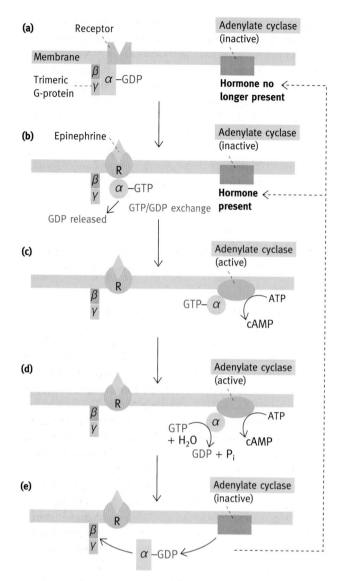

Fig. 27.19 The control of adenylate cyclase activity by a hormone such as epinephrine. The geographical locations of the subunits shown are hypothetical, the essentials being that the α-GTP complex activates adenylate cyclase. The steps (a)–(e) are referred to in the text.

The hormone thus switches on cAMP production using the G-protein as an intermediary. Activation of cAMP production by the α-GTP complex must be limited in time – it must be switched off, for otherwise a single hormone stimulus would last indefinitely, long after the response would be appropriate.

The α subunit of the G protein in its GTP-bound form has enzyme activity – it is a GTPase. It hydrolyses its attached GTP molecule to GDP (Fig. 27.19(d)). The activity is low so that the hydrolysis occurs only after a delay. As soon as GDP is formed, the α subunit reverts to its original state; it detaches from the adenylate cyclase which is now inactivated. The α subunit rejoins its old partners, the β and γ subunits, and the

G-protein–GDP trimeric complex is reassembled, in contact with the receptor (Fig. 27.19 (e)). If the latter still has bound hormone, the whole cycle can start again (back to Fig. 27.19(b)). If the receptor no longer has hormone attached (back to Fig. 27.19(a)), the process comes to a halt. Destruction of cAMP by phosphodiesterase in the cell completes the reversal (see Fig. 27.18(b)). Thus, to continue cAMP production, the α subunit runs back and forth from receptor to adenylate cyclase, the duration of its stay on the adenylate cyclase being that required for the hydrolysis of its attached GTP molecule. The situation is rather like that of a timed light switch on a staircase where you press a button, the light goes on, and the button slowly comes back out and switches off the light after a minute or two. You have to keep on pressing the button at intervals for the light to stay on. The G-protein is a timing device to limit the period of activation of adenylate cyclase. The system has an amplifying effect, since one molecule of hormone, on binding to a receptor, causes the synthesis of many molecules of cAMP, each of which activates further.

The importance of the GTP hydrolysis switch-off is seen in **cholera**. Gut mucosal cells secrete Na^+ into the intestinal lumen and this is activated by cAMP. The cholera toxin inactivates the GTPase activity of the G-protein α subunit. Thus, once a hormone stimulus activates the adenylate cyclase, it cannot be switched off; it is frozen into the α-GTP state. The prolonged cAMP production results in massive loss of Na^+, accompanied by water molecules, causing massive diarrhoea and possible death from fluid and electrolyte loss.

GTPase-activating proteins (GAPs) regulate G-protein signalling

Earlier we described GAPs acting on Ras-type signal transduction proteins. GAPs controlling the heteromeric G-proteins also exist, acting on the α subunit, whose GTPase activity can be enhanced more than 2000-fold.

Different types of G-protein receptor

In the case given in Fig. 27.19 the GTP-α subunit stimulates adenylate cyclase activity and is called G_s (s for **stimulatory**). Another type of receptor for epinephrine (known as the α_2 receptor) operates similarly, except that the GTP-α subunit (called G_i **for inhibitory**) inhibits adenylate cyclase. Examples of G_i receptors are those for **angiotensin** and **somatostatin**. Thus, a hormone can exert different effects on different cells, according to the type of receptor present. It illustrates the way in which the use of different G-proteins associated with particular receptors can control different signal transduction pathways. Since there are multiple forms of each of the G-protein heterotrimer subunits, different combinations are possible giving flexibility of control options.

How does cAMP control gene activities?

In Chapter 16 (page 250), we described how cAMP activates **protein kinase A (PKA)**. PKA, as well as being involved in metabolic control (page 249), is part of a pathway of gene control,

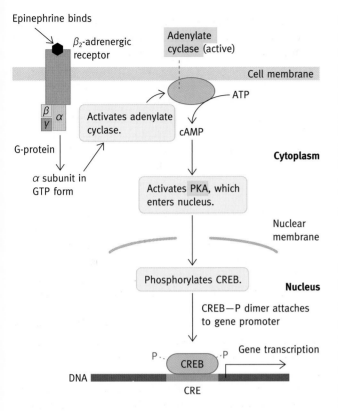

Epinephrine binds

β_2-adrenergic receptor

Adenylate cyclase (active)

Cell membrane

ATP

Activates adenylate cyclase.

cAMP

β α γ

G-protein

α subunit in GTP form

Cytoplasm

Activates PKA, which enters nucleus.

Nuclear membrane

Phosphorylates CREB.

Nucleus

CREB—P dimer attaches to gene promoter

P CREB P

Gene transcription

DNA

CRE

Fig. 27.20 The β_2-adrenergic receptor function. Binding of epinephrine to the receptor causes the heterotrimeric G-protein to convert to the GTP form, detachment of the α subunit – GTP, which activates adenylate cyclase. The cyclic AMP (cAMP) produced activates protein kinase A (PKA), which is transported into the nucleus. There, the PKA phosphorylates cAMP-response-element-binding protein (CREB), which dimerizes and in this form is a transcription factor for specific genes. This activates the genes, resulting in the appropriate responses to the epinephrine. CRE, cAMP-response element of gene promoter.

as shown in Fig. 27.20. Activated PKA is transported into the nucleus. The promoters of several cAMP-inducible genes contain **cAMP-response elements (CREs)**. We remind you that a response element is a section of DNA promoter to which a transcription factor binds and activates the gene. A **CRE-binding protein (CREB)**, when phosphorylated by PKA, dimerizes and becomes an active transcription factor. Mutations that permanently activate the α subunit lead to excessive PKA activation and may be oncogenic.

Desensitization of the G-protein receptors

Typically, cellular responses to extracellular signals diminish with prolonged exposure to the signal molecule. In some cases synthesis of the receptor diminishes, and/or the receptors are endocytosed and destroyed in lysosomes as happens for example with the insulin receptor. The reduction in receptor

numbers is known as **downregulation**. An alternative strategy is desensitization in which the receptor is inactivated when the signal is prolonged. Many of the G-protein-coupled receptors are desensitized by a family of enzymes called **G-protein receptor kinases (GRKs)**, which phosphorylate the receptors inactivating them. Then an inhibitory protein, β-arrestin, binds to the phosphorylated site. This is not to be confused with the tyrosine kinase receptors in which activation is coupled with phosphorylation. The GRKs phosphorylate only *activated* G protein receptors to which their cognate signal molecules are bound – they only inactivate receptors that are in the activated state. There is evidence that the GRKs are themselves subject to controls.

The phosphatidylinositol cascade: another example of a G-protein-coupled receptor, which works *via* a different second messenger

Other G-protein-coupled receptors exist that use quite different signalling pathways from that described for the adrenergic receptors. In the signalling system we will now describe, a different enzyme is activated by the α-GTP (α subunit in the GTP form), leading to the formation of a second messenger other than cAMP. Examples of signals that use the mechanism include acetylcholine and vasopressin (ADH).

The start of this story is the phospholipid membrane component, **phosphatidylinositol-4,5-bisphosphate (PI(4,5)P$_2$)**, the structure of which is given in Fig. 27.21. *This signalling pathway is different from the PI 3-kinase one already described, which also starts with PI(4,5)P$_2$* (Fig. 27.15). Once again a G-protein links the receptor to the intracellular signal pathway. Binding of the hormone to a receptor causes the GTP-α subunit to exchange its GDP for GTP; it then migrates to, and activates, a membrane-bound enzyme, **phospholipase C (PLC)**, which splits PI(4,5)P$_2$ into **inositol trisphosphate (IP$_3$)** and **diacylglycerol (DAG)** (Fig. 27.21). The IP$_3$ leaves the membrane and causes release of Ca^{2+} from the lumen of the endoplasmic reticulum (ER) where the ion is stored in high concentration relative to that in the cytoplasm. It opens IP$_3$-gated Ca^{2+} channels in the membrane, allowing the ion to escape back into the cytoplasm. The signal is reversed when hormone is no longer attached to the receptor due to GTP hydrolysis, destruction of the IP$_3$ and return of Ca^{2+} to the lumen by a Ca^{2+}/ATPase pump. Thus, the combination of a hormone, or other signal, with a receptor associated with the phosphatidylinositol cascade results in increases of intracellular DAG and Ca^{2+} (Fig. 27.22).

DAG is the physiological activator of **protein kinase C (PKC)**, which is also activated by Ca^{2+}. Cytoplasmic PKC is attracted to the membrane by DAG and activated. It is a protein kinase with multiple target proteins. In fact, multiple versions of PKC exist. As one example, it is involved in phosphorylating transcription factors in the nucleus. It also has a controlling role in cell division regulation as is illustrated by the tumour-promoting effect of **phorbol esters**. These are analogues of DAG

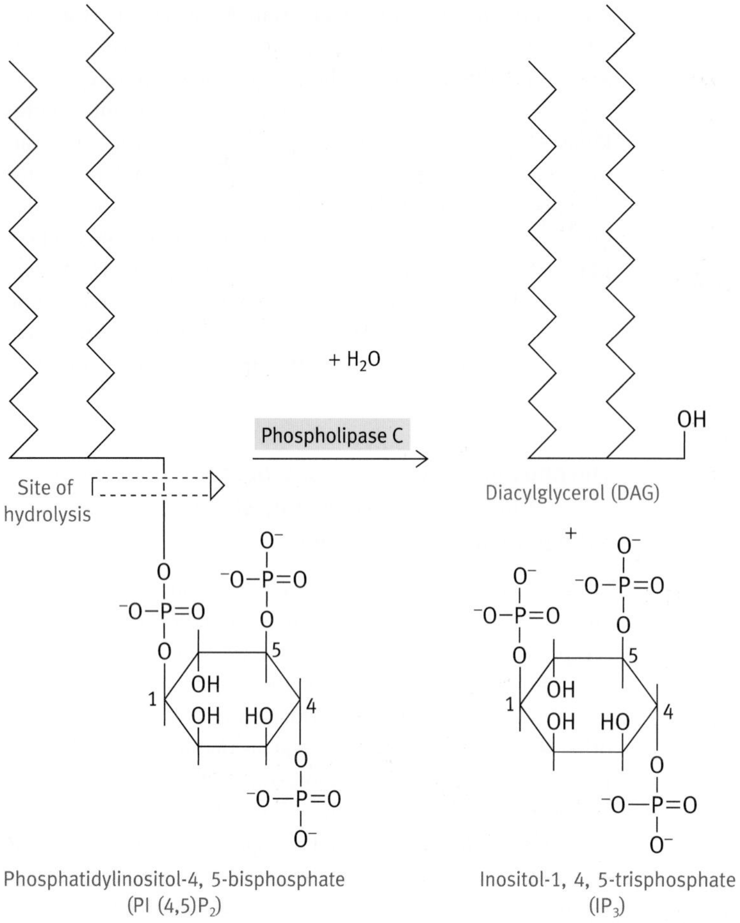

$+ H_2O$

Phospholipase C

Site of hydrolysis

Diacylglycerol (DAG)

Phosphatidylinositol-4, 5-bisphosphate (PI (4,5)P$_2$)

Inositol-1, 4, 5-trisphosphate (IP$_3$)

Fig. 27.21 Hydrolysis of phosphatidylinositol-4,5-bisphosphate (PI (4,5) P$_2$) to diacylglycerol (DAG) and inositoltrisphosphate (IP$_3$).

(Fig. 27.23) capable of activating PKC, leading to cell division. It may seem incongruous that DAG, a normal cellular signalling molecule, has the same effect in activating PKC as does a promoter of tumour formation, but DAG is rapidly destroyed and activates PKC only when required, whereas phorbol esters are longer-lived and deliver an inappropriately prolonged signal. DAG and Ca^{2+} are both needed for maximal activation of PKC but quite apart from this, Ca^{2+} is an important second messenger on its own.

Other control roles of calcium

Ca^{2+} ions control a wide variety of cellular processes. The human body contains about 1 Kg Ca – about 99% of it as a structural component of bones and teeth, and about 1% in the blood and extracellular fluid; only a tiny proportion is intracellular. The strategy is to keep the cytoplasm of cells very low in Ca^{2+} concentration by Ca^{2+}/ATPases, which pump it either to the outside, the mitochondria, the ER lumen, or in the case of skeletal muscle into a special sac called the **sarcoplasmic reticulum** (page 131). Appropriate signals open gated calcium channels (page 114), which release the ion back into the cytoplasm where it acts as a regulator. The steep concentration gradient across the membranes means that there is an instant delivery of Ca^{2+} to the cytoplasm.

A general second-messenger role of the ion involves combination with a widely distributed protein called **calmodulin**. It has four sites for binding Ca^{2+} with high affinity, causing a conformational change in the protein. Calmodulin is sometimes found in association with the enzymes it controls, or it may be free and attach to enzymes in its Ca^{2+}-bound form, depending on the enzyme. The Ca^{2+} causes a conformational change in the calmodulin which alters the activity of the enzyme it is associated with. A number of calmodulin-Ca^{2+}-activated protein kinases exist and Ca^{2+} can, via this route, exert multiple cellular effects. The target proteins of the calmodulin-activated kinases include glycogen phosphorylase, and myosin light chains (page 132); the full list is much longer than this.

Vision: a process dependent on a G-protein-coupled receptor

The versatility of G-protein signalling pathways using different specific G-proteins is shown by this example, in which light is

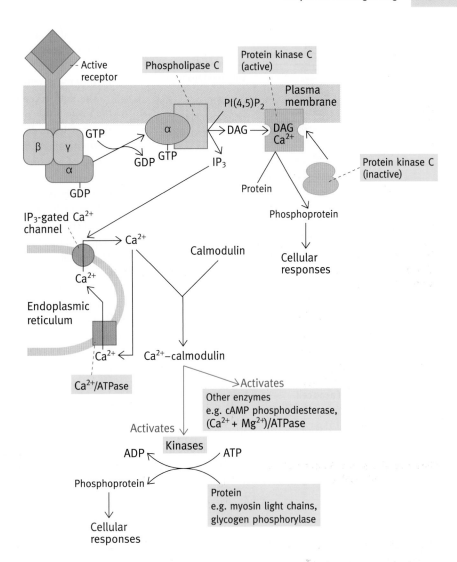

Fig. 27.22 The phosphatidylinositol cascade: interactions of diacylglycerol (DAG), inositoltrisphosphate (IP$_3$), and Ca^{2+} as second messengers. Note that the G$_\alpha$ protein binds to phospholipase C (PLC) only in the GTP-complexed form. On hydrolysis to GDP the process reverses. This illustrates the versatility of G-protein-associated receptors. Different receptors are associated with different G-proteins whose α subunits, when in the GTP form, control the activities of different enzymes – in this case G$_\alpha$ activates phospholipase C. See text page 447 for description of what happens. (Compare with Fig. 27.19, where the G-protein α subunit activates adenylate cyclase.)

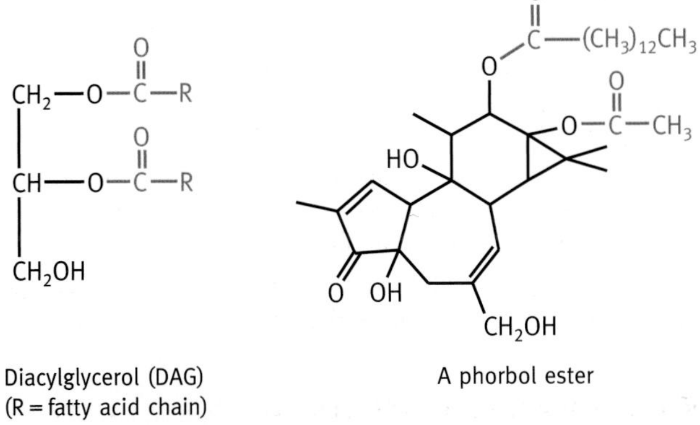

Diacylglycerol (DAG)
(R = fatty acid chain)

A phorbol ester

Fig. 27.23 Phorbol esters are analogues of diacylglycerol, the natural activator of protein kinase C. (The complete structure of the phorbol ester is given only to illustrate this point.)

the signal. The basic problem of vision is to convert the stimulus of light photons into chemical changes, which result in impulses in the optic nerve carrying signals to the brain.

The vertebrate retina has two types of cells for light detection: **rods** for black-and-white and dim-light vision, and **cones** for colour vision. The rod cell (which is the one that we will discuss) has three sections. The middle part has the mitochondria, nucleus, etc. One end of the cell makes a synapse with a bipolar cell that connects with the optic nerve. At the other end is a cylindrical rod-shaped section in which there is a stack of membranous discs (as many as 2000) embedded in the cytoplasm (Fig. 27.24). These contain the light-detection machinery.

Transduction of the light signal

The process of light detection is complex in detail; here we deal only with the essential principles. In the dark, the rod cells have a relatively high level of **cyclic GMP** (**cGMP**), analogous to cAMP, synthesized by a **guanylate cyclase** (see Fig. 27.28). In this instance,

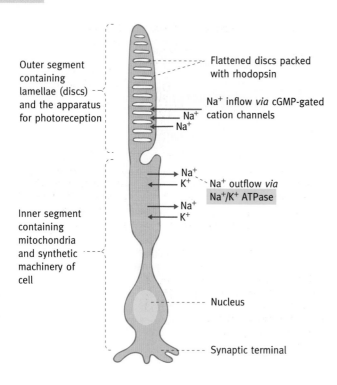

Fig. 27.24 Structure of a rod cell.

cGMP is not strictly a second messenger, since it is not produced as a result of receptor activation. In the cell membrane there are ligand-gated cation channels (page 113), which are kept open by cGMP binding to them as their controlling ligand.

In the dark, the constant inflow of Na^+ through these channels gives the cell membrane a potential across it at equilibrium of -30 mV (see page 118 if you want to refresh your memory on membrane potentials).

The light receptor in the discs is **rhodopsin**, a complex of the protein opsin and the visual pigment, **11-*cis*-retinal**, synthesized from dietary retinol (vitamin A) and beta-carotene (pro-vitamin A). The 11-*cis*-retinal is linked to the $\varepsilon-NH_2$ amino group of a lysine residue in the rhodopsin. Absorption of a photon converts rhodopsin to meta-rhodopsin II. As with other G-protein-associated receptors it has seven transmembrane helices. Light causes activation of rhodopsin due to a conformational change in the visual pigment to become **all-*trans*-retinal** (Fig. 27.25). This causes a conformational change in the cytoplasmic domain of rhodopsin. The associated heterotrimeric G-protein, called **transducin**, exchanges its bound GDP for GTP. The α-GTP complex detaches and activates a membrane-bound enzyme, in this case cGMP phosphodiesterase, which destroys cGMP (Fig. 27.26). This enzyme has a similar action to the cAMP phosphodiesterase shown earlier in Fig. 27.18(b). *Lowering* the cGMP level results in closure of the cation channels, the inflow of Na^+ ceases and the membrane potential increases to -70 mV – it becomes hyperpolarized. This triggers a nerve impulse in the optic nerve to flow to the visual centre in the brain and all in the time it takes you to see a light spot appear.

The recovery of the cell after illumination is complex. A primary event is the inactivation of the α-GTP subunit by its GTPase activity, which results in the reassembly of the trimeric

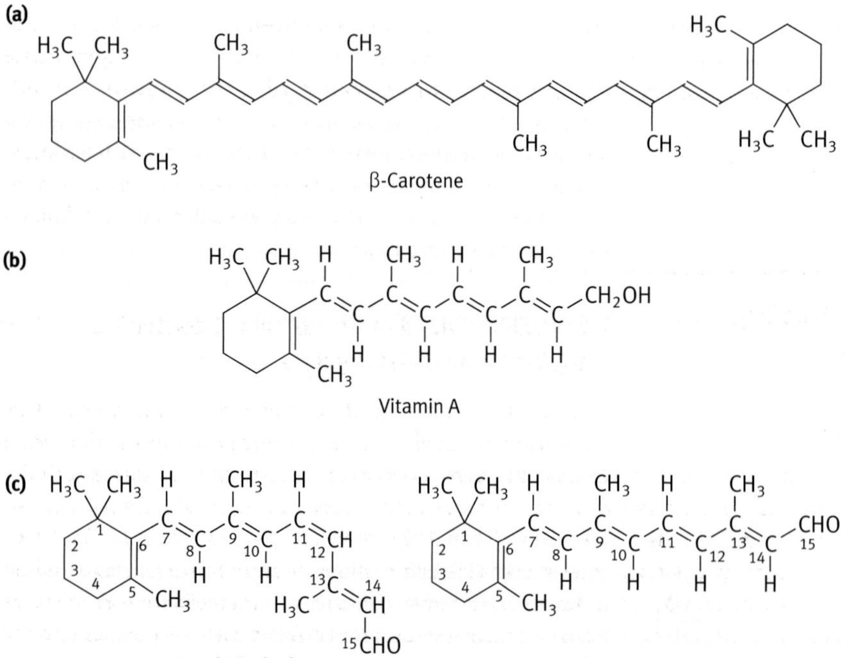

Fig. 27.25 (a) Structure of β-carotene, which is the principal accessory photosynthetic pigment in plants and a precursor of vitamin A in animals. **(b)** Structure of vitamin A. **(c)** Structures of 11-*cis*-retinal and all-*trans*-retinal. These structures are given for reference purposes only.

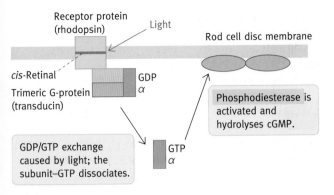

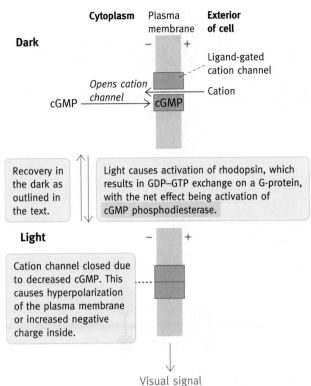

Fig. 27.26 The G-protein-coupled receptor involved in vision. The receptor consists of seven transmembrane helices with 11-*cis*-retinal as the chromophore. The cytoplasmic domain on light stimulation causes the trimeric G-protein transducin to exchange its GDP for GTP. The α-GTP subunit migrates to the membrane and activates a phosphodiesterase which hydrolyses cyclic GMP (cGMP). This results in closure of cation channels and cessation of Na⁺ inrush. The consequent increase in membrane polarization is converted into a signal in the optic nerve.

Fig. 27.27 Simplified diagram of the visual process. The action of light on rhodopsin results in the activation of cyclic GMP (cGMP) phosphodiesterase. Decreased cGMP results in the closure of a cation channel, hyperpolarization of the cell membrane, and an optic nerve impulse.

transducin. Recovery of the cell involves inactivation of the activated rhodopsin, lowering of the Ca^{2+} level in the cell, which stimulates cGMP synthesis, and recycling of the rhodopsin by a complex pathway.

To summarize the whole process (Fig. 27.27), in light, cGMP levels fall, cation channels close, hyperpolarization of cell membrane leads to visual signal in the optic nerve; after illumination, cGMP levels are restored, Na⁺ channels open, and the system is restored to be ready for the next photon.

Colour vision in the cone cells is due to the presence of three different visual pigments, proteins (all with 11-*cis*-retinal attachments), each more than 95% homologous in amino acid sequences to rhodopsin. The three pigments are responsible for red, green and blue vision respectively.

We now come to two different pathways where cGMP is acting as a true second messenger.

Signal transduction pathway using cGMP as second messenger

Membrane receptor-mediated pathways

The heart produces a neuropeptide hormone which regulates salt balance and affects blood pressure. It acts via cGMP as a second messenger. The reaction for the formation of cGMP is shown in Fig. 27.28. The hormone, **atrial natriuretic peptide**, produced by endothelial cells, combines with its membrane receptor on kidney cells whose *inner domain is a guanylate cyclase* and this is activated by a conformational change result-

ing from the attachment of the neuropeptide to the receptor (Fig. 27.29). The raised cGMP level mediates cell responses by activating protein kinases. This results in the appropriate cellular effects, including increased Na⁺ excretion by the kidneys. Hydrolysis of cGMP by a phosphodiesterase reaction analogous to that for cAMP (Fig. 27.18(b)) confers reversibility.

There is a second control system producing cGMP as a second messenger. (cGMP is, as stated, not definable as a second messing in the visual process.)

Nitric oxide signalling – activation of a soluble cytoplasmic guanylate cyclase

The second guanylate cyclase is present in the cytosol. This has a haem molecule as its prosthetic group to which binds a signalling molecule of surprising simplicity – **nitric oxide (NO)**. The haem molecule functions as a detector of NO at concentrations as low as 10^{-8} M and transduces the signal into cGMP synthesis from GTP. NO is produced from the arginine guanidino group by **nitric oxide synthase**, in endothelial cells lining parts of the vascular system and elsewhere. The NO diffuses into the smooth muscle of blood vessels, causing cGMP production,

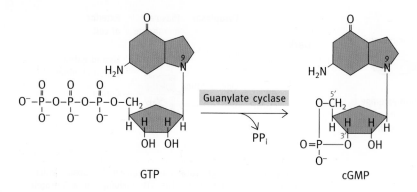

Fig. 27.28 Formation of 3′, 5′-cyclic GMP by guanylate cyclase.

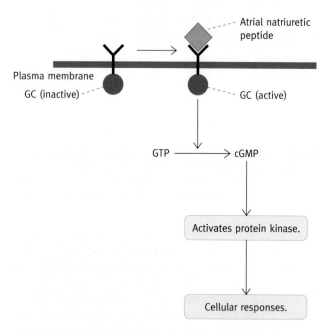

Fig. 27.29 Production of the second messenger cyclic GMP (cGMP) by a membrane receptor (R), activated by the atrial natriuretic peptide (blue). It has a single polypeptide chain with a membrane-spanning sequence and a guanylate cyclase catalytic site (GC) on the cytoplasmic face. On attachment of the signal peptide to the receptor, it undergoes a conformational change responses. which activates the cyclase and leads to the production of cGMP from GTP. This, *via* the activation of a protein kinase, leads to the cellular responses.

which, in turn, causes muscle relaxation and vessel dilatation. The NO has a lifetime of a few seconds. NO is also produced in response to shearing forces exerted by blood flow on the endothelial cells lining the vessels. This results in vasodilatation. Since NO is oxidized to higher oxidized states of nitrogen in seconds, it is a very locally acting (paracrine) hormone; being lipid soluble it escapes from cells producing it and enters adjacent cells. Trinitroglycerine, a drug long used in the treatment of angina, slowly produces NO thereby relaxing blood vessels (including coronary vessels) and reducing the workload of the heart. NO is part of a complex regulatory system with multiple physiological effects. The phosphodiesterase which destroys cGMP is inhibited by the drug sildenafil (**Viagra®**). This potentiates the effect of NO, production of the latter being increased by sexual stimulation. Dilatation of blood vessels in the penis aids erection.

It has been suggested recently (**Carey** *et al.* in Further reading) that control of the soluble guanylate cyclase is more sophisticated than hitherto believed. It seems that, in resting states, smooth muscle tone is maintained by a base low level of NO that combines with the haem prosthetic group and partially activates the cyclase. When there are bursts of NO production, such as caused by liberation of acetylcholine, the higher concentration of NO combines at a non-haem site and gives a transitory full activation of the cyclase. This results in immediate smooth muscle relaxation.

Summary overview

Figure 27.30 gives a simplified overview summary of the signalling pathways dealt with in this chapter with emphasis on the protein kinases and second messengers involved.

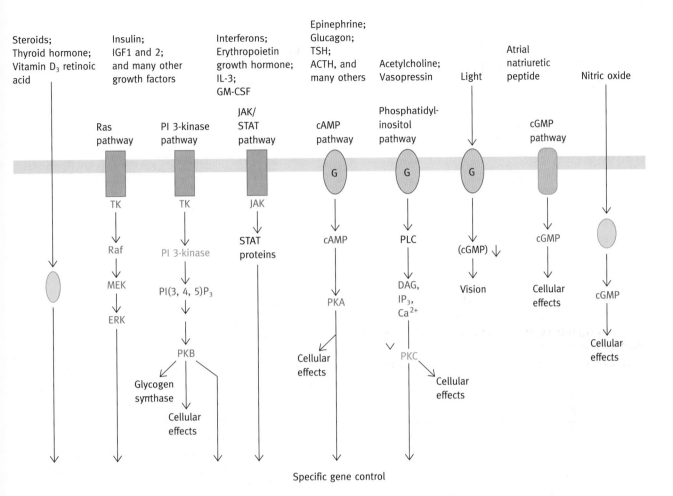

Fig. 27.30 Simplified summary diagram of the signal tranduction pathways. The figure shows only the kinases (red), which are activated by the receptors, and second messengers formed (green). Yellow receptors are intracellular. Receptors shown in red are phosphorylated on binding of the signal (TK, tyrosine kinase; JAK, Janus kinase). G indicates that the receptor is of the heterotrimeric G-protein-coupled type. The type of receptor shown in orange is unusual in that the guanylate cyclase which forms the second messenger is part of the receptor itself. Note that the pathways are of extraordinary complexity with multiple parallel pathways. Note also that the light receptor is an exception in that its effect is to reduce the level of cGMP which is not a second messenger in this context. EGF, epidermal growth factor; IGF, insulin-like growth factor; IL-3, interleukin 3; GM-CSF, granulocyte–macrophage-colony stimulating factor; TSH, thyroid-stimulating hormone; ACTH, adrenocorticotrophic hormone. The lists of growth factors, cytokines and hormones using the different pathways are illustrative examples only, not exhaustive lists.

 Summary

Cells have receptors that receive signals from other cells. In unicellular organisms such as yeast and bacteria this is limited in scope, but in animal cells such as mammalian ones, the number of signals needed to coordinate their activities is large. They control the most fundamental aspects of gene control, cell survival, programmed cell death, and cell division as well as metabolism. Cancer usually involves malfunction of signalling pathways.

The signals are variously proteins, peptides, steroids, and other lipid-related molecules, and nitric oxide. They are hormones, neurotransmitters, growth factors, and cytokines.

The signalling molecules bind to specific receptors of target cells and activate signalling pathways, which result in gene-control events in the nucleus and/or more direct effects on metabolism.

Steroids enter the cell directly and attach to intracellular receptors but water-soluble signals, the predominant class, combine with membrane-bound receptors exposed on the cell exterior. These span the membrane and are exposed also on the inside where they cause an allosteric change in their cytoplasmic domains. This leads to activation of signalling pathways within the cell.

There are two main classes of membrane receptor, tyrosine kinase-associated and G-protein-linked. The tyrosine kinase-associated receptors dimerize in the membrane on signal binding and tyrosine —OH groups of the cytoplasmic domains are phosphorylated. Protein phosphorylation is the predominant process in this class of signalling. Adaptor proteins bind to tyrosine phosphate groups by specific domains such as SH2 and link the receptors to other proteins of signal transduction pathways, again by specific domains such as SH3. Variable forms of these domains are found on large numbers of different proteins. Chains of signal transducing proteins are assembled into pathways that transmit the signal to the nucleus where they control gene activities.

The Ras pathway, universally found, is the prototype tyrosine kinase-associated pathway. It transmits signals *via* a serine/threonine kinase-amplifying cascade to gene-control proteins (transcription factors), which enter the nucleus. The Ras protein, which is one of the components of the signalling pathway, has a GTP/GDP switch. Ras is activated only when the receptor has a signal molecule attached to it. It involves a bound molecule of GDP being exchanged for GTP. However, the bound GTP molecule is slowly hydrolysed to GDP thus inactivating Ras and blocking the signal transduction pathway. This is a timing device to limit the effect of a signal activation event. It can be reactivated again only if the signal is still attached to the receptor. The signal is destroyed so that the pathway remains activated only so long as the signal is being produced, usually by other cells.

When the signalling molecule is no longer present, phosphatases deactivate the phosphorylated proteins. Overactivity of the Ras protein due to mutation impairing the GTP/GDP switch off is associated with a number of human cancers, a fact that is true of many of the signalling pathways (Chapter 30).

In another type of tyrosine kinase pathway, the JAK/STAT pathway, activated receptors cause phosphorylation of cytoplasmic STAT proteins that enter the nucleus and act as gene controllers. This is a direct pathway in contrast to the multi-component Ras pathway. Controls exist to reverse the signal so that runaway activation of the pathway does not occur. Insulin operates via another pathway involving PI 3-kinase present in the membrane.

G-protein-linked receptors are many and versatile; over half of the pharmaceutical drugs are targeted to them. The classical G-protein-associated pathway is that activated by epinephrine (adrenaline). The membrane receptors are linked on the cytoplasmic face to a heterotrimeric protein with α, β, and γ subunits. On receipt of a signal, the α subunit exchanges GDP for GTP and migrates to, and activates, a membrane-bound adenylate cyclase that produces the second messenger, cAMP. (Please see Chapter 16 for a fuller treatment of second messengers.) The signal is automatically cancelled after a brief period because the α subunit slowly hydrolyses the GTP to GDP. The subunit then detaches and rejoins its partners in the G-protein. The GTP/GDP switch is a molecular timing mechanism. In cholera, GTP hydrolysis is inhibited, thus inappropriately indefinitely extending the formation of cAMP, which causes the intestinal symptoms of the disease.

Other G-protein-associated receptors activated by their cognate signalling molecule cause the formation of different second messengers, the phosphatidylinositol cascade being an example. Here the second messengers are inositol triphosphate (IP_3), which causes Ca^{2+} release into the cytoplasm and diacylglycerol formation, which activates protein kinase C. The Ca^{2+} is important in many control systems.

Visual receptors in which photons are the signal also are G-protein-linked systems but in this case the signal causes a reduction in cGMP. This results in a signal in the optic nerve to the visual centre in the brain.

cGMP production is the second messenger of a different type of membrane receptor which when activated is itself a guanylate cyclase.

Nitric oxide signalling is different again in that it causes cGMP production but the receptor is located in the cytoplasm. Signals are cancelled by the hydrolysis of cGMP by a phosphodiesterase (inhibited by Viagra®).

Further reading

Struhl, K. (1998). Fundamentally different logic of gene regulation in eukaryotes and prokaryotes. *Cell*, **98**, 1–4.
A minireview looking in broad terms at the mechanisms of gene transcription control in the two classes.

Hunter, T. (2000). Signalling – 2000 and beyond. *Cell*, **100**, 113–27.
A major review that comprehensively discusses all aspects of cell signalling in clearly defined sections. Probably more suitable for the advanced student.

Various authors (2002). Mapping cellular signaling. *Science*, **296**, 1634–57.
Multiple, succinct coverages of G-protein pathways, T cell receptors, integrins, JAK/STAT, and PI 3-kinase pathways.

G-proteins

Neer, E. J. (1997). Intracellular signalling: turning down G-protein signals. *Curr. Biol.*, **7**, R31–3.
Reviews the family of regulators of G-protein signalling (RGS) which modulate the GTPase activity of the α subunits of G-proteins.

Dhanasekaran, N. *et al.* (1998). Regulation of cell proliferation by G proteins. *Oncogene*, 17, 1383–94.

Ross, E. M. and Wilkie, T. M. (2000). GTPase-activating proteins for heteromeric G proteins: regulators of G protein signaling (RGS) and RGS-like proteins. *Annu. Rev. Biochem.*, 69, 795–827.

Receptors

Pawson, T. (1994). Look at a tyrosine kinase. *Nature*, 372, 726–7.
A News and Views item on the three-dimensional structure of the insulin receptor.

Pitcher, J. A., Freedman, N. J., and Lefkowitz, R. J. (1998). G protein-coupled receptor kinases. *Annu. Rev. Biochem.*, 67, 653–92.
A comprehensive review of the densensitization (downregulation) of receptors. More suitable for advanced students.

Inflammatory response

Ghosh, S., May, M. J., and Kopp, E. B. (1998) NF-κB and Rel proteins: evolutionarily conserved mediators of immune responses. *Ann. Rev. Immunol.*, 16, 225–60.

Buckbinder, L. and Robinson, R. P. (2002). The glucocorticoid receptor: molecular mechanism and new therapeutic opportunities. *Curr. Drug Targets Inflamm. Allergy*, 1, 127–36.

Chen, G. and Goeddel, D. V. (2002). TNF-R1 signaling: a beautiful pathway. *Science*, 296, 1634–5.
Short review summarizing the engagement of tumour necrosis factor with its receptor and the pathway leading to the downstream events, apoptosis and NF-kB activation.

Yamamoto, Y. and Gaynor, R. B. (2004). IκB kinases: key regulators of the NF-κB pathway. *Trends Biochem. Sci.*, 29, 72–9.

Perkins N. D. (2007). Integrating cell signalling pathways with NF-kB and IKK function. *Nature Reviews. Mol. Cell Biol.*, 8, 40–62.
Comprehensive account of role in inflammatory response. NF = nuclear factor; IKK = NF-kB kinase.

Nitric oxide

Mayer, B. and Hemmens, B. (1997). Biosynthesis and action of nitric oxide in mammalian cells. *Trends Biochem. Sci.*, 22, 477–81.
A succinct review.

Signalling pathways

Nishizuka, Y. (1994). Protein kinase C and lipid signalling for sustained cellular responses. *FASEBJ.*, 9, 484–96.
Summary of the signalling pathways using inositol phospholipid hydrolysis, and the degradation of various other membrane lipid constituents.

Karin, M. and Hunter, T. (1995). Transcriptional control by protein phosphorylation: signal transmission from the cell surface to the nucleus. *Curr. Biol.*, 5, 747–57.
Compares the Ras pathway and the JAK pathway.

Quilliam, L. A., Khosravi-Far, R., Huff, S. Y., and Der, C. J. (1995). Guanine nucleotide exchange factors: activators of the Ras superfamily of proteins. *BioEssays*, 17, 395–404.
A fairly advanced but readable review of the control of Ras activity.

Burgering, B. M. T. and Bos, J. L. (1995). Regulation of Ras-mediated signalling: more than one way to skin a cat. *Trends Biochem. Sci.*, 22, 18–22.
Extensively reviews interaction of the Ras pathway with other signalling routes.

Elion, E. A. (1998). Routing MAP kinase cascades. *Science*, 281, 1625–6.
A 'perspectives' article briefly and clearly summarizing scaffold proteins. See also the two similar following articles (pages 1668 and 1671) by H. J. Schaeffer et al. and A. J. Whitmarsh et al., respectively.

Irvine, R. (1998). Inositolphospholipids: translocation, translocation, translocation. . . . *Curr. Biol.*, 8, R557–9.
Pleckstrin homology (PH) domains interact with membrane inositol phospholipids.

Mochly-Rosen, D. and Gordon, A. S. (1998). Anchoring proteins for protein kinase C: a means for isozyme selectivity. *FASEBJ.*, 12, 35–2.
Describes isozymes of the PKC family.

Whitmarsh, A. J. and Davis, R. J. (1998). Structural organisation of MAP kinase signalling modules by scaffold proteins in yeast and mammals. *Trends Biochem. Sci.*, 23, 481–5.
Reviews the role of scaffold proteins in the organization of different MAP kinase pathways.

Evans, D. R. H. and Hemmings, B. A. (1998). What goes up must come down. *Nature*, 394, 23–4.
A News and Views article dealing with the physical association of MAP kinases and phosphatases and the switch-off of signalling pathways.

Sceffzek, K., Ahmadian, M. R., and Wittinghofer, A. (1998). GTPase-activating proteins: helping hands to complement an active site. *Trends Biochem. Sci.*, 23, 257–62.
Reviews the GAPs or proteins that regulate the GTPase activities of small GTP-binding proteins involved in signal transduction.

Vanhaesebroek, B., Leevers, S. J., Panoyotou, G., and Waterfield, M. D. (1998). Phosphoinositide 3-kinases: a conserved family of signal transducers. *Trends Biochem. Sci.*, 22, 267.
Describes several families of PI 3-kinases and their roles in signalling pathways.

Krugman, S. and Welch, H. (1998). PI 3-kinase. *Curr. Biol.*, 8, R828.
A single-page quick guide giving salient features of the signalling pathways activated by this enzyme.

Alessi, D. R. and Cohen, P. (1998). Mechanism of action and function of protein kinase B. *Curr. Opin. Genet. Dev.*, 8, 55–62.
Deals with the activation of the kinase and its role in insulin signalling.

Cohen, P. (1999). The Croonian Lecture 1998. Identification of a protein kinase cascade of major importance in insulin signal transduction. *Phil. Trans. Ser. B. Roy. Soc. Lond.*, **354**, 485–95. *A readable review of this work.*

Williams, J. G. (1999). Serpentine receptors and STAT activation: more than one way to twin a STAT. *Trends Biochem. Sci.*, **24**, 333–4. *A concise summary of the way in which membrane receptors activate STAT proteins.*

Misra, S., Miller, G. J., and Hurley, J. H. (2001). Recognising phosphatidylinositol 3-phosphate. *Cell*, **107**, 559–62. *Minireview on signalling role of PI(3)P in membranes.*

Brazil, D. P., Park, J., and Hemmings, B. A. (2002). PKB binding proteins: getting in on the Akt. *Cell*, **111**, 293–303. *Reviews diverse roles of PKB (also known as Akt).*

Ceulemans, H., Stalmans, W., and Bollen, M. (2002). Regulator-driven functional diversification of protein phosphatase-1 in eukaryotic evolution. *BioEssays*, **24**, 371–81.

O'Shea, J. J., Gadina, M., and Schreiber, R. D. (2002). Cytokine signaling in 2002: new surprises in the Jak/STAT pathway. *Cell*, **109**, S121–31. *Reviews the field and covers negative regulation.*

Hammes, S. R. (2003). The further redefining of steroid-mediated signaling. *Proc. Natl.Acad. Sci. U.S.A.*, **100**, 2168–70. *A new development – membrane steroid receptors.*

Park, S.-H., Zarrinpar, A., and Lim, W. A. (2003). Rewiring MAP kinase pathways using alternative scaffold assembly mechanisms *Science*, **299**, 1061–4.

Pawson, T. and Nash, P. (2003). Assembly of cell regulatory systems through protein interaction domains. *Science*, **300**, 445–52. *Advanced research-level review of modular domains and signalling proteins.*

Ptashne, M. and Gann, A. (2003). Imposing specificity on kinases. *Science*, **99**, 1025–7.

Parker, P. J. and Murray-Rust, J. (2004). PKC at a glance. *J. Cell Sci.*, **117**, 131–2. *Survey of the various members of the PKC family, phylogenetic relationships, and domain structures. Poster and interactive version online at **www.jcs.biologists.org/***

Hancock, J. F. and Parton, R. G. (2005). Ras plasma membrane signalling platforms. *Biochem. J.*, **389**.

Meder D. and Simons., K. (2005). Ras on the roundabout. *Science*, **307**, 1731–3. *A perspectives article.*

Carey, P. L., Winger, J. A., Derbyshire, E. R. and Marietta, M. A. (2006). Nitric oxide signalling: no longer simply on or off. *Trends Biochem. Sci.*, **31**, 239.

Moscat, J. *et al.* (2007). Signal integration and diversification through p62 scaffold protein. *Trends Biochem. Sci.*, **32**, 95–100

Vogt, P. K. *et al.* (2007). Counter-specific mutations in PI 3-kinase. *Trends Biochem. Sci.*, **32**, 342–9. *PI 3-kinase mutation frequencies exceeding 30% are found in diverse human cancers.*

Haemopoietic growth factors

Metcalf, D. (1991). The 1991 Florey Lecture. The colony-stimulating factors: discovery to clinical use. *Phil. Trans. Ser. B Roy. Soc. Lond.*, **333**, 147–73. *An in-depth review of the progression from unidentified factors to their widespread clinical use.*

Interferons

Stark, G. R. *et al.* (1998). How cells respond to interferons. *Annu. Rev. Biochem.*, **67**, 227–64. *A comprehensive review.*

Calcium homeostasis

Bootman, M. D. and Berridge, M. J. (1995). The elemental principles of calcium signalling. *Cell*, **83**, 675–8. *Minireview on the sources of Ca^{2+} for generating signals: Ca^{2+} release from intracellular stores and Ca^{2+} entry across the plasma membrane.*

Monteith, G. R. and Roufogalis, B. D. (1995). The plasma membrane calcium pump – a physiological perspective on its regulation. *Cell Calcium*, **18**, 459–70. *The role of the plasma membrane Ca^{2+}/Mg^{2+}-dependent ATPase in cellular signalling and in intracellular calcium homeostasis.*

Berridge, M. A., Bootman, M. D., and Lipp, P. (1998). Calcium – a life and death signal. *Nature*, **395**, 645–7. *A News and Views feature which puts into perspective the role of calcium in its universal signalling roles.*

Hoeflich, K. P. and Ikura, M. (2002). Calmodulin in action: diversity in target recognition and activation mechanisms. *Cell*, **108**, 739–42. *Structural studies on calmodulin complexes with adenylate cyclase and Ca^{2+}-activated K^+ channel.*

Taylor, C. W. (2002). Controlling calcium entry. *Cell*, **111**, 767–9. *Reviews Ca^{2+} channels and their control.*

Problems

1 How is cAMP production controlled? In particular, describe the role of GTP in the process. What is the relevance of the latter to cholera?

2 How does cAMP exert its effect as second messenger?

3 How does nitric oxide exert its controlling effect?

4 cAMP and cGMP are not the only second messengers. Describe another system.

5 Make a general comparison of the activation mechanism of three types of receptor: (a) the epinephrine receptor activating adenylate cyclase; (b) the EGF receptor of the Ras pathway; and (c) the interferon receptor.

6 What are the classes of signalling molecules between cells in the mammalian body?

7 Lipid-soluble signalling molecules directly enter cells but lipid-insoluble ones do not. Does this mean that the two types act in totally different ways? Explain your answer.

8 If a protein is found to have an SH2 domain, what is its likely function in the cell?

9 In a general way, without detail, compare the salient features of a Ras signalling pathway with a JAK/STAT pathway.

10 G-protein-associated receptors are tremendously versatile in the signals the different receptors respond to. Explain the principle of G-protein signalling and why it can be so versatile.

11 Can cGMP be properly regarded as a second messenger in the visual system? What about the nitric oxide signalling system?

12 Most tyrosine kinase-associated receptors dimerize on receipt of a signal. What is the exception to this?

13 What is the role of GTP/GDP switches? Illustrate your answer with examples from cell signalling, protein targeting and protein synthesis respectively.

14 Ras is a GTPase but it is not called a G protein. Explain why.

Chapter 28

Manipulating DNA and genes

DNA-manipulation techniques are a major part of a revolution in medical and biological sciences in general. Areas referred to as **recombinant DNA technology, gene technology, genetic engineering,** and **biotechnology** depend on DNA manipulation. At one time, doing anything with DNA seemed to verge on the impossible because of the colossal size of the molecules and because it appeared to be an endless string of nucleotides with few distinctive characteristics. It has turned out, surprisingly, that DNA can be manipulated more easily than other macromolecules such as proteins or complex polysaccharides. Despite the almost awe-inspiring biological roles of DNA, the chemistry of the molecule is relatively simple. It has a straightforward structure with only four different nucleotide units. The major property of DNA that makes it susceptible to manipulation is that it can combine accurately with other pieces of DNA by specific Watson–Crick base pairing. There also exists a wealth of different enzymes that nature uses in defense, DNA repair and replication that seem almost tailor-made for recombinant DNA technology. This means that DNA can be cut, the pieces joined together, duplicated, sequenced, and detected by labelled complementary DNA probes. The technology is reminiscent of the cut, paste, copy, and find procedures on a computer.

When two molecules of DNA of different origin are covalently linked together, end to end, the resultant molecule, which does not occur naturally, is known as a **recombinant DNA** molecule. This is different from natural genetic recombination in cells.

DNA manipulations are very often the most powerful approach available in biological and medical sciences. To give a few examples, the technologies make it possible to isolate individual genes, to determine their base sequences, to manipulate the base sequences in any desired way, to transfer genes back to their original source or from one species to another, and to detect genetic abnormalities. It is possible to produce proteins such as human hormones or other factors using bacteria, or other cells loaded with extra genes, as protein factories. The proteins can be produced in virtually unlimited amounts even though they may occur in the body only in very small amounts. Amino acid sequences of proteins may be modified by changing the DNA coding for them. Minute amounts of DNA pieces can be rapidly amplified. This is used in forensic science in which human DNA is 'fingerprinted' for identification purposes. The base sequence of the coding region of genes permits deduction of the amino acid sequence of proteins (and is often the easiest way to determine them). Evolutionary relationships can be traced from DNA sequences and since DNA has some degree of stability under favourable circumstances, studies on ancient Egyptian mummies and even extinct fossilized organisms are possible. The general biological applications are almost unlimited. This applies to the therapeutic industry where, together with protein technologies, it plays a crucial role, and is the basis of much of the biotechnology industries. The new technology of DNA chips, or microarrays as they are more properly called, enables the expression of large numbers of genes to be studied simultaneously.

The volume of information on DNA now available would be of little avail unless there were mechanisms for storing, retrieving, and analysing it. This brings us to the international DNA databases in which information of all kinds related to DNA is recorded as it becomes available from research. The databases are freely available on the Internet, together with software programs for retrieving and interrogating the data in a wide variety of ways. In fact, 'mining the databases' has become an important research activity in its own right. Computational skills as well as a knowledge of molecular biology are required for this. The DNA databases complement the protein databases already described (page 81). Use of the information in databases is referred to as **bioinformatics** and increasingly there are specialized courses in this subject. More information on this aspect is given at the end of this chapter.

Basic methodologies

Some preliminary considerations

It is usual to talk of a fragment or a piece of DNA being sequenced, cloned, or whatever. In general manipulations require multiple copies of such pieces. Although minute actual quantities of DNA are handled, 1 mole of any chemical compound contains 6.03×10^{23} molecules. Procedures such as DNA sequencing will be incomprehensible unless this is kept in mind. If we consider a single typical gene or other piece of DNA in which we are interested, there will be two copies of it per diploid cell among another approximately 25,000 genes in human samples and these constitute only about 20% of the total DNA of a eukaryotic genome. This means that it is impracticable to isolate a gene or other piece of DNA directly from a cell lysate as one does for proteins. Therefore, recombinant DNA technology involves amplifying (multiplying) the piece of interest. Cloning is the main method of doing this but a newer technology (the polymerase chain reaction) can often achieve it with greater ease and rapidity.

Most of the manipulations of DNA are performed using enzymes. These have the required specificity and can be applied to minute amounts of material handled. A wide range of enzymes are readily available (polymerases of various types to synthesize DNA, ligases to join pieces, kinases to add phosphoryl groups, phosphatases to remove them, nucleases of various types, reverse transcriptase to copy RNA into DNA). The list is large. An important exception is the nonenzymic chemical synthesis of DNA pieces with sequences made to order by automatic machines.

Genes and other pieces of DNA of interest are parts of gigantic molecules, the chromosomes, and a first step is to obtain reproducible fragments of a size that can be handled.

Cutting DNA with restriction endonucleases

The only thing that distinguishes a gene or other DNA section from all the other DNA is its base sequence. The discovery of a group of enzymes in different bacteria (see below) which hydrolyse DNA at specific sequences was the vital first development that opened the way to manipulating DNA.

The class of enzymes called **restriction enzymes** *recognize short sequences of bases* so as to make cuts at precise points in both strands of a DNA molecule. Different bacteria have different restriction enzymes that recognize different base sequences in the DNA and therefore have different cutting sites. The bacteria have these enzymes as protection against bacterial DNA viruses (bacteriophages). When these infect a cell, they insert their own DNA, which takes over the targeted cell, and directs its biochemical activities to synthesizing new virus particles, which are then released. The cell's restriction enzymes cut the invading DNA, which destroys the phage. There will be large numbers of short base sequences in the infected cell's DNA identical to the sequence targeted by the restriction enzyme. So why doesn't the latter destroy the cell's own DNA? They protect themselves by methylating A or C bases in the restriction sites in their own DNA; the enzymes no longer recognize the methylated sites. The sequences in the DNA of an invading virus that has been produced in a strain of cell with different restriction enzymes will not be protected by methylation against those of

the target cell it is invading and will be attacked. The infection will be restricted. Over 100 restriction enzymes cutting at specific base sequences are now known so that we now have great control over where we cut DNA. To illustrate this point, an enzyme from *Escherichia coli* cuts double-stranded DNA at the sequence:

$$5'G{\downarrow}AATTC3'$$

$$3'CTTAA{\uparrow}G5'$$

and that from *Bacillus amyloliquefaciens* cuts at the sequence:

$$5'G{\downarrow}GATCC3'$$

$$3'CCTAG{\uparrow}G5'$$

These two enzymes make staggered cuts in the two DNA strands but others make straight-through cuts. *HAE*III is an example, producing 'blunt-ended' fragments:

$$5'GG{\downarrow}CC3'$$

$$3'CC{\uparrow}GG5'$$

Different enzymes recognize sequences four to eight base pairs in length. The longer-sequence-specific enzymes have fewer sites in a given DNA molecule and therefore cut the DNA into longer fragments than do the enzymes specific for shorter sequences.

The enzymes are named after the bacterium (and bacterial strain) of their origin and a Roman numeral where more than one enzyme occurs in that species. Thus the three enzymes mentioned above are called *Eco*RI, *Bam*HI, and *Hae*III, respectively. *Eco*RI was the first to be isolated from *E. coli* strain R.

Restriction enzymes make it possible to cut DNA at base sequences with precision, producing defined fragments that can be characterized in a preliminary way simply by their size. With so many different restriction enzymes now available commercially, each cutting at a different DNA sequence, so that you can always find one suitable for your purpose.

Separating DNA pieces

DNA pieces can be separated by **gel electrophoresis**. For this purpose, the pores in the gel must be sufficiently large to allow the DNA molecules to migrate through them. For shorter pieces of DNA, up to about 1000 base pairs in length, **polyacrylamide** gels are suitable, and the apparatus is basically the same as used for proteins (page 74). DNA molecules differing by a single nucleotide are separated. Each phosphate group in DNA carries a negative charge so that molecules migrate to the anode, smaller molecules moving faster than large ones due to molecular sieving.

For longer pieces of DNA, **agarose** gels are used, because it is not possible to create stable polyacrylamide gels with sufficiently large pores. Agarose is a passive uncharged seaweed polysaccharide. Each phosphate group in DNA carries a negative charge so that molecules migrate to the anode. The regular repeat of the phosphate groups means that all DNA molecules have the same charge to mass ratio. As a consequence, the rate or speed of migration depends wholly on the size of the molecule, such that smaller molecules move faster than the larger molecules due to molecular sieving.

Visualizing the separated pieces

If all you want to do is to locate any DNA it may be detected by staining with fluorescent DNA-binding dyes such as ethidium bromide (which intercalates between the bases) and viewing the fluorescence in UV light. However, if the mixture was derived from chromosomal DNA there will be large numbers of pieces spread out all over the gel and staining in this way will give a meaningless blur.

Detection of specific DNA fragments by nucleic acid hybridization probes

On page 317 we described the self-annealing properties of DNA strands. If a mixture of different pieces of double-stranded DNA is heated, the hydrogen bonds holding the Watson–Crick base pairs together are disrupted and the strands separate. On cooling, the pieces find their original partners again by base pairing, so that only correctly base-paired double strands are reformed. This theme of specific base pairing is central to recombinant DNA work.

The specificity of hybridization makes it a sensitive tool for identifying specific sequences irrespective of how many other molecules are present, so that a single gene sequence for example can be detected in the DNA from an entire genome. To do this a **hybridization probe** is needed; a probe is a piece of DNA complementary to the one you wish to find, and with some sort of label incorporated in it, such as radioactivity or fluorescence. A probe optimally needs to be at least 20 nucleotides long to give sufficient hydrogen-bonding attachment and is often hundreds of bases long. Short probes can be synthesized chemically by an automatic machine or a longer probe may be a cloned fragment (see below). The DNA being probed is first rendered single stranded by heat or NaOH. The hybridization is carried out in defined conditions of temperature and ionic strength. The sensitivity of hybridization to detect DNA sequences can be varied. If you want only perfect base pair matching to occur, a temperature just below the melting point of a double helix is used where a single mismatch will prevent hybridization (these conditions are known as high stringency). At lower temperatures hybridization will occur despite imperfect matching. The degree of stringency becomes important in using single nucleotide polymorphisms (SNPs) as genetic markers for locating disease-causing genes (page 472).

The probe may be labelled with radioactivity. This may be achieved by enzymically adding a radioactive phosphoryl group

(^{32}P) or synthesizing the probe from radioactive nucleotides. Fluorescent tags are often now preferred to avoid using radioactivity. Methods for obtaining a suitable probe will vary with the individual experiment.

Southern blotting

Hybridization probing is often used to visualize DNA molecules that have been separated by electrophoresis. Probing can not be done conveniently on either agarose or polyacrylamide gels, so it is necessary to first transfer the DNA fragments to an alternative supporting medium, typically a nylon membrane, while retaining their pattern of distribution. This technique is known as Southern blotting after the discoverer (Edwin Southern). The DNA in the gel is first made single-stranded by exposure in dilute alkali, and then transferred to the nylon membrane giving an exact print of the gel. In the case of a radioactive probe, the membrane is autoradiographed by exposing it to an X-ray film, which is then developed. A band on the film indicates the position of the piece of DNA of interest.

The blotting method can also be used to detect mRNA using a labelled DNA hybridization probe. It has been used as a semiquantitative assay method. The method is known as **Northern blotting** as a play on the name of Southern blotting.

A further application is called **Western blotting**. This is used for proteins, not DNA, by using a specific antibody.

Sequencing DNA

Determination of the base sequence is of central importance. The original method is based on enzymic replication of DNA. It is called the **dideoxy method** or **chain-termination method**. It was developed by Fred Sanger of Cambridge, UK. He was awarded a second Nobel Prize for DNA sequencing in 1980.

The principle of DNA sequencing by the chain termination method

The sequencing procedure requires that the piece of DNA is copied, *in vitro*, by DNA polymerase. This requires (apart from the four deoxynucleoside triphosphates (dNTPs)) that the DNA to be copied is single stranded, and that a primer is hybridized to the start site because DNA polymerase cannot initiate new chains. Duplex DNA is rendered into single-stranded form by treatment with NaOH or heat. The single-stranded DNA to be sequenced is incubated with primer, a suitable DNA polymerase, dATP, dGTP, dCTP, and dTTP. The products are separated by polyacrylamide gel electrophoresis to separate DNA molecules. Each addition of a nucleotide alters the migration of a DNA piece so that chains form separate bands, each being one nucleotide different in length from the next.

The next point is the crucial one; it concerns **dideoxy** derivatives of nucleoside triphosphates (**ddNTPs**), because if one of

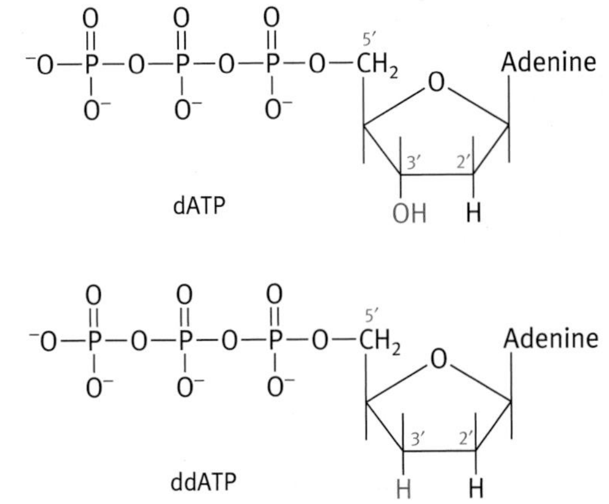

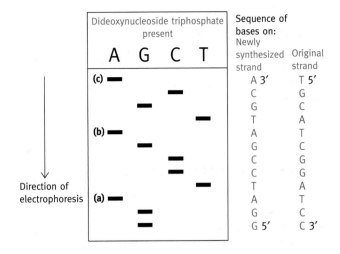

Fig. 28.1 The structures of deoxyATP (dATP) and dideoxyATP (ddATP). The absence of the 3′-OH group means that, when a ddNTP is added to a growing DNA chain, the chain is terminated.

Fig. 28.2 An autoradiograph sequencing gel. The sequence is read from the bottom to the top. **(a)**, **(b)**, and **(c)** are the bands produced in the presence of dideoxyATP. See text for explanation. Manual sequencing has been largely replaced by the automated method described below, but the basic principle remains the same.

these molecules is added to a growing DNA chain, synthesis stops. This is because DNA polymerase adds a nucleotide to the 3′-OH of a growing DNA chain and ddNTPs lack the 3′-OH group (Fig 28.1). Hence although they can still be added to a chain *via* their 5′-phosphate, the chain is then terminated.

We remind you that when we talk of a 'piece' of DNA being sequenced, multiple copies of that piece are involved in the experiments. Even a minute amount of DNA contains a large number of individual molecules. Suppose that we have in the copying process all four dNTPs plus a small amount of *one* ddNTP. If we take ddATP as an example, every time addition of an A is specified by the template strand, most of the new chains will have a normal adenine nucleotide added, and will continue to grow, but a fraction will, by chance, have a dideoxy form of the adenine nucleotide added, thus terminating those particular chains. (The fraction terminated will depend on the relative proportions of dATP and ddATP.) This ratio is adjusted so that the terminated chains are sufficient in number to be detected as a separate band on an electrophoretic gel by autoradiography. The rest of the molecules will go on being added to until another A is due to be added and the same will happen again.

How is this interpreted as a base sequence?

Suppose the piece of DNA being sequenced has a sequence with Ts placed as shown:

3′ X—X—T—X—X—X—X—T—X—X—T 5′

If, during the copying ddATP is present (along with dATP) such that, at the addition of each A, a small proportion of the growing chains are terminated, the copying will produce the following population of chains (attached to the primer):

(a) 5′X—X—ddA 3′

(b) 5′X—X—A—X—X—X—X—ddA3′

(c) 5′ X—X—A—X—X—X—X—A—X—X—ddA3′.

On a sequencing gel these bands will be seen as the bands in Fig. 28.2 (left-hand column).

If a second incubation contains ddTTP instead of the ddATP, an analogous set of chains terminating in T will be produced and, similarly, terminating in C and G if ddCTP or ddGTP, respectively, are present in duplicate incubations. All four incubations are run side by side on a gel, giving bands shown in Fig. 28.2. From this the base sequence of the piece of DNA can be read off from the bottom of the gel upwards as the sequence of the newly synthesized chain (and therefore of the partner to the template strand being sequenced) rather than of the template strand itself. This gives the sequence in the 5′ → 3′ direction, since synthesis always proceeds in that direction.

Automated DNA sequencing

At first sequencing was done manually with radioactive labels. Now an automated procedure is used but the basic principle is the same. In this the four ddNTPs are labelled each with a differently coloured fluorescent label. Electrophoretic separation of the chains is done in fine capillaries containing a fluid matrix rather than a gel. Resultant separations are automatically scanned for the different colours, and a computer prints out the base sequence. Figure 28.3 shows a section of a sequencing run from an automated sequencer. It is now usual for research workers to send a sample of DNA with a primer in a small plastic tube to a sequencing service and wait for the answer.

The development of automated sequencing has made it possible to sequence the entire 3.2 billion bases in human DNA by the international Human Genome Project consortium in 2003. Many other genomes including *E. coli*, yeast, *Drosophila*, mouse, the rice plant, *Arabidopsis* (a favourite plant for genetic studies), and others are also available. Today, because of the Human Genome Project and other programmes, DNA work is

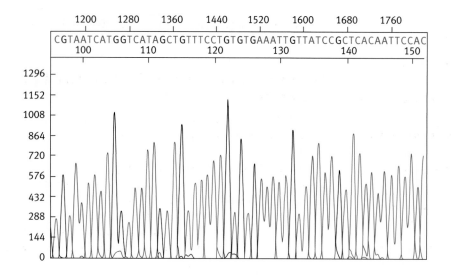

Fig. 28.3 Graph from automatic sequence analyser. Fluorescence detection of fragments produced by the dideoxy method, the fluorescence being of a different colour for each of the four ddNTPs. Kindly provided by Arthur Mangos and Dr Z. Rudzki, Molecular Pathology, Institute of Medical and Veterinary Sciences, Adelaide, Australia.

therefore often performed with sequence information already available in databases.

Emulsion DNA sequencing

This is a recent advanced method that is capable of rapidly sequencing whole genomes, which may be sequenced in a few weeks. It is not designed for the relatively short pieces of DNA used in laboratory cloning experiments. The whole genome is fragmented randomly by enzymic digestion or sonication and all of the pieces sequenced together. The process starts in an oil–water emulsion in which each minute droplet acts as a separate compartment for polymerase chain reaction (PCR) amplification (see below). Each reaction product becomes bound to a minute DNA-binding bead and sequenced by non-Sanger procedures so that large numbers of sequences are determined simultaneously. The sequences are assembled into the complete genome electronically by detecting overlaps in the sequences.

Amplification of DNA by the polymerase chain reaction

This technique has assumed enormous importance in DNA studies of all kinds because it enables a stretch of DNA to be picked out and quickly amplified exponentially. It is so sensitive that a few molecules of DNA hidden among millions of others can be detected selectively and amplified. The essential requirement is that the sequences flanking each end of the selected section are known so that complementary DNA primers can be obtained. The specificity of selecting a particular section of DNA from all the others depends on this. The principle of the method is that the selected section, primed on both strands of the DNA, is enzymically replicated from the four dNTPs and then this is repeated with the old and the new strands all being replicated from the added primers giving an exponential increase of the section. The incubation mixture contains a heat stable DNA polymerase and excess primers, as well as the dNTPs, so that no more components need be added throughout the cycles. Progression through the cycles is determined solely by the temperature changes.

To describe the method let us consider a stretch of DNA that you want to amplify. This is shown in green in Fig. 28.4(a). The parent strands are separated by heating (Fig. 28.4(b)). For simplicity we will illustrate in Fig. 28.4 only what happens to one of the two parent strands since the same applies to both.

Two chemically synthesized DNA primers, about 20 nucleotides in length, complementary to the 3′ flanking sequences on the two template strands are added in large excess (because each round incorporates the primers into the new strand). These attach on cooling to priming sites (one shown in Fig. 28.4(c)). Replication to the end of the piece occurs, producing a 'long' product (Fig. 28.4(d)). *That is the end of the first cycle.*

The next cycle is started by heating to separate the strands (Fig. 28.4(e)) and after cooling, primers attach to both strands and replication from these occurs. The *Taq* polymerase used is heat stable and does not need to be added again. The original strand once again gives rise to a long product but the *first* long product will be copied in the opposite direction to produce a 'short product' (Fig. 28.4(f)) *which is the wanted copy.* At the end of the second cycle we therefore have the original strand and two long products and one short product. *That is the end of the second cycle.*

The third cycle is started by again heating to give strand separation; again primers attach on cooling (Fig. 28.4(g)). Replication produces another short strand giving a duplex of the latter (Fig. 28.4(h)). From now on the short strands increase exponentially as the number of cycles increases but the long

a few minutes. Amplification of many millions of fold is easily achieved (2^n fold where n = number of cycles).

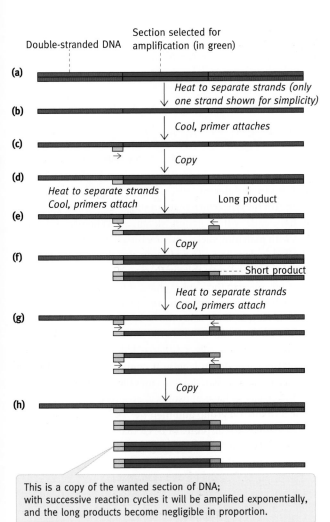

Double-stranded DNA

Section selected for amplification (in green)

(a)

Heat to separate strands (only one strand shown for simplicity)

(b)

Cool, primer attaches

(c)

Copy

(d)

Heat to separate strands
Cool, primers attach — Long product

(e)

Copy

(f)

Short product

Heat to separate strands
Cool, primers attach

(g)

Copy

(h)

This is a copy of the wanted section of DNA; with successive reaction cycles it will be amplified exponentially, and the long products become negligible in proportion.

Fig. 28.4 The basic principle underlying amplification of a DNA section by the polymerase chain reaction (PCR). Incubations contain four dNTPs, a heat-stable DNA polymerase, and primers. **(a)** The green bars represent the section chosen for amplification by the use of appropriate primers (arrows), each complementary to the 3′ flank of the section in one strand to be amplified. **(b)** Heating separates the strands and, from this point, to keep the diagram to a manageable size, amplification of only one strand is shown; the process amplifies both strands. Cool so that the primer appropriate for this strand binds. DNA is replicated (in this case beyond the end of the desired piece). Heating separates the strands. **(f)** The new strand is primed. **(g)** The new strand is replicated. **(h)** The separated wanted section. You can see that we now have reproduced our selected piece. With about 25 rounds of replication, heating, cooling, priming, and synthesis, the piece can be amplified a millionfold. Note that, every time priming and copying occurs, all of the original and new molecules are so primed and copied that a detailed diagram of what happens becomes rather involved. The essential point is that an enormous amplification of the selected piece occurs.

products increase only arithmetically. The latter therefore becomes a minute proportion of the total and can be ignored.

The entire process is automated with the heating and cooling performed in as many cycles as you wish, each cycle taking only

Analysis of multiple gene expression in cells using DNA microarrays

One of the newest additions to the armoury of molecular biology tools are the **DNA microarrays**, colloquially referred to as **DNA chips**.

Northern blotting can be used to detect the transcription of single genes by hybridizing mRNA to single-stranded DNA probes of the gene from which they were transcribed. Genomics however is concerned with *the degree of* transcription (and expression) of many genes, even thousands, at any one time and in any one tissue so that overall patterns of gene expression, or large parts of it, can be studied.

The interest stems from the realization that the interactions of many genes determine the life of cells and organisms. If the degree of expression of many genes could be determined it could for example throw new light on gene control in normal and diseased situations. As an example, comparing total gene transcription in normal and cancer cells of the same tissue might lead to better classification of the cancers and more specifically tailored therapeutic regimes. A human has about 25,000 genes whose expression might be looked at. How could this be done for large numbers of them at any one instant of time? DNA microarray technology performs this seemingly impossible task.

The principle is that an array of pieces of DNA which are specific targets are spotted on to a DNA-binding surface called a chip, most often of glass. (*Note that these are referred to as the probes.*) The spots are minute in size and are applied robotically or synthesised *in situ*, the location of each spot being identifiable and each sequence corresponding to one in a known specific gene. Many thousands of such sequences can be placed on a postage-stamp-sized chip. The specific target sequences in the spots may be cDNA copies of genes or parts of them amplified by PCR or synthetic oligonucleotides, synthesized on the glass *in situ* by robotic means. The availability of whole genome sequences means that the sequence of any identifiable gene is known. Ready-made microarrays specific for groups of genes such as those relevant to cancer are available commercially.

Suppose your aim is to compare gene expression in cancer cells with that in normal cells of the same type, using an appropriate DNA microarray. mRNA is isolated from the two cell samples and cDNA libraries (page 468) made from them. The cDNAs are labelled with fluorescent dyes, red in the case of the cDNA made from cancer cells and green for the normal control cell samples. The strands are separated by heat and allowed to hybridize with the target probe sequences on the microchip. The chip is washed and scanned automatically so that the intensity of the different-coloured fluorescence on each spot is quantified. Genes whose expression is increased in the cancer cells as compared with normal ones appear as red fluorescent

Fig. 28.5 Gene expression analysis using a DNA microarray (or DNA chip). A library of about 19,000 oligonucleotide probes for human gene coding regions was hybridized with complementary DNA from two cell lines, labelled with a green and red fluorescent dye respectively. A small section of it was enlarged for this figure. See text for explanation provided. We are grateful to Mark Van der Hoek, of the Adelaide Microarray Facility, Institute of Medical and Veterinary Sciences, University of Adelaide, Australia, who kindly supplied this photograph.

spots because more cDNA has hybridized to the gene as compared to that from the normal cells. In the reverse situation where expression of a gene in the cancer cells is decreased relative to the normal control, the spot is green and if there is no difference between the two types of cells the spot is yellow. Dark spots indicate little or no expression of either type. A computer records the results in terms of amount of specific mRNAs in each of the spots in the microarray. Since each of these corresponds to a known gene the method gives a global picture of the expression of all the genes at any one time in the cancerous and normal tissues.

Figure 28.5 shows a comparison of expression when a cDNA library was probed with reverse-transcribed labelled cDNA generated from RNA extracted from different cell lines. Different patterns of expression are found.

Joining DNA to form recombinant molecules

One of the central techniques is to create new DNA molecules by joining together pieces that are not found as such in nature. The new molecules are recombinants. There are various ways of doing this and which is used will depend on the actual experiment. The simplest is known as sticky-end hybridization. A restriction enzyme that makes a staggered cut such as *Eco*RI produces the cohesive or sticky ends; these are small complementary overhangs which can re-anneal under suitable conditions.

$$
\begin{array}{l}
-X-X-G\!\downarrow\!A-A-T-T-C-X-X-\\
-X-X-C-T-T-A-A\!\uparrow\!G-X-X-
\end{array}
$$

$$Eco\text{RI}\downarrow$$

$$
\begin{array}{ll}
-X-X-G & A-A-T-T-C-X-X-\\
-X-X-C-T-T-A\!=\!A & G-X-X-
\end{array}
$$

Sticky ends

If DNA from different sources is cut with the same restriction enzyme (of the type which makes a staggered cut) the fragments will have ends with matching stickiness. When two pieces of DNA with matching sticky ends are mixed, they join together as indicated in Fig. 28.6. The nicks in the chains are covalently sealed by a DNA ligase; this synthesizes a bond between the 3′-OH of one piece to the 5′-phosphate of the other.

Sometimes it is necessary to join pieces of DNA with blunt ends. This can be done using the ligase enzyme from T4 phage. Another way is to add linkers to the blunt ends you wish to join. A linker is a chemically synthesized duplex DNA sequence about 10 nucleotides in length. It contains sequences that are specifically cut by restriction enzymes that produce sticky ends. The linker has to have phosphoryl groups enzymically added to the 5′-hydroxyl group and then the T4 ligase can blunt-end ligate the linker to the pieces of DNA that you wish to join. The linkers are then cut with an appropriate restriction enzyme and the pieces joined by sticky-end hybridization and ligation.

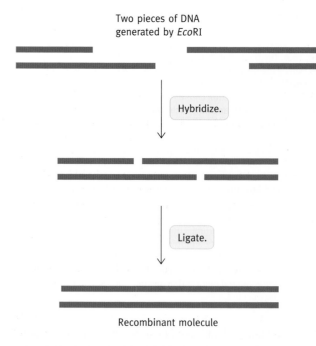

Two pieces of DNA
generated by *Eco*RI

Hybridize.

Ligate.

Recombinant molecule

Fig. 28.6 Principle of construction of recombinant DNA molecules by sticky-end hybridization.

Virtually any two pieces of DNA can be joined into a recombinant molecule by available technologies.

Cloning DNA

It is necessary to have large numbers of individual molecules of the DNA sequence of interest so that there are sufficient to be detected, sequenced, or whatever. One very efficient method is to amplify it by PCR. However, cloning the piece in a vector is used for many purposes. In DNA technology, the word 'cloning' usually refers to the process of multiplying pieces of DNA in a living cell. To do this the DNA is covalently attached to the DNA of a **cloning vector** which the host cell can replicate.

Cloning in plasmids

There are several different vectors to choose from. They differ mainly in the size of the piece of DNA that they can accommodate but some have other useful characteristics. The most commonly used vectors are bacterial plasmids for they are the easiest to handle and have long been the backbone of recombinant DNA technology. They can handle DNA inserts to be cloned, up to approximately 10 kb pairs in length.

The *E. coli* cell has a single major circular chromosome that constitutes most of the cell's genetic make up. In addition, however, there are tiny separate minichromosomes or plasmids, often in multiple copies in the cytoplasm – they are circular DNA molecules carrying a handful of genes that usually have a protective role in the cell. Typically they carry genes conferring antibiotic resistance on the cell. Each plasmid has an origin of replication to provide for duplication of the plasmid in the cell. Cloning not only amplifies pieces of DNA but at the same time separates each into a bacterial cell of its own so that a pure clone of a specific fragment selected out of a large number of different pieces can be obtained in unlimited amounts. *E. coli* cells can be separated by conventional plating on agar, each cell growing up into a colony containing large numbers of identical cells, each carrying multiple copies of the same plasmid with its specific DNA insert. Provided you can identify a colony that is carrying the piece of DNA in which you are interested, it is then a routine matter to grow the cells in liquid culture and produce as many copies of the plasmids with its DNA insert as you want. The plasmids are easily isolated because of their small size compared with the main chromosome. Your wanted piece of DNA can be released from the plasmid by cutting with the same restriction enzyme that was used to construct the recombinant plasmid in the first place.

Fig. 28.7 shows the general principle of constructing a recombinant plasmid. The plasmid is cut at a specific cloning site using a restriction enzyme giving sticky ends. The DNA piece to be cloned is generated by cutting with the same enzyme so that

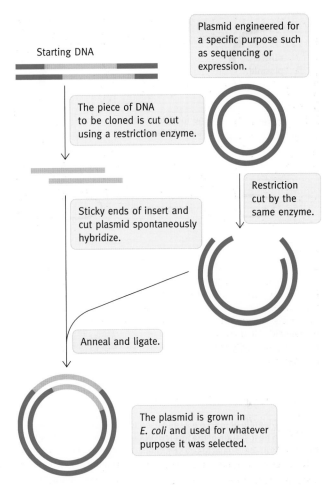

Fig. 28.7 Method of constructing a recombinant plasmid for cloning a DNA insert.

the pieces and the cut plasmid have complementary sticky ends which anneal together. The cut plasmids and the pieces are hybridized and enzymically ligated (covalently joined) together. This forms a **library** of recombinant plasmids each carrying a section of the genome. The next step is to introduce the plasmids into *E. coli* cells, which we'll take as our example.

Plasmids, naturally, are weakly infectious but *E. coli* cells can be made 'competent' to take up plasmids more readily, a process known as **transformation**. At its best, transformation is inefficient; it is rare for one plasmid to enter a bacterial cell so only a very small proportion of cells acquire one. For two to enter one cell is so rare that it can be ignored as a possibility. This ensures that a cell will almost never acquire more than one plasmid.

Specially engineered plasmids are used for particular purposes. pBR322 was a much-used one engineered for cloning purposes but a series of later cloning vectors have been derived from it. The one we will describe is pUC18. Its special feature is

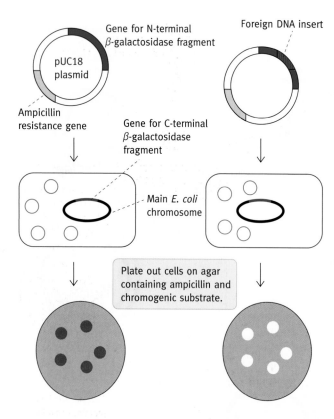

Fig. 28.8 Diagram of pUC18 plasmid used to clone a piece of foreign DNA (red). The insertion of the foreign DNA inactivates the gene coding for one part of the β-galactosidase protein. The *Escherichia coli* cells used have been engineered to contain the gene for the rest of the β-galactosidase protein. The plasmids without the insert allow the synthesis of the fragment that complements the latter so that active enzyme is produced. This hydrolyses a chromogenic substrate taken up from the medium and turns the bacterial colonies blue. The plasmids with the foreign insert do not produce active enzyme and the colonies appear white. The white colonies are screened for the wanted insert by a hybridization probe or other means (see text).

that it is possible to tell by visual inspection whether a bacterial colony has received a plasmid with an insert (known as blue/white selection). Fig. 28.8 shows the construction of the plasmid. It has a gene for ampicillin resistance and a gene for the N-terminal 146 amino acids of the enzyme β-galactosidase. The gene also has several restriction enzyme sites which can be used as cloning sites, The one that we are interested in is an *Eco*R1 restriction site situated in the gene for the N-terminal fragment of β-galactosidase. The first task is to get rid of the *E. coli* cells that do not have a plasmid by culturing them all in a medium containing ampicillin. Cells with a plasmid are protected because of the ampicillin resistance gene but the ones that have not received a plasmid do not survive. Most of the incorporated plasmids will not have received the DNA insert and again we do not want those cells. A visual selection method

has been engineered into the plasmid. This is where the gene for the N-terminal fragment of β-galactosidase come in. If it joins to the missing C-terminal portion of the enzyme the two reconstitute the active enzyme. An engineered *E. coli* strain is used for the cloning into whose main chromosome the gene for the missing C-terminal section of β-galactosidase has been inserted. Any plasmid which had the DNA fragment inserted into it cannot synthesise the N-terminal β-galactosidase fragment because the insertion disrupts the coding region. A plasmid with a successful insert will therefore not give rise to active β-galactosidase. On the other hand cells with plasmids lacking the DNA insert will produce both the N-terminal and the C-terminal enzyme fragments and generate active enzyme.

The cells are plated on a medium containing ampicillin and a chromogenic (colour-generating) substrate which enters the cells and gives rise to a blue colour if hydrolysed (by β-galactosidase if it is present). Colonies of cells carrying the plasmid without a DNA insert produce the enzyme and turn blue. Cells carrying a recombinant plasmid do not produce active enzyme and the colonies are white (Fig. 28.8).

These white colonies are screened for the desired fragment. If the sequence of the DNA fragment sought is known, a hybridization probe can be constructed by chemical synthesis. The screening involves overlaying the colonies by a DNA-binding membrane and taking a 'print' of the colonies (marking both plate and membrane so that the positions of colonies on the two can be cross identified). The print will contain some of the cells from each colony; the DNA in them is released by NaOH and fixed to the membrane in single stranded form by UV irradiation and/or baking. Nonspecific DNA-binding sites are swamped with random DNA, washed and the print is then subjected to hybridization with the probe, washed and autoradiographed. Cells hybridized with the probe will give a black spot; the colony can then be identified on the original plate and cells from it grown up in culture on any desired scale. The plasmids are prepared from disrupted cells and the cloned DNA fragment released by *Eco*R1 digestion.

A collection of *E. coli* cells which collectively contain all of the genome of the organism from which the starting DNA was obtained is known as a **genomic library**. From these it is possible to clone any gene or DNA section that can be identified. Nowadays, workers often obtain such libraries from other laboratories or commercial sources for they can be amplified simply.

Cloning using bacteriophage λ as vector

Plasmids, as mentioned, can be used for cloning DNA fragments up to 10 kb pairs in length, which is convenient for many purposes but it may be necessary to clone longer pieces than that. Lambda phage (bacteriophage λ can be used for pieces up to 20 kb pairs. When would this be needed? Some genes are very large and plasmids could not handle the whole length.

Bacteriophage λ is a bacterial virus which infects *E. coli*. It is comprised of a shell of protein molecules (the head) enclosing the single molecule of DNA and a tail which in effect is a device for injecting the DNA into the cell (Fig. 2.8) where it replicates. One of the remarkable properties of lambda phage is that the DNA is automatically packaged into the heads by the self-assembly principle and can be done *in vitro* using 'packaging kits' with the necessary protein components. Packaging requires that the ends of the phage DNA molecule, known as **cos sites**, are the correct distance apart but the middle section can be replaced by a piece of foreign DNA to be cloned. The foreign DNA must be 15–20 kb in length or the **cos** sites will not be the critical distance apart and will not be packaged. An engineered phage called Charon, after the boatman in ancient Greek mythology who ferried souls across the River Styx to the underworld is used. (Lambda ferries DNA pieces into *E. coli*.) It has two *Eco*RI restriction sites placed so that the correct-size piece of DNA is excised from the centre, and used as cloning sites for insertion of the DNA pieces to be cloned. The phage so assembled *in vitro* is infective and is replicated in the bacterial cell. *E. coli* cells are mixed with enough recombinant phage to infect only a very small proportion so that a cell will receive only a single one. Here it is replicated and new phage produced. The cells are grown as a lawn on an agar plate. Each new phage causes infected cells around it to lyse and the released phage infect neighbouring cells so that clear plaques are created in the opaque lawn of bacteria. The plaques are screened for the wanted insert as described in Fig. 28.9. As in the case of plasmids the procedure creates a genomic library from which wanted pieces of DNA can be obtained by the cloning procedure described. Workers usually obtain ready-made libraries from other laboratories for this purpose.

Cloning very large pieces of DNA

Cosmids are hybrids of lambda phage and plasmids used for cloning pieces 40–50 kb pairs in length. As described, the terminal arms of lambda are essential for packaging DNA into the phage heads. If the entire portion of the phage DNA between the cos sites is removed and replaced by the piece of DNA to be cloned the two cos sites are still the correct distance apart for packaging. However, it no longer has the genes needed for replication. To cope with this, the recombinant phage DNA has been joined to a piece of DNA essential for reproduction of a plasmid (the origin of replication). Hence the name **cos**- for the cos sites and -**mid** for the plasmid. After infection of the cells by the assembled phage particles the recombinant molecule is replicated as a plasmid but packaged as a phage.

For cloning pieces of DNA 500 kb pairs or longer, **yeast artificial chromosomes** (YACs) are used. These are artificial DNA chromosomes with telomeres, centromeres, and origins of replication and are reproduced in the host cells. **Bacterial artificial chromosomes** (BACs) are also used.

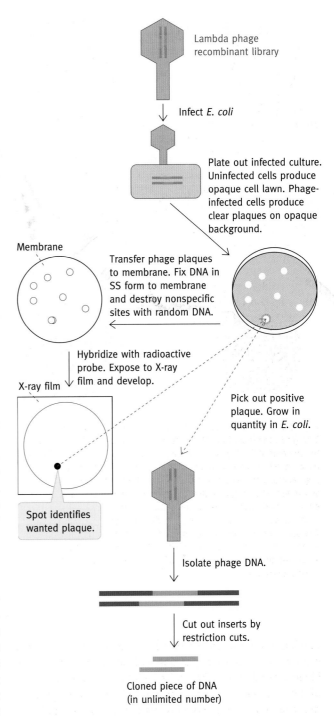

Fig. 28.9 Illustration of screening a bacteriophage *λ* library for a wanted DNA piece. A genomic library of lambda phage consists of *Escherichia coli* cells each carrying a different recombinant DNA molecule. Screening is used to pick out individual cells with the desired cloned insert. The cells are plated out as a lawn. Those cells harbouring a phage lyse and the released phage infects other cells causing clear plaques on the lawn. A membrane print is made and probed for the wanted insert. This identifies a wanted phage plaque which can then be propagated in quantity, and the DNA insert released from the isolated phage by the restriction enzyme used to form the recombinant. SS, single-stranded.

Applications of recombinant DNA technology

Production of human and other proteins

This has advantages over their isolation from cells in which they occur naturally. A protein of therapeutic value such as human growth hormone, which is the therapy for dwarfism, used to be isolated from human pituitary glands obtained from cadavers. Quite apart from the difficulty of obtaining the source material, there was the risk of transferring to the patient an infectious agent such as a prion (page 397). Insulin previously obtained from beef pancreas sometimes produced an immunological reaction which is less of a risk if human insulin produced in bacteria (or more commonly today in yeast) is used. A major advantage of the technology is that it can produce large amounts of proteins which may occur in the body only in small amounts. Tissue plasminogen activator is an example. The reason is, of course, that the amount of a protein produced naturally is a function of the promoter of the gene. In recombinant DNA technology the gene is placed under a powerful promoter.

To produce eukaryotic proteins in *E. coli*, DNA copies of the mRNA are used, known as **complementary DNA (cDNA)** because bacterial genes are not interrupted by introns and *E. coli* cannot splice (page 360) the RNA transcripts as in eukaryotes. In a eukaryotic cell, an mRNA has already had the introns spliced out so that cDNA can be copied from it and can be translated into protein in *E. coli* provided the necessary bacterial translational signals are added. A starting point is a cDNA library from which the cDNA gene of interest may be isolated. We will describe the preparation of such a library. Note that a population of mRNA from which cDNA is copied varies from tissue to tissue and from time to time depending on which genes are being expressed at the time.

Preparation of a cDNA library

mRNA is only a small fraction of the total RNA of a cell and, in the case of human cells, has (with a few exceptions) polyA tails (page 371). The RNA preparation is passed down a column of inert support carrying synthetic oligo-dT ligand (short, single-stranded 'DNA with only T bases). The ribosomal and transfer RNAs run through while mRNA binds, due to A—T hydrogen bonding of the tails to the ligand; the mRNA is then eluted using solutions of low ionic strength.

The isolated mRNA mixture is now copied in the test-tube into DNA using the polymerase domain of viral reverse transcriptase. This gives an RNA—DNA duplex. The RNA strand is destroyed with NaOH (page 316) and the single-stranded DNA is copied to give double-stranded DNA by an exonuclease-free DNA polymerase I (Fig. 28.10).

The procedure now is to clone the cDNA molecules into a library of plasmids from which the one carrying the cDNA of

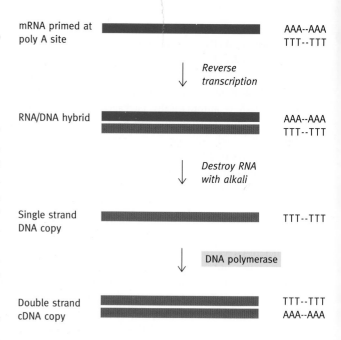

| mRNA primed at poly A site | AAA--AAA TTT--TTT |

Reverse transcription

| RNA/DNA hybrid | AAA--AAA TTT--TTT |

Destroy RNA with alkali

| Single strand DNA copy | TTT--TTT |

DNA polymerase

| Double strand cDNA copy | TTT--TTT AAA--AAA |

Fig. 28.10 Preparation of a complementary DNA (cDNA) library from eukaryotic messenger RNA (mRNA) (green). The reverse transcriptase is primed with oligo-dt and a DNA copy (red) is synthesized. The RNA is destroyed by alkali and the single DNA strand duplicated by DNA polymerase.

interest is selected by screening, much as was described on page 647. If a hybridization probe is not available and the cDNA was cloned into an expression vector (see below) then the clone of interest may be identified using a radioactive antibody against the protein.

Expressing the cDNA in *E. coli*

To produce eukaryotic proteins in *E. coli*, plasmids engineered so that production of the wanted protein is maximized, are available commercially. Built into such a plasmid are a strong bacterial promoter and also a sequence which transcribes into a ribosome-binding (Shine–Dalgarno) site on the mRNA (page 387). The vector has a polylinker site. (A polylinker is a synthetic short piece of DNA, incorporated into the plasmid, containing multiple cloning sites specific for different restriction enzymes.) The ends of the cDNA are made blunt-ended using a suitable exonuclease and a linker is added to each end using T4 ligase. The linker is cut to give a sticky end enabling the cDNA to be inserted into the plasmid. The latter is transfected into bacterial cells where mRNA is produced and translated into the wanted protein (Fig. 28.11).

Bacteria do not glycosylate proteins but this may not affect their function. If glycosylation is required, eukaryotic cells and appropriate expression vectors may be used. A wide variety of human proteins are now produced in *E. coli*, yeast, and other

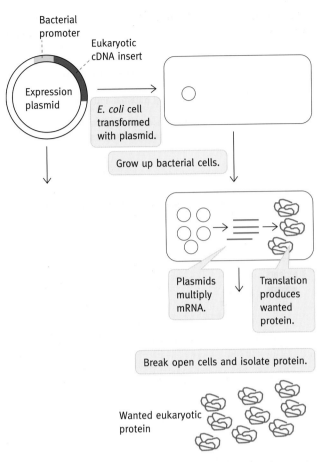

Fig. 28.11 Expression of a eukaryotic protein in *Escherichia coli* using an expression plasmid. See text for explanation.

Site-directed mutagenesis

This enables one or more selected amino acid residues in a protein to be altered and makes it possible to investigate protein structure and mechanism of action. For example in studying the mechanism of an enzyme reaction an amino acid in the active site can be changed to see what effect the change has on activity.

We start with a bacterial plasmid carrying a cloned cDNA gene for the enzyme. Suppose that you wish to change a selected serine residue to a cysteine whose coding triplets in the DNA are, for purposes of illustration, AGA and ACA respectively (the code for these amino acids is degenerate). The two strands of the duplex are separated by heating and a synthetic primer about 20 nucleotides in length containing the mismatched nucleotide (shown in red in Fig. 28.12) at the serine codon site is added. The primer is long enough to hybridize despite the mismatch. The primer is extended by DNA polymerase I to form a double-stranded plasmid and ligation completes the plasmid which is transfected into a bacterial cell where it replicates. Two versions of plasmids will be produced after replication from the two strands. One will contain a normal gene and the other the mutated gene. The mutated gene can be isolated and the gene expressed to produce the mutated protein. An alternative is to replace a section of DNA with a 'cassette' carrying the desired sequence. A section of the gene in an expression plasmid is

cells. Examples include insulin, tissue plasminogen activator (used for blood-clot removal), and human growth hormone. The viral hepatitis B protein used as a vaccine against the disease is also produced in this way. They carry no danger of infecting the patient with a disease.

When you transform *E. coli* with the expression plasmid, you want to produce as much protein as possible so a strong promoter is used. However, the cell cannot both grow rapidly and produce lots of foreign protein at the same time. To cope with this, an inducible promoter is used. The bacteria grow rapidly without inducer and then when sufficient cells are formed you switch on synthesis of the wanted protein with an inducer. (If you need reminding what an inducible promoter is, see page 359.)

When eukaryotic proteins are being produced in bacteria, they may not fold up properly and produce so much of the wanted protein that it precipitates out as 'denatured' inclusion bodies. Procedures have been developed by which, in some cases, these can be dissolved and refolded properly *in vitro*, though larger proteins may present difficulties.

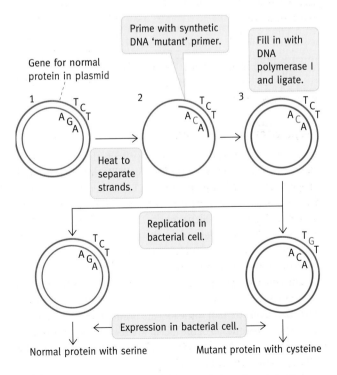

Fig. 28.12 Steps in site-directed mutagenesis to replace a single amino acid of a protein. See text for explanation.

excised by two restriction enzyme cuts and replaced by a synthetic DNA piece coding for the wanted amino acid sequence.

PCR in forensic science

An application in forensic science is now widely accepted as evidence in criminal trials and paternity determinations. Each human being has a large number of polymorphic loci (recognizable sequences in the DNA) in their chromosomes with short repeated sequences known as **microsatellites** and also as **short tandem repeats** or **STRs** (see Box 28.1). These markers can be readily detected by PCR. In forensic work the microsatellites with tetra- or pentanucleotide-repeat units (such as GATA, GATA, GATA, GATA, etc.) are usually used. We will use the example of a GATA repeat in the subsequent account. The number of repeats in different chromosomal loci typically varies from 4 to 40 in number and there are different alleles at the same locus amongst individuals in the population. In a pair of homologous chromosomes, one derived from the father and one from the mother, the repeat number in a given locus is by chance likely to be different in the two chromosomes (Fig. 28.13). If you select, say, nine microsatellite loci in a sample of DNA and examine the repeat number in both chromosomes of homologous pairs, you will have a pattern of (2×9) 18 numbers. (In some places it has become routine to use 14 loci rather than 9 in DNA profiling). *(Note that in Fig. 28.13 we show only three loci being used for the purpose of clarity in illustrating the principle.)* This pattern is not quite unique to each individual but the chance of another *unrelated* individual having the same pattern is one in billions (though identical twins will have the same pattern as each other). Near certainty exists if a pattern is found in an individual to be different from the forensic sample for then they are beyond reasonable doubt not the same. Therefore DNA profiling is an extremely decisive method for excluding suspects in a case.

How is this put into practice? From genetic chromosome mapping studies and the human genome sequence, the location and flanking sequences of large numbers of microsatellite loci are known. Using a pair of synthetic primers that flank a given locus the repeat section is amplified by PCR. Note that primers that would hybridize to the core GATA sequence itself would be useless – they must be complementary only to the *flanking* sequences, as these are locus specific. The length of the product from each locus will depend on the number of GATA repeats and so the products from the two chromosomes will usually give rise to two bands on an electrophoresis gel. From the three different loci, the combined PCR products will give a pattern of up to 6 bands as shown in Fig. 28.13 and up to 18 from nine loci. (It will not be 6 or 18 if some loci are homozygous.) In current profiling technique, all of the loci of interest are PCR'd together in one mix, known as a '**multiplex**'. The pattern obtained from the DNA of a suspect is compared with that obtained from a forensic sample. DNA profiling is decisive. Because of the sensitivity of PCR, minute traces of DNA in bloodstains, semen

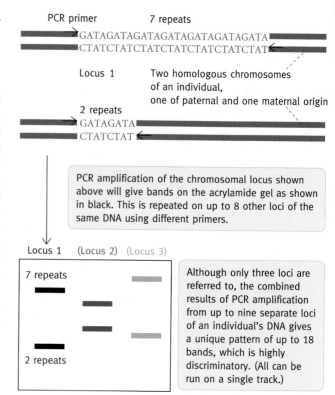

Fig. 28.13 Diagram illustrating the principle of DNA fingerprinting in forensic science. A microsatellite locus on homologous chromosomes of an individual is shown. Polymerase chain reaction (PCR) products give two bands. This process is repeated on up to eight more selected loci on the DNA of the same individual so that altogether a pattern of up to 18 bands will result that is unique to the individual. The process is repeated on the forensic sample of DNA to see if the patterns match. (Note that loci 2 and 3 are not illustrated but the bands that might be produced from them are shown. Also, although we show only three locus results, in practice up to nine may be used to increase the certainty of identification. Note also that the smaller pieces run faster than the larger ones, so that different repeat numbers are separated into discrete bands.)

samples, swabs wiped on the steering wheel of a car, or a single hair follicle for example is sufficient. The procedure can also establish paternity since by chance the mother and father will have different alleles at many loci, and these are inherited by the rules of classical genetics.

Locating disease-producing genes

In situations where nothing is known about a single gene genetic disease, apart from the phenotypic symptoms, it can be extremely valuable if the gene responsible for the disease can be located and isolated. Studies of the gene may open the way to a better understanding of the disease and of devising therapies for it. A strategy known as **positional cloning** makes this possible in some cases; it has indeed led to the isolation of the genes for

BOX 28.1 Repetitive DNA sequences

About half the DNA of the genome is made up of repetitive sequences of different types. See Chapter 21, page 328, for long interspersed repeated sequences (LINES) and short interspersed repeated sequences (SINES).

Tandemly repeated sequences are the other main category of repetitive DNAs. These consist of sequences of short nucleotide units arranged one after the other. An example is five bases in tandem, TTCCA/ TTCCA /TTCCA (with their complementary strands) repeated dozens or thousands of times. Repeats of this style are found particularly in the tightly packed heterochromatin of centromeres, in patches near the centromeres and near the ends of the chromosomes. They constitute perhaps 10% of the total DNA. Nothing is known of their function, if any, though it has been suggested that they may originate from transposons.

Most tandem repeated sequences are highly polymorphic. This gives them some important practical applications in human genetics. The term 'polymorphism' refers to the situation in which, within a population, there are several or many variants of a sequence at a given locus. In other words the sequences at this locus is highly likely to be different between individuals. Within an individual there will be two versions, one each on the maternal and the paternal chromosomes. A common form is where the number of repeats of a tandem repeated sequence unit at a given chromosomal locus varies between two homologous chromosomes. The number is usually different between the maternal and paternal chromosomes. As an example, at a particular locus there could be 5 repeated units on one chromosome and 30 on its homologous partner. Geneticists have identified large numbers of such polymorphisms throughout the human genome. These are important as genetic markers; for example to locate a disease producing gene and as the basis for DNA fingerprinting in forensic science. Thee longer ones are known as variable number of tandem repeats (VNTRs) but for experimental convenience the shorter tandem repeats known as STR polymorphisms (STR = short tandem repeats) are most used. They are also known as microsatellites.

Genetic markers are very important in locating disease-producing genes and in forensic science. VNTRs have been commonly used, but now single nucleotide polymorphisms have replaced them, except in the case of DNA finger-printing where, for special reasons, STRs are preferred.

---

cystic fibrosis and Huntington's disease. In the latter case, study of the gene led to the discovery that the disease is associated with a different type of mutation. The gene codes for a protein known as **huntingtin** of unknown function but widespread in brain cells. The coding region in normal people contains a CAG triplet repeat which codes for a stretch of polyglutamine in the protein. In patients with the disease, which causes intellectual decline and ultimate death, there is an expansion of the number of CAG triplets so that the protein has a longer polyglutamine stretch in it. The number of repeats is variable with the likelihood of the disease occurring, increasing with increase of the repeat number. The number in a normal person is up to 26, while the disease always occurs if the number exceeds 39. This apparently causes abnormal aggregates to form in the cell, though the link between this and expression of the disease is not understood. Another disease causing mental retardation is the fragile X-syndrome. This is associated with an expansion of the CGC triplet which is not in a coding region of the gene. It is associated with the occurrence of a fragile site in the X chromosome. The gene is believed to be involved in control of the expression of many genes in the brain. The inheritance patterns in Huntington's disease and the fragile X-syndrome have idiosyncratic features and for a fuller understanding of these diseases, genetics treatises should be consulted.

The principle of the positional cloning strategy is based on genetic linkage. If two loci are close together on a chromosome they tend to be inherited together – they are tightly genetically linked – while if far apart they tend to be inherited independently as illustrated in Fig. 28.14. This is due to genetic recombination in meiosis in which pairs of homologous chromosomes

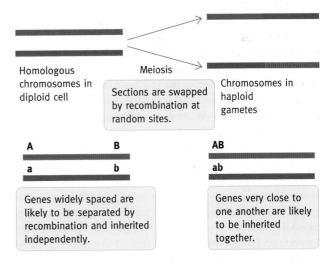

Fig. 28.14 Diagram to illustrate the principle of genetic linkage.

align, and two of the four chromatids swap sections generating (natural) recombinants; the probability of a recombination event occurring in the DNA section between genes is greater the further they are apart.

Previous genetic studies have identified polymorphic genetic markers. While more than 99% of the bases in human DNA are the same in two individuals, there are many sites in which variation occurs in regions of the genome where they have no deleterious effects. The positions of vast numbers of these have been mapped on the human genome and are used as genetic markers since they can be detected experimentally by DNA methodologies. If it is found that a disease is inherited tightly co-inherited with a genetic marker then it is likely that the disease gene is close to the genetic marker on the chromosome.

Several different types of genetic markers have been used, one replacing the other as techniques for their detection became easier. The earliest were **restriction fragment length polymorphisms** (**RFLPs**). These are based on the fact that a single nucleotide change can change the restriction sites on a piece of DNA so that a different pattern of restriction fragments are produced on enzyme digestion, and detectable by Southern blot analysis. The method was successful in isolating the genes for cystic fibrosis and Huntington's disease. However, restriction analysis takes several days and the amount of work in gene isolation was enormous. Newer markers, particularly the microsatellites already described above, have largely replaced it. These markers can be detected in hours using PCR amplification followed by electrophoresis. The latest type of genetic markers of increasing importance are **single nucleotide polymorphisms** (**SNPs**) (though microsatellites are still preferred for forensic DNA fingerprinting). These are, as the name implies, changes in one nucleotide of a base pair so that A—T becomes A—G or C—T for example. They have several advantages. They occur very frequently in the genome so that several millions, and their positions, are known in the human genome and the sequence in which they occur can be read off the human genome sequence. The next important thing is that they can be detected by DNA hybridization even though only a single base change is involved. For this a probe is made to the sequence containing the SNP and the hybridization done at high stringency. DNA microarrays are used so that huge numbers can be studied simultaneously and automatically.

We will give a simplified outline of the method of searching for a disease gene. First the general location of the gene in the genome is identified. For this about 300 widely spaced SNPs are selected spread over the entire genome. DNA fragment preparations from multiple patients with the disease and from many normal people are separately labeled with red and green dyes respectively (page 463) mixed and applied to microarrays containing probes for all of the SNPs. A statistical excess of red spots at a locus indicates a locus with high linkage to the disease gene, green a negative association and yellow no association. This gives a broad indication of the area of DNA housing the gene in a particular chromosome. Concentration on this by using the same technique, but with more closely spaced markers gives a narrower localization. When the gene containing locus has been narrowed down to a piece of only perhaps 10–20 genes, computer methods may be used to look for open reading frames (sequences long enough to code for proteins without any stop codons), and start sites to identify genes. Because of introns it may be difficult to identify open reading frames of genes in eukaryotes.

There are several other established methods but the results of the human genome project often greatly simplify the task of establishing candidate genes in the isolated piece of DNA. A computer programme known as a gene browser is used to interrogate databases about what is known of genes and their expression in the section of interest. The nature of candidate genes may be usefully determined by searching the relevant DNA sequences (obtained from the human genome project) for domains such as those characteristic of kinases, transmembrane helices, SH2, SH3, and pleckstrin homology domains associated with signaling pathways, or multiple zinc fingers characteristic of transcription factors. A genome wide search for known homologous genes may also give clues on the nature of the genes and possibly which one is likely to be related to the disease. Final identification of the particular gene of interest is done by whatever methods are possible for the particular gene; it can then be cloned and studied. In the case of cystic fibrosis the final localisation was by RNA–DNA hybridization using mRNA obtained from skin pores where disease symptoms are expressed.

The microarray method of detecting SNPs, using genome-wide microarrays, is assuming great medical importance. Individual patients can be screened for large numbers of SNPs so that it may become possible to use the presence of individual SNPs as genetic markers to indicate susceptibility to many of the common diseases. The commentary paper by **Topol** *et al.* (in the Further reading list) refers to identification of DNA markers associated with susceptibility to complex diseases including acute lymphoblastic leukaemia, type 2 diabetes, and coronary heart disease. They point out however that identification of an associated marker does not give information on the biological reason for the susceptibility to a disease so that a large research effort lies ahead on this area of medicine.

Knockout mice

Mutants have long played an important role in biochemistry and molecular biology. For decades prokaryotic mutants have provided information on the function of genes and of the proteins they code for. Mutants are comparatively easy to obtain in bacteria because huge numbers of cells can be chemically mutated at once and the desired mutant selected by screening procedures. Obtaining mammalian mutants is a much more difficult task for obvious reasons but procedures have been developed for obtaining strains of mutant animals that specifically lack a known gene. Such animals are useful medically in that they can act as models of human diseases caused by the lack of a given gene. Mice are the favourite experimental mammal for such work and the term **knockout mice** is an accepted one. (Mice with an inserted gene are called knockin mice.) Libraries of knockout mice as models of different human diseases are being developed.

As an example a mouse model of the human disease familial hypercholesterolaemia has been made by knockout of the gene for the low-density lipoprotein receptor. The production of these strains requires that every cell in the body should carry the mutation and this requires that the mutation is present in germ line cells which makes the method of obtaining them more complex.

There is a quite different reason for wanting to knock out specific genes and that is in the hope of therapeutic treatment of certain human diseases. For example a cancer may be due to the presence of an oncogene, and silencing it may be a useful treatment. The new phenomenon of RNA interference (RNAi) described in Chapter 25 offers the potential of doing this relatively easily though its actual application to human therapy has not yet been made.

Method of obtaining a specific gene knockout

In this, a specific gene in the mouse is selected and a **targeting vector** constructed using the methods of recombinant DNA technology. One method is to partially or fully replace the selected gene with a piece of foreign DNA (see below). The basis of the targeting is that at both ends of the section to be incorporated there are regions of homology with the section it is to displace. Homologous recombination at each end of the DNA to be inserted specifically targets the selected gene. The technique of gene targeting is only half of the problem, because there follows the question of how you arrange that *all* the cells of a mouse have the mutation. **Embryonic stem (ES) cell** technology (as described below) achieves this.

The embryonic stem (ES) cell system

There are various groups of **stem cells** in the body known as **adult stem cells**, which provide replacement cells for a limited range of cell types. Thus there are adult stem cells for blood cells (see Fig. 2.7), sperm, skin, bone, and epithelial cells.

Embryonic stem (ES) cells are **pluripotent** – they can give rise to all types of cells. (Some **ES-like**, pluripotent, **cells** have also been found in some other adult somatic tissues.) After fertilization, a mammalian egg divides to form a solid ball of cells, which develops into a **blastocyst**. This is a sphere of cells containing a cavity and an inner cell mass (Fig. 28.15), the ES cells. The latter will give rise to all cells of the mature animal, but at this stage individual cells are not committed to becoming any particular type in the adult. They can be propagated in long-term culture, and under appropriate conditions they do not differentiate but retain their pluripotency. The cultured cells can be injected back into the cavity of a blastocyst and when the latter is re-introduced into a mouse foster mother it develops into progeny in which the introduced (foreign) cells contribute to all tissues, including germline cells. In any one tissue of progeny mice some of the cells will be derived from the cells of the 'natural' blastocyst and some from the (foreign) injected ES cells which were obtained from a different blastocyst. This provides a route for gene targeting described below.

Gene targeting

The first step is to construct a **targeting vector**. In the example that we will use here, the target gene is inactivated by replacing it with a **neomycin-resistance gene (NRG)**. In the vector, the NRG has flanking sequences homologous to the targeted gene (Fig. 28.16) so that recombination at each end of the replacement section causes the new piece of DNA with NRG to replace the targeted gene. The procedure for obtaining knock-out mice is given in Fig. 28.15. We suggest that you follow the steps in the figure.

ES cells are obtained from a mouse blastocyst and cultured *in vitro*.

- The cells are **transfected** by the targeting vector, usually by **electroporation**. This involves giving an electric shock to the cells, which transitorily forms holes in the cell membrane and allows the vector to enter. Most frequently, in mouse cells, the introduced DNA integrates into the host chromosomes in a random fashion, which is not what is wanted, but a low but useful percentage (0.5–10%) integrate by homologous recombination at the target site.

- Stably transfected cells are now resistant to the neomycin due to expression of the neomycin gene so that these cells can be selected by growing them in the presence of the antibiotic. Cells without the neomycin-resistance gene fail to grow. After that, it is necessary to select those cells in which integration has occurred by homologous recombination rather than by random insertion. A rather clever trick facilitates this. In constructing the targeting vector, a gene for **thymidine kinase**, derived from the herpes simplex virus, is inserted at one end of the construct, well outside the region of homology to the target gene. If the DNA is incorporated randomly, the thymidine kinase gene goes in with it and is expressed; if the incorporation is by homologous recombination the kinase gene is not incorporated (because it is outside the homology region). If the kinase gene is expressed, it makes the cell sensitive to **gancyclovir**, since the latter has to be phosphorylated to be toxic to the cell and the viral thymidine kinase is needed for this. Thus, in the **positive-negative selection**, as it is called, only those cells with the neomycin gene but without the thymidine kinase gene will grow, and these are the wanted homologous recombinants.

- The cells that survive this test are separately grown up as clones, and individual clones of cells are examined to confirm that the replacement gene has been correctly inserted by homologous recombination. This is done using recombinant DNA techniques previously done with Southern blotting or more recently by PCR and sequencing.

- Cells in which the targeted gene has been replaced are grown up and injected into the cavity of a mouse blastula derived from an animal whose coat colour is different from that of the mouse from which the injected ES cells originated (see below).

- The hybrid blastula is placed in a pseudopregnant foster mother (pseudopregnancy results from mating with a sterile male). The injected, mutated, cells participate in

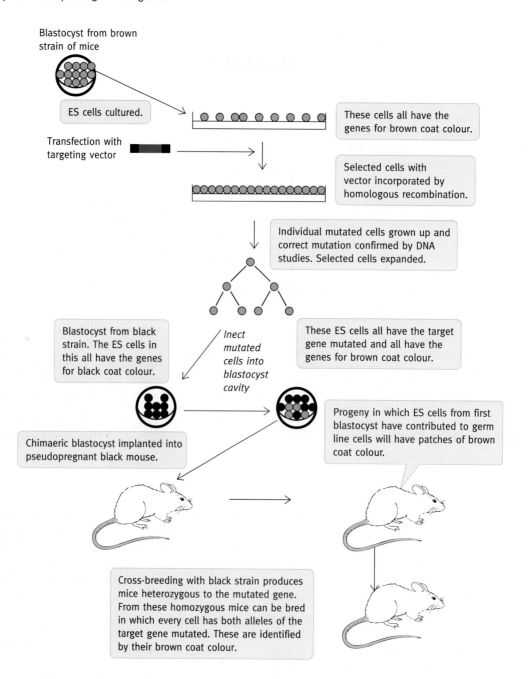

Blastocyst from brown strain of mice

ES cells cultured.

These cells all have the genes for brown coat colour.

Transfection with targeting vector

Selected cells with vector incorporated by homologous recombination.

Individual mutated cells grown up and correct mutation confirmed by DNA studies. Selected cells expanded.

Blastocyst from black strain. The ES cells in this all have the genes for black coat colour.

Inect mutated cells into blastocyst cavity

These ES cells all have the target gene mutated and all have the genes for brown coat colour.

Chimaeric blastocyst implanted into pseudopregnant black mouse.

Progeny in which ES cells from first blastocyst have contributed to germ line cells will have patches of brown coat colour.

Cross-breeding with black strain produces mice heterozygous to the mutated gene. From these homozygous mice can be bred in which every cell has both alleles of the target gene mutated. These are identified by their brown coat colour.

Fig. 28.15 Outline of procedure for obtaining knockout mice by gene-targeted mutation. The diagram illustrates how coat colour provides a convenient guide in selecting progeny. The brown progeny of the first mating will be +/− and +/+ for the relevant gene. Male and female mice are selected by DNA studies and inbred. The progeny will be +/+, +/− and −/−. The latter, which are selected by DNA studies, are homozygous for the knockout gene. ES, embryonic stem cells.

formation of tissues as the embryo develops so that in each tissue of the progeny mouse, some cells will have originated from the original blastula cell mass and some from the injected transfected cells. In other words, the mouse will be a chimera.

- As a ready guide to what is occurring genetically, coat colour is used. Let us suppose that the original blastula

from which the ES cells for manipulation were obtained had a brown coat colour and the recipient blastula into which modified cells were injected came from a black mouse strain. The mouse resulting from implantation of the latter into the foster mother will be a patchy black/brown mixture as shown in Fig. 28.15. These are not yet the end of the story for only some of their cells are mutated.

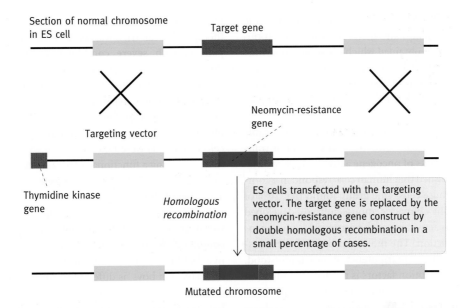

Section of normal chromosome
in ES cell

Target gene

Targeting vector

Neomycin-resistance
gene

Thymidine kinase
gene

*Homologous
recombination*

ES cells transfected with the targeting
vector. The target gene is replaced by the
neomycin-resistance gene construct by
double homologous recombination in a
small percentage of cases.

Mutated chromosome

Fig. 28.16 One method of gene targeting by homologous recombination in embryonic stem (ES) cells. ES cells are obtained from a mouse embryo at the blastocyst stage and cultured. They are then transfected with a targeting vector constructed by recombinant DNA methods. In this, the target gene is replaced by a neomycin-resistance gene flanked by sequences homologous to the normal gene. Homologous recombination replaces the target gene with the neomycin-resistance gene construct in a small number of cases. Stably transfected cells are cultured and cells which have incorporated the foreign gene by homologous recombination are identified by recombinant DNA methods and cultured. The mutant cells are used to obtain knockout mice as described in the text and illustrated in Fig. 28.15. The yellow rectangles represent normal genes flanking the target gene. The blue rectangle represents the addition of a viral thymidine kinase gene which is placed outside of the region of homology between the vector and targeted chromosome. This is to permit negative selection of cells in which the vector is incorporated randomly rather than by homologous recombination.

- What is needed are mice in which every cell in the body is a knockout (null) mutant with respect to your target gene including the germline cells. To obtain these, the chimaeric mice are crossed with a black mouse of the strain from which the host blastula was derived. This will usually give rise to some intermediate coloured progeny, indicating that they are heterozygotes derived from germline cells and thus all cells have one mutated gene and one normal gene.

- Inbreeding of these results in progeny half of which are brown and half black. Appropriate cross breeding of these produce knockout mice in which all the cells in the body of the brown mouse lack the targeted gene (see explanation in Fig. 28.15).

Stem cells and potential therapy for human diseases

We have already described the nature of stem cells in Chapter 2 (page 22) and in the previous section the production and properties of embryonic mouse stem cells are described. The hope that it might be possible to cure human degenerative disease and restore damage caused by certain traumatic injuries by the introduction of stem cells to replace lost or damaged ones. As an example a heart attack causes sections of cardiac muscle to die and, in principle, stem cells could replace these.

There are impediments. First, embryonic stem cells would be immunologically different from the patients' cells and therefore subject to immunological rejection. Secondly, to obtain embryological stem cells involves the destruction of a human embryo and this has raised ethical considerations. The ideal situation would be if a patient's own differentiated adult somatic cells could be reprogrammed from the differentiated state back to the pluripotent state of embryonic stem cells. This would overcome both objections. It was once thought that differentiation of cells in which pluripotency was lost was irreversible but the famous Dolly the sheep work (page 22) showed that the differentiated nucleus could be reprogrammed to the stem cell state by injecting it into an egg which had had its own nucleus removed. This is called **somatic cell-nuclear transfer (SCNT)**. The resultant embryo would produce autologous stem cells (that is cells immunologically the same as the patient's cells). However, it still requires a human donor for the egg, which raises both practical and ethical objections. It perhaps is worth emphasizing that stem cell cloning technology is different from cloning of animals such as Dolly the sheep.

Recently advances have been reported in obtaining autologous stem cells in mice by reprogramming skin fibroblasts. Three separate groups (see **Rossant** reference in Further reading for a summary) have generated cells with stem cell-like characteristics by introducing transcription factors into the cells by retroviral transfection and then selecting for pluropotency. Cell lines have been obtained which resemble embryonic stem cells almost identical to embryonic stem cells including the ability to generate all cell types when injected into blastocysts of early embryos which are then allowed to develop in pseudopregnant mice.

Much remains to be done before the technique is applicable to human therapy. The use of retroviral transfection of transcription factors poses risks of oncogene activation (page 521). It will have to be established that human cells respond in a comparable way to the mouse cells. One group found that 20% of offspring derived from chimaeric mouseembryos developed tumours attributed to activation of the retroviral c-*myc* oncogene used in the induction. The molecular basis of the transformation of adult cells to pluripotency is not understood. Expression of the transcription factors is needed for induction of the pluripotency but not for maintenance of that state.

A different potential source of autologous stem cells has been discovered recently by a different group (see the **Seandel** *et al.* reference in Rurther reading). These identified a cell marker for a subset of testicular ES-like stem cells which made it possible to efficiently isolate and culture them as cell lines that resembled embryonic stem cells. They passed the crucial test of being able to contribute to the formation of chimaeric embryos when injected into blastocysts. These cells were found to be able to regenerate cardiac tissue *in vitro* and functional blood vessels *in vivo* in mice. The discovery appears to have the potential to generate autologous cells for treatment of male patients, though the authors add that the cells should be used therapeutically only after extensive preclinical experimentation has been done.

Transgenic animals and gene therapy

Genes can be directly inserted into animals and plants to give rise to different genotypes, and thus phenotypes. This is referred to as **transgenesis**. It has been found that pieces of DNA injected into cells may be incorporated into the cell's DNA in variable degrees of efficiency where it may be expressed. The first example of this was achieved by injecting a cloned gene for growth hormone into the nucleus of a fertilized mouse egg. The egg was then implanted in a mouse to develop. It was found that some cells of the progeny contained multiple copies of the growth hormone gene. As a result it grew to about twice the size of a normal mouse. Moreover, some of the germ line cells contained the extra genes so that it is possible to breed transgenic animals with the extra genes.

This has given rise to hopes that it might permit genetic human diseases to be remedied by **gene therapy**, which involves correction of a gene deficiency by inserting an untargeted normal gene or genetic material into cells.

Initially, gene therapy was considered primarily for the treatment of diseases in which there is a known single-gene deficiency, such as β-thalassaemia (page 68), cystic fibrosis, or Duchenne's muscular dystrophy (page 130). The first gene-therapy clinical trial was performed in 1990. This was on patients with **severe combined immunodeficiency disease (SCID)** in which the gene for **adenosine deaminase** (page 304) is deficient in lymphocytes. This leads to a build up of adenosine, the metabolic repercussions of which are to cause a toxic build up of dATP and inhibition of DNA synthesis in the lymphocytes. This impairs the immune response. Bone marrow cells of patients were treated *in vitro* with a crippled retroviral vector carrying a functional adenosine deaminase gene and the cells returned to the patient. This gave encouraging results with a number of children.

In 2001, a French group used gene therapy to cure children with a very severe and fatal immune defect called **X-linked, severe combined immunodeficiency** (then treatable only by isolation in a 'bubble' to prevent infection). This disease, found almost exclusively in boys, is different from that caused by adenosine deaminase deficiency (above). It is caused by a mutation in a lymphocyte protein that is common to a number of interleukin receptors. These are necessary for the immune cells to signal each other to raise appropriate immune responses (page 500). The researchers used a retroviral vector carrying a corrective functional receptor protein gene. This was inserted *in vitro* into the patients' bone marrow haematopoietic stem cells. They were then transplanted back into the patients. Remarkably, 10 out of 11 patients benefited significantly from the therapy with most appearing to have been cured. Unfortunately, within 3 years of the treatment, two of the patients had developed leukaemia, which appears to have been due to insertional mutagenesis. That is, the insertion of the retroviral gene vector activates a protooncogene or disrupts a tumour-suppressor gene (page 521). Because of this danger, gene therapy is probably considered only in cases where there is no other hope. Despite the successes in a few cases, gene therapy has unfortunately proved more difficult than was anticipated.

In plants, foreign genes can be inserted successfully into the chromosomes by using as a cloning vector a naturally occurring (but suitably altered) plasmid (the Ti, or tumor-inducing, plasmid). This is found in the pathogenic soil bacterium *Agrobacterium tumefaciens*. This has been important in agriculture. Alternatively, DNA molecules may be literally shot into plant cells, with a gun-type instrument that fires a cloud of fine shot loaded with DNA. In this way, crop plants are being engineered with specified phenotypic characteristics – for example, to be resistant to herbicides. The purpose of this is to control weeds by blanket spraying with the herbicide – only the resistant crop plant survives.

DNA databases and genomics

DNA databases are the essential partners to the protein databases described in Chapter 5. Genomics refers to the study of large numbers of genes. It is the partner to **proteomics** (page 79) in which large numbers of proteins are studied together. The computational use of the protein and DNA databases is collectively known as **bioinformatics**. The databases are now of great importance in biochemistry and molecular biology, medicine, and virtually all biological sciences.

When a gene or other section of DNA has been isolated and sequenced, the information about it is recorded in one of the international databases in the public domain. Free usage of software is available to interrogate the bases and analyse the data in ways appropriate to the type of questions being asked. When an unidentified gene or other DNA sequence is isolated and fully or partially sequenced, a search of the databases for matching sequences will reveal whether information on that piece of DNA, or a closely related one, already exists. The complete sequence may then be available and its location in the chromosomes relative to other genes identifiable. The search might also, for example, reveal that it is part of a family cluster of related genes; its function may be known or perhaps an asso-

ciation of it with a disease may have been determined from other work. It is also possible to analyse a DNA sequence for the presence of open reading frames (page 387) or potential splicing patterns (page 361) to give but two examples.

The databases resulting from the Human Genome Project (see references in the Further reading at the end of this chapter) are now revealing, depending on the particular case, information of importance to medicine, biotechnology, or basic science. This can speed up a research project to an almost unimaginable degree. The uses are almost limitless and 'mining the databases' is an important branch of research in many areas of biological science.

Use of the DNA databases is mainly a research activity for it requires computational skills and considerable knowledge of biochemistry and molecular biology. Specialist courses on bioinformatics are given in many departments. Further information is given in the bioinformatics overview section on page 81. Some relevant website addresses are given in Box 5.1 on page 80.

Summary

The technology of DNA manipulations has become the most powerful approach to many biological and medical problems. It permits the isolation of genes, determination of their nucleotide sequences, detection of abnormal genes, and production of human and other proteins in unlimited amounts in hosts such as yeast and bacteria. It can also produce proteins with specific amino acid substitutions.

The techniques are many and varied but a number of basic principles apply.

- DNA can be cut with precision at known sequences using a battery of restriction enzymes.

- DNA sequences can be identified by hybridization with probes obtained by isolation or synthesis.

- Recombinant DNA molecules in which different pieces of DNA are joined together can be produced in a number of ways.

- Recombinant molecules or individual sections of DNA can be amplified and purified by cloning techniques.

- The polymerase chain reaction can amplify a selected stretch of DNA in a chromosome, millions of folds in an hour or so. The selection of pieces to be so amplified can be done using sequence information from the human-genomed-project maps.

- DNA pieces can be sequenced using the dideoxy method, which is now fully automated.

The availability of the human genome sequence has made available markers called microsatellites that can be amplified by the polymerase chain reaction. This is the basis of DNA fingerprinting. The microsatellite markers can also be used to locate disease-producing genes. Single nucleotide polymorphisms (SNPs) have been identified in large numbers, and increasingly are the genetic markers of choice. Genome-wide linkage between SNPs and diseases is a rapidly developing field. The use of microarrays (DNA chips) permits the study of the expression of large numbers of genes at once and promises to be important in genomics in general and in the study of cancer and other complex diseases.

Gene therapy involves the insertion of normal genes into patients to correct abnormal ones. Success has been recorded only in limited cases, but the concept has medical potential and is being investigated. In plants, transgenesis has had wide application in genetically engineering new varieties with favourable characteristics.

Animal models of diseases due to faulty genes (called knockout mice) can be made using stem cell technology. More recently a simpler method of gene silencing called RNA interference (RNAi) has been developed. In principle it involves a specific short RNA piece targeted to a particular mRNA. It causes the destruction of the messenger or inhibition of its translation. The method (see Chapter 25) is attracting enormous attention.

Essential parts of DNA technology are the DNA databases. These, with the help of appropriate software for retrieving and analysing the vast amount of information, are a vital part of the relatively new science of bioinformatics, which is assuming great importance. Mining the databases is now an important part of molecular biology.

Further reading

The human genome

The Human Genome Project. (1999). *Curr. Biol.*, **9**, R908–9.
A summary of the quick-guide type summarizing the project in one and a half pages.

Dennis, C. and Gallagher, R., Eds. (2001). *The Human Genome.* Nature Publishing Group.

The polymerase chain reaction

Saiki, R. K., *et al.* (1988). Primer-directed enzymatic amplification of DNA with a thermostable DNA polymerase. *Science*, **239**, 487–91.
Detailed account of the technology of the PCR reaction.

Gene targeting

Melton, D. W. (1994). Gene targeting in the mouse. *BioEssays*, **16**, 633–8.
Describes the technology of obtaining knockout mice and discusses its application to study of human diseases.

DNA microarrays

Knight, J. (2001). When the chips are down. *Nature*, **410**, 860–1.
A cautionary tale on DNA microarrays.

Stoughton, R. B. (2005). Application of DNA microarrays in biology. *Annu. Rev. Biochem.*, **74**, 53.

Topol, E. J., Murray, S. H. and Frazer, K. A. (2007). The genomics goldrush. *J. Amer. Med. Assoc.*, **298**, 218–21.
Deals with the medical potential of using microarray analysis of single nucleotide polymorphisms to make genome-wide profiles of individual patients. The goldrush refers to the enormous activity in this field.

Williams, D. A. and Baum, C. (2003). New challenges ahead. *Science*, **302**, 400–1.

Stem cells

Karnoub, A. E., Dash, A. B., Vo, A. P., Sullivan, A., Brooks, M. W., Bell, G. W., Richardson, A. L., Polyak, K., Turbo, R. and Weinberg, R. A. (2007). Mesenchymal stem cells within a tumour stroma promote breast cancer metastasis. *Nature*, **449**, 557–65.

Okita, K., Ichisaka, T. and Yamanaka, S. (2007). Generation of germline – competent induced pluripotent stem cells. *Nature*, **448**, 313–7.

Rossant, J. (2007). The magic brew. *Nature*, **448**, 260–2.
The brew is the cocktail of transcription factors used to induce adult skin cells to become plutipotent. A News and Views article summarizing three reports on this subject.

Seandel, M. *et al.* (2007). Generation of functional multipotent adult stem cells from GPR125+ germline progenitors. *Nature*, **449**, 346–50.
Reports of the isolation from testis of pluripotent stem cells with potential for male therapy.

Wernig, M. *et al.* (2007). In vitro reprogramming of fibroblasts into a pluripotent ES-cell-like state. *Nature*, **448**, 318–24.

Gene therapy

Morgan, R. A. and Anderson, W. E. (1993). Human gene therapy. *Annu. Rev. Biochem.*, **62**, 191–217.
Deals with the candidate diseases, gene-transfer methods, treatment of adenosine deaminase deficiency, and concludes with safety and ethical questions.

Capechi, M. R. (1994). Targeted gene replacement. *Sci. Am.*, **270**(3), 34–41.
Describes gene replacement by homologous recombination and transgenesis techniques.

Anderson, W. E. (1995). Gene therapy. *Sci. Am.*, **273**(3), 96–8B.
Includes a list of 14 diseases being tested in clinical trials of gene therapy.

Sunyaer, S. R., *et al.* (2000). SNP frequencies in human genes: an excess of rare alleles and differing modes of selection. *Trends Genet.*, **16**, 335–7.
(SNP = single nucleotide polymorphism.)

Gene cloning

Gurdon, J. B. and Colman, A. (1999). The future of cloning. *Nature*, **402**, 743–6.
A special News and Views article describing how Dolly was cloned by nuclear transplantation including discussion of the potential use of stem cell technology for growth of organ transplants, the ethical and legal implications, and the future of cloning.

Vogel, G. (2000). In contrast to Dolly, cloning resets telomere clock in cattle. *Science*, **288**, 586–7.

Moore, D. W. (2001). *Bioinformatics: sequence and genome analysis.* Cold Spring Harbor Laboratory Press.

Imarisio, S. et al. (2008). Huntington's disease: from pathology and genetics to potential therapies. *Biochem. J.* **412**, 191–209.
General review of the disease caused by a trinucleotide expansion.

Problems

1 How does a restriction enzyme differ from pancreatic DNase?

2 *Eco*RI cuts at a hexamer base sequence; such a sequence must occur many times in *E. coli* DNA. Why does the enzyme not destroy the cell's own DNA?

3 What is meant by the term 'sticky ends', as applied to DNA molecules?

4 What is a genomic clone? What is a cDNA clone? How do they differ? Refer to eukaryotes in your answer.

5 What are the steps in making a genomic library in bacteriophage λ?

6 What is a dideoxynucleoside triphosphate? What is the precise function of these in the Sanger technique of DNA sequencing?

7 Explain in a few sentences the importance and principle of the polymerase chain reaction, specifying the requirements for it to be used.

8 What is a bacterial plasmid expression vector?

9 What is restriction analysis?

10 What is a stem cell and why are they of great current interest?

11 Describe in conceptual outline how mammalian genes can be targeted to produce specific mutants, generally known as knockout mice.

12 In the polymerase chain reaction (PCR):
(a) Why are two primers added?
(b) Why are the primers added in large excess?
(c) Which enzyme is used and what are its special characteristics for use in the PCR?

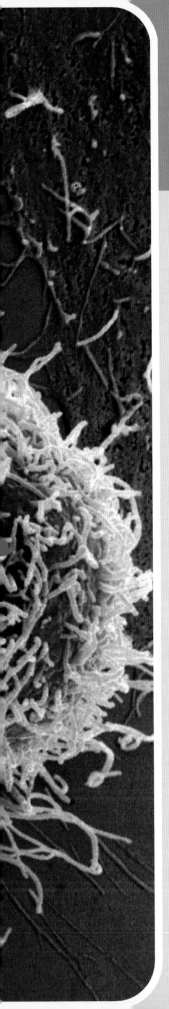

Part 5

Molecular biology in health and disease

Cervical cancer cell, SEM
Coloured scanning electron micrograph (SEM) of a cervical cancer cell. Most cervical cancer arises from the
flattened cells that cover the cervix. It may take years to develop but will then spread rapidly to nearby tissues and
other organs. Regular smear tests allow the dentification of pre-cancerous cells, which can then be removed or
destroyed.
STEVE GSCHMEISSNER/SCIENCE PHOTO LIBRARY

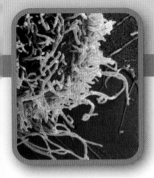

Chapter 29

Special topics: blood clotting, xenobiotic metabolism, reactive oxygen species

At the molecular level, life is a dangerous business, particularly in organisms as complex as mammals. In many areas there is a delicate balance between the necessity of having chemical mechanisms, without which life would be impossible, and the danger inherent in the mechanisms. We are in some respects in the situation of riding on the back of a tiger, which might be an effective way of moving but has dangers if it does not go strictly to plan. It is efficient to circulate oxygen to all the cells of the body by a high-volume flow of blood but this means that the body has to be ready to instantly form a clot at the site of a wound to prevent death by bleeding. Unfortunately, inappropriate blood clotting (thrombosis) also constitutes a life-threatening hazard, and this must be guarded against without impairing the ability of rapid wound response.

It is also efficient to generate useable energy from foodstuffs by transferring electrons to oxygen, and provided the oxygen molecule is reduced to water the process is benign. No process is perfect, however, and incompletely reduced **reactive oxygen species** (ROS), often **free radicals**, are generated, which are destructive to biological molecules. There are also other causes of ROS generation such as ionizing radiation and certain chemicals.

We need to take in a wide variety of foods to obtain essential nutrients but this means the intake of a variety of potentially toxic molecules, which if not disposed of would be dangerous.

In this chapter we have collected together the enzymic mechanisms by which the body protects itself from such hazards. It is more usual to find these attached to other chapters where they have mechanistic relevance, but here they are presented as biological topics in their own right.

Other protective mechanisms fit into specific chapters. The immune system is the major protective mechanism against attack by disease-causing pathogens but this is a subject in its own right and is dealt with in Chapter 30. DNA repair to cope with damage caused by ionizing radiation, ultraviolet (UV) light,

and mutations is also a protective mechanism but this has already been covered in Chapter 22 in which DNA synthesis is dealt with. Similarly, tumour suppression genes, which protect us from cancer, are best left to Chapter 33 where cancer is discussed.

Blood clotting

Coagulation of blood is needed to form blood clots, which plug holes in damaged blood vessels and prevent bleeding. The response has to be rapid and relatively massive, while the initial signal in chemical terms is small. A massive amplification of the signal is needed – the response must, in quantitative terms, be vastly greater than the signal.

We have seen earlier that biochemical amplification is achieved by means of a **reaction cascade**. In this, an enzyme is activated that then activates another enzyme and so on. The fact that the enzymes activated are themselves catalysts means there is an amplification at each step. In case this is not clear, if a single enzyme molecule activates 1000 molecules of the next enzyme in 1 minute and each molecule of the second enzyme does the same for the next enzyme and so on, in a cascade of four steps, in a very short time vast numbers of active molecules of enzyme number four are created. This enzyme at the end of the cascade can rapidly catalyse a massive response.

Blood clotting can conveniently be divided into two parts. There is the cascade resulting in the activation of the enzyme that forms the clot, and there is the mechanism of clot formation itself by that enzyme. The cascade process is based on **proteolysis** – hydrolysis of peptide bonds of inactive precursor proteases activates them (*cf.* trypsinogen activation to trypsin in digestion). All of the necessary inactive precursor proteins are present in the blood waiting for the signal of a damaged blood vessel. They all have to be normally inactive.

What are the signals that clot formation is needed?

When a wound occurs, the endothelial cell layer lining the blood vessels is damaged, exposing the structures underneath, such as collagen fibres. These have a negatively charged or 'abnormal' surface. The blood-clotting response is a localized reaction around the site of damage. Initially, a temporary plug is formed by the aggregation of blood platelets around the hole. Liberation of ADP and thromboxane A_2 activates other blood platelets to aggregate on the wound. For the formation of a clot, a small group of proteins in the blood absorb to the abnormal surface, the net result of which is that two proteases mutually activate each other. One of these is called factor XII. (The nomenclature is slightly confusing in that some 'factors' are enzymes and some are cofactors for enzymes and they are not numbered according to the sequence of their appearance in the process.) Factor XII activates a cascade of three steps resulting in active factor X (another protease). Factor X activates **prothrombin** to the active protease, **thrombin**. (The prefix 'pro' or the suffix 'ogen' refer to the inactive molecules which are converted to the active form.) Thrombin causes clotting (or thrombus formation). (We usually expect enzyme names to end in 'ase' but several of the classic proteases end with 'in' – for example thrombin, pepsin, trypsin, chymotrypsin, and others.)

This is called the **intrinsic pathway** because, if blood is put into a glass vessel, the glass surface triggers clotting. Nothing has to be added; therefore the process is **intrinsic**.

There is also an **extrinsic pathway**. This is triggered by the release of a protein complex called tissue factor from damaged cells and tissues. Since something has to be added to blood, it is called the *extrinsic* pathway, which is shorter than the intrinsic one. A protease is activated and this activates the same factor X as occurs in the intrinsic pathway, resulting again in active thrombin formation. The two pathways are set out in Fig. 29.1.

The intrinsic pathway, being longer, is slower to cause clot formation when measured *in vitro*, than the extrinsic pathway, also measured *in vitro*. However, in the genetic disease haemophilia A, in which blood clotting fails to occur, it is the intrinsic pathway that is deficient due to the absence of factor VIII required for factor X proteolytic activation. However, it seems that, for normal physiological clotting, both pathways function as one, both being essential. Interactions between the two pathways have been identified to be involved in physiological clotting .

How does thrombin cause thrombus (clot) formation?

In the circulating blood there is a protein called **fibrinogen**. The basic molecular unit consists of short rods made up of three polypeptide chains; two of these rods are joined together by S—S bonds near their N-terminal ends, forming the fibrinogen monomer. As shown in Fig. 29.2, at their joining points two of

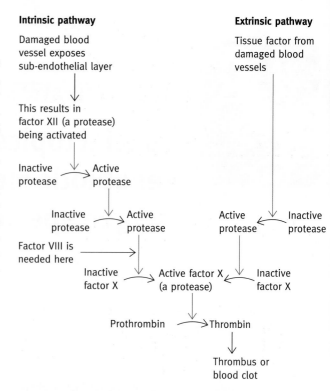

Fig. 29.1 Simplified diagram of intrinsic and extrinsic pathways of blood clotting. The names of the various proteases and other factors involved are omitted for simplicity. (Thirteen factors, numbered I-XIII, are in fact known.) The proteases listed are specific for their particular substrate. Factor VIII is the protein missing in patients with **haemophilia A.**

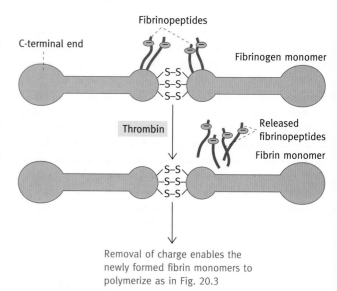

Fig. 29.2 Fibrinogen monomer and its conversion to fibrin monomer. Each half of the fibrinogen monomer is composed of three polypeptide chains, two of which terminate in the negatively charged fibrinopeptides.

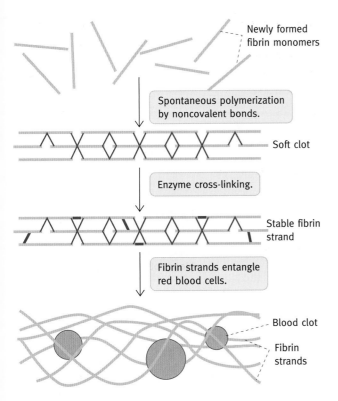

Fig. 29.3 Spontaneous polymerization of fibrin monomers and their enzymic cross-linking to form a stable fibrin strand. The fibrinogen cannot polymerize until thrombin hydrolyses off the fibrinopeptides because (1) the negative charges prevent association and (2) the central sites with which the ends of the fibrin monomers associate are masked by the fibrinopeptides. Noncovalent bonds are shown in blue; the covalent cross-links, shown in red, are arbitrarily represented in position and number.

the three chains in each short rod project as negatively charged peptides, called **fibrinopeptides**. The negative charges on the monomers mutually repel and prevent association.

Thrombin cleaves the fibrinopeptides off, giving a **fibrin monomer**. The fibrin monomer is now able to polymerize spontaneously by noncovalent bond formation. The sites at the end of the fibrin monomer are complementary to sites at the centre of adjacent molecules so that a staggered arrangement forms from the polymerization (Fig. 29.3). This gives a so-called 'soft clot'. A more stable 'hard clot' is formed by subsequent covalent cross-linking between the side chains of adjacent fibrin molecules.

The covalent cross-links are curious in that a glutamine side chain on one monomer is joined to a lysine side chain on the next in an enzymic transamidation reaction:

$$-CONH_2 + H_3N^+ - \longrightarrow -CO-NH- + NH_4^+.$$

(Glutamine   (Lysine   (Cross-link)
side chain)   side chain)

The fibrin strands entangle blood cells, forming a blood clot.

Keeping clotting in check

Blood clotting is potentially dangerous unless limited to local sites of bleeding. Once started, there is the danger of an autocatalytic process like this getting out of hand, causing inappropriate clotting. An elaborate series of safeguards exists. Proteinase inhibitors (for example, antithrombin) in blood 'dampen down' and prevent the clotting reactions from spreading; heparin, a sulphated polysaccharide (a glucosaminoglycan; page 59) present on blood vessel walls, increases this inhibitory effect; another protease, **plasmin**, dissolves blood clots. Plasmin itself is formed from inactive **plasminogen**; activation of this occurs by a protein, **tissue plasminogen activator** (**TPA**, or **t-PA**), released from damaged tissues. Tissue plasminogen activator is one of the therapies for coping with blood clotting. Although present in tissues in minute amounts, its gene and the corresponding cDNA have been isolated and used for commercial production of the protein for injection (the technology is described in Chapter 28). This, and other inhibitory mechanisms, limit the reaction to the damaged surface area, the latter being required for initiation of the process. Blood clotting control is complex; inappropriate clotting is responsible for large numbers of deaths.

Blood clotting requires aggregation of blood platelets. This is encouraged by thromboxane A_2 (page 232), which is synthesized by cells lining blood vessels. The synthesis is selectively inhibited by low doses of aspirin (100 mg/day – see Box 14.2 for action). This is commonly used medically to help prevent vascular problems.

Rat poison, blood clotting, and vitamin K

The widely used rat poison, **warfarin**, kills by preventing blood clotting so that the rodents die from unchecked internal bleeding from minor lesions that continually occur. Warfarin is used clinically, for example, in patients with atrial fibrillation, in which areas of stagnant blood due to defective atrial contractions raise the risk of clotting. It is structurally similar to **vitamin K** (K for the German *Koagulation*) and acts as a competitive inhibitor. It competes with the vitamin for an enzyme site and inactivates it (have a quick look at the two structures in Fig. 29.4). Vitamin K is needed for **prothrombin** conversion to **thrombin**; it acts as a cofactor in an unusual enzyme reaction that adds an extra $-COOH$ group, using CO_2, to several glutamic acid side chains of prothrombin. The carboxyglutamate is needed to bind Ca^{2+}, which is essential in the prothrombin → thrombin activation process. Half the human body's vitamin K is from the diet, and half formed from gut bacteria.

$$\begin{array}{ccc} | & & | \\ CH_2 & & CH_2 \\ | & & | \\ CH_2 & + CO_2 \longrightarrow & HC-COO^- \\ | & & | \\ COO^- & & COO^- \end{array}$$

This group efficiently binds Ca^{2+}

A glutamic acid   Carboxyglutamate
side chain of
prothrombin

(a)

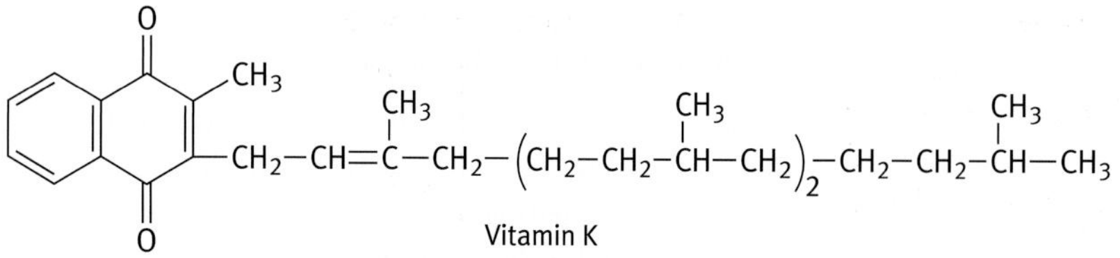

Vitamin K

(b)

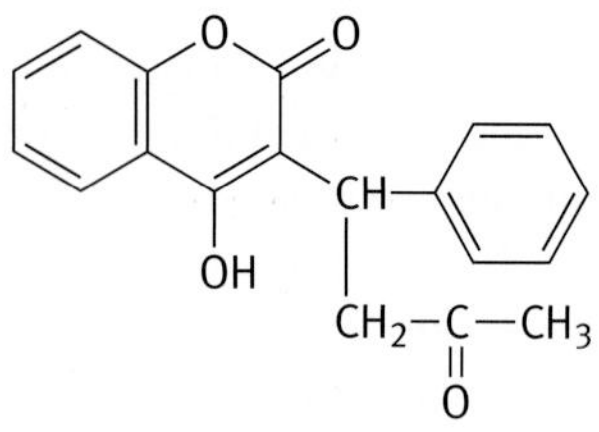

Warfarin

Fig. 29.4 Comparison of the structures of (a) vitamin K and (b) warfarin. Vitamin K is needed for blood clotting. Warfarin antagonizes vitamin K and prevents blood clotting.

Protection against ingested foreign chemicals (xenobiotics)

Large foreign molecules such as proteins are dealt with by the immune system (Chapter 30). Small foreign molecules (which are not indicative of invasion by a living pathogen) are coped with by different systems, enzymic in nature.

Human beings ingest large numbers of different foreign chemicals, collectively referred to as **xenobiotics** (*xeno* meaning foreign). These include pharmaceuticals, pesticides, and herbicides, as well as complex structures of plants. Many of these are relatively insoluble in water, but soluble in fats, and they therefore tend to partition into the hydrocarbon layer of membranes and the fat globules of adipose cells rather than being excreted in the urine or bile. Unless they are rendered more polar and therefore more water soluble, they will accumulate in the body with deleterious consequences. To facilitate their excretion, foreign chemicals are metabolized in several ways, but the **cytochrome P450 (P450)** system is of central importance. P450 is a haem–protein complex. The name comes from P for pigment and 450 from the absorption maximum (in nanometres) of the complex formed with carbon monoxide. (CO is not involved in the reaction – it just happens to give a complex with a spectrum that makes measurement of the amount of P450 easy.) A typical reaction of the P450 system is to add a hydroxyl group to an aliphatic or aromatic grouping. This is known as **phase I** of xenobiotic metabolism. In **phase II** different enzymes add various highly polar groups to the hydroxyl groups, of which glucuronate is the major one, thus increasing their water solubility and facilitating excretion in the urine.

We now need to look a little more closely at these processes. First the P450 system.

Cytochrome P450

There are multiple isoforms or isozymes of P450s in the body. A nomenclature system has been adopted based on amino acid sequence homologies. All are given the name **CYP** followed by a number indicating families (>40% homology), then a capital letter indicating subgroups (>55% homology), and then a number defining an individual isoform. An example is CYP2B4. The nomenclature is relevant because of the medical importance of the enzymes but for the purposes of this discussion only the term P450 will be used.

P450 enzymes have two roles. Some are concerned with normal metabolic roles. A number are needed for the conversion of cholesterol to steroid enzymes so that the adrenal glands are rich in them. The second role is in xenobiotic metabolism.

The remarkable thing about the P450 system is the large number of different compounds that it attacks, including some that living organisms could not have encountered before. It may be that the collection of compounds such as terpenes, alkaloids, etc. found in plants were developed as protection for the plants against attacks (for example, grazing) by animals. The latter, therefore, it may be speculated, evolved a 'detoxifying' system that could cope with almost anything present in plants and other food. The basis of this versatility is that a P450 enzyme has a wide specificity – it attacks a variety of related structures – and different P450 enzymes exist with different but overlapping specificities.

P450 enzymes are found in most tissues, the liver having the largest amounts. They are anchored into the smooth endoplasmic reticulum (ER) facing the cytoplasm. P450s collectively can bring about a surprising variety of reactions, including dehalogenations and desulphurations, but the most important are hydroxylations. In hydroxylations, a foreign compound, AH, is attacked according to the reaction:

$$AH + O_2 + NADPH + H^+ \rightarrow A\text{—}OH + H_2O + NADP^+$$

It is called a **monooxygenase reaction** because it uses only one atom of oxygen from each O_2 molecule. NADPH is used to reduce the other oxygen atom to water. It is also called a **mixed-function oxygenase** because it both hydroxylates AH and reduces O to H_2O. The electrons from NADPH are transferred, one at a time, to the Fe^{3+} in the haem of P450 by a P450 reductase enzyme also present in the smooth ER membrane.

Secondary modification – addition of a polar group to products of the P450 attack

This is known as **phase II** of xenobiotic metabolism. The purpose is to convert the products of the P450 attack into a more water-soluble form for excretion in the urine. There are several reactions for doing this but we will describe the most important ones only. These are **glucuronidation** and conjugation with **glutathione**.

The glucuronidation system

This is present in the smooth ER and it transfers the highly hydrophilic **glucuronate** group from UDP-glucuronate to the hydroxyl group generated on a foreign chemical by P450 (Fig. 29.5). UDP-glucuronate is produced by oxidation of UDP-glucose. Glucuronidation facilitates excretion in the urine and bile by increasing water solubility of the chemical. The same system is used for the excretion of endogenous products such as bilirubin arising from haem degradation (page 291).

The glutathione S-transferase system

Glutathione is a tripeptide of glutamic acid, cysteine, and glycine, the peptide link between glutamate and cysteine being on the γ-carboxyl (Fig. 29.6). It is present in large amounts in liver, muscle and other tissues. Its main function appears to be to maintain a reducing situation in the cells by virtue of its —SH group. For this reason it is abbreviated to GSH.

Glutathione S-transferases add the sulphur to xenobiotics, which include halogenated molecules, epoxide metabolites of carcinogens, and others. The transferases are present in the smooth ER. The reaction catalysed is as follows:

$$RX + GSH \rightarrow RSG + HX$$

where R is an electrophilic xenobiotic.

The reaction is an important defence against reactive carcinogenic molecules for it prevents them reacting with DNA and causing genetic damage. The product, RSG, of the transferase reaction is modified before excretion. The glutamyl and cysteinyl groups are hydrolysed off and the cysteinyl amino group acetylated to form a **mercapturic acid**, which is excreted.

Glutathione is involved in quite separate important protective reactions against peroxides.

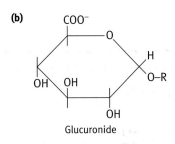

(a)

Glucose-1-phosphate
— UTP
→ PP$_i$
UDP-glucose
H_2O — 2 NAD$^+$
→ 2 NADH + H$^+$
UDP-glucuronate
— R–OH
UDP-glucuronosyl transferase
R–glucuronate + UDP

(b)

COO$^-$
OH OH
OH
O–R
H
Glucuronide

Fig. 29.5 (a) The glucuronidation system. **(b)** Structure of a glucuronide.

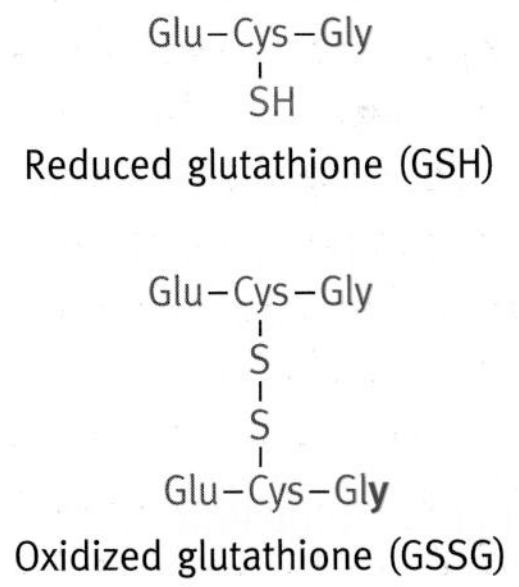

Glu–Cys–Gly
SH
Reduced glutathione (GSH)

Glu–Cys–Gly
S
S
Glu–Cys–Gly
Oxidized glutathione (GSSG)

Fig.29.6 Structures of reduced glutathione (GSH) and of the oxidized form (GSSG). Glu, Cys, and Gly are abbreviations for glutamate, cysteine, and glycine, respectively, using the three-letter system.

Medical significance of P450s

Many pharmaceutical drugs are attacked by P450 so that the half-life of a drug in the body is related to the rate it is attacked and excreted. The level of a given enzyme may vary from one individual to another because of genetic variations so that the correct dose of a drug may vary from person to person. Another aspect is that many P450s are induced by drugs. Thus if barbiturates are fed to a rat there is a massive proliferation of the smooth ER and of the P450s in it. Once the inducing drug is disposed of, the smooth ER returns to normal. This can complicate drug therapy for if a patient on a correct dose of one drug is given another that induces the P450 that attacks the first one, then the dose of the latter may now be inadequate. An illustrative case is warfarin the anticlotting agent, the dose of which is carefully calibrated. If the patient is also given a P450-inducing drug the warfarin dose could now be inadequate.

The P450s are not always beneficial in their attacks on xenobiotics. An ironic twist to the story is that the oxidation of some substances by the P450 increases their carcinogenic effect.

Multidrug resistance

Another form of protection of cells against toxic chemicals is the reduction of their accumulation inside cells. Many cells, including those of human tissues, express a P glycoprotein (P for permeability) in their cell membranes that is an ATP-driven multidrug transporter of drugs out of the cell. It is one of a very large family of ABC (ATP-binding cassette) transporters with common structural features (page 112). They are found in bacteria and eukaryotes. A remarkable range of chemicals is transported, among them several of the anticancer drugs used in chemotherapy. This has caused much interest in the phenomenon but a much wider range of chemicals reacts with the transporter. These include many pharmacological agents and cytotoxic chemicals. Multidrug resistance can occur after prolonged administration of a drug due to P glycoprotein(s) being induced. The acquired resistance might include drugs other than to the original one causing the resistance in the first place.

Since steroids are also transported and the transporter protein is prevalent in steroid-secreting adrenal cortical cells, the system may have this as a primary biological role. It has recently been found to transport cholesterol out of cells (page 173) and participate in the formation of HDL particles and so is antiartherogenic. The molecules transported have no chemical similarities but all are amphipathic compounds preferentially soluble in lipids.

Protection against reactive oxygen species

As stated in the introduction, oxidation of foodstuffs may sometimes lead to incompletely reduced reactive oxygen species (ROS). ROS may be compounds such as hydrogen peroxide (H_2O_2) and they may also be free radicals, which have an unpaired electron, such as superoxide (O_2^-) and the highly reactive hydroxyl radical ($OH^{\cdot}$).

Formation of the superoxide anion and other reactive oxygen species

As described in Chapter 12, oxygen is an ideal electron sink for the energy-generating electron transport system. Its position on the redox scale means that, in energy terms, electrons from NADH have a long way to 'fall', meaning that the negative free-energy change of the overall oxidation of NADH to produce water is large. O_2 accepts four electrons and four protons to give H_2O, the final product of the electron transport chain:

$$O_2 + 4e^- + 4\,H^+ \rightarrow 2H_2O$$

O_2 is relatively unreactive and therefore itself does no chemical damage, and the product is H_2O. During evolution, the switch from energy generation by anaerobic metabolism to the use of oxygen as the electron sink was one of the most important events. There is, however, a darker side to the story, for O_2 has the potential to be dangerous in the body.

One way in which the danger occurs is when a single electron is acquired by the O_2 molecule to give the **superoxide anion**, a free radical that is a reactive corrosive chemical agent:

$$O_2 + e^- \rightarrow O_2^-$$

Although free radicals such as the superoxide anion are present in minute amounts, they set up chain reactions of chemical destruction in the body. Unpaired electrons are usually very reactive and seek a partner. The unpaired electron of O_2^- acquires a partner by attacking and destroying a covalent bond of some cell constituent. In doing so it acquires a partner electron, and so solves its own problem, so to speak, but also this generates a new free radical species with an unpaired electron from the attacked molecule that, in turn, attacks yet another molecule of cell constituent producing yet another free radical, and so on. The destruction initiated by the free radicals is thus a self-perpetuating chain of destructive reactions.

Superoxide is formed in the body in several ways. In the electron transport chain, the final enzyme that donates electrons to oxygen, cytochrome oxidase, does not release partially reduced oxygen intermediates in any significant amounts – it ensures that an O_2 molecule receives all four electrons resulting in H_2O formation. However, in the respiratory chain it is inevitable that small amounts of superoxide are formed. Moreover, mutations in mitochondrial DNA may block electron transport pathways and deflect electrons into formation of reactive oxygen species. Mitochondria have few or no DNA-repair systems, so that mutations accumulate in them.

In addition, there are other oxidation reactions in the body that produce small amounts of dangerous oxygen species. Spontaneous oxidation of haemoglobin (Hb) to methaemoglobin (Fe^{3+} form) is another source; as a rare event, the oxygen in oxyhaemoglobin (HbO_2) instead of leaving as O_2 and leaving behind haemoglobin in the Fe^{2+} form, leaves as O_2^- with the formation of methaemoglobin, the Fe^{3+} form.

There are other sources of ROS. Ionizing radiation, by its interaction with water generates ROS, which can lead to free radical attack on biological components. When phagocytes ingest a bacterial cell there is a rapid increase in oxygen consumption; this is used to oxidize NADPH *via* a mechanism that deliberately generates superoxide anions. These are shed into the vacuole and converted to H_2O_2, which helps to destroy the contained bacterial cell. Excess neutrophils attracted to irritated joints may lead to superoxide release and contribute to arthritic damage.

Some oxidases that directly oxidize metabolites using O_2 and quite distinct from the respiratory pathways we have described, generate H_2O_2.

Flavoprotein oxidases, with FAD as their prosthetic group, generally catalyse reactions of the type:

$$AH_2 + O_2 \rightarrow A + H_2O_2$$

Xanthine oxidase, involved in purine metabolism (Chapter 20), is of this type. H_2O_2 is potentially dangerous because, in the presence of metal ions such as Fe^{2+}, it can generate the highly reactive hydroxyl radical (OH·), not to be confused with the hydroxyl anion (OH^-).

$$H_2O_2 + Fe^{2+} \rightarrow Fe^{3+} + OH^· + OH^-$$

(H_2O_2 is the result of two electrons being added to O_2; when three electrons are added, OH· and OH^- are formed.) The hydroxyl radical can attack DNA and other biological molecules.

The biological injuries caused by free radical damage are not precisely established but it has been suggested they contribute to ageing, cataract formation, the pathology of heart attacks, and other problems.

Basically, there are two protective strategies – one chemical and the other enzymic. The two protective enzymes, catalase and superoxide dismutase, are important. Another enzyme that destroys H_2O_2 is glutathione peroxidase, described below. Since brain has little catalase (the enzyme that decomposes H_2O_2 to H_2O – see below), this latter enzyme possibly is the main one for protection against H_2O_2 in that organ.

Box 29.1 Red wine and cardiovascular health

Epidemiological studies in the 20th century revealed what became known as the French Paradox – despite high consumption of saturated fat, France was found to have relatively low rates of coronary heart disease

Comparing figures for heart disease in men aged 55–64 years in Europe, North America, and Australasia, it was also found that the highest number of deaths were in traditional beer and spirit-drinking countries, while France had the lowest number, and the highest wine consumption.

Recent epidemiological studies have shown that it is moderate consumption of red wine specifically that reduces the risk of death from cardiovascular disease.

The substances in red wine that have this effect are polyphenols found in the grape skin and seeds. The polyphenol content of wine varies with grape variety, cultural conditions, and the wine making process. The most abundant polyphenols in red wine are the procyanidins, made up of repeating units of 2 to 6 molecules of the polyphenol, catechin (oligomeric procyanidins – OPCs). Some other sources of OPCs are cocoa solids such as found in dark chocolate, and in several fruits, particularly cranberries.

Procyanidins are antioxidants. They quench oxygen free radicals and prevent damage by these agents. In blood vessels they appear to protect low-density lipoproteins (LDLs, page 171) from free radical-mediated oxidation. Plaque formation in arteries involves LDL oxidation, and heart attacks ensue when plaques in arteries burst; blood clots form and cause blockages.

Procyanidins may have an additional protective effect by increasing the synthesis and release of nitric oxide (NO, page 451) by the vascular endothelium, thus promoting vasodilatation, and lowering the blood pressure.

(+)-Catechin monomer

Procyanidins

n = 0 (dimer) to 4 (hexamer)

Experiments with cultured endothelial cells identify the most potent vasoactive polyphenols in red wine as OPCs. A wide range of red wines across the world was tested and the OPC content of each wine correlated with the suppression of synthesis of endothelin-1, a powerful vasoconstricting peptide.

See references in Further reading.

Mopping up oxygen free radicals with vitamins C and E

The *chemical strategy* is to dampen or **quench** the chain reactions, initiated by superoxide, by using **antioxidants**. The requirement for a quenching reagent is that it should itself be attacked by a free radical such as the superoxide anion but generate a free radical insufficiently reactive to perpetuate the chain reaction. The main biological quenching agents are **ascorbic acid** (vitamin C) and **α-tocopherol** (vitamin E). The former is water soluble, the latter lipid-soluble, and so, between them, they protect in both phases of the cell. (Note that there are medically undesirable side effects of excessive doses of either.) These two vitamins are not the only antioxidants – some normal metabolites such as uric acid are effective; **β-carotene** (Fig. 27.25) is also an antioxidant. Certain polyphenolic compounds (**procyanidins**) found in red wine and other food sources have antioxidant properties (see Box 29.1)

Bilirubin, produced from haem breakdown by haem oxygenase, is also an effective antioxidant.

Enzymic destruction of superoxide by superoxide dismutase

Most or all animal tissues contain the enzyme **superoxide dismutase** and, most appropriately, it occurs in mitochondria. It also occurs in lysosomes (Chapter 26) and peroxisomes (Chapter 14) as well as in extracellular fluids such as lymph, plasma, and synovial fluids. Superoxide dismutase catalyses the reaction:

$$2O_2^- + 2H^+ \rightarrow H_2O_2 + O_2$$

It is an oxido-reduction reaction between two superoxide anions. The hydrogen peroxide is destroyed by **catalase**:

$$H_2O_2 \rightarrow 2H_2O + O_2$$

The glutathione peroxidase–glutathione reductase strategy

Glutathione we have already met. It is a thiol tripeptide (γ-glutamyl-cysteinyl-glycine), which is abbreviated to GSH (Fig. 29.6). As stated, it is found in most cells where, because of its free thiol group, it functions as a reducing agent, for example to keep proteins with essential cysteine groups in the reduced state. This reaction with proteins is nonenzymic, but another protective action of GSH is to inactivate peroxides via the action of **glutathione peroxidase**, producing **oxidized glutathione (GSSG)**:

$$H_2O_2 + 2GSH \rightarrow GS—SG + 2H_2O$$

Organic peroxides (R—O—OH) generated by free radical attack on membrane lipid are also destroyed in this way. The GSSG is subsequently reduced by NADPH, the reaction being catalysed by **glutathione reductase**:

$$GSSG + NADPH + H^+ \rightarrow 2GSH\ NADP^+$$

Red blood cells depend for their integrity on GSH, which reduces any ferrihaemoglobin (methaemoglobin) to the ferrous form, as well as destroying peroxides. This explains why red blood cells have the pentose phosphate pathway (Fig. 17.1) to supply the NADPH needed for the reduction of GSSG. Usually, patients with defective glucose-6-phosphate dehydrogenase, the first enzyme in the pentose phosphate pathway, have enough enzymic activity for normal function. However, when extra stress is placed on the cell, for example by the accumulation of peroxides due to the action of the antimalarial drug **pamaquine** (a glycoside found in fava beans) and other drugs, the supply of NADPH can no longer be maintained (see Box 17.1). The integrity of the cell membrane is impaired and haemolysis results in such patients from taking the drug. As with sickle cell anaemia, lack of glucose-6-phosphate dehydrogenase may have some protective effect in areas where lethal forms of malaria exist.

 Summary

There are number of processes in the body which are essential protective devices against different hazards.

Blood clotting involves two separate pathways of proteolytic enzymes, which converge at active factor X. This activates pro-thrombin to thrombin, a proteolytic enzyme that converts fibrinogen to fibrin. Fibrin is a fibrous complex that entraps blood cells into a soft clot, which is then stabilized by cross-link formation between the strands.

One pathway is triggered by a wound exposing an 'abnormal' surface, such as collagen. This triggers the activation of a long proteolytic cascade in which each activated component activates the next. The purpose is to amplify a minute signal into a massive response. It is called the intrinsic pathway.

The shorter extrinsic pathway results from release of a factor from damaged cells. Both pathways are needed for efficient clotting.

Vitamin K is needed for clotting. It is involved in the carboxylation of a number of glutamate side chains of prothrombin. The carboxyglutamate binds Ca^{2+} efficiently, which is needed for the activation of prothrombin to thrombin. The same applies to several other clotting factors. The agent warfarin antagonizes vitamin K and prevents clotting. It is used therapeutically, in carefully graded amounts, as a guard against inappropriate clotting and also as a rat poison.

A complex system is important in protecting against inappropriate clotting since all the components are present in blood; inappropriate clotting causes many deaths.

Xenobiotics are foreign chemicals, including pesticides, pharmaceuticals, and many others. They would accumulate in the body unless excreted. Since most are lipid soluble, they must be rendered more hydrophilic for excretion in the urine.

The cytochrome P450 family of enymes typically hydroxylate a wide variety of the compounds, followed by a secondary addition of polar attachments. Multiple P450s exist, each with wide overlapping specificities so that a vast range of chemicals can be attacked.

The P450 systems are of great medical importance since they can interfere with drug therapy regimes.

Reactive oxygen species are formed in the body by several mechanisms and are destructive agents. Superoxide radicals are converted to hydrogen peroxide by the enzyme superoxide dismutase. The enzyme catalase destroys the peroxide. Another system involving glutathione protects red blood cells against peroxides. Vitamins C and E are antioxidants. They protect against free radicals as quenching agents – they terminate the destructive reaction chains initiated by free radicals (but can have side effects if ingested excessively).

Further reading

Blood clotting

Scully, M. F. (1992). The biochemistry of blood clotting: the digestion of a liquid to form a solid. *Essays Biochem.*, **27**, 17–36.
Review of clotting, limiting of the process, and medical aspects.

Hooper, N. M. (Ed.) (2002). Proteases in biology and medicine. *Essays Biochem.*, **36**, 1–167.
A complete issue with 12 reviews on many aspects of proteases, including caspases, cancer, proteasomes, and blood clotting.

Sadler, J. E. (2004). K is for Koagulation. *Nature*, **427**, 493–4.
Short News and Views article about identifying a key component of vitamin K metabolism that is the target of the anticoagulant drug, warfarin.

Cytochrome P450

Coon, M. J., Ding, X., Pernecky, S. J., and Vaz, A. D. N. (1992). Cytochrome P450; progress and predictions. *FASEB J.*, **6**, 669–73.
A crisp overview of different aspects of this important field.

Eastabrook, R. W. (1996). The remarkable P450s: a historical review of these versatile hemoprotein catalysts. *FASEB J.*, **10**, 202–4.
Summarizes the metabolic roles of P450s in broad terms.

Reactive oxygen species

Babior, B. M. (1987). The respiratory burst oxidase. *Trends Biochem. Sci.*, **12**, 241–3.
Reviews the deliberate production of superoxide by phagocytes.

Rusting, R. L. (1992). Why do we age? *Sci. Am.*, **267**(6), 130–41.
Contains a section on the evidence that free radicals may be a significant factor in ageing.

Harris, E. (1992). Regulation of antioxidant enzymes. *FASEB J.*, **6**, 2675–83.
A review of how the levels of enzymes destroying oxygen free radicals are adjusted to needs.

Rose, R. C. and Bode, A. M. (1993). Biology of free radical scavengers: an evaluation of ascorbate. *FASEB J.*, **7**, 1135–2.
Discusses the properties required for free radical scavengers and looks at ascorbate (vitamin C) in detail.

Fridovich, I. (1994). Superoxide radical and superoxide dismutase. *Annu. Rev. Biochem.*, **64**, 97–112.
Concise review of the field.

Uddin, S. and Ahmad, S. (1995). Dietary antioxidants protection against oxidative stress. *Biochem. Edu.*, **23**, 2–7.

Lenaz, G. (2001). The mitochondrial production of reactive oxygen species: mechanisms and implications of human pathology. *IUBMB Life*, **52**, 159–64.

Di Castelnuovo, A. *et al.* (2002). Meta-analysis of wine and beer consumption in relation to vascular risk. *Circulation*, **105**, 2836–44.

Fitzpatrick, D. F., Bing, B., Maggi, D. A., Fleming, R. C. and O'malley, R. M. (2002). Vasodilating procyanidins derived from grape seeds. *Ann. N. Y. Acad. Sci.*, **957**, 78–89.

Dell'Agli, M., Busciala, A. and Bosisio, E. (2004). Vascular effects of wine polyphenols. *Cardiovasc. Res.*, **63**, 593–602.

Corder, R., Mullen, W., Khan, N. Q., Marks, S. C., Wood, E. G., Carrier, M. J. and Crozier, A. (2006). Red wine procyanidins and vascular health. *Nature*, **444**, 566.

Problems

1 Blood clotting involves cascades of enzyme activations. What is the rationale for such a long process?

2 Explain how thrombin triggers the formation of a blood clot.

3 The spontaneous polymerization of fibrin monomers forms a 'soft clot'. How is this converted into a more stable structure?

4 Why is vitamin K needed for blood clotting?

5 What is the function of cytochrome P450?

6 Why should NADPH be involved in an oxygenation reaction?

7 What is the function of UDP-glucuronate in disposing of water-insoluble compounds?

8 What is multidrug resistance?

9 What is superoxide?

10 What mechanisms exist for guarding against the deleterious effects of superoxide?

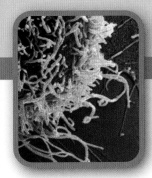

Chapter 30

The immune system

Overview

The immune system is a large subject area. Apart from its medical importance, it is one of immense relevance to molecular biology. In this chapter we have attempted to give an outline of the basic molecular mechanisms involved and to give sufficient cell biology background of the subject to put them into a meaningful context.

The body is constantly under the threat of invasion by pathogenic organisms such as bacteria and viruses. The immune system is the main protection against this. Its necessity is illustrated by the results of an impaired immune system due to genetic defects such as adenosine deaminase deficiency (page 304) or to infection by HIV (the AIDS virus). It is also necessary as a protection against abnormal cells being produced within the body. Immunological surveillance against cancer cells is believed to be an important role.

There are two separate immune systems. One is the **innate system**, which is mainly a protection against infection by microorganisms. It is less protective than the **adaptive immune system** but has the advantage that it is instantaneous, while the adaptive response usually takes a week or more to become fully effective. Bacterial infections can multiply very rapidly so that this is important. The innate response is also necessary for the triggering of the adaptive response.

This chapter is mainly about the adaptive response and we shall give only a brief outline of the innate system.

The innate immune system

Phagocytic white cells known as macrophages, present in tissues, and neutrophils in blood, engulf and destroy pathogens, and an inflammatory process ensues. They secrete chemokines, which attract more white cells to the site of infection, and also cytokines, which cause cells to produce an inflammatory response (see page 434 for more on this). The phagocytes recognize components of the infective prokaryotes that never occur in the body. Typical examples are the bacterial cell wall components such as lipopolysaccharides – carbohydrates anchored into the membrane lipid bilayer by fatty acid hydrocarbon chains. There are no counterparts to this in an animal and so the phagocytes can safely attack them as foreign. Prokaryote protein synthesis (page 338) is initiated by N-formylmethionine while eukaryotes do not have formyl groups attached. Such formylated proteins are powerful chemokines which attract macrophages to the site of infection. The innate immune system also has natural killer cells (NK cells) which also destroy pathogen-infected cells, and possibly also some types of cancer cells, but the recognition procedure is different from that adopted by the killer cells in the adaptive immune system (page 499).

The innate immune response does not confer any protection against subsequent infection by the same pathogen – there is no lasting immunity as happens with the adaptive response.

The adaptive immune response

This will be referred to simply as the immune response. In evolutionary terms it is a relatively recent development confined to vertebrates. It depends on the recognition of one or more specific macromolecules, mainly proteins, recognized as foreign to the body. A foreign macromolecule in the body is a warning signal that a defensive response is needed against an invader. A molecule that elicits a response is known as an **antigen** (antibodies are **gen**erated). A patient recovering from a disease such as measles is protected, often for a long time, from a further measles infection but not from other pathogens.

The adaptive response requires the help of the innate system to activate macrophages. Vaccination involves injection of killed pathogens or harmless proteins derived from pathogens; to trigger the immune response it is necessary to stimulate an innate immune response by injecting an immunostimulant to start the process. This is usually given in the form of an adjuvant, a preparation of microbial origin which mimics infection. It stimulates the activity of phagocytes in engulfing foreign material.

The immune system responds to macromolecules. Proteins are the most important class. Carbohydrate attachments to proteins are also often antigenic, as are some polysaccharides themselves. The immune system does not protect against small foreign molecules such as drugs though antibodies can develop targeted to small molecules attached to proteins. Other enzymic protection systems deal with unattached small foreign molecules (page 486).

The problem of autoimmune reactions

The body has thousands of proteins differing from foreign proteins only in the detail of amino acid sequences and yet the distinction between 'self' and foreign proteins must be made because otherwise the immune system would attack the components of the body in **autoimmune reactions**. The avoidance of autoimmune attack is of overriding importance, and much of the elaboration of immune systems is because of this requirement. The protection cannot be absolutely perfect, as evidenced by the existence of **autoimmune diseases**. An example of an autoimmune disease is **myasthenia gravis** in which an autoimmune reaction destroys the acetylcholine receptors of muscle, thus preventing nervous stimulation of contraction. **Insulin-dependent diabetes** is the result of an autoimmune attack on insulin-producing cells of the pancreas. **Rheumatic fever** is an autoimmune attack on heart cells caused by *Streptococcal* infections. (Antibodies against bacterial antigens happen to cross-react with a heart muscle protein.)

The cells involved in the immune system

All blood cells have their origin in bone marrow stem cells which continually divide (see Fig. 2.7). Two of the principle players in the immune system are **B cells** and **T cells**. They are known as **lymphocytes**. Although B and T cells both originate from the same stem cells in the bone marrow, there is a major difference; B cells multiply and undergo their primary maturation in the bone marrow, but those cells committed to becoming T cells migrate to the thymus gland, a small structure behind the breastbone, where they multiply and mature – hence the name T (for thymus) cell.

B cells are responsible for producing antibodies, but only after an elaborate preparation. To do this they require the cooperation of a class of T cells known as **helper T cells** – they help the B cells to do their job. Two separate classes of T cells are generated in the thymus – helper T cells and **cytotoxic or killer T cells**.

A group of phagocytic cells are also involved in immunity. **Dendritic cells** are the most important of these. They engulf foreign material, an activity usually stimulated by the invading organism. The engulfed proteins (antigens) are hydrolysed into short peptides, which are placed in grooves of special proteins that are then displayed on the outside of the phagocytic cell. Since they present the peptides derived from the antigens for inspection, they are known as **antigen-presenting cells (APCs)**. The T cells are activated only by peptides presented by an APC – they do not react directly with a foreign antigen. The binding results in the activation of these cells, whose function will be described shortly.

There are two arms to the adaptive immune response

First, there is the production of **antibodies**. These are soluble proteins in the blood that combine with the foreign **antigens**. As stated, an antigen is a macromolecule, normally foreign to the body, which leads to **anti**body **gen**eration. The attachment of the antibody to antigen results in protective events described later. This type of immunity is called **humoral immunity**, for, in times past, body fluids were referred to as humors.

The second type of immunity is known as **cell-mediated immunity**. In this, the **cytotoxic T cells** or **killer T cells** (the names are interchangeable) recognize abnormal cells in the body. For instance, the abnormality might result from a viral infection in the cell. The killer cell attaches to the cell and destroys it. This stops a infection from spreading, since cell destruction aborts virus replication. Direct contact is made between the killer cell and its target – hence the term, cell-mediated immunity. Thus the two mechanisms complement each other – to use virus infection as the example, the humoral system is extracellular and attacks a virus in the blood or in the mucous membrane before it infects a cell. The antibodies cannot attack the virus once it is inside a cell so a different system is needed. The cell-mediated immune system destroys the host cell after it has become infected. The killer T cell delivers a 'death signal' to the outside of the infected cell, which results in events inside the targeted cell that lead to its destruction.

Where is the immune system located in the body?

There is no single organ for the immune system. Instead there are vast numbers of separate cells which, *en masse,* would be equivalent to a large organ. They are distributed in spleen, lymph nodes, intestinal, and mucosal lymphoid tissue, and about 10% in blood and lymph circulation. Lymph is a solution of soluble blood proteins and small molecules – it is essentially a blood 'filtrate' lacking red blood cells that leaks out of the blood and bathes cells. It is drained into thin-walled lymphatic vessels, which return the lymph to the blood *via* two main ducts. The lymphocytes are able to squeeze through certain capillary walls and so can migrate from blood to lymph. In being carried in the lymph, they encounter foreign material in all parts of the body. The lymph is propelled by the movements of the body. On the route of the lymph returning to the blood, the lymphatic vessels expand at places into **lymph nodes**, known as **secondary lymphoid centres** (bone marrow and thymus being primary), in which high concentrations of B and T cells are found. The nodes are where cells undergo the interactions and the differentiation processes that occur when an immune response is being mounted. They are strategically placed in the armpit, groin, tonsils, adenoids, and intestine so that lymphocytes and foreign antigens draining from the tissues are likely to encounter nodes. Dendritic cells (or other APCs) displaying foreign peptides on their surface encounter T cells in the lymph nodes and activate any that are capable of binding to the displayed peptide. Activation of helper T cells and cytotoxic cells are crucial steps in the generation of both types of immune response by mechanisms to be described shortly.

With that general introduction we will deal with immunity under two main headings – humoral or antibody-based immune protection, and then cell-mediated immunity.

Antibody-based or humoral immunity

Structure of antibodies (immunoglobulins)

Antibodies are **immunoglobulins**. There are several different classes of these. Immunoglobulin G (**IgG**) is the one produced in largest amounts as an immune response develops. We will look at its structure first. It is a Y-shaped protein made up of two identical **light (L) polypeptide** chains and two identical **heavy (H) chains** held together by disulphide bonds (Fig. 30.1).

The ends of the Y arms have hypervariable regions on both the heavy and light chains so that in a collection of IgG molecules they will all have the same constant regions (see Fig. 30.1) but molecules will have a different specific version of the variable regions. It is these hypervariable regions of the two chains that form the antigen-binding site on each of the Y arms. On each arm the light and heavy variable regions form a single binding site to which an antigen might fit. Because of the variability, vast numbers of binding-site specificities exist, one per immunoglobulin molecule, the sites on the two arms being identical on each individual molecule. The binding to the antigen is by noncovalent bonds. An antibody thus has two identical antigen-binding sites, which means that it can cross-link antigen molecules. At the fork of the Y there is a flexible 'hinge' region, which increases the cross-linking ability of the molecule.

Although foreign molecules need to be large to produce an immune response, a given antibody binds to only a small part of the antigen – in the case of a protein antigen only a few amino acids. The specific part of the antigen that is recognized by an antibody is called an **epitope**. Thus a given protein antigen provokes the production of a large number of different antibody molecules, each combining with a specific epitope.

What are the functions of antibodies?

The serum of an immunized animal contains multiple antibodies recognizing different epitopes on a given antigen so that an aggregate of antigens and antibodies forms, which activates

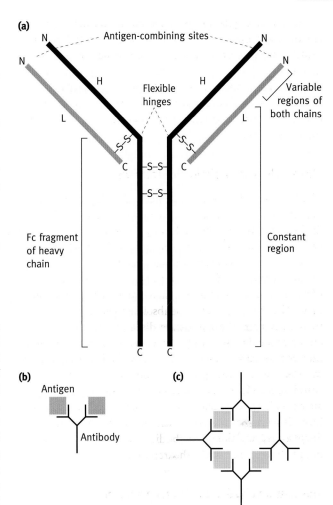

Fig. 30.1 (a) An immunoglobulin G (IgG) molecule. IgM and IgA molecules differ in the Fc fraction, but are similar in the variable part. H, heavy chain; L, light chain. N and C refer to the N-terminal and C-terminal ends of the polypeptides, respectively. The Fc fragment is the C-terminal half of the two H chains, bound covalently into a single fragment by —S—S— bonds. The name arises from crystallizable (c) fragment (F), one of the products of papain hydrolysis of the antibody molecule at the two peptide bonds at the flexible hinges. **(b)** Cross-linking of antigen with single specific epitopes. **(c)** Antigens with multiple antigenic determinants form a cross-linked insoluble cluster with antibody.

the **complement system**. This is a group of proteins in blood; antibody-antigen complexes, when assembled on a bacterial cell, attract complement proteins. The complement-coated bacterium or other antigen is engulfed by **phagocytes** which digest it.

The different classes of antibodies

There are several classes of immunoglobulins differing in the **constant regions** of their H chains – they are constant only within each class. These differences have nothing to do with the antigen-binding sites but are related to the precise role of the antibody (see below).

When the body responds to an antigen, the antibody produced first is **IgM**, a multisubunit, or polymeric, form of antibody with 10 antigen-combining sites. It is particularly efficient at binding viruses and bacteria into a network, because of its multiple combining sites, and in activating complement and promoting phagocytosis. More prolonged challenge by the same antigen results in production of **IgG**. The **class switching** changes the constant region of heavy chains but not the variable sections of the H and L chains. The antigen-binding specificity therefore does not change but the cell switches the class it makes. (The mechanism of gene segment rearrangements is discussed below.)

IgA, another class, is of importance as a first line of defence. It is transported through the membranes of epithelial cells that line the intestine and lungs, by combination with a special secretory polypeptide, into the mucous layer of the intestine and respiratory tract. For example, it protects against the cholera bacillus establishing infections in intestinal cells in humans previously exposed to the disease. It is also a component of breast milk. The baby absorbs the IgA from its gut into its blood and so gets some immediate protection against infection.

IgE binds to receptors on mast cells (a type of white blood cell) of the body and triggers the release of cytokines and histamine. These help to combat infections partly by dilating blood vessels but also are responsible for allergies such as **hay fever**.

Generation of antibody diversity

The body is potentially exposed to vast numbers of different antigens and it is potentially capable of producing an adequate number of different antibodies to combine with these. The principle is that the immune system produces B cells each differing in the genes coding for antibodies. Each newly developed B cell can produce, in terms of antigen specificity, only a single antibody species, but each cell, as released from the bone marrow, produces a different one. Given the vast numbers of cells involved, whatever antigen comes along, there will be an antibody that by sheer chance combines with it.

What is the source of this variability? Human cells contain around 25,000 genes to code for all the proteins of the body, whereas the body has the potential of making about 10^{12} different immunoglobulins (estimates vary), each requiring a different gene to code for it. Clearly, something special must arrange for the enormous diversity of immunoglobulin genes. We are used to cells attempting to ensure that their genes are constant in their protein-coding capacity but here we have exactly the opposite – a mechanism to ensure that there is tremendous variability in genes coding for the antigen-binding sites of antibody proteins.

To understand how this generation of diversity is achieved, let us look first at the L chain of immunoglobulin molecules. As outlined, it has two sections – a terminal variable region that participates in the antigen-binding site and a constant region that is identical in all immunoglobulin molecules of a particular class. Consider a stem cell in the bone marrow *before* it has become committed to becoming a B cell. At this stage it has not assembled its gene for the immunoglobulin L chain but it has sections of DNA from which a gene coding for an antibody will be assembled, a very exceptional process. One section codes for the constant (C) domain; there are four separate sections each of which codes for different short peptide sequences used to join the constant to the variable gene section, called **joining** (**J**) **sections**. And, finally, there are about 300 **variable** (**V**) **sections** each of which codes for an end of an L chain. The 300 V sections are all different from one another, as are the four J sections, and they will thus code for different peptide sequences.

When the bone marrow stem cell is committed to becoming a B lymphocyte, a rearrangement of these DNA sections occurs so that a functional L chain gene is produced by site-specific or somatic recombination (explained below). In the assembly of the gene, one C section is joined by one of the four J sections to one of the 300 V sections. This results in a new composite gene different in each of the B lymphocytes, as shown in Fig. 30.2. *The DNA recombination is a completely random process.* Thus, in the final mRNA coding for an IgG L chain, any one of 300 V sections is joined to any one of the four J sections giving 1200 different combinations. The recombination, known as somatic, or site-specific, recombination, involves excision of a piece of DNA and rejoining of the remaining sections. The recombination is brought about by two proteins which cleave the DNA at specific sites, thus excising a section of DNA. The double-stranded breaks so created are end-joined by other enzymes in a sloppy, imprecise way creating more diversity. Yet further diversity is created by an enzyme which adds random nucleotides to the cut ends before joining. The number of possible L chain variants is greatly increased by this joining variability.

The H chain gene is assembled in a similar random manner from variable sections giving large numbers of possible H chain gene variants. The variability is greater in the H chain assembly due to the greater diversity in joining events than occurs with L chain assembly. Since the antigen-binding site is made from the combination of the variable parts of the H and L chains, and there are large numbers of different H and L genes, the number of different binding sites coded by the assembled L and H genes is sufficient to account for the immune system's ability to produce vast numbers of different antibodies. In summary, each developing B cell assembles, at random, a gene coding for one antibody with its specific antigen-binding site.

Since the B cells are diploid there will be both maternal and paternal DNA sections so that production of more than a single gene coding for antibodies might be expected. However, a process of allelic exclusion ensures that only one version of H and L chains is assembled from maternal or paternal sections so that a B cell is monospecific in its antibody-binding.

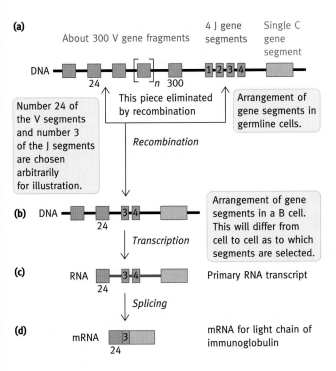

Fig. 30.2 The process of rearrangement leading to a functional L chain immunoglobulin gene. **(a)** Arrangement of gene segments in the stem cell where the immunoglobulin genes are not being expressed. **(b)** The randomly chosen V gene (V_{24} in this example) is joined to one of the J genes (J_3 in this example), the intervening DNA being excised. **(c)** Transcription begins at the V_{24} gene segment, the J genes (J_3 and J_4) also being transcribed. **(d)** After messenger RNA splicing to remove transcripts of J_4 and introns, those corresponding to V_{24}, J_3, and C now make up the mRNA. Allelic exclusion ensures that only one of the pair of alleles becomes a functional immunoglobulin gene so that a given cell produces only one immunoglobulin, not two.

Activation of B cells to produce antibodies

This involves an elaborate sequence of events. A preview of each step, without going into detail, may help you keep track of what is happening in the subsequent description.

- In the bone marrow, each B cell produces a limited number of copies of its own particular antibody molecules, the binding specificity of which is determined by the random gene assembly process described above. These antibody molecules are not released at this stage but are fixed in the membrane with the antigen-binding site displayed on the outside.

- In the bone marrow, any B cell that binds an antigen (which will be a self-component) is eliminated (clonal deletion).

- The survivors, which are likely to be specific only for foreign antigens, are released into the circulation.

- If a released B cell now meets and binds its antigen, it multiplies into a clone of cells. These cells cannot produce secreted antibodies until activated by a helper T cell which itself has been activated by meeting a peptide derived from the same antigen (which has to be presented to it by an **APC**, usually a dendritic cell).

- After activation by an activated helper T cell, the B cell multiplies and differentiates into an antibody-secreting plasma cell.

We will now describe how these events happen.

Deletion of potentially self-reacting B cells in the bone marrow

The immature B cells formed in the bone marrow have between them the potential to produce antibodies to attack just about every macromolecular component in the body. Any B cell whose antibody, if produced, would be against a self-component must be eliminated to avoid the risk of autoimmune diseases. This is done during primary maturation in the bone marrow. There the B cells are exposed to the self-antigens they are ever likely to meet.

Each immature B cell in the bone marrow produces about 10^5 copies of its antibody and displays them on its surface (fixed into the membrane) inviting, as it were, recognition by an antigen. It does not release the antibody at this stage; in this role, the antibody is a **receptor** or recognition 'aerial'. If, during *primary maturation* of a B cell in the bone marrow, an antigen is encountered which combines with its displayed antibody, it is assumed (as it were) that it is a self-component which must be tolerated (not attacked) by the immune system. *At this stage of development the* B cell, on binding to an antigen, dies. After this selective killing, the surviving B cells should respond only to foreign antigens. They are now released into circulation. Peter Medawar in the UK and Macfarlane Burnet in Australia shared the 1960 Nobel Prize for the theory explaining the basis of immunological tolerance to self-components, acquired before birth.

The theory of clonal selection

When the B cell survivors of the killing process in the bone marrow emerge into the circulation, they include a vast number of cells each with its own randomly designed antibody receptor exposed so that, when a foreign (unprocessed) antigen is encountered, a few of them will by sheer chance combine with it. The number of cells happening to do so will be small, but the body has to mount a large-scale response requiring a large number of B cells specific for that antigen. The answer to this problem is **clonal selection**, a mechanism proposed by Macfarlane Burnet. The principle is simple. An antigen, by binding,

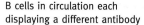
B cells in circulation each
displaying a different antibody

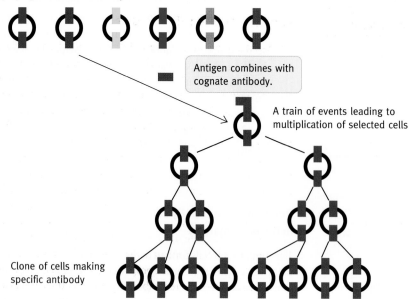

Antigen combines with
cognate antibody.

A train of events leading to
multiplication of selected cells

Clone of cells making
specific antibody

Fig. 30.3 The principle of clonal selection. The released population of B cells contains a vast array of cells, each of which makes a different antibody which at this stage is displayed as a receptor in the membrane (coloured blocks). An antigen binds to its cognate receptor and triggers a train of events leading to multiplication of the selected cell forming a clone. The antigen thus engineers its own destruction by selecting for amplification the clone of cells most suited to attack it. This binding of the antigen is only the initial signal leading to multiplication, which requires a complex set of events to be described. Note that at this stage the antibody is not released but is an integral membrane protein with its binding sites displayed as receptors. See text.

selects the B cells carrying a displayed antibody specific for itself and initiates a train of events which leads to a multiplication of those particular cells. The displayed antibody is an antigen receptor, and binding activates signalling pathways that ultimately leads to clonal expansion (increase in the numbers of that particular cell). A foreign invader (or other antigen), in short, arranges for its own destruction – which has a rather satisfying poetic justice about it. The principle is illustrated in Fig. 30.3.

B cells must be activated before they can develop into antibody-secreting cells

The selection process described above results in a clone of B cells which recognize an antigen. However, it is not yet ready to produce an antibody against the antigen. The B cell internalizes the bound antigen, hydrolyses it into peptides and displays these on the outside of the cell (how this is done is explained shortly). The B cell is now functioning as an antigen-presenting cell. Fig. 30.4.

You can regard the B cell at this stage as being on a 'standby' basis waiting for a further stimulus. This is given by a **helper T cell** produced in the thymus. Each helper cell has on its surface multiple copies of a receptor for a specific peptide if it is displayed on the surface of a B cell. In the thymus any receptor that would bind to such a displayed peptide from a 'self' antigen is destroyed (since it could cause an autoimmune disease). Released helper cells should recognize only peptides from foreign antigens.

The released helper T cells themselves require activation before they can 'help' a B cell. This is done by **dendritic cells**; these **antigen-presenting cells** (APCs) are widely distributed in the body. They phagocytose foreign pathogens, hydrolyse them into peptides and display these on their external surface. If the helper T cell recognizes and binds to its specific target peptide on the dendritic cell, this means that that there is a foreign invader which needs to be tackled. The binding causes the dendritic cell to produce cytokines (Fig 30.5), which stimulate the helper cell to produce a self-stimulating cytokine (autokine) that induces multiplication of the helper T cell into a clone of activated cells.

The final stage is that if now an *activated* helper cell recognizes its specific peptide on the surface of the B cell, cytokines are produced which cause the B cell to multiply and differentiate into a clone of antibody producing cells. The antibody is released to search for its target antigen. Initially the B cell, before activation by a helper cell, displayed antibody molecules fixed in the cell membrane. The fully activated **plasma cell**, as the secreting B cells are called, release the antibody because differential splicing (page 362) at this stage eliminates two exons from the mRNA, which code for a polypeptide segment that fixes the antibody into the membrane at the earlier stage (Fig 30.6).

The requirement for an activated helper cell to activate the B cells in effect is a double check that the target of the B cell is foreign and not a self-protein (since *both* B and T cells are survivors from a process that eliminated those cells responding to self-antigens).

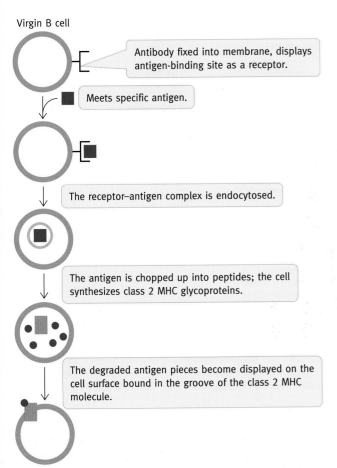

Virgin B cell

Antibody fixed into membrane, displays antigen-binding site as a receptor.

Meets specific antigen.

The receptor–antigen complex is endocytosed.

The antigen is chopped up into peptides; the cell synthesizes class 2 MHC glycoproteins.

The degraded antigen pieces become displayed on the cell surface bound in the groove of the class 2 MHC molecule.

The B cell is now in an activated state; it does not produce antibodies unless it binds to a helper cell that has been activated by an antigen-presenting cell displaying the same antigen. If this happens, the B cell matures into an antibody-secreting plasma cell and also multiplies due to cytokine release by the helper cell. (This is illustrated in Fig. 30.6.)

Fig 30.4 Conversion of a virgin B cell into an activated state but not yet to an antibody producing cell.

There are cases where helper T cells are not required for the production of antibodies. Polysaccharides of the type found typically in the lipopolysaccharides attached to the outer membrane of some bacteria are antigenic. The antigens in this case have multiple recognition sites in the carbohydrates occurring one after another. Possibly because of this they give signals to the B cells that are sufficient to cause their differentiation into antibody-secreting cells without helper T cells being involved. It is believed that a special class of B cells is involved. This type of response does not, however, produce memory cells so there is no lasting immunity.

Fig 30.7 summarizes the process of antibody production.

Affinity maturation of antibodies

The B cell(s) were initially selected for multiplication by an antigen that combined with the cell's displayed antibody. However, the latter's binding site was generated by random variation and the antigen just by chance happened to fit to it. The fit is not likely to be particularly good, so the antibody produced is of low binding efficiency and therefore of low protective effect. During B cell multiplication, after antigen binding, rapid site-specific mutations occur that produce yet more variations in the antibody-binding sites of the multiplying B cells. This is coupled with a selection of those cells with improved binding affinity for the relevant antigen. Cells with poor binding ability that do not capture an antigen molecule die. As the number of antigen molecules falls, owing to their removal by phagocytosis, the competition for them increases. The result is that during an immune response there is a progressive Darwinian evolutionary improvement in the antibody quality, as only the best B cells are able to capture an antigen molecule and survive. The process of **affinity maturation** takes place in special centres (germinal centres) in secondary lymphoid organs. The antibody class-switching that occurs as the immune response progresses (page 496) is also an important aspect of antibody quality improvement.

Memory cells

When B cells are activated and they proliferate, not all of the resultant clone of cells mature into antibody-secreting plasma cells. A proportion become long-lived memory cells that may circulate for years or an animal's lifetime. They are the basis of long-term immunity from repeat infections. If an appropriate antigen is encountered at a subsequent time, the immunological response is very rapid. Hence, vaccination can often give long-term protection against an infective disease.

Cell-mediated immunity (killer T cells)

This is the second major arm of the adaptive immune response. It is called cell-mediated because the effector cells, known as **killer T cells** or **cytotoxic cells**, must make physical contact with their target cells. This is in contrast to the humoral response, where the effector agent is an antibody that can attack its antigen target, and cell contact is not involved in the actual attack process.

Why is a second arm of the adaptive immune response needed? The humoral response protects against pathogens in the blood before they have infected host cells. Once in the cells

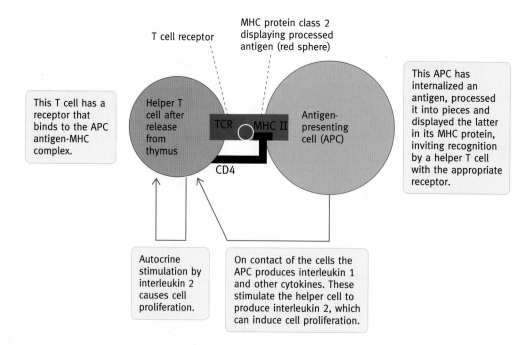

Fig. 30.5 Activation of a helper T cell by an antigen-presenting cell (APC), a dendritic cell. The autocrine stimulation of the helper T cell allows cell multiplication to occur after separation of the cells. CD_4 is a glycoprotein on the T cell that interacts with the MHC 2 protein of the APC. This interaction is necessary in addition to the binding of the T cell receptor (TCR) with the APC MHC-antigen complex. The CD_4 protein is the receptor by which the AIDS virus (HIV) infects the helper T cell.

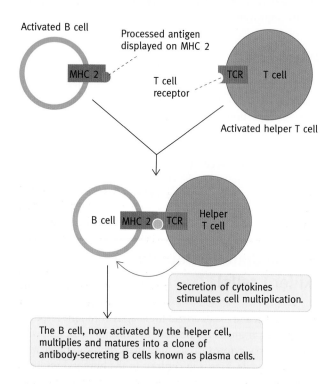

Fig. 30.6 A helper T cell activating a B cell to become a clone of plasma cells secreting antibody. The cytokines are growth factors that stimulate local cells to divide (paracrine action). On activation of a B cell to become a plasma cell, differential processing of RNA transcripts eliminates the membrane-anchoring section of the antibody. TCR, T cell receptor.

the antibodies cannot reach them so a different strategy is needed. The killer T cells attack and trigger the destruction of the infected cell, and this aborts multiplication of the pathogen. But how does the killer T cell 'know' which cells are infected? A remarkable mechanism achieves this. Somatic cells continually 'sample' their cytoplasmic proteins, hydrolyse them into peptides in proteasomes (page 339) and display these peptides on their surface for killer T cells to inspect. The killer T cells develop in the thymus and undergo the same selective process as occurs with B cell in the bone marrow. The cells each have a randomly assembled receptor to recognize a single specific peptide displayed on the outside of cells; any cell that recognizes 'self' peptides are eliminated so that released killer T cells should recognize only peptides derived from foreign proteins. The cells have to be activated (described below) and then they inspect host cells for displayed foreign peptides. Each killer T cell is specific for a single peptide. Displayed self-peptides are not recognized by the killer cell since any cell with the potential to do so will have been destroyed in the thymus. Binding of the killer T cell to a host cell results in its destruction of the infected cell.

How are killer T cells activated after release from the thymus? This again is the job largely of the dendritic cells. The dendritic cells engulf pathogens, hydrolyse them into peptides and display these on their exterior surface (see below). If a killer T cell recognizes its target peptide among these it binds to it. Cytokines are released that cause the killer T cell to multiply into a clone of active cells. These now 'know', as it were, that

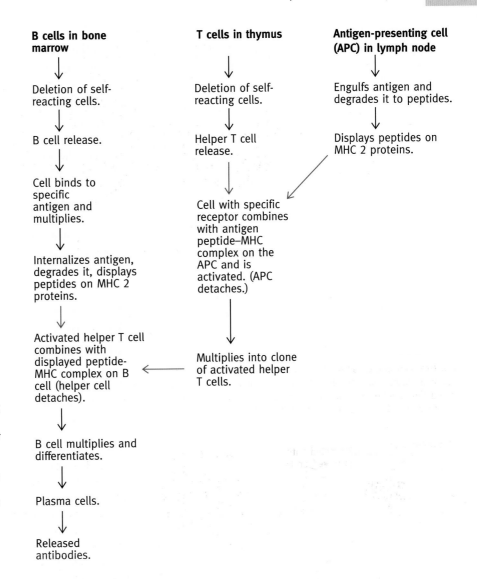

B cells in bone marrow

↓

Deletion of self-reacting cells.

↓

B cell release.

↓

Cell binds to specific antigen and multiplies.

↓

Internalizes antigen, degrades it, displays peptides on MHC 2 proteins.

↓

Activated helper T cell combines with displayed peptide-MHC complex on B cell (helper cell detaches).

↓

B cell multiplies and differentiates.

↓

Plasma cells.

↓

Released antibodies.

T cells in thymus

↓

Deletion of self-reacting cells.

↓

Helper T cell release.

↓

Cell with specific receptor combines with antigen peptide–MHC complex on the APC and is activated. (APC detaches.)

↓

Multiplies into clone of activated helper T cells.

Antigen-presenting cell (APC) in lymph node

↓

Engulfs antigen and degrades it to peptides.

↓

Displays peptides on MHC 2 proteins.

Fig. 30.7 Summary scheme of antibody production by B cells. The crucial point to note is that a B cell cannot proceed to release antibodies until it is given the signal to do so by a helper T cell. The helper cell is activated by combining with an antigen-presenting cell (APC) that has engulfed the *same* antigen that was endocytosed by the B cell. The APC displays the antigen peptides on its MHC proteins and this is recognized by the specific helper T cell receptor. The helper T cell thus activated has to see the identical MHC-peptide complex on the B cell. You will notice that in this sense the B cell is acting as an antigen-presenting cell. Cytokines are liberated at activation steps and play an essential part in the process (see Fig. 30.5). The red arrows are to draw attention to cell–cell interactions.

there is a target pathogen around and seek out infected host cells.

Mechanism of action of killer T cells

There are two mechanisms by which target cells are killed. The killer T cell makes contact with the cell and either delivers an apoptopic death signal (page 514), instructing the cell to self destruct, or it liberates **perforin**, which renders the cell permeable and allows entry of proteases, which the T cell also secretes. These also trigger cell death.

The role of the major histocompatibility complexes (MHCs) in the displaying of peptides on the cell surface

The peptides to be displayed on the outside of antigen presenting cells (B cells, helper cells, and dendritic cells) are placed in a groove of newly synthesized MHCs, which are then transported to the cell surface. The receptors on helper cells and killer T cells recognize the complex of their target peptides in MHC grooves, not the peptide alone or the MHC alone. There are two main classes of MHCs called **class 1** and **class 2**.

Most somatic cells of the body have class 1; killer T cells recognize only peptides held in class 1. Therefore most somatic cells of the body are subject to killer T cell surveillance for infection. The somatic cells display peptides produced by proteasomes in the cytoplasm on newly synthesized class 1 MHCs.

The antigen presenting cells are different. They engulf antigens and fragment them into peptides in endosomes (page 416), not proteasomes as in class 1. Peptides produced in this way are routed to be displayed on newly synthesized class 2 MHCs. There was a puzzle in that dendritic cells were known to activate both killer T cells and helper T cells, which require display on different classes of MHCs. An important recent discovery (see **Dudziac** *et al.* reference in Further reading) has revealed

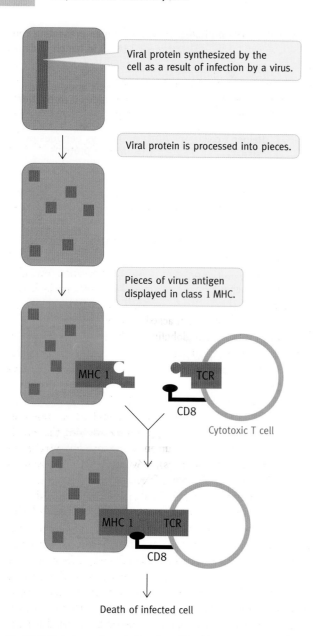

Viral protein synthesized by the cell as a result of infection by a virus.

Viral protein is processed into pieces.

Pieces of virus antigen displayed in class 1 MHC.

MHC 1

TCR

CD8

Cytotoxic T cell

MHC 1   TCR

CD8

Death of infected cell

Fig. 30.8 Sequence of events in cell-mediated immune reaction to a foreign antigen synthesized inside cells (for example, as a result of virus infection). A proportion of the clone of cells develops into memory cells. The infected cell may be killed by perforation of its cell membrane due to the release of the protein perforin, or death may be due to apoptosis. TCR, T cell receptor.

that there are two different subsets of dendritic cells one specific for Class 1 MHC display of peptides and the other for class 2.

The use of the different MHCs for peptide display in antigen presenting cells means that killer T cells do not attack antigen presenting cells and helper cells do not recognize somatic cells.

Some viruses try to elude killer T cell attack by inhibiting the display of MHC-peptides. Remarkably, the natural killer (NK) cells of the innate immune system recognize cells with deficient display and attack them.

CD proteins reinforce the selectivity of T cell receptors for the two classes of MHCs

The binding of T cell receptors to their targets is primarily determined by recognition of peptide-MHC complexes, but this is reinforced by additional proteins. Killer T cells express a protein called CD 8, which binds to a *constant* region of MHC 1 proteins; helper T cells instead express a protein, CD4, which binds to class 2 MHCs. (CD means cluster of differentiation; they are cell surface markers.) This extra binding is necessary for effective binding of the different cells to their targets and so confines the two types of T cells to their proper targets (Figs. 30.5, 30.8). The CD4 is the route by which the AIDS virus enters and destroys the helper T cells thus removing humoral protection of the body with disastrous effects.

Why does the human immune system reject transplanted human cells?

At first sight it is puzzling why the body should have a mechanism for immune attack on cells surgically transferred from one individual to another. Tissue grafts and organ transplants are recent developments, so there was no known evolutionary reason to guard against them. The main cause of rejection is the MHC proteins on cells. Individuals have several genes coding for variants of the MHC proteins. In addition the genes for these are highly polymorphic (page 328) within a population – they vary from one individual to the next. It is unlikely, therefore, that the MHC proteins of one individual are exactly the same as those of the next. Those that are different will therefore be antigenic if transferred to another person. The evolution of such variation in the MHC molecules of the population is probably a protection for the species as a whole. Viruses and other pathogens employ devices to outwit the immune system. Suppose that, by chance, a virus evolves so that its processed antigen is not displayed on a host MHC protein or not recognized by the cytotoxic T cells. The cell is not labelled as virally infected and is not attacked. The virus might then multiply and kill its host. However, the next individual it infects is unlikely to have exactly the same MHC genes and the chance of the same evasive device being successful again is reduced. The disease is therefore less likely to sweep through the whole population.

Monoclonal antibodies

If an animal is injected (immunized) with an antigen, antibodies appear in the blood against the antigen and from this, an immune serum can be obtained and used for a variety of purposes such as experimentally to detect a particular protein.

However, such sera are, in the molecular sense, extremely crude for there are many different antibodies specific for a wide variety of epitopes of different proteins – really to whatever antigens the animal has been exposed to in the preceding period. The antibody of interest is only a very small proportion of the total. It was realized that a 'pure' antibody, in which every immunoglobulin molecule is specific for the antigen of interest, would be a powerful tool in biochemistry and medicine.

The achievement of this earned Niels Jerne, Georges Kohler, and Cesar Milstein a Nobel Prize in 1984. A given B cell, as explained, produces one antibody in terms of binding specificity and one only (though many copies of it). If a B cell that produces the wanted antibody could be isolated and grown in culture, a pure antibody would be produced. It is possible to isolate such a specific B cell, but B cells will not grow in culture for more than perhaps a few days. Only transformed cancerous cells will grow indefinitely in culture.

Pure clones of cancerous B cells are available in tumours called **multiple myelomas**; these are derived from single B cells which multiply uncontrollably and produce large amounts of a pure immunoglobulin and will grow indefinitely in culture. The immunoglobulin it produces will bind to whatever antigen for which the parent cell was specific. This could be anything, so its immunoglobulin is of no practical use. The solution was to isolate a B cell that produced the wanted specific antibody (but would not grow in culture) and fuse it with a multiple myeloma cancer cell, which does not produce the wanted antibody but does grow easily. This can produce an immortalized B cell that secretes the wanted antibody.

The technique, briefly, is to immunize mice with an antigen and a few weeks later to prepare B cells from the spleen (a secondary lymphoid organ). These are a mixed population of B cells producing different antibodies but including some specific for your injected antigen. They are fused with myeloma cells (this can be done simply by using a common chemical, polyethylene glycol). The fused cells are called **hybridoma cells**, but there will also be large numbers of unwanted cells which have not fused. These are eliminated by culturing the mixed population in a selective medium (not described here), which does not allow unfused cells to survive. The hybridomas are collected and inoculated into small wells in a plastic block. By appropriate dilutions it is arranged that only a single hybridoma cell is placed in a small volume of culture fluid in each well. After suitable growth, the wells are tested for the presence of antibody reacting to the antigen of interest. Once identified, cells can be grown up in unlimited quantities producing the monoclonal antibody also in unlimited amounts. The cell culture can be stored indefinitely in liquid nitrogen and grown up whenever more of the antibody is wanted.

Monoclonal antibodies have become the basis of a biotechnology industry, so powerful are they as a tool. Almost any protein can be detected by developing the appropriate monoclonal antibody. They are used in clinical laboratories for measuring specific human proteins or for HIV or other virus detection.

This is the basis of **ELISA (enzyme-linked immunosorbent assay)**, widely used in clinical medicine; polyclonal antibodies are also used in this in some cases but monoclonal ones have greater specificity. Suppose you want to test for the presence of a hormone in urine or of a virus in some biological fluid. The sample is adsorbed to the bottom of a plastic well. The excess is washed away and any remaining sites on the plastic well that could adsorb proteins are swamped by nonspecific protein. An antibody specific for the antigen of interest that has had conjugated with it an enzyme, such as horseradish peroxidase, is added to the test well. If the virus or hormone is present, the antibody will attach. The excess is washed away. A colourless substrate for the peroxidase is added which the attached enzyme converts to a coloured compound. This is quantitated colorimetrically and indicates the amount of the antigen of interest present in the test sample. Monoclonal antibodies give absolute specificity. There are several variations which can increase the sensitivity. In one of these, the initial (rabbit) antibody used has no enzyme attached. A second antibody, such as goat anti-rabbit immunoglobulin, specific for the first antibody (with attached enzyme in this case) binds to the first antibody; this increases the amount of enzyme fixed since multiple copies bind, and so increases the sensitivity of the assay. For example, to detect antibodies to HIV the patient's blood is added to a well containing adsorbed HIV antigen. If an antibody is present in the blood it will bind. After appropriate washing the well is treated with an antibody to human immunoglobulin (raised in another species, often in goats), to which peroxidase has been bound. Production of a colour from added substrate indicates that there is an antibody to the HIV in the patient's blood.

Experimentally, monoclonal antibodies can be labelled with fluorescent compounds or with radioactivity and used to identify specific protein production, for example in cells of embryos (Fig. 8.13 on page 133 was obtained in this way). They also have the potential to target therapeutic agents to specific sites in the body, such as tumours, an area of great interest because cancerous cells may have antigenic 'markers' on them to which specific antibodies could attach. The antibody could for example be loaded with a therapeutic agent against the cancer.

Humanized monoclonal antibodies

The use of antibodies to combat diseases has been used for about a century. Tetanus infection was treated by injection of horse serum from animals that had been injected with the tetanus toxoid. Subsequent injections carried the risk of potentially fatal immune responses to the horse serum proteins.

Mononoclonal antibodies specific for a given target protein do not have the contaminating proteins but since monoclonal antibodies are generated in mice, they still have the drawback of a possible immune response to the mouse proteins, which at least could diminish the effectiveness of the treatment.

To overcome this there have been developments of 'humanized' monoclonal antibodies, in which the variable region of the mouse antibody is retained (the specific 'business ends' of the molecule) while other constant protein components are either removed or replaced by human versions of the proteins. The latter is achieved by genetic engineering mice so that their immune system has human genes replacing the relevant mouse ones. This approach has had considerable success, and a number of diseases have been therapeutically treated with such monoclonal antibodies. The partially humanized antibodies reduces the risk of unwanted immune reactions.

Recently this has culminated in the development of a strain of mice (known as the xenomouse) by genetic engineering that produce totally human monoclonal antibodies. Such an antibody produced in this way has received FDR approval in 2006 for use in humans (see article by **C. T. Scott** in Further reading).

 ## Summary

The immune system of animals protects against pathogenic invaders. There is an innate immune system in which phagocytes engulf invading pathogens. This gives immediate protection, and helps to trigger the adaptive response

The adaptive immune system is the most important and the one mainly discussed in this chapter. It responds to macromolecules, mainly proteins and complex polysaccharides, but not to small foreign molecules unless attached to large ones. Foreign macromolecules are indicative of an invader. They are called antigens because antibodies are generated in response to them. There are two arms of the adaptive immune systems, the humoral in which antibodies are produced in the blood and the cell-mediated immunity in which cytotoxic or killer T cells destroy abnormal cells such as virus-infected ones.

Cells of the immune system, known as lymphocytes, originate in the bone marrow. B cells multiply and undergo their primary maturation there and when released are the generators of antibodies. T cells are produced in the bone marrow but migrate to the thymus gland where they multiply and undergo their primary differentiation. When released and activated they become either killer T cells or helper T cells. Helper T cells are essential to 'help' B cells to form antibodies.

Antibodies are proteins; several classes exist for different roles but immunoglobulin G illustrates essential features of all. There are two heavy and two light polypeptide chains forming a Y-shaped molecule. The N-terminal ends of each arm are hypervariable and make the specific antibody-binding sites, both identical on one molecule but different on individual molecules. Each B cell makes a single antibody in terms of binding specificity and individual cells make different ones.

Each B cell in the bone marrow assembles a different gene (from pre-existing sections – a unique situation) for the variable regions of the heavy and light chains randomly, and displays the resultant antibody on the outside of the cell. If it combines with a protein in the bone marrow it is likely to be self-reacting and undergoes self-programmed death (apoptosis; Chapter 30), because it could cause an autoimmune disease.

Survivors after release, on binding to a (presumptively foreign) antigen, multiply. The antigen, thus selects cells capable of protecting against itself, a process known as clonal selection.

The bound antigen is engulfed by the B cell and processed into peptides, which are displayed, on class 2 major histocompatibility complex (MHC) proteins, on the cell surface. The selected B cells must combine with an activated helper T cell specific for the same antigenic peptide before it differentiates into an antibody-secreting cell. The helper T cell is activated by an antigen-presenting cell, usually a dendritic cell, displaying the same foreign peptide in its class 2 MHC proteins as displayed on the B cell it will activate. The activation processes involve liberation of cytokines, which are of pivotal importance in the entire system of immunity. The AIDS virus destroys helper T cells, thus impairing antibody production. The combination of the activated helper cell and its cognate B cell causes cytokine release, which complete the signal for the B cell to multiply and secrete antibodies.

The humoral response protects against, say, a virus, in the blood but once the virus infects a cell the antibodies cannot reach it. To cope with this, the cell-mediated immune response destroys the infected cell. Killer T cells maturing in the thymus have receptors for foreign antigen peptides displayed on class 1 MHC proteins. Those that would attack self cells are deleted.

The killer cells must be able to recognize which cells are infected. For this purpose, all cells sample proteins in their cytoplasm (normal and foreign) and hydrolyse them into peptides. These are displayed on class 1 MHC proteins. The killer cell receptors recognize foreign peptides displayed in this way and destroy the cell. Before this can occur, however, the killer cell itself must be activated by combining with the same foreign peptide displayed on class 1 MHC proteins of dendritic cells. The use of two classes of MHC proteins is a form of compartmentation. It ensures that killer cells do not attack B cells displaying the same foreign peptide, since in this case it is on a class 2 MHC protein.

In organ transplantation, immune rejection results from the foreign MHC proteins. These vary between individuals and constitute a protection for the species against an infection.

Monoclonal antibodies are preparations obtained from a clone of a single type of activated B cell so that the antibody molecules are all identical, as opposed to antibodies isolated from blood in which a wide variety if immunoglobulins are present.

Monoclonal antibodies are used for detection of specific proteins in cells and may have application in targeting chemotherapeutic agents specifically to cancer cells. They, and polyclonal antibodies, have an important use in the ELISA method for clinical detection of antigens and antibodies.

Mononoclonal antibodies generated in mice may have an immune response to the mouse. 'Humanized' monoclonal antibodies in which the variable region of the mouse antibody is retained while other constant protein components are either removed or replaced by human versions of the proteins is achieved by genetic engineering mice so that their immune system has human genes replacing the relevant mouse ones.

Further reading

Various authors (1993). *Sci. Am.*, **269**(3), 20–108.
A complete issue of Scientific American *devoted to all aspects of immunity.*

Generation of gene diversity

Sekiguchi, A. and Frank, K. (1999). V(D)J recombination. *Curr. Biol.*, **9**, R835.
A one-page quick guide to the essential facts of how the immune system generates gene diversity.

Monoclonal antibodies

Yelton, D. E. and Scharff, M. D. (1981). Monoclonal antibodies: a powerful new tool in biology and medicine. *Annu. Rev. Biochem.*, **50**, 657–80.
General review of techniques and applications.

Chien, S. and Silverstein, S. C. (1993). Economic impact of applications of monoclonal antibodies to medicine and biology. *FASEB J.*, **7**, 1426–31.
Despite the title, includes a summary of how monoclonal antibodies were developed; this is followed by their practical application and monetary value to industries.

Scott, C. T. (2007). Mice with a human touch. *Nature Biotech.*, **25**, 1075–7.
Deals with humanized monoclonal antibody production.

Zola, H. (2001). Monoclonal antibodies: therapeutic uses. *Encyclopedia of Life Sciences*, Nature publishing group.
A comprehensive description of the development of the field, which summarizes the history of humanized antibody development.

Antigen presentation

Engelhard, V. H. (1994). How cells process antigens. *Sci. Am.*, **271**(2), 44–51.
An account of antigen processing and display of fragments on MHC proteins.

Robertson, M. (1999). Antigen presentation. *Curr. Biol.*, **9**, R829–30.
A 'primer' article describing how antigens are presented on MHC 1 and MHC 2 proteins. Includes an account of how viruses sabotage the process for their own ends.

Miller, J. F. A. P. (2001). Ruby anniversary: forty years of thymus immunology research. *Nature Immunol.*, **2**, 663–4.
A fascinating short historical perspective, very readable.

Werlen, G., et al. (2003). Signaling life and death in the thymus: timing is everything. *Science*, **299**, 1859–63.
Describes differentiation of T cells so that receptors do not react with self.

Dudziac, D., et al. (2007). Differential antigen processing by dendritic cell subsets in vivo. *Science*, **315**, 107.

LuDeng and Mariuzza, R. A. (2007). Recognition of self-peptide-MHC complexes by autoimmune T-cell receptors. *Trends Biochem. Sci.*, **32**, 500–8.
Describes how T cell receptors distinguish between self and foreign peptides displayed on MHC 1 protein.

Pathogens and the immune system

Goodenough, U. W. (1991). Deception by pathogens. *Am. Sci.*, **79**, 344–55.
Describes how, at the molecular level, bacteria and viruses attempt to escape immunological detection.

Felgner, P. L. (1998). DNA vaccines. *Curr. Biol.*, **8**, R551–3.

Haemopoiesis and haemopoietic growth factors

Metcalf, D. (1991). The 1991 Florey Lecture. The colony-stimulating factors: discovery to clinical use. Phil. Trans. Ser. B Roy. Soc. Lond., **333**, 147–73.
An in-depth review by the discoverer of these important proteins.

Problems

1   Describe the structure of an IgG molecule.

2   Explain how the body can have genes to code for so many different antibodies.

3   What are the different classes of lymphocyte involved in immune protection and their function?

4   What is an antigen-presenting cell?

5   When a host cell displays, for example, a viral antigen on its surface which class of MHC molecule is it displayed on? Which class of MHC molecule does a cytotoxic T cell recognize?

6   What is meant by the term clonal selection theory?

7   What is the principle of avoidance of autoimmunity, or how does the immune system become self-tolerant?

8   Production of an antibody by a B cell after stimulation (to become a plasma cell) by a helper T cell does not change the antigen-binding site of the antibody, but the antibody has to be released instead of residing in the membrane. How is this achieved?

9   When a B cell (plasma cell) starts to secrete antibody, the latter may have a relatively poor affinity for its antigen but this is rapidly improved. How is this achieved?

10 Which immune systems have the glycoprotein CD4 and which CD8? What are their roles in the immune reaction? What is the relevance of CD4 to the AIDS virus?

11 Why are there two classes of MHCs (major histocompatibility antigens)?

12 What are humanized monoclonal antibodies?

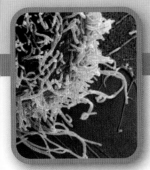

Chapter 31

The cell cycle and its control

The eukaryotic cell cycle

All life depends on the self-reproduction of cells, which has gone on continuously for billions of years, since the first primordial self-replicating 'unit' appeared. Cells multiply by dividing into two and, before the next cell division, the daughter cells grow and double in size so that normal cell size is maintained. Due to mutual signaling by cells (in complex eukaryotes such as mammals), rapid somatic cell division occurs when increased numbers are needed, but then this ceases and cell division is restricted to replacing dead cells. Each cell division requires the total DNA in the nucleus to be replicated exactly so that there are two sets of chromosomes, no less and no more. The two sets are then segregated into the two daughter cells at cell division. This must be done with absolute precision or both daughter cells would be genetically abnormal. It is not surprising that the eukaryotic cell cycle has an elaborate system of controls and checkpoints. The process has been tightly conserved, so that the cycle is the same in all eukaryotes.

The cell cycle is divided into separate phases

In a typical replicating animal cell in culture, from the completion of one cell division to the next takes about 24 hours depending on the cell type (Fig. 22.4). The process of mitosis itself, in which the duplicated chromosomes are separated and cell division completed, takes about an hour. During mitosis (described on page 323) the chromosomes are in a highly condensed phase and are inactive – they are not transcribed into messenger RNA. When the daughter cells are formed by cell division the cells come out of mitosis and the chromosomes are unraveled from the highly compacted state to the more extended form pervading the nucleus. In the period between mitoses, known as **interphase**, the genes can be active and the cell more or less continuously synthesizes most of the proteins and other components needed for cell growth. This, however, does not apply to DNA synthesis, which

is confined to the **S phase** of perhaps 7 hours' duration out of a 24-hour cycle in mammalian cells. Prior to the S phase there is a G_1 **phase** (G for the gap in DNA synthesis) lasting somewhere around 10 hours, in which the cell prepares to enter S phase. After completion of S phase, in which the genome is completely duplicated, the cycle progresses to the second gap phase or G_2 in which the cell prepares for mitosis. Interphase therefore consists of $G_1 + S + G_2$. The latter leads into the mitosis (**M**) phase. The replicated DNA condenses into compact chromosomes and mitosis occurs, culminating in the separation of the two daughter cells in a process known as **cytokinesis**, as described earlier.

The cell cycle phases are tightly controlled

The cell cycle must be controlled in a decisive manner to avoid genetic abnormalities. For this reason progression through the phases of the cycle is subject to controls and checkpoints to ensure that everything is in order. There are several vital requirements to be met.

- If the DNA of a cell is damaged and not repaired, the cell must not be allowed to proceed to mitosis to avoid the production of genetically abnormal, potentially cancerous, daughter cells.

- The DNA must be completely replicated before mitosis for the same reason. Equally important, and again for much the same reason, the DNA must be replicated once and once only. This means that the replicons, the sections of DNA under the control of single centres of origin (page 333) must fire only once per cell cycle.

- At the metaphase stage of mitosis the two sets of chromosomes must be correctly positioned at the equator attached to spindle fibres. If this were not so, genetically abnormal daughter cells would result from the imperfect segregation of the chromosomes between the two daughter cells.

- Each phase of the cycle must be completed before the next phase is initiated.

- The next requirement is in a different category. Replication of the DNA must not proceed in a mammalian cell unless a mitogenic signal to proceed to cell division is received by that cell from other, neighbouring cells. In an animal as complex as a mammal, individual cells are part of a vast community that exists for the welfare and reproduction of the organism as a whole. They must fit in with the needs of other cells, and paramount among these is the obligation not to proceed to cell multiplication independently. This is what happens in cancer. To this end there is a system of cell signalling pathways to coordinate cell multiplication with requirements of the body as a whole. The effectors of this control are intercellular signals, proteins known as **growth factors** and **cytokines** (see page 432).

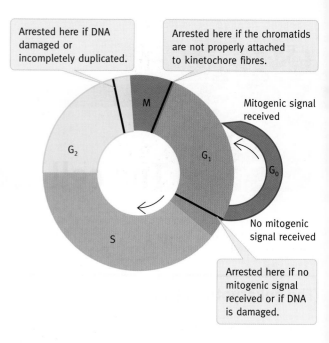

Fig. 31.1 The eukaryotic cell cycle, showing checkpoints in G_1, G_2, and M phases.

Cell cycle controls

Cytokines and growth factor control in the cell cycle

The cytokines and growth factors are signalling molecules which bind to cell-surface receptors and activate pathways that control genes. Deficiencies in this control can lead to cancer and to chaotic cell multiplication in general.

In the present context, it is their mitogenic (mitosis-stimulating) effects that are important, which, as it were, give permission to a cell to divide. If such a mitogenic signal is not received by a cell in early G_1 phase, the cycle does not proceed, and it enters a quiescent G_0 **phase** in which it metabolizes normally but does not undergo mitosis. Most somatic cells are in G_0 most of the time. On receipt of a mitogenic signal, however, the G_0 cell can re-enter the cycle again at G_1.

The intercellular cytokine/growth factor signals mainly come from neighbouring cells. These signals maintain correct organ cell numbers, for once adult size is reached, cell multiplication largely ceases except to replace dead cells and for wound healing. Although much is known of the mechanisms of individual signal transduction pathways, how these collectively add up to coordination of the mass of cells in tissues is unclear. The mitogen control of cell division operates mainly in the G_1 phase of the cell cycle as described later.

Cell cycle checkpoints

In addition to the control by mitogenic factors, there are other checks related to safety requirements. Towards the end of G_1, G_2, and M phases, there are checkpoints at which the cycle is halted if one of the safety requirements is not met. The arrest gives an opportunity for the defect to be rectified, in which case the cycle can proceed to the next phase. If the fault is not corrected, the cell is instructed to self-destruct by activating a pre-programmed chain of events leading to **apoptosis**

(page 512), as the self-sacrifice is called. Figure 31.1 shows the checkpoints in mammalian cell cycles. We will come to the nature of these checkpoints shortly.

Cell cycle controls depend on the synthesis and destruction of cyclins

As in so many cellular controls, protein kinases are of overwhelming importance. The cycle kinases are however of a unique kind; they are without activity on their own and require the binding of other proteins known as **cyclins** before they have activity and are known as **cyclin-dependent protein kinases** or **Cdks**. There is a complex variety of Cdks operating in a mammalian cell cycle, each operating in its own designated zone in the cell cycle. The Cdks remain essentially constant in amount in the cell during the cycle but the cyclins are destroyed at the end of each phase and new ones are synthesized to enable progression to the next phase; this is the basis of the main control of the cell cycle. The sequence of cyclin involvement is quite complex but Fig. 31.2 shows the generally accepted Cdk/cyclin combinations that operate in the various phases. (The regions of the cycle they operate in have not been precisely defined but they are associated with the occurrence of definite cycle phases). The diagram is mainly to give some idea of the complexity and most of subsequent figures will refer to the cyclins simply as G_1, S, G_2 and M cyclins.

In some cases different cyclins activate different Cdks, but in others the same Cdk is activated to operate in different phases by different cyclins. As shown in Fig. 31.2, Cdk2 acts in G_1

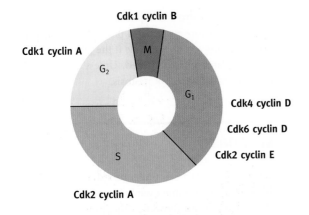

Fig. 31.2 Diagram showing the proposed Cdk/cyclin complexes and their appropriate sites of operation in the mammalian cell cycle. Other complexes are believed to be involved. (Cdk3 cyclin C is believed to regulate the re-entry, shown in Fig. 31.1, of G_0 cells into G_1.)

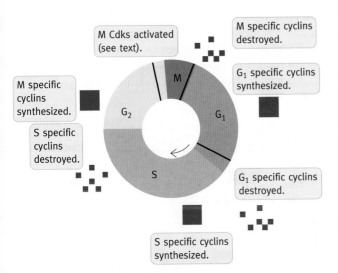

Fig. 31.3 Cyclin–Cdk cycle. Simplified diagram of cell cycle control by synthesis and destruction of phase-specific cyclins that activate phase-specific Cdks. The M phase Cdks are inactive until just prior to M phase when they are rapidly activated by dephosphorylation (see text). Note that more than one cyclin and Cdk occur in a given phase but this is omitted for simplicity.

attached to cyclin E, but in S phase attached to cyclin A. The mechanism of cyclin activation is known. Cdks in the absence of cyclins are inactive because their active site is blocked due to the conformation of the protein. Combination with a cyclin causes a conformational change which partly unblocks the site. In the case of Cdk1, complete activation requires phosphorylation of the Cdk on a threonine residue. This is done by a Cdk activating kinase (CAK). The activated Cdk/cyclin complexes are themselves regulated further. In the case of Cdk1 operating in the M phase, further control exists in progression to M phase

(see below). There are also Cdk inhibitory proteins (Cdks); a family of these exists specific for different ones.

At the start of each cycle phase, genes have to be activated so that the appropriate cyclins are synthesized. If this does not happen, the cycle cannot proceed through that phase. At the end of each phase, the cyclins are ubiquitinated (page 399) and destroyed by proteasomes and new cyclin synthesis specific for the next phase is needed (Fig. 31.3). This may seem an expensive way to achieve control but it is a decisive procedure leaving no room for partial inactivation or reversibility. Fig. 31.2 shows that cyclin D is needed for the progression of G_1 into the S phase. Cyclin D synthesis requires the presence of a mitogenic signal from growth factors. If this is absent, cyclin D synthesis does not occur, and the cycle is shunted into the G_0 phase. It is known that several human cancers are associated with unregulated cyclin D synthesis which allows G_1 to progress into S phase when it is not appropriate.

Controls in G_1 are complex

The progression through G_1 to S phase involves multiple gene controls. At the end of the preceding M phase, all cyclins are destroyed so that at the start of G_1 there are no active Cdks. For G_1 cyclin synthesis to occur, the cell must receive a mitogenic signal, which activates cyclin synthesis and this in turn results in activation of Cdks needed to activate genes necessary for the cycle to progress to S phase. Synthesis of G_1 cyclins is delayed after entry into G_1, probably to give time for mitogenic signals to operate.

If this requirement is satisfied, the cell can progress to the G_1 checkpoint. If not, it is shunted into G_0 (as happens most of the time with somatic cells).

The G_1 checkpoint

This is an important control for, if the cycle is allowed through it, the cell is committed to proceed right through to M phase. In mammals it is known as the **restriction point**. The best-known check is that if the DNA of a cell is damaged, it is not allowed through the checkpoint. The chief player is the protein p53. This is described more fully later in Chapter 33 (page 522) because it is a tumour suppressor but, briefly, in the presence of damaged DNA, p53 increases in amount in the cell and is activated. It is a transcription factor that sets up a train of events; the cell cycle is halted so that DNA repair can be attempted. If this fails, the cell is signalled to self-destruct by apoptosis to avoid the risk of genetically abnormal cells being produced.

Another protein of importance in the G_1 checkpoint control is the **retinoblastoma** protein. This is covered on page 522 in the Cancer chapter. This halts the cycle in the absence of a mitogenic signal.

How is DNA damage detected?

This is a truly remarkable phenomenon. The mammalian genome is inevitably subject to damage by ionizing radiation, reactive oxygen species, and other agents. Fortunately, the repair mechanisms described earlier (page 344) detect and repair many types of lesions, for otherwise complex life would be impossible. If they are not repaired, as stated, the cycle must not be allowed to proceed. Damaged DNA involves exposure of single-stranded sections. For example, a replicative fork stalled for some reason will have stretches of single-stranded DNA. Double-stranded DNA breaks may also have some terminal single-stranded DNA. A protein known as **replicative protein A (rpA)** attaches to the single-stranded DNA and this attracts a complex of protein kinases to assemble. This is the signal that is detected by p53, which halts the cycle to enable the DNA lesion to be repaired. If this is not done, the p53 initiates apoptopic destruction of the cell. One of the proteins in the complex (ATM) is mutated in the human disease **ataxia telangiectasia** in which, among other things, the person has increased sensitivity to radiation and risk of cancer.

ATM protein is a kinase associated with the DNA of the genome. It is activated by double strand breaks, which lead to blocking of the G → S phase transition.

Progression to S phase

Once past the G_1 checkpoint, the G_1 cyclins are destroyed by ubiquitin-dependent proteolysis, S phase cyclins are synthesized, and the cycle enters the S phase. These result in initiation of DNA replication. The initiation of duplication in each replicon is effected by a complex of proteins by a mechanism that ensures that each fires only once per cell cycle. Once committed to S phase, the cycle advances to the checkpoint in M phase at the end of G_2, the S phase cyclins being destroyed at the end of S phase.

Progression to M phase

The mitotic-related cyclins are synthesized and accumulate in the cell during S and G_2 phases, and combine with the relevant Cdks. Cyclin-Cdk complexes, as stated, require addition of a phosphoryl group for activation, but in the case of M phase Cdk1, when this is done, the enzyme is not immediately active because other kinases add two additional inhibitory phosphate groups. Just before mitosis, the two phosphate groups are removed causing activation of the Cdk. The reason for this convoluted process is probably that it permits build up of the triply phosphorylated, inactivated, form and then a very rapid

activation by simple hydrolytic dephosphorylation just prior to mitosis. However, before entering M phase, the G_2 checkpoint must be passed. This arrests the cycle if the DNA has not been completely replicated or is damaged. Unless the fault is corrected, the cell destroys itself by apotosis.

Mitosis phase

In the final M phase, the cell undergoes dramatic changes in which the nuclear membrane disappears, the DNA condenses into compact mitotic chromosomes, the spindle fibres develop and the chromosomes become positioned at the spindle equator ready for segregation in anaphase. A single mitotic Cdk initiates all of these cellular events. In metaphase, the final checkpoint before anaphase ensures that all chromosomes are correctly positioned at the equator. Any chromosome not attached to the kinetochore fibres (page 324) is a signal to halt the cycle. Once past these checks, the spindle is disassembled, the nuclear membrane reforms, and the cell undergoes mitosis and cytokinesis, and the mitotic cyclins are destroyed. The cell is now ready to commence a new cell cycle.

A summary of the Cdk-cyclin control of the cell cycle is given in Fig. 31.4.

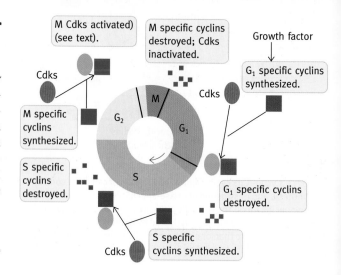

Fig. 31.4 Simplified diagram of control mechanism. In mammalian cells the cell cycle cannot pass the restriction point in G_1 unless a mitogenic signal is received, which activates the synthesis of G_1-specific cyclins. The latter are required to activate kinases (Cdks) necessary for activation of genes involved in the progression into S phase at which stage the cyclins are degraded. Activation of the same kinases by mitosis-specific cyclins is required for progression to the M phase. The different cyclins target the Cdks to different substrates as appropriate for the different phases of the cell cycle. Red represents inactive Cdks and green the activated forms. Note that in the interests of clarity the figure does not show the multiplicity of Cdks or of the activating cyclins involved. The details of these do not affect the essentials of the control scheme.

Summary

The eukaryotic cell cycle is divided into sections. These are the first gap (gap in DNA synthesis) phase (G_1), the DNA synthesis phase (S), the second gap phase (G_2), and the mitotic phase (M), occurring in that sequence.

All DNA synthesis occurs in S phase. Progression through the phases depends on the synthesis of cyclin proteins specific for different phases, and at the end of each phase they are destroyed by ubiquitin-associated proteolysis. The cyclins are required to activate protein kinases specific for each phase. They also determine which substrate a given kinase attacks.

Cyclin synthesis in G_1 requires the receipt by the cell of a mitogenic signal from a growth factor or cytokine; in its absence the cell enters a quiescent G_0 phase, which is the state of most somatic cells. A mitogenic stimulus reverts it to G_1.

To advance from the G_1 to S phase, the cycle passes a checkpoint. This halts the cycle if DNA is damaged and, if not repaired, the cell is given an apoptopic signal to self-destruct. Once in S phase the cycle is committed to proceed to mitosis.

Entry to M phase has to pass a checkpoint at the end of G_2. If the DNA has not been replicated completely or is damaged the cycle is halted.

After entering M phase a further check is made to establish that all of the chromosomes are correctly placed on the mitotic spindle. If not, the cycle is halted since it could result in abnormal daughter cells. Cytokinesis separates the daughter cells.

Further reading

Nigg, E. A. (1995). Cyclin-dependent protein kinases: key regulators of the eukaryotic cell cycle. *BioEssays*, **17**, 471–80.
Describes the role of phosphorylation by cyclin-dependent protein kinases in the regulation of the cell cycle.

Breeden, L. L. (2000). Cyclin transcription: timing is everything. *Curr. Biol.*, **10**, R586–8.
Concise review of cell-cycle control.

Cleveland, D. W., Mao, Y., and Sullivan, K. F. (2003). Centromeres and kinetochores: from epigenetics to mitotic checkpoint signaling. *Cell*, **112**, 407–21.
A comprehensive review of most aspects of mitosis and cell-cycle control.

Lowndes, N. E. and Murguia, J. R. (2000). Sensing and responding to DNA damage. *Curr. Opin. Genet. Dev.*, **10**, 17–25.

Bartek, J. and Lukas, J. (2003). Damage alert. *Nature*, **421**, 486–7.
A News and Views summary of how cells recognize DNA damage.

Zou, L. and Elledge, S. J. (2003). Sensing DNA damage through ATRIP recognition of RPA-ssDNA complexes. *Science*, **300**, 1542–8.

Problems

1 The restriction point in the G_1 phase of the cell cycle is of major importance. Discuss this with reference to the need for a growth factor signal for cell division to proceed.

2 Discuss the role of cyclins in the eukaryotic cell cycle.

3 If a mitogenic (growth factor) signal is not received in G_1 the cell enters the G_0 phase. Discuss this.

4 What checks are made at the G_1 restriction checkpoint and the mitotic checkpoint?

Chapter 32

Apoptosis

Overview

The name **apoptosis** is derived from the Greek prefix 'apo' meaning detached, and 'ptosis' meaning falling, with allusion to falling leaves. Leaf fall is a form of programmed death essential to the tree. Apoptosis is programmed cell death, essential to the life of multicellular animals.

Apoptopic death is not the only form of cell death within living animals; there is also **necrosis**. This occurs as a result of crude mechanical damage or oxygen deprivation to cells, which stop ATP production. The cell contents leak out of ruptured membranes and can set up inflammation in the surrounding tissue. It is a messy, potentially dangerous, method of getting rid of unwanted or damaged cells. When severe damage occurs, such as long-term oxygen deprivation to cells, pores open up in cell membranes, there is a large influx of calcium ions, the cells swell and burst. Necrotic destruction of cells is not a regulated process; it is essentially an accidental one not deliberately entered into by the body. This is a fundamental difference from apoptosis, which is a deliberate process brought about by a highly regulated elaborate multipathway process. In apoptosis the cell is killed but does not rupture the membrane. The cell is systematically degraded to a shrunken remnant by fragmentation of the DNA, and blebbing off of small vesicles, which are phagocytosed by white cells. The phagocyte recycles the degraded material. The method avoids the problem of inflammation that disrupted necrotic cells cause.

What is the purpose of apoptosis?

The importance of deliberate programmed cell death in the life of an animal, and its scale and complexity, has in the past decade probably come as a big surprise to many. It has two broadly different roles. In embryonic development it is essential that, at the appropriate stage, certain cells are removed. Obvious examples are the disappearance of the tail of the tadpole on development into a frog. Development of the nervous system involves death of neurons that fail to make profitable connections. In adult animals there is a constant vast amount of cell death occurring. For example, in the bone marrow and thymus, development of cells of the immune system involves the destruction every day of large numbers, probably billions, of B and T cells that would cause autoimmune attacks if allowed to survive. These are essential normal processes in which the cells self-destruct on receipt of signals. There is another reason. In an organism as complex as a human the potentiality for things to go wrong become great. An obvious example is that a single cell which escapes normal controls on cell division can result in a cancer that could be fatal and cause the demise of the trillions of cells making up the human. Similarly, if the DNA of a cell becomes damaged beyond repair, to allow it to continue to divide could result in a cancer that could destroy the animal Another major reason for programmed self-destruction is that one of the main strategies of the immune system to combat virus or other infection of cells (page 499) is for activated killer T cells of the cellular immunity system to deliver a signal to the cell to self-destruct thus aborting virus replication.

The complexity of cell growth regulation is such that virtually all cells have an elaborate mechanism in them ready to cause self-destruction should such irreparable trouble in the life cycle become apparent. The protein p53, the tumour suppressing protein (see page 522), plays a central role in this by triggering self-destruction if damage to the genome is detected that cannot be repaired. It is striking that p53 is deficient in about 50% of all human cancers, and probably is the reason that apoptosis has not been initiated, so that the 'corrupt' cell is allowed to survive and divide, when it should have been destroyed.

Cells thus contain a self-destruction 'kit' poised to be activated, a very delicate situation that has to be counter-balanced by a system to prevent accidental activation. There is a constant fine balance between proteins that promote apoptosis and others that prevent it. If the balance of these goes wrong, disease can result. For example, if a damaged genome requires apoptosis of the cell to avoid cancer and the excessive amounts of anti-apoptopic proteins prevent it, then cancer can develop. It has become apparent in the past decade that the self-destructive mechanisms present in most animal cells are of extraordinary complexity, to the point where they begin to look almost as complicated as the elaborate mechanisms that constitute the living process.

There are two broad methods of initiating apoptosis

The two methods of initiating apoptosis are called **intrinsic** and **extrinsic**. The intrinsic pathway follows from a wide variety of events that happen inside the cell, although what induces the apoptosis may come from the outside, such as radiation, antibiotics, and agents that damage DNA. Short-term oxygen deprivation can also induce apoptosis. (Long-term may cause necrosis.) It may be initiated by the p53 protein detecting damaged DNA. However, in the intrinsic induction of apoptosis, there is no deliberate signal delivered to the cell instructing it to destruct. In simplistic terms it is a cell response to a wide variety of stressful situations.

The extrinsic method of apoptopic induction on the other hand depends on the existence of special protein **death receptors** produced by cells and displayed on their outside surface. When specific ligands combine with these receptors present on cytotoxic cells (killer T cells) of the immune system, the cell self-destructs.

Mechanism of an intrinsic pathway of apoptosis

Given the central benign role of mitochondria in producing most of the cell's ATP, the idea that they become deadly killers in intrinsic apoptosis is almost difficult to accept, yet such is the case. We remind you that one of the electron transfer proteins in oxidative phosphorylation (page 202) is cytochrome c. Unique among the cytochromes, it is not an integral membrane protein, but only loosely attached by noncovalent bonds to the inner mitochondrial membrane. It is small, soluble and can be relatively easily experimentally extracted from the mitochondria. This protein, vital to ATP production, if released into the cytoplasm, is a death sentence to the cell.

A separate form of stress that can induce apoptosis is called endoplasmic reticulum (ER) stress, in which proteins cotranslationally transported into the ER lumen fail to fold properly so that the unfolded polypeptides accumulate. These are transported into the cytoplasm and lead to apoptosis (see **Rutkowski** in Further reading).

Mechanism of cytochrome C release

The initial 'insult' to the cell which, as mentioned, can be a wide variety of things like radiation, drugs, or a damaged genome, first results in the formation of pores in the mitochondrial membranes. P53 plays a role in this, probably by activating genes that produce proteins needed in the formation of pores. This is not completely understood. Cytochrome c leaks out into the cytoplasm. There it forms a complex with the protein **apoptotic protease-mediating factor** (**apaf-1**) and an inactive cytoplasmic proteolytic enzyme called **procaspase-9**. dATP, being present in the cell for DNA synthesis, is a required cofactor. The complex, apaf-cytochrome c-procaspase-9-dATP, polymerizes into a larger complex called an **apoptosome** (Fig. 32.1). In humans the apoptosome probably has 7 apaf-1 and 7 cytochrome c molecules. The apaf-1 subunits have a domain to which the procaspase-9 binds. As a result of the binding, procaspase-9 is activated as a proteolytic enzyme. It is not certain what the mechanism of this activation is. Possibly the procaspase has low level activity, and clustering them together causes mutual activation, but activation by conformational change is another possibility. The active **caspase-9** then activates three other procaspases also present in the cytoplasm in a proteolytic cascade, Caspases are so named because they have a **c**ysteine residue at their active site and attack proteins, cleaving after an **asp**artate residue (c-aspases). They target large numbers of proteins for cleavage. Destruction of a cellular DNase inhibitor protein activates the Dnase, which then takes the cell's DNA apart. The cell, as already described, blebs off degraded material as vesicles leaving a shrunken remnant, all of which are removed by phagocytosis.

Regulation of the intrinsic pathway of apoptosis

In this knife-edge situation, with the cell constantly delicately balanced between life and death, very tight control is needed. Biochemical controls in delicate situations are most often of the push–pull type in which opposing controls give stability. In cells, there are two groups of proteins, one favouring apoptosis and the other inhibiting it. They all belong to the **Bcl-2** family. **Bax** and **Bad** are two of the former family (*pro-apoptopic*) and believed to be involved in the initial formation of pores in the mitochondrial membrane. **Bcl-2** and **Bcl-X2** are important anti-apoptotic proteins that prevent pore formation and cytochrome c release. Bax and Bad block the anti-apoptotic activity of Bcl-2 and Bcl-X2. Whether a stimulus, such as a stress factor, produces cell destruction will therefore depend on the balance between the pro-apoptopic and anti-apoptotic proteins. As already mentioned, it is known that the protein p53 is important in the induction of apoptosis so that if p53 is deficient, then for example a cell with damaged genome, which should be destroyed by apoptosis, may survive and result in a cancer (discussed further on page 522).

There are also more direct controls on apoptosis. There is a family of proteins called **IAP** (**inhibitor of apoptosis**) that prevent caspase activation and inhibit caspase activity.

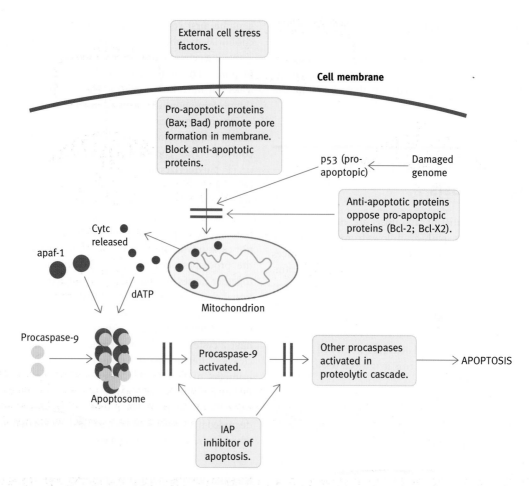

Fig. 32.1 Highly simplified diagram of the main features of the intrinsic pathway of apoptosis induction. The intrinsic pathway is a response to a variety of cell stress events, which include radiation and toxic chemicals. The stress stimulus leads to the formation of pores in the mitochondrial membrane and the release of cytochrome c. In the presence of dATP and apoptopic protease mediating factor (apaf1), cytochrome c forms a complex to which procaspase-9 attaches and is activated. The structure of the complex, called an apoptosome, is not fully known and varies in different species. It is not known how many cytochrome c molecules are attached, nor their location in the complex. The process is controlled by opposing proteins of the Bcl-2 family. Bcl-2 and Bcl-X2 inhibit apoptosis. Bax and Bad promote it; they have a domain which combines with a site on the Bcl proteins and block their action. The role of p53 is greatly simplified in the diagram. Green indicates pro-apoptopic pathway; red indicates anti-apoptopic pathway.

Mechanism of the extrinsic pathway of apoptosis

The extrinsic pathway is quite different in its induction from that of the intrinsic pathway. It is based on deliberate signals being delivered to protein death receptors which somatic cells produce on their surface. Several of these are known but we will use the one called **Fas** for illustration. Fas protein is a member of the **TNF** (**tumour necrosis factor**) receptor superfamily. The Fas receptor is the one that killer T cells of the immune system use to deliver a self-destruct signal to cells that it has recognized to be infected with a virus against which it has been activated. (We have described this system in Chapter 30.) The other death receptors have the same mechanism of action. Such killer T cells produce on their surface a protein called the **Fas ligand**, which

specifically recognizes, and binds to, the Fas receptor on the infected cell. The binding triggers the cell to self-destruct by apoptosis and in this way aborts virus production in that cell.

The Fas receptor is a transmembrane protein. Attachment of Fas ligand to the external domain causes the cytoplasmic domain to recruit an adaptor protein (**FADD**) to the cytoplasmic domain of the receptor (Fig. 32.2). Procaspase-8 then binds as a cluster to the FADD–receptor complex and is activated possibly due to the procaspases having a low level activity, and the clustering causing mutual activation (as may be the case for the intrinsic pathway). Once an active caspase is formed, a proteolytic cascade would ensue in which the family of procaspases are activated, as already described leading to cell destruction.

An alternative extrinsic delivery of an apoptopic signal to cells has been discovered. Killer T cells attach to target cells and

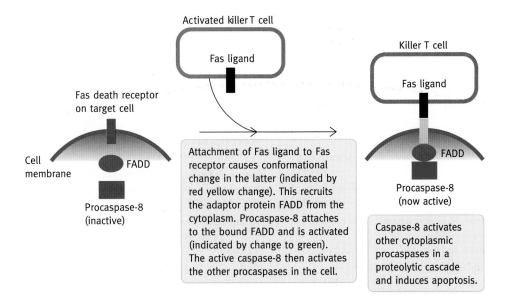

Fig. 32.2 The extrinsic pathway (described in the text).

release proteases from granules in their cytoplasm which the target cell takes up. They activate cellular procaspases directly by proteolysis. The enzymes have been christened **granzymes**. The killer cells also release **perforin**, which causes holes to form in the target cell membrane by which the granzymes enter.

Understanding of the complex mechanism(s) of apoptosis and its controls has opened up the hope that targeted drugs to induce apoptosis may attack cancer cells. Inhibitors of apoptosis may possibly also be used to treat or manage degenerative diseases.

Summary

Apoptosis is a normal way of disposing of unwanted cells and occurs on a large scale. The cell takes itself to pieces without lysis, the remnant body being phagocytosed. This avoids necrotic lysis that could cause inflammation and toxic effects.

Eukaryotic cells contain inactive proteolytic enzymes called caspases that are activated to destroy the cells. The activation may arise basically in two ways; in one, internal events such as DNA damage cause the release of cytochrome *c* from mitochondria. This causes aggregation and activation of caspases. Alternatively a killer T cell delivers a death signal by combining with a death receptor on the cell surface, which again results in aggregation and activation of caspases.

Further reading

Nicholson, D. W. and Thornberry, N. A. (1997). Caspases: killer proteases. *Trends Biochem. Sci.*, **22**, 299–306.

Depraetere, V. (1998). Fas. *Curr. Biol.*, **8**, R704.
One-page note about this cell-surface receptor, one of the main triggers of death in cells of the immune system.

Strasser, A., O'Connor, L., and Dixit, V. M. (2000). Apoptosis signaling. *Annu. Rev. Biochem.*, **69**, 217–45.

Nicholson, D. W. (2001). Baiting death inhibitors. *Nature*, **410**, 33–4.
News and Views summary of caspase control of apoptosis.

Chen, G. and Goeddel, D. V. (2002).TNF-R1 signaling: a beautiful pathway. *Science*, **296**, 1634–5.
Short review summarizing the engagement of tumour necrosis factor with its receptor and the pathway leading to the downstream events, apoptosis and NF-KB activation.

Hooper, N. M., Ed. (2002). Proteases in biology and medicine. *Essays Biochem.*, **36**, 1–167.
A complete issue with 12 reviews on many aspects of proteases, including caspases, cancer, proteasomes, and blood clotting.

Johnstone, R. W., Ruefli, A. A., and Lowe, S. W. (2002). Apoptosis: a link between cancer genetics and chemotherapy. *Cell*, **108**, 153–64.

Chakravati, D. and Hong, R. (2003). SET-ting the stage for life and death. *Cell*, **112**, 589–91.

Apoptosis by granzyme release

Downward, J. (2003). Metabolism meets death. *Nature*, **424**, 896–7.
A News and Views article on a protein that controls cell death found in complex with a protein involved in glucose metabolism.

Liu, C. Y. and Kaufman, R. (2003). The unfolded protein response. *J. Cell Sci.*, **116**, 1861–2.
Brief account of apoptosis mechanisms.

Newmeyer, D. D. (2003). Mitochondria: releasing power for life and unleashing the machineries of death. *Cell*, **112**, 481–90.

Ravichandran, K. S. (2003). 'Recruitment signals' from apoptotic cells: invitation to a quiet meal. *Cell*, **113**, 817–20.
Discusses phagocytosis of apoptotic cell 'corpses'.

Varfolomeev, E. E. and Ashkenazi, A. (2004). Tumor necrosis factor: an apoptosis juNKie? *Cell*, **116**, 491–7.

Halestrap, A. (2005) A pore way to die. *Nature*, **434**, 578–9.
A News and Views article comparing apoptosis and necrosis.

Lavrik, I., Golks, A., and Krammer P. H. (2005). Death receptor signalling. *J. Cell Sci.*, **118**, 265–7.
A 'Cell Science at a Glance' article.

Arian, C., Brumatti, G., and Martin, S. J. (2006). Apaptosomes: protease activation platforms to die from. *Trends Biochem. Sci.*, **31**, 243–7.

Rutkowski, D. T. and Kaufman, R. J. (2007). That which does not kill me makes me stronger: adapting to chronic ER stress. *Trends Biochem. Sci.*, **32**, 400–6.
Deals with response to accumulation of unfolded proteins in ER.

Problems

1   What are caspases? Describe their role and the mechanisms by which they become involved in it.

2   What is the purpose of apoptosis?

3   What initiates apoptosis in cells (apart from the killer T cell death signal)

4   What is the broad mechanism by which cytochrome C causes apoptosis?

5   What regulates apoptosis in cells?

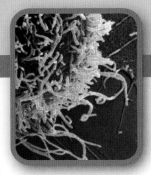

Cancer

Chapter 33

General concepts

We will first deal with the general cell biology of cancer and then with some of the basic molecular aspects.

Development of cancer involves loss of control of a number of processes; above all, it involves the uncontrolled multiplication of cells. Rapid cell division occurs normally throughout life in stem cells, in tissues such as those producing blood cells in the bone marrow and germline cells that produce gametes (though they have been found slowly dividing or quiescent in many other adult somatic tissues). Somatic cells for the most part stop dividing except for the small numbers of divisions needed to replace dead cells, and many retain the ability to divide rapidly if they are given the requisite signals to do so. Wounding of the skin, for example, fires off cell replication to heal the wound, but when this is achieved *normal cell division ceases*. If two-thirds of a rat liver is surgically removed, cell division restores the original size in a week or two but then the *division ceases*. The signalling pathways from mitogenic growth factors and cytokines, as referred to earlier, control the replication process so that it is appropriate to the needs of the body as a whole.

A cancer cell is aberrant in that it replicates without mitogenic signals from other cells or with a diminished requirement for them. Cell division may occur without 'permission' from other cells, or it may be that they fail to self-destruct by apoptosis after DNA damage, or other cell cycle mishap, has occurred. This difference between normal and cancerous cell division can be illustrated by tissue-culture experiments. Adherent mammalian cells of epithelial and mesenchymal origin will divide on nutritive medium in plastic dishes if they are supplied with growth factors, usually provided by foetal calf serum. The cells multiply and spread out to cover the surface of the plate in a single layer until they are in contact with each other and (with some cell types) they then stop dividing. This has been called **contact inhibition**. Unlike normal cells, cancer cells keep on dividing after the plate is completely covered, the cells piling up on one another to form a solid mass, which is essentially the equivalent of a tumour (Fig. 33.1).

Most normal cells can divide only a limited number of times

Normal somatic cells cannot be repeatedly subcultured on nutritive plates indefinitely but cancer cells can. In an earlier chapter (page 345) we described how a linear chromosome is shortened after each round of DNA replication, now believed to be a molecular cell-division 'counter' that limits the number of divisions a given cell line can undergo. To cope with this DNA loss, each eukaryotic chromosome has excess DNA in the form of **telomeres** at its ends. These are extensions of each DNA

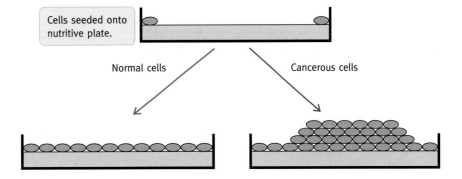

Fig. 33.1 Growth characteristics of normal and some cancerous animal cells in culture. Normal cells show contact inhibition – they stop dividing when they are in contact with each other. Cancer cells do not.

strand consisting of repeating short units of nucleotides put on by the enzyme **telomerase**. The telomeric DNA has two roles; first, it protects the DNA ends from the attention of repair enzymes and nucleases but it is also there to be progressively lost on chromosome replication and, in being sacrificed, protects the 'real' chromosomal DNA carrying the genes. In rapidly dividing stem cells, the telomerase replenishes the lost telomeric DNA so that no matter how many times the chromosome is replicated it is protected from shortening. However, somatic cells do not contain telomerase so that once the telomere is reduced to a critical length cell division ceases. It would seem that somatic cells have a 'ration' of telomeric DNA for so many cell divisions and that is it. Nonetheless, cancer cells derived from those same normal somatic cells are immortal.

Cancer cells have no limitation on the number of cell divisions they can make

There is no intrinsic limitation on the ability of cancer cells to divide, as is the case with normal cells. Cancerous human fibroblasts and many other cancers have been found to have telomerase activity, unlike the normal cells from which they developed, and this is true for most but not all cancer cells. About 20% of cancer cell types (more in the case of some bone cancers) do not switch on telomerase synthesis. They nevertheless are immortal because of the adoption of the alternative method of telomere elongation. They use ALT (alternative lengthening of telomeres) in which the telomere of one chromosome acts as the template for another one.

This does not imply that telomerase activity causes cancer, but that it is a corequisite for the immortality characteristic of cancer cells. One cancer cell line, known as HeLa (derived from the name of the patient), originating from a cervical cancer has been continuously cultured in countless laboratories for well over half a century.

Types of abnormal cell multiplication

There are two broad classes of abnormal cell growth in the body. Some cells grow and form a solid tumour, which simply gets bigger but remains as a single 'lump'; this is a **benign tumour**, so called because it does not spread to adjacent tissues. It may not pose a threat to life other than from the effects of physical compression or its own biochemical activity, if it, for example, is a hormone-producing tumour, and may often be relatively easy to cure surgically. An ordinary mole is a benign tumour. The other sort are **malignant cancer cells**; they invade neighbouring tissues and ultimately may break off and migrate to lymph nodes and elsewhere and set up further cancer centres. This spread is known as **metastasis**. Such metastasizing cancers are the dangerous ones; the cells have properties over and above their immortality, which enables them to spread into vitally important organs.

Cancers arising from epithelial cells (the most common) are called **carcinomas**, and from muscle cells, **sarcomas**. Leukaemias are 'cancers of the blood' in which large numbers of abnormal white blood cells continue to multiply and circulate without giving rise to, and often at the expense of, normal functional white blood cells.

Malignant Darwinism: cancer development involves an evolutionary progression of mutations

A cancer is a clone of cells originating from a single somatic cell that outgrows its neighbours. A cancer cell is the phenotypic expression of a corrupted genome that has Darwinistic advantages over its parent cells. It is initiated by **genetic mutation**, which may be a factor in why the incidence of cancer increases with age. There has been more time for initiating mutations to have accumulated. However, single genetic changes are not sufficient to cause cancer. Additional multiple genetic events have to occur, often over years. Cancerous cells undergo cytological changes that are diagnostically important; they tend to de-differentiate; that is to say, they lose the characteristic phenotype of the original cell from which they were descended.

Uncontrolled growth alone does not constitute a cancer, which usually starts as benign cell growths due to one or a few genetic changes. During this abnormal cell division the clone of cells goes through a natural selection process in which further genetic changes accumulate in the multiplying cells. The comprehensive review by **Hanahan and Weinberg** (see Further reading) discusses the cellular abnormalities that are found in most human cancers. The changes do not occur in a fixed progressive series so that individual cancers of the same type may have different histories of mutations. Some of the genetic changes give variants growth advantages over neighbouring tumour cells. Tumours can promote angiogenesis (development of blood vessels which sustain development of the tumour mass so that the cells are not restricted in growth by oxygen deprivation). A tumour cell may develop the ability to break loose from its starting clone of cells and escape to colonize other parts of the body. This metastasis requires several changes in a cell, incompletely understood, but which include cell-surface changes enabling it to detach from other cells in the tumour, and secretion of enzymes to break down connective tissue basal lamina layers (page 56) so that the cell enters the lymphatics or bloodstream.

A major recent development has been the discovery that a specific microRNA (called microRNA-10b) is highly expressed in metastatic breast cancer cells, and that non-metastatic breast cancer cells can be induced to metastasize by overexpression of microRNA-10b (**Li Ma** *et al.* in Further reading). It seems that it inhibits a homeotic gene whose expression restrains cell migration during embryogenesis. See Chapter 25 for more on microRNA genes.

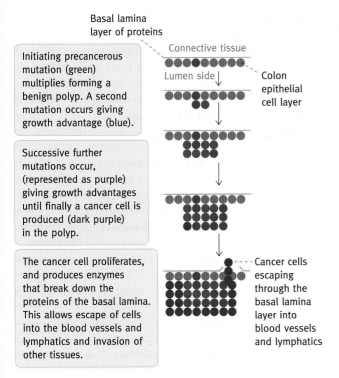

Basal lamina layer of proteins

Connective tissue

Lumen side

Colon epithelial cell layer

Initiating precancerous mutation (green) multiplies forming a benign polyp. A second mutation occurs giving growth advantage (blue).

Successive further mutations occur, (represented as purple) giving growth advantages until finally a cancer cell is produced (dark purple) in the polyp.

The cancer cell proliferates, and produces enzymes that break down the proteins of the basal lamina. This allows escape of cells into the blood vessels and lymphatics and invasion of other tissues.

Cancer cells escaping through the basal lamina layer into blood vessels and lymphatics

Fig. 33.2 Cartoon of development of a colorectal cancer. An initial mutation in an epithelial cell causes it to proliferate into a benign polyp. A series of further mutations occur giving the mutant cells growth advantages until the accumulation of mutations result in the development of a cancer cell (deep purple). This multiplies and metastasises by invading the epithelial layer and breaking down the basal lamina proteins. The cancer cells escape into the lymphatics and blood vessels and colonize other tissues of the body. Only four mutations are depicted here to keep the figure to a reasonable size. The progression from initial mutation to malignancy usually takes several years. Large numbers of cells are obviously formed at each stage but could not be shown here. The basal lamina layer is described on page 56.

Development of colorectal cancer

A classic example of the progression of cancer development is the colorectal cancer. This originates from a single cell in the gut epithelial layer causing the formation of a precancerous mass of cells known as a polyp (Fig. 33.2). If not removed (as is possible relatively simply during colonoscopy), the polyp cells go on multiplying and cell division provides the opportunity for accumulation of further mutations, which might lead to the evolution of a corrupted cancer genome over a period, usually of several years. The polyps often leak blood into the gut, and this can be detected by a simple widely available faecal test. Removed polyps can be checked cytologically to see whether any malignancy had developed. Figure 33.2 is a cartoon depiction of the genetic changes conferring growth advantages, ultimately resulting in cells that invade adjacent tissues and metastasize to other locations in the body.

Mutations cause cancer

The causes of cancer are basically at the genome level and they include the following:

- Carcinogenic chemicals that react with and covalently modify the DNA. The cytochrome P450s activate some of the procarcinogens to actual carcinogens. An example of a natural carcinogen is the **aflatoxins**, which are liver carcinogens. These may be present in foods such as peanuts that have been fungus-infected due to improper storage.

- Ionizing radiation causes chromosome breakages and rearrangements.

- UV light is well known to cause melanomas. Other skin lesions are frequent in patients with a certain type of defective DNA-repair mechanism (page 347).

- In addition to the above external agents, reactive oxygen radicals, superoxide, and particularly hydroxyl radicals generated in cells covalently modify DNA, and set up destructive chain reactions (page 488).

- Point mutations may be produced in any cell division due to faulty DNA replication (or mutations giving predisposition to cancer may be inherited *via* germline cells). The mutation rate is greatly increased if DNA-repair mechanisms are deficient.

- Viruses can also produce cancers in animals. **Papilloma virus**, which causes cervical cancers, is one of a small group that cause human cancers. **RNA retroviruses** are of great interest from a molecular biology viewpoint (see page 348) in that they are known to cause cancer in animals (see below), and, in one case, a human leukaemia. **HIV** is associated with the development of cancers but this arises from the suppression of the immune response. It seems that continuous immunological surveillance is important to destroy cancer cells presumably because of their surface changes. Without this, cancers would occur more frequently than they do.

Tumour promoters

In experiments in which mouse skin is painted with a carcinogen, such as benz(a)pyrene, to cause mutations, induction of a cancer also requires several applications of a substance known as a **tumour promoter** (not to be confused with a gene promoter). Phorbol esters are frequently used; these mimic the second messenger, diacylglycerol, which activates protein kinase C (page 447). Promoters do not cause cancer on their own and it is not totally understood how they work, but probably by stimulating cell division they both increase the chance of a cell with an initiating mutation leading to progeny cells, which acquire the further mutations leading to cancer. Different types of promoters presumably occur naturally.

The types of genetic change involved in cancer

As stated, cancers result from multiple mutations acquired possibly over years and accumulating gradually in successive generations derived from an initial clone of cells. Some of the corrupt genes may have little to do with the generation of cancer and be incidental to it while others may be indirectly involved in conferring a selective growth advantage. Much is yet to be discovered about the mechanisms by which cancer cells acquire their particular characteristics such as the capacity to metastasise, though there has been a recent breakthrough in understanding of this (page 408). However there are two broad classes of genes, the mutation of which is known to be of critical importance and about which, biochemically, a good deal is known.

One broad class is the **oncogenes**. These give abnormal control signals leading to uncontrolled cell division (*onco* = Greek for mass or tumour). Their effect is positive; they do something that actively causes or facilitates the development of cancer. Put colloquially, they are the 'bad' genes. They are most often derived from normal, essential, genes in control pathways, described below.

The second broad category of genetic changes that contribute to cancer development is due to the loss of functional genes that protect cells against uncontrolled cell division, the **tumour-suppressor genes.** They are in colloquial terms the 'good' genes. Mutation of these genes does not itself lead to uncontrolled cell growth, but removal of their protective effect means that if oncogenes become activated, cells are more likely to make the progression to the cancerous state. These are also described below.

DNA-repair genes are also protective but are not conventionally described as 'tumour-suppressor' genes. In Chapter 22 (page 343) we described the mechanism of methyl-directed mismatch repair involving Mut H, S, and L proteins in *Escherichia coli.* It has been found that proteins corresponding to Mut S and L (but not H) are present in eukaryotes, mutations of which are associated with a form of heritable bowel cancer in humans. Since mismatched bases (if not corrected) can lead to mutations and possibly to the formation of oncogenic forms of normal genes, it follows that DNA-repair systems are protective. We will now describe the **oncogenes** and the **tumour-suppressor genes** in turn.

Oncogenes

An oncogene is an abnormal, dominant gene that can initiate uncontrolled cell divisions or cause inappropriate cell survival,

as in the case of Bcl-2 (see below). They may be detected in cancers by isolating and fragmenting the DNA and introducing the pieces into normal cells (transfection), where they cause abnormal cell division. There are, however, a great many genes that are not known to have any potential to cause cancer when mutated. What makes particular genes liable to initiate cancer if mutated?

This is the point at which several major topics come together. In Chapter 27 we discussed cell signalling in which the predominant emphasis was on gene control and the way in which cytokines, growth factors, and hormones from outside the cell control specific genes in the nucleus to produce the required responses. If you consider the Ras pathway (page 437) as an illustrative example, signals from other cells attach to receptors on the outside of the target cell and a signalling pathway conducts the instruction to the nucleus via a series of intermediate proteins. Mitogenic growth factors act *via* Ras or other signalling pathways. Mutation of genes in such pathways are associated with cancer development. Cancer in fact typically involves at least one signalling pathway becoming faulty.

How can a signalling pathway become faulty? Consider a cascade of events such as shown below in which A–D are components of the signalling pathway (arrows represent activations, not conversions, please note).

$$\text{Mitogenic factor} \rightarrow \text{cell receptor} \rightarrow A \rightarrow B \rightarrow C \rightarrow D$$
$$\rightarrow \text{cell division}$$

If any of the protein components of the chain were to be altered, as a result of gene mutation, so that it is in a permanently active state, it will be the same as if a mitogenic growth factor or cytokine is present, even in the absence of the latter (Fig. 33.3). The nucleus is erroneously given the signal to activate genes leading to cell division. The pathway fires off its signal without a mitogenic signal having been received by the cell receptor so that, to revert to the cell-cycle controls, the cell is permitted to proceed through G_1 in the absence of 'permission'. This is in contrast to a *normal* cell which goes into G_0 phase in the absence of a signal (page 508). A *gene in a signalling pathway of this type has the potential to become an oncogene, and is therefore called a **protooncogene**.* Oncogenes arise from normal essential genes; it is the *importance* of the protooncogenes in control that makes them potentially oncogenic.

A totally different type of oncogene is that such as that for anti-apoptotic Bcl-2 proteins (overexpressed in many cancers) which favour cancer progression. The excess Bcl-2 protein blocks protective apoptosis signals from tumour-suppressor genes (page 513) and thus allows cells (e.g. with damaged DNA) to survive and become cancerous. If the apoptotic signal were not blocked as in normal cells, the cell would self-destruct and prevent progression to cancer.

How are oncogenes acquired?

Protooncogenes maybe converted to oncogenes in several ways. The simplest is that a point mutation in a protooncogene may

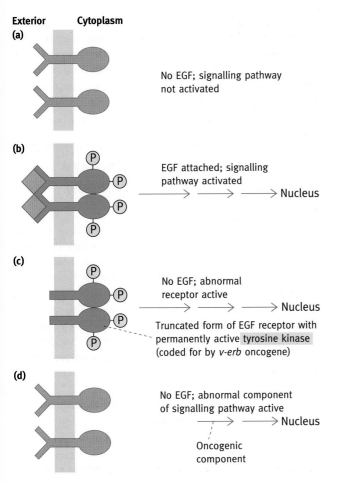

Exterior   Cytoplasm

(a)   No EGF; signalling pathway not activated

(b)   EGF attached; signalling pathway activated → Nucleus

(c)   No EGF; abnormal receptor active → Nucleus

Truncated form of EGF receptor with permanently active tyrosine kinase (coded for by *v-erb* oncogene)

(d)   No EGF; abnormal component of signalling pathway active → Nucleus

Oncogenic component

Fig. 33.3 Oncogene products as components of signalling pathways from the membrane to the nucleus – the epidermal growth factor (EGF) receptor is used as an example: **(a)** normal receptor – not activated; **(b)** normal receptor – activated; **(c)** permanently active truncated EGF receptor; **(d)** normal receptor – pathway component permanently active. Examples of the latter in human cancers are oncogenic forms of Ras and Raf in the Ras signalling pathway.

result in the production of an abnormal hyperactive protein that causes abnormal cell cycle control. The Ras protein gene, one of the early components of the signalling pathway, is a protooncogene. It has the automatic GTP/GDP switch, which limits the period of its activation; the Ras protein has, in effect, to continually ask the receptor whether the signalling molecule is still there. If it is not, in normal cells, the whole pathway is switched off, but in the oncogenic form the protein lacks the ability to hydrolyse GTP and so cannot operate the **GTP/ GDP-off switch**. Deficiency of the GTPase-activating protein, described on page 446, may also be involved in this. The pathway therefore fires off uncontrollably once activated. Many human cancers have an abnormal Ras protein, in some cases due to a point mutation in the gene. It is also known that the next component in the pathway, **Raf**, occurs as an oncogene in many human cancers in which the protein lacks the normal regulatory domain. It continually activates the pathway. This

concept of intermediate signalling pathway components being abnormally active is illustrated diagrammatically in Fig. 33.3.

Chromosomal translocations can result in oncogene formation by fusing a section of a protooncogene gene with a different gene. The resultant fusion protein may be overproduced and/ or be hyperactive. Alternatively the coding region of a normal protooncogene may be placed under the control of a strong promoter which results in overexpression of the gene. **Burkitt's lymphoma** is a cancer of human B cells of the immune system. It is due to the *c-myc* gene being put under the control of a strong immunoglobulin gene promoter by a chromosomal breakage and rearrangement, resulting in the formation of excessive amounts of the transcription factor Myc. Since Myc is a transcription factor that activates genes involved in cell growth and cell replication, this leads to excessive gene activation and to uncontrolled multiplication of the lymphocytes.

Retroviruses are RNA viruses known to cause cancers in animals and in one human leukaemia. When a retrovirus infects a cell, its reverse transcriptase copies the RNA into DNA, which then randomly inserts itself into the host DNA as a permanent genetic passenger. The retrovirus may insert upstream of protooncogenes, and, *via* their strong promoter elements, over-express it. Alternatively, the retrovirus may insert into tumour-suppressor genes (see below) and disrupt normal production of protein coded for by that gene. In some cases, a retrovirus acquires an oncogene from the host genome as it is reproduced in a cell which becomes inserted into a new host DNA on infection.

A note on nomenclature of viral oncogenes

Retroviral oncogenes are named (in italics) after the retroviruses carrying them. Thus, *ras* was found in the rat sarcoma virus. The oncogene is identified by the letter 'v' (for virus) and the protooncogene by 'c' (for cellular) so that we have v-*ras* and c-*ras* etc. The expressed proteins are written as Ras, Raf, etc. (not in italics). As an example, the v-*erb* oncogene codes for an abnormal truncated form of the receptor for epidermal growth factor (EGF; Fig. 33.3). It is permanently in the active state. If this gene is inserted into a host cell by a virus, the abnormal receptor is produced and is present in the cell membrane. It remains permanently in the phosphorylated, active state even though there is no EGF. It thus continually, and improperly, activates the Ras pathway which causes cell division in the absence of EGF which normally would be required for this to occur.

Tumour-suppression genes

We have emphasized that oncogenes commonly affect signal transduction pathways and gene control. When we come to tumour-suppressor genes the emphasis shifts to the cell-cycle control and particularly to the restriction point in G_1 phase of the cycle (page 508), which determines whether the cycle may proceed into S phase. Several tumour-suppressor genes are known, mutation of which has been found in particular groups of cancers. The two best-known tumour suppressors are **p53** and the **retinoblastoma** genes. The retinoblastoma genes are so named because a mutant form of this was first discovered in a rare retinal cancer. It has since been found to be present in a number of other cancers. Tumour-suppressor genes are a protection against the development of cancer.

Mechanism of protection by the p53 gene

It is the *absence* of a normal protective p53 protein that favours cancer development, rather than p53 mutants promoting cancer. p53 is described as the guardian of the genome. Defective p53 genes are involved in the development of over half of all human cancers. If only one p53 allele is deficient there is still functional p53 protein coded for by the other allele, so that mutation of both alleles is needed to lose the protective effect. The mutations are recessive.

Consider a normal cell in early G_1 phase. If appropriate mitogenic factors are present, the latter will activate genes leading to the production of the G_1-specific cyclins, which activates the Cdks. In the *normal* cell, the amount of p53 protein is low and inactive and the cycle can proceed. If the DNA is damaged however, the p53 gene is activated and more p53 protein is produced. p53, in the presence of damaged DNA, becomes phosphorylated on a serine residue which both activates it and decreases its rate of destruction. p53 is a transcription factor that activates several other genes including p21 whose protein product inactivates the G_1-specific Cdk activity. This arrests the cycle at the restriction point, giving time for the DNA to be repaired. If this is done, the p53 gene is no longer activated and the level of p53 protein returns to its low level and the cycle continues normally. If the DNA damage is not repaired, the p53 causes the cell to self-destruct by the apoptopic mechanism. In the absence of functional p53 genes, the cycle is not arrested and the apoptosis signal is not delivered, so a cell with abnormal DNA is allowed to replicate, thus increasing the chance of cancer developing.

Mechanism of protection by the retinoblastoma gene

This is another well-known tumour-suppressor gene. The protein coded for by the **retinoblastoma** gene (**Rb**) also blocks the cycle at the G_1 checkpoint. It combines with a transcription factor (**E2F**) and prevents this from activating genes required for synthesis of proteins involved in the progression into S phase. It therefore restrains the cell cycle. In *normal* cells Rb is inactivated; in such cells, when a mitogenic stimulus is received, the genes for G_1 cyclin synthesis are activated so that the Cdk activity increases. The Cdk phosphorylates and inactivates the Rb protein, preventing it combining with E2F so that the cycle can advance to S phase. However if the cycle is proceeding in an aberrant way due to an oncogene bypassing the need for a mitogenic signal, in the absence of the normally required Cdks, the Rb is not inactivated and it arrests the cycle. Retinoblastoma gene defects are known to be associated with certain types of cancer such as retinoblastoma, osteosarcoma, and small cell lung cancer.

Molecular biology advances have potential for development of new cancer therapies

Advances in understanding the molecular biology of the cell may provide new and more specific targets for attacking the causes of cancer. The membrane receptor tyrosine kinases which are of central importance in signalling pathways are attracting much attention as potential therapeutic targets. There are many approaches being investigated such as to target toxins to specific cancer cells using immunological methods based on the fact that the cancer cell surface is changed from the normal. Strategies to boost the immune system to increase immunological surveillance are also being attempted.

There are other approaches such as agents aimed at preventing vascularization of tumours that could help lead to their demise since the mass of cells become short of oxygen without this.

The discovery of RNA interference (RNAi, page 408) and its applicability to mammalian cells is very recent, but it raises the possibility of it being used against cancer. The small interfering RNAs (siRNAs) are easily synthesized chemically. Their sequence specificity means that, in principle, any messemger RNA can be targeted to silence (or knockdown) expression of specific genes. Oncogenes are an obvious target. Much work has already been done on mammalian cell cultures, including cancer cells. The silencing was 80–90% effective in these experiments. An interesting advance has been to use a viral vector to deliver to cultured cells a piece of DNA which is transcribed into a specific siRNA.

The completion of the human genome project facilitates the identification of disease-producing genes (and just about everything else connected with things genetical). A problem in understanding and treating cancer is that individual cancers have different characteristics which makes their classification difficult and this complicates devising specifically tailored therapeutic regimes. The development of DNA microarrays (page 463), which make it possible to study the simultaneous expression of large numbers of genes, has the potential to classify cancers in a way that may help the establishment of improved therapy regimes. By comparing the expression of blocks of genes in normal and cancerous tissues new understanding of the disease may emerge. A similar promise is inherent in the related technology of proteomics (page 79).

Summary

Cancer cells grow uncontrollably when they should not, often due to abnormalities of cellular signalling systems. They become independent of mitogenic signals and proceed to S phase in the absence of such a signal. They evade signals to self-destruct and because they acquire the ability to lengthen their telomeres can divide indefinitely. They progressively mutate, which gives them growth advantages and finally may metastasize to invade other tissues and spread throughout the body, a process which itself requires multiple mutations.

Development of cancer is a process of malignant Darwinian evolution with successive mutations conferring growth advantages on individual cells in a tumour. While a single initiating mutation may cause cell division, development of an invasive cancer requires a series of undefined mutations, which may take several years. The mutations do not occur in a fixed order, and cancers of the same type may have different histories and mutations.

A typical example is colorectal cancer in which first a benign polyp develops. Usually only after several years does it become a cancer. Removal of the polyp in colonoscopy will prevent it developing into a malignant form.

Major causes of mutations are carcinogenic agents, such as chemicals, certain viruses, and ionizing radiations. It may result also from faulty DNA replication or abnormal translocations of DNA sections.

A major class of cancer-causing mutations result in the formation of oncogenes ('bad genes') from normal protooncogenes. Protooncogenes are genes coding for proteins in signalling systems that control replication.

Alternatively the mutations may affect the activities of tumour-suppressor genes ('good genes'), which guard against cancer; p53 and the retinoblastoma gene are prime examples. About half of all human cancers are associated with deficient p53 protein. It works by halting the cell cycle if the genome is damaged. If this is not repaired the p53 somehow delivers an apoptopic signal for the cell to self-destruct, thus avoiding development of a potentially cancerous cell. If both p53 genes are deficient this safeguard is not present and an oncogenic mutation is more likely to progress to a cancer. The retinoblastoma protein can prevent the cell cycle from progressing from G_1 into the S phase if appropriate mitotic signals are not present.

Many new targets for possible cancer drug therapies are arising from understanding the molecular biology of cancer.

Further reading

Fearon, E. R. (1999). Cancer progression. *Curr. Biol.*, **9**, R873–5.
A 'primer' article giving essential information very briefly.

Hanahan, D. and Weinberg, R. A. (2000). The hallmarks of cancer. *Cell*, **100**, 57–70.
Includes succinct statement of six essential alterations in cancer development.

Bernards, R. and Weinberg, R. A. (2002). A progression puzzle. *Nature*, **418**, 823.
A one-page comment on cancer progression to metastasis.

Lu, J. *et al.* (2005). MicroRNA expression profiles classify human cancers. *Nature*, **435**, 834–8.

Ma, L., Teruya-Feldstein, J., and Weinberg, R. A. (2007). Tumour invasion and metastasis initiated by microRNA-10b in breast cancer. *Nature*, **449**, 682–8.

Karnoub, A. E., *et al.* (2007). Mesenchymal stem cells within a tumour stroma promote breast cancer metastasis. *Nature*, **449**, 557–63.

Telomeres

Johnson, F. B., Marcniak, R. A., and Guarente, L. (1998). Telomeres, the nucleolus and aging. *Curr. Opin. Cell Biol.*, **10**, 332–8.

Discusses the relationship between telomere lengthening and replicative senescence. Includes discussion of the relationship to human ageing.

Oncogenes

Cavanee, W. K. and White, R. L. (1995). The genetic basis of cancer. *Sci. Am.*, **272**(3), 50–7.
Describes the accumulation of genetic defects that can cause normal cells to become cancerous.

Diffley, J. F. X. and Evan, G. (1999). Oncogenes and cell proliferation. Cell cycle, genome integrity and cancer – a millennial view. *Curr. Opin. Genet. Dev.*, **10**, 13–16.
An informative editorial overview of the whole field. It also refers to subsequent reviews in the same issue of the journal.

Tumour-suppressor genes

Brehm, A. and Kouzrides, T. (1999). Retinoblastoma protein meets chromatin. *Trends Biochem. Sci.*, **24**, 142–5.
Discusses the effect of the tumour-suppressor retinoblastoma protein on chromatin remodelling.

Macleod, K. (2000). Tumor suppressor genes. *Curr. Opin. Genet. Dev.*, **10**, 81–93.

p53 tumour suppressor

Elledge, R. M. and Lee, W.-H. (1995). Life and death by p53. *BioEssays*, **17**, 923–30.
Summarizes the many functions of the P53 gene product, including its role in cancer suppression.

p53 tumour suppressor (1998). *Curr. Biol.*, **8**, R476.
A single-page quick guide giving salient facts of this important gene mentioning that at the time of publication there were 11,226 papers on the subject.

Lozano, G. and Elledge, S. J. (2000). P53 sends nucleotides to repair DNA. *Nature*, **404**, 24–5.
A News and Views article describing a new aspect of the tumour suppressor p53, namely that it activates a gene coding for a subunit of ribonucleotide reductase. This produces the deoxyribonucleotides required for DNA replication and repair.

Caspari, T. (2000). Checkpoints: how to activate p53. *Curr. Biol.*, **10**, R315–7.
Activation and stabilization involves protein kinase ATM.

Sharpless, N. E. and DePinho, R. A. (2002). P53: good cop/bad cop. *Cell*, **110**, 9–12.
A minireview.

DNA mismatch repair

Prolla, T. A. (1998). DNA mismatch-repair and cancer. *Curr. Opin. Cell Biol.*, **10**, 311–16.
Mutations of mismatch repair proteins are associated with the development of tumours in mice and in humans.

Problems

1 Explain what a protooncogene is. What are the mechanisms by which a protooncogene can become an oncogene?

2 How does the p53 gene protect against the development of cancer?

3 What are oncogenes and tumour-suppressor genes?

4 When somatic cells become cancerous they often develop telomerase activity. Explain the significance of this.

Figure acknowledgments

Molecular modelling

Atomic coordinates used are from the Protein Data Bank Research Collaboratory for Structural Bioinformatics (PDB). See Berman, H. M., Westbrook, J., Feng, Z., Gilliland, G., Bhat, T. N., Weissig, H., Shindyalov, I. N., and Bourne, P. E. (2000). The Protein Data Bank. *Nucleic Acids Res.*, **28**, 235–42; http://www.pdb.org/

References prior to 30 June 1999 are from the Brookhaven National Laboratory PDB. See Bernstein, F.C., Koetzle, T. F., Williams, G. J. B., Meyer Jr, E. F., and Brice, M. D.

Rodgers, J. R., Kennard, O., Shimanouchi, T., and Tasumi, M. (1977). The Protein Data Bank: a computer-based archival file for macromolecular structures. *J. Mol. Biol.*, **112**, 535–42.

Illustrations taken from PDB Molecule of the Month installations and illustrations were prepared by the author and illustrator, David S. Goodsell of The Scripps Research Institute.

Figures prepared with the program MOLMOL: Koradi, R., Billeter, M., and Wüthrich, K. (1996). *J. Mol. Graphics*, **14**, 51–5.

Fig. 1.5 Deoxy human hemoglobin; PDB code 1A3N. J. Tame and B. Vallone. Primary citation: not available.

Fig. 4.7 (a) Oxymyoglobin; PDB code 1MBO. Phillips, S. E. (1980). Structure and refinement of oxymyoglobin at 1.6 A resolution. *J. Mol. Biol.*, **142**, 531.

Fig. 4.7 (b) Staphylococcal nuclease; PDB code 1A2T. Wynn, R., Harkins, P. C., Richards, F. M., and Fox, R. O. (1996). Mobile unnatural amino acid side chains in the core of staphylococcal nuclease. *Protein Sci.*, **5**, 1026.

Fig. 4.7 (c) Triosephosphate isomerase; PDB code 1AG1. Verlinde, C. L., Noble, M. E., Kalk, K. H., Groendijk, H., Wierenga, R. K., and Hol, W. G. (1991). Anion binding at the active site of trypanosomal triosephosphate isomerase. Monohydrogen phosphate does not mimic sulphate. *Eur. J. Biochem.*, **198**, 53.

Fig. 4.7 (d) Pyruvate kinase; PDB code 1A3W. Jurica, M. S., Mesecar, A., Heath, P. J., Shi, W., Nowak, T., and Stoddard, B. L. (1998). The allosteric regulation of pyruvate kinase by fructose-1,6-bisphosphate. *Structure*, **6**, 195.

Fig. 4.9 Insulin; PDB code 3INS. Wlodawer, A., Savage, H., and Dodson, G. (1989). Structure of insulin: results of joint neutron and X-ray refinement. *Acta Crystallogr. B*, **4**, 99.

Fig. 4.22 Deoxyhemoglobin; PDB code 1A3N. Tame, J., and Vallone. B. Primary citation: not available.

Fig. 4.23 (a) Oxymyoglobin; PDB code 1MBO. Phillips, S. E. (1980). Structure and refinement of oxymyoglobin at 1.6 A resolution. *J. Mol. Biol.*, **142**, 531.

Fig. 4.23 (b) Oxyhemoglobin; PDB code 1HHO (B chain). Shaanan, B. (1983). Structure of human oxyhaemoglobin at 2.1 A resolution. *J. Mol. Biol.*, **171**, 31.

Fig. 6.3 (a) Yeast hexokinase A (unbound); PDB code 1HKG. Steitz, T. A., Shoham, M., and Bennett Jr., W. S. (1981). Structural dynamics of yeast hexokinase during catalysis. *Phil. Trans. R. Soc. Lond. B Biol. Sci.*, **293**, 43.

Fig. 6.3 (b) Yeast hexokinase A with o-toluoylglucosamie; PDB code 2YHX. Anderson, C. M., Stenkamp, R. E., and Steitz, T. A. (1978). Sequencing a protein by X-ray crystallography. II. Refinement of yeast hexokinase B co-ordinates and sequence at 2.1 A resolution. *J. Mol .Biol.*, **123**, 15.

Fig. 7.9 Porin membrane protein; PDB code 1BH3. Schmid, B., Maveyraud, L., Kromer, M., and Schulz, G. E. (1998). Porin mutants with new channel properties. *Protein Sci.*, **7**, 1603.

Fig. 7. 16 Potassium channel protein; PDB code 1BL8. Doyle, D. A., Morais Cabral, J., Pfuetzner, R. A., Kuo, A., Gulbis, J. M., Cohen, S. L., Chait, B. T., and MacKinnon, R. (1998). The structure of the potassium channel: molecular basis of K+ conduction and selectivity. *Science*, **280**, 69.

Fig. 7.17 Potassium channel protein; PDB code 1K4C. Zhou, Y., Morais-Cabral, J. H., Kaufman, A., and Mackinnon, R. (2001). Chemistry of ion coordination and hydration revealed by a K+ Channel-Fab Complex at 2.0 A resolution. *Nature*, **414**, 43. Picture created in RasMol for Molecule of the Month, 2003.

Fig. 24.9 (a) Yeast tRNAPhe; PDB code 4TNA. Hingerty, B., Brown, R. S., and Jack, A. (1978). Further refinement of the structure of yeast tRNAPhe. *J. Mol. Biol.*, **124**, 523.

Fig. 24.9 (b) EF-G in complex with GDP; PDB code 1DAR. al-Karadaghi, S., Aevarsson, A., Garber, M., Zheltonosova, J., and Liljas, A. (1996). The structure of elongation factor G in complex with GDP: conformational flexibility and nucleotide exchange. *Structure*, **4**, 555.

Fig. 24.9 (c) Ternary complex of Phe-tRNAPhe, EF-Tu, and a GTP analog; PDB code 1TTT. Nissen, P., Kjeldgaard, M., Thirup, S., Polekhina, G., Reshetnikova, L., Clark, B. F., and Nyborg, J. (1995). Crystal structure of the ternary complex of Phe-tRNAPhe, EF-Tu, and a GTP analog. *Science*, **270**, 1464.

Fig. 24.16 (a) and (b) GroEL and GroES complex; PDB code 1AON. Xu, Z., Horwich, A. L., Sigler, P. B. (1997). The crystal structure of the asymmetric GroEL-GroES-(ADP)7 chaperonin complex. *Nature*, **388**, 741.

Figures prepared with the program PyMOL: DeLano, W. L. (2002). The PyMOL Molecular Graphics System. World Wide Web **www.pymol.org**.

Cover picture A1C12 subcomplex of F_1F_o ATP synthase; PBD code 1C17. Rastogi, V. K., and Girvin, M. E. (1999). *Nature*, **402**, 263–8.

Fig. 12.24 The conformation of the epsilon- and gamma-subunits within the *E. coli* F_1 ATPase; PBD code 1JNV. Hausrath, A. C., Capaldi, R. A., and Matthews, B. W. (2001). *J. Biol. Chem.*, **276**, 47227–32.

Fig. 21.5 Structure of a B DNA dodecamer: conformation and dynamics; PDB code 1BNA. Drew, H. R., Wing, R. M., Takano, T., Broka, C., Tanaka, S., Itakura, K., and Dickerson, R. E. (1981). *Proc. Natl. Acad. Sci. USA*, **78**, 2179–83.

Other figures

Fig. 2.1 Harry, E. J., Callister, H., and Wake, R. G. (1981). J. *Bacteriol.*, **145**, 1042–51, with permission of the American Society of Microbiology and R. G. Wake. Photograph courtesy Dr E. J. Harry, School of Molecular and Microbial Biosciences, University of Sydney.

Fig. 2.7 After Fig. 1 in Metcalf, D. (1991). *Phil. Trans. Roy. Soc., Lond.* B **333**, 147. Reproduced (modified) by permission of D. Metcalf and the Royal Society.

Fig. 2.8 After Fig. 1.1 in Ptashne, M. (1992). *A Genetic Switch: Phage l and Higher Orhanisms* (2nd edn), Cell Press, and Blackwell Scientific Publications, with permission of Blackwell Publishing and M. Ptashne.

Fig. 6.10 Slightly modified from Fig. 1 of Dodson, G. and Wlodawer, A. (1998). *Trends Biochem. Sci.*, **23**, 347–52, with permission from Elsevier Science.

Fig. 8.8 From Figs 7 and 8 in Geeves, M. A. and Holmes, K. C. (1999). *Annu. Rev. Biochem.*, **68**, 687–728, reproduced with permission from Annual Reviews, Inc., and M. Geeves.

Fig. 8.13 (b) After Fig. 2.13 in Lewin, B. (1994). *Genes V*, Oxford University Press, Oxford.

Fig. 8.18 After Fig. 1(e) in McNiven M. A. and Ward, J. B. (1998). *J. Cell Biol.*, **106**, 111, with permission from the Rockefeller University Press.

Fig. 12.23 Reproduced from Fig. 1 of Hutcheon, M. L., Duncan, T. M., Ngai, H., and Cross, R. L. (2001). *PNAS*, **98**, 8519–24., with permission from the National Academy of Sciences.

Fig. 12.24 Reproduced from Fig. 2(c) of Abrahams, J. P., Leslie, A. G. W., Lutter, R., and Walker, J. E. (1994). *Nature*, **370**, 621–8, with permission from Nature Publishing Group, and J. Walker.

Fig. 21.3(c) and (b) adapted from Figs 2.4(a) and 3.1 respectively in Calladine, C. R. and Drew, H. W. (1992).

Understanding DNA, Academic Press, with permission of Elsevier Science.

Fig. 21.4 Reproduced from Fig. 2.1 in Ptashne, M. (1992). *A Genetic Switch: Phage l and Higher Organisms* (2nd edn), Cell Press, and Blackwell Scientific Publications, with permission of Blackwell Publishing and M. Ptashne.

Fig. 21.8 Fig. 28.19 in Lewin, B. (1994), *Genes V*. Oxford University Press, Oxford. (The electron micrograph was provided to Lewin by Barbara Hamkalo.)

Fig. 22.14 Fig. 3(a) from Krishna, T. S. R., Kong, X.-P., Gary, S., Burgers, P. M., and Kuriyan, J. (1994). *Cell*, **79**, 1233–43. Copyright by Elsevier Science. Photograph kindly provided by Dr J. Kuriyan.

Fig. 22.15 Based on Fig. 1 of Kelman, Z. and O'Donnell, M. (1995). *Annu. Rev. Biochem.*, **64**, 171, with permission of Annual Reviews.

Fig. 23.17 Reproduced from Fig. 29.10 in Lewin, B. (1994). *Genes V*, Oxford University Press, Oxford.

Fig. 23.25 Reproduced from Fig. 1 in Bjorklund, S. and Gustafsson, C. M. (2005). *Trends Biochem. Sci.*, **30**, 240–4, with permission from Elsevier Science.

Fig. 23.29 Adapted from Fig. 4 in Ross, J. (1995). *Microbiol. Rev.*, **59**, 423, with permission from the American Society for Microbiology and J. Ross.

Fig. 23.30 (b). Fig. 3(a) in Ohlendorf, D. H., Anderson, U. F. and Matthews, B. W. (1983). *J. Moc. Evol.*, **19**, 109. Reproduced with copyright permission of Springer-Verlag, and B. W, Matthews.

Fig. 24.5 Fig. 9.5 from Lewin, B. (1994). *Genes V*, Oxford University Press, Oxford.

Fig. 24.10 From Fig. 1 of Noller, H. F. (1991). *Annu. Rev. Biochem.*, **60**, 193. Reproduced with copyright permission of Annual Reviews, Inc.

Fig. 24.11 Reproduced from Fig. 1(a) Mankin, A. S. (2006). *Trends Biochem. Sci.*, **31**, 11–3, with permission from Elsevier Science.

Fig. 24.17 Reproduced from Fig. 5(b) of Baumeister, W., Walz, J., Zühl, F., and Seemüller, E. (1998). *Cell*, **92**, 367–80, with permission from Elsevier Science and W. Baumeister.

Fig. 26.17 Based on Fig. 2 of Mattaj, I. W. and Englmeier, L. (1998). *Annu. Rev. Biochem.*, **67**, 265, with permission from Annual Reviews, Inc., and I. Mattaj.

Fig. 26.18 Based on Fig. 2 of Mattaj, I. W. and Englmeier, L. (1998). *Annu. Rev. Biochem.*, **67**, 265, with permission from Annual Reviews, Inc., and I. Mattaj.

Fig. 26.19 Modified from Fig. 1 of Jans, D. A. (1998). *Australian Biochemist*, **29**, 5–10, with permission from D. Jans.

Fig. 27.18 After Fig. 2 in Dohlman, H. G., Caron, M. G., and Lefkowitz, R. J. (1987). *Biochemistry*, **26**, 2660, with copyright permission of the American Chemical Society.

Answers to problems

Chapter 1

1 A high entropy level means a relatively low energy level and a low entropy level a relatively high energy level.

2 It is difficult to define what it is though everyone feels that they know what it is. What one knows is that there are different types of energy, which can be converted into one another.

3 There is potential energy, which may be gravitational potential energy like a rock about to fall, chemical potential energy of a molecule about to react and liberate energy. There is the kinetic energy of a moving object. Gravitational potential energy becomes kinetic energy when the rock falls. There is heat energy, but this can do work only if there is a temperature gradient.

4 Molecules such as starch are long strings of glucose units linked together like a string of the same letter. Proteins and DNA are made of different units linked together so the sequence of these can vary and convey a message or instruction. They are more like meaningful sentences. The information in the sequences of these molecules is the basis of life.

5 It is an RNA molecule that delivers the message of genes from the DNA to the ribosomes, which translate the message into proteins. The message is in the form of a copy of the sequence of bases that make up the DNA of the gene.

6 The structures of both DNA and RNA are such that by hydrogen bonding capability they can direct their own replication. This would have been essential to establish life in the primitive environment before there were any of the elaborate replication mechanisms of the modern cell had been developed. Without accurate self-replication there can be no life.

Chapter 2

1 Antibody protection against influenza is directed at the haemagglutinin protein. This is continually mutated but, because there are many epitopes on the molecule, the immune protection loss is partial and gradual. This antigenic drift leaves people susceptible to infection but residual protection means that it is mild. If by a recombination between different strains a totally new haemagglutinin is present (antigenic shift), a lethal pandemic can result.

2 Somatic cells are the 'ordinary' cells of most tissues. They are differentiated into liver cells, muscle cells, etc., and at cell division give rise only to their specific type. Their ability to divide is limited due to telomere shortening and they only divide to repair damage and replace dead cells. Stem cells are of two broad types, embryonic and adult. The former are pluripotent – they give rise to all types of cells in the body. At cell division the two daughter cells can either differentiate into somatic cells or remain as stem cells and thus are constantly renewed. Telomeres are maintained and cell division is unlimited. Adult stem cells are multipotent but not pluripotent. They renew themselves but can give rise also to some somatic cells. An example is bone marrow stem cells, which can give rise to any type of blood cell.

3 Prokaryotes are the bacteria. They are very small and are surrounded by a rigid cell wall. They have no internal membranes and thus lack a defined nucleus. Their DNA is in contact with the cytoplasm. They have a single circular main chromosome but in addition small circular plasmids of DNA may also be present which replicate independently. Prokaryotes in general have specialized in rapid cell division as their survival strategy. Prokaryotes are typically haploid; their DNA is segregated at cell division in a relatively simple way without obvious physical changes. Eukaryotic cells are those of plants and animals. They are typically about 1000 times larger in volume than bacteria; they have no cell wall outside of their plasma membrane (plants excepted). The most important characteristic is the possession of a 'true' nucleus surrounded by a membrane and containing the chromosomes. They also have a variety of internal membranes in the form of other organelles. Eukaryotic chromosomal DNA molecules are linear and, except for stem cells and cancer cells, mammalian cells are limited in the number of cell divisions they can undergo. Eukaryotic cells are typically diploid (gametes excluded); segregation of DNA involves the elaborate physical events of mitosis.

4 A living cell has to take in molecules which it needs from outside and get rid of waste molecules. This means that there is a high volume of traffic across the membrane usually requiring transport mechanisms in the membrane. The traffic has to be adequate to sustain the needs of the cell; the bigger the cell the more traffic is required. The ratio of membrane area to volume of the cell diminishes with size so that cells must remain small

enough for this to remain adequate for the needs of the cell. Also molecules must reach all parts of the cell. Diffusion is slow so that small size is also required for this.

5   There has to be sufficient room to accommodate the molecules needed for life. If you were building a doll's house but had to use full size bricks this would set a lower limit to how small you could make it.

6   It is best to use an organism that gives the best chance of getting an answer, usually the simpler the better. The system works because the fundamental processes of life are basically the same in all life forms.

Chapter 3

1   The amount of ATP that can be synthesized using 5000 kJ of free energy is 5000/55, which equals 90.91 mol. The weight of disodium ATP produced is thus 551×91 g daily, or 50, 141 g, which is equal to 72% of the man's body weight. The reason why this is possible is that ATP is being continuously recycled to $ADP + P_i$ and back again to ATP.

2   The $\Delta G^{0'}$ value refers to standard conditions where ATP, ADP, and P_i are present at 1.0 M concentrations. In the cell the concentrations will be very much lower, the actual ΔG value for ATP synthesis will be different from the $\Delta G^{0'}$ value according to the relationship

$$\Delta G = \Delta G^{0'} + RT 2.303 \log_{10} \frac{[ADP][P_i]}{[ATP]}.$$

3   ATP and ADP are high-energy phosphoric anhydride compounds, whereas AMP is a low-energy phosphate ester. The factors that make hydrolysis of the former more strongly exergonic include the following.

Release of phosphate relieves the strain caused by the electrostatic repulsion between the negatively charged phosphate groups. The released phosphate ions fly apart. A factor also contributing to the exergonic nature of the hydrolyses is the resonance stabilization of the phosphate ion which exceeds that of the phosphoryl group in ATP. Hydrolysis of AMP causes little increase in resonance stabilization.

4   (a) The nonpolar molecule cannot form hydrogen bonds with water molecules so that those of the latter surrounding the benzene molecule are forced into a higher-order arrangement in which they can still hydrogen bond with each other (for the total bonding does not change). This increased order lowers the entropy and increases the energy of the system; insertion of the benzene molecules into the water is therefore opposed and they are forced into a situation with minimal benzene/water interface as spherical globules and then as a separate layer. The effect is known as hydrophobic force.

(b) The polar groups on glucose can form hydrogen bonds with water.

(c) The Na^+ and Cl^- ions become hydrated. The energy release as a result of this exceeds that of the ionic attraction between them. The separation also has a large negative entropy value.

5   The enzyme AMP kinase transfers a phosphoryl group from ATP to AMP by the reaction

$$ATP + AMP \rightarrow 2ADP$$

Hydrolysis is not involved so that there is no significant $\Delta G^{0'}$ change in the reaction.

6   In the cell, PP_i will be hydrolysed to $2P_i$, whereas with a completely pure enzyme there will be no inorganic pyrophosphatase present. In the former case, the $\Delta G^{0'}$ of the total reaction will be $-32.2 - 33.4 + 10$ kJ mol^{-1} = -55.6 kJ mol^{-1}. For the completely pure enzyme, the $\Delta G^{0'}$ will be -22.2 kJ mol^{-1}.

7   Ionic bonds, hydrogen bonds, and van der Waals forces with average energies of 20, 12–29, and 4–8 kJ mol^{-1}, respectively. The energies of activation for formation of weak bonds are very low and so can occur without catalysis. Weak bonds in large numbers can confer definite structures on molecules, but nonetheless can be broken easily, resulting in flexibility of structures.

8   (a)

9   The binding of the molecules is *via* noncovalent bonds. Since these are weak, several are needed to make the binding effective. Noncovalent bonds are short range and, in the case of hydrogen bonds, highly directional. Put these considerations together and only molecules that exactly fit together can make the required number of bonds. This is the basis of biological specificity. The fact that the bonds are weak (low ΔG) means that they can form spontaneously, without catalytic assistance, and importantly can dissociate to make the associations reversible, which is also essential in many situations. In the

(b)

$$pH = pK_a + \log \frac{[salt]}{[acid]}$$

The relevant pK_a is 7.2.

$$pH = 7.2 + \log \frac{[Na_2HPO_4]}{[Na_2HPO_4]}$$
$$= 7.2 + \log 1$$
$$= 7.2.$$

(c) Histidine with a pK_a value of 6.5.

case of antibody-antigen binding in which irreversibility is needed for immune protection, very large numbers of bonds are involved. Many structures in the cell, in some cases very elaborate ones, depend on the same principle of specific protein-protein interactions to bring about their self-assemblage.

10 In chemistry, a high-energy bond is one requiring a large amount of energy to break it, which is the reverse of the intended biological concept. Most importantly, however, it is incorrect to envisage the energy of a molecule to be located in a particular bond but rather, it resides in the molecule as a whole. It is the free energy fall in converting ATP to its products that provides the energy for work. A fall in free energy is a property of a complete reaction, not only of the breakage of a particular bond. This does not alter the fact that Lipmann's concept was important in that ATP behaves *as if the* energy is located in the bond. For this reason it can still be found in textbooks.

11 It is the relentless drive of the universe to achieve maximum stability or lowest energy state. The most stable forms of atoms are when the outer valence electron shell is fully occupied by 8 electrons (the octet rule). Atoms with this structure are the inert gases such as helium. Atoms on earth cannot change their electron numbers but by sharing electrons such as occurs in covalent bond formation they can mimic the or approach the noble gas structure and thus achieve greater stability.

12 Nothing can be exempt. To assemble a living cell from smaller environmental molecules, energy is needed. This is obtained by the breakdown of food molecules such as glucose to CO_2 and water, which leave the cell. In doing so this increases entropy to a greater degree than the assembly of the cell lowers entropy. Therefore the total entropy of the universe is increased and the second law is obeyed. Life is thus an ordinary process mechanistically and not magical.

13 Free energy is a term which applies to chemical reactions not a property of individual molecules. One can talk of the free energy of formation of a molecule because that refers to reactions. In a chemical reaction there is a liberation of energy but only part of this is available to do work such as being coupled to the synthesis of molecules with a higher energy content. This is the fraction known as free energy. It is expressed as ΔG – that is the change in free energy occurring in a reaction. This is the all-important thermodynamic term in considering the reactions of life. The part of the total energy change which is not available for utilisation in work goes in satisfying the second law of thermodynamics which specifies that any happening must increase the total entropy of the universe.

14 Entropy is the degree of randomness in any system and in the universe. A system of high entropy is at a lower energy than one at a lower entropy so that increasing entropy level increases stability. The 'aim' of the universe is to achieve maximum entropy and this appears to be the driving force of everything. The second law of thermodynamics demands that all happenings raise the entropy of the universe. It is a somewhat elusive concept to grasp. One common way to raise total entropy is to release heat from a chemical reaction which increases the random movement of gas molecules in the air. All reactions must contribute to entropy increase, otherwise they would be defying the second law; this part is not available for work. What is left is free energy.

Chapter 4

1 It refers to the amino acid sequence of the polypeptide chain. The amino acids arelinked together by peptide (—CO—NH—) bonds. The structure of such a chain is given on page 49.

2 The polypeptide chain of a protein is folded into a specific, usually compact shape, the precise folding being dependent on noncovalent bonds and covalent disulphide bonds between the amino acid residues. Heat disrupts the former bonds, causing the polypeptide chains to unravel and become entangled into an insoluble mass, devoid of biological activity. This is known as protein denaturation.

3 Examples of all are given on pages 46 and 47.

4 (a) Around 4; (b) around 10.5–12.5; (c) around 6.5.

5 Aspartic and glutamic acids, lysine, arginine, and histidine. The carboxyl and amino groups of all other amino acids are bound in peptide linkage (excluding the two end ones).

6 Primary, secondary, tertiary, and quaternary. Illustrated in Fig. 4.2.

7 It lies in the four-way desmosine group shown in Fig. 4.12.

8 Collagen fibres are designed for great strength. To achieve this, three polypeptides are associated in a closely packed triple superhelix. The pitch of the supercoil is such that there are three amino acid residues per turn. This arranges that where the polypeptides touch, there is always a glycine residue. The side group of a single hydrogen atom permits close association whereas bigger side groups would prevent this.

9 (a) Proteins are functional only in their native state, which means their correct three-dimensional folded state. This, in most proteins, is due to noncovalent bonds, which are easily disrupted by mild heat. The polypeptides unfold and become irreversibly tangled up into aggregates.
(b) A number of proteins such as insulin have a few disulphide bonds (three in the case of insulin) which, being covalent, are not destroyed by heat. Although noncovalent bonds are still important in insulin structure, the disulphide bonds confer a degree of increased stability. The extreme case is keratin where large numbers of disulphide bonds produce a very stable structure. You can boil hair as long as you like without producing any permanent structural changes.

10 Particularly in the larger proteins, when the three-dimensional structure is determined, separate regions of the protein are often seen (see Fig. 4.7(d)) separated by unstructured polypeptide. The impression is that these are essentially independent globular sections, which could exist on their own. Indeed, in a number of cases, domains have been isolated from multi-functional enzymes and found to have partial catalytic activity on their own. This is sometimes of great practical use in that it permits the preparation of an enzyme fragment with a wanted activity, but is free from other activities of the complete enzyme. It is essential for a protein domain to be formed from a given section of a chain; it cannot leave the domain and then double back into it, because then the structure would not be stable on its own if separated from the rest of the protein. Domains are of very great interest because it is likely that they have played an important role in evolution by a process of domain shuffling. This is described in detail in later chapters but for the moment it is often found that domains are coded by sections of genes which are believed to be shuffled about to form new genes. When expressed, these give rise to new functional proteins. Evolution of new enzymes by this means would be much faster than gradually altering proteins by point mutations.

11 Phenylalanine and isoleucine are hydrophobic and are to the maximum possible extent shielded from water in the hydrophobic interior of globular proteins. Aspartic acid and arginine have highly charged side chains at physiological pH and therefore need to be on the outside exposed to water. Inside a protein molecule these groups would tend to have a destabilizing effect because they cannot have their bonding capabilities satisfied in the hydrophobic environment. Formation of bonds releases energy so that only when bonding is maximized do you get the most stable structure.

12 The charges on the GAGs cause the molecule to adopt an extended conformation and the GAG units permit the formation of very large molecules so that the total volume of water trapped is large.

13 The polypeptide backbone has hydrogen-bonding potential which must be satisfied if structures of maximum stability are to be achieved. The backbone cannot avoid crossing the hydrophobic interior of proteins and the hydrophobic side chains of the amino acid residues in these regions offer no possibility of hydrogen-bond formation. The solution is for backbone groups to hydrogen bond to other backbone groups and this is what happens in both the a helix and the β-pleated sheet structures.

14 As shown in Fig. 4.20, myoglobin has the higher affinity and the curve is hyperbolic as compared with the sigmoid curve of haemoglobin. Myoglobin functions purely as an oxygen reserve in muscle. The high affinity means that it maximizes the storage. It is released only when the O_2 tension in the muscle falls. Haemoglobin has to take up the maximum amount of O_2

in the lungs and release the maximum amount in the capillaries. The sigmoid curve helps this. It releases O_2 most rapidly in the capillaries because the O_2 tension there corresponds to the steep part of the curve.

15 On binding of oxygen to the haem iron, Fe moves into the plane of the porphyrin ring. This requires the slightly domed tetrapyrrole to flatten. The Fe is attached to the F8 histidine of the protein and the molecule re-arranges itself. This allows the tetrapyrrole to flatten. The movement affects subunit interactions causing the tetramer to change its conformation from the T to the R state (see Fig. 4.27).

16 Adult haemoglobin in the de-oxygenated state has a cavity that binds 2:3-bisphosphoglycerate (BPG) by positive charges on the protein side chains. Binding of BPG is possible only in the de-oxygenated state and therefore binding increases unloading of oxygen – it reduces affinity of the haemoglobin for oxygen. Foetal haemoglobin has a γ subunit instead of the adult β one. It lacks one of the BPG charge-binding groups. BPG therefore binds to foetal haemoglobin less tightly, thus raising its affinity for oxygen above that of the maternal haemoglobin.

17 It is necessary for the major transport of CO_2 as HCO_3^-, as illustrated in Fig. 4.29.

18 The statistical chance of a polypeptide with a random primary sequence folding up into a functional domain is vanishingly small. It is observed that many different proteins have folded structures which are strikingly similar but performing different roles. This can be explained by domain shuffling in which a polypeptide primary sequence that folds up to form a functional entity is used over and over again in different domain combinations and with variations that adapts it to new roles. This would lead to more rapid evolution than depending on single amino acid residue changes due to point mutations.

19 The disease was the first understood case of a 'molecular disease'. In this a single amino acid residue change in haemoglobin from glutamate to valine results in deoxyhaemoglobin crystallizing out as long rods which distort the red blood cell leading to serious vascular problems. The disease has apparently been positively selected in those areas with a vicious form of malaria, against which the abnormal haemoglobin gives a protection, which outweighs the deleterious effects of the disease.

20 Neutrophils secrete elastase in the mucous lining of lung tissue that destroys elastin, the elastic structural material of lungs. This, unchecked, converts the minute alveoli into much larger structures with a greatly reduced surface area for gas exchange. α_1-Antitrypsin in the blood diffuses into the lung and keeps elastase in check and thus prevents damage. Smoking has two effects: (a) it inactivates α_1-antitrypsin by converting a crucial methionine side chain to a sulphoxide ($S \rightarrow S = O$); (b) by irritating the lungs it attracts neutrophils resulting in more elastase being liberated.

21 The peptide bond is a hybrid between two structures.

The link is approximately 40% double bonded. This restricts rotation of the peptide bond.

22 Proline. It is an imino acid rather than an amino acid. It also is an α-helix breaker while the others are good α-helix formers.

Chapter 5

1 The original Sanger technique was to label the N-terminal amino group of a peptide derived from partial hydrolysis of the protein. The label is coloured or fluorescent for ease of detection and the labelling is stable to acid hydrolysis. The peptide is hydrolysed and the products separated by chromatography. This identifies the first (N-terminal) amino acid residue. Sequencing a protein by this procedure is very laborious, because the exercise has to be performed on many peptides and the complete sequence determined from overlapping peptides. This method would now not be used for anything but very small peptides, if that. The Edman procedure is to label the peptide with a reagent, which under appropriate conditions causes the N-terminal amino acid to detach, still labelled. This exposes the next amino acid residue to be so labelled. The procedure is automated and can be used for sequences up to about 30 residues. The labelled amino acids detached in turn are identified by chromatography. The method is laborious if a complete protein has to be sequenced and requires the disruption of the protein into oligopeptides which are sequenced separately. Mass spectrometry can determine the sequence of oligopeptides very rapidly and promises to be of increasing importance. It should not be forgotten that often the quickest method of sequencing a protein is to isolate the DNA responsible for coding the amino acid sequence of the protein. The understanding of this statement may have to wait until later sections of the book have been studied.

2 It refers to the total collection of proteins present in cells. This varies from time to time and from cell to cell as proteins are synthesized and destroyed.

3 Proteomics refers to the simultaneous study of large numbers of proteins at once. For example it may be a study of the changing protein profile during development or between normal and cancerous tissues. To do this work it is necessary to be able to identify proteins very rapidly and from small amounts such as in single spots on two-dimensional gels. Mass spectrometry has been the main development in doing this. Proteins may be sufficiently characterized by the method to search for them in protein databases in minutes.

4 The SDS molecules insert into the proteins by their hydrophobic tails and denature them. The large numbers of SDS molecules that become inserted with their strong negative charge outside swamp all charges of the native protein. This results in separations based purely on molecular sieving and therefore of molecular size. Nondenaturing gels by contrast separate proteins both by molecular sieving and the charges on the proteins. SDS can also solubilize proteins.

5 Molecular sieving depends on packing matrices with different pore sizes. Entry of a protein into the beads retards it as compared with a larger protein that flows around the beads. Ion-exchange chromatography involves the use of packings with different ionic groups. Proteins are absorbed selectively and may be eluted selectively by buffers. Reverse-phase chromatography uses a hydrophobic packing to which proteins absorb selectively and can be eluted selectively by solutions of different hydrophobicities or ionic strengths. Affinity chromatography involves attachment to the column packing of groups which are known to specifically bind to the wanted protein. It can be eluted by a solution of the affinity molecule.

6 The protein can be digested with, say, trypsin and a peptide mass analysis by MALDI-TOF spectrometry performed on it. The pattern is usually sufficient to identify the protein in a database particularly if about five peptides are obtained. An alternative is to obtain limited sequence data by tandem mass spectrometry which unambiguously identifies a protein.

7 Mass spectrometry involves ionized molecules in the gas phase which suited many organic molecules but not proteins, which are large and nonvolatile. In 1988 two methods for obtaining suitable ions from proteins were developed. MALDI (matrix-assisted laser-desorption ionization) involves a UV-light-absorbing matrix mixed with the protein. Laser light causes formation of charged-protein-derived ions. Electrospray, involving spraying a solution at a high electrical potential, also achieves this.

8 Protein databases have been established in various international centres to store information on proteins as it is reported. They contain vast amounts of information on all aspects of protein structure and function. This includes amino acid sequences, tertiary structures and much else. The databases are freely available as well as software for analysing the information in many ways. Mining the databases is itself a research activity. They can enormously speed up research because if a protein can be identified with one in a database information is then immediately available. They are of constant use in proteomics in which large numbers of proteins have to be identified.

9 Tandem mass analysers separated by a collision chamber are used. The first analyser selects the peptide to be sequenced. It enters the middle chamber and collides with argon gas molecules which fragments the peptides in a predictable way.

The fragmentation products are separated in the second mass analyser and the results displayed as a spectrum from which the amino acid sequence can be deduced.

10 The first method is X-Ray diffraction of proteins labelled by heavy metal isomorphous replacement. A newer version of this is to use synchrotron radiation. For this proteins labelled with incorporated selenomethionine can be used. These are conveniently made in *Escherichia coli* or other expression system of cDNA molecules. A different method, currently applicable for small proteins is nuclear magnetic resonance (NMR). This requires high protein concentrations but crystallisation is not needed. However, synchrotron radiation can use extremely small crystals, which are easier to obtain than those needed for laboratory X-ray diffraction.

Chapter 6

1 Very little of significance because the plot will be arbitrary depending on the time period selected for the measurement of activity. This is because, at higher temperatures, destruction of the enzyme is likely to occur and, at any given temperature, the amount of destruction will be proportional to the time the enzyme is exposed to that temperature. If information is sought on the heat stability of the enzyme it is much better to expose it for a standard time at the different temperatures and, after cooling, to measure the activity in each sample at the normal incubation temperature.

2 The $\Delta G^{0'}$ value of a reaction determines whether a reaction may proceed, but says nothing about the rate at which it does proceed (if at all). The latter is determined by the energy of activation of the reaction and the rate at which the transition state is formed.

3 This can involve several factors:
(a) the active site binds the transition state much more firmly than it does the substrate and, in doing so, lowers the activation energy;
(b) it positions molecules in favourable orientations;
(c) it can exert general acid-base catalysis on a reaction;
(d) it may position a metal group which facilitates the reaction.

4 A double reciprocal plot (Lineweaver–Burke plot) is needed. A noncompetitive inhibitor will reduce the V_{max} at infinite substrate concentration (the intercept at the vertical axis) but will not alter the K_m. A competitor inhibitor on the other hand has no effect on the V_{max} at infinite substrate concentration but does change the K_m. This is illustrated in Figs 6.6 and 6.8.

5 An affinity constant for a substrate binding to an enzyme represents the equilibrium $E + S \rightleftharpoons ES$. A K_m, which is a measure of the concentration of S at which the enzyme is working at half maximum velocity, is derived from a measure of the rate of product formation. This involves the catalytic rate at which the enzyme converts ES to product. For a K_m to represent a true affinity constant, it is necessary that the rates of the reaction $E + S \rightarrow ES$ and the reverse $ES \rightarrow E + S$ are very large compared with the rate of removal of ES by conversion to product, so that the latter can be ignored in the substrate-binding equilibrium.

6 A transition state binds more tightly to the active centre of an enzyme than does the substrate molecule. This is the fundamental basis of the mechanism of enzyme catalysis. The affinity in one recorded case was a thousand or more times greater than that of the substrate. Therefore, there is considerable interest in the potential of stable transition state molecules as specific enzyme inhibitors.

7 It is essential to make sure that you are measuring initial velocities in which the amount of reaction catalysed is linear with time. Otherwise other limiting factors may occur such as inhibition by product, exhaustion of substrate or denaturation of enzyme. In these situations the activity measured is not a true reflection of the amount of enzyme.

8 The serine —OH is perfectly positioned next to a histidine group, which readily accepts the proton thus freeing the oxygen atom to make a bond with the target carbon atom.

9 The aspartate —COO⁻ group forms a strong hydrogen bond with the histidine side chain and holds the imidazole ring in the tautomeric form and most favourable orientation to accept the proton from serine (see Fig. 6.11). Amidation of the carboxyl group greatly reduces its hydrogen-bonding potential.

10 The enzymes have different active sites for binding their respective substrates. Chymotrypsin has a large hydrophobic pocket to accommodate the large aromatic amino acid residues. Trypsin has an aspartate residue at the binding site to which the basic amino acid residues characteristic of trypsin substrates attaches. Elastase accepts only peptides with small amino acid residues because a pair of valine and threonine residues restricts entrance to the site.

11 A thiol protease is similar to a serine protease except that cysteine replaces the active serine and the intermediate acyl enzymes are thiol esters. Papain is such an enzyme. Aspartyl proteases have a pair of aspartyl residues which alternate in acting as proton donors and acceptors in the catalytic mechanism. One such enzyme is essential for replication of the AIDS virus.

Chapter 7

1 The basic structure of all biological membranes is the lipid bilayer. It is formed from components which have amphoteric structures, comprised of a hydrophilic head and two long hydrocarbon tails which are hydrophobic. The molecules arrange themselves so a double layer is formed producing a two-dimensional sheet, with the hydrophilic heads on the outside and the hydrophobic tails pointing inwards, towards each other, sandwiched between them. The centre of the lipid bilayer is therefore hydrophobic. The structure is held together by non-covalent forces. The adjacent heads are exposed to water, and the arrangement maximizes van der Waals forces between the hydrophobic tails. The arrangement is thermodynamically optimal and therefore stable.

2 It acts as a fluidity buffer by preventing the polar lipids from associating too closely towards the centre of the bilayer, but also acts as a wedge at the surface layer.

3

$$
\begin{array}{l}
CH_2\text{---}O\cdot CO\text{---}R \\
CH\text{---}O\cdot CO\text{---}R \\
\qquad\quad O \\
\qquad\quad \| \\
CH_2\text{---}O\text{---}P\text{---}O^- \\
\qquad\quad | \\
\qquad\quad O^-
\end{array}
$$

4 Lecithin (choline); cephalin (ethanolamine); phosphatidyl-serine (serine).

5 They put a kink into the tail which prevents close packing of the fatty acyl tails of lipid bilayers and thus make the membrane more fluid.

6 To do so would require stripping the molecules of H_2O molecules, an energetically unfavourable process.

7 It involves a membrane protein forming a hydrophilic channel which permits solutes to traverse the membrane in either direction, according to the direction of the concentration gradient. No energy input is involved. The anion transport of red blood cells is an example; it allows Cl^- and HCO_3^- ions to pass through in either direction.

8 A triacylglycerol is a neutral molecule with no hydrophilic groups. The molecule is therefore water insoluble and to minimize contacts with water the molecules are forced to become spherical globules or form a separate layer. A polar lipid is amphipathic with a charged group at one end and hydrophobic tails at the other. It can therefore assemble into a bilayer structure with hydrophobic tails in contact with each other and shielded from water. The charged group can interact at the surface with water giving the bilayer a stable minimum free energy arrangement of the molecules.

9 It is simply another way of constructing an amphipathic molecule of the same type as a glycerophospholipid – one that can participate in the formation of a lipid bilayer. Sphingosine-based lipids are prevalent in myelinated nerves, but exactly why this particular structure is selected for this is not known nor is there a clear understanding of why such a variety of polar lipids is employed in different membranes.

10 This is done by symport and antiport systems. In the case of glucose active transport (a symport), the glucose molecule is cotransported into the cell with sodium ions whose concentration is much higher outside than inside. The latter are ejected from the cell by the Na^+/K^+ ATPase that maintains the ion gradient. Thus ATP indirectly supplies the energy for glucose transport. An example of an antiport system is the pumping of Ca^{2+} from a cell, driven by the co-transport of Na^+ into the cell. The energy once again comes from the ATP-dependent Na^+/K^+ ATPase.

11 It is a membrane channel formed by proteins, which opens and closes on receipt or cancellation of a signal respectively. A ligand-gated channel opens on binding of a specific molecule. The acetylcholine-gated channel in nerve conduction is one. A voltage-gated channel opens in response to a change in membrane potential. The Na^+ and K^+ voltage-gated channels in neuron axons are examples.

12 The drug 'freezes' the Na^+/K^+ pump in the phosphorylated form when applied to the outside face of the protein in the presence of K^+. This inactivates the pump and results in the intracellular concentration of Na^+ rising, which lowers the steepness of the gradient of this ion across the cell membrane. This causes an increase in the intracellular Ca^{2+} concentration, which increases the strength of the heart muscle contraction. The increase in Ca^{2+} is due to the fact that an antiport system driven by the Na^+ gradient exports calcium from the cell. Lowering the Na^+ gradient reduces the outflow of Ca^{2+}.

13 A shell of water molecules surrounds ions in solution. The hydrated ions are too large to pass through the channel. Eight molecules of water surround a potassium ion; normally there is an energy barrier to removal of these but the selectivity filter is lined by peptide carbonyl groups which bind to the ion exactly mimicking, in a thermodynamic sense, the water molecules. The potassium ion is able to slip from solution into the channel without any thermodynamic barrier. It can move through the pore from one binding site to the next and as it leaves the pore is once again hydrated. A sodium ion is smaller but in its hydrated form too large to traverse the pore. The dehydrated ion however too small to bind to the carbonyl groups lining the pore and so there is a thermodynamic barrier to it doing so.

Chapter 8

1 Acetylcholine stimulation of the muscle receptor leads to liberation of Ca^{2+} from the sarcoplasmic reticulum into the myofibril. On the actin filaments are tropomyosin molecules and, in turn, these have a Ca^{2+}-sensitive troponin complex attached to them. When Ca^{2+} binds, a conformational change occurs; it was believed that the tropomyosin blocked the attachment of myosin heads to actin. This is now questioned. A Ca^{2+} ATPase returns the Ca^{2+} to the sarcoplasmic reticulum and terminates the contraction.

2 In smooth muscle myosin heads there is a p-light chain that inhibits the binding of the myosin head to the actin fibre, preventing contraction. Neurological stimulation opens channels allowing Ca^{2+} to enter the cell. The Ca^{2+} combines with calmodulin regulatory protein, which activates a myosin light chain kinase. Phosphorylation of the light chain abolishes its inhibitory effect and contraction occurs.

3 In one case ATP is synthesized from ADP + P_i. The synthesis on the enzyme surface involves little or no free energy change, but energy is required for release of ATP. This is mediated by a conformation change in the proteins of the ATP synthase head. In the case of myosin, ATP is split into ADP + P_i but on the enzyme surface there is little free energy change involved. It is on the release of the P_i and ADP that the power stroke of contraction occurs. This is mediated by a conformational change in the myosin head. In neither case is there a covalent intermediate in the synthesis or breakdown of the ATP.

4 It used to be thought that the myosin head swings at its point of attachment to the coiled-coil rod of myosin so that the angle of the head to the actin filament changed. This is now known to be incorrect. The angle of attachment of the head to the actin filament does not change but an α helix of the head known as the lever arm swings to cause the power stroke. The diagram in Fig. 8.7 provides an illustration of this.

5 Contraction may occur in such cells. Actin filaments anchored in the cell membrane provide the means for small bundles of myosin molecules to exert a contractile force. A second role is for actin filaments to form a transport track along which special minimyosin molecules move. The latter have a myosin head but the rod-like structure is replaced by a short tail to which vesicles may be attached.

6 Hollow tubes formed by the polymerization of tubulin protein subunits. They have a definite polarity with (+) and (−) ends. The tubulin monomers have a bound molecule of GTP. The end GTP-monomer added protects the tubule from collapse. However, the tubulin has low GTPase activity and hydrolyses the GTP to GDP + P_i after an interval. Unless GTP-monomers are added before this, the tubule collapses.

7 The microtubule-organizing centre (MTOC) protects the (−) end. The (+) ends *of growing* tubules are protected by a tubulin-GTP cap. The GTP is hydrolysed slowly, removing the protection. Unless a new tubulin-GTP molecule is added before this, the microtubule collapses. When the microtubule reaches a 'target' structure it is then protected. The assembly of the cytoskeleton is complex, and proteins known as filamins and formins are involved.

8 They are molecular motors that move along microtubule tracks and can pull a load with them. Kinesin and dynein travel in opposite directions (in terms of microtubule polarity) on the microtubule track. Myosins move by the swinging-lever

mechanism as in muscle. Kinesin and dynein move by a step-wise movement involving swivelling of the two heads.

9 No, they cannot contract. It is believed that the shortening is due to depolymerization but precisely what causes chromosome movement is not certain.

10 Filaments 10 nm in diameter which are intermediate in this respect between microtubule filaments (20 nm) and actin filaments (6 nm). In specialized cases, they form the structural basis of hair. They may be associated with conferring toughness on structures such as neurofilaments and the Z discs of sarcomeres. However, mutant cultured cells survive without them. Lamins of the nuclear membrane are disassembled in cell division – phosphorylation is involved in this.

11 They are both monomers with a definite polarity and both have an associated molecule of nucleotide triphosphate, ATP in the case of actin and GTP in the case of tubulin. They both polymerize into fibres, actin filaments and microtubules respectively. In both cases the ATP or GTP is hydrolysed to the diphosphate after polymerization. The triphosphate state in both cases favours polymerization and the diphosphate state, depolymerization. In a growing actin filament or microtubule, it is necessary to add a new subunit before the previously added one is converted to the diphosphate state in order to prevent polymer collapse.

Chapter 9

1 Pepsin as pepsinogen; chymotrypsin as chymotrypsinogen; trypsin as trypsinogen; elastase as proelastase; carboxypeptidase as procarboxypepidase.

The proteolytic enzymes are potentially dangerous in that they could attack proteins lining the ducts. There are no components that amylase might attack. Mucins coating the gut cell lining of the intestine protect cells against proteolytic attack.

2 In the stomach the acid pH causes a conformational change in pepsinogen that activates molecules to self-cleave the extra peptide in pepsinogen that inactivates the enzyme. Once some pepsin is formed, it activates more pepsinogen so that an autocatalytic activation cascade occurs. In the small intestine, enteropeptidase, an enzyme produced by gut cells, activates trypsinogen to trypsin. This, in turn, activates the other zymogens.

3 Pancreatitis, or inflammation of the pancreas, due to proteolytic damage of pancreatic cells.

4 Because the gut cells can absorb only monomers (plus monoacylglycerols). Fat, protein, polysaccharides, and disaccharides cannot be absorbed.

5 Milk contains lactose, which must be hydrolysed to glucose and galactose by the enzyme lactose. After infancy, many individuals lose the capacity to produce this. The lactose is not absorbed and is fermented in the large intestine. It attracts water into the gut by its osmotic effect, producing diarrhoea.

6

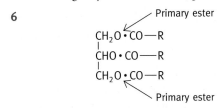

Primary ester

Primary ester

TAG or triacylglycerol (sometimes called triglyceride, but this is chemically incorrect). The central bond is a secondary ester, the other two are primary.

7 The fat is emulsified by intestinal movement and by the monoacylglycerol and free fatty acids produced by initial digestion together with bile salts, so that the lipase can attack the emulsified substrate. The products of digestion (free fatty acids and monoacylglycerol) are carried to the intestinal cells as disc-like mixed micelles with bile salts that have a high carrying capacity for such products. Probably at the cell surface the micelle breaks down.

8 The free fatty acids and monoacylglycerol are resynthesized into TAG. The TAG itself cannot traverse cell membranes but it is incorporated into lipoprotein particles, called chylomicrons. These have a shell of stabilizing phospholipid, some proteins, and a centre of TAG and cholesterol and its ester. The chylomicrons are released by exocytosis into the lymphatics and eventually discharge into the blood as a milky emulsion via the thoracic duct.

9 The osmotic pressure of glucose precludes this. The osmotic pressure of a solution is related to the number of particles of solute in it. By polymerizing thousands of glucose molecules into a single glycogen molecule, the osmotic pressure is correspondingly reduced.

10 Glycogen stores in the liver are sufficient to last through 24 hours of starvation only, but there may be sufficient TAG for weeks. TAG is a more concentrated energy source – it is more highly reduced and is not hydrated. If the energy in fat were present as glycogen, the body would have to be much larger.

11 Yes – glucose can be converted to acetyl-CoA and thence to fat.

No – acetyl-CoA cannot be converted into pyruvate.

No – there are no special amino acid storage proteins (excluding milk) in the body.

12 (a) It is the glucostat of the body. It stores glucose as glycogen in times of plenty; it releases glucose in fasting to keep a constant blood sugar level which the brain needs.

(b) During prolonged starvation it synthesizes glucose for the same reason; it converts fats to ketone bodies which the brain (and other tissues) can use for part of its energy supplies and so conserve glucose.

(c) It synthesizes fat and exports it to other tissues.

13 No.

14 They store fat in times of plenty; they release fatty acids in fasting.

15 They are completely dependent on glucose, which they convert to lactate. Since mature red blood cells lack mitochondria, they are unable to produce ATP by any metabolic route except glycolysis.

16 When blood glucose is high, insulin is released. This is the signal for tissues to store food. As the blood glucose levels fall, insulin levels fall and glucagon levels rise. The latter is the signal for the liver to release glucose and for adipose cells to release free fatty acids for the tissues to utilize. The brain is not insulin-dependent for its uptake of glucose, so this can proceed even in starvation, when insulin levels are extremely low. In addition, epinephrine can override all controls, causing massive turnout of glucose and fatty acids to cope with an emergency requiring the 'fight-or-flight' reaction.

17 Ketone bodies are produced in excess in uncontrolled type I diabetes. However, in prolonged fasting, when glycogen sources are exhausted, they supply muscles with a ready supply of metabolizable substrate, which is preferentially utilized as energy source. This minimizes the utilization of glucose. In starvation, keeping up the blood glucose level assumes top priority because the brain must be supplied with it. Nonetheless, during starvation the brain can adapt to use ketone bodies for perhaps half of its energy needs, thus economizing on glucose. Preserving glucose in this situation is important. To provide pyruvate for the liver to synthesize glucose, muscle proteins are broken down, stimulated by cortisol and it is clearly advantageous to keep muscle wasting to the minimal level consistent with keeping the brain supplied with glucose.

18 Leptin is produced by adipose cells at a rate inversely related to the level of fat reserves. It inhibits appetite and has been found to be effective in controlling obesity in mutant mice and in a small number of children who lack the hormone. Unfortunately, obese humans are not usually deficient in the hormone and indeed its levels may be high. This might be theoretically due to resistance but also may be a consequence of evolutionary factors.

19 Ghrelin, produced by the empty stomach, stimulates appetite via a centre in the brain. Leptin and adiponectin, produced by fat cells (greatest when fat reserves are high) and a peptide produced by the intestine in the presence of food have the reverse effect. Insulin to a lesser degree has a leptin-like effect. Cholecystokinin produced by an extended stomach gives a feeling of satiety.

Chapter 10

1

$$G\text{-}1\text{-}P + UTP \rightarrow UDPG + PP_i.$$

$$PP_i + H_2O \rightarrow 2P_i.$$

$$UDPG + \text{glycogen }(n) \rightarrow UDP + \text{glycogen }(n+1).$$

The hydrolysis of PP$_i$ makes the overall reaction strongly exergonic and irreversible.

2 The reaction named never occurs in the cell because the required substrate, inorganic pyrophosphate, is immediately destroyed. It is the reverse reaction that occurs in the cell but, for systematic nomenclature reasons, the enzyme is named for the forward reaction.

3 Only the liver (and the less quantitatively important kidney) does so. These alone possess the enzyme glucose-6-phosphatase.

$$\text{Glucose-6-phosphate} + H_2O \rightarrow \text{glucose} + Pi$$

4 Glucose phosphorylation. Glucokinase has a much higher K_m than hexokinase. During starvation the liver turns out glucose into the blood, primarily to supply the brain (and red blood cells). The first reaction in the uptake of glucose into brain and liver is phosphorylation of glucose. The lower affinity of glucokinase for glucose, as compared with hexokinase, means that the liver does not compete with brain for blood sugar. It efficiently takes up glucose only when blood sugar levels are high. Since it is not as sensitive to inhibition by glucose-6-phosphate as is hexokinase, it can take up glucose and synthesize glycogen even when cellular levels of glucose-6-phosphate are high.

5 Many glycolipids and glycoproteins contain galactose. Removal of galactose from the diet does not prevent synthesis of these because UDP-glucose can be epimerized to UDP-galactose.

6 In the capillaries, lipoprotein lipase splits free fatty acids from TAG, and these immediately enter adjacent cells.

7 VLDL are lipoproteins, resembling chylomicrons, that carry cholesterol and TAG from the liver to peripheral tissues.

8 Conversion to bile salts in the liver. Excess cholesterol is a risk factor in causation of heart attacks. Removal of cholesterol is therefore of importance.

9 The enzyme lecithin: cholesterol acyltransferase (LCAT) transfers fatty acyl groups from lecithin to cholesterol.

10 TAG is *not* released; a hormone-sensitive lipase, stimulated by glucagon (or epinephrine in emergency situations), hydrolyses off free fatty acids that are carried in the blood to the tissues attached loosely to serum albumin. By contrast, fat transport from the liver to other tissues occurs *via* VLDL.

11 In familial hypercholesterolaemia the patient lacks LDL receptors and so cannot remove cholesterol-rich LDL from the blood.

12 Adipose cells liberate free fatty acids, which are carried attached to serum albumin. This dissociates from the latter and enters cells. Liver exports fat as VLDL. In extrahepatic tissues, lipoprotein lipases release free fatty acids from the VLDL and these enter the adjacent cells to be metabolized.

13 Cholesterol is transported to peripheral cells from the liver in the form of VLDL and taken up by extrahepatic cells in the form of LDL formed by fat depletion of VLDL. Cholesterol in excess of requirements is transported out of cells by the ABC1 transport system. It is taken up by HDL and converted to cholesterol ester, which is transferred to LDL. Part of the latter are taken up by the liver resulting in the delivery of cholesterol back to the liver. This is the reverse flow. It is important because the liver converts part of the cholesterol to bile acids which are secreted into the intestine and is one route of avoiding cholesterol excess in the body. HDLs are associated with protection against arterial blockages and heart attacks.

Chapter 11

1 Glycolysis, the citric acid cycle, and the electron transport chain; cytoplasm, mitochondrial matrix, and the inner mitochondrial membrane, respectively.

2 Structures on page 178.

$$AH_2 + NAD^+ \rightleftharpoons A + NADH + H^+$$
$$B + NADH + H^+ \rightleftharpoons BH_2 + NAD^+$$

3 FAD is another hydrogen carrier; it is reduced to FADH$_2$. It is not a coenzyme but a prosthetic group attached to enzymes.

4 In aerobic glycolysis, glucose is broken down to pyruvate. The NADH produced is reoxidized by mitochondria. In anaerobic glycolysis, the rate of glycolysis exceeds the capacity of mitochondria to reoxidize NADH. This can occur, for example, in the 'fight-or-flight' reaction. Since the supply of NAD$^+$ is limited, if NADH were not reoxidized sufficiently rapidly to NAD$^+$, glycolysis and its ATP production would cease. An emergency mechanism reoxidizes NADH by reducing pyruvate to lactate.

$$CH_3COCOO^- + NADH + H^+ \rightarrow CH_3CHOHCOO^- + NAD^+$$

Pyruvate   Lactate   Lactate
                              dehydrogenase

5 Structures on page 18.
−31 as compared with −20 kJ mol^{-1} for the carboxylic ester – that is, the thiol ester is a high-energy compound.

The vitamin, pantothenic acid, does not participate in the reactions involving CoA, unlike the situation with other coenzymes, where the vitamin is the 'active part' of the molecule.

6 Pyruvate + CoA—SH + NAD$^+$ → Acetyl—S—CoA + NADH + H$^+$ + CO$_2$

$\Delta G^0 = -33.5$ kJ mol^{-1}

7 The acetyl group is fed into the citric acid cycle.

8 They are re-oxidized by the electron transport chain producing H$_2$O and generating ATP in the process (though this is achieved indirectly via a proton gradient).

9 The Nernst equation relates $\Delta G^{0'}$ and ΔE_0 values as follows. $\Delta G^{0'} = -nF\Delta E_0$ (F is the Faraday constant = 96.5 kJ V^{-1} mol^{-1})

and ΔE_0 is the difference between the redox potentials of the electron donor and acceptor. In the example given ΔE_0 equals -1.035 V $(-0.219 - 0.816$ V$)$.

Therefore, $\Delta G^{0'} = -2(96.5$ kJ V^{-1} mol$^{-1})(-1.035$ V$) = -193$ $(-1.035) = 194.06$ kJ mol^{-1}.

10 Breakdown of fatty acids to acetyl-CoA.

11 (a) Yes. Glucose is converted to pyruvate and pyruvate dehydrogenase converts this to acetyl-CoA. The latter can be used to synthesize fat.

(b) No. Fatty acids are broken down to acetyl-CoA. To synthesize glucose, pyruvate is needed. The pyruvate dehydrogenase reaction is irreversible. There can, in animals, be no net conversion of acetyl-CoA (and therefore of fatty acids) to glucose.

Chapter 12

1 The splitting of the C_6 molecules into two C_3 molecules occurs by the aldol split catalysed by aldolase. This requires the structure:

$$
\begin{array}{c}
R \\
| \\
C=O \\
| \\
R'-C-R'' \\
| \\
H-C-OH \\
| \\
R
\end{array}
$$

The conversion of glucose-6-phosphate to fructose-6-phosphate produces the aldol structure capable of splitting in this way. In straight-chain formulae they are as follows.

$$
\begin{array}{ll}
CHO & CHOH \\
| & | \\
CHOH & CO \\
| & | \\
CHOH & CHOH \\
| & | \\
CHOH & CHOH \\
| & | \\
CHOH & CHOH \\
| & | \\
CH_2OPO_3^{2-} & CH_2OPO_3^{2-} \\
\\
\text{Glucose-6-phosphate} & \text{Fructose-6-phosphate}
\end{array}
$$

2 It occurs when a 'high-energy phosphoryl group' transferable to ADP is generated in covalent attachment to a substrate. Oxidation of glyceraldehyde-3-phosphate by the mechanism given on page 191 is an example.

By contrast, electron transport in mitochondria generates a proton gradient which drives ATP synthesis. No phosphorylated intermediates interpose between ADP + P$_i$ and ATP.

3 Pyruvate kinase works in the reverse reaction but the nomenclature of kinases always derives from the reaction using ATP. The reaction is irreversible because the product of the reaction, enolpyruvate, spontaneously isomerizes into the keto form, a reaction with a large negative $\Delta G^{0'}$.

4 Two and three, respectively. Phosphorolysis of glycogen using P$_i$ generates glucose-1-phosphate, convertible to glucose-6-phosphate. To produce the latter from glucose, a molecule of ATP is consumed.

5 Cytoplasmic NADH itself cannot enter the mitochondrion; instead the electrons must be transported in by one of two alternative shuttles, and which is used affects the outcome. The malate-aspartate shuttle (Fig. 12.29) reduces mitochondrial NAD$^+$, but the glycerol phosphate shuttle reduces FAD attached to glycerol phosphate dehydrogenase of the inner mitochondrial membrane. The redox potential of the former is more negative than that of the latter and the two enter the electron transport chain at different respiratory complexes. The yield of ATP is lower from the glycerol phosphate shuttle.

6 Oxidation forms a β-keto acid that readily decarboxylates.

7 The reaction is catalysed by pyruvate carboxylase.

$$
\begin{array}{c}
COO^- \\
| \\
C=O \quad + \quad HCO_3^- \quad + ATP \\
| \\
CH_3
\end{array}
$$

$$
\downarrow
$$

$$
\begin{array}{c}
O \\
|| \\
C-COO^- \\
| \\
H_2C-COO^-
\end{array}
\quad + \quad ADP + P_i + H^+
$$

Acetyl-CoA cannot be involved. Although this participates in citric acid formation, both carbon atoms are lost as CO_2 in one turn of the cycle. Therefore there can be no increase in cycle acids. Also acetyl-CoA cannot be converted to pyruvate in animals. (In bacteria and plants, acetyl-CoA can produce C_4 acids via the glyosylate cycle; see Chapter 15.)

8 Biotin, a B-group vitamin. Using ATP, it forms a reactive carboxybiotin that can donate a carboxy group to substrates. The reaction is:

9 This is shown in Fig. 12.17.

10 They are both mobile electron carriers. Ubiquinone connects respiratory complexes I and II with III and cytochrome c connects complexes III and IV. The former exists in the hydrophobic lipid bilayer; the latter in the aqueous phase loosely attached to the outside of the inner mitochondrial membrane.

11 To pump protons from the mitochondrial matrix to the outside of the inner mitochondrial membrane and so generate a proton and charge gradient that can be used to drive ATP synthesis.

12 In the eukaryote, NADH generated in the cytoplasm has to have its electrons transferred to the electron transport pathway. If the glycerol-3-phosphate shuttle is employed for this we lose one ATP generated per NADH that has to be so handled. Since there are two NADH molecules generated per glucose glycolysed, this leads to a potential loss of two ATP molecules. In *E. coli* there is no such problem. In addition, in *E. coli* there is no expenditure of the energy used in eukaryotes to exchange ATP and ADP across the mitochondrial membrane.

13 In one turn of the cycle, one acetyl group is disposed of, the products being nine reducing equivalents (effectively nine high-energy hydrogen atoms) in the form of three $NADH + H^+$, one $FADH_2$, and two molecules of CO_2. A further proton is needed for generation of CoA—SH making a total of nine hydrogen atoms. The acetyl group supplies three hydrogen atoms and one oxygen atom, leaving a deficit of six hydrogens and three oxygens. These partly come from the input of two water molecules; in addition, if you examine the breakdown of succinyl-CoA, it effectively also splits a further water molecule. This ingenious thermodynamic device is indirect and easily missed but it is important in appreciating the energetics of the metabolism of the acetyl group in the citric acid cycle.

14 The proton flow back into the matrix causes a rotary motion in the F_0 unit in the inner mitochondrial membrane. This drives a shaft inside the F_1 unit, which somehow causes conformation changes in the F_1 subunits. The energy stored in the conformation changes brings about the synthesis of ATP from $ADP + P_i$ but it is important to note that it is the release of ATP which requires the energy. The reaction forming ATP from $ADP + P_i$ at the enzyme surface involves little free energy change (a concept which takes a little getting used to). The diagram in Fig. 12.21 would be suitable in this question.

15 There are three sites capable of synthesizing ATP in each F_1 unit but they progress through different stages referred to as open (O state), loose binding (of ADP and P_i – the L state), and tight binding (T state). ATP synthesis occurs at the last stage. Cooperative interdependence means that at any one time only one of the three is in the O state, one in the L state, and one in the T state. The individual sites change simultaneously. Thus the L state can change to the T state only if the preceding O state

changes to the L state and so on. Figure 12.27 would be useful here.

16 There are four complexes I–IV. I accepts electrons from NADH and transports them to ubiquinone (Q). Q carries electrons to III. II accepts electrons from $FADH_2$ and also transports them to Q. The FAD is the prosthetic group of succinate dehydrogenase and the fatty oxidation pathway and of the fatty acyl-CoA dehydrogenase. III transports the electrons to IV (via cytochrome c) which finally delivers them to oxygen to form water.

I pumps protons to the outside in a manner not yet understood. II does not pump protons; the free energy fall in the transference of electrons from $FADH_2$ to Q is insufficient for this. III pumps protons by the Q cycle in which Q is reduced using protons from the matrix and oxidized on the opposite face of the membrane, liberating protons to the outside. IV pumps protons but the mechanism is not fully elucidated. Figure 12.20 would be useful here.

17 It depends on the aspartate residues in the ring of c subunits. In the unprotonated charged state it is thermodynamically obliged to reside in a hydrophilic environment. When protonated, if free to do so, it will move to a hydrophobic environment in the lipid bilayer. In short the principle is that the minimum free-energy state is when charged groups are in a hydrophilic environment and protonated uncharged ones in a hydrophobic environment, and if free to do so will move to this situation.

18 Acetyl-CoA. The rest are involved in the citric acid cycle. Acetyl-CoA enters the cycle but is not part of the cycle itself.

Chapter 13

1 (a) From free fatty acids carried by the serum albumin in the blood; these originate from the fat cells.
(b) From chylomicrons – lipoprotein lipase releases free fatty acids and TAG.
(c) From VLDL produced by the liver – released in the same way as (b).

2 The brain and red blood cells (the latter have no mitochondria).

3 (a) Activation to convert them to acyl-CoA derivates.
(b) On the outer mitochondrial membrane.
(c) In the mitochondrial matrix.
(d) As carnitine derivatives, as shown in Fig. 13.1.

4 The reactions in Fig. 13.2 are analogous to the succinate → fumarate → malate → oxaloacetate reactions in the citric acid cycle, both in the reactions and in the electron acceptors involved.

5 One molecule of palmitic acid produces eight acetyl-CoAs and generates seven FADH$_2$ and seven NADH molecules. NADH oxidation is estimated to produce 2.5, and 1.5, molecules of ATP per molecule of NADH and FADH$_2$ respectively. If you add to this the yield of ATP from the oxidation of eight molecules of acetyl-CoA (ten per molecule), the total is 108 molecules of ATP (counting the GTP from the citric acid cycle as ATP). From this must be subtracted two used in the activation reaction, giving a net yield of 106 molecules.

6 After two rounds of β-oxidation, the *cis*-Δ^3-enoyl-CoA is isomerized to become the *trans*-Δ^2-enoyl-CoA.

7 In situations of rapid fat release from adipose cells such as occurs in starvation or diabetes type 1, the liver converts acetyl-CoA to ketone bodies that are released into the blood. These are preferentially utilized by muscle, thus conserving glucose; most importantly, the brain can obtain about half its energy needs from ketone bodies.

8 Acetoacetate synthesis occurs in the mitochondrial matrix and cholesterol synthesis in the cytoplasmic compartment, the process occurring on the endoplasmic reticulum membrane.

9 These are membrane-bounded vesicles in the cytosol that contain oxidases. These enzymes oxidize a variety of substrates, using oxygen, and generate H$_2$O$_2$.

$$R'H_2 + O_2 \rightarrow R' + H_2O_2$$

Catalase destroys the hydrogen peroxide.

Substrates include those not metabolized elsewhere; for example, very-long-chain fatty acids are shortened. Oxidation of the cholesterol side chain to form bile salts is believed to occur here. The essential role of peroxisomes is underlined by a fatal genetic disease in which some tissues lack peroxisomes.

Chapter 14

1 Fatty acids are synthesized two carbon atoms at a time, but the donor of these is a three-carbon unit, malonyl-CoA. Acetyl-CoA is converted to the latter by an ATP-dependent carboxylation. The subsequent decarboxylation results in a large negative ΔG$^{0'}$ value. In other words, the point of the carboxylation and decarboxylation is to make the process of adding two-carbon-atom units to the growing fatty acid chain irreversible.

2 This is given in Fig. 14.1.

3 In eukaryotes, all of the enzyme reactions are organized into a single protein molecule with the enzymic functions catalysed by separate domains. The functional unit is a dimer with the two molecules cooperating as a single entity. In *E. coli* the different activities are catalysed by separate enzymes. The advantage of the eukaryotic situation is that the intermediates are transferred

from one active centre to the next. In *E. coli* the components must diffuse to the next enzyme so that the process is slower.

4 See pages 178 and 225 for structures. NAD$^+$ is used in catabolic reactions – it accepts electrons for oxidation and energy generation. NADP$^+$ is involved in the reverse – in reductive syntheses. The existence of the two is a form of metabolic compartmentation that facilitates independent regulation of the processes.

5 The main site in the body is the liver.

6 The acetyl-CoA in the mitochondrion is converted to citrate; the latter is transported into the cytosol where citrate lyase cleaves citrate to acetyl-CoA and oxaloacetate. This is an ATP-requiring reaction that ensures complete cleavage.

$$Citrate + ATP + CoA\text{—}SH + H_2O \rightarrow acetyl\text{-}CoA + oxaloacetate + ADP + P_i$$

7 The oxaloacetate from the citrate lyase reaction is reduced to malate by malate dehydrogenase, an NADH-requiring reaction. The malate is oxidized and decarboxylated to pyruvate by the malic enzyme, an NADP$^+$- requiring reaction. This scheme effectively switches reducing equivalents from NADH to NADPH. The pyruvate generated returns to the mitochondrion, as illustrated in Fig. 14.4.

This generates only one NADPH per malonyl-CoA produced, while the reduction steps in fatty acid synthesis require two. The rest is generated by the glucose-6-phosphate dehydrogenase system, described in Chapter 16.

8 The scheme is shown in Fig. 14.5. Glycerol phosphate and fatty acyl-CoAs are used.

9 In the synthesis of glycerol-based phospholipids there are two routes. In one, phosphatidic acid is joined to an alcohol such as ethanolamine (see Fig.14.6). For this the alcohol is activated; in phospholipid synthesis, the activated molecule is always a CDP-alcohol. For some glycerophospholipid syntheses, the diacylglycerol component is activated (see Fig. 14.6). Again, this is by formation of the CDP – diaclygylcerol complex. The situation is reminiscent of the use of UDP-glucose whenever an activated glucose moiety is wanted.

10 (a) Eicosanoids have 20 carbon atoms. They include prostaglandins, thromboxanes, and leukotrienes.
(b) All are related to, and synthesized from, polyunsaturated fatty acids.
(c) Prostaglandins cause pain, inflammation, and fever. Thromboxanes affect platelet aggregation. Leukotrienes cause smooth muscle contraction and are a factor in asthma, by constricting airways.
(d) Aspirin inhibits cyclooxygenase, an enzyme involved in their synthesis, and thus it can suppress pain and fever, and also inhibit blood clotting (see Box 14.2).

11 Mevalonic acid is the first metabolite committed solely to cholesterol synthesis. Structural analogues of mevalonic acid have been found to inhibit HMG-CoA reductase, which is the enzyme responsible for mevalonate production. The drugs act in the body as competitive inhibitors of HMG-CoA reductase.

Chapter 15

1 The brain cannot use fatty acids; it must have glucose. So must red blood cells, which have no mitochondria and can generate energy only from glycolysis.

2 The substrate of pyruvate kinase is the enol form of pyruvate, but the keto/enol equilibrium is overwhelmingly to the keto form and hence the enzyme has no substrate. The solution lies in a metabolic route in which two high-energy phosphate groups are expended.

$$\text{Pyruvate} + \text{ATP} + \text{HCO}_3^- \rightarrow \text{oxaloacetate} + \text{ADP} + \text{P}_i$$

(Pyruvate carboxylase)

$$\text{Oxaloacetate} + \text{GTP} + \text{H}_2\text{O} \rightarrow \text{PEP} + \text{GDP} + \text{CO}_2$$

(PEP-CK)

3 Fructose-1:6-bisphosphatase, producing fructose-6-phosphate, and glucose-6-phosphatase.

4 No. Only the liver and kidney produce free glucose.

5 In normal nutritional situations (nonstarvation) strenuous muscular activity can generate lactate by anaerobic glycolysis. This travels in the blood to the liver where it is converted back to glucose. Release of glucose into the blood and its uptake by muscle completes the cycle. See Fig. 15.4.

6 Glycerol kinase of liver is required to convert glycerol to glucose by the route shown in Fig. 15.5. Glycerol release occurs in starvation where a prime concern is to produce blood glucose. Since only the liver can do this it makes sense for the glycerol to have to travel to the liver rather than being metabolized by adipose cells that cannot release blood glucose.

7 *Via* the glyoxalate cycle shown in Fig. 15.6. The principle is that the two decarboxylating reactions of the citric acid cycle are bypassed.

8 In animals, pyruvate carboxylase synthesizes oxaloacetate (C4) from pyruvate (C3). However, in organisms with the glyoxylate cycle, an extra molecule of acetyl-CoA is converted in the net sense to malate and therefore no topping-up reaction is needed.

9 The liver must be supplied with a suitable substrate for gluconeogenesis. Muscle wasting produces free amino acids, which are converted largely to alanine. The alanine migrates to the liver where it is converted to pyruvate.

10 In starvation, the liver must synthesize glucose to supply the brain since glycogen stores are rapidly exhausted. The pyruvate for gluconeogenesis arises from lactate produced by red blood cells and from alanine coming from muscle. Alcohol raises the ratio of reduced to oxidized NAD^+ in the liver; this can impair conversion of lactate to pyruvate because the equilibrium between these is easily shifted by increased NADH levels and also may cause reduction of pyruvate formed from alanine to lactate. Thus the liver may be deprived of pyruvate needed for gluconeogenesis.

Chapter 16

1 One, by allosteric control; two, by covalent modification of the enzyme, the chief mechanism for this being phosphorylation.

2 (a)

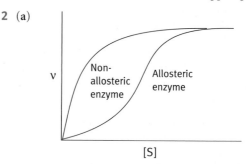

(b) Allosteric modulators (with a few exceptions in which V_{max} is changed) work by changing the affinity of an enzyme for its substrate. This requires that the substrate concentrations for such enzymes are subsaturating, which is usually the case. A positive allosteric modulator moves the sigmoid substrate/velocity curve to the left and a negative effector to the right (Fig. 16.4). A sigmoidal relationship amplifies the effect of such changes on the reaction velocity and so increases the sensitivity of control. This can be seen in Fig. 16.3.

3 They would have no effect since the modulators change the affinity of the enzyme for its substrate. At saturating concentrations this cannot be used to increase the rate of the reaction.

4 It is that an allosteric modulator need have no structural relationship to the substrate of the enzyme. This means that completely distinct metabolic systems can interact in a regulatory fashion.

5 Intrinsic regulation is usually allosteric and can apply to a single cell. It keeps the metabolic pathways in balance. However, it cannot determine the overall direction of metabolism of a cell – whether, for example, it will store glycogen and fat or release these. These are determined by extrinsic controls (hormones, etc.), which direct the activities of cells to be in harmony with the physiological needs of the body.

6 See Fig. 16.19. AMP activates glycogen phosphorylase and phosphofructokinase while ATP inhibits the latter. The salient feature is that high ATP/ADP ratios stop glycolysis while a lower ATP level (resulting in increased AMP) speeds it up. High citrate levels also logically slow the feeding of metabolite via glycolysis into the citric acid cycle. High acetyl-CoA levels may be an indication that oxaloacetate is low – hence activation of pyruvate carboxylase to perform its anaplerotic reaction. At the same time high acetyl-CoA levels indicate that the supply of pyruvate by glycolysis is adequate and inhibition of glycolysis at the PEP step is therefore logical.

7 Direct allosteric controls, controls on a kinase that phosphorylates and inactivates PDH, and controls on protein phosphatase that reverses the phosphorylation (Fig. 16.25).

8 High blood glucose levels stimulate insulin release; low glucose levels stimulate glucagon release.

9 A hormone such as epinephrine is a first messenger; in combining with a cell receptor it causes an increase in a second molecule that exerts metabolic effects. This is a second messenger. For the two hormones mentioned, this is cAMP. It allosterically activates a protein kinase, PKA, whose activity has diverse metabolic effects.

10 By mobilizing glucose transporters into a functional position in the cell membrane (Fig. 16.12).

11 It causes an amplifying cascade of activation, starting with PKA activation as shown in the scheme on page 253.

12 The differences are summarized in Fig. 16.20. In liver glycolysis is inhibited; in muscle it increases in rate.

13 Different cells have receptors for different hormones. Thus, cell A has a receptor for hormone X; cAMP in cell A has effects appropriate to hormone X. Cell B does not have a receptor for hormone X but does for hormone Y. cAMP in cell B elicits responses appropriate to hormone Y but not those appropriate to hormone X.

14 Fructose-2:6-bisphosphate. cAMP decreases its level in liver. See Fig. 16.22. The mechanism is complex, involving a second phosphofructokinase, PFK2.

15 Gluconeogenesis is switched on by glucagon whose second messenger is cAMP. The latter activates a kinase that phosphorylates and inactivates pyruvate kinase. In muscle, epinephrine produces cAMP as second messenger. The aim of epinephrine here is to maximize glycolysis so that inactivation of pyruvate kinase here would be inappropriate.

16 cAMP activates a hormone-sensitive lipase that hydrolyses TAG.

17 Such reactions are irreversible in the cell and act as one-way valves in the pathway; they also ensure that the pathway goes to completion. The carboxylation of acetyl-CoA to malonyl-CoA in fatty acid synthesis is an example in which the reaction serves no apparent purpose other than to confer irreversibility due

to the subsequent decarboxylation reaction. However, many pathways have to be reversed; an example is glycolysis, which has to operate in the reverse direction for gluconeogenesis. In such situations it is necessary to reciprocally control the different directions or the pathway would be simultaneously breaking down glucose units and synthesizing them. To separately control a pathway in the two directions, different enzymes are required. At the irreversible steps it is necessary to have a different reaction for the reverse direction and this is often a control site. PFK is a typical such step in glycolysis. The reverse step is catalysed by fructose-1:6-bisphosphatase, the two steps being reciprocally controlled.

18 (a) Insulin is the signal to store food, in the case of glucose as glycogen. It is therefore logical that insulin should activate glycogen synthase.
(b) The default state is that glycogen synthase is inactivated by GSK3. There are several kinases phosphorylating the synthase in different positions but insulin activation involves removal of the —P groups added by GSK3 and these are the sites mainly involved in synthase control. To activate the synthase, the GSK3 has therefore to be inhibited and, in addition, the protein phosphatase which dephosphorylates the synthase and thereby activates it is activated by insulin. PKB is the enzyme which inactivates GSK3 by phosphorylating the latter. PKB is activated by a signal pathway activated by binding of insulin to its cell membrane receptor. Figure 16.17(b) summarizes this complicated sequence.

19 It has an important role in the control of lives phosphorylase. If there is plenty of glucose around there is no point in breaking down glycogen. Glucose binds to phosphorylase a (the active phosphorylated form) and induces a conformational change which makes the enzyme more susceptible to attack by the protein phosphatase 1. This converts the phosphorylase a into the relatively inactive 'b' form.

20 Malonyl-CoA, the first metabolite committed solely to fat synthesis, inhibits the conversion of fatty acyl-CoAs to the carnitine derivative. The latter is essential for transporting the acyl-CoAs into mitochondria where fatty acid oxidation takes place.

21 The strongly charged phosphate group is very effective at producing a conformational change in a protein, which in most cases is responsible for altering the activity of the enzyme.

22 This is a broadly acting control which shuts down ATP-consuming synthesis reactions and activates catabolic ATP-generating pathways. The rationale is that AMP is a sensitive signal of reduced phosphorylation charge – that ATP supplies are low. There is only a relatively small amount of ATP in a cell and life depends on its very rapid regeneration. ATP shortage would be much more dangerous than shutting down anabolic reactions temporarily. The AMPK achieves this.

Chapter 17

1 It supplies ribose-5-phosphate for nucleotide synthesis; it can supply NADPH for fat synthesis; and it provides for the metabolism of pentose sugars.

2 Glucose-6-phosphate is converted to ribose-5-phosphate + CO_2 and $NADP^+$ is reduced. The reactions are given in Fig. 17.1.

3 Transaldolase and transketolase are the main enzymes whose reactions are given in Fig. 17.2. (Other enzymes of the glycolytic pathway may also participate.)

4 Part of the ribose-5-phosphate is converted to xylulose-5-phosphate, a ketose sugar (since transaldolase and transketolase must have a ketose sugar as donor). The following manipulations now occur.

(1) $2C_5 \rightarrow C_3 + C_7$ (transketolase)

(2) $C_7 + C_3 \rightarrow C_4 + C_6$ (transaldolase)

(3) $C_5 + C_4 \rightarrow C_3 + C_6$ (tranketolase)

The $2C_5$ in reaction 1 are ribose-5-phosphate and xylulose-5-phosphate; the final C_3 compound is glyceraldehyde-3-phosphate. This is converted to glucose-6-phosphate with loss of P_i. The net effect is that six molecules of ribose-5-phosphate are converted to five molecules of glucose-6-phosphate + P_i. Thus the cell can produce NADPH with no net increase in ribose-5-phosphate.

5 The oxidative part of the pathway converts glucose-6-phosphate to ribose-5-phosphate and CO_2. If you consider that six molecules of the former, and the six molecules of the latter are recycled to five glucose-6-phosphate, then, on a balance sheet, six molecules of glucose-6-phosphate have produced six molecules of CO_2 which, on paper, looks like the oxidation of a glucose molecule. It is not; one molecule of glucose-6-phosphate is not converted to six molecules of CO_2.

6 NADPH is required to reduce glutathione, a molecule necessary for the protection of the red blood cell. Patients lacking the enzyme are sensitive to the antimalarial drug, pamaquine, resulting in a haemolytic anaemia.

Chapter 18

1 Light reactions involve the splitting of water using light energy, and the reduction of $NADP^+$ to NADPH. Dark reactions mean the utilization of that NADPH to reduce CO_2 and water to carbohydrate. The term 'dark' implies that light is not essential, not that it occurs only in the dark. In fact, dark reactions will occur maximally in bright sunlight.

2 Photosystems contain many chlorophyll molecules. When excited by a photon, one of the electrons in the latter is excited to a higher energy level. Resonance energy transfer permits this excitation to jump from one molecule to another until it becomes trapped in special reaction-centre molecules called antenna chlorophyll molecules. The excitation is insufficient for resonance energy transfer but sufficient for an electron to enter the electron transport chain shown in Fig. 18.6.

3 (a) Electrons passing along the carriers of photosystem II result in ATP synthesis by the chemiosmotic mechanism as they traverse the cytochrome *bf* complex (see Fig. 18.6).
(b) When all of the $NADP^+$ is reduced, electrons passing along the carriers of photosystem I are diverted to the cytochrome *bf* complex and so generate more ATP (see Fig. 18.7).

4 It is the chlorophyll $P680^+$; that is, the reaction-centre pigment of photo system II that has been excited by resonance energy transfer from antenna chlorophyll molecules to donate an electron to pheophytin, the first component of the photo system II electron transport chain. $P680^+$ is an electron short and has a strong tendency to accept an electron; that is, it is a strong oxidizing agent.

5 Thylakoids are formed by invagination of the inner chloroplast membrane (*cf.* inner mitochondrial membrane), which explains the apparent opposite orientation of proton pumping.

6 The enzyme Rubisco (ribulose-1:5-bisphosphate carboxylase) splits ribulose-1:5-bisphosphate into two molecules of 3-phosphoglycerate, a molecule of CO_2 being fixed in the process (see reaction on page 277).

7 As in Fig. 18.10.

8 If CO_2 is fixed first into 3-phosphoglycerate by Rubisco, the plant is known as a C_3 plant. These are subject to full competition between oxygen and CO_2 in the Rubisco reaction. However, in high-temperature, high-sunlight areas, in C_4 plants, CO_2 is first fixed into oxaloacetate by pyruvate phosphate dikinase and PEP carboxylase (see Fig. 18.11). This is reduced to malate which is transported into the underlying bundle sheath cell where the Calvin cycle occurs. Decarboxylation of the malate occurs so that the ratio of CO_2/O_2 in the cell is greatly increased – the concentration of CO_2 in bundle sheath cells may be raised 10–60-fold by this means. The whole scheme is given in Fig. 18.11. Variations in detail of the scheme given occur in different C_4 plants but all are devoted to the same basic strategy.

9 In animals this cannot happen directly – pyruvate kinase cannot form PEP from pyruvate. However, the enzyme in plants is pyruvate-phosphate dikinase in which two phosphoryl groups of ATP are used, making the reaction thermodynamically feasible.

Chapter 19

1 Oxidation results in the formation of a Schiff base, hydrolysable by water.

$$\underset{\mid}{\overset{\mid}{C}}HNH_2 \xrightarrow{-2H} \underset{\mid}{\overset{\mid}{C}}=NH \xrightarrow{H_2O} \underset{\mid}{\overset{\mid}{C}}=O + NH_3$$

2 Glutamic acid.

3 Transdeamination is the most prevalent mechanism. The amino group of many amino acids are transferred to α-ketoglutarate, forming glutamate. The latter is deaminated by glutamate dehydrogenase. As an example:

1. Alanine + α-ketoglutarate → pyruvate + glutamate.

2. Glutamate + NAD^+ + H_2O → α-ketoglutarate + NADH + NH_4^+

Net reaction: Alanine + NAD^+ + H_2O → pyruvate + NADH + NH_4^+.

4 Pyridoxal phosphate (Fig. 19.2). The mechanism of transamination is given in Fig. 19.3.

5 By removal of H_2O, to result in a Schiff base.

6 A glucogenic amino acid is one that, after deamination, can give rise to pyruvate (or phosphoenolpyruvate). This may be indirect – any acid of the citric acid cycle is glucogenic. A ketogenic amino acid is one that cannot give rise to the above but rather gives rise to acetyl-CoA. Only leucine and lysine are purely ketogenic but some, such as phenylalanine, are mixed. Ketogenic amino acids produce ketone bodies only in circumstances appropriate to this such as starvation. Otherwise the acetyl-CoA is oxidized normally.

7 Phenylalanine is not normally transaminated; it is converted to tyrosine and then metabolized (see Fig. 19.5). If the phenylalanine conversion to tyrosine is defective, phenylalanine does transaminate, producing phenylpyruvate, which causes irreparable brain damage to babies and early death.

8 To supply the reducing equivalent for the formation of H_2O from one atom of the oxygen molecule used in the hydroxylation reaction.

9 By formation of 5-adenosylmethionine (SAM); this has a sulphonium ion structure conferring a strong leaving tendency on the methyl group. The formation of SAM is illustrated in Fig. 19.6.

10 5-Aminolevulinate synthase (ALA) carries out the reaction shown in Fig. 19.9 and ALA dehydratase produces the pyrrole, porphobilinogen, by the reaction in Fig. 19.10.

11 This is shown in Fig. 19.14.

12 Both situations call for high rates of deamination of amino acids (in starvation muscle proteins break down to allow glucose synthesis). The amino nitrogen must be converted to urea.

13 (a) Ammonia is converted to glutamine, which is transported to the liver and hydrolysed.

$$\text{Glutamate} + \text{ammonia} \xrightarrow[\text{ATP} \quad \text{ADP} + P_i]{} \text{glutamine}$$

$$\text{Glutamine} + H_2O \longrightarrow \text{glutamate} + \text{ammonia}$$

(b) Amino nitrogen is transported from muscle as alanine. The alanine cycle is shown in Fig. 19.16.

14 Any blockage in the cycle leads to ammonia toxicity. Deficiencies are known to occur in the level of the enzyme which synthesizes N-acetyl glutamate, in argininosuccinate synthetase and lyase. Usually the deficiency is only partial but in severe cases mental retardation and death can occur. Attempts to cause supplementary excretion of ammonia are used to treat the condition. Feeding of benzoate or phenylacetate in large amounts causes excretion of the glycine and glutamine conjugates respectively. In the case of argininosuccinate lyase deficiency, feeding of excess arginine and a low-protein diet results in excretion of argininosuccinate. The reason is that the arginine is converted to urea (which is excreted) forming ornithine which is converted to argininosuccinate, using up a molecule of ammonia.

15 Increased activity in liver is associated with acute intermittent porphyria (maybe variegate phoryia), though the connection between 5-aminolevulinic acid (ALA) production and the neurological effect is not understood.

Chapter 20

1 PRPP or 5-phosphoribosyl-1-pyrophosphate is the universal agent. Its formation from ribose-5-phosphate and ATP and its mode of action are given on page 300.

2 Tetrahydrofolate (FH_4).

Formyl-FH_4 =

3 Serine. Serine hydroxymethylase transfers —CH_2OH to FH_4 leaving glycine and forming N^5,N^{10}-methylene FH_4. This is oxidized to N^5,N^{10}-methylene FH_4 by an $NADP^+$-requiring reaction. Hydrolysis of this produces formyl-FH_4. The reactions are shown on page 303.

4 It ribotidizes guanine and hypoxanthine in the purine salvage pathway.

5 HGPRT is missing so that purine salvage cannot occur. However, the brain has the direct pathway and it may be that overproduction of purine nucleotides *de novo* occurs due to increased levels of PRPP (because it is not used by salvage). In such patients, excess uric acid is formed but prevention of this by allopurinol does not relieve the neurological symptoms. And gout patients do not suffer the latter, so these symptoms are unexplained.

6 It mimics the structure of hypoxanthine. It is a potent xanthine oxidase inhibitor. This inhibition results in xanthine and hypoxanthine formation rather than that of uric acid.

7 As in Fig. 20.8. It is based on feedback inhibition.

8 No. Thymidylate synthetase transfers a C_1 group from methylene FH_4 to dUMP and at the same time reduces it to a methyl group. The H atoms for this are taken from FH_4 so that FH_2 is a product as shown on page 306.

9 It inhibits reduction of FH_2, generated by thymidylate synthase, and so prevents dTMP production essential for cell multiplication. FH_4 is essential for the thymidylate synthase reaction.

10 Vitamin B_{12} is required for the methylation of homocysteine to methionine, the methyl donor being methyl FH_4. Lack of vitamin B_{12} causes FH_4 to be 'trapped' as the methyl compound and unavailable for the other folate-dependent reactions.

Chapter 21

1 This is shown on page 316.

2 Genetic material needs to be as chemically stable as possible. DNA is more stable than RNA. This is because, as shown on page 317, the 2'-OH of RNA can make a nucleophilic attack on the phosphodiester bond making RNA less chemically stable than DNA.

3 The phosphodiester bond is longer than the thickness of a base; a straight chain structure of DNA would leave hydrophobic faces of the bases exposed to water. Because of hydrophobic forces the bases are collapsed together by sloping the phosphodiester bond as shown in Fig. 21.3(c).

4 B DNA; right-handed; 10 base pairs.

5 One chain runs $5' \to 3'$ in one direction and the other $5' \to 3'$ in the other direction. Thus in a linear DNA molecule each end has a 5' end of one chain and a 3' end of another chain.

6 A $5' \to 3'$ direction means that you are moving from a terminal 5'-OH group to a 3'-OH group of the polynucleotide chain.

7   5' CATAGCCG 3'

              3' GTATCGGC 5'

Watson–Crick base pairing explains the complementary sequence. By convention a single sequence is written with the 5'end to the left.

8 **(a)** The DNA of a eukaryotic chromosome is wound around an octamer of histone proteins, two turns per nucleosome occupying 146 base pairs. Successive nucleosomes are connected by linked DNA.
(b) When chromatin is digested with DNase, the 146-base-pair sections of DNA are protected; this observation was important in the discovery of nucleosomes (see Fig. 21.7).

9 Repetitive DNA; an Alu sequence is a few hundred bases long but repeated almost exactly hundreds of thousands of times; scattered throughout the human chromosome.

10 The statement is true to a large extent but not completely so. It has been known for a long time that a few genes such as those for ribosomal and transfer RNAs do not code for protein but very recently it has been found that in so-called junk DNA there are large numbers of 'micro genes', coding only for small RNA sequences, which in some way seem to play a part in the expression of conventional genes and may determine characteristics of eukaryotic organisms.

11 In mitosis, chromosomes are replicated and a copy of each chromosome is segregated into each of the two daughter cells, which like the parent cell are diploid ($2n$). In meiosis, haploid sperm and egg cells are produced (n). Two cell divisions are involved but only in the first is DNA replicated. At the first cell division each daughter cell receives two copies (as paired chromatids) of one chromosome of each of the parental homologous pairs, randomly assigned. In the second cell division DNA synthesis does not occur and the chromatids of the chromosome duplexes are pulled apart and segregated into the two daughter cells, which each thus receives single copies of one of each pair of the parental homologous chromosomes. The copy has been derived either from the mother or the father of the organism. The gametes are haploid (n). During meiosis the duplex pairs swap sections of the chromatids leading to greater genetic diversity

12 Uracil. The others are components of DNA. Uracil occurs in RNA but not DNA.

13 The prokaryotic genome is haploid and circular and not enclosed by a membrane. It does not have the defined nucleosome structure found in eukaryotes. The eukaryotic genome consists of linear chromosomes, terminating in telomeres and is surrounded by a nuclear membrane. Eukaryotic cells are typically diploid. At cell division the chromosomes condense into a tightly packed form unlike prokaryotic cell division where this does not occur.

Chapter 22

1 A section of DNA whose replication is initiated at a single origin of replication.

2 Ahead of the replicative fork positive supercoils occur.

3 In *E. coli* gyrase, a topoisomerase II, introduces negative supercoils. In eukaryotes a topoisomerase I relaxes positive supercoils.

4 As shown in Figs 22.6 and 22.7.

5 In winding DNA around a nucleosome, a local negative supercoil is introduced, but, since no bonds are broken, there cannot be a net negative supercoiling. The local negative super-coil is compensated for by a local positive supercoiling else-where. This is relaxed by topoisomerase I, leaving a net negative supercoiling in the DNA.

6 dATP, dGTP, dCTP, and dTTP.

7 Cytosine readily deaminates to uracil; if uracil were a normal constituent of DNA it would be impossible for repair enzymes to recognize and correct the mutation. Use of T instead of U removes the problem.

8 (a) No. An RNA primer is required. DNA polymerase cannot initiate new chains.
(b) Synthesis proceeds in the $5' \rightarrow 3'$ direction; by this it is meant that the new chain is elongated in the $5' \rightarrow 3'$ direction, new nucleotides being added to the free $3'OH$ of the preceding nucleotide. It does not refer to the template strand, which runs antiparallel.

9 Hydrolysis of inorganic pyrophosphate; base pairing of the nucleotides and breaking of the high-energy phosphate bond of dNTPs, liberating pyrophosphate.

10 Polymerase I is the enzyme that processes Okazaki frag-ments into a continuous lagging strand. It has a $5' \rightarrow 3'$ exonu-clease and a $3' \rightarrow 5'$ exonuclease, and is a DNA polymerase with low processivity. Its activities are summarized in Fig. 22.16.

11 Correct base pairing of the incoming nucleotide triphos-phate with the template nucleotide is the primary essential. The enzyme also has a proofreading activity. It has a $3' \rightarrow 5'$ exonuclease activity that removes the last incorporated nucleotide if this is improperly base paired.

12 The methyl-directed mismatch repair system is described in Fig. 22.18. There is evidence that proteins similar to *E. coli* proteins exist in humans and that, where these are deficient, there is an increased risk of cancer.

13 Thymine dimers are formed when DNA is subjected to UV irradiation. Two adjacent thymine bases become covalently linked. Repair can either be direct by a light-dependent repair system that disrupts the bonds and restores the separate bases; or it can be subject to excision repair (Fig. 22.19).

14 As shown in Fig. 22.21, removal of the $3'$ Okazaki fragment primer leaves an unreplicated section.

15 The answer lies in telomeric DNA synthesized as described in Fig. 22.22. The telomerase is a reverse transcriptase.

16 Processivity is the ability of a DNA polymerase to duplicate the coding strand of DNA for the very long stretches involved in DNA replication without falling off. The sliding clamps achieve this. These are annular proteins that are placed around the DNA to be copied, at the site of an RNA primer; the polymerase attaches to the clamp and so is prevented from falling off. DNA polymerase I has low processivity because its job is to remove the RNA primer from Okazaki fragments. If it had high pro-cessivity it would remove the newly synthesized DNA of the fragments and replace them unnecessarily. It is best that soon after it has reached the DNA part, it detaches to allow the ligase to heal the nick.

17 DNA. All the rest are the precursors of RNA.

Chapter 23

1 RNA polymerase uses ATP, CTP, GTP, and UTP as substrates (cf. dATP, dCTP, dGTP, and dTTP). Most importantly, RNA polymerase can initiate new chains; DNA polymerase requires an RNA primer. RNA synthesis never includes proofreading.

2 This is illustrated in Fig. 23.4. In more detail the promoter has a -10 (Pribnow) box and a -35 box.

3 RNA polymerase attaches nonspecifically to the DNA, but, when joined by a sigma factor molecule, it binds firmly to a promoter. The Pribnow box and the -35 box position orientate the polymerase correctly. The enzyme synthesizes a few phos-phodiester bonds and then the sigma protein flies off and the polymerase progresses down the gene transcribing mRNA.

4 One is by the stem loop structure shown in Fig. 23.7, formed by G—C base-pairing. In this, internal G—C base pair-ing in the mRNA prevents bonding of the mRNA to the tem-plate DNA strand. Following this, a string of uracil nucleotides further weakens the attachment (only two hydrogen bonds in a U—A pair), facilitating detachment. There is still some uncertainty as to why this is a reliable terminator. The second method depends on the Rho factor, a helicase that unwinds the mRNA–DNA hybrid. At the termination site, the polymerase pauses (possibly due to a region in the DNA rich in G—C base pairs with triple bonds to be broken) and the Rho factor catches up and detaches the mRNA.

5 The precise base sequence of the −10 and −35 boxes, their distance apart, and the bases in the +1 to +10 region.

6 The operon is blocked by a repressor protein. In the presence of lactose, the repressor detaches (see Fig. 23.10).

7 The primary transcript in eukaryotes has introns that must be spliced out; the 5′ end is capped; the mechanism of termination is unknown; the mRNA is, with few exceptions, polyadenylated at the 3′ end.

8 The mechanism is shown in Fig. 23.13. It is based on a consensus sequence that determines where splicing occurs and on a trans-esterification reaction involving little change in free energy. Split genes may have facilitated evolution by promoting exon shuffling. Differential splicing can also result in a single gene giving rise to different proteins.

9 The eukaryotic polymerase II does not attach to the DNA directly but rather attaches to a protein complex that assembles on the DNA (Fig. 23.23). In addition, any number of transcription factors may be associated with the complex (Fig. 23.24).

10 In the default condition, eukaryotic genes are shut down. This is due to nucleosomes blocking the promoter sites. Chromatin remodelling unblocks the site by removal of the nucleosomes. This is done by the enzyme histone acetyltransferase (HAT) which acetylates the tails of the histones that constitute the nucleosome octamer. This has the effect of removing the positive charge on the histone lysine side chains. Somehow this results in chromosome remodelling.

11 In insect salivary glands, the formation of 'chromosome puffs' at the time of localized gene transcription can be seen in the microscope. Secondly it is known that genes at the time of activation become hypersensitive to added DNase, indicating that the enzyme has increased access to the DNA. This has been demonstrated for the haemoglobin gene among others.

12 The enzyme histone deacetylase is believed to reverse the activation process started by the histone acetyltransferase.

13 In the case of HAT, the enzyme is part of the coactivator. The latter binds to the activating transcription factor(s) and brought into proximity to the nucleosomes on the promoter selected by the transcription factors. Negative control may be effected in essentially the same way, a repressing transcription factor binding to the promoter and attracting the deacetylase.

14 The 'introns early' view is that they have always been present in modern genes. The latter are postulated to have been formed from primitive mini-genes which fused together, the introns being the fused nontranslated regions. On this view, uninterrupted prokaryotic genes are due to all excess DNA having been discarded in the interests of rapid cell division.

The alternative view is that introns are late additions to prokaryotic genes which are regarded as primitive, and the

introns may almost be regarded as parasitic DNA. It is generally accepted that introns may have been important in evolution by facilitating exon shuffling. The argument on intron origins still rages.

15 Helix–turn–helix proteins, leucine zipper proteins, helix–loop–helix proteins, zinc fingers proteins, and homeodomain proteins.

16 The proteome is the complete collection of proteins present in a cell at any one time. The genome is the complete collection of genes. The proteome varies according to which genes have been activated so that the number of proteins varies in cells from different tissues and from time to time according to physiological controls. The genome is essentially fixed if we exclude occasional examples of gene amplification.

17 No; alternative splicing may produce multiple versions of proteins from a single gene transcript.

Chapter 24

1 If only 20 codons were used, there would be 44 not coding for any amino acid. Any mutation in a gene-coding region would then be highly likely to inactivate the gene by prematurely introducing a stop codon. Instead, by using 61 codons for amino acids, a base change would either cause no change or would substitute a different amino acid. Because of the arrangement of the genetic code, many of these substitutions would be conservative and cause minimal change to the change to the protein structure.

2 The answer lies in the wobble mechanism described on page 384.

3 The convention is that RNA is shown with the 5′ end to the left. When mRNA is shown in this manner, a tRNA anticodon is base paired in an antiparallel fashion. The tRNA molecule is therefore shown with the 5′ end to the right.

4 In the case of some amino acids, at the stage of attaching the amino acid to tRNA. In all cases at the stage of elongation; a pause in GTP hydrolysis on the EF-Tu gives time for incorrectly paired aminoacyl-tRNAs to leave the ribosome.

5 In general GTP hydrolysis to GDP and P_i is believed to result in conformational changes in the relevant proteins. GTP is involved in assemblage of the *E. coli* initiation complex. It is involved in the delivery of aminoacyl-tRNAs to the ribosome by EF-Tu. It is involved in the translocation step.

6 Figure 24.8 is necessary for this explanation. The straddling has the advantage that in moving from one site to the other the tRNA is never completely detached. It also means that the peptidyl group does not have to physically move relative to the ribosome.

7 Figure 24.12 explains this. The scanning mechanism of selecting the start AUG means that there can be only one start site.

8 They bind to nascent polypeptides and prevent premature, improper folding associations. Appropriate release of chaperones facilitates correct folding though the process is not fully understood. Other chaperonins provide a suitable folding box.

9 The prion diseases described on page 397. In these, the faulty protein is due to different folding of the identical polypeptide. They are typified by mad-cow disease.

10 (a) Yes, you can deduce the amino acid sequence of the protein that an mRNA will code for because each codon unambiguously specifies an amino acid.
(b) No, you cannot deduce the base sequence of an mRNA from the amino acid sequence of the protein it codes for. This is because of the degenerate nature of the genetic code, or redundancy. Since several codons may code for a given amino acid it is impossible to deduce which of the alternatives was actually used by the cell in translating the messenger.

11 If one or two bases were deleted the reading frame would be shifted so that improper codons would be read thereafter. The result of this would be meaningless polypeptide after the first 99 amino acids. Or, if the new reading frame encountered a termination codon, the polypeptide would be prematurely terminated. If all three bases were deleted, the only effect would be to delete one amino acid residue and it is possible that the protein would be functional, depending on the effect of the deletion on the structure of the protein.

12 The peptidyl transferase has been shown to be active even after extraction of proteins from the ribosome. It is in fact not an enzyme but a ribozyme. Also, deletion of a single adenine base from one of the ribosomal RNAs by ricin inactivates the ribosome.

13 Transcription is the synthesis of mRNA coded for by the gene; translation is the synthesis of polypeptide coded for by the mRNA. The terms derive from the fact that transcription implies copying in the same language (DNA and RNA are both polynucleotides). Translation implies copying in a different language (mRNA is a polynucleotide, proteins are polypeptides).

14 It is based on an iron-dependent inhibition of mRNA translation as shown in Fig. 24.16.

15 Proteasomes are protein organelles consisting of a central core made of rings of protein subunits with caps at both ends. Inside the cavity of the central core are proteolytic enzymes which hydrolyse proteins into peptides and amino acids. The caps act as the gateway for entry of proteins destined for destruction. The proteasomes selectively destroy individual protein molecules, which are targeted into the cavity. The vital role is selective destruction of proteins such as regulatory proteins in cell cycle control. They also play a vital part in the immune system by producing peptides, for example from viruses to be displayed of the outside of cells inviting attack by killer T cells of the immune system. The importance of proteasomes is underlined by the fact that they have been highly conserved throughout evolution. Mutations causing non-functional proteasomes in yeast are lethal. In very recent times it has been realized that protein breakdown in cells especially by proteasomes has very great fundamental importance, so the field has become one of the most intensively studied – a revolution in the image of the field.

16 The entry ticket is the attachment of the small protein ubiquitin to target proteins. Polyubiquinated proteins are allowed to enter the proteasome cavity for destruction. The ubiquitin is removed during the entry process and recycled. There is still much to learn about what determines the selection of proteins for ubiquitination but certain N-terminal sequences are known to be a factor.

17 No. The free-energy release that occurs with a correct base pairing and an incorrect one are not sufficiently different to give the required discrimination. However the ribosome participates in the process in that a triplet of bases in the *E. coli* small subunit RNA checks that the first two anticodon-codon base pairing involves genuine Watson–Crick hydrogen bonds before peptide synthesis is allowed to occur.

18 Hsp 70 with a bound molecule of ATP binds to exposed hydrophobic groups on nascent polypeptide chains emerging from ribosomes and prevents unprofitable (improper) hydrophobic associations which could hinder correct folding. The chaperone detaches on ATP hydrolysis and allows the polypeptide to fold.
   Hsp means heat-shock protein. These are formed when cells such as *E. coli* are subjected to sudden temperature increases. They are in fact chaperones whose function is to assist refolding of proteins denatured by the heat. In this their role is the same as helping to fold the (unfolded) nascent polypeptide chains emerging from the ribosomes.

Chapter 25

1 It is an RNA transcript which is not translated into a protein, as is messenger RNA. Several types have been known for decades. Examples are ribosomal RNAs, transfer RNAs, SnoRNAs. These act as 'infrastructure' supporting the expression of protein coding genes. The new type is microRNAs (miRNAs) produced from microgenes, about 75 nucleotides long.

2 They are very widely distributed including most of the areas formerly known as junk DNA. This includes the introns of protein-coding genes. The ENCODE project has shown that at least 80% of the bases in the genome of eukaryotes are transcribed into RNAs. The protein coding section of conventional genes occupies only a small percentage of the genome. Microgenes exist in very large numbers.

3 The ENCODE project and other studies have shown that microgenes are conserved across different species such as humans and mice. Such conservation indicates that they must have a function naturally selected.

4 There are several. Overall it is believed that they were used in the evolution of complexity in eukaryotes by exerting epigenetic control over protein coding genes rather than by increasing the number of the latter. There is no clear correlation between complexity and conventional gene number. Prokaryotes have remained relatively simple; there is little 'junk' DNA and microgenes appear to be comparatively unimportant so far as is known.

It is also believed that miRNAs by making it possible to silence protein coding genes protected eukaryote s from excessive transposon proliferation.

5 Very little is known of this. At present the only action of miRNAs mechanistically understood is to make gene silencing possible by RNA interferance.

6 Microgenes are transcribed into miRNAs about 75 bases long which adopt a hairpin form. These are processed into small interfering double stranded RNAs (siRNAs) which attach to RISC complexes. One strand is selected and guides the RISC complex to complementary sequences in mRNAs which are then either destroyed or silenced.

7 It lies in the possibility of specifically silencing protein-coding genes. Theoretically a wide variety of diseases could be treated. For example cancer might be treated by targeting an oncogene. Such silencing has been demonstrated in tissue culture. Or a virus could be targeted. The beauty of the system would be that siRNAs can be cheaply made and potentially specific. However the field is in its infancy and problems of delivering the siRNAs to target tissues or cells is an obvious problem.

Chapter 26

1 In cotranslational transport, the polypeptide is transferred through the target membrane as it is synthesized. This occurs in the transport of proteins into the ER. Posttranslational transport is where the protein is fully synthesized in the cytoplasm, released, and then transported through its target membrane. Examples are mitochondrial proteins, nuclear proteins, and peroxisomal proteins.

2 (It will probably be necessary to refer to Fig. 26.6 to follow this explanation.) In almost every case, GTP hydrolysis occurs to produce a conformational change in a protein. Commonly the hydrolysis occurs only after a slight delay, this essentially being a timing device to permit other happenings to occur. An example of this is in the transport of proteins through the ER. It is postulated that, after the SRP in its GDP-bound form joins the docking protein on the ER, it has to lose its grip on the signal peptide so that the latter can insert into the translocon. To do this the GDP is exchanged for GTP; this reduces the affinity of the SRP for the signal peptide and increases its affinity for the docking protein. Since the SRP now has to leave the latter and return to the cytoplasm, it hydrolyses the GTP back to GDP which reduces its affinity for the docking protein. However, in its GDP form it would also bind to the signal peptide again if the latter was not safely esconced in the translocon. This is thepostulated point of the delay in hydrolysing the GTP – it may be to allow time for the signal peptide to enter the channel. There are other examples (see Q. 7).

3 To destroy unwanted molecules and structures imported into the cell by endocytosis and to destroy components of the cell destined for destruction.

4 They are vesicles produced by the Golgi sacs containing an array of hydrolytic enzymes. The enzymes are acid hydrolases with an optimum pH of 4.5–5.0. Proton pumps in the membrane maintain this pH inside the vesicle. They fuse with endosomes to form the lysosomes.

5 A variety of genetically determined lysosomal storage disorders exist in which the absence of a specific hydrolytic enzyme causes the lysosomes to become overloaded with material normally disposed of.

6 In Pompe's disease, a fatal genetically determined one, there is a lack of a lysosomal α-l:4-glycosidase that normally degrades glycogen. The lysosomes become overloaded with glycogen. However, it is not clear why disposal of glycogen by this means should be needed. It is not part of the usual accounts of glycogen metabolism.

7 (It will probably be necessary to refer to Figs. 26.17 and 26.18 to follow this explanation.) The Ran-GDP/Ran-GTP exchange enzyme exists only inside the nucleus. The function of Ran-GTP is to effect the release of the cargo from the importin-cargo complex arriving from the cytoplasm. The Ran–GTP–importin complex migrates back to the cytoplasm. It is now necessary for the GTP to be hydrolysed to allow the Ran-GDP to dissociate and release the importin. The Ran–GTPase, however, must be activated by GAP, the GTPase-activating protein. This is found only in the cytoplasm, not in the nucleus (the Ran-GDP itself must return to the nucleus). The released importin picks up another cargo and carries it into the nucleus to start the whole cycle again.

8 They are cluster of hydrophobic amino acids in the proteins lining the nuclear pore. (F and G are the one-letter abbreviations for phenylalanine and glycine respectively). The repeats are binding sites for importins which move through the pore from one repeat to the next.

9 Proteins destined for secretion are transported cotranslationally into the endoplasmic reticulum and from there to the cell membrane by transport vesicles. Proteins destined for the mitochondria are delivered in an extended form to receptors on the mitochondrial membrane. Nuclear proteins are delivered to the nuclear pore by importins. Peroxisomal proteins are delivered to receptors on the organelle in a fully folded form.

Chapter 27

1 The production is described in Fig. 27.19. GTP hydrolysis has a timing function. In cholera the GTPase is inactivated so that cAMP production remains activated once stimulated.

2 It allosterically activates protein kinase A (PKA). This can have metabolic effects depending on phosphorylation of key enzymes but, in addition, there are genes with cAMP-response elements (CREs). Inactive transcriptional factors can be activated by phosphorylation by PKA. Therefore cAMP can have extensive gene-control effects.

3 It activates a cytoplasmic guanylate cyclase that produces cGMP. The latter is a second messenger in some systems.

4 The phosphatidylinositol cascade involves the receptor-mediated activation of membrane-bound phospholipase C. This releases inositol trisphosphate (IP_3) and diacylglycerol (DAG). The former increases cytoplasmic Ca^{2+}; the latter activates a protein kinase, PKC. Thus, IP_3 and DAG are second messengers. The scheme is shown in Figs 27.21 and 27.22.

5 (a) Involves an allosteric change of the cytoplasmic domain on binding of the hormone.
(b) Involves receptor dimerization and activation of self-phosphorylation of tyrosine residues by an intrinsic kinase.
(c) Involves receptor dimerization and association with a separate tyrosine kinase that phosphorylates the receptor (see Figs 27.7 and 27.16).

6 Neurotransmitters; homones, cytokines, and growth factors; vitamin D_3 and retinoic acid; nitric oxide is in a class of its own.

7 In both cases the signal combines with a receptor protein and this triggers a response which in terms of gene control usually causes a protein to enter the nucleus and effect gene control. In the case of a lipid-soluble signal such as glucocorticoid it combines with a cytoplasmic receptor protein which enters the nucleus and functions as a transcription factor. In the case of a water-soluble signal such as EGF, it combines with a membrane receptor, and this results ultimately in a kinase entering the nucleus and activating a transcription factor. The two systems are fundamentally similar, the differences being in the detail. (In some cases the lipid signal receptor resides in the nucleus, but the same general principle applies.)

8 It is likely to be a component of a signalling pathway which specifically binds to an activated membrane receptor of the tyrosine kinase type.

9 Both have tyrosine kinase-associated receptors. The Ras-associated receptors are themselves tyrosine kinases, whereas the JAK/STAT pathways depend on cytoplasmic kinases associating with the receptors. The Ras pathway is activated by many hormones whereas the other pathway is best known for its association with cytokine signals such as interferon. The Ras receptor activates a long cascade of proteins terminating in a kinase that migrates into the nucleus and activates transcription factors. The JAK/STAT receptors bind the JAKs which phosphorylate tyrosine groups of the receptor. The STAT proteins are attracted from the cytoplasm to bind to these and are themselves phosphorylated by the double-headed JAKs. The phosphorylated STAT proteins migrate into the nucleus where they activate genes. Thus the latter pathway is a much more direct system from receptor to gene whereas the Ras pathway is very long. This is probably to allow amplification of the signal. Less certainly, it may also give more opportunity for cross-talk between Ras and other pathways and provide opportunities for extra controls.

10 All G-proteins are heterotrimeric proteins associated with membrane receptors. The latter do not become phosphorylated but on the binding of the cognate ligand undergo a conformational change. This in turn causes a change in the conformation of the attached G-protein such that the GDP attached to the α subunit is exchanged for GTP. This α-GTP subunit detaches from its partner β and γ subunits and migrates to a target enzyme located on the membrane. In the case of a typical G-protein, the one associated with the β-adrenergic receptor, it activates adenylate cyclase which produces a second messenger cAMP. The activation is terminated by the slow intrinsic GTPase activity of the α subunit which acts as a timing device. The α subunit in its GDP form migrates back to the receptor to reform the heterotrimeric G-protein. If the receptor is still activated, the process is repeated. The versatility is due to the fact that different receptors have different G-proteins whose activated α subunit activates or inhibits different enzymes involved in production of second messengers. Thus stimulation of receptors by epinephrine may activate or inhibit cAMP production depending on the nature of the particular G-protein subunit associated with the receptor; in the phosphatidylinositol system, the G-protein subunit activates phospholipase C as one example of the different target enzyme specificities that exist. In this case the second messengers are DAG and $PI(3,4,5)P_3$. (Note that other GTPase proteins exist, but the term G-protein is restricted to the type described above.)

11 Not really; cGMP is continually produced in the rod cells to keep the cation channels open. The effect of light is to cause the activation of the enzyme which destroys cGMP. In the case of the nitric oxide signalling system cGMP is a second messenger for the signal activates either a membrane receptor or an intracellular receptor to produce the cGMP.

12 The insulin receptor does not dimerize but it structurally resembles a covalently permanently dimerized receptor of the EGF type.

13 The breakdown of GTP to GDP + P$_i$ may seem to have no purpose in that no chemical bonds are formed or obvious work done. However the breakdown is often a timing device. The G protein activation of cAMP synthesis is an example in which the α subunit–GTP activates adenyl cyclase but GTP hydrolysis switches this off. In protein targeting, it is postulated that GTP hydrolysis by the signal receptor protein reduces its affinity for the peptide and allows the latter to insert into the translocon channel. In protein synthesis GTP hydrolysis allows the EF-Tu to leave the ribosome but only after a slight delay to allow incorrect amino acyl–tRNAs to detach. There are other examples such as the uncoating of COP-coated vesicles and the auto inactivation of Ras.

14 Ras is called a small monomeric GTPase. The term G-protein refers only to the heterotrimeric signalling proteins such as the adrenalin receptor.

Chapter 28

1 Pancreatic DNase randomly cuts DNA. A restriction enzyme cuts only at specific short sequences of bases.

2 The adenine bases in all of the relevant hexamer sequences of *E. coli* strain R DNA are methylated so that the enzyme does not recognize it. Invading foreign DNA will not be so protected.

3 It refers to ends resulting from a staggered cut made by many restriction enzymes.

The overhanging ends will automatically base pair and 'stick' together.

4 A genomic clone refers to a cloned section of DNA identical to the sequence of DNA in a chromosome. A cDNA clone means complementary DNA obtained from a eukaryotic mRNA. It refers to cloned DNA identical to the mRNA for a gene. The cDNA lacks introns; eukaryotic genomic clones have them.

5 These are outlined in Fig. 28.9.

6 The nucleotide lacks a 3′OH and therefore, when added by DNA polymerase to a growing DNA chain, terminates that chain. The application of this to sequencing is described on page 460.

7 A specific section of DNA on a chromosome can be amplified logarithmically. It involves copying the section of DNA and copying the copies *ad infinitum*. Its importance is that a minute amount of DNA, too small for any studies, can be amplified at will. The essential, apart from enzymes and substrates, is that you must have primers for copying the section in both directions and this means knowing the base sequences at either end of the piece to be amplified.

8 It is an engineered plasmid with a convenient insertion site for, say, a cDNA clone for a specific protein, but which also contains appropriate bacterial DNA transcriptional signals (a promoter) and translational signals. The DNA is transcribed and the mRNA translated in a bacterial cell. It can be used to produce large quantities of specific proteins.

9 It is a technique for detecting gene abnormalities. If DNA is cut by a pattern of restriction enzymes and the resultant pieces separated on an electrophoretic gel, a pattern of bands can be visualized by hybridization methods. Mutations may produce or remove restriction sites so that the pattern may be altered. The method was important, but is largely replaced by PCR.

10 Stem cells divide and the progeny cells can either proceed to develop into mature terminally differentiated cells or divide into more stem cells. This can occur indefinitely, unlike somatic cells which have a limited potential to divide. Thus a stem cell population keeps up a continual supply of replacement cells. Adult stem cells are committed to become a certain class of cells; bone marrow stem cells differentiate into many types of blood cells. Embryonic stem cells found in blastocysts are pluripotent; they can give rise to any type of the cell in the body. They are of especial interest because they can be cultivated *in vitro* and retain this pluripotency. Apart from their use in producing knockout mice, they have the potential to be of enormous medical importance in that they might be developed in controlled *in vitro* conditions into replacement cells for the human body. Production of nerve cells to repair neurological damage is but one possible example.

11 (The procedure is complex and you will probably have to follow Figs 28.15 and 28.16 to understand this answer.) The principle is to inactivate a specific gene by homologous recombination. For this a targeting vector is constructed so that the normal gene is replaced by an inactivated one. The method uses embryonic stem cells obtained from a mouse blastocyst. These are mutated by the vector and cells with the targeted gene correctly replaced are grown up and injected into the blastocyst of another mouse. The blastocyst is implanted into a mouse to develop. The offspring contains cells derived from both the mutated cells and the recipient blastocyst cells. By appropriate cross-breeding, mice which are homozygous to the mutated gene can be obtained. In the procedure, the targeted stem cells are taken from a blastocyst from a mouse of a coat colour which is different from that from which the recipient blastocyst is taken. This enables the process to be followed by coat colour of the progeny.

12 (a) To replicate the DNA, synthesis in both directions is required and a different primer is needed for each direction.
(b) Each time a section of DNA is replicated the primer becomes incorporated into the new strand. Since an enormous amplification is required sufficient primers must be available to support the indefinite numbers of replications.
(c) DNA polymerase; since at each cycle the mixture has to be heated to separate strands a heat-stable enzyme preparation from a thermophilic bacterium (*Taq* polymerase) is used.

Chapter 29

1 The initial stimulus for blood clotting is, in quantitative terms, very small. To obtain a sufficiently rapid response amplification is needed. Cascades are the biological method of amplification. The final enzyme activation is that of prothrombin to thrombin, an active proteolytic enzyme (see Fig. 29.1).

2 Fibrinogen monomer proteins are prevented from spontaneous polymerization by negatively charged fibrinopeptides (Fig. 29.2) that cause mutual repulsion. Removal of these by thrombin permits the association of fibrin monomers, illustrated in Fig. 29.3.

3 Covalent cross-links are formed between monomers in the polymer by an enzymic transamidation process between glutamine and lysine side chains:

$$-CONH_2 + H_3N^+ - \rightarrow -CO-NH- + NH_4^+$$

4 In the conversion of prothrombin to thrombin, a glutamic acid side chain is carboxylated; the carboxyglutamate binds Ca^{2+}. Vitamin K is a cofactor in the carboxylation reaction.

$$\begin{array}{c} | \\ CH_2 \\ | \\ CH_2 \\ | \\ COO^- \end{array} + CO_2 \longrightarrow \begin{array}{c} | \\ CH_2 \\ | \\ HC-COO^- \\ | \\ C-COO^- \end{array}$$

Other conversions in the proteolytic cascade involve this conversion.

5 It is involved in the conversion of hydrophobic xenobiotics to more soluble ones. A typical reaction is:

$$AH + O_2 + NADPH + H^+ \rightarrow AOH + H_2O + NAD^+$$

6 To reduce one oxygen atom to H_2O. It is known as a mixed-function oxygenase.

7 The glucuronyl group is donated to an —OH on the foreign molecule, rendering it much more polar (see Fig. 29.5).

8 It involves ATP binding cassette transporters (ABC transporters) that remove a variety of compounds from cells. It transports steroids in steroid-secreting adrenal cortical cells, but all transported compounds (including anticancer drugs used in chemotherapy) are lipid-soluble amphipathic compounds, indicating a more general role. It has recently been shown to transport the cholesterol to the outside of the cells to be removed to form high-density lipoprotein (HDL).

9 It is an oxygen molecule that has acquired an extra electron.

$$O_2 + e^- \rightarrow O_2^-$$

Superoxide is extremely reactive, causing damage.

10 Most eukaryotic cells have superoxide dismutase and catalase:

$$2O_2^- + 2H^+ \rightarrow H_2O_2 + O_2 \text{ (dismutase)}$$
$$2H_2O_2 \rightarrow 2H_2O + O_2 \text{ (catalase)}$$

Secondly, there are antioxidants such as ascorbic acid (vitamin C) and α-tocopherol (vitamin E). Superoxide is dangerous because it attacks a molecule and generates another free radical, causing a self-perpetuating chain of reactions. Antioxidants are quenching agents. When attacked by superoxide, the free radical produced from these is insufficiently reactive to perpetuate the chain of reactions.

Chapter 30

1 As in Fig. 30.1.

2 The principle is that each B cell assembles its own immunoglobulin genes randomly from a selection of DNA sections as described in Fig. 30.2.

3 B cells produce antibodies (after activation and maturation into plasma cells). Helper T cells are (in most cases) required for B cells to do this. Cytotoxic or killer T cells bind to host cells displaying a foreign antigen and kill them by perforating their membrane or inducing apoptosis.

4 It is a type of phagocyte (usually a dendritic cell) that engulfs a foreign antigen, processes it into pieces, and displays it in combination with its MHC molecules. If an inactive helper T cell combines with the antigen–MHC complex it is activated (see Fig. 30.6). B cells after their initial encounter with an antigen also are antigen-presenting cells.

5 In both cases MHC class 1.

6 Each B cell produces a different antibody. For any given antigen there will be very few B cells specific for it, but, when the antigen binds to the displayed antibody, the B cell proliferates into a clone. Thus the antigen automatically selects which B cells are to proliferate.

7 During *primary* maturation of B cells in the bone marrow or of T cells in the thymus, if an antigen binds, the cell is eliminated, the principle being that such antigens will be 'self'. After release from the bone marrow or thymus, the binding to an antigen activates the cells to multiply, but further activation by dendritic cells is required.

8 Differential splicing of introns; at the 3′ end of the immunoglobulin gene is an exon that codes for an anchoring polypeptide sequence. At the onset of secretion a switch in the splicing eliminates this during mRNA formation.

9 The onset of secretion is accompanied by rapid somatic mutation in the cells, which modifies the variable site. Since binding of the antigen causes cell proliferation this constitutes a progressive selection mechanism for those cells producing as 'better' antibody. This is known as affinity maturation.

10 CD4 is present on helper T cells. The protein binds to a constant protein component of MHC class 2 molecules and thus confines interactions to B cells, helper T cells, and antigen-presenting cells. Cytotoxic T cells have CD8 which binds to MHC class 1 molecules; this restricts the interaction of the killer T cells to host cells. The CD4 protein is the receptor by which the AIDS virus infects helper T cells.

11 It is a form of compartmentation which directs immune cells to their correct targets. Most somatic cells have class 1, for which killer T cells are specific. Helper T cells and B cells are class 2. This means that killer T cells will not attack them and helper T cells will not combine with somatic cells.

12 They are monoclonal antibodies produced in mouse systems in which the constant mouse protein groups are replaced by their human counterparts. They are used for therapeutic injection into humans to eliminate or reduce immunological reactions generated by mouse proteins. A new strain of mice has been engineered to produce completely human versions of the antibodies.

Chapter 31

1 In G_1, a growth factor signal is necessary to activate genes that code for the G_1-specific cyclins needed to activate the kinases which push the cell cycle through the restriction point. In the absence of growth factor signal the cycle is arrested in the G_0 phase awaiting the arrival of a signal.

2 Cyclins are proteins which are necessary for the cyclin dependant kinases (Cdks) to act. They not only activate Cdks but also direct their activity to be appropriate to the needs of the different phases of the life cycle. Cyclins are specific to particular phases of the life cycle and are completely destroyed at the end of their appropriate phase of the cycle. New cyclins are synthesised as the cycle progresses from one phase to the next.

3 At the end of cell division the cyclins are destroyed and new G_1 specific cyclins synthesised. Without this the cell cannot progress to the G_1 checkpoint. If no mitotic signal is received the cell enters Go in which it functions normally in interphase but does not divide. Most somatic cells are in this state most of the time. If a signal is received it can enter the G_1 phase again.

4 If the DNA is damaged in any way the cell is not allowed to proceed to the S phase. Damage is detected by the protein P53 which halts the cycle progression. In addition the retinoblastoma protein also can halt the cycle at this point (described in chapter 33). At the mitotic checkpoint each chromosome (chromatid at this point) must be properly aligned on the mitotic spindle. If each is not attached to a kinetochore fibre the cycle is halted.

Chapter 32

1 They are proteolytic enzymes with a cysteine residue their active centre which attack a wide variety of proteins at sites adjacent to an aspartate residue. They exist in multiple copies in all cells in an inactive state and become activated in response to an apoptopic signal for the cell to destroy itself. They exist in cells as inactive procaspases. One form of apoptopic stimulus releases cytochrome c from mitochondria; another is a death signal from a killer T cells. In both instances the procaspases are caused to aggregate probably resulting in mutual activation by partial proteolysis. The cascade of proteolysis which ensues destroys the cell.

2 It has a vital role in destroying unwanted cells. An obvious example is the destruction of huge numbers of B and T cells of the immune system to eliminate those that would cause an autoimmune reaction. The importance of apoptotic destruction is that it is a neat non-messy way of cell disposal without necrosis that could cause problems. It also has a vital protective role in eliminating potentially cancerous cells. Another role is that killer T cells of the immune system destroy infected cells by delivering a signal for the cell to self-destruct.

3 A wide variety of 'insults' to cells can induce it including radiation or a damaged genome. P53 plays a vital part in this. If a cell is damaged to the point where an abnormal potentially cancerous state is present P53 triggers its destruction. In fact about half of human cancers are associated with P53 deficiency. It is not totally understood how P53 induces apoptosis but and early event is the release of cytochrome C from mitochondria into the cytoplasm which is a death sentence.

4 Cytochrome C causes procaspases which are inactive proteolytic enzymes to aggregate. Activation of these occurs and the caspases then destroy the cell. The activation process is not fully understood, but it may be that the procaspases have a low activity, and aggregation results in mutual activation.

5 Cells are delicately balanced between life and death. There are opposing controls of pro- and anti-apoptosis. Broadly there are two groups of proteins within the BC12 family. Bax and Bad are pro-apoptopic and are believed to initiate cytochrome C release. Bel-2 and Bcl-X2 are important anti-apoptopic proteins which prevent cytochrome C release from mitochondria. It is the balance between the two groups which will determine the fate of a cell. There are additional direct controls inhibiting caspases.

Chapter 33

1 A protooncogene is a normal gene which can become converted into an oncogene. It is a gene coding for a protein which is involved in cellular control processes most commonly those which affect cell replication. They were discovered when it was found that oncogenes of retroviruses, known to be cancer-producing, have almost identical counterparts in normal cells. Conversion of a protooncogene to an oncogene usually causes excessive signalling – either the gene is over-expressed or the gene product (the signalling molecule) has inappropriately prolonged life. The change may be caused by a single base mutation or by the gene being placed under the control of an excessively active promoter. In the case of Burkitt's lymphoma chromosomal rearrangements places a protooncogene under the control of an immunoglobulin gene promoter. A very active retroviral promoter may be inserted at a point in a chromosome so as to control the expression of the protooncogene.

Alternatively a mutation may produce an excessively long-lived transcription factor. In the case of the *Ras* oncogenes, mutations eliminate the GTPase activity so that the signalling pathway cannot be switched off.

2 The p53 gene is normally expressed at a low level but this increases if the DNA is damaged. The resultant high levels of p53 protein activate genes that cause the inhibition of the cyclin-dependent kinases which are needed for the cell to proceed through the restriction point in G_1 The delay allows time for the DNA to be repaired. When this is done, the p53 expression is no longer stimulated, the p53 protein diminishes and the cell cycle proceeds. If the DNA is not repaired the p53 gene can also induce apotosis of the cell.

3 An oncogene is a gene involved in the control of cell division which has become mutated so that it gives inappropriate mitogenic signals leading to uncontrolled cell division. They are the 'bad' genes. Tumour-suppressor genes protect against cancer. The best known is p53. It is activated by damaged DNA and arrests the cell cycle at the G_1 checkpoint. If the damage is not rectified it signals the cell to die by apoptosis. Mutation of both alleles of a tumour-suppressor gene makes a cell with an oncogene more likely to progress to cancer.

4 A somatic cell shortens its telomeres at every round of DNA replication and this limits the number of cell divisions it can undergo. A cancer cell has to be able to undergo unlimited numbers of divisions and the continual lengthening of the telomeres by telomerase or ALT permits this.

Index of diseases and medically relevant topics

Index